DEVELOPMENTS IN GEOTECHNICAL ENGINEERING
FROM HARVARD TO NEW DELHI 1936-1994

PROCEEDINGS SYMPOSIUM ON DEVELOPMENTS IN GEOTECHNICAL ENGINEERING
BANGKOK/THAILAND/12-16 JANUARY 1994

Developments in Geotechnical Engineering

From Harvard to New Delhi 1936-1994

Edited by
A.S.BALASUBRAMANIAM
S.W.HONG, D.T.BERGADO, N.PHIEN-WEJ & P.NUTALAYA
Asian Institute of Technology, Bangkok, Thailand

A.A.BALKEMA / ROTTERDAM / BROOKFIELD / 1994

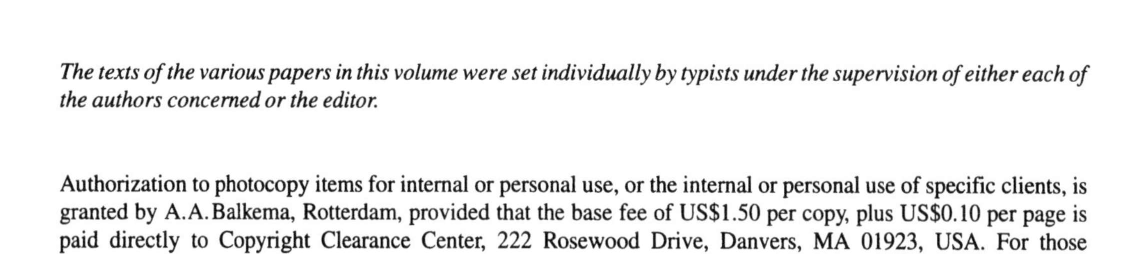
The texts of the various papers in this volume were set individually by typists under the supervision of either each of the authors concerned or the editor.

Authorization to photocopy items for internal or personal use, or the internal or personal use of specific clients, is granted by A.A. Balkema, Rotterdam, provided that the base fee of US$1.50 per copy, plus US$0.10 per page is paid directly to Copyright Clearance Center, 222 Rosewood Drive, Danvers, MA 01923, USA. For those organizations that have been granted a photocopy license by CCC, a separate system of payment has been arranged. The fee code for users of the Transactional Reporting Service is: 90 5410 522 4/94 US$1.50 + US$0.10.

Published by
A. A. Balkema, P.O. Box 1675, 3000 BR Rotterdam, Netherlands (Fax: +31.10.4135947)
A. A. Balkema Publishers, Old Post Road, Brookfield, VT 05036, USA (Fax: 802.276.3837)

ISBN 90 5410 522 4
© 1994 A. A. Balkema, Rotterdam
Printed in the Netherlands

Developments in Geotechnical Engineering, Balasubramaniam et al. (eds) © 1994 Balkema, Rotterdam, ISBN 90 5410 522 4

Preface

This tenth volume in a series of Balkema Publications on the Annual Geotechnical Symposium sponsored by the Asian Institute of Technology and the Southeast Asian Geotechnical Society is devoted to the Developments in Geotechnical Engineering (from Harvard to New Delhi 1936-1994). The previous volumes dealt with the Prediction versus Performance in Geotechnical Engineering (1992), Geotechnical Aspects of Restoration Works on Infrastructures and Monuments (1990), Computer and Physical Modeling in Geotechnical Engineering (1989), Environmental Geotechnics and Problematic Soils and Rocks (1988), Geotechnical Aspects of Mass and Material Transportation (1987), Recent Developments in Laboratory and Field Tests and Analysis of Geotechnical Problems (1986), Recent Developments in Ground Improvement Techniques (1985), Geotechnical Aspects of Coastal and Offshore Structures (1983), and Geotechnical Problems and Practices of Dam Engineering (1982).

Fifty-two papers of international significance are included in this volume which are contributed mostly by well-known and senior engineers who have spent a lifelong period in the geotechnical engineering profession. These papers review the developments that have taken place in the field of geotechnical engineering since the first international conference on Soil Mechanics and Foundation Engineering was held in Harvard University in 1936 until the January 1994 conference in New Delhi, India. For ease of reference, the papers are categorized into six sections based on certain subject matters. The first section covers the *Developments in the Theory and Practice in Geotechnical Engineering*. It contains thirteen valuable contributions from across the globe, i.e. Australia, Canada, USA, France, Germany, Thailand, Indonesia, Japan and Korea, related to new and unconventional methods on *in-situ* undisturbed sampling, testing and performance monitoring, and evaluation of liquefaction resistance of sands, horizontal mass permeability of clay, state parameter, plasticity analysis, studies on consolidation theories and a personal account on the contributions in geotechnical engineering after more than a score of professional life, among others.

The second section discusses the *Engineering Behaviour of Soils*. It contains six papers that deals on the strength of residual soils with inclusions, expansive, swelling and collapsing behaviours, schistous sands and a study on geomorphology as determinant of engineering problem in a river basin in Australia. Section three deals on *Natural Hazards and Environmental Geotechniques*. It has ten papers from Japan, Taiwan, Singapore, Australia, Canada, USA, and Austria that deal with seismic deformations, industrial waste reuse, waste containment, decontamination and rehabilitation, seepage studies, risk assessment, mitigation, and satellite application to environmental geotechnology. Section four incorporates topics on *Embankments, Excavations and Buried Structures* contributed by twelve prominent authors coming from the six continents whereas section five deals on *Soil-Structure Interactions* with five papers. Finally, section six is on *Ground Improvement Techniques* with six papers.

The release of this volume was made possible through the efforts, help and support received from many individuals and organizations or institutions, and are, therefore, gratefully acknowledged. First of all, acknowledgement is due to the following members of the General Committee of the Southeast Asian Geotechnical Society: Dr Ooi Teik Aun (President), Dr C. D. Ou, Dr Za-Chieh Moh, Dr S. B. Tan, Dr E. W. Brand, Dr W. H. Ting, Prof. S. L. Lee, Dr A. W. Malone, Prof. Sambhadharaksa, Dr K. Y. Yong, Dr J. C. Li, Dr S. F. Chan and Dr Clive Franks.

At the Asian Institute of Technology, the editors would also like to express their sincere thanks, among others, to Prof. Alastair North, President and to Dr Pisidhi Karasudhi, former Dean of the School of Civil Engineering and now Vice-President for Development, for their wholehearted support in our professional activities. The meticulous and untiring work of Mr Elmer Bandalan, Ms Vilma C. Chandraratne and Mr U. P. L. Lokuge in checking and proofreading the manuscripts, not to mention the page-by-page scrutiny of Dr S. W. Hong, have made possible the production of this volume in its present form.

A. S. Balasubramaniam

Developments in Geotechnical Engineering, Balasubramaniam et al. (eds) © 1994 Balkema, Rotterdam, ISBN 90 5410 522 4

Table of contents

1 *Developments of theory and practice in geotechnical engineering (including soil sampling, laboratory and field testing, design methods and construction works)*

2 *Engineering behaviour of soils (including collapsible and expansive as well as other problematic soils)*

3 *Natural hazards and environmental geotechniques*

4 *Embankments, excavations and buried structures*

5 *Soil-structure interactions*

6 *Ground improvement techniques*

1 Developments of theory and practice in geotechnical engineering (including soil sampling, laboratory and field testing, design methods and construction works)

Developments in Geotechnical Engineering, Balasubramaniam et al. (eds) © 1994 Balkema, Rotterdam, ISBN 90 5410 522 4

Contributions in geotechnical engineering: Soil mechanics and foundation engineering

A. S. Balasubramaniam (Bala)
Geotechnical Engineering Program, School of Civil Engineering, Asian Institute of Technology, Bangkok, Thailand

ABSTRACT: This paper summarizes the author's contribution to Soil Mechanics and Foundation Engineering while at the Asian Institute of Technology (AIT) for over twenty years. It describes the research work carried out on coastal deposits of soils, and in particular the Bangkok sub-soils, in relation to infra-structure development. It draws extensively upon thesis research carried out by graduate students in the field of Soil Engineering at AIT.

The Engineering behavior of the Bangkok sub-soils are analysed in terms of Critical State Soil Mechanics concepts and the results of analyses were then used in numerical solutions related to Highway Embankments, Deep Foundations and Deep Excavations. Deterministic numerical analyses were carried using finite element methods with inputs of the soil properties as required by Critical State Soil Mechanics. These data were then compared with the values computed from the traditional analysis. Higher order statistics and simple probabilistic analysis were also combined with the finite element methods to obtain more meaningful parameters that would allow engineers to make better judgments than using traditional safety factors.

Various types of ground improvement techniques are described and the use of polymer grids in shallow reinforcements and the use of surface and deep seated chemical stabilisations which involve lime and cement additives are recommended.

The use of proper in-situ tests such as the piezo-cone, the pressuremeter and the dilatometer is strongly advocated in comparison to the traditional dominant role played by the Standard Penetration Tests. Promising areas for research and developments are also touched upon: these include the use of the Centrifuge in pollution control; liquefaction studies and accelerated tests of a week duration in time scale, realising the effects of several hundred years in the field. Non-traditional, problematic soils, natural mixed gradational soil deposits, and soft rocks could dominate the research work in the next decade to re-examine the empirical conclusions which appear in routine soil mechanics text based on the experiences of saturated and well behaved soils.

1 INTRODUCTION

Looking at the developments in Geotechnical Engineering since the First International Conference on Soil Mechanics and Foundation Engineering was held in Harvard University in 1936 on the occasion of the three hundredth year anniversary celebration of Harvard University, to the New Delhi conference in January 1994, one could conclude that Soil Mechanics and Foundation Engineering is still a young discipline, only about 50 years old. What then are the Asian Institute of Technology's contributions to these developments, being also a young institution (only 31 years old), founded only in 1967?

This paper thus reviews the significant contributions to these developments while the author was involved in research and developments at AIT over the last twenty years.

1.1 *Soil Mechanics and Foundation Engineering at AIT*

Soil Mechanics and Foundation Engineering started at AIT in 1967 as a field of study concurrent with the founding of the Southeast Asian Geotechnical Society (SEAGS). Research and development in Geotechnics embrace Engineering Geology, Soil Engineering, Rock Mechanics and Engineering Geophysics and can broadly be presented under three major sections, namely: Infra-structure Developments, Resources Developments and Mitigation of Natural Hazards. Selected contributions in these areas as drawn out mainly from the research activities of our graduate students and the professional activities form the most relevant material for this paper.

My interest in Soil Mechanics emerged when I became a post-graduate student in 1965, after only

29 years of advancement of the subject, a period that had been interrupted for several years by the Second World War. Thus, true contributions in Soil Mechanics emerged only from the early fifties. My earlier training at Cambridge University and at the Norwegian Geotechnical Institute equipped me well I guess in the activities that relate to Deltaic Deposits and their developments (BALASUBRAMANIAM, 1969 to 1975). Our involvement in the developments of Geotechnics related to Deltaic Deposits was the most appropriate choice of research at AIT and the impact in this area over the last two decades has been significant.

1.2 *Developments in Asia: AIT's Contributions*

The last two decades have witnessed a so-called modernisation in many Asian countries which have resulted in the construction of very large urban development projects. These include a dramatic expansion of cities and towns, highways and expressways, large bridges and tunnels, and port facilities and airports.

Of the many factors that play an important role in development, the geological and soil conditions of the terrain in which the construction is carried out, seem to have a very high priority when viewed in terms of the cost of the project. Asia covers a land area of several million square kilometers and its physiography and geology are most complex. The most successful economic developments in Asia have taken place in the low, flat and deltaic plains as shown in Fig. 1.

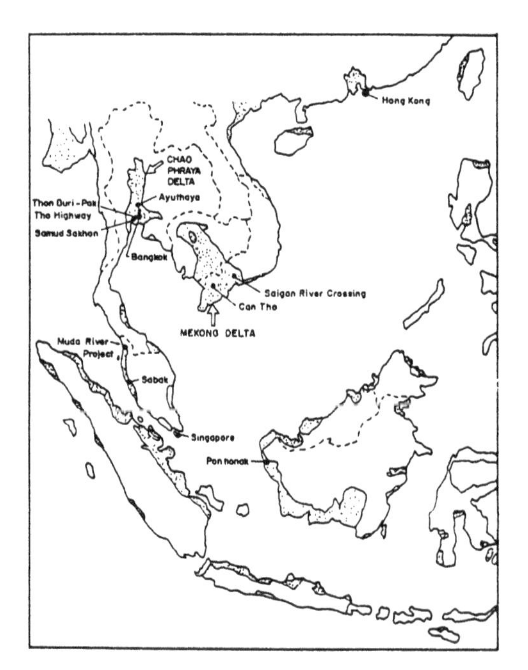

Fig. 1 Distribution of recent clays in Southeast Asia.

The Asian Institute of Technology was most wise in selecting this landform as the study area, since these regions cover a major portion of the population than the others. The reason for this is that civilization in Asia, like many other civilizations, first began on the deltaic plains as the soils in these regions are more fertile and are amenable to easy irrigation through the annual flooding. However, this is an unfortunate situation from a development point of view as the deltaic soils which are most excellent for agricultural purposes are the worst soil conditions one could imagine for civil engineering construction. As a result, these deltaic deposits are an excellent terrain for research and development in Soil Mechanics and Foundation Engineering. Thus, the Asian Institute of Technology has rightly exploited this situation and there has been extensive research activities in particular on the Bangkok sub-soils and Aquifer systems.

The engineering properties of the soils in Southeast Asia are documented in the Golden Jubilee Proceedings of the Southeast Asian Geotechnical Society as edited by BALASUBRAMANIAM, BERGADO, and CHANDRA, (1985b).

2 BANGKOK SUB-SOILS AND AQUIFER SYSTEMS

The traditional developments in teaching and research in Soil Mechanics and Foundation Engineering have been on the behavior of saturated clays. Those who are trained in Soil Mechanics and Foundation Engineering will be quite familiar with the research work on the normally consolidated soft Boston Blue Clay and others in the United States of America. Equally well known are the studies on the heavily overconsolidated and stiff London Clay and Weald Clay. Impressive work on the sensitive soft clay in Scandinavia and Canada is also well documented. The Bangkok clay is one of the most well researched natural deposit in Asia and it is the only deposit on which the Critical State Soil Mechanics has been so elaborately studied and applied. Before these studies are described, a brief description of the Bangkok sub-soils will be given.

The Chao Phraya Plain consists of a broad basin filled with sedimentary soil deposits which form alternate layers of sand, gravel and clay. The profile of the surface of the bedrock is still undetermined, but its level in the Bangkok area is known to vary in ranges of 550 to 2000 m (Fig. 2). The available geological, hydrological and geophysical evidence suggests that there exist eight confined aquifers which are somewhat inter-connected but separated by impervious strata of clay. A typical shallow soil profile and some index properties of the soil along the coastal line at Pom Prachul in Bangkok is shown in Fig. 3.

2.1 *Engineering Behavior of Bangkok Sub-Soils*

The engineering problems related to the Bangkok

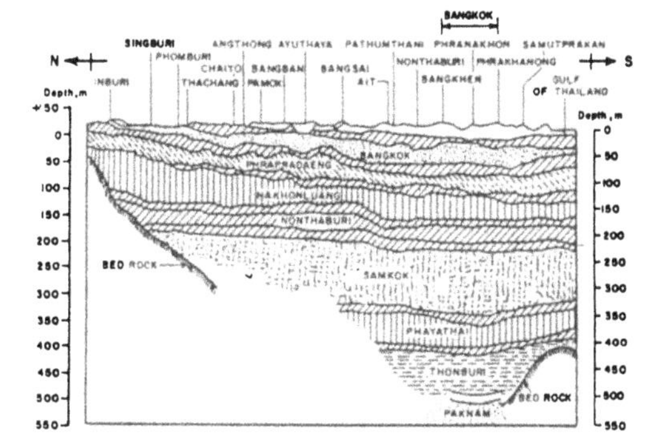

Fig. 2 Systems of aquifers under the Chao Phraya plain.

sub-soils are of three types. The first one deals with construction activities that impose a small loading or an unloading. This can be in the form of an embankment for highway construction; a development project; a shallow excavation for an irrigation canal; a basement for building; or pipe laying for water supply, sewerage or natural gas consumption. For such engineering purposes, an understanding of the engineering properties up to a depth of about 20 to 25 m above the first sand layer is sufficient. Then comes a class of problems related to deep foundations and deep excavations for tall buildings. Currently, the tallest buildings in Bangkok are founded in the second sand layer located around 50 to 60 m deep while the deep foundations require an understanding of the engineering properties of sub-soils comparable to even as deep as 100 to 120 m. The third class of problems relates to the subsidence of Bangkok as a result of deep well pumping. It is important that any site investigation, laboratory and field testing program recognises these three classes of problems.

When it comes to the soil mechanics analysis of structures, two types of analysis are often carried out. One is a deformation analysis to evaluate the movement of the structure under external loads and the other is the stability analysis to evaluate the safety of the structure against an outright failure. For the first class of problems, the engineering behavior of the soft Bangkok Clay is crucial. The soft Bangkok Clay in the lower Chao Phraya Plain extends from 200-250 km in the East-West direction and 250-300 km in the North-South direction. The thickness of the soft to medium stiff clay in the upper layer varies from 12 to 20 m while that of the total clay layer including the lower stiff clay is about 15 to 30 m. Thicker deposits are found close to the Gulf of Thailand and the thickness decreases towards the north.

The one-dimensional consolidation and compressibility characteristics of Bangkok subsoils have been studied in detail up to the depths of 100 m (YAU, 1976; ADIKARI, 1976; TOWAN, 1976; JIANN, 1977; TSAI, 1981; PARENTELA, 1982; KERDSUWAN, 1984). The studies include natural water contents, index properties, accurate determination of relationship between void ratio and pressure, and determination of the primary and secondary consolidation parameters. K_o measurements were carried out by WANG (1975) and TANTIKOM (1981).

One of the significant contributions made by the Asian Institute of Technology is teaching and research on Critical State Soil Mechanics. Comprehensive work carried out on Bangkok

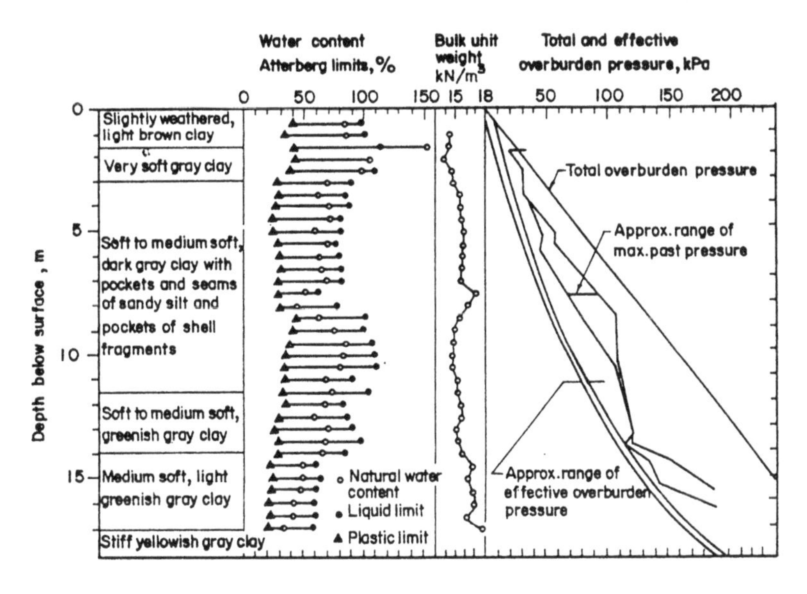

Fig. 3 A typical Bangkok subsoil profile (Pom Prachul).

sub-soils include compression and extension tests under a variety of applied stress paths both at in-situ stress levels and at higher pre-shear consolidation stresses. These results are discussed in relation to the critical state concepts and the associated stress strain theories (HWANG, 1975; WAHEED, 1975; CHAUDHURY, 1975; LI, 1975; AHMED, 1976; HASAN, 1976; SARKER, 1979; WIJEYAKULASURIYA, 1986; SUROSO, 1987). Thus, extensive studies were made on the application of the critical state concepts for the behavior of undisturbed samples of Bangkok sub-soils. Quasi-static repeated loading tests were also conducted (CHUI, 1975; LO, 1976) and the engineering properties of sand in the First Sand Layer were studied by SUDTHIKANONG (1976), MOLLAH (1977) and JIANN (1977). The other dissertations related to the Engineering Behavior of Soils are TJANDRAWIBAWA (1978), TONYAGATE (1978), NILAWEERA (1978), LAM (1982), HAMID (1985), TABE (1987), KOK (1987), CHENG (1987), MAMPITIYA (1987), ZAMALUDIN (1988), LIM (1988), MONOI (1989), DANGOL (1989) and UDAKARA (1989).

2.2 *Critical State Concepts on Bangkok Clays*

The predictions from the critical state theories are often compared with the behavior of remolded specimens of Kaolin prepared under controlled conditions in the laboratory. BALASUBRAMANIAM (1969) has carried out a series of extensive stress-controlled triaxial tests on normally and overconsolidated specimens of Kaolin under a large variety of applied stress paths and compared the experimental behavior with theoretical predictions. As a logical extension of the extensive study that has already been made on the behavior of remoulded specimens of Kaolin, the research program at the Asian Institute of Technology is directed towards extending the study to the behavior of undisturbed specimens of natural deposit of Soft Clay which is extensively found in Southeast Asia, particularly in Bangkok.

2.2.1 *Stress Paths and State Boundary Surface*

Figure 4 shows the stress paths for compression and extension tests on Weathered Clay at low pre-shear consolidation pressures which are less than the maximum past pressure. The behavior thus corresponds to overconsolidated specimens. The stress paths are found to be approximately sub-parallel to the q-axis, indicating that the mean normal stress, p, do not vary much during shear. These results are in good agreement with the elastic wall concept used extensively in the critical state theories. According to these theories, only elastic volumetric strains take place inside the state boundary surface.

Since the volumetric strain is zero in an undrained test, there would not be any change in the mean normal stress, thus, the stress paths rise parallel to the q-axis in the (q,p) plot until failure is reached. The dotted lines in Fig. 4 correspond to extension tests under unloading conditions. It is noted that the effective stress paths in compression and in extension under loading and unloading conditions are similar for a first degree of approximation.

The effective stress paths followed by the specimens under undrained condition at higher pre-shear consolidation pressures for Weathered Clay and Soft Clay were found to be virtually similar. The results are presented in the (q/p_e, p/p_e) plot, where p_e is the mean equivalent pressure (Fig.5).

2.2.2 *State Paths for Normally Consolidated States*

The state paths followed by all the test specimens in the normally consolidated states are shown in Figs. 5 (a) to (j) for each type of test condition. Also, for each type of applied stress path, the state paths are found to be independent of the pre-shear consolidation pressure and also the state boundary surface is virtually the same. Using this unique state boundary surface, the volumetric strain for any stress-increment in the normally consolidated state can be determined. The state paths shown in Fig. 5 also include those corresponding to anisotropic consolidation. Since one dimensional consolidation in the oedometer is also a special type of anisotropic consolidation with zero lateral strain, the volumetric strain and hence the axial strain during the one-dimensional consolidation can be also predicted from the state boundary surface presented in Fig. 5.

2.2.3 *Shear Strain Contours in Undrained Tests*

The shear strain contours for Weathered Clay sheared under compression and extension conditions are shown in Figs. 6 (a) to (c). Under compression conditions, the shear strain contours are found to be nearly subparallel to the p-axis. These constant shear strain contours were called as the constant q yield loci and were used in the prediction of shear strains on specimens of clay sheared along stress paths which lie below the state boundary surface.

2.2.4 *Stress Ratio - Strain Relationships*

For any one type of applied stress path, the stress ratio - strain relationships are found to be unique for normally consolidated specimens. The stress ratio - strain relationships for undrained and drained tests are presented in Figs. 7 and 8 for all the tests carried out on Weathered Clay and Soft Clay in the normally consolidated states. It is noted that for all the undrained tests, the shear strain ϵ is a unique function of the stress ratio, ($\eta = q/p$). Also, for the undrained tests the ratio of pore pressure to mean normal stress was found to be a unique function of the stress ratio. In the case of drained tests, both the volumetric strain and the shear strain are found

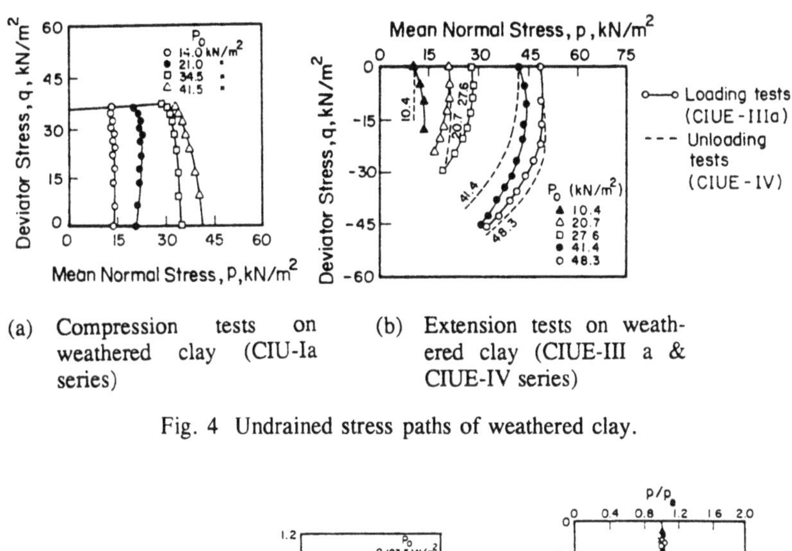

(a) Compression tests on weathered clay (CIU-Ia series)

(b) Extension tests on weathered clay (CIUE-III a & CIUE-IV series)

Fig. 4 Undrained stress paths of weathered clay.

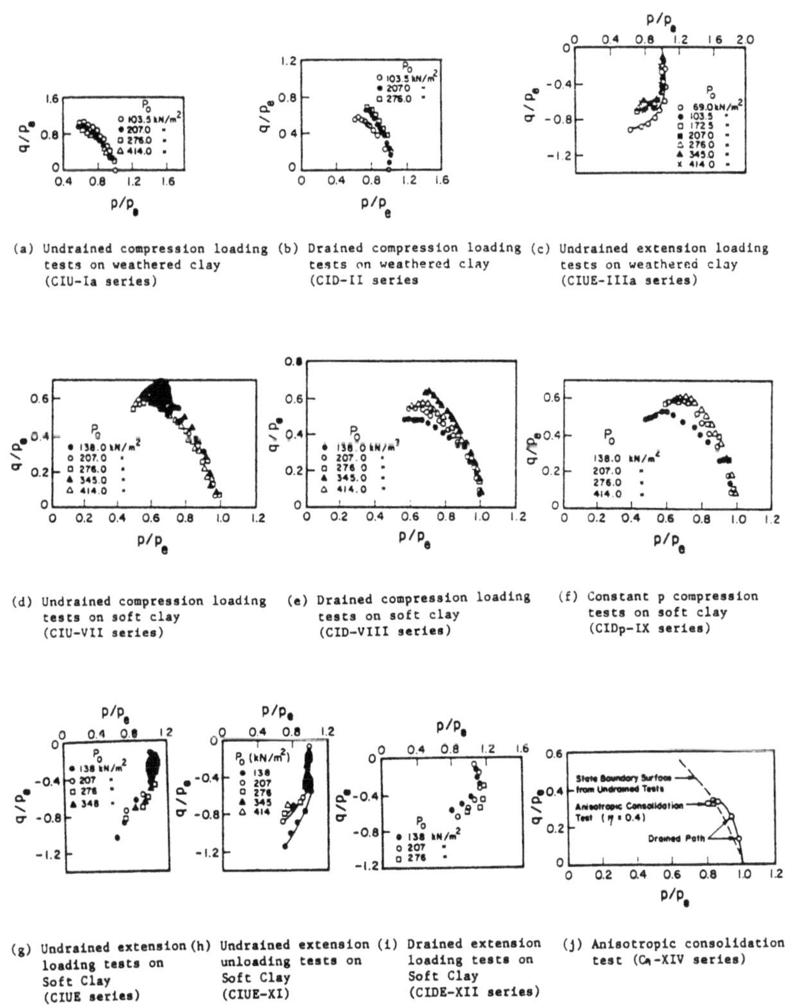

(a) Undrained compression loading tests on weathered clay (CIU-Ia series)

(b) Drained compression loading tests on weathered clay (CID-II series)

(c) Undrained extension loading tests on weathered clay (CIUE-IIIa series)

(d) Undrained compression loading tests on soft clay (CIU-VII series)

(e) Drained compression loading tests on soft clay (CID-VIII series)

(f) Constant p compression tests on soft clay (CIDp-IX series)

(g) Undrained extension loading tests on Soft Clay (CIUE series)

(h) Undrained extension unloading tests on Soft Clay (CIUE-XI)

(i) Drained extension loading tests on Soft Clay (CIDE-XII series)

(j) Anisotropic consolidation test (C_A-XIV series)

Fig. 5 State paths followed by triaxial specimens in normally consolidated states.

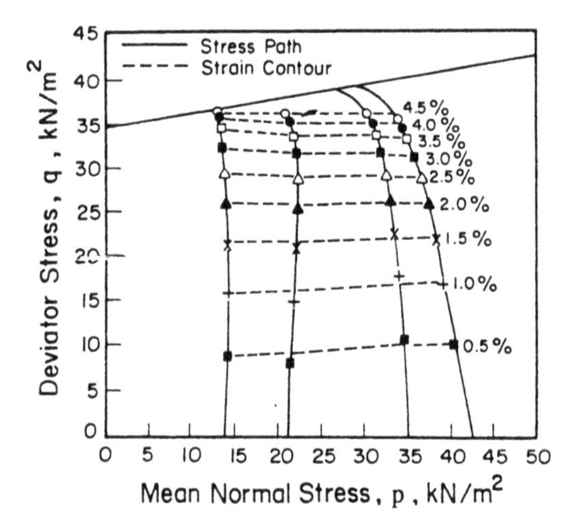

(a) Compression loading tests on weathered clay (CIU-Ia series)

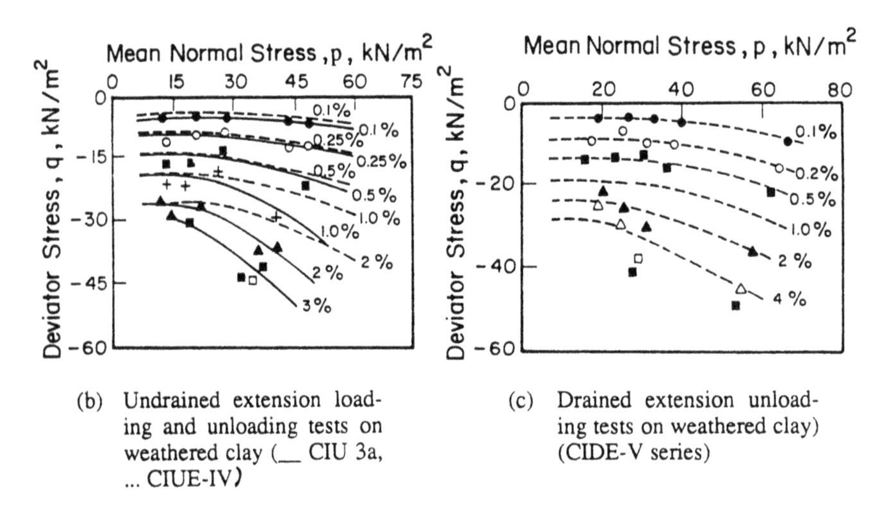

(b) Undrained extension loading and unloading tests on weathered clay (__ CIU 3a, ... CIUE-IV)

(c) Drained extension unloading tests on weathered clay) (CIDE-V series)

Fig. 6 Constant shear strain contours for specimens sheared from pre-shear consolidation pressure less than the maximum past pressure.

to be unique functions of stress ratio η.

2.2.5 *Anisotropic Consolidation*

During anisotropic consolidation, the stress ratio, σ_1/σ_3 was maintained constant (where σ_1 is the major principal stress and σ_3 is the minor principal stress). Figure 9 illustrates a possible experimental procedure for such a test in the (q, p) plot. A specimen is sheared under undrained condition from a pre-shear consolidation pressure, p_o. The undrained stress path is denoted by AB. From B, the specimen is subjected to anisotropic consolidation along the path BB'. During anisotropic consolidation q/p (= η) is maintained constant. Anisotropic consolidation tests were carried out on both Weathered Clay and Soft Clay. The strain paths are linear for Soft Clay. However, for Weathered Clay the strain paths corresponding to any particular stress-ratio are found to consist of two straight lines. During the initial phase $(dv/d\epsilon)\eta$ is higher than in the final phase (Fig. 10). This would imply that in the initial phase of shear, the distortional strain is small in magnitude than in the subsequent normally consolidated state. The (e, log p) relationships during the anisotropic consolidation in the normally consolidated states are found to be linear and the slope is equal to 0.51.

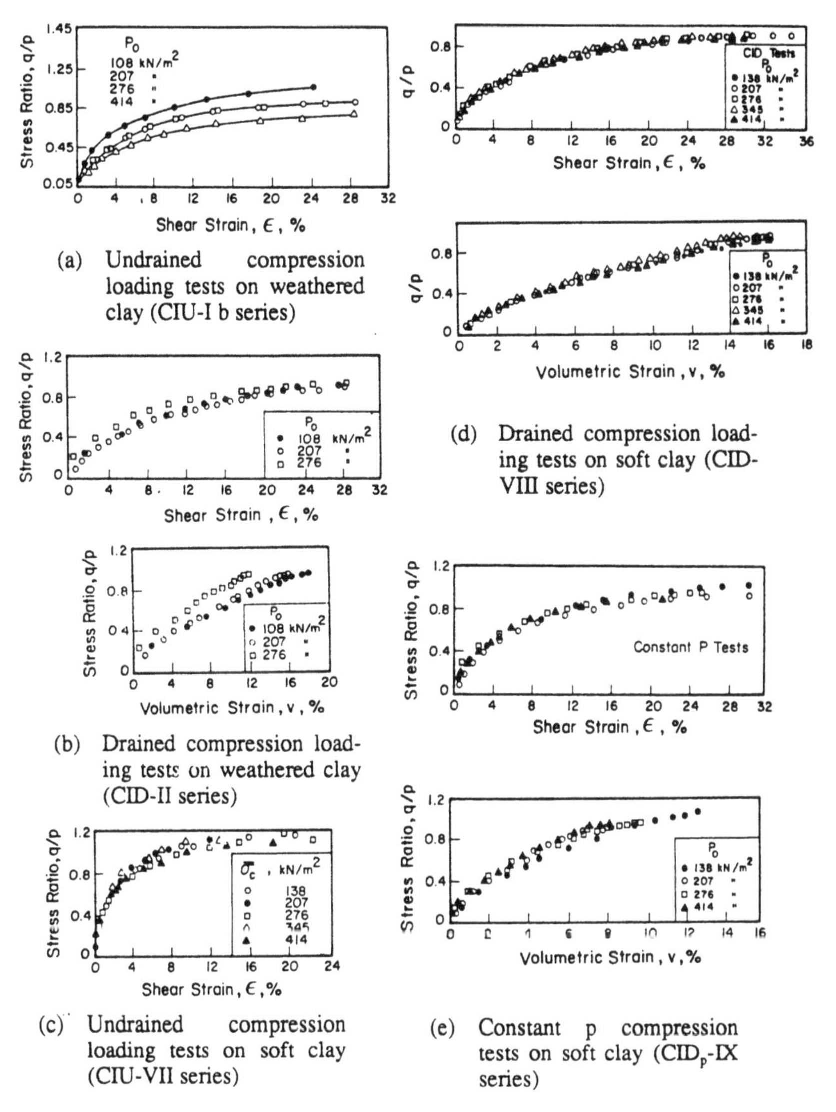

Fig. 7 Stress ratio strain relationships for compression tests in the normally consolidated states.

2.2.6 *Critical State Parameters and Strength Envelopes*

The end points of the specimens in (q, p), (w, log p) and (w, log q) plots are shown in Figs. 11 (a) to (d). For Weathered Clay in compression under undrained conditions at low pre-shear consolidation pressure, the ϕ value is found to be considerably low. Also, the (w_f, log p_f) relationship and the (w_f, log q_f) relationships are found to be curved. Similar results are also noted for the specimens of Weathered Clay sheared under extension conditions. For these specimens the (q_f, p_f, w_f) relationships under different types of applied stress paths are found to be the same for a first degree of approximation.

For Weathered Clay tested in a normally consolidated state in compression and in extension, the end points are found to coincide with the critical state line. The projection of the critical state line in (q, p) plot is linear. Similarly, linear relationships are noted in (w, log p) and (w, log q) plots. The end points of the specimens of Soft Clay sheared in compression under undrained, fully drained and constant p conditions are also found to lie on linear projections in (q, p), (w, log p) and (w, log q) plots. However, some differences are noted in all plots especially between the undrained tests and the other tests.

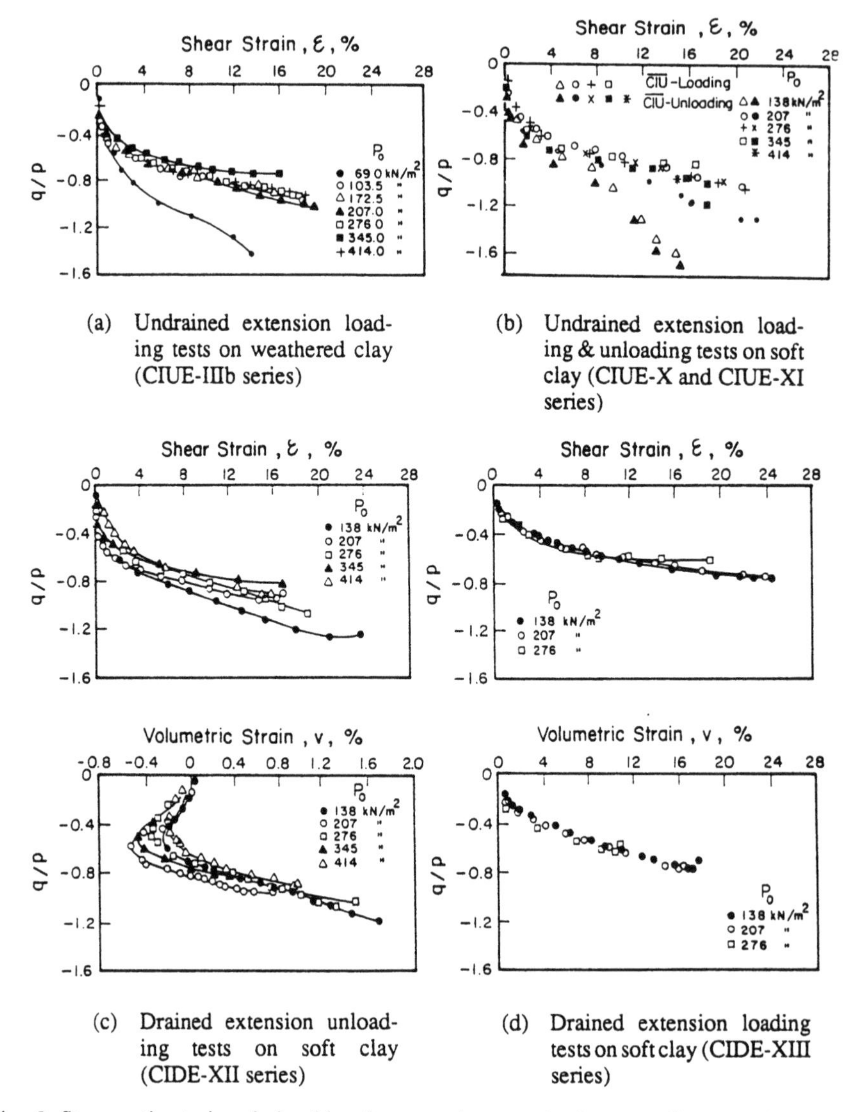

(a) Undrained extension loading tests on weathered clay (CIUE-IIIb series)

(b) Undrained extension loading & unloading tests on soft clay (CIUE-X and CIUE-XI series)

(c) Drained extension unloading tests on soft clay (CIDE-XII series)

(d) Drained extension loading tests on soft clay (CIDE-XIII series)

Fig. 8 Stress ratio strain relationships for extension tests in the normally consolidated states.

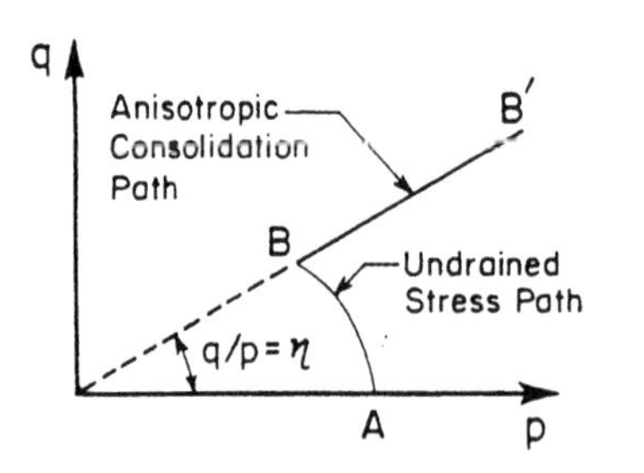

Fig. 9 Anisotropic consolidation path in (q,p) plot.

The results of the extension tests carried out on Soft Bangkok Clay are found to lie on straight lines in (q_f, p_f) plot, but the straight lines do not pass through the origin. It thus appears that Soft Bangkok Clay exhibits a small degree of cohesion under extension condition. The (w_f, log p_f) and the (w_f, log q_f) relationships are similar to those exhibited by normally consolidated clay specimens.

2.2.7 *Prediction of Strains Using Critical State Theories*

In this section, the experimentally observed strains

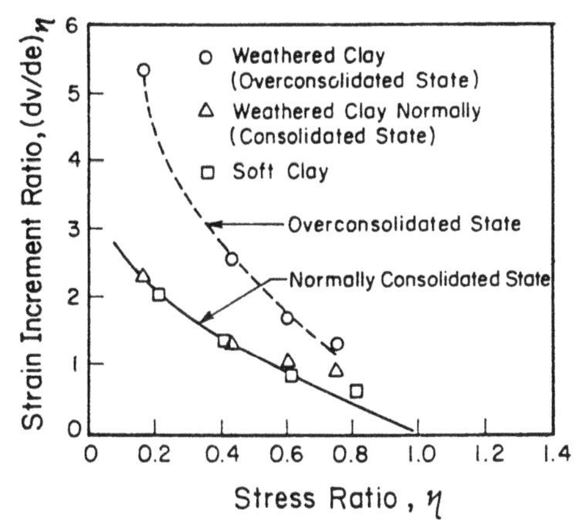

Fig. 10 Strain increment ratio during anisotropic consolidation.

are compared with the strains predicted from the Critical State Theories. Three theories are employed and these are the Cam Clay theory, by ROSCOE, SCHOFIELD & THURAIRAJAH (1963), SCHOFIELD & WROTH (1968); the Revised Theory by ROSCOE & BURLAND (1968); and the Incremental Stress-Strain theory of ROSCOE & POOROOSHASB (1963).

The fundamental soil parameters used in the Critical State Theories are, λ, κ and M. λ is the slope of the isotropic consolidation line in the (e, log p) plot, κ is the slope of the isotropic swelling line in the (e, log p) plot. Isotropic consolidation and swelling tests carried out on Soft Bangkok Clay indicate that the value of λ is 0.51 and that of κ is 0.091. Also the critical state parameter M is taken as 1.0. In the Revised Theory of ROSCOE and BURLAND, corrections are made for the shear strain from the contributions due to the constant q yield loci. The contributions from the constant q

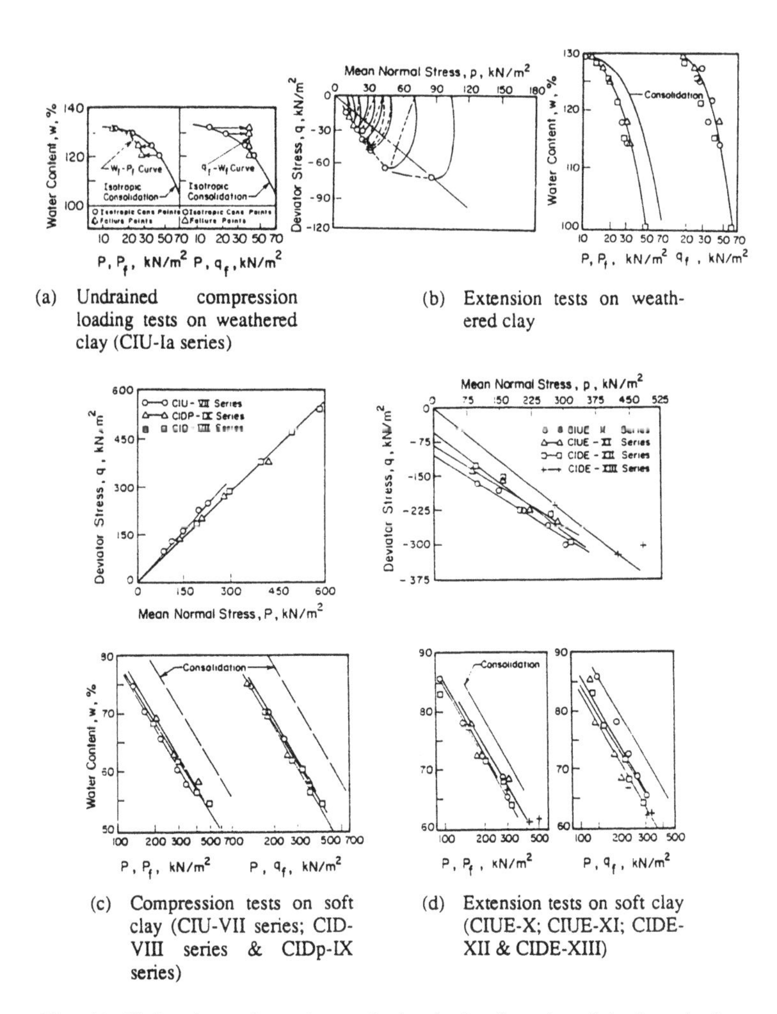

(a) Undrained compression loading tests on weathered clay (CIU-Ia series)

(b) Extension tests on weathered clay

(c) Compression tests on soft clay (CIU-VII series; CID-VIII series & CIDp-IX series)

(d) Extension tests on soft clay (CIUE-X; CIUE-XI; CIDE-XII & CIDE-XIII)

Fig. 11 End points of specimens in (q,p), (w, log p) and (w,log q) plot.

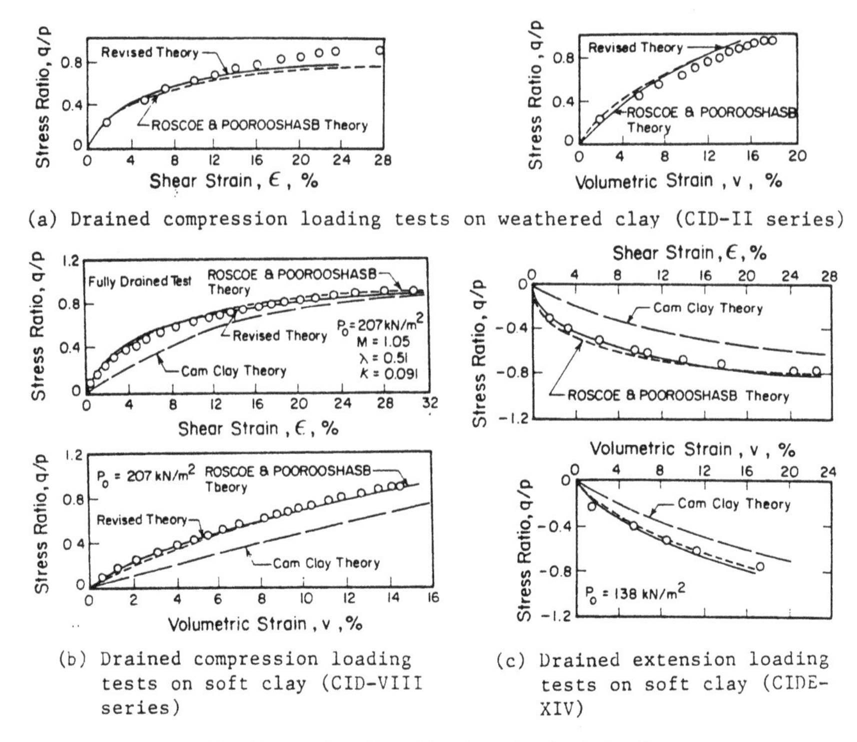

(a) Drained compression loading tests on weathered clay (CID-II series)

(b) Drained compression loading tests on soft clay (CID-VIII series)

(c) Drained extension loading tests on soft clay (CIDE-XIV)

Fig. 12 Observed and predicted strains in drained tests.

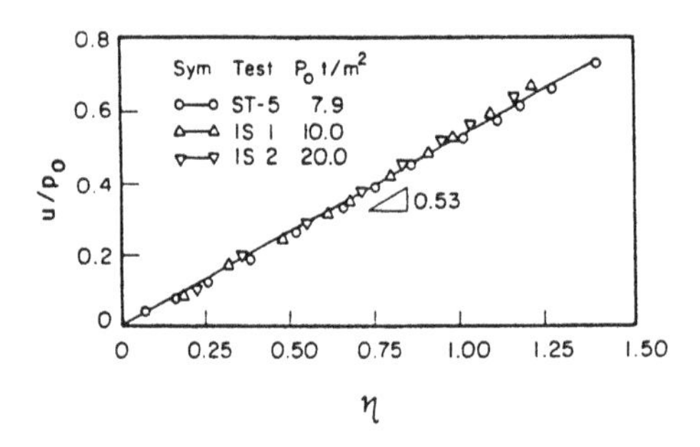

Fig. 13 (u/p_o, η) plot for normally consolidated samples under monotonic loading.

loci were approximately the same as the shear strain obtained from undrained tests in (q/p, ϵ) plot. The experimentally observed strains and those predicted from the Critical State Theories are shown in Fig. 12. The Cam Clay theory is found to overpredict the strains while the predictions from the Revised theory are good.

2.2.8 *Concluding Remarks from the Critical State Concepts*

A series of comprehensive triaxial compression and extension tests were carried out on Weathered and Soft Bangkok Clay and the results were compared with the predictions from a number of stress-strain theories. The following conclusions have been reached:

(i) For Weathered Clay sheared under compression and extension conditions from pre-shear consolidation pressures less than the maximum past pressure, the stress paths under the undrained conditions are nearly sub-parallel to the q-axis in a (q, p) plot.

(ii) For both Weathered and Soft Clay in the normally consolidated state, the state paths are found to be somewhat the same in the (q/p_e, p/p_e) plot.

(iii) For Weathered Clay sheared under compression and extension conditions from pre-shear consolidation pressures less than the maximum past pressure, the constant shear strain contours are nearly sub-parallel to the p-axis in the (q, p) plot.

(iv) For any one type of applied stress path, unique stress ratio strain relationships are observed both for Weathered Clay and Soft Clay, in the normally consolidated states.

(v) For each type of clay under similar applied stress paths, unique water content-strength relationships are noted. These relationships are found to coincide with the critical state line for specimens sheared from the normally consolidated states.

(vi) The incremental stress-strain theory of ROSCOE and POOROOSHASB (1963) and the Revised Critical State Theory are found to predict successfully the experimentally observed strains in the drained tests.

2.3 *Pore Pressure - Stress Ratio Relationships*

More recent study is confined to the analysis of the pore pressure developments under monotonic and cyclic loadings. In the case of monotonic loading under in-situ stress conditions, three types of undrained behavior were considered, i.e. isotropic, anisotropic and K_o-consolidation cases. Also, cyclically loaded samples were subsequently sheared under undrained conditions. Normally and overconsolidated behaviors were studied and simple relationships were established for the trend in pore pressure developments. These results are presented in Figs. 13 to 16 as pore pressure stress-ratio relationships while further details are given in HANDALI (1987), MONOI (1989) and BALASUBRAMANIAM et al. (1989a). Accurate determination of the in-situ yield loci of soft clays are now under progress as an extension of the work of BALASUBRAMANIAM & HWANG (1980) on the yielding of Weathered Bangkok Clay. Also, an extension of the Cambridge stress-strain theories are now being sought by KIM (1991) for samples sheared from in-situ stress conditions in the lightly overconsolidated state.

2.4 *Concluding Remarks on the Engineering Behavior of Bangkok Subsoils*

The material given here is only a very brief presentation of the extensive work done on Bangkok subsoils. Most of the work done up to now refer to axi-symmetric triaxial stress conditions. Even with such stress conditions, the behavior at in-situ stress levels needs additional work to completely understand the stress path-dependent behavior as needed for the solutions of engineering problems. Plane strain test data should also supplement the axi-symmetric data for a proper understanding on the stress-strain behavior.

3 HISTORICAL LANDMARKS IN SOIL MECHANICS

It would perhaps be useful for us to look back on the historical developments in Soil Mechanics and Foundation Engineering, and this is refered to with a scientific bias. Over the last two hundred years, one could count on ones fingers, the number of researchers who made significant contributions to the

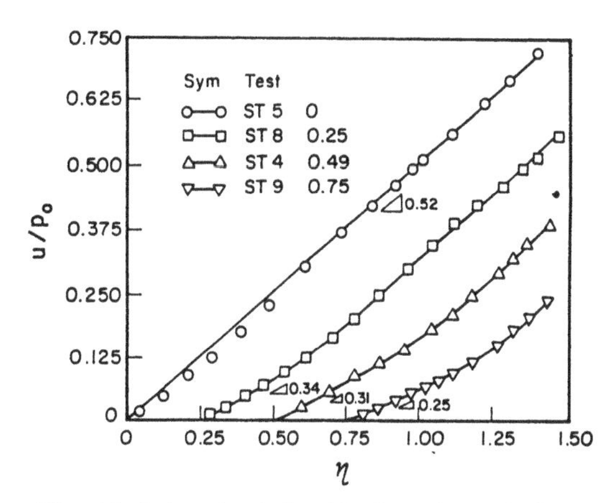

Fig. 14 (u/p_o, η) relationship for anisotropically consolidated samples.

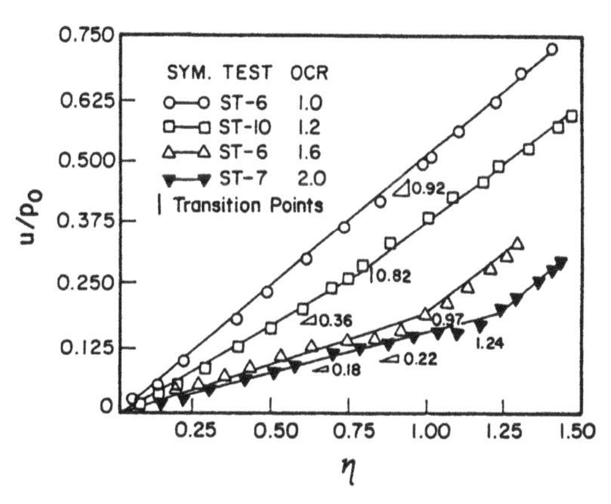

Fig. 15 (u/p_o, η) plots for overconsolidated samples under monotonic loading.

field from a fundamental point of view. The historical background on the development of Soil Mechanics is never complete without referring to the contributions of the Shear Strength of Soils by COULOMB in 1773. It will come as no great surprise to learn that developments in Soil Mechanics and Foundation Engineering up until 1773 was through a succession of experimentations without any real scientific character. Quoting KERISEL (1985); "Indeed it could have not been otherwise when at the beginning of the seventeenth century, the most advanced minds of the age discussed concepts without providing a definition". GALILEO and DESCARTES both make reference to the ideas of speed and of distance without firstly stating what is meant by the words used. At this time, we find only a set of empirical rules without any mathematical equations or Soil Mechanics formulae.

Following Coulomb's law of shear strength, one may remember the DARCY's law for the flow of fluids through porous media which dates back to

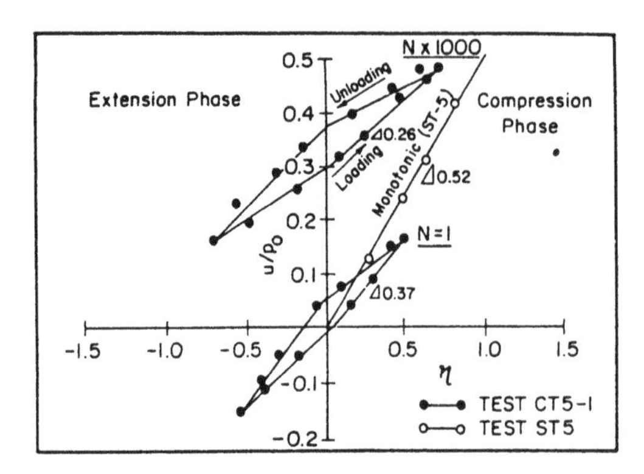

Fig. 16 (u/p_o, η) plot for normally consolidated sample CT5-1.

1856. Often modern Soil Mechanics is considered to have developed after the contributions of the principle of effective stress by TERZAGHI in 1924. Yet an important contribution often ignored by leading Soil Mechanics specialists is OSBORNE's (1986) Concepts and Experiments on Dilatancy, which show that a granular material when sheared changes in volume. Such a Law when recast in the theory of Plasticity as a stability criterion by DRUCKER (1959) and combined with the energy balance concept of ROSCOE, SCHOFIELD & WROTH (1958), ROSCOE & POOROOSHASB (1963), ROSCOE, SCHOFIELD & THURAIRAJAH (1963), SCHOFIELD & WROTH (1968) and ROSCOE & BURLAND (1968) gave us a simple expression for the first time for the dilatancy rate under radial paths.

SCHOFIELD (1980) presented an ingenious Rankine Lecture to illustrate the effect of natural water content and Liquidity Index on the strength and deformation characteristics of saturated clays. It is his approach which when pursued in a systematic manner, will give a proper understanding of the failure mechanisms in Slurry Mechanics in Dredging works (which is time dependent); ground improvement work in soft clays which flow and yield; the rupture mechanism in medium stiff clays; and, the fracture mechanism in very stiff clays.

3.1 *Revision of Soil Mechanics Curricula*

Truly speaking, Soil Mechanics as a basic discipline should be completely reorganised in the undergraduate Civil Engineering curricula in most universities in Asia, so that students are taught the elements of Critical State Soil Mechanics. This would allow them to see the whole spectrum of consolidation, shear, deformation prior to and during failure as well as strength and failure criterion. Indeed this is an aim of the curriculum of Soil Mechanics at the Asian Institute of Technology. It is hoped that, gradually, teachers in Soil Mechanics in Asia would justifiably be replaced by those who have an understanding of the fundamental concepts. A completely new text must be prepared for undergraduate Soil Mechanics which would bridge the gap between an average engineer and those who dwell around the tensor space with mathematics.

4 LABORATORY AND FIELD EQUIPMENT AND INSTRUMENTATION

4.1 *Laboratory Equipment*

Most of the major developmental projects in Sedimentary Soils and Residual Soils demand reliable geotechnical parameters obtained by laboratory testing. In the last two decades, excellent laboratory facilities for soil testing have been established at the Asian Institute of Technology. These include equipment for classification and index tests, compaction, consolidation, triaxial and direct shear tests. Equipment are also available for cyclic loading and large scale triaxial and oedometer tests on rockfill materials. Soil Mechanics and Foundation Engineering as a subject can never be taught or researched properly without good laboratory testing facilities. Over the years, AIT has trained numerous graduates who have returned to their home countries and have improved their own laboratory and field testing facilities in a commendable manner.

Plane strain equipment for large strains and a Centrifuge of an appropriate size should be added to the facilities so that such phenomena as environmental pollution, liquefaction of soils and time effects of several hundred years can be properly modelled in the laboratory.

4.2 *Field Equipment and Instrumentations*

Field tests can be classified into two groups: small scale tests and large scale tests. The vane tests, the Dutch cone tests and the Standard Penetration Tests fall under the category of small scale tests used commonly in South-east Asia. Vane tests are used to determine the shear strength of soft clays while Dutch cone tests are used for measurements at varying depths depending on the capability of the machine. Substantial developments have taken place in the use of the cone penetration tests. Standard Penetration Tests are used extensively in Stiff Clay, Residual Soils and Sand Formations.

It is in the development of the small scale field tests where AIT has failed to achieve its objective. The small scale field testing facilities at AIT, and especially in Thailand, need to be improved remarkably to keep abreast with the recent developments.

Standard Penetration Tests have been used extensively in the Bangkok subsoils and could be intelligently replaced by other sophisticated and accurate field testing devices such as the piezo-cones, the pressuremeter tests and the dilatometer tests.

4.3 *Large Scale Field Tests & Instrumentations*

Large scale field tests and instrumentations are now being carried out extensively for embankments, excavations and ground improvement techniques. The first such series of fully instrumented tests were conducted by the Asian Institute of Technology in 1973. This has been a catalysis for subsequent tests to be carried out in Singapore, Hong Kong, Taiwan, Indonesia and recently for a very important and magnificent series of such tests in Malaysia in the soft clay formations. A state-of-the-art lecture on this subject was given by BALASUBRAMANIAM (1980a) in the Sixth Southeast Asian Conference on Soil Engineering in Taipei.

Authoritative lectures on developments in laboratory and field tests and the analysis of Geotechnical problems were given in several international symposia organized by the Asian Institute of Technology and subsequently published as books through Balkema Printers (BALASUBRAMANIAM, BERGADO & CHANDRA, 1985a; and BALASUBRAMANIAM et al., 1989c).

5 EMBANKMENTS FOR HIGHWAYS AND EXPRESSWAYS

Thousands of kilometers of roads and highways in Asia and Southeast Asia are constructed on soft clays, and often the pavements are placed on shallow embankments only a few meters high. Thus, the stability and settlement characteristics of these embankments become important considerations. Extensive studies have been conducted at the Asian Institute of Technology on the stability and settlement characteristics of these low and high embankments (HO, 1976a; SIVANDRAN, 1975; SIRIWARDANE, 1976; LEE, 1978; AMERATUNGA, 1979; KAMPANANONDE, 1984; PARNPLOY, 1985; KUNTIWATTANAKUL, 1986). These studies include traditional stability and settlement analyses, and finite element analysis both with a deterministic and probabilistic approach. A comprehensive presentation on the performance of these embankments was given as a Guest Lecture in the Third International Conference on Numerical Methods in Geomechanics in Aachen, Germany in 1979 (BALASUBRAMANIAM et al., 1979b) and also in the Symposium on Geotechnical Aspects of Soft Clays (BALASUBRAMANIAN & BRENNER, 1981a). The other dissertations related to stability of slopes and subsidence are SAMOON (1978), HASNAIN (1985) and YAO (1986).

In the analysis of the test embankments as presented in many of our publications, isotropically and K_o-consolidated triaxial compression test data with a variety of applied stress paths were used and often the stress-strain behavior was modelled through concepts of Critical State Soil Mechanics.

Two types of embankment materials are often used. One category is sand which is frictional in nature and the second is cohesive and frictional residual soil. Recent studies carried out in Bangkok, Malaysia and Indonesia indicate that in the design of sand embankments, the strengths of fill materials have virtually no effect on the failure height on the embankments, while for residual soils the fill materials offer substantial resistance in increasing the failure height of the embankments.

5.1 *Embankments in the Bangkok Plains*

The excess pore pressure developed during the embankment construction loading at Pom Prachul site is shown in Fig. 17, wherein the pore pressure build up is gradual at low stress levels and increases at a much higher rate after a certain critical stress state is reached (BALASUBRAMANIAM et al., 1985d). The pore pressure parameter, μ, in excess of the critical stress is plotted in Fig. 18 with respect to the overconsolidation ratio. The magnitude of μ decreases with an increase in OCR. The estimated value of the Skempton's pore pressure parameter A is plotted versus the OCR in Fig. 19. Typical effective stress paths using pore pressure on the center line beneath the embankment are shown in Fig. 20. These stress paths are similar to those for both lightly and heavily overconsolidated clays. Normally, settlements during construction of embankments are separated into three components:(i) the elastic settlement at constant volume due to the imposed shear stresses (ii) the plastic settlement due to the induced shear stresses exceeding the shear strength, and (iii) consolidation settlement due to dissipation of the pore pressure. The consolidation settlement during construction as calculated by the method proposed by LEROUEIL et al. (1978) is found to be comparable to actual measured values. Post construction settlements were determined by dividing the entire compressible layer into a series of thin layers. The following methods were used: (i) a conventional one-dimensional method, (ii) a three dimensional method, (iii) the modified method of LEROUEIL et al. (1978), and (iv) the graphical method of ASAOKA (1978). The modified method of LEROUEIL et al. (1978) was found to be most suitable for the evaluation of post construction settlements as well.

5.1.1 *Stability of Test Embankments*

BOONSINSUK (1974) and HO (1976a) evaluated the stability of test embankments using both effective stress and total stress analysis methods. Also, shear strengths as obtained from a variety of tests were employed in the undrained analysis. MESRI (1975) proposed a relationship $c_u = 0.22\sigma_p$ which correlated well with the average field vane strength when corrected with the correction factor, μ of BJERRUM (1972), where σ_p is the preconsolidation pressure. The comparison between BJERRUM's correction factor, μ, and MESRI's correction factor $0.22\sigma_p/c_u$

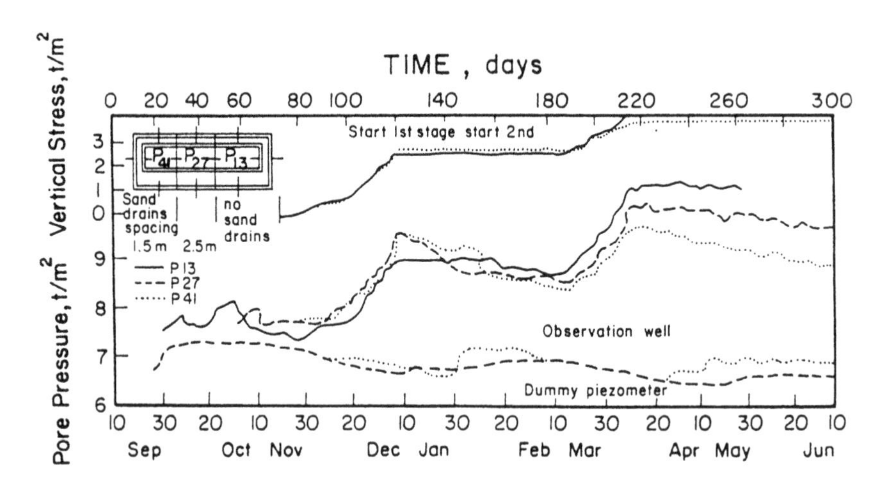

Fig. 17 Typical piezometric reading.

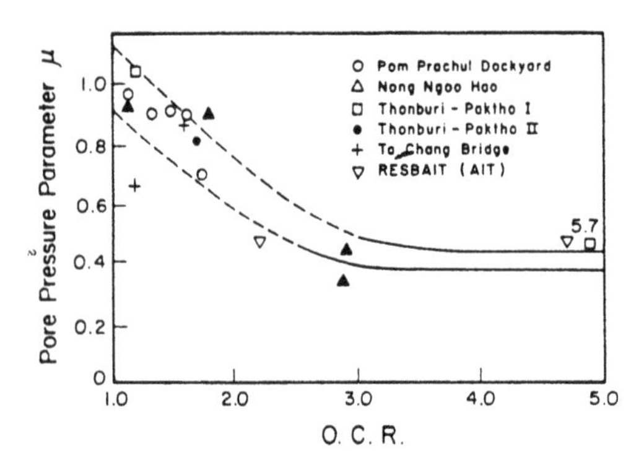

Fig. 18 Relationship between pore pressure parameter, μ, and overconsolidation ratio (OCR) for soft Bangkok clay.

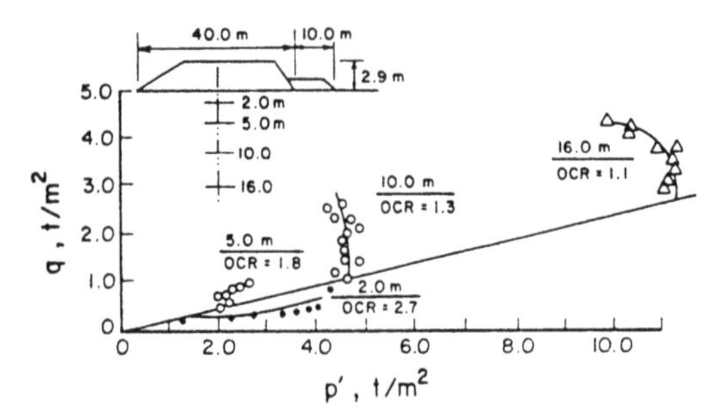

Fig. 20 Effective stress path due to embankment loading at each depth, Nong Ngo Hao test embankment.

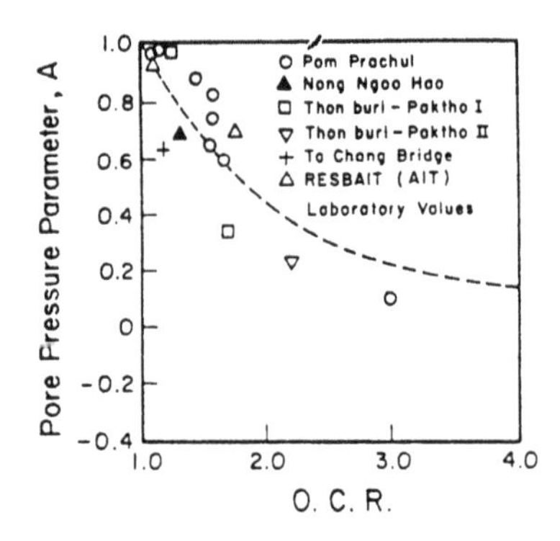

Fig. 19 Pore pressure parameter, A vs OCR beneath the centerline of embankment in soft Bangkok clay.

for vane strength were computed and the average ratios of c_u (vane)/σ_p for a number of sites in the Bangkok Plain are plotted versus the plasticity indices in Fig. 21. The factor of safety using MESRI's relationship for the test embankment in the Bangkok Clay ranged from 0.93 to 1.06. The undrained strength is thus independent of the plasticity index and is a unique function of the pre-consolidation pressure.

5.1.2 *Probabilistic Analysis*

Probabilistic techniques were introduced by SIVANDRAN (1979), ANANDA (1979) and GANESHANANTHAN (1979) to evaluate the stability and settlement of embankments and excavations and the carrying capacity of piled

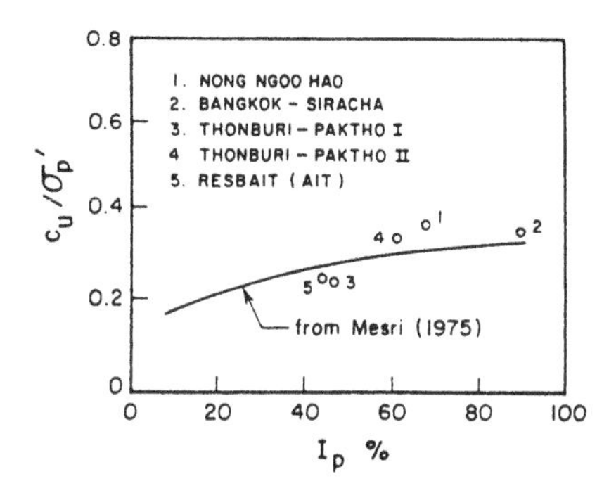

Fig. 21 The ratio c_u/σ_p with plasticity index in soft Bangkok clay.

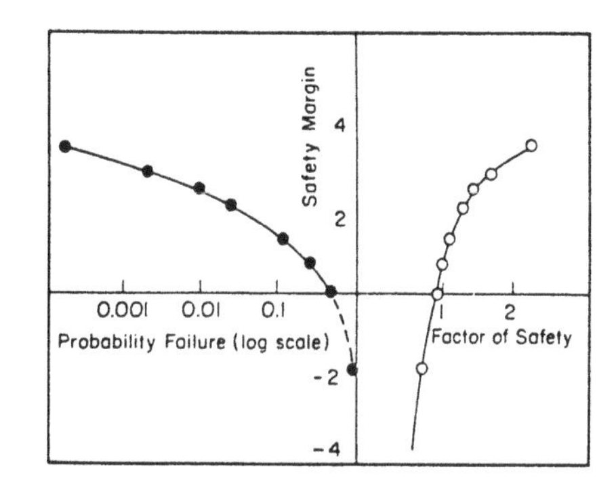

Fig. 23 Safety margin vs probability of failure and safety factor.

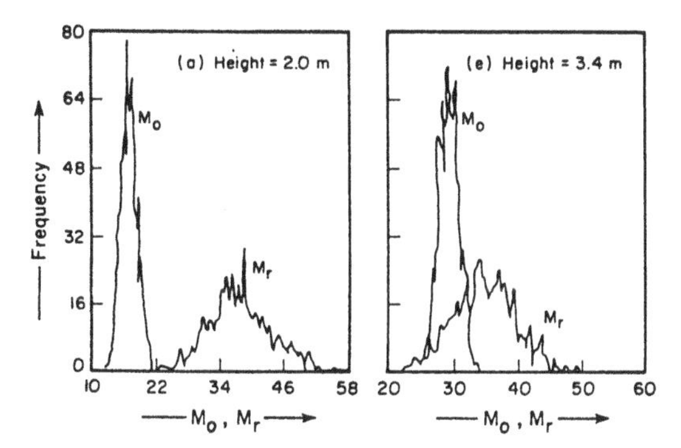

Fig. 22 Frequency distribution for overturning and resisting moments for embankment heights of 2.0 m and 3.4 m.

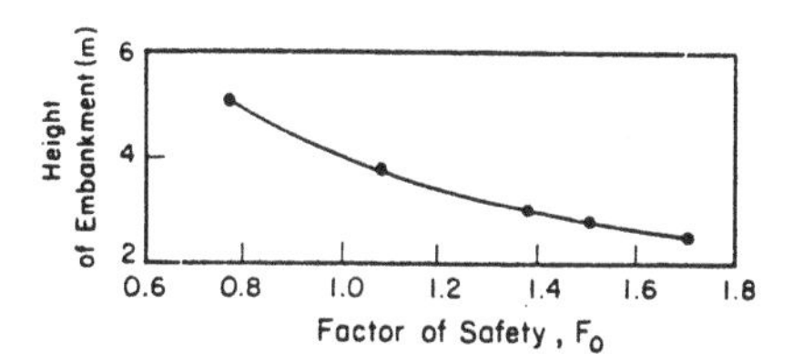

Fig. 24 Height of embankment vs safety factor.

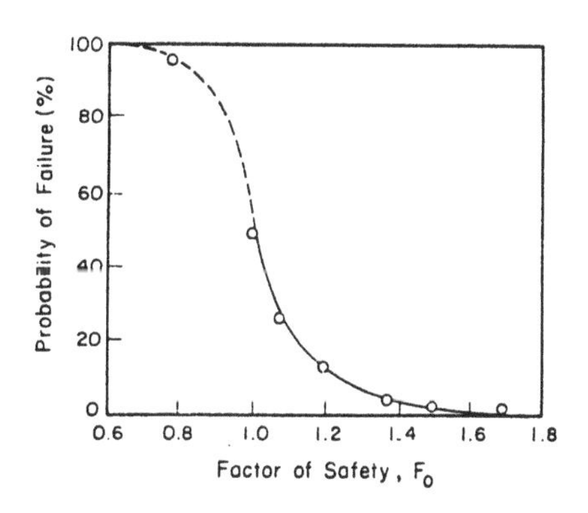

Fig. 25 Probability of failure vs safety factor.

foundations. In the stability analysis of full scale test embankments, the probability distribution of the factor of safety and the safety margins were obtained as a direct consequence of the uncertainty of the soil properties using the Monte Carlo simulation technique. The aim was to develop a functional relationship to describe the actual probability of failure in terms of the conventional safety factors for soft Bangkok Clay. These results are presented in Figs. 22 to 25. The probability of failure decreases asymptotically with the factor of safety. The probability of failure reached values less than 0.01 for a safety factor of 1.5. Because the soft Bangkok clay exhibits a slight overconsolidation, the e - log σ_v was defined to be made up of two linear portions corresponding to the over consolidated and the normally consolidated ranges. Histograms for the compression index C_c and C'_c in the normally consolidated as well as the overconsolidated states were plotted. The C_c was modelled as a normally distributed variable with a coefficient of variation (COV) of 0.18. The C'_c in the overconsolidated range has COV of 0.15. Probability distributions for consolidation settlements were constructed using simulation techniques. In the evaluation of immediate settlements, the probabilistic analysis was applied to an advanced analytical model, the finite element technique. Frequency distributions of immediate settlements were constructed. Using soil properties from probability distributions, the immediate settlements were evaluated using bilinear and nonlinear analysis resulting in a COV of 0.112 and 0.067, respectively.

5.2 *Prediction versus Performance*

Even though studies on the behavior of embankments on soft clay have now been made for

Table 1 Actual and Predicted Condition at Failure

Predictor	Fill Thickness (m)	Fill Height (m)	Slip Surface Depth (m)	Maximum Embankment Settlement (m)	Maximum Surface Heave (m)	Maximum Lateral Movement (m)	Excess Pore pressure in Piezometer P2 (m)
Bala	5.0	4.35	5.0	0.65	0.18	0.35	1.0
Nakase	3.5	3.20	4.6	0.30	0.25	0.50	• 6.2
Poulos	3.8	3.45	5.9	0.35	0.05	0.16	5.7
ACTUAL	5.4	4.70	8.2	0.70	0.15	0.37	9.3

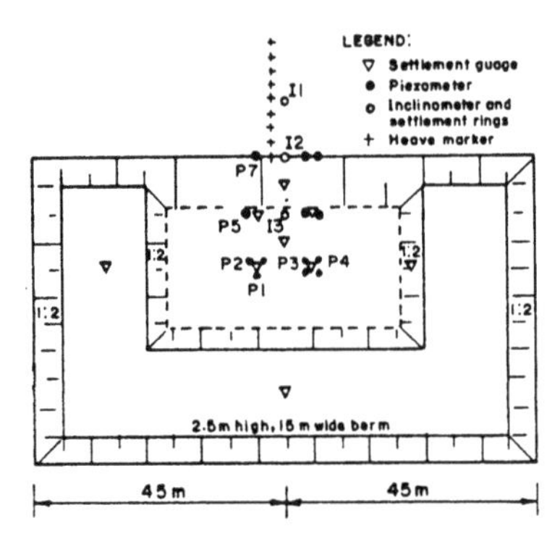

Fig. 26 Plan of Muar test embankment showing positions of key instruments.

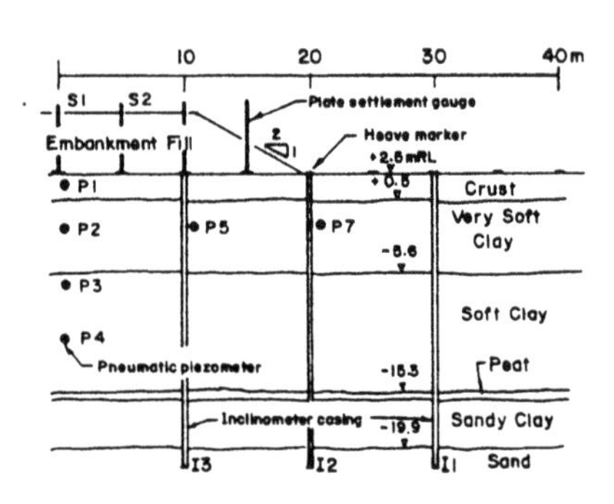

Fig. 27 Cross-section through center of Muar test embankment showing instrumentation.

more than twenty five years, the realistic predictions of the failure height, pore pressures and deformations are still a challenge. Some reasons for this include case studies from full scale tests in which all the parameters were well documented but are still lacking in Southeast Asia. Although there are excellent case records in Bangkok Clay and other deposits, the reliability of the field measurements, especially the pore pressures is sometimes questionable. Also, lateral movements are seldom measured and thus there is an element of doubt on the mode of settlements, whether it is the consolidation type or of the undrained yielding type. Further, the same degree of uncertainty arises as to whether a prolonged settlement with time is a consequence of secondary consolidation or due to undrained creep. In many instances, the embankment dimensions are much wider than the thickness of the compressible clay layer, and yet the settlements during the construction period indicate a very high proportion of immediate settlement under undrained conditions.

5.3 *Test Embankment in Muar Flat Site, Malaysia*

In a recent excellent prediction symposium held in Kuala Lumpur, organized by the Institution of Engineers, Malaysia, and the Malaysian Highway Authority, international experts from different locations made wide ranging predictions on the failure heights, the pore pressure developments and the settlement and the lateral movements of an embankment (Fig. 26 and 27) tested to failure. These predictions are summarised in Table 1 and Figs. 28 to 33. It is worthwhile that full scale instrumented field tests are carried out as far as practicable.

The related publications on the performance of embankments are given in the references. The Asian Institute of Technology has indeed extended their research work on Embankments, extensively for the Malaysian Soil conditions.

5.4 *Concluding Remarks on Embankments on Soft Clays*

(i) It is important that proper geotechnical parameters be evaluated from stress path dependent laboratory tests and the current computer software available be modified to take into account such path dependent material properties.

(ii) In the case of compacted residual soil fill material, the tensile mode of failure be further investigated.

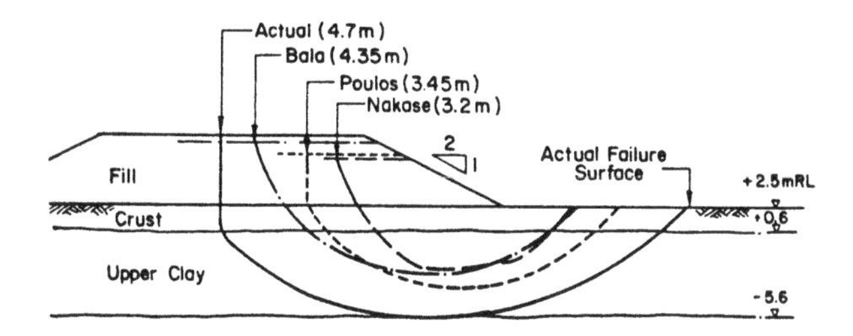

Fig. 28 Actual and predicted failure heights and failure surfaces for Muar test embankment.

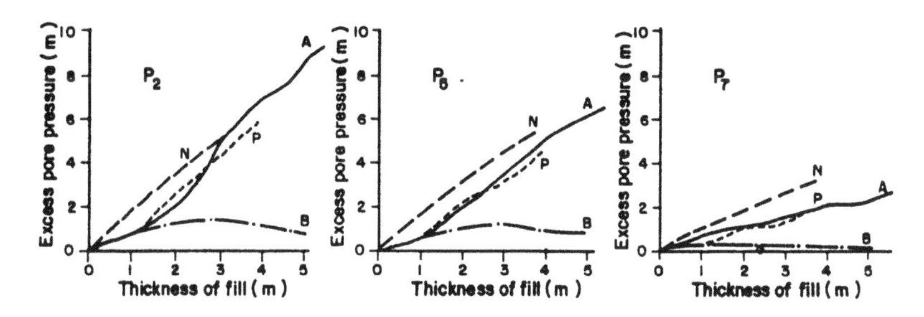

Fig. 29 Actual and predicted excess pore pressures at P2, P5 & P7.

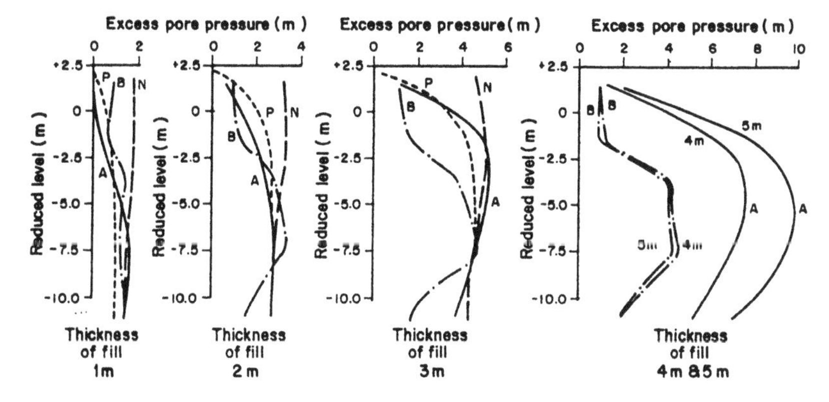

Fig. 30 Actual and predicted excess pore pressures profiles under embankment center.

6 DEEP FOUNDATIONS

Deep foundations particularly piled foundations has been a subject of research at the Asian Institute of Technology for a long time (KOBAKIWAL, 1978; LIM, 1978; GANESHANANTHAN, 1979; PHOTO-YANUVAT, 1979; KHOO, 1980; CHEN, 1981; CHUNG, 1983; KHATRI, 1983; CHIN, 1984; SELVANAYAGAM, 1984; TAN, 1988; BUENSUCESO, 1985; LIN, 1985; YU, 1987; HANIFAH, 1988 and CHIA , 1990). Piled foundation studies were conducted in Bangkok, Taipei, Hong Kong, Manila, Kuala Lumpur and Singapore. However, the most comprehensive study has been on the Bangkok Plain.

Driven piles of various sizes and depths are used extensively in the Bangkok Plain to support important structures (BALASUBRAMANIAM, PHOTO-YANUVAT, GANESHANANTHAN & LEE, 1981e). The design practice of piled foundation for these structures are largely empirical and load tests is always carried out in projects to confirm the design loads. Extensive research has been conducted at the Asian Institute of Technology to study the bearing capacity of driven piles where comparisons were made between the ultimate test load value and those computed by several methods. The study included the analysis of the bearing capacity of piled foundations using the total stress as well as the effective stress methods. In the total

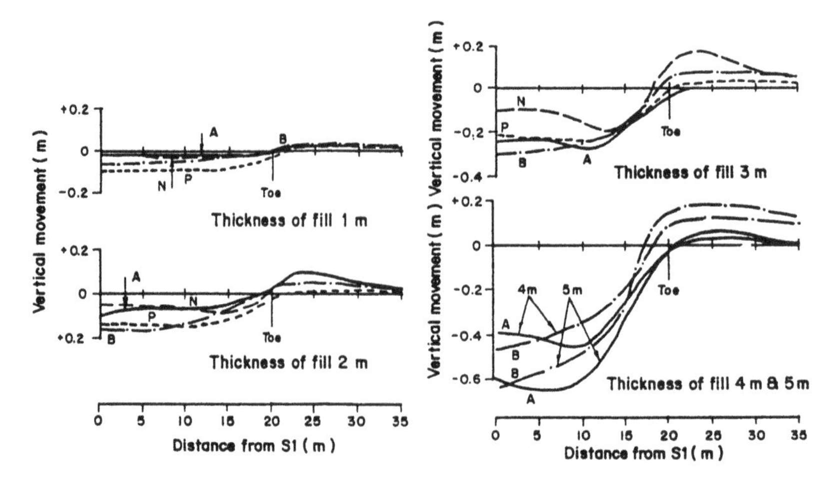

Fig. 31 Actual and predicted surface settlements.

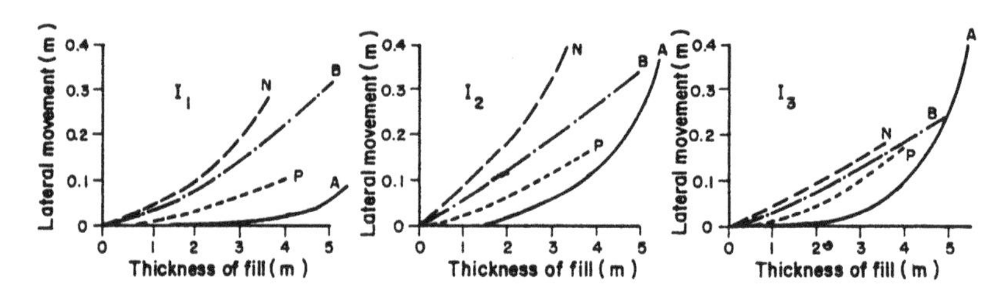

Fig. 32 Actual and predicted lateral displacements at I1, I2 & I3.

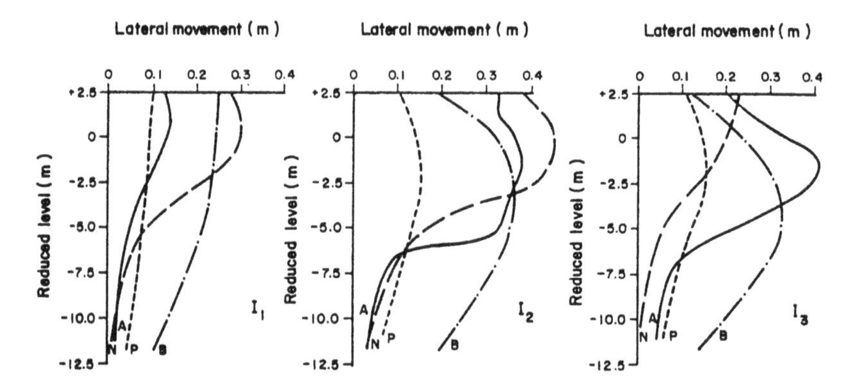

Fig. 33 Actual and predicted lateral displacement profiles at I1, I2 & I3.

stress method, undrained strength as obtained from field vane tests on soft and medium clay and UU tests on stiff clay are typically used. For piles extending into the sand layer, Standard Penetration Tests are used to evaluate the shaft friction and end bearing in the sand stratum. It is also common to carry out Dutch cone tests in the Bangkok subsoils to evaluate the bearing capacity of driven piles. Due to the lack of data on the effective stress strength parameters of the Bangkok subsoils, it has been somewhat difficult to carry out accurate analysis of the bearing capacity of pile foundations using the effective stress method. However, with the data acquired over the years, an attempt was made to employ the effective stress analysis to determine the bearing capacity of driven piles. Wave equation

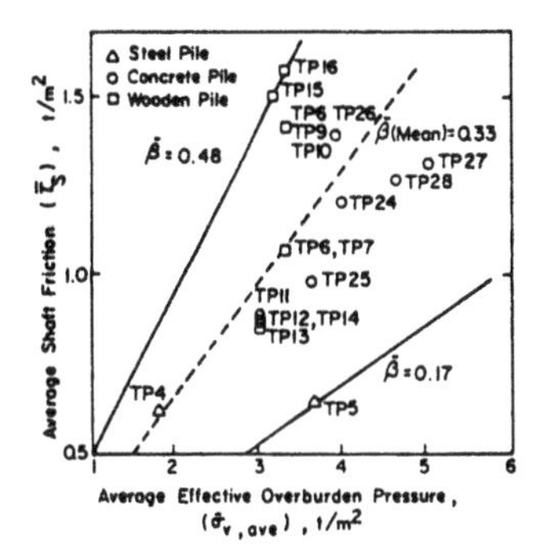

Fig. 34 Estimation of pile load test in Bangkok clay.

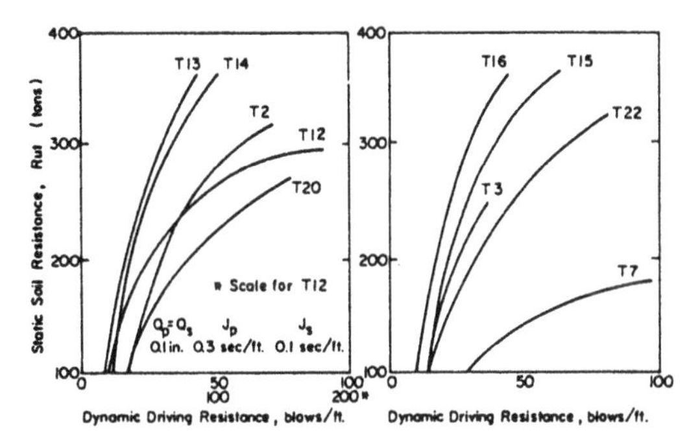

Fig. 35 Static soil resistance vs dynamic resistance for ten test piles embedded entirely in Bangkok clay.

analysis is now increasingly used in many countries. A parametric study using the wave equation analysis has also been made and it enabled the selection of appropriate soil parameters for the estimation of the bearing capacity of driven piles in Bangkok subsoils. Some of the results obtained from the effective stress analysis and the wave equation methods are shown in Figs. 34 to 37 and full details can be found in BALASUBRAMANIAM et al. (1981e).

Bored piles are now being used extensively for tall buildings and bridges in the Bangkok Plain. These piles often extend to the first or second sand layer. In doing so, the piles are constructed for a substantial length in the upper soft clay layer. The major construction problems relating to large diameter bored piles are casing, boring and concreting, borehole stability, cleaning of boreholes and vertical alignment. One of the major difficulties is the squeezing of soft clay into the pile shafts as the tubes are withdrawn forming a waisted or necked shaft. The soft clay near the surface is easily remoulded especially in the rainy season. Where ground water table is high, poor access conditions and poor working surface usually occur due to soft clay bedding. The soft clay tends to cave into the

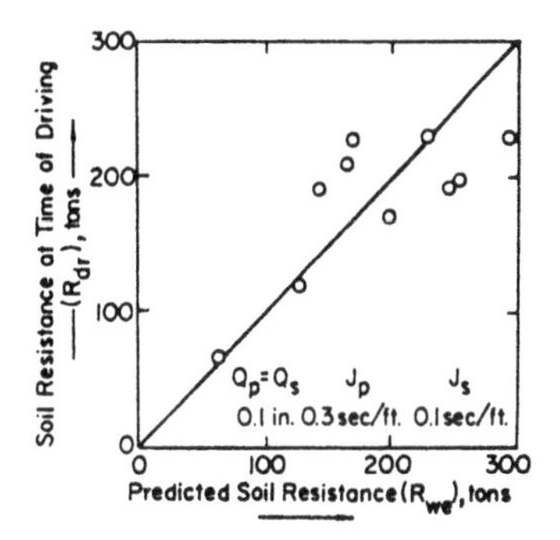

Fig. 36 Relationship between soil resistance and predicted soil resistance for piles in Bangkok subsoil.

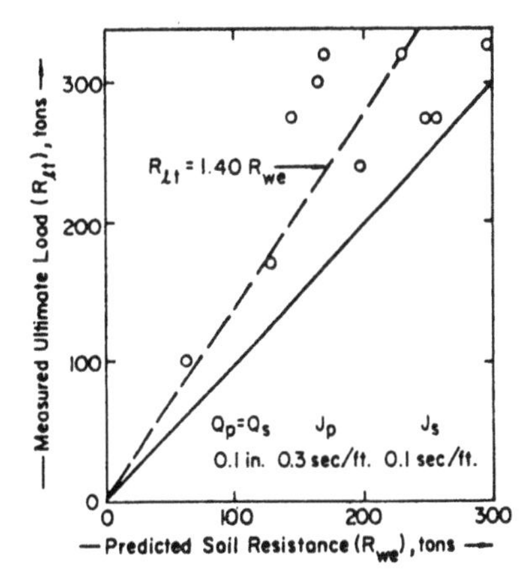

Fig. 37 Correlation between measured and ultimate load and predicted soil resistance.

borehole and casing is usually needed, often with bentonite slurry. Due to the compressibility of the soft clay, negative skin friction appears to be important in many of the bored piles. Normally the soft clay is highly plastic and low strength with organic silt and frequent pockets of sands. In addition, it consists of slickensides of fissures and is heterogeneous in nature. The problem of obtaining parameters for design purposes arises due to difficulties of estimating undrained shear strength both in the laboratory and in the field.

As Bored Piled Foundations are now used extensively in Bangkok for very tall buildings there is currently a continuous research programme at the Asian Institute of Technology (Fig. 38).

6.1 *Negative Skin Friction in Instrumented Piles*

Negative skin friction arises when the soil surrounding a pile settles faster than the pile. Such

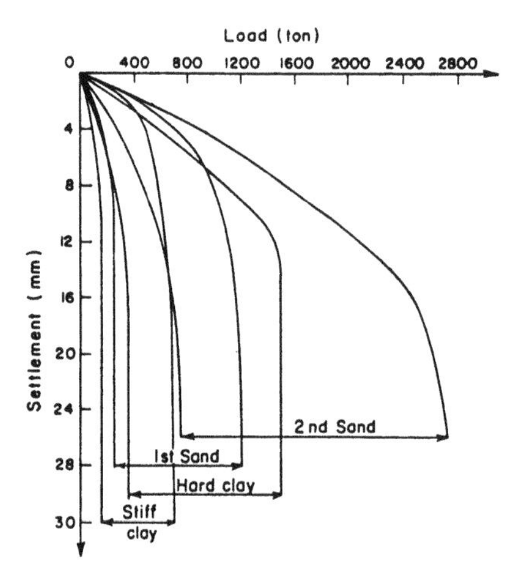

Fig. 38 Comparison of load-settlement curves with pile tip at different layers.

negative skin friction reduces the carrying capacity of a pile as well as increasing the compressive stress in the pile and the pile tip.

Driven piles are used as foundation elements for structural purposes in the Bangkok Plain. The soft Bangkok clay which ranges in thickness from about 10 to 15 m can be completely remoulded during pile driving and the subsequent consolidation of the remoulded clay can cause settlement, resulting in possible negative skin friction. Also, the extensive deep well pumping in the Bangkok Plain causes piezometric draw-down in the sub-soils which in turn brings about in the consolidation and compression of the clay and sand layers resulting in negative skin friction. Such negative skin friction can extend to great depths especially in bored piles which are found around 50 to 60 m.

In addition to the above two phenomena, most of the area in the Bangkok Plain is low lying and is close to a meter above mean sea level. Thus, these areas are heavily prone to floods and as such, most development projects for housing and industries require the ground to be filled with about 1.5 to 2 m of fill. This fill causes settlement in the upper soft clay and results in negative skin friction in piles.

6.1.1 *Pile and Field Instrumentation*

In a comprehensive study by PHAMVAN (1990), negative skin friction was stimulated around two driven piles by a fill of 2 m height. The two tension piles were instrumented and their performance under the embankment surcharge was recorded for a period of about 9 months. One of the two tension piles (pile T2) was coated with a bitumen slip coat layer, while pile T1 was left uncoated. The pore water pressure due to pile driving as well as under the embankment surcharge was also monitored in addition to the surface and sub-surface settlements. An extensive laboratory and field testing program was also carried out to evaluate the necessary geotechnical parameters to be used in the interpretation of the results.

The two tension test piles were hollow, prestressed, precast and spun concrete type; the outside and inside diameters are 0.4m and 0.25m, respectively. Each instrumented test pile was divided into six segments, five of which were 4 m in length. The last section was 6 m long. Load cells were placed at the pile tip as well as at the connection joint of each segment. A schematic diagram of the strain gauge type of load cell used is shown in Fig. 39. A tell-tale system was adapted for the measurement of the pile compression and its movements. A total of twelve tell- tale rods were used in each instrumented pile. They ran from the pile tip to the top of the pile at intervals of 4 m. During the pulling tests, LVDT gauges were used to measure the deformations and were directly connected to the computer-controlled data logger unit. The pile section modulus was determined from the load-deformation characteristics of two segments of piles each 1m long.

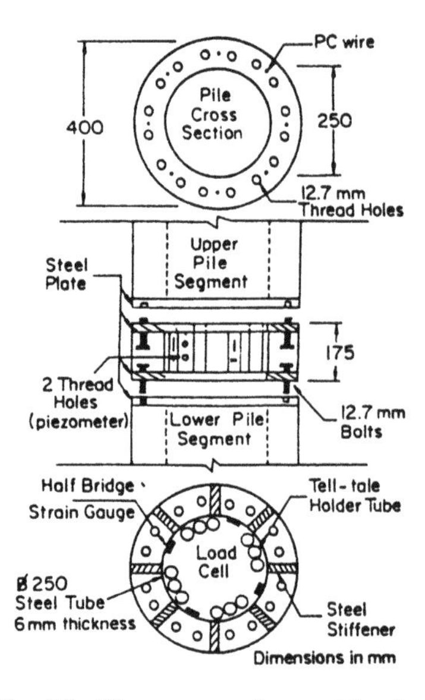

Fig. 39 Pile cross-section and load cell.

The variation in pore pressure on the surface of the pile was determined with a closed hydraulic system piezometer installed at each load cell, except at the pile tip. The active piezometers located around the two instrumented piles were of the closed hydraulic system type with mercury; they were laid 0.5m and 1m away from the center of the pile. The settlement at the site was monitored by simple surface settlement plates and screw type deep settlement points. The locations of the instrumentations are given in Fig. 40.

All the piles were driven by means of a standard drop hammer (6 tons). The two instrumented piles were driven in stages. The first two pile segments were connected on the ground to form a 8m long pile. The 8m segment was then lifted to stand over the designated position and was pushed slowly at an approximate rate of 1 m/minute by its own weight and the hammer weight. When the pile tip reached a depth of 8 m, the hammer was lifted out and another load cell was connected to the top of the pile. Tell- tale rods were lowered through the encased tubes to seat at the required depth. The pile was left for 8 or 9 days and then subjected to the pull-out test. When the first pull-out test was completed, tell-tale rods and encased tubes were extended. The next pile segment was then connected. Fine sand and glue were filled into the piezometer cell of the top load cell. The driving took place the next morning, after overnight curing of steel-filler epoxy and porous material glue. The driving was carried out continuously until the pile tip reached a depth of 12 m. This process was then repeated for the next two pile segments.

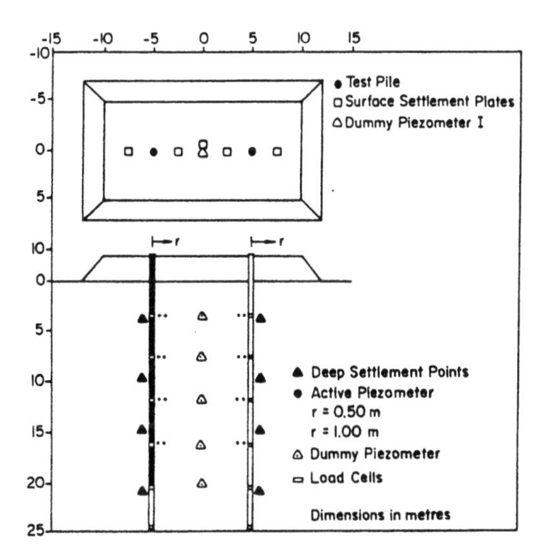

Fig. 40 Ground monitoring systems, instrumented piles and embankment surcharge load.

6.1.2 *Pile Pullout Tests*

Quick maintained load tests (QL) were adopted for the tension tests, in accordance with ASTM Standard D1143-81 and D3689-83. The final pile segment, which was the last 6m long segment, was connected to the existing pile after the fourth pull-out test had been completed. The pile was then driven to a depth of 25m, leaving about 2m of the pile length above the ground. A 24m x 14m embankment with side slope of 2:1 was then constructed 2m above the test area. The embankment provides a surcharge load which causes excessive settlement of ground resulting in negative skin friction in the test piles. Continuous observations of the loads and settlements were taken for nearly nine months.

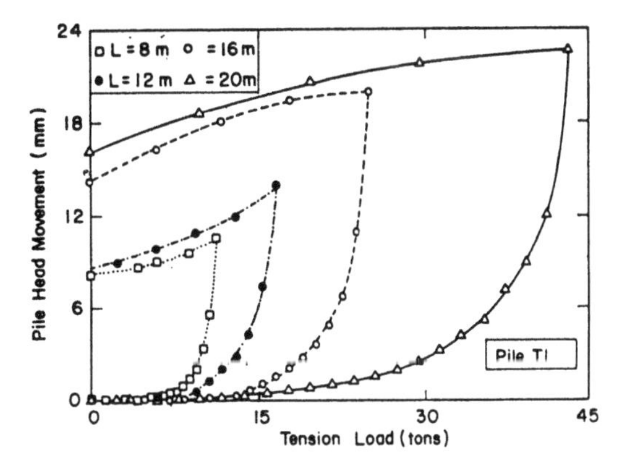

Fig. 41 Load-uplift curve of pile T1.

6.1.3 *Short Term Pullout Test Results*

The tension load-pile head uplift curves for pile T1 are shown in Fig. 41 while Fig. 42 shows similar load-pile movement characteristics for pile T2. Although the pile capacities were substantially smaller, the failure loads occurred at nearly the same magnitude of pile head movement as for pile T1.

The so-called "total stress" method of predicting the shaft capacity of driven piles in clay correlates the limiting skin friction, τ_s, with the undrained shear strength of the soil, s_u, the relationship being: $\tau_s = \alpha\ s_u$. The calculated values of α based on the short-term pull-out tests as carried out in this study are listed in PHAMVAN (1990). Figure 43 shows that at the time of each pull-out test, the excess pore water pressure developed at the pile surface due to pile driving has not completely dissipated.

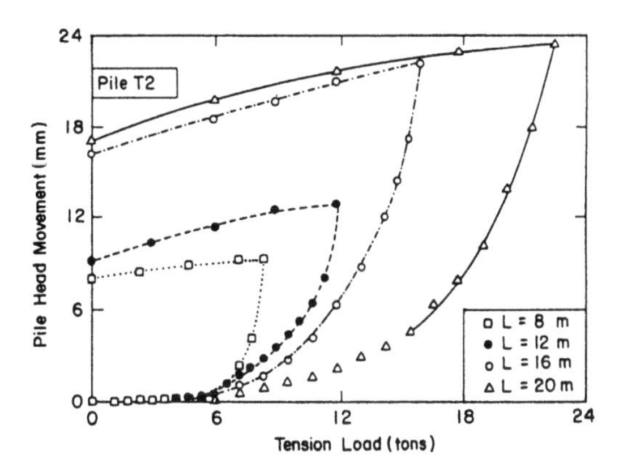

Fig. 42 Load-uplift of pile T2.

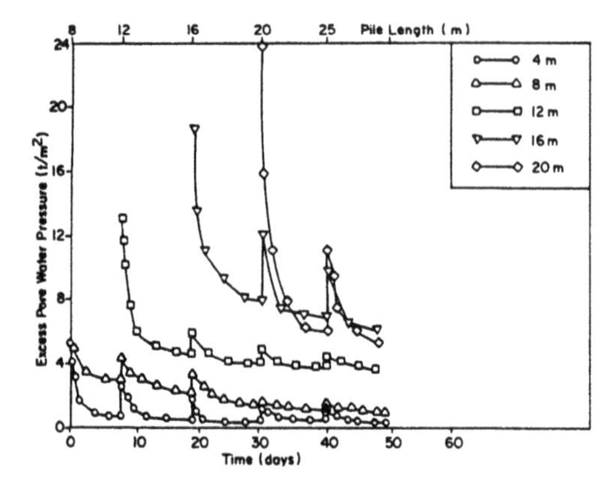

Fig. 43 Monitored time history of excess pore water pressure due to pile driving at pile surface (r = 0.20 m).

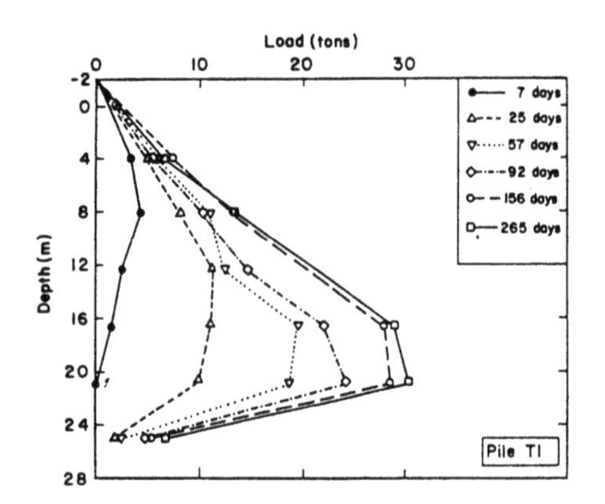

Fig. 44 Load distribution along pile shaft (pile T1).

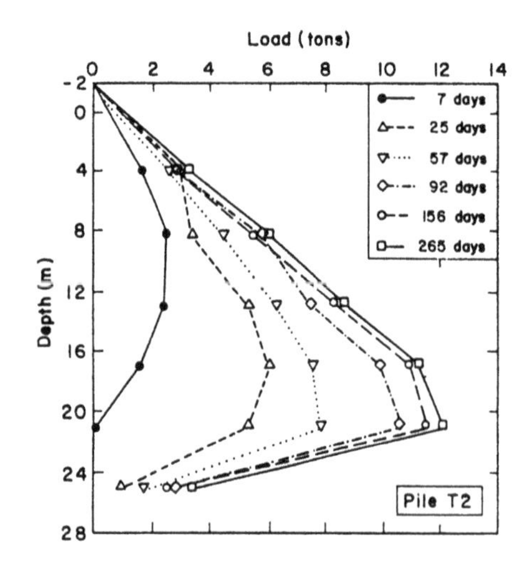

Fig. 45 Load distribution along pile shaft (pile T2).

Consequently, predictions made using the total stress approach must be treated with caution because values of τ_s are largely influenced by excess pore water pressure; in case the excess pore pressure has not completely dissipated, values of α will be considerably small.

6.1.4 *Long Term Measurements*

The axial load distribution along the length of piles T1 and T2 was monitored using load cells and tell-tale rods for a period of up to 9 months after the placement of the surcharge load. The changes in the load distribution curves with time are shown in Figs. 44 and 45.

The load distribution curves also show that the point at which the maximum axial load reading recorded changed with time. For both piles, the maximum load was observed at the -8.3m measurement point at 7 days, and moved downwards with an increase in time. This phenomenon can be explained by the distribution of the settlement along the pile length shown in Fig. 46. At the top 10m length of the pile, the relative movement (or settlement of the soil after deducting the pile settlement) was large enough to mobilise negative skin friction a few days after placement of the embankment. At depths greater than 10m, it was about 2 months before the required relative movement took place. This confirms that the development of negative skin friction (which provides the axial load on the pile) is dependent on the relative movement between the pile and the consolidating soil layer.

Long term analysis for negative skin friction can be made with the effective stress approach where the limiting frictional stress is considered in terms of the estimated effective stresses acting around the pile. The stress most easily determined is the effective overburden pressure, σ_{vo} and τ_s is expressed as $\tau_s = \beta\sigma_{vo}$ where β depends on the friction angle between the pile and the soil and the ratio of the horizontal to vertical stress. The calculated values of β within the soft clay layer, based on the results of the long-term monitoring were very stable and were between 0.15-0.2 for the uncoated pile and 0.05-0.1 for the coated pile.

6.1.5 *Estimation of Negative Skin Friction*

The pile load distribution curves from long term measurements can be used to obtain a profile with the depth of the skin friction at the soil-pile interface, as illustrated in Fig. 47. In this figure, the average undrained shear strength properties of the test site and the skin friction calculated from the short term pull-out tests are also shown. The negative skin friction developed on the uncoated pile is quite low in the weathered clay layer, about 1 t/m², and increases to about 1.5 t/m² in the soft clay layer. The adhesion factor was found to be close to unity in the middle portion of the soft clay zone but was only 0.3 in the weathered clay layer. On the other hand, the negative skin friction on the bitumen-coated pile was quite uniform for the whole subsoil profile; the average negative skin friction was 0.5 t/m²and the adhesion factor was 0.15 in the weathered clay and 0.35 in the soft clay. Comparing the skin friction for the coated and uncoated piles, it can be seen that the bitumen coating material was able to reduce the negative skin friction to about one third of the uncoated pile. A positive skin friction acting on the uncoated pile was found to be very close to the average undrained shear strength of the medium stiff clay. This implies that full mobilisation of the positive skin friction has taken place in the uncoated pile. For practical purposes, the estimation of negative skin friction based on the undrained shear strength might be applied as an upper limit for Bangkok sub-soils. It may also be noted that for the bitumen coated pile, the bottom 4 m segment was left uncoated, and in this uncoated section, an adhesion factor of 0.3 was obtained.

The comparison between the short term pull-out skin friction and the negative skin friction gave reasonable agreement in the soft clay region for the uncoated pile. The short term skin friction was higher in the weathered zone. The negative skin friction on the bitumen-coated pile was nominally smaller than the skin friction evaluated from the short term pull-out tests. Based on the results of this study, the negative skin friction on driven piles may be predicted with some degree of accuracy using results obtained from the short term pull-out test.

Negative skin friction could also be evaluated using the effective stress approach. The plot of the parameter along the soil profile is shown in Fig. 48. It is significant to note that both short term pull-out and long term measurements gave very close values of ß within the soft clay layer. The maximum negative skin friction that was developed is shown to be about $0.25\sigma_{vo}$ with an average value of $0.2\ \sigma_{vo}$ for the soft clay layer. The application of bitumen coating meanwhile, caused a reduction of negative skin friction to only $0.1\ \sigma_{vo}$. Again, it seems that the results of short term pull-out tests may be used to predict the long term behavior.

6.1.6 *Concluding Remarks on Negative Skin Friction*

The results of the investigations revealed that

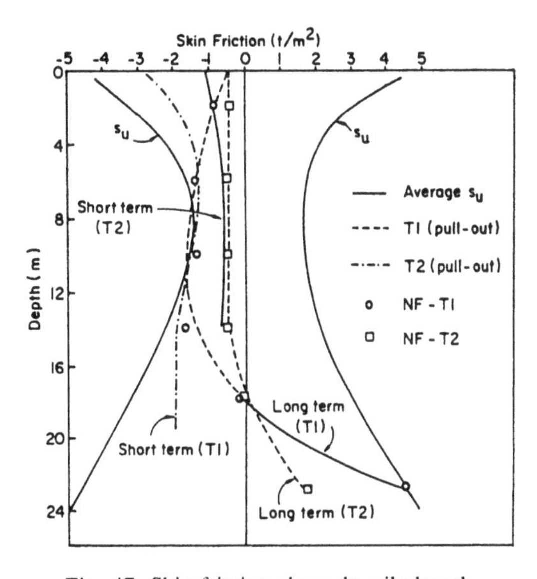

Fig. 47 Skin friction along the pile length.

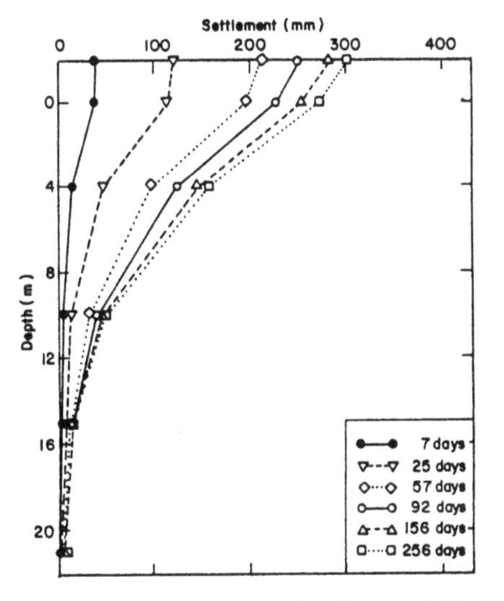

Fig. 46 Distribition of ground settlement (pile T1).

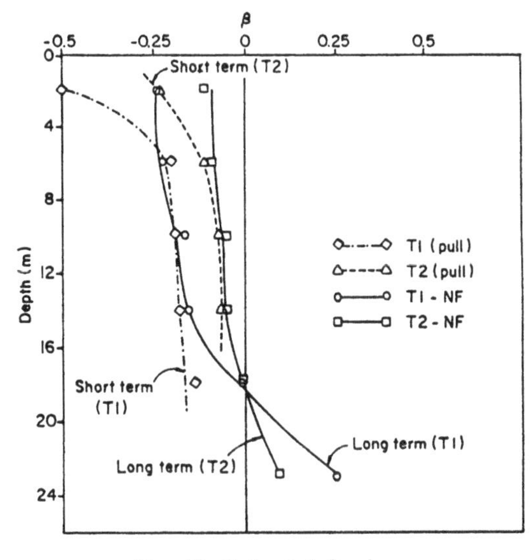

Fig. 48 Estimated β values.

predictions of the negative skin friction acting on driven piles in the Bangkok sub-soils in the long term condition may be made using parameters obtained from the results of short term pull-out tests. For both total and effective stress methods, the α and β parameters obtained from short term tests agreed closely with the results of long term measurements, specifically for the soft clay layer. This finding was found to be valid for both uncoated and bitumen coated piles.

The load distribution curves along the pile length were shown to increase with time. This behavior was found to correlate well with observations of the changes with time of the settlement-depth distribution curves for the consolidating clay layer. The maximum negative skin friction was found to have been mobilised about 3 and 6 months after the placement of the embankment load, for the coated and uncoated piles, respectively. The application of bitumen coating was also found to cause a reduction in the negative skin friction to about 40% of that acting on an uncoated pile.

Although pore pressure measurements have been reported to have been successfully used to predict the development with time of negative skin friction along the pile shaft, the results of this study suggest that a significant scatter in pore pressure- time history data exists due to the variation of pore pressure readings with piezometer location and time. Furthermore, the presence of undissipated pore pressures due to pile driving presented some complications in the monitoring of the actual pore pressure within the consolidating clay layer. Hence, when using pore pressure measurements in calculating the effective stresses within the consolidating clay strata, some cautions have to be exercised. This is particularly relevant when applying the Finite Element Method to study the development of negative skin friction on driven piles.

6.2 *Concluding Remarks on Deep Foundation in Bangkok Subsoils*

The Standard Penetration Test may be supplemented with other sophisticated in-situ tests for the estimation of in-situ soil properties; namely the pressuremeter, the dilatometer and piezo-cone. Also, instrumented piles may be tested for long term load transfer estimation and the effect of negative skin fiction which arises from a number of sources. Non-destructive pile load tests of the ultra sonic type and the vibration measurement type (PDA) should be used more in practice.

7 GROUND IMPROVEMENT TECHNIQUES

Research work on Ground Improvement Techniques began at the Asian Institute of Technology as early as 1975 in collaboration with the Royal Thai Navy Dockyard in Bangkok (OOI, 1978; MALLAWAARATCHY, 1979; GHO, 1979; HOSSAIN (1981); KOO, 1984; SINGH, 1985; IBRAHIM, 1988; RAMLE, 1988; YONG, 1988; YANG, 1988; LAW, 1989 and KHAN, 1990). An excellent International Symposium held on Ground Improvement techniques in 1982 at the Asian Institute of Technology had a substantial impact on the use of these techniques in Southeast Asia and elsewhere (BALASUBRAMANIAM, 1982c; and BALASUBRAMANIAM, CHANDRA & BERGADO, 1984b).

The various types of ground improvement techniques which have been commonly adopted in many countries for more than two decades have not yet found their way into actual construction activities in the Bangkok Plain. The practitioners are too fond of using driven piles for deep foundations and are only reluctantly giving into the introduction of bored piles. Even other modes of patent piles are less favoured in the construction industry.

7.1 *Performance of Sand Drains in Bangkok Test Embankments*

Pilot studies on the performance of sand drains, sandwicks and vacuum drainage have been carried out. The performance of sandwicks in accelerating the consolidation of soft Bangkok clay was studied in a full scale test embankment at Pom Prachul Dockyard (BALASUBRAMANIAM, BRENNER, MALLAWARATCHY and KUVIJITJARU, 1980b). The embankment was 90 m long, 33 m wide, divided into three sections, namely, a section with no drain, a section with drains of 2.5 m spacing, and a section with drains of 1.5 m spacing. The plan and elevation of this test embankment are shown in Fig. 49. The sand drains consisted of small diameter (5 cm) sandwicks and were installed by the displacement method. The finished sandwicks extended to a depth of 17 m below the ground surface. The embankment was built in two stages, firstly to a height of 1.8 m, and secondly to final height of 2.35 m. The settlement time records obtained from this study are presented in Fig. 50 and indicate that sandwicks are not substantially effective in accelerating consolidation settlements. A recent study conducted by MOH et al. (1987) indicated that vacuum drainage is effective provided the drains are not extended below the depth from which the natural piezometric drawdown occurs due to deep well pumping.

7.1.1 *Lime and Cement Column Techniques*

Among other ground improvement techniques, especially for road embankments and bridge approach roads, the lime column technique seems to possess several advantages and merits over other methods such as dynamic consolidation, granular columns and deep chemical mixing.

A comprehensive research program on lime and cement column techniques is currently underway at the Asian Institute of Technology and the results being interpreted within the Critical State Soil

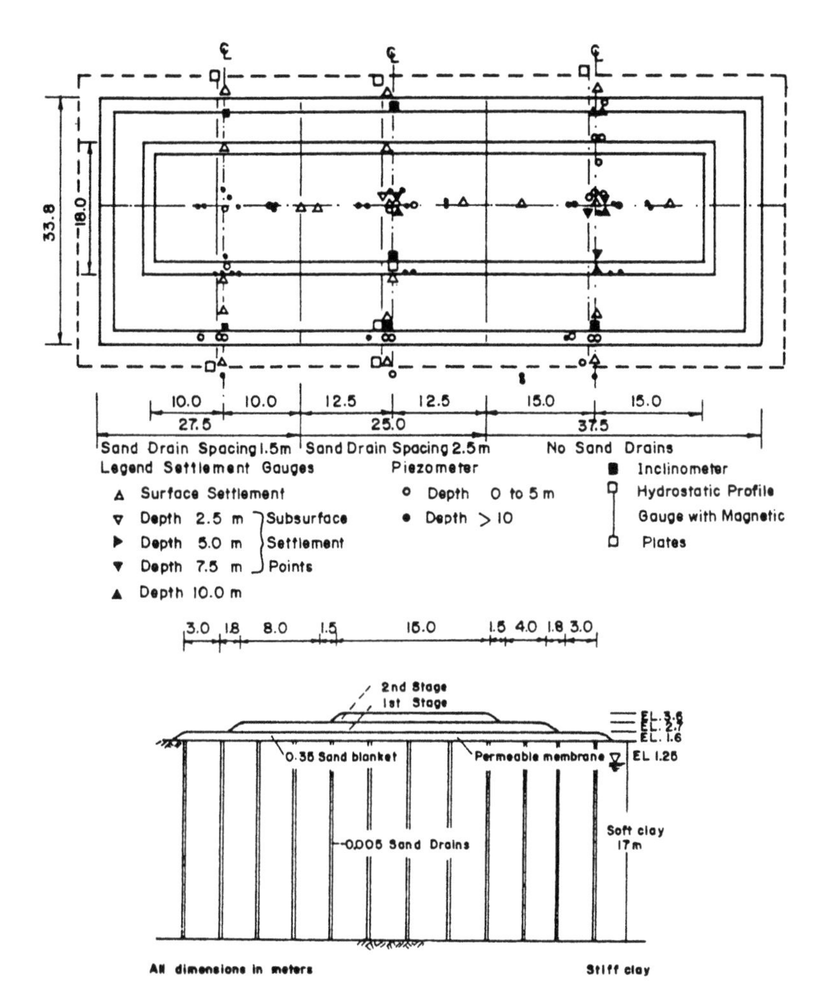

Fig. 49 Plan and elevation of Pom Prachul test embankment.

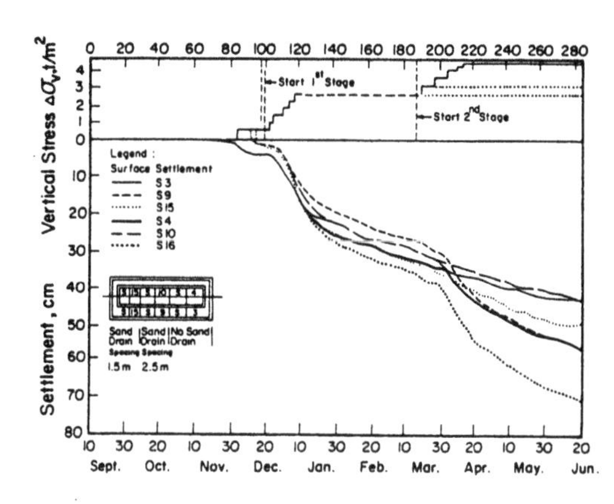

Fig. 50 Typical settlement point readings at Pom Prachul site.

Mechanics framework (BUENSUCESO, 1990). Excellent research work on the concepts and applications of reinforced earthworks are now in progress by BERGADO and his colleagues.

The work on Lime Treated Clay was undertaken by BUENSUCESO (1990) for a proper understanding on the fundamental behavior under Oedometer type of Stress conditions as well as in triaxial shear. The data were to be used both in shallow and deep stabilisations. BUENSUCESO (1990) carried out a comprehensive series of oedometer tests to study the $e - \ln \sigma_v$ relations for lime treated clay under various percentages of lime treatment. He also conducted triaxial tests with the undrained, fully drained and anisotropic consolidation stress conditions. The undrained stress paths at two pre-shear consolidation stresses of 5 t/m² and 60 t/m² are shown in Figs. 51 (a) and (b). At low pre-shear consolidation pressure, even 5

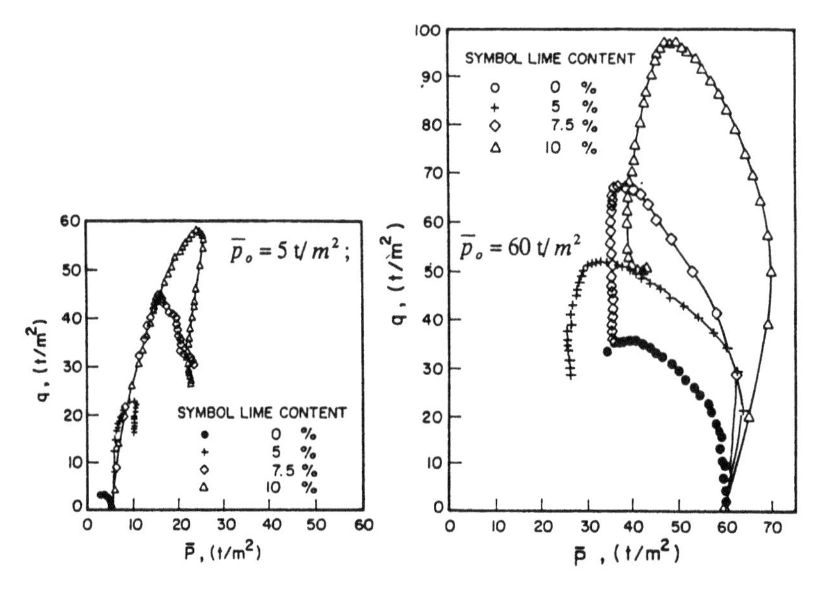

Fig. 51 Undrained stress paths (2 months curing time).

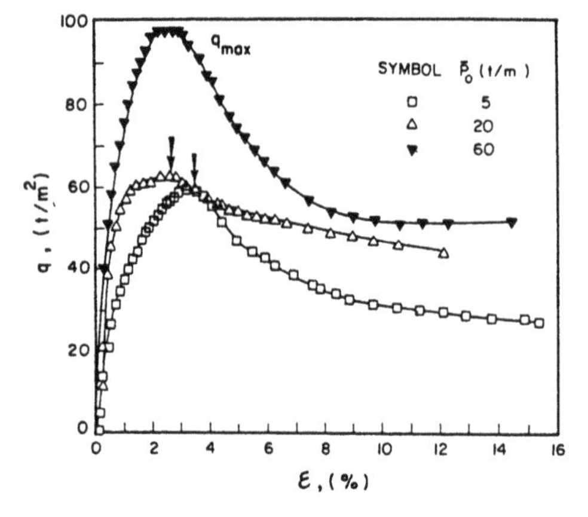

Fig. 52a (q,ε) plot for lime treated clay (10% lime content, 2 months curing time).

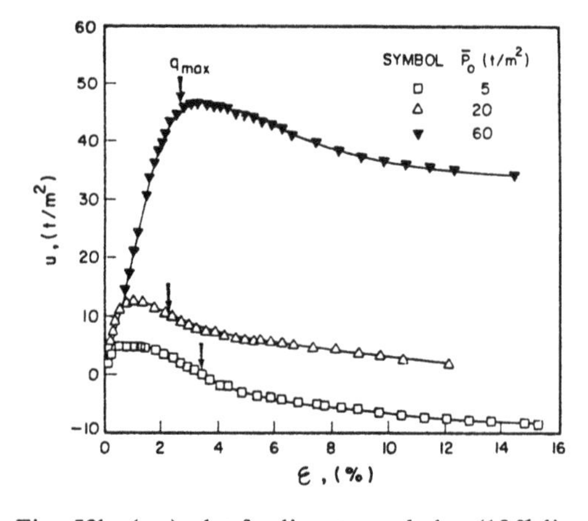

Fig. 52b (u,ε) plot for lime treated clay (10% lime content, 2 months curing time).

percent lime addition is sufficient to convert the undrained stress path from a normally consolidated type to a heavily overconsolidated type. With 10 percent lime and two months curing period, a ten fold increase in the strength is noted. For specimens sheared from 60 t/m² pre-shear consolidation pressure, the deviator stress at failure reached as high as 100 t/m². Typical (q, ε) and (u, ε) relations under undrained conditions are shown in Figs. 52 (a) & (b), respectively. Similarly, the (q, v) and (v, ε) relations during triaxial tests are shown in Figs. 53 (a) and (b). The undrained stress paths of lime treated clay are presented in Figs. 54 (a) and (b). In these figures, it can be clearly seen that with the optimum lime content and the proper curing period, the undrained stress paths are virtually vertical in the (q, p) plot. This remains true even beyond the critical state line of the untreated clay due to the development of small pore pressure when compared to corresponding specimens sheared under untreated conditions. The peak strength envelope also reaches a limiting upper bound. However, with the large strains, the lime treated clay showed substantial reduction in the strength from the peak value and approaching lower values at the critical state (Fig. 54). These results are very promising in having a proper understanding on the fundamental behavior of shallow and deep treated soils with lime and cement additives.

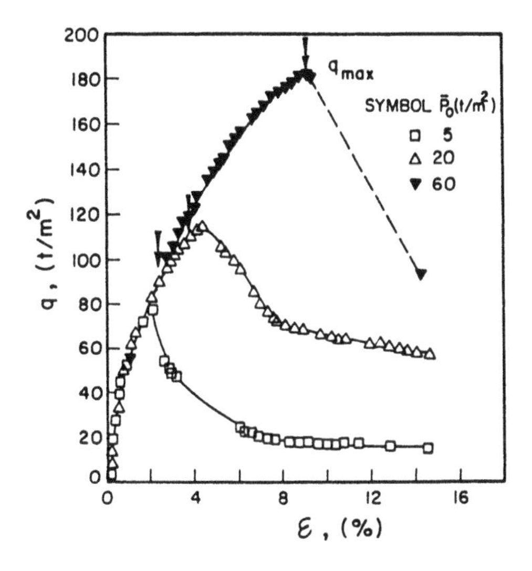

Fig. 53a (q-ε) plot for lime treated clays from CID tests (10% lime content; 2 months curing).

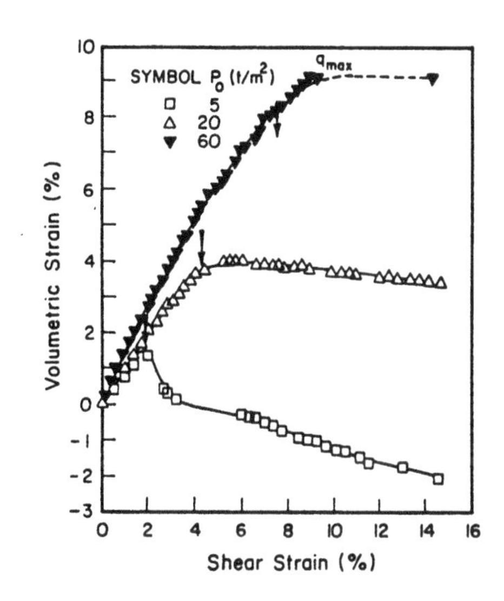

Fig. 53b (v-ε) plot for lime treated clays from CID tests (10% lime content; 2 months curing).

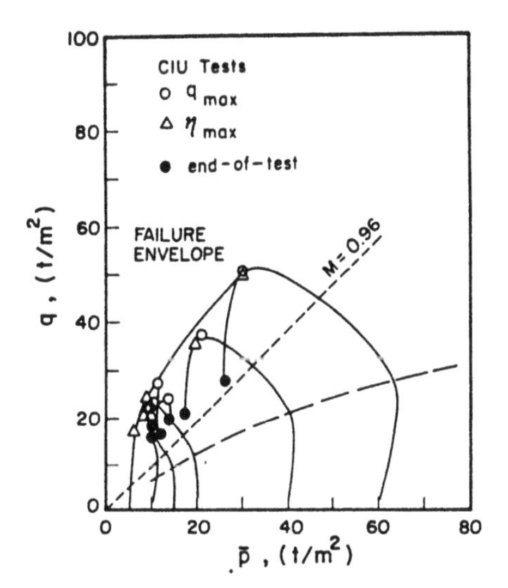

Fig. 54a Effective stress paths for lime treated clay (5% lime content; 2 months curing).

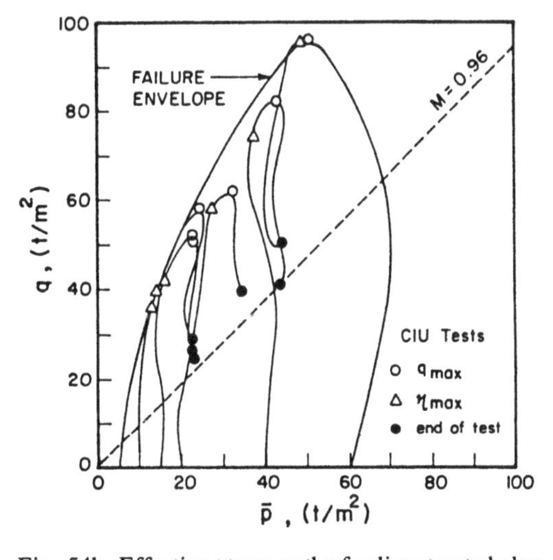

Fig. 54b Effective stress paths for lime treated clay (10% lime content; 2 months curing).

7.2 *Concluding Remarks on Ground Improvement*

In the next decade extensive use of current ground improvement techniques which include the use of deep chemical mixing, lime and cement column and reinforcements such as the polymer geogrids should be encouraged instead of the traditional piling methods.

8 SHALLOW AND DEEP EXCAVATIONS

Excavations in soft clay deposits are frequently made for a number of purposes namely, basements for tall buildings, canals for water supply and drainage, dry docks and for pipe-laying in the transmission of natural gas, and domestic and industrial use of water. Comprehensive research activities on shallow and deep excavations were conducted at AIT by SIVANDRAN, 1976; HO, 1976b; SIRIWARDANE, 1976; LEE, 1978; ANANDA, 1979; CHIA, 1986; SAEED, 1979. Deep excavations are often either supported with struts and bracings, or unsupported with the use of bentonite slurry or jet grouted piles and cement piles. The two excavations which were classical in

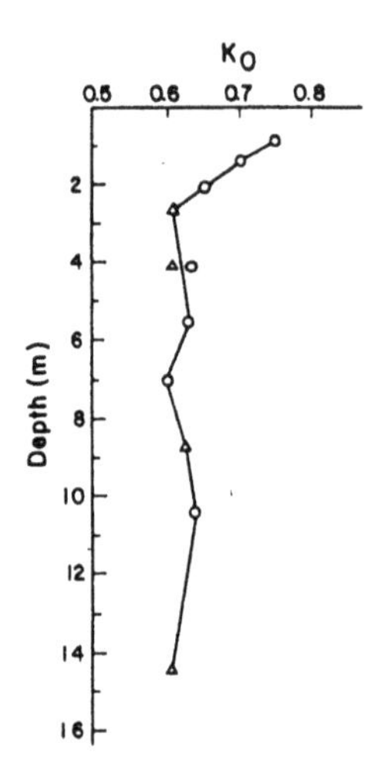

Fig. 55 Variation of K_o with depth for Nong Ngoo Hao site.

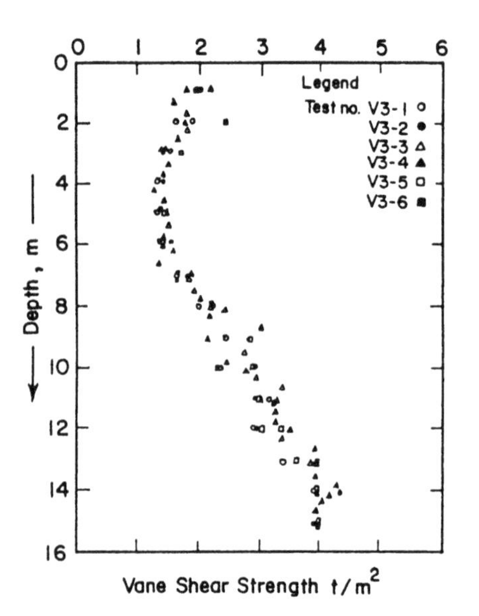

Fig. 56 Vane strength profile (Nong Ngoo Hao).

nature are: (i) the fully documented full scale field excavation carried out at the Nong Ngoo Hao site in Bangkok; and (ii) the techniques adopted for the construction of the gateway, the open basin and the dry docks at the RTN Dockyard site in Pom Prachul, Thailand.

The laboratory tests carried out for each excavation works are isotropically and K_o-consolidated triaxial compression and extension tests under in-situ type of stress. These triaxial tests are of the unloading types wherein the axial and lateral stresses are maintained

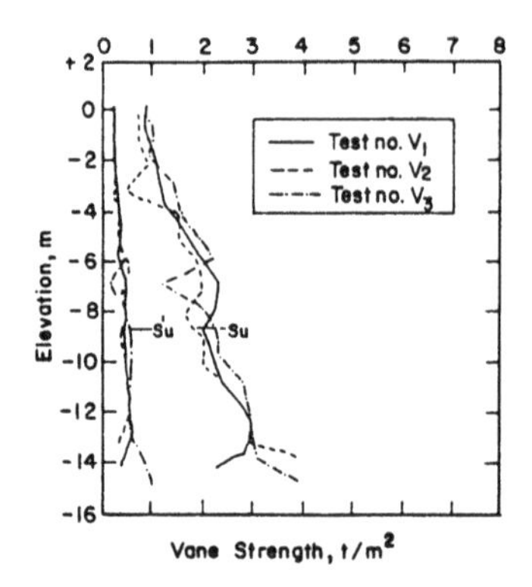

Fig. 57 Vane strength profile (Pom Prachul).

constant or lowered to simulate unloading compression and extension conditions (WAHEED, 1976; LI, 1976; AHMED, 1976; HO, 1976b; KIM, 1991). The stability conditions of excavations are such that the negative pore pressure developed results in higher safety factors just after the excavation is completed and the safety factors are critical under long term conditions, when the negative pore pressures gradually rise to neutral conditions. Both small scale field tests and large scale instrumented field tests are common to evaluate the necessary geotechnical parameters in the safe design of these excavations. The vane tests are the commonly carried out field tests on a small scale while fully instrumented large scale excavations are also conducted in a limited number. Both total and effective stress stability analysis are carried out using the vane strength and the triaxial strength parameters. Numerical analysis involving the finite element method needs to be employed to accurately estimate the deformations of excavations under short term and long term conditions. Both deterministic and probabilistic analysis were carried out to evaluate the safety and the deformations of such excavations.

The coefficient of earth pressure at rest as obtained from the triaxial laboratory tests under zero lateral strain conditions are shown in Fig. 55 for the Nong Ngoo Hao clay. The only field measurements available for Bangkok clay are from hydraulic fracturing tests done with the piezometers. The K_o values obtained from the hydraulic fracture tests are comparable to the values obtained from laboratory measurements. In fact, in-situ K_o measurements could have been determined more accurately using earth pressure cells or the self boring pressuremeter. But such equipment are not yet available in Bangkok.

Figures 56 and 57 illustrate the vane strength profile at the test sites; namely Nong Ngoo Hao and Pom Prachul. Tables 2,3,4,5 and 6 include the strength parameters c and ϕ, and the undrained strength values for the Nong Ngoo Hao site.

Table 2 Consolidated Undrained Triaxial Tests Strength Parameters in the Overconsolidated Range (Nong Ngoo Hao Clay).

Depth (m)	Shear Strength Parameters		Test Conditions	Stress Conditions
	$\bar{c}$(kN/m^2)	$\bar{\phi}$(deg)		
	11.7	19.0	Strain Controlled	CAU
1.1-1.3	11.5	16.7	Stress Controlled	Compression
	10.1	13.1	Strain Controlled	CIU
	11.4	15.5	Stress Controlled	Compression
	11.7	20.1	Strain Controlled	CAU
	13.1	16.8	Stress Controlled	Compression
	11.7	19.0	Strain Controlled	CIU
2.5-3.0	13.1	17.1	Stress Controlled	Compression
	0	51.4	Strain Controlled	CIU Extention σ_1 decreasing
	0	62.1	Strain Controlled	CIU Extention σ_3 increasing
	0	68.8	Strain Controlled	CID Extension σ_1 decreasing
4.0-4.5	8.8	15.3	Stress Controlled	CAU Compression σ_3 decreasing
	3.9	16.1	Strain Controlled	CIU Compression
	16.8	18.6	Strain Controlled	CAU
5.5-6.0	8.9	21.3	Stress Controlled	Compression
	9.2	13.8	Stress Controlled	
	9.9	10.1	Strain Controlled	CIU Compression
7.0-7.5	13.3	16.6	Strain Controlled	CAU Compression

Table 3 Summary of Unconsolidated Undrained Triaxial Unloading Tests (σ_1 decreasing test on 76mm ϕ specimen)

Depth (m)	Test No.	Water Content (%)	Cell Pressure (kN/m^2)	s_u (kN/m^2)
1.6	UU1-1	125	24.0	16.5
2.8	UU1-2	120	42.7	10.5
4.2	UU1-3	120	65.7	9.6
5.8	UU1-4	126	88.3	10.6

Table 4 Summary of Unconsolidated Undrained Triaxial Unloading Tests (σ_1 decreasing test on 35mm ϕ specimen)

Depth (m)	Test No.	Water Content (%)	Cell Pressure (kN/m^2)	s_u (kN/m^2)
1.4	UU2-1	127.1	24.0	12.8
2.1	UU2-2	128.3	33.3	18.1
2.8	UU2-3	120.0	42.7	11.1
4.7	UU2-4	124.6	65.7	11.4
5.8	UU2-5	120.0	88.3	11.8

8.1 *Trial Excavation at Nong Ngoo Hao*

The trial excavation was 80 m long with a slope of 1:2.5 (Fig. 58). It took 30 days to acquire the required dimensions. Excavation was then stopped at a depth of 4 m to observe its performance in the long term. The excavation did not fail after construction to a depth of 4 m, showing that the factor of safety at that time was greater than unity. However it was anticipated that the excavation would fail due to the dissipation of negative pore pressure set up due to stress relief. In view of the anticipated failure, the excavation was kept empty of rain water for a few weeks following the end of

Table 5 Summary of Unconsolidated Undrained Triaxial Unloading Tests
(σ_3 decreasing test on 76mm ϕ specimen)

Depth (m)	Test No.	Water Content (%)	Cell Pressure (kN/m^2)	s_u (kN/m^2)
1.4	UU3-1	130.0	24.0	15.8
3.2	UU3-2	120.8	42.7	15.5
4.3	UU3-3	125.2	65.7	14.1
5.8	UU3-4	125.3	88.3	14.7

Table 6 Summary of Unconsolidated Undrained Triaxial Unloading Tests
(σ_3 decreasing test on 35mm ϕ specimen)

Depth (m)	Test No.	Water Content (%)	Cell Pressure (kN/m^2)	s_u (kN/m^2)
1.4	UU4-1	120.0	24.0	16.7
2.0	UU4-2	129.0	33.3	25.2
2.8	UU4-3	121.0	42.7	18.6
4.7	UU4-4	119.3	65.7	14.7
5.8	UU4-5	121.0	88.3	17.1

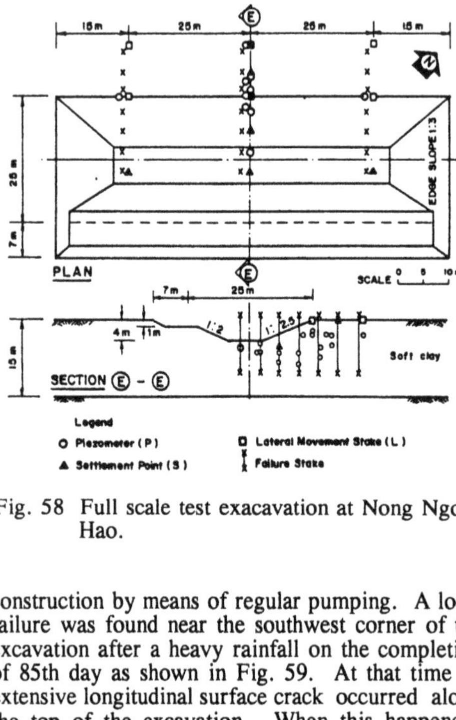

Fig. 58 Full scale test exacavation at Nong Ngoo Hao.

construction by means of regular pumping. A local failure was found near the southwest corner of the excavation after a heavy rainfall on the completion of 85th day as shown in Fig. 59. At that time an extensive longitudinal surface crack occurred along the top of the excavation. When this happened, there were only 2.2 m of water in the bottom of the excavation, which probably prevented the failure of the excavation. On the 93rd day, the pumping operation was improved and water level in the excavation began to fall rapidly. After continuous pumping for more than 24 hours, the depth of water was brought down to 0.7 m when the excavation failed on the morning of the 94th day.

Only five failure stakes were recovered along the failure surface but the co-ordinates of failure surface obtained from these stakes could not make a realistic slip surface. Thus vane shear tests were carried out at three different locations along the length of the excavation to determine the depth of the remoulded zone and thereby obtain the shape and position of the failure surface. The vane tests indicate that the failure slip was approximately circular. Furthermore, at two locations the failure surfaces were almost coincident (Fig. 60), but at the third location, a quite different slip surface was obtained. Theoretical analysis showed that the slip surfaces obtained from the two locations were close to the computed critical failure surface. This slip surface was therefore adopted as an actual failure circle in the analysis.

8.1.1 *Strength Characteristics*

Six vane tests were carried out at every meter depth and the results are summarized in Fig. 56. The results of cone resistance and local friction are presented in Fig. 61. The shear strength from direct shear tests, unconfined compression tests and also the results of CIU and CK_oU tests are presented in Fig. 62.

8.1.2 *Total Stress Stability Analysis*

The total stress analysis is used to find the factor of safety at the end of construction. The shear strength introduced in the stability analysis is thus the in-situ undrained shear strength of the clay which existed prior to construction. For excavations, the shear strength decreases with time due to the dissipation of the negative pore pressure so that the total stress analysis at the end of construction will not be the critical case. However, the total stress analysis is relatively easier to perform than the effective stress analysis both in experimental investigation and analysis, and the factor of safety obtained can be an indicative reference to the long term safety factor. The end-of- construction safety factors as determined by Swedish circle method with undrained strengths are shown in Table 7.

The tension crack actually occurred at failure and the effect of the tension crack as shown in Table 8 is to reduce the safety factor by about 8% when compared with the corresponding safety factor without the effect of the tension crack. When the tension crack was introduced, the average vane strength yielded a safety factor of 1.59 as against

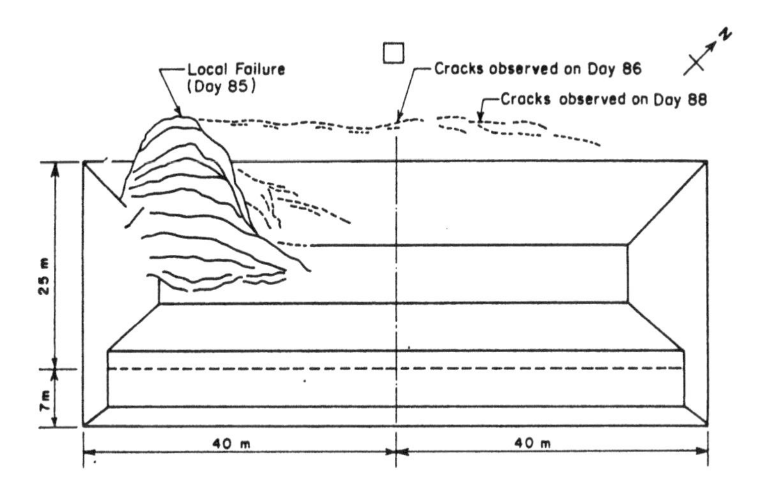

Fig. 59 Plan of visible cracking prior to failure.

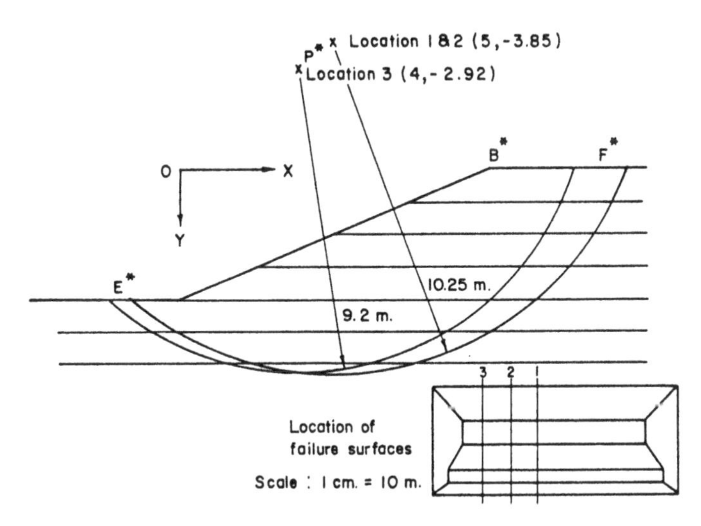

Fig. 60 Actual failure surface as confirmed by vane test.

1.74. Similarly, the safety factors obtained from the direct shear tests are 1.68 and 1.73 with and without the tension crack, respectively. The tension crack also alters the position of the critical failure surface. The effect of water pressure on the tension crack is shown in Table 8. The average vane strength yielded safety factors of 1.41 and 1.74 with and without hydrostatic water pressure, respectively, giving an error of 14% in the safety factor which is the maximum error when compared with the case without the tension crack. The critical surface computed with water pressure acting in the tension crack is also closer to the actual failure surface. The end-of-construction safety factors are also summarized in Table 8, which shows the results from 11 types of undrained strength values obtained from different tests. Safety factor values obtained using eight types of unconsolidated undrained strengths are given. For the direct shear strengths, average vane strength and the strength from UU tests with σ_3 decreasing, the safety factors are 1.59, 1.41 and 1.46, respectively. These safety factors will approach unity when BJERRUM's correction factor is used. The CK_0U strength gave too high a safety factor.

8.1.3 *Effective Stress Stability Analysis*

Strength parameters measured by CIU, CK_0U, CK_0D (strain controlled) and CK_0U stress controlled tests are used in the analysis with both measured and predicted pore pressures. The selected values of the pore pressure parameter A are given in Table 9.

Table 7 End-of-Construction Safety Factors without Tension Cracks (Total Stress Analysis)

Shear Strength	Safety Factor by Swedish Circle Method
Average Vane strength	1.72
Direct Shear Strength	1.73
Unconfined Shear Strength	1.13
Unconsolidated Undrained Strength	1.33

Table 8 End-of-Construction Safety Factors by Swedish Circle Method (Total Stress Analysis)

Strength	Factor of Safety					
	Without tension cracks		With tension cracks		With water pressure on tension Cracks	
	Actual Slip Surface	Critical Failure Circles	Actual Slip Surface	Critical Failure Circles	Actual Slip Surface	Critical Failure Circles
Vane Strength Average	1.85	1.74	1.76	1.59	1.54	1.41
Direct Shear Strength	1.84	1.73	1.81	1.68	1.73	1.59
UU-1 σ_1 decreasing 76mm ϕ sample	1.09	1.03	1.04	0.98	1.01	0.95
UU-2 σ_1 decreasing 35mm ϕ sample	1.29	1.15	1.24	1.11	1.21	1.08
UU-3 σ_3 decreasing 76mm ϕ sample	1.73	1.63	1.68	1.55	1.60	1.46
UU-4 σ_3 decreasing 35mm ϕ sample	1.84	1.69	1.81	1.63	1.73	1.55
UU-5 Compression 76mm ϕ sample	1.44	1.31	1.38	1.24	1.33	1.19
UU-6 Compression 35mm ϕ sample	1.46	1.33	1.41	1.28	1.36	1.23
UU-7** unloading 76mm ϕ sample	1.24	1.09	1.16	0.99	1.10	0.95
UU-8** unloading 35mm ϕ sample	1.36	1.20	1.30	1.14	1.25	1.11
CK_oU Strength	2.28	2.10	2.27	2.07	2.06	1.83

Table 9 Selected Values of pore pressure Parameter A

Depth (m)	A
0.0 - 2.0	0.19
2.0 - 3.5	0.16
3.5 - 4.8	0.24
4.8 - 6.5	0.18

GRAY's formulae are used for the calculation of the reduction in stresses due to excavation. The methods of stability analysis considered are the Swedish circle and the BISHOP's simplified methods. Table 10 summarises the theoretical safety factor of the excavation at the end of construction, failure and long term, respectively. In all cases the long term safety factor is the most critical one for the stability of an excavation. The safety factors at failure are higher than the long term values showing that the excavation is too deep to stand permanently. Different strength parameters gave different safety factors. The safety factors under long term conditions with the effect of water pressure in the tension crack are 1.21, 1.40, 1.27 and 1.31 from CIU, CK_oU (strain controlled), CK_oU (stress controlled) and CK_oD strength parameters, respectively, with the BISHOP's simplified method. In all cases, the differences in the safety factors are particularly high between the CIU and CK_oU strength parameters.

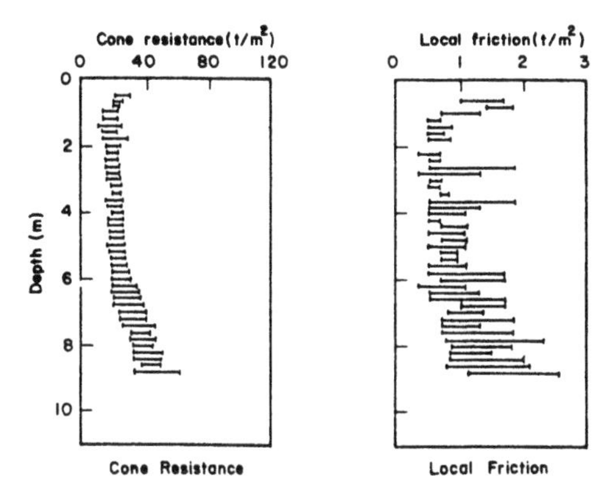

Fig. 61 Dutch cone test results

8.2 *Recent Research at AIT on Excavations*

Currently, a continuous research program is underway at AIT on the performance of various types of construction techniques utilized in excavations for basements of tall buildings. These include the use of sheet piled braced excavations, the use of cement grouted piles and the use of diaphragm walls with and without bracing (SRICHAIMONGKOL, 1991).

9 AIRPORTS AND SEAPORTS

9.1 *Airports*

The Asian Institute of Technology was fully involved in the comprehensive geotechnical studies related to the proposed and controversial site at the Nong Ngoo Hao for an international airport. The geotechnical data collected from these studies have now become a valuable source of information for all types of projects in Soft Clays. Two major airports were constructed in similar terrain in the last two decades. They are the Changi International Airport and the International Airport in Jakarta. HSU (1979) conducted his thesis research on the improvements of the soil properties at the Changi site in Singapore. Prof. S.L. LEE and his colleagues at the National University of Singapore have done extensive work on the ground improvement works at the Changi Airport Site as well as other sites in Singapore. The Changi project in Singapore was on a 645 hectare of reclaimed land and about 40 million cubic meters of sand fill was dredged from the seabed adjacent to the site. Dynamic consolidation combined with vertical drainage was used as a ground improvement technique to improve the underground soil formation with the soft clay layer. This is probably the only site in Southeast Asia where Dynamic consolidation has been used on a large scale, although the technique has been used on a smaller scale in Thailand, Malaysia, and Indonesia. The two

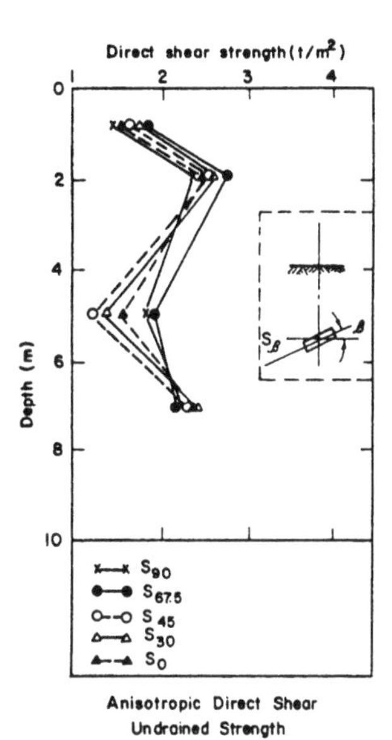

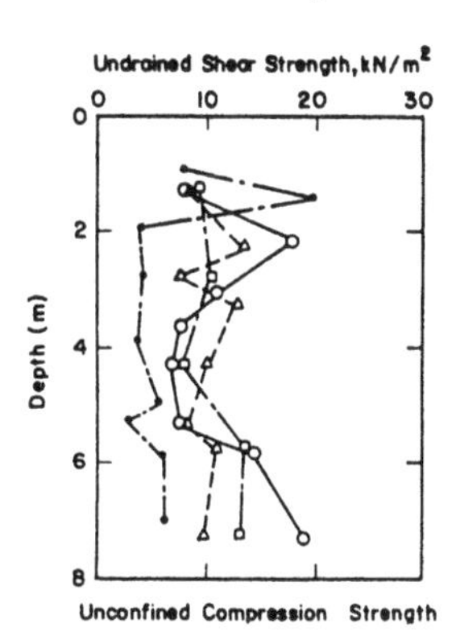

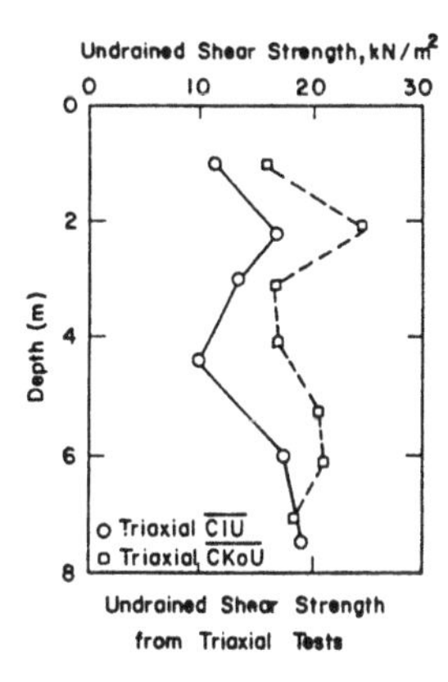

Fig. 62 Strength profile at Nong Ngoo Hao test excavation site.

Table 10(a) Effective Stress Analysis by Bishop's Simplified Method

Shear Strength Parameters	Factor of Safety							
	Predicted pore pressure end of construction		Measured pore pressure end of construction		Measured pore pressure at failure		Long term conditions	
	Without Tension Cracks	With Tension Cracks & Water Pressure	Without Tension Cracks	With Tension Cracks & Water Pressure	Without Tension Cracks	With Tension Cracks & Water Pressure	Without Tension Cracks	With Tension Cracks & Water Pressure
CIU	1.73	1.61	1.72	1.60	1.51	1.39	1.40	1.21
CAU (Strain Controlled)	2.26	2.13	2.17	2.04	1.85	1.71	1.67	1.40
CAU (Stress Controlled)	2.28	2.12	2.19	2.04	1.80	1.66	1.51	1.27
CAD	-	-	-	-	-	-	1.47	1.31

Table 10(b) Effective Stress Analysis by Swedish Circle Method

Shear Strength Parameters	Factor of Safety							
	Predicted pore pressure end of construction		Measured pore pressure end of construction		Measured pore pressure at failure		Long term conditions	
	Without Tension Cracks	With Tension Cracks & Water Pressure	Without Tension Cracks	With Tension Cracks & Water Pressure	Without Tension Cracks	With Tension Cracks & Water Pressure	Without Tension Cracks	With Tension Cracks & Water Pressure
CIU	1.54	1.46	1.52	1.42	1.34	1.22	1.17	1.05
CAU (Strain Controlled)	1.99	1.92	1.97	1.84	1.64	1.50	1.36	1.21
CAU (Stress Controlled)	2.40	1.97	2.01	1.86	1.60	1.45	1.32	1.08
CAD	-	-	-	-	-	-	1.22	1.14

international symposia held in AIT on Geotechnical Aspects of Mass and Material Transportation (BALASUBRAMANIAM, CHANDRA, BERGADO, & RANRUCCI, 1985c) and Recent Developments in Ground Improvement Techniques (BALASUBRAMANIAM, CHANDRA & BERGADO, 1984b) contain authoritative lectures and papers related to ground improvement techniques as applied to airport constructions. GUO (1979) reviewed the design and performance criteria for highways and airfields.

9.2 *Sea Ports*

Two major seaports were constructed in Thailand; one in Bangkok in a soft clay deposit in Pom Prachul for the Royal Thai Navy Dockyard, and the other is the Eastern Sea Board Development Project in Lamchabang.

9.3 *RTN Dockyard Construction in Pom Prachul, Bangkok*

The RTN Dockyard involved the construction of some heavy marine works supported on approximately 20,000 piles. The works included two dry docks, a holding basin and basin entrance, basin gauges, riverside jetties, workshop service buildings, etc. (Fig. 63). These services needed excavations up to 12 m in depth and all excavations were of the supported type. The construction sequence for the structures like the dry docks, basin entrance, basin quays and riverside jetties has been dealt with in this paper along with the discussion of general features and structural loads for these works. Details of the construction works are discussed below.

9.3.1 *Steel H-Piles*

The steel H-Piles are used to take up tension,

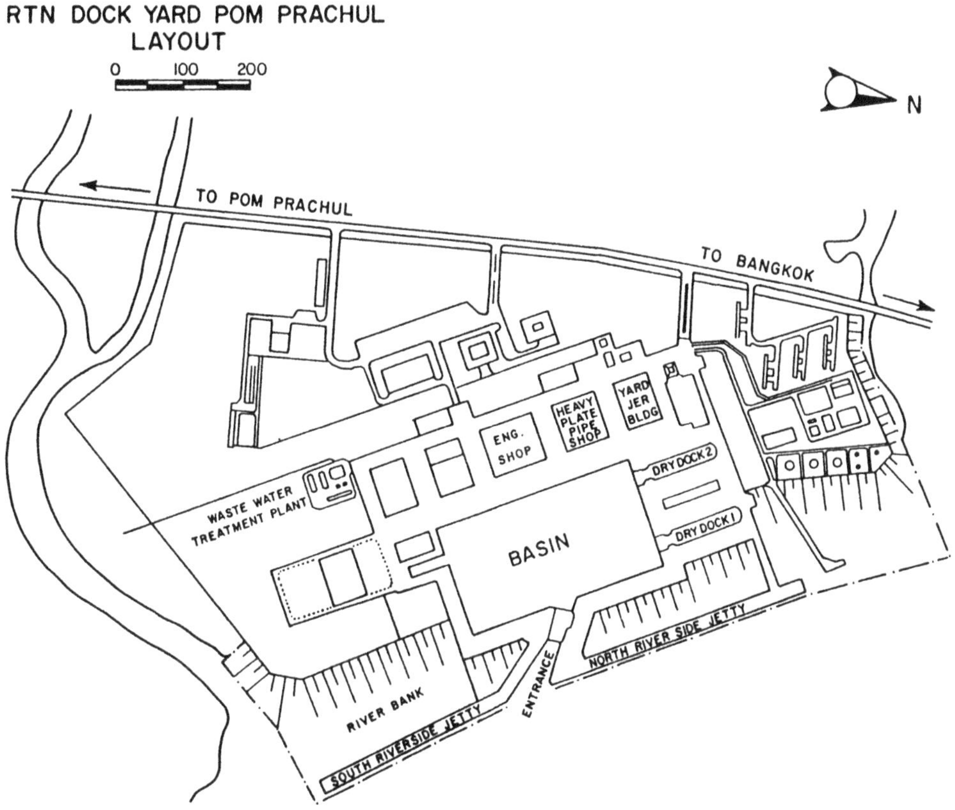

RTN DOCK YARD POM PRACHUL
LAYOUT

Fig. 63 Site plan for Pom Prachul Dockyard.

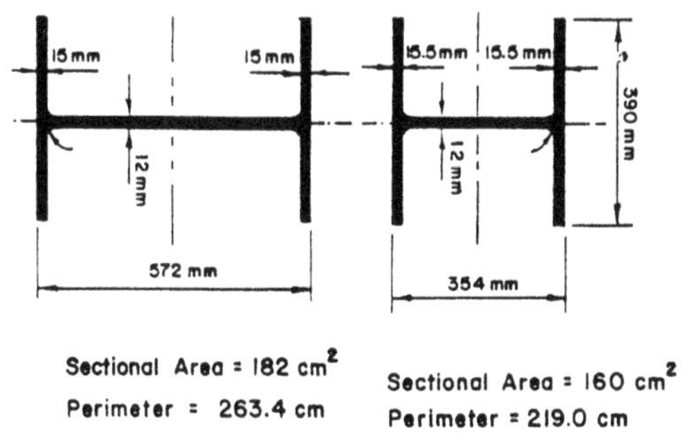

(a) Steel pile KZ 600 PS (b) Steel pile HZ 360 PS

Fig. 64 Steel sheet pile sections.

compression and lateral loads. The steel bearing piles beneath the Dry Dock floor and the Basin Entrance had been designed to take uplift when there is no water, and to take compression when there is water. Also, the steel H-Piles serve as king piles for the combined sheet piled walls in Dry Docks and quays and for the box piled wall in the Basin Entrance. The steel piles used consist of the ARBED steel HZ section as shown in Fig. 64. The steel grade of the piles was St, Sp, and Ss which have yield strength of 40,000 t/m^2. The bearing piles beneath the floor of the Dry Dock and Basin Entrance were driven prior to the excavation. This was made possible by the use of a dolly.

9.3.2 *Steel Piles*

Steel piled structures constituted an important geotechnical aspect in the design and construction of the Naval Dockyards. Sheet piling was used both in temporary and permanent works. Around the walls of the Dry Docks, a sheet piled wall was employed as temporary wall for retaining the soft clay during the construction of the Dry Dock.

Along the quays, sheet piling formed part of the permanent structure and served as cut-off against the river as well as a retaining wall for the soft clay. In the Basin Entrance, box piling was employed to provide a dry area during the construction period. Later on, part of the box piling was incorporated into permanent works.

New developments in ship building demand port facilities of great depth and size. This necessitated

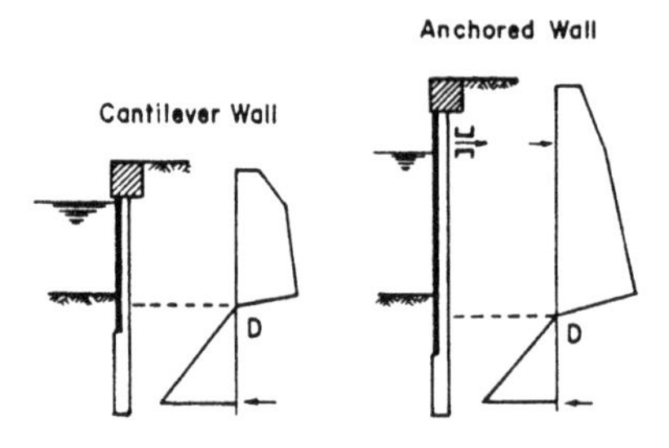

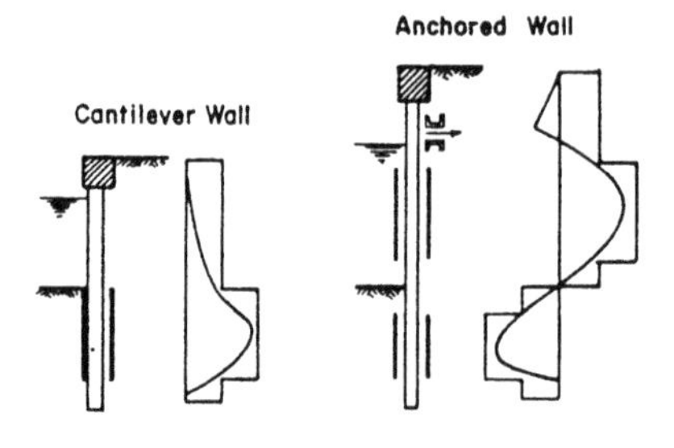

Fig. 65 Typical pressure and moment diagrams.

the use of a high section modulus and moment of inertia in the steel walls formed out of sheet piling without a concomitant increase, in the weight per unit square area of the wall. Taking these considerations into account, the ARBED HZ steel sheet piling system was selected.
This system is comprised of the following sections:

(i) special H-sections (beams with a bulb at each end of the flanges);

(ii) a new Z-sheet piling section similarly provided with a bulb at each end;

(iii) an interlocking section to pin the above described elements together.

By connecting the H-sections, or inserting double Z-sections between the H-sections driven at a certain distances from one another, it was possible to obtain a steel wall allowing a variety of combinations. One significant feature of the ARBED system is that the intermediary Z-sections do not require the same driving depth as the H-sections. This resulted in a reduction in the weight of the wall.

The active thrust of the soil was taken up by the H-pile and the intermediary Z-sheet piling while the reactions of the sub- soil i.e. the passive pressures acted only on the H-piles. Figure 65 shows typical earth pressure and bending moment diagrams.

The combined wall along the periphery of the quays employs larger HZ-600 LS sections along with the BZ 9.5 intermediary section and RH-16 interlocks as shown in Fig. 66. The strong box piled wall in the Basin Entrance consists of continuous HZ-600 LS sections linked together with RH-16 (Fig. 67). Table 11 gives parameters of different combinations of sheet pile sections employed at Pom Prachul.

9.3.3 *Load Tests on Sheet Piles*

The load tests on sheet piles were, in fact "snap through" tests performed on intermediary sheet piles i.e. BZ 9.5 sections with RH-16 interlocks. The objective of these tests was to verify that intermediary sheet piles would remain in their locks and would not deform excessively so as to lose flexural strength. The conclusion arrived at from these tests was that there is no danger of intermediary sheet pile "snap through" or displacement from interlocks under design pressure as long as the king piles and intermediate piles are driven accurately in position and are vertical.

9.3.4 *Sheet Piling in Dry Docks*

As stated earlier, the sheet piling forms a temporary retaining structure around the boundary of the Dry Docks during the construction phases (Fig. 68). In the permanent structure, the sheet piled wall will be retained i.e. it will not be withdrawn though it will not be taking up any load.

This sheet piled wall was supported laterally by a system of steel struts at two levels, i.e. at -0.28 m and at -5.75 m. At level -5.75 m, these struts were connected to a reinforced concrete ring beam that serves as a waling. In the dock entrance area, this ring beam took up the lateral thrust on the sheet piled wall by its arching action. At higher elevation i.e. at - 0.28 m, the struts were in level with the floor of the service gallery except in the dock entrance area where they were connected to an outer ring beam. Figure 69 illustrates the sheet pile profiles in different works at the dockyard. In dry docks, the sheet piled wall was retaining about 8.5 m of soft clay.

9.3.5 *Sheet Piling in Quays*

Combined sheet piled walls ran along the periphery of the quays around the Basin. Unlike in the Dry Docks, this sheet piled wall is a member of the permanent structure and has to support about 7 m height of soil. The anchoring system of this sheet piled wall is illustrated in Fig. 70. The top of the king piles was cast into the floor of the service gallery which, in turn, was tied to a number of sloping anchor beams. These anchor beams were connected to a capping beam of raker piles.

9.3.6 *Box Piling in Basin Entrance*

The Box piled wall employed in the Basin Entrance

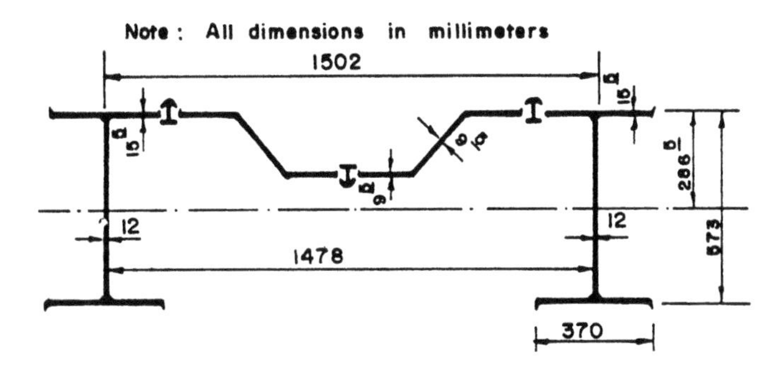

Fig. 66a Combined sheet piled wall formed by HZ 600A and BZ 9.5 sections.

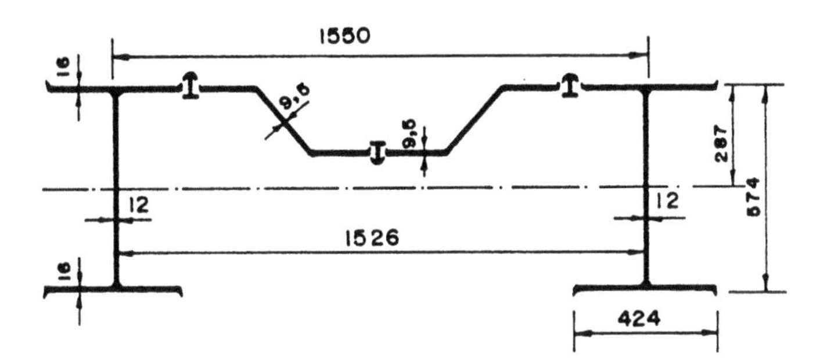

Fig. 66b Combined sheet piled wall formed by HZ 600 LS and BZ 9.5 sections.

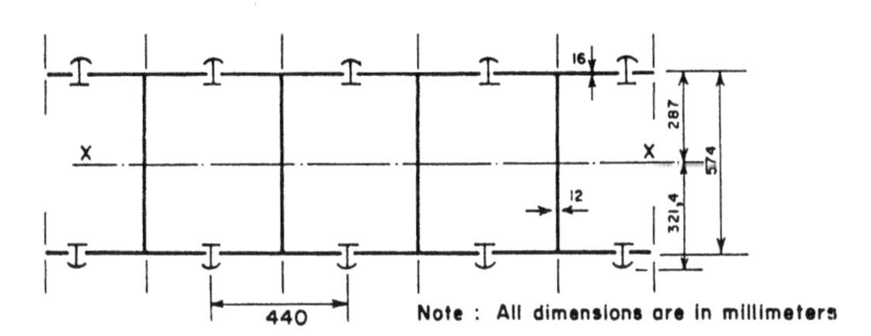

Fig. 67 Box piled wall formed by double HZ 600 LZ sections.

had been designed as both temporary and permanent works. As temporary works, the box piled wall served to provide a dry working area thereby retaining earth on the northern and southern flanks of the Basin Entrance and providing a cut-off from the river on the eastern side. During the construction period, the level of earth retained was +0.5 m and the dredge level was -10.0 m. Hence the box piled wall will be retaining about 10.5 m of earth. During this period, the box piled wall was supported by a system of bracing in the form of a reinforced concrete ring beam at level -2.0 m. This ring beam acted as an arch to take up the lateral thrust from all the three sides of the piled wall. As a permanent system, the box piled wall was designed to retain earthfill (level -2.0 m) on one side and water on the other side (level varies from -1.7 m to +2.2 m). In the final structure, the head of the box piles was cast in a capping beam which was tied back to a number of anchor beams. The anchor beams transferred the horizontal load to raker piles.

The development of coastal structures is of great importance in Southeast Asia and in Asia. Excellent lectures were presented in the International Symposium on Coastal and Offshore Structures (YUDHBIR & BALASUBRAMANIAM, 1983) and the Kozai Club Seminars held in 1988 and 1989. A comprehensive paper on the coastal developments in soft clay deposits is given by BALASUBRAMANIAM, PHIEN-WEJ and KUHANANDA (1988d).

Table 11 Statistical Properties of Sheet Pile Wall

Type of Sheet Pile Wall	System Width (m)	Sectional Area (cm^2/m)	Weight (kg/m)	Moment of Inertia (cm^4/m)	Section Modulus (cm^3/m)
Combined wall formed by HZ 600A & BZ 9.5	1.50	257.50	202.1	87640.0	3059.0
Combined wall formed by HZ 600 LS & BZ 9.5	1.55	383.00	210.4	176409.0	9527.3
Continuous wall formed by HZ 600 LS (Box pile wall)	0.44	591.40	401.4	393192.4	1370.0

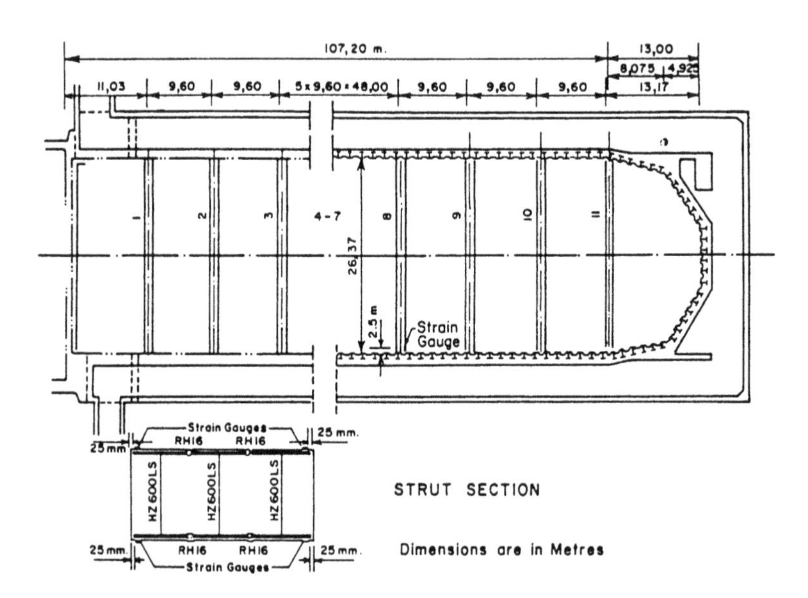

Fig. 68 Temporary retaining structure formed by sheet piling around the boundary of dry docks.

9.4 *Concluding Remarks on Excavations*

With the increasing need for deep basements in buildings for underground storage and parking facilities, deep excavations will increase in number in most major cities. Up to now very little field instrumentation and in-situ measurements have been done in deep excavations in the Bangkok subsoils as compared to those in Singapore, Hong Kong or Taipei.

10 RESOURCES DEVELOPMENTS

The geotechnical aspects of resources developments deal with dams and reservoirs for water resources and hydro-power as well as energy resources related to natural gas, oil and lignite.

10.1 *Water Resources*

Water is invariably the most important of all resources. The situation of water resources developments in Asian countries and elsewhere is crucial, primarily because of the population explosion. In the last two decades, developments in the areas of hydro-power, irrigation, and water resources have been most active. Accordingly, several projects of various sizes have been completed and many are still in the construction, planning and investigation. Geotechnical studies relate to engineering geology, site investigations, construction materials, foundation treatments, and construction problems. The latter relates to power houses, dams and tunnels. Another important research area is the evaluation of the safety of old dams in operation and remedial steps for unsafe structures. My

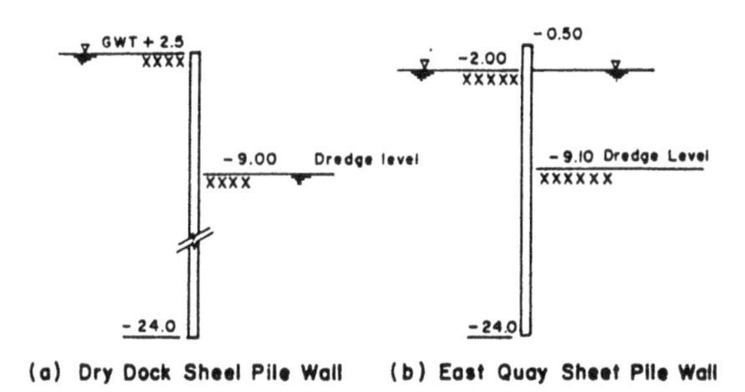

Fig. 69 Sheet pile profile.

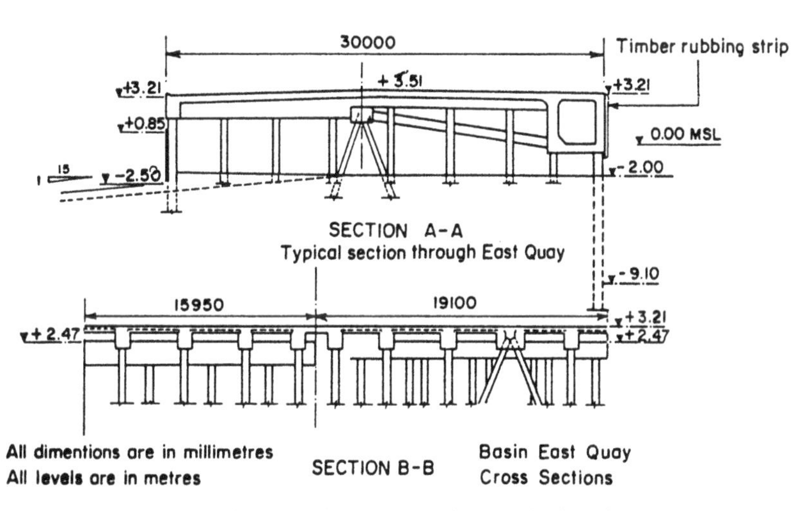

Fig. 70 Anchoring system of sheet piled wall.

involvement in the geotechnical aspects of dams has been very limited except for initiating a comprehensive research program at AIT on the strength and compressibility characteristics of rockfill materials which was initiated at a time when several rockfill dams were built in Thailand and in Southeast Asia (BALGUNAN, 1984; HSU, 1984; LIN, 1985; LIEW, 1987; & LEE, 1987). An incremental stress-strain theory for Rockfill materials was proposed by BALASUBRAMANIAM, LEE and WIJEYAKULASURIYA (1987b). Authoritative presentations on the Engineering Geological Aspects of Dams are the domain of our Colleague Prof. PRINYA NUTALAYA and the vast area of the applications of rock mechanics in Asia are the specialty of Drs. INDRARATNA and PHIEN-WEJ. Worthy of mention here is the most successful International Symposium on Large Dams held in AIT in 1980 (BALASUBRAMANIAM, 1981d; and BALASUBRAMANIAM et al., 1981f). Some aspects related to the geotechnics in dam engineering is included in my special lecture at the Eighth Southeast Asian Geotechnical Conference held in Kuala Lumpur in 1985 on Geotechnical Aspects of Resource Development (BALASUBRAMANIAM et al., 1985f).

10.2 *Natural Gas and Oil*

In response to the energy crisis in 1973 and in subsequent years, the onshore and offshore exploration of oil and gas increased exponentially. Indeed, in a most successful Coastal and Offshore Symposium held at AIT in 1981, experts from various parts of the world concentrated on the Gravity Platforms for oil explorations. The natural gas project in the Gulf of Thailand alone involves the installation of over 20 fixed offshore platforms and several hundred kilometers of offshore pipe laying. Fixed offshore platforms are used in Thailand and in Southeast Asia.

A General Report on Geotechnical Aspects of Waterfront and Offshore Structures given by BALASUBRAMANIAM & BERGADO (1983) at the Seventh Asian Regional Conference on Soil Mechanics and Foundation Engineering held in Haifa, summarises such aspects as the site investigation methods, evaluation of geotechnical parameters, offshore piling practice, design principle of gravity platforms, installation of concrete gravity structures, use of centrifugal models in offshore and nearshore waters, instrumentation and data acquisition, and the development of grouting works, scour protection and the theory behind high capacity anchors. At the Asian Institute of Technology, CHIN (1984) has reviewed the Piled Foundations for Offshore Platforms in Southeast Asia.

11 NATURAL HAZARDS

Southeast Asia comprises one of the greatest concentrations of natural hazards in the world. The natural hazards can be grouped as earthquake hazards, volcanic hazards, hazards from ground failures and hydrologic hazards. Professor Prinya Nutalaya has initiated a research program on Natural Hazards and has documented more than 400 earthquakes in Indonesia alone with magnitudes higher than 4 on the Richter scale. Similar hazards are experienced elsewhere in the region.

Engineering for protection from Natural Disasters was the subject of an International Conference held at AIT in 1980 (KARASUDHI, BALASUBRAMANIAM and KANOK-NUKULCHAI, 1980); yet the research at AIT in this direction has not been pursued as vigorously as one would have expected. An equal concern could be expressed for research works on hazards from ground failure which include landslides, problematic soils and subsidence.

Many of the topics in natural hazards can also be covered under the umbrella of Environmental Geotechniques, while the latter can also be ballooned to include geotechnical control and environmental protection; environmental geotechnical aspects of major infra-structure projects; sanitary wastes and their use and disposals; as well as problematic soils and rocks (BALASUBRAMANIAM, CHANDRA, BERGADO, and NUTALAYA, 1988). Under the broad theme of Environmental Geotechnics, restoration work and the preservation of historical structures and ancient monuments (BALASUBRAMANIAM et al., 1990) should also form important aspects for future research.

12 FUTURE DIRECTIONS OF DEVELOPMENTS

Critical State Soil Mechanics was the catchword in the sixties in Soil Mechanics and Foundation Engineering. However, in the nineties, and the decades to follow, Soil Mechanics alone is not in a critical state; our whole profession will be in danger if adequate precautions are not taken for proper training of our undergraduates and the graduate students. It seems that there is a lowering of standards in undergraduate education in most universities in Asia, a truly detrimental trend. Most National Universities including those in the NICs (Newly Industrialised Countries) are suffering from severe inbreeding. There is a need for updated laboratory and field testing facilities and teaching methods in the national universities in Asia. Many of them cannot afford to even subscribe to the journals in our discipline because of sky rocketing prices.

Traditional teaching and research in Soil Mechanics have concentrated on the behavior of saturated soils and the time has now come to extend this knowledge to partially saturated soils and problematic soils and rocks in the Asian context. The impact of computers has been both positive and harmful. More software development is needed and software should be freely available for teaching, research and professional practices. In addition, concerted efforts need to be placed on research related to rural development, resources development and the mitigation of natural hazards. Continuing education of the highest quality must constantly inform institutes of the developments of other Research Institutes both within and beyond the region.

13 CONCLUSIONS

My concluding remarks look ahead at the future directions in research and developments rather than re-emphasising past successes. At the Asian Institute of Technology, graduate students are admitted from countries with very wide and diverse scientific and professional backgrounds. As such, Soil Mechanics and Foundation Engineering should be considered in terms of the poorest and the least developed countries in Asia as well as the highly developed and industrialised countries.

In the last five decades remarkable progress has been made in the fundamental, analytical and construction techniques needed for large urban development projects. What has been neglected are the rural development projects which cover a very large area and utilise low cost, less mechanical and automated equipment and techniques. Also, in these underprivileged countries, the soil deposits are somewhat different from those in the low land plains. Therefore for rural development, one can foresee a vast potential in terms of research and developments of residual soils and other problematic soils. Included herein is the work associated with resources developments and the prevention of natural hazards on small scale projects. The successful implementation of this task would demand considerable revisions in the current teaching and research methods as needed to develop the vast areas in Asia which do not obey the classical Soil Mechanics as traditionally taught in schools.

For both the rural and urban developments, there is a need:

(a) to prepare proper geotechnical maps with the available geotechnical information well documented and readily accessible in computer data bases.

(b) for proper Laboratory and Field Testing facilities to handle non-text book soil types and for the evaluation of the associated geotechnical parameters.

(c) for inter-disciplinary field of activities which involve

(i) Soil Mechanics and Fluid Mechanics say Slurry Mechanics in dredging and reclamation works;

(ii) Soil Mechanics and Environmental Studies such as Environmental Geotechnics to deal with pollution problems;

(iii) Soil Mechanics and Structural Mechanics in Soil Structure Interaction Problems;

(iv) Soil Mechanics and Soil Physics in terms of Soil Erosion studies and Landslides;

(v) Soil Mechanics and Chemistry in terms of developing new construction materials as well as in ground improvement works.

Thus, the next decade will take us more and more into non-traditional type of activities which embrace a cosmopolitan type of combinations in Science, Engineering and Social Study disciplines.

ACKNOWLEDGEMENTS

It is virtually impossible to make a full list of acknowledgement to a large number of colleagues and friends from all parts of the world who have helped me for the last eighteen years. Prof. Alastair North, the current President of AIT, and the former Presidents Dr. R.B. Banks and Dr. Milton E. Bender have contributed significantly to our advancement and are hereby acknowledged for their valuable support. The Division of Geotechnical and Transportation Engineering originated with the efforts of Dr. Za-Chieh Moh, who was the first Chairman, followed by Dr. E.W. Brand, Prof. John Hugh Jones, Prof. Prinya Nutalaya and Dr. Y. Honjo; their administrative wisdom in maintaining the reputation of Geotechnical Engineering be recorded. Sincere appreciation is also given to the many learned colleagues of the past and the present in the Geotechnical and Transportation Engineering Division: Dr. R.P. Brenner, Dr. G. Rantucci, Dr. D.T. Bergado, Dr. A.M. Richardson, Dr. A. Kazi, Prof. T. Onodera, Mr. S. Holmberg, Dr. A. Tomiolo, Dr. F. Prinzl, Dr. S. Chandra, Dr. R. H. Whiteley, Dr. Y. Yamada, Dr. I. Towhata, Dr. J. Kuwano, Dr. N. Phien-wej, Dr. B. Indraratna, Prof. T. Akagi, Prof. H. Ohta and Prof. Yudhbir. Appreciation is also extended to the Division Secretaries Mrs. Vatinee Chern, Mrs. Uraivan Singchinsuk and Mrs. Vannee Sithikosoljit.

In the preparation of the material, Prof. Jean Kerisel, Mr. Pierre Londe, Prof. Andrew Schofield, Dr. Za-Chieh Moh, Mr. Ove Eide, Prof. A. Nakase, Prof. T. Kimura, Mr. T. Hosoi, Dr. M.F. Randolph, Mr. S. Balachandran, Mr. N.K. Ovesen, Dr. Chin Der Ou, Dr. Woo Siu Mun, Dr. Ting Wen Hui, Dr. Ooi Teik Aun, Dr. Chan Sin Fatt, Prof. K.R. Massarsch, Prof. R.G. Campanella, Mr. J.F. Corte, Mr. Rodrigo Murillo and numerous others have made very valuable help and assistance.

Special thanks also to Dr. R.A. Hawkey, our Academic Secretary for his kind and strong support in the preparation and arrangement of the Inaugural Lecture.

Both the written and verbal presentation would not have materialised if not for the untiring work of our Research Associates Mr. Manoj Kumar Panda, Mr. Hilario Carvajal, Mr. Shahadat Hossain, Mrs. Kenneth Enriquez and Miss P. Ratnayake.

The preparation of this manuscript was done by Mr. Elmer P. Bandalan, Research Associate.

REFERENCES

Adikari, G.S.N. (1976). Statistical evaluations of strength and deformation characteristics of Bangkok clays, M.Eng. Thesis, AIT, Bangkok, Thailand.

Ahmed, M.A. (1976). Stress-strain behaviour and strength characteristics of stiff Nong Ngoo Hao clay during extension tests, M.Eng. Thesis, AIT, Bangkok, Thailand.

Ameratunga, J.J.P. (1979). Finite element analysis of consolidation and subsidence problems, M.Eng. Thesis, AIT, Bangkok, Thailand.

Ananda, J.M.T. (1979). Probabilistics approach to the stability and deformation of excavations and natural slopes, M.Eng. Thesis, AIT, Bangkok, Thailand.

Asaoka, A. (1978). Observational procedure of settlement prediction, Soils and Foundations, Vol. 18, No. 4, pp. 87-101.

Balasubramaniam, A.S. (1969). Some factors influencing the stress-strain behaviour of clays, Ph.D. Thesis, Cambridge University, 3 Volumes.

Balasubramaniam, A.S. (1973). Stress history effects on the stress-strain behaviour of a saturated clay, Geotechnical Engineering Journal, Vol. 4, pp. 91-111

Balasubramaniam, A.S. (1974a). Critical study of the uniqueness of state boundary surface for saturated specimens of kaolin, Geotechnical Engineering Journal, Vol. 5, pp. 21-38.

Balasubramaniam, A.S. (1974b). Local strains in cylindrical specimens of a saturated clay during consolidation, Geotechnical Engineering Journal, Vol. 5, pp. 89-107.

Balasubramaniam, A.S. (1975a). Recoverable strains in triaxial specimens of a saturated clay, 1st Baltic Conference on Soil Mechanics and Foundation Engineering, Gdansk, Poland, pp. 45-57.

Balasubramaniam, A.S. (1975b). Histograms of local strains in cylindrical specimens of a saturated clay, 2nd Int. Conf. on Application of Statistics and Probability of Soils & Structural Engineering, Aachen, Germany, pp. 231-248.

Balasubramaniam, A.S. (1975c). Effects of repeated loading on the stress-strain behaviour of a saturated clay, Istanbul Conference on Soil Mechanics and Foundation Engineering, Istanbul, Turkey, pp. 1-8.

Balasubramaniam, A.S. (1975d). Behaviour of a normally consolidated clay in stress ratio strain space, Symposium on Recent Developments in the Analysis of Soil Behaviour and Their Applications to Geotechnical Structures, University of New South Wales, Australia, pp. 275-287.

Balasubramaniam, A.S. (1975e). A critical re-appraisal of the incremental stress-strain theory for normally consolidated clays, Geotechnical Engineering Journal, Vol. 6, No. 1, pp. 15-32.

Balasubramaniam, A.S. (1975f). Stress-strains behaviour of a saturated clay for states below the state boundary surface, Soils and Foundations, Vol. 15, No. 3, pp. 13-25.

Balasubramaniam, A.S. (1975g). Strain increment ellipses for a normally consolidated clay, 5th Pan-American Conference on Soil Mechanics and Foundation Engineering, Buenos Aires, Argentina, Vol. 1, pp. 249-258.

Balasubramaniam, A.S. (1975h). Instrumented piles tested in fine sand, 5th Asian Regional Conference on Soil Mechanics and Foundation Engineering, Bangalore, India, pp. 153-160.

Balasubramaniam, A.S. (1980). *In-situ* tests and instrumentation in soft clay deposits, State-of-the-Art Report, Sixth Southeast Asian Conference on Soil Engineering, Taipei, Taiwan, R.O.C., Vol. 2, pp. 225-266.

Balasubramaniam, A.S. (1981). General report on the symposium on large dams, Ground Engineering.

Balasubramaniam, A.S. (1982). Ground improvement techniques in civil engineering and resources development, Vol. 13, Bo. 1, pp. 97-114.

Balasubramaniam, A.S. & Bergado, D.T. (1983). General report on geotechnical aspects of waterfront and offshore structures, Theme Paper, 7th Asian Regional Conference, Haifa, Israel, Vol. 2, pp. 153-168.

Balasubramaniam, A.S. & Brenner, R.P. (1981). Compressibility and settlement characteristics of soft clays, Chapter 10, Geotechnical Aspects of Soft Clays, Elsevier Publishing Co., Ltd., pp: 481-566.

Balasubramaniam, A.S., Bergado, D.T. & Chandra, S. (1985a). Recent developments in laboratory and field tests and analysis of geotechnical problems, A.A. Balkema Publishers, The Netherlands, 621 p.

Balasubramaniam, A.S., Bergado, D.T. & Chandra, S. (1985b). Geotechnical engineering in Southeast Asia, A Commemorative Volume for the 1985 Golden Jubilee International Conference, Southeast Geotechnical Society, A.A. Balkema Publishers, The Netherlands, 342

Balasubramaniam, A.S., Bergado,D.T., Lee, Y.H., Chandra, S. & Yamada, Y. (1985). Stability and settlement characteristics of structures in soft Bangkok clay, Proc. 11th Int. Conf. in Soil Mech. and Foundation Engineering, San Francisco, U.S.A., Vol. 3, pp: 1641-1648.

Balasubramaniam, A.S., Brenner, R.P., Mallawarachy, R.V. & Kuvijitjaru, S. (1980). Performance of sand drains on bangkok clays, Sixth Southeast Asian Conference on Soil Engineering, Vol. 1, pp. 447-468.

Balasubramaniam, A.S., Chandra, S. & Bergado, D.T. (1984). Recent developments in ground improvement techniques, A.A. Balkema Printers, The Netherlands, 587 p.

Balasubramaniam, A.S., Chandra, S., Bergado, D.T. & Nutalaya (1988). Environmental geotechniques and problematic soils and rocks, A.A. Balkema Publishers, The Netherlands, 564p.

Balasubramaniam, A.S., Chandra, S., Bergado, D.T., & Rantucci, G. (1987). Geotechnicals aspects of mass and material transportation, A.A. Balkema Publishers, The Netherlands, 533 p.

Balasubramaniam, A.S., Handali, S., Phien-wej, N., & Kuwano, J. (1989). Pore pressure stress ratio relationship, 12th International Conference on Soil Mechanics and Foundation Engineering, Rio de Janerio, Brazil, Vol. 1, pp: 11-14.

Balasubramaniam, A.S., Honjo, Y., Rantucci, G., Bergado, D.T., Phien-wej, N., Indraratna, B., & Nutalaya, P. (1990). Geotechnical aspects of restoration work on infrastructures and monuments, Balkema Printers, 365 p.

Balasubramaniam, A.S., Lee, Y.H. & Wijeyakulasuriya, C.V. (1987). Stress-strain behaviour and strength characteristics of rockfills, 8th Asian Regional Conference, Kyoto, Japan, Vol. 1, pp: 21-24.

Balasubramaniam, A.S., Phien-wej, N. & Kuhanandha, N. (1988). Coastal developments in soft clay deposits, Seminar on Engineering for Coastal Developments, The Kozai Club, Bangkok, Thailand.

Balasubramaniam, A.S., Phota-Yanuvat, C., Ganeshananthan, R. & Lee, K.K. (1981). Performance of friction piles in Bangkok subsoils, 10th Int. Conf. on Soil Mech. and Foundation Engineering, Stockholm, Sweden, pp. 605-610.

Balasubramaniam, A.S,, Sivandran, C, & Ho, Y.M. (1979). Prediction of deformation under embankment loading in soft Bangkok clay using probability analysis, Proc. Int. Conf. on Computer Applications in Civil Engineering, Roorkee, India.

Balasubramaniam, A.S., Yudhbir, Tomiolo, A. & Younger, J.S. (1981). Geotechnical problems and practice of dam engineering, A.A. Balkema Publishers, The Netherlands, 392 p.

Balgunan, N. (1984). Engineering properties of shales and soft rocks, M.Eng. Thesis, AIT, Bangkok, Thailand.

Boonsinsuk, P. (1974). Stability analysis of a test embankment on Nong Ngoo Hao clay, M.Eng. Thesis, AIT, Bangkok, Thailand.

Buensuceso, B.R. (1985). Foundation engineering practice in Manila, M.Eng. Thesis, AIT, Bangkok, Thailand.

Buensuceso, B. (1990). Engineering behaviour of lime treated soft Bangkok clay, D.Eng. Thesis, AIT, Bangkok, Thailand.

Chaudhury, A.R. (1975). Effects of applied stress paths on the stress-strain behaviour and strength characteristics of soft Nong Ngoo Hao clay, M.Eng. Thesis, AIT, Bangkok, Thailand.

Chen, C.S. (1981). Review and study of foundation for tall buildings in Hong Kong, M.Eng. Thesis, AIT, Bangkok, Thailand.

Cheng, T.Y. (1987). Geotechnical characteristics of Sungshan formation within Taipei City, M.Eng. Thesis, AIT, Bangkok, Thailand.

Mampitiya, A.A.S. (1987). Engineering behaviour of peat deposits in urban areas of Colombo, Sri Lanka, M.Eng. Thesis, AIT, Bangkok, Thailand.

Chia, B. (1990). Load settlement characteristics and bearing capacity of triangular reinforced concrete piles, M.Eng. Thesis, AIT, Bangkok, Thailand.

Chin, M.C. (1984). Pile foundations for offshore platforms in Southeast Asia, M.Eng. Thesis, AIT, Bangkok, Thailand.

Chui, K.F. (1975). Effects of repeated loading on the stress-strain behaviour of soft Bangkok Clay under undrained conditions, M.Eng. Thesis, AIT, Bangkok, Thailand.

Chung, C.H. (1983). Deep foundation in the cities of Southeast Asia, M.Eng. Thesis, AIT, Bangkok, Thailand.

Drucker, D.C. (1959). A definition of stable inelestic material, Journal Appl. Mech. Trans., ASME, 26, pp. 101-106.

Ganeshananthan, R. (1979). Probabilistic approach to carrying capacity of pile foundation, M.Eng. Thesis, AIT, Bangkok, Thailand.

Gho, Beng-de (1979). Review and studies related to rock and soil anchors, M.Eng. Thesis, AIT, Bangkok, Thailand.

Hamid, R.M. (1985). Mapping of very soft layer in AIT Campus, using dutch cone tests, M.Eng. Thesis, AIT, Bangkok, Thailand.

Handali, S. (1987). Cyclic behaviour of clays for offshore types of loading, D.Eng. Thesis, AIT, Bangkok, Thailand.

Hanifah, H.M.A.A. (1988). Under-reamed bored piles in Bangkok clay, M.Eng. Thesis, AIT, Bangkok, Thailand.

Hasan, Z. (1976). Stress-strain behaviour and strength characteristics of stiff Bangkok clays during triaxial compression tests, M.Eng. Thesis, AIT, Bangkok, Thailand.

Hasnain, V.N. (1985). Instrumentation for subsidence measurements in AIT Campus, M.Eng. Thesis, AIT, Bangkok, Thailand.

Ho, Y.M. (1976a). A re-analysis of the stability of Nong Ngoo Hao test embankment, M.Eng. Thesis, AIT, Bangkok, Thailand.

Ho, Y.M. (1976b). A re-analysis of the stability of Nong Ngoo Hao test excavation, M.Eng. Thesis, AIT, Bangkok, Thailand.

Hossain, M.M. (1980). Uses of mini piles and compaction grouting in underpinning works, M.Eng. Thesis, AIT, Bangkok, Thailand.

Hsu, W.B. (1984). Large oedometer tests on rockfill materials, M.Eng. Thesis, AIT, Bangkok, Thailand.

Hwang, Z.M. (1975). Stress-strain behaviour and

strength characteristics of weathered Nong Ngoo Hao clay, M.Eng. Thesis, AIT, Bangkok, Thailand.

Ibrahim, H.B. (1988). Laboratory and field tests of lime column techniques in Bangkok soft clay, M.Eng. Thesis, AIT, Bangkok, Thailand.

Jiann, D.M. (1977). Geotechnical observations from a deep bore hole at Rangsit, M.Eng. Thesis, AIT, Bangkok, Thailand.

Karasudhi, P., Balasubramaniam, A.S. & Nukulchai, W.K. (1980). Proc. of Int. Conf. on Engineering for Protection from Natural Disasters, AIT, Bangkok, Thailand.

Kerdsuwan, T. (1984). Basic properties and compressibility characteristics of first and second layers of Bangkok subsoils, M.Eng. Thesis, AIT, Bangkok, Thailand.

Kerisel, J. (1985). The history of geotechnical engineering up until 1700, Proc. of the Tenth Int. Conf. on Soil Mech. and Foundation Engineering, San Francisco, Golden Jubilee Volume, pp. 3-94.

Khatri, M.A. (1983). Review of the recent developments in the field of deep and shallow foundation and settlements, M.Eng. Thesis, AIT, Bangkok, Thailand.

Kampananode, N. (1984). Settlement predictions and performance of railway embankments at Cacheongsao, M.Eng. Thesis, AIT, Bangkok, Thailand.

Khoo, S.K. (1980). Geotechnical problems related to foundations for tall buildings - Case Studies, M.Eng. Thesis, AIT, Bangkok, Thailand.

Kim, S.R. (1991). Strength and deformation characteristics of lightly overconsolidated clay, D.Eng. Thesis, AIT, Bangkok, Thailand.

Kobakiwal, A.Q. (1978). Bearing capacity and settlement characteristics of driven piles in sand, M.Eng. Thesis, AIT, Bangkok, Thailand.

Kok, K.N. (1987). Geotechnical effects on some of the engineering properties of soft Bang Phli clay, M.Eng. Thesis, AIT, Bangkok, Thailand.

Kou, M.I. (1984). Recent developments of grouting in rocks and soils, M.Eng. Thesis, AIT, Bangkok, Thailand.

Kuntinawattanakul, P. (1986). Use of microcomputers in predicting settlements of embankments and effects of pile installations, M.Eng. Thesis, AIT, Bangkok, Thailand.

Lam, W.K. (1982). Recent developments in collapsible and dispersive soils, M.Eng. Thesis, AIT, Bangkok, Thailand.

Law, K.H. (1989). Strength and deformation characteristics of cement treated clays, M.Eng. Thesis, AIT, Bangkok, Thailand.

Khan, M.J. (1990). Effects of fly ash with lime cement additives on the behaviour of soft Bangkok clay, M.Eng. Thesis, AIT, Bangkok, Thailand.

Lee, K.K. (1978). Carrying capacity of driven and bored piles in bangkok subsoils, M.Eng. Thesis, AIT, Bangkok, Thailand.

Lee, Y.H. (1987). Strength and deformation characteristics of rockfills, D.Eng. Thesis, AIT, Bangkok, Thailand.

Lee, Y.N. (1978). Finite element analysis of earth retaining structures, M.Eng. Thesis, AIT, Bangkok, Thailand.

Leroueil, S. Tavenas, F., Mieussens, C. & Peignand, M. (1978). Pore pressure in clay foundation under embankment, Part II, Canadian Geotechnical Journal, Vol. 15, No. 1, pp.66-82.

Li, Y.G. (1975). Stress-strain behavior and Strength Characteristics of Soft Nong Ngoo Hao Clay Under extension conditions, M. Eng. Thesis, AIT, Bangkok, Thailand.

Liew, S.K. (1987). Correlation between drained and undrained triaxial tests for rockfill materials, M.Eng. Thesis, AIT, Bangkok, Thailand.

Lim, C.S. (1988). Engineering properties of Brunei soft clay, M.Eng. Thesis, AIT, Bangkok, Thailand.

Lim, S.Y. (1978). Bearing capacity and settlement characteristics of driven piles in clay, M.Eng. Thesis, AIT, Bangkok, Thailand.

Lin, W.P. (1985). Triaxial and large scale oedometer tests on rockfill materials, M.Eng. Thesis, AIT, Bangkok, Thailand.

Ling, S.Y. (1985). Review and study of geotechnical aspects of shallow and deep foundations in Taipei, R.O.C., M.Eng. Thesis, AIT, Bangkok, Thailand.

Lo, W.B. (1976). Repeated loading tests on Bangkok subsoils, M.Eng. Thesis, AIT, Bangkok, Thailand.

Mallawaaratchy, G.V. (1979). Performance of sand drains in soft clay, M.Eng. Thesis, AIT, Bangkok, Thailand.

Mesri, G. (1975). Discussion of new design procedure for stability of soft clays, Journal of the Geotechnical Engineering Division, ASCE, Vol. 103, No. GT5, pp. 417-430.

Moh, Z.C. & Woo, S.M. (1987). Preconsolidation

of soft Bangkok clay by nondisplacement sand drain and surcharge, Proc. of the ninth Southeast Asian Geotechnical Conference, bangkok, Thailand, Vol. 2, pp. 8.171-8.184.

Mollah, M.A. (1977). Strength and deformation characteristics of bangkok sands in triaxial compression, M.Eng. Thesis, AIT, Bangkok, Thailand.

Monoi, Y, (1989). Deformation of heavily overconsolidated clays, M.Eng. Thesis, AIT, Bangkok, Thailand.

Nilaweera, R.B.W. (1979). Modulus of elasticity and Poisson's ratio for triaxial specimen of kaolin under different applied stress paths, M.Eng. Thesis, AIT, Bangkok, Thailand.

Ooi, H.K. (1978). Stabilized soil as construction materials, M.Eng. Thesis, AIT, Bangkok, Thailand.

Parentela, E. (1982). Engineering properties of stiff Bangkok clays, M.Eng. Thesis, AIT, Bangkok, Thailand.

Parnploy, U. (1985). Deformation analysis and settlement prediction of Bangsue Bangpakon highway - Section 1, M.Eng. Thesis, AIT, Bangkok, Thailand.

Phamvan, P. (1990). Negative skin friction on driven piles in bangkok subsoils, D.Eng. Thesis, AIT, Bangkok, Thailand.

Photo-Yanuvat, Chukiat (1979). Carrying capacity of driven piles in bangkok subsoils, M.Eng. Thesis, AIT, Bangkok, Thailand.

Ramle, A.B. (1988). Strength and compressibility characteristics of soft muar clay stabilized with lime and lime-rice husk ash, M.Eng. Thesis, AIT, Bangkok, Thailand.

Roscoe, K.H. & Burland, J.B. (1968). On the generalized stress-strain behaviour of wet clay, engineering plasticity, Cambridge University Press, pp. 535-609.

Roscoe, K.H. & Poorooshasb, H.B. (1963). A theoretical and experimental study of strains in triaxial tests on normally consolidated clays, Geotechnique, Vol 13, pp. 12-38.

Roscoe, K.H., Schofield, A.N. & Thurairajah, A. (1963). Yielding of clays in states wetter than critical, Geotechnique, Vol. 13, pp. 211-240.

Samoon, M.I. (1978). Stability analysis of natural and cut slopes, M.Eng. Thesis, AIT, Bangkok, Thailand.

Sarker, M.A.R. (1979). Strength and deformation of Bangkok subsoils under *in-situ* stress conditions, M.Eng. Thesis, AIT, Bangkok, Thailand.

Schofield, A.N. & Wroth, C.P. (1968). Critical state soil mechanics, McGraw-Hill, London.

Selvanayagam, A.N. (1984). Piled foundation problems in Kuala Lumpur limestone, M.Eng. Thesis, AIT, Bangkok, Thailand.

Singh, J.P. (1985). Ground improvement techniques for road embankments in Bangkok area, M.Eng. Thesis, AIT, Bangkok, Thailand.

Siriwardane, T.H.J. (1977). Finite element analysis of embankments and slurry trenches, M.Eng. Thesis, AIT, Bangkok, Thailand.

Sivandran, C. (1976). Finite element analysis of an embankment and an excavation at Nong Ngoo Hao, Bangkok, M.Eng. Thesis, AIT, Bangkok, Thailand.

Sivandran, C. (1979). Probabilistics analysis of stability and settlement of structures on soft Bangkok clay, D.Eng. Thesis, AIT, Bangkok, Thailand.

Sivandran, C., Chiev, K. and Balasubramaniam, A.S. (1979). Application of probability theory to the finite element method in prediction settlements in soft Bangkok clay, Proc., 3rd Int. Conf. on Numerical Methods in Geomechanics, Aachen, Germany, Vol. 3, pp. 1025-1032.

Srichaimongkol, Wanchai (1991). Performance of supported and unsupported excavations in Bangkok subsoils, M.Eng. Thesis, AIT, Bangkok, Thailand.

Sudthikanung, Suebsant (1976). Some aspects of the engineering properties of Rangsit sand, M.Eng. Thesis, AIT, Bangkok, Thailand.

Suroso (1987). Behaviour of natural deposits of medium stiff clay under in-situ stresses, M.Eng. Thesis, AIT, Bangkok, Thailand.

Tabe, G. (1987). Cyclic behaviour of overconsolidated clays, M.Eng. Thesis, AIT, Bangkok, Thailand.

Tan, S.F. (1985). Engineering properties and the evaluation of driven piled foundations of the coastal subsoils in Penang Island, Malaysia, M.Eng. Thesis, AIT, Bangkok, Thailand.

Tantikom, Supachai (1981). K_o measurements on Bangkok clays, M.Eng. Thesis, AIT, Bangkok, Thailand.

Tjandrawibawa, Soebianto (1978). Strength and deformation characteristics of composite soils, M.Eng. Thesis, AIT, Bangkok, Thailand.

Tonyagate, Werapong (1978). Geotechnical properties of Bangkok subsoils for subsidence analysis, M.Eng. Thesis, AIT, Bangkok, Thailand.

Saeed, S.A. (1979). Geotechnical problems associated with the construction of RTN dockyard in Pom Prachul, M.Eng. Thesis, AIT, Bangkok, Thailand.

Towan, Chana (1976). Consolidation characteristics of stiff Nong Ngoo Hao Clay, M.Eng. Thesis, AIT, Bangkok, Thailand.

Tsia, C.Y. (1981). A monograph on the engineering properties of bangkok subsoils, M.Eng. Thesis, AIT, Bangkok, Thailand.

Udakara, D.D.S. (1989). Estimation of yield loci for natural deposits of clays, M.Eng. Thesis, AIT, Bangkok, Thailand.

Waheed, U. (1975). Stress-strain behaviour and strength characteristics of weathered bangkok clays under extension conditions, M.Eng. Thesis, AIT, Bangkok, Thailand.

Wang, W.T. (1975). K_o determination by hydraulic fracturing, M.Eng. Thesis, AIT, Bangkok, Thailand.

Wijeyakulasuriya, C.V. (1986). Shear behaviour of soft clays with particular reference to soft Bangkok clay", M.Eng. Thesis, AIT, Bangkok, Thailand.

Yang, C.W. (1988). Effect of lime on strength and compressibility of soft clays, M.Eng. Thesis, AIT, Bangkok, Thailand.

Yao, C.W. (1986). Landslide studies in Taiwan, R.O.C., M.Eng. Thesis, AIT, Bangkok, Thailand.

Yau, H.L. (1976). An analysis of consolidation tests data for settlement computations, M.Eng. Thesis, AIT, Bangkok, Thailand.

Yong, C.C. (1988). Ground improvement techniques suitable for approach roads and embankments, M.Eng. Thesis, AIT, Bangkok, Thailand.

Yu, T.C. (1987). A study of the behaviour of driven piles at Hsin-ta power plant and Kaoshiung harbor, Southern Taiwan, M.Eng. Thesis, AIT, Bangkok, Thailand.

Yudhbir & Balasubramaniam, A.S. (1983). Geotechnical aspects of coastal and offshore structures, A.A. Balkema Publishers, The Netherlands, 280 p.

Zamaludin, A. (1988). Consolidation characteristics of soft Muar clay with the effects of vertical sand drains, M.Eng. Thesis, AIT, Bangkok, Thailand.

Developments in Geotechnical Engineering, Balasubramaniam et al. (eds) © 1994 Balkema, Rotterdam, ISBN 90 5410 522 4

Plasticity analysis in geotechnical engineering: From theory to practice

W.F.Chen
School of Civil Engineering, Purdue University, West Lafayette, Ind., USA

T.K.Huang
Department of Civil Engineering, National Chung-Hsing University, Taichung, Taiwan

ABSTRACT: The paper describes several readily understandable and soluble plasticity models for soils, shows how these models can be used to obtain solutions of practical geotechnical engineering problems with either the limit analysis method or the finite element method, and leads to the generalization of the classical limit analysis method to the modern computer-based finite block method. Directions of further applications of these methods are also indicated.

1 INTRODUCTION

1.1 *Plasticity in Soil Mechanics*

The stress-strain behavior of soils is not linearly elastic for the entire range of loading of practical interest. In fact, actual behavior of soils is much complicated and they show a great variety of features when subjected to different loading conditions. A number of constitutive models of soils have been developed for practical applications. No one mathematical model can completely describe the complex behavior of real soils under all conditions. The successful application of a soil model in engineering practice needs only to address the soil behaviors that are significant in a considered problem and disregard what is minor effects in that class of phenomena (Chen, 1980).

The incremental formulations of elasticity-based soil models have been popularly used in recent years by geotechnical engineers. Since these soil models are all reversible, they fail to identify plastic deformations when unloading occurs. This shortcoming can, to some extent, be rectified by introducing the theory of plasticity.

The development of the modern *theory of soil plasticity*, as a new field, was strongly influenced by the well-established theory of metal plasticity. Soil mechanics specialists have been preoccupied with extending these concepts to answer the complex problems of soil behavior. Tresca's yield condition, used widely in *metal plasticity*, provides the basis on which the important concept of the *limit equilibrium* of a soil media had been firmly established in soil mechanics(Terzaghi, 1943). In the theory of limit equilibrium, the introduction of stress-strain relations was obviated by the restriction to the consideration of equations of equilibrium and a yield condition. However, the key to obtain a valid solution for kinematically restrained cases requires the basic knowledge of the stress-strain relations. Otherwise, a so-called solution is merely a guess.

The general *theory of limit analysis*, developed in early 1950's, considers the stress-strain relation of a soil in an idealized manner. This idealization, termed *normality* or the *associated flow rule*, establishes the *limit theorems* on which limit analysis is based. There have been an enormous number of practical geotechnical problems available (Chen, 1975; Chen and Liu, 1990). Many of the solutions obtained by the method are remarkably good when comparing with the existing results for which satisfactory solutions already exist. Although the limit analysis method can provide some very important results, such as ultimate load, stability factor, etc., the deformational behavior during the loading process prior to failure is not available.

Two major advances in the development of soil plasticity were reported in the following two classical papers "*Soil Mechanics and Plastic Analysis or Limit Design*" by Drucker and Prager (1952) and "*Soil Mechanics and Work-Hardening Theories of Plasticity*" by Drucker et al. (1957). In the former, the authors extended the Coulomb criterion to three-dimensional soil mechanics problems. This is now known as the *Drucker-Prager model*. In the later, the concept of work-hardening plasticity was introduced into soil mechanics. The innovative ideas in these two papers have led in turn to the generation of many soil models, leading to the development of the *critical state soil mechanics* at Cambridge University, U.K. In recent years, these new soil models have grown increasingly complex as additional experimental data are collected, interpreted, and matched. This extension marks the beginning of the modern development of a consistent *theory of soil plasticity* (Chen, 1975,1982; Chen and Baladi, 1985; Chen and Mizuno, 1990).

1.2 *Perfect Plasticity*

The fundamental difference between the elasticity-based and the plasticity-based models lies in the treatment of loading and unloading. In the plastic theory, the concept of *loading criterion* is introduced to treat separately the different behaviors of the materials in loading and unloading. For relating the increment of plastic strain $d\varepsilon_{ij}^{p}$ to the stress state σ_{ij} and stress increment $d\sigma_{ij}$, the simple *flow theory* (or *incremental theory of plasticity*) is generally used in the soil plasticity. The flow theory is based on three fundamental assumptions: (1) the existence of *initial yield surface*; (2) the evolution of subsequent loading surfaces (*hardening rule*); and (3) the formulation of an appropriate *flow rule*. The difference between the *perfect plasticity* and the *work-hardening* plasticity is that the yield surface is fixed in stress space and a stress state σ_{ij} is not permitted to move outside the yield surface. For soils, as for metals, perfect plasticity is an excellent design simplification, while more complex stress-strain behavior of soil may be approximated by the more sophisticated work-hardening plasticity models. This will be described in the following section.

The fixed yield function f can be in general described as :

$$f(\sigma_{ij}) = f_c \tag{1.1}$$

For a perfectly plastic material, the yield surface is fixed in stress space, and therefore plastic deformation occurs only when the stress path moves on the yield surface (see Fig. 1).

Thus, the loading condition for plastic flow is given by :

$$f = f_c, \quad df = 0 \tag{1.2}$$

On the other hand, elastic behavior occurs if, after an increment of stress, the new stress state is within the elastic domain, that is:

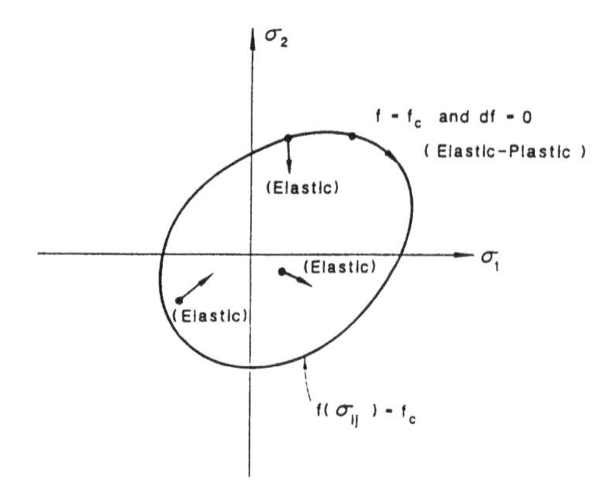

Fig. 1. Yield surface for perfectly plastic material.

$$f < f_c \tag{1.3}$$

or for the case of a stress path moving towards the elastic domain originally on the yield surface is expressed as:

$$f = f_c \text{ and } df < 0 \tag{1.4}$$

The plastic strain increment is defined by the plastic potential g in the form:

$$d\varepsilon_{ij}^{p} = d\lambda \frac{\partial g}{\partial \sigma_{ij}} \tag{1.5}$$

where $d\lambda$ is a positive scalar of proportionality dependent on the stress state and loading history. If the potential and yield surfaces coincide with each (f=g), it is called *associated flow rule*. Otherwise, it is the *non-associated type*.

The classical models such as Tresca, von Mises, Coulomb and Drucker-Prager types all belong to the category of perfect plasticity. The Tresca and von Mises failure criteria might be used to describe approximately the response of soils under undrained conditions. The Coulomb and Drucker-Prager types are applicable for soils under general conditions coupled with a suitable pore pressure model adopted.

1.3 *Work-Hardening Plasticity*

The incremental theory of plasticity for a hardening material is based on the same fundamental assumptions as the perfect plasticity. The yield surface is permitted to move outside the yield surface. Several hardening rules for different work-hardening plasticity models are proposed to describe the evolution of the subsequent yield surfaces for strain-hardening materials.

In general, the loading function is expressed in terms of the stress state σ_{ij}, the plastic strain ε^p_{ij}, and the hardening parameter k, that is:

$$f = f(\sigma_{ij}, \varepsilon^p_{ij}, k) \quad (1.6)$$

As shown in Fig. 2, the elastic-plastic behavior occurs if the stress state is originally on the yield surface and moves beyond the present boundary. In this case, the loading condition for plastic flow is defined by:

$$f = f_c, \; df > 0 \quad (1.7)$$

The elastic behavior occurs under the following condition as :

$$f < f_c \quad (1.8)$$

$$\text{or } f = f_c \text{ and } df < 0 \quad (1.9)$$

However, note that the neutral loading occurs when the initial and subsequent stresses fall on the same yield surface ($f=f_c$ and $df=0$). In this case of *neutral loading*, deformations are assumed to be purely elastic.

The growth of subsequent yield surfaces is described by the *hardening rule*. The choice of a specific hardening rule depends primarily on the ease with which it can be applied and its ability to represent the hardening behavior of a particular material. In general, there are three commonly used types of work-hardening models. These are *isotropic*, *kinematic*, and *mixed hardening models* (Chen and Saleeb, 1982 and Chen, 1994). In the isotropic hardening model, the yield surface expands or contracts uniformly as plastic flow continues. In the kinematic hardening model, the yield surface translates as a rigid body in stress space without changing its initial shape and orientation. A mixed hardening rule combines the isotropic and kinematic behaviors and provides more flexibility in describing the complicated behaviors of soils during hardening or softening.

The *cap model* (DiMaggio and Sandler, 1971) and the *Lade model* (Lade, 1979) are two commonly used isotropic hardening models. The cap model is

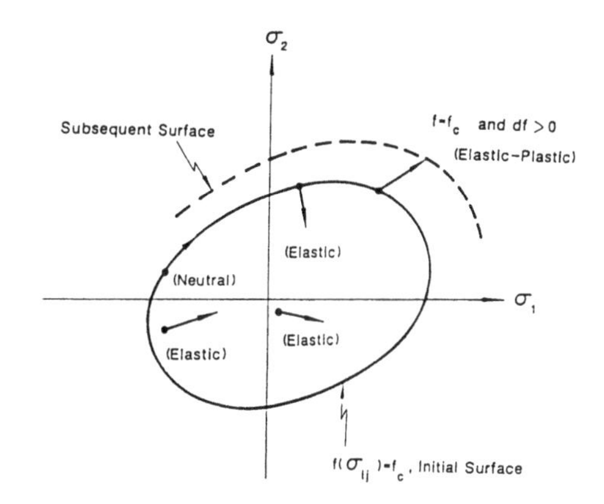

Fig. 2. Yield surface for a hardening material.

suitable for clays and the Lade model is appropriate for sands under monotonic loading condition. In a strict sense, the isotropic hardening model can not reproduce the hysteretic behavior under unloading-reloading path. Only elastic deformation is allowed in the isotropic hardening models during unloading. However, it is generally observed from many soil tests that during unloading, both elastic and plastic deformations occur before the stress is fully reversed.

The kinematic hardening model is applicable for approximately describing the behaviors of metals under unloading-reloading condition. However, the soils exhibit an apparent anisotropy in the tensile and compressive strength. The mixed hardening model may provide a more realistic representation of soil behavior under reverse, and particularly cyclic loading condition. Among the mixed hardening models, the Mroz(1967), Prevost(1978 a,b) *multiple surface models* and the *bounding surface model* are usually used in the application of geotechnical engineering problems under cyclic load. The *multiple surface models* have the versatile ability in describing the complicated behavior of soils. However, much more parameters are needed to describe the work-hardening modulus, size and position for each yield surface etc. (Prevost, 1982). Therefore, more storages and calculations are required for updating all the yield surfaces in a finite element analysis program for engineering applications.

The *boundary surface model* introduced earlier for metals (Dafalias and Popov, 1975) was extended to analyze the behavior of clay under monotonically triaxial load (Dafalias et al., 1980) and cyclic load (Dafalias et al., 1981). In the bounding surface model, two surfaces (one for elastic, the other for ultimate state) are used. The hardening modulus of

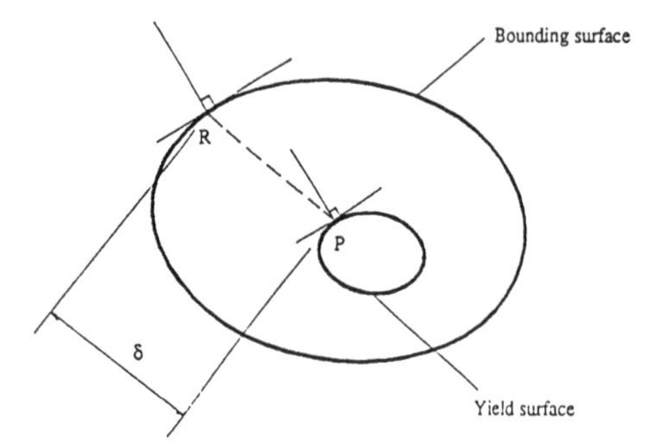

Fig. 3. Yield and bounding surfaces in stress space.

the stress point on the yield surface is assumed to be a function of the relative distance between the current stress point P on the yield surface and its conjugate point R on the bounding surface, as shown in Fig.3.

1.4 *Plasticity Analysis in Geotechnical Engineering*

Various methods exist for incorporating the appropriate soil plasticity models into finite element computer programs for solving practical geotechnical problems that require nonlinear stress analyses involving both material and geometric nonlinearities. Iterative computations based on linearly incremental formulation can be treated with comparative ease within the framework of the finite element software. Consequently, practical problems can be analyzed and reasonable data for use in the design can be generated using the finite element calculation.

Since geotechnical engineering problems are so complex, however, moderate idealizations in the use of soil plasticity models should be performed with great care to obtain reliable solution by the finite element analysis. In recent years, the development of more sophisticated and realistic plasticity soil models has rapidly advanced with the aid of finite element computational work. Consequently, the range of application for the finite element method has been extended to more general geotechnical engineering problems under static as well as dynamic loading conditions. Although plasticity analysis in geotechnical analysis has been widely used, there are some analytical and practical difficulties. These are:

(1) Selection of soil plasticity models,
(2) Evaluation of model parameters, and
(3) Assessment of analytical results.

In item (1), the selection of an appropriate soil model depends on the characteristics of the considered problems, the accuracy for the analysis, and the loading condition. Item (2) should be considered that the related model parameters should be evaluated through a simple and practical manner. Otherwise, the use of this soil model will not receive wide acceptance. As for item (3), results of a finite element method may be checked against those obtained by conventional methods, such as limit equilibrium and limit analysis methods.

When an elaborate idealization of the geotechnical engineering problem is made, the finite element plasticity analysis becomes meaningful and powerful. It should be noted that the numerical results through finite element method are only response to the idealized conditions and may not be taken as the real behavior. Such obtained results must be assessed by experimental data from laboratory and test data in situ. Then, the applicability of the soil plasticity models used and their numerical implementation procedures adopted in plasticity analysis can be re-evaluated.

2 LIMIT ANALYSIS OF PERFECT PLASTICITY

2.1 *Basic Concepts*

For an elastic-plastic material, there is as a rule a three-stage development in a solution, namely the initial elastic response, the intermediate contained plastic flow and finally the unrestricted plastic flow. The complete solution by this approach is likely to be cumbersome for all but the simplest problems, and methods are needed to furnish the load-carrying capacity in a direct manner. Limit analysis is the method which has been used widely in conventional soil mechanics for evaluating the *plastic collapse load* without carrying out the step-by-step elastic-plastic analysis.

The limit analysis method considers the stress-strain relationships of the soil in an idealized manner. For assessing the limit load, the soil is idealized as an elastic-perfectly plastic medium and the changes in geometry of structures are neglected. These idealizations lead to the limit load condition at which displacements can increase continuously without limit while the load remains constant.

The *limit analysis method* is based on the establishment of the upper and lower bound solutions. The upper bound solution obtained by equating the rate of external work to the rate of internal dissipative energy in an assumed deformation mode satisfies the velocity boundary conditions and compatibility conditions. The lower bound solution determined from an assumed distribution of stress field satisfies the equilibrium equations , the stress boundary conditions and nowhere violates the yield condition. By suitably choosing the stress and velocity fields, the two solutions can be made close enough for evaluating the required *plastic limit load.* Applications to classical and extended problems in soil mechanics using the limit analysis method can be found in the books by Chen(1975) and Chen and Liu (1990).

2.2 *The Classical Applications*

In this section, two examples of classical applications will be illustrated using the basic techniques of the upper-bound theorem. The first one is the limit load of a punch indentation problem for a Tresca material and the second is the critical height of a vertical cut for a general Coulomb material.

Figure 4 shows a simple rigid-body rotational mechanism about O for a punch indentation problem. By equating the rate of external work to the rate of internal energy dissipation, the upper load limit is evaluated as :

$$P^{u} = 6.28\ cb \tag{2.1}$$

where c is the cohesion of the Tresca material. The rotational mechanism of Fig.4 may be generalized by taking the radius and the position of the rotating center as two independent variables (see Fig.5). By equating the rates of external work to the internal energy dissipation and minimizing the solution with respect to the two variables r and θ, a lower upper load limit becomes:

$$P^{u} = 5.53\ cb \tag{2.2}$$

The failure mechanism involving only rigid-body translation as shown in Fig. 6 can be considered for evaluating the upper bound limit load. Through the computation of energy dissipation rate between the rigid blocks and the external work rate, the upper-bound solution obtained from this translated mechanism is:

$$P^{u} = 5.78\ cb \tag{2.3}$$

From the above results, it indicates that different layouts of the failure mechanism may be employed to reduce the upper load. However, the limit load corresponding to these mechanisms is seen not sensitive to the particular layout of the mechanism. In other words, the minimum is a rather flat curve, and a wide variety of geometrical layouts within the

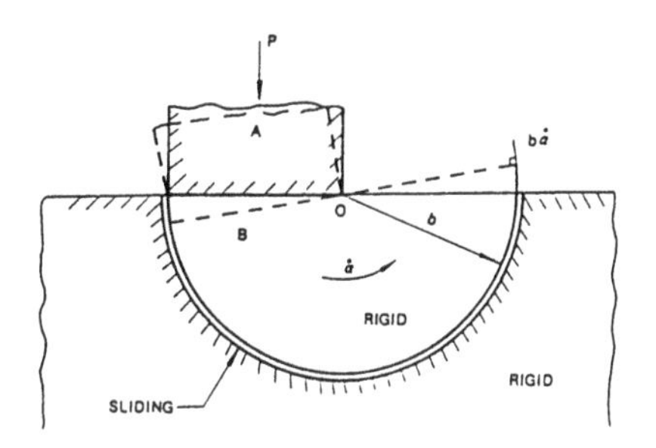

Fig. 4. Rotational mechanism

$P^{u} = 5.53\ cb$

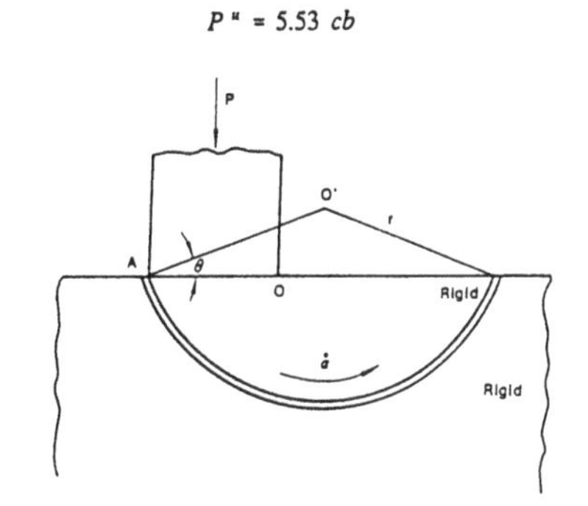

Fig. 5. Rotating block with center at O' for purpose of minimizing upper bound

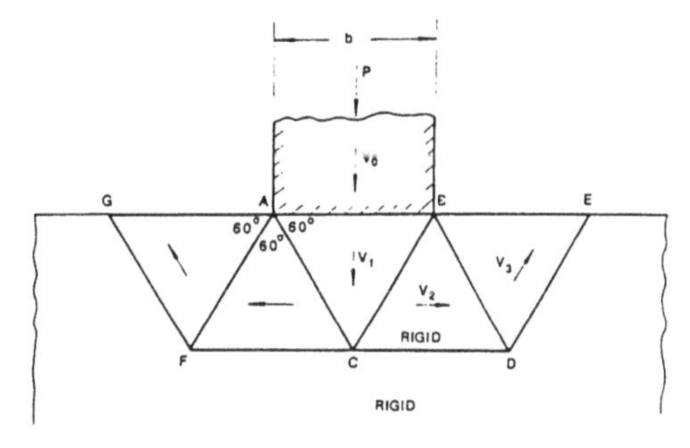

Fig. 6. Simple rigid-block translation for a rough punch.

same mechanism would be expected to give only slightly different upper bounds.

The upper load of the punch indentation problem can be improved by introducing radial shear zones to the failure mechanism (see Fig. 7). After equating the rates of external work to the internal dissipative energy, the upper bound load yields:

$$P^{u} = (2 + \pi)\ cb \tag{2.4}$$

This result is the correct value of limit load and the mechanism corresponds to the *slip-line solution*

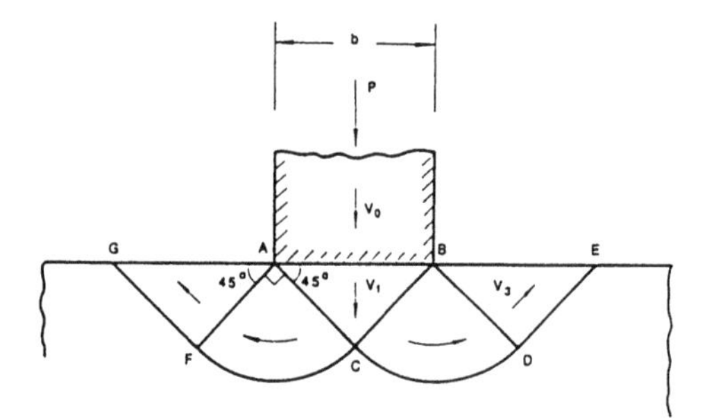

Fig. 7. Velocity field with a radial shear zone for a rough punch (Prandtl mechanism)

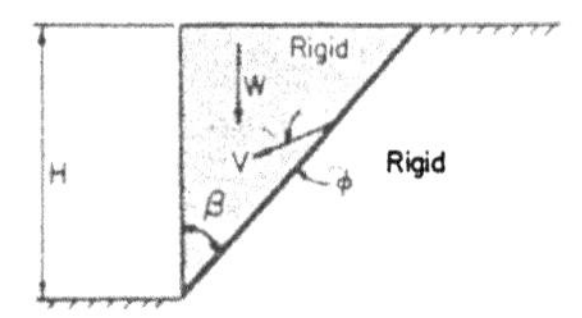

Fig. 8. Critical height of a vertical cut.

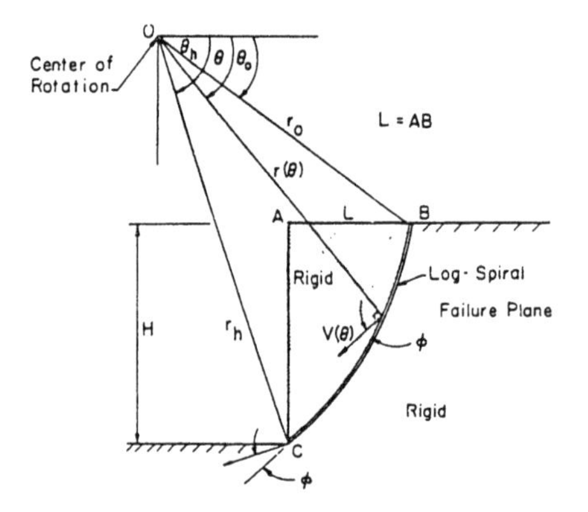

Fig. 9. Rotational failure mechanism for the critical height of a vertical cut.

proposed by Prandtl in 1921.

The other example of the classical application using the upper-bound theorem is shown in Fig. 8. In the figure, two failure mechanisms are assumed, translational and rotational. For translational mechanism, we assume first that the failure occurs by sliding along a plane making an angle β with the vertical. A limiting condition is reached when the rate at which the gravity load is doing work equals the rate of energy dissipation along the surface of sliding. Minimizing the solution with respect to the variable β, the critical height of a vertical cut gives:

$$H^{cr} = \frac{4c}{\gamma} \tan\left(\frac{\pi}{4} + \frac{1}{2}\phi\right) \qquad (2.5)$$

where ϕ and c are, respectively, the frictional angle and cohesion. γ is the unit weight of soil. This value of critical height is the same value obtained by the conventional Rankine analysis (limit equilibrium method).

An improved upper-bound solution may be obtained by considering a logspiral discontinuity (see Fig. 9). The slope angles θ_0 and θ_h of the chords OB and OC, respectively and the height H of vertical cut are selected for convenience in the solution of the critical height. After equating the rates of work done due to gravity load and the internal dissipative energy along the logspiral sliding surface, the height H of a vertical cut becomes a function of the slope angles θ_0 and θ_h. Taking the derivative of the solution with respect to θ_0 and θ_h, the height of a vertical cut has a minimum value as:

$$H^{cr} = \frac{3.83c}{\gamma} \tan\left(\frac{\pi}{4} + \frac{1}{2}\phi\right) \qquad (2.6)$$

The value 3.83 of upper-bound in Eq.(2.6) is an improvement of the previous solution 4.0 as given in Eq.(2.5). The lower value 3.83 is the same as that obtained by Fellenius(1927) using the conventional limit equilibrium method.

2.3 *Some Recent Developments*

The normality assumption required for the upper-bound technique of limit analysis method leads to a much too large dilation for frictional soils during plastic flow than that can be explained experimentally. This has been the center of the dispute. Previous investigators, such as Chen(1975), have concentrated on how the techniques of limit analysis can be applied to solve *soil stability problems*. However, little work has been done on why these techniques are applicable to soils, especially for cohesionless soils. Chen and Chang(1981) later examined the applicability of the upper-bound limit analysis to soil medium and discussed the range of validity of the basic assumptions in the use of limit analysis techniques.

On the other hand, most of the early applications of limit analysis of perfect plasticity to soil mechanics have been limited to soil statics. Recent works attempt to extend this method to soil dynamics, especially for earthquake-induced stability problems. Recent results show convincingly that the upper-bound analysis method can be applied to soils for obtaining reasonably accurate solutions of slope failures and lateral earth pressure subjected earthquake excitation. Different aspects of these advances made in recent years were reported in several books, theses, conference proceedings, and state-of-the-art reports. This includes the books by Bazant(1985), Desai and Gallagher(1981), and Dvorak and Shield(1984); the theses by Chang(1981), Saleeb(1981), Chang(1984), Chan(1980), Mizuno(1981), and McCarron(1985); the Conference Proceedings by ASCE(Yong and Ko, 1981, Yong and Selig, 1982), and the state-of-the-art reports by Chen(1980, 1984), and Chen and Chang(1981), among others.

Although the upper-bound limit analysis method can be applied to solve stability problems with any type of failure criterion, almost all solutions that are at present known, are based on the well-known linear Mohr-Coulomb failure criterion. However, in many practically geotechnical problems, such as the frozen gravel embankments used in offshore acrtic engineering, experimental data have shown that the frozen gravel follows a highly nonlinear failure criterion. Wang and Liu(1988) successfully combined the upper-bound limit analysis method with the conventional limit equilibrium method and developed a realistic and practical method for the solution of a class of stability problems in nonlinear soil mechanics (see Chen and Liu, 1990).

2.4 *Some Engineering Applications*

2.4.1 Seismic Stability of Slopes

In current seismic stability analysis of slopes, there are two basic approaches. One is *stress-strain analysis* and the other is the conventional *pseudo-static analysis*. A complete *progressive failure analysis* of a slope can be fulfilled through the former approach, but it is too complicated for practical application(Makdisi and Seed, 1979). On the other hand, the traditional pseudo-static approach is too crude to predict the seismic response of a slope. In this traditional analysis, the pseudo-static force is applied as a permanent force, whereas in reality the reduction in the stability of a slope exists only during the short period of time for which the inertia force is acting. Thus, during the earthquake, the factor of safety may drop below unity a number of times which will induce some movements of the failure section of a slope, but not cause the collapse of a slope. Therefore, the stability of a slope should depend on the cumulative displacements developed during an earthquake. Newmark(1965) first proposed the important concept that the seismic stability of a slope should be evaluated in terms of the *displacement accumulated* rather than the traditional *minimum factor of safety*.

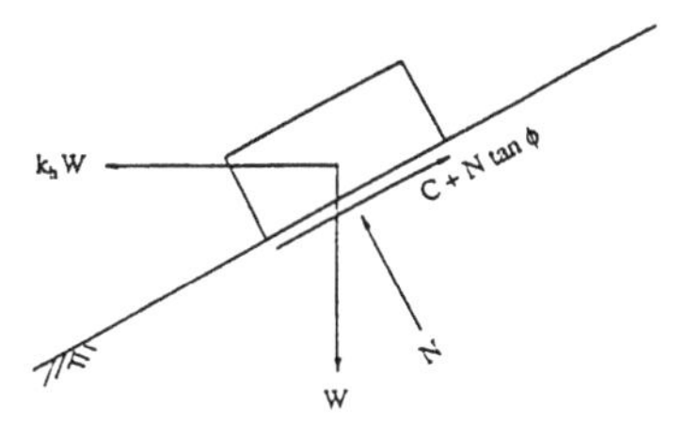

Fig. 10. Rigid block on an inclined plane.

In the following, an extended pseudo-static procedure based on the average acceleration data within the sliding mass(Chang, et al., 1984; Chen and Liu, 1990) is used to evaluate the accumulated displacement of a slope during an earthquake. The upper-bound limit analysis method is adopted to determine the slope failure mechanism, and the Newmark's concept(1965) is used to evaluate the movement of a slope subjected to earthquake excitation.

Newmark (1965) developed a procedure for evaluating the potential displacements of an embankment due to earthquake excitation. He illustrated his concept by considering a block sliding downward on an inclined plane (Fig.10), and comparing it with the downhill movement of a soil mass along a failure surface. The movements of the slope occur if the inertia force induced by earthquake on a potential sliding mass exceeds the *yield acceleration*. The movements stop when the inertia effect is reversed, i.e., a negative velocity or velocity heading uphill is not allowed in this analysis.

By searching the accelerations of the earthquake loading history relative to the yield acceleration and integrating the effective velocity on the sliding mass, the accumulated displacements during earthquake can be evaluated. Using the accumulated displacement, the safety of the slope can then be evaluated. The steps for performing this seismic displacement analysis are outlined as follows.

(1) Assume a failure mechanism and find its corresponding yield acceleration factor at which the downward movement occurs.
(2) From the acceleration time history of an earthquake, apply the calculated pseudo-static forces to the slope.
(3) When the applied acceleration exceeds the yield acceleration, downward movement starts to occur. The accumulated displace-

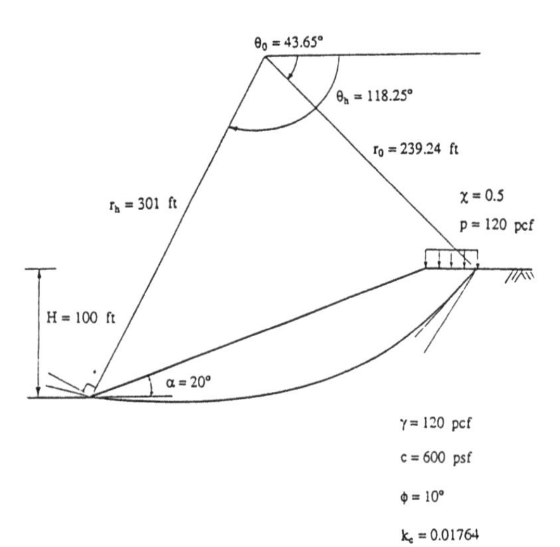

Fig. 11. Example problem.

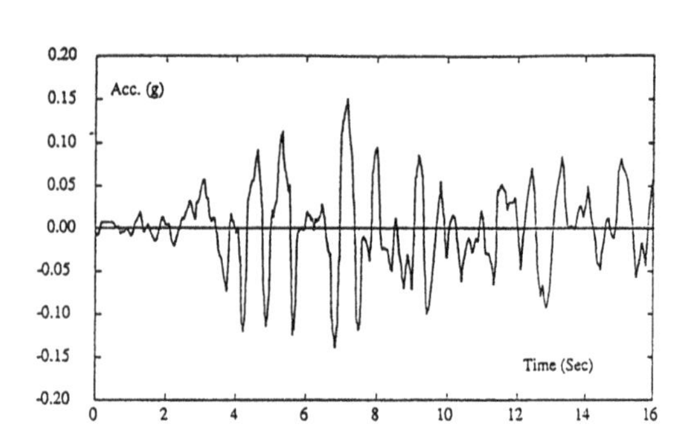

Fig. 12. Time history of acceleration history of Pasadena earthquake.

ment can be evaluated by integrating all the positive velocities throughout the time history.

A computer program for evaluating the accumulated displacement of a slope subjected to earthquake excitation based on the above procedure has been developed. The illustrated problem is shown in Fig.11. The design earthquake record selected is Pasadena earthquake (CALTECH, 1971) scaled to 0.15g (Fig.12). After solving the equilibrium problem, the yield acceleration factor for the logspiral failure surface is 0.017. For simplicity, this yield acceleration factor is assumed constant during earthquake. Horizontal and vertical displacements at the top of slope are shown in Fig.13. The displacement increases sharply close to the positive peak acceleration, and stops moving at some periods in which lower acceleration is continuously applied. In fact, the displacement at any point along the

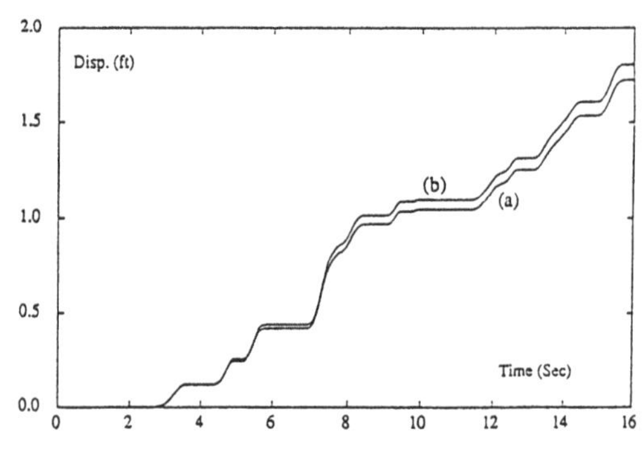

Fig. 13. (a) Horizontal and (b) Vertical displacement at the top of the failure zone.

failure surface can be obtained by simply introducing the corresponding angle θ.

2.4.2 Design Consideration of Rigid Retaining Structures

In the design of retaining structures, either ultimate-load-based or displacement-based, both the magnitude and the distribution of lateral earth pressure acting on the structure is of great importance. Most of the methods for assessing the earth pressure are developed based on translational wall movement and may not be suitable for other modes of wall movement. A method for assessing the earth pressure, taking into consideration the mode of wall movement, is therefore needed.

A so-called *modified Dubrova method* developed by Chang(1981) is developed for attempting to assess the acting point of resultant lateral earth pressures and to investigate the effect of wall movement including excitation force. Dependence of strength mobilization on wall movement is carefully studied. Proper distributions for strength parameters ϕ (the frictional angle), and δ (wall friction angle), which completely control the analyzed results, are suggested for three types of wall movement based on the consideration of change in state of stress in the backfill during wall yielding.

Figures 14 and 15 show the comparisons of static active and passive earth pressures based on the Dubrova's distribution and the recommended distribution considering the possibly rotational modes of movement. Also shown in the figures are Rankine solutions which represent the solutions of modified Dubrova's method for translational wall movement. It indicates that the calculated lateral earth pressure distribution is significantly influenced by the assumed distribution of wall friction mobilization. The

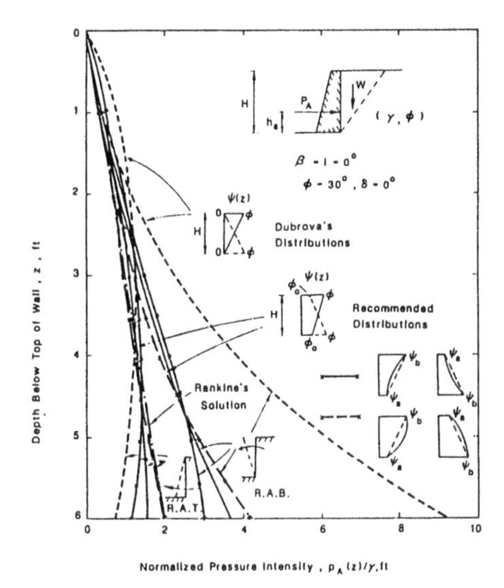

Fig. 14. Distribution of static active earth pressure based on different distributions of mobilized strength.

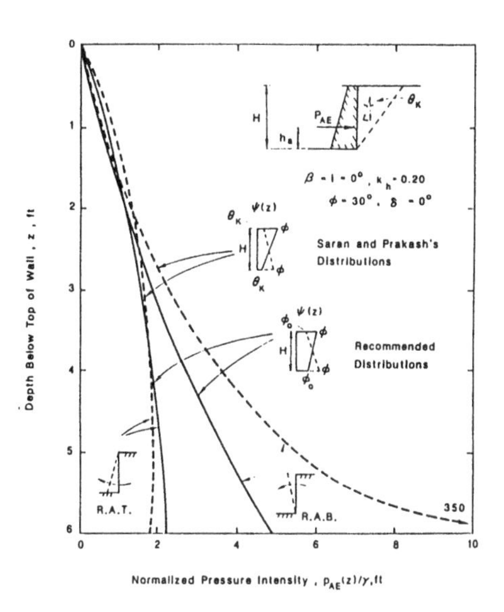

Fig. 16. Distribution of seismic active earth pressure based on different distributions of mobilized strength.

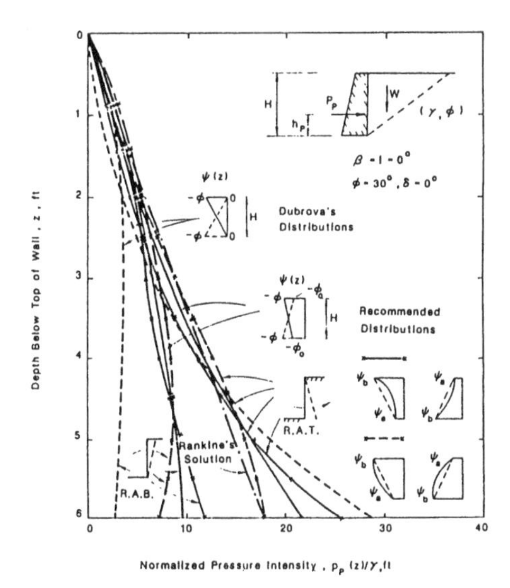

Fig. 15. Distribution of static passive earth pressure based on different distributions of mobilized strength.

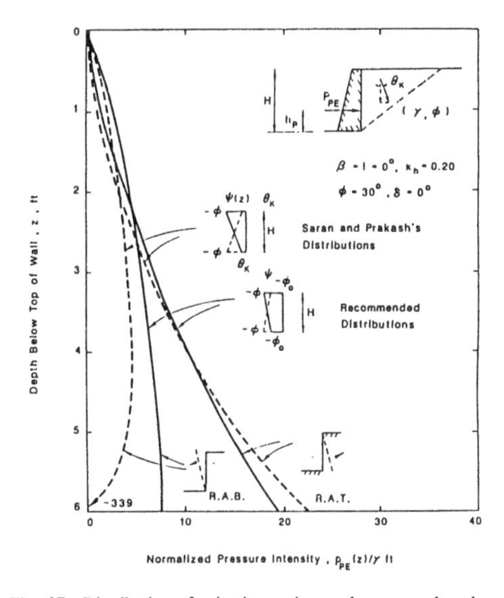

Fig. 17. Distribution of seismic passive earth pressure based on different distributions of mobilized strength.

corresponding acting points for those distribution can also be obtained from the earth pressure distribution. An apparent difference in the location of acting points for different modes of wall movement can be expected.

For seismic case with k_h = 0.2, the active and passive earth pressure corresponding to different distributions are shown in Figs. 16 and 17. The differences in both the resultant pressure and the acting points, similar to the active case, are larger for the case of rotation about the toe.

3 FINITE ELEMENT ANALYSIS OF CAP PLASTICITY

3.1 *The Cap Plasticity Model*

Drucker et al. (1957) first suggested that the soils might be modeled as a work-hardening plasticity

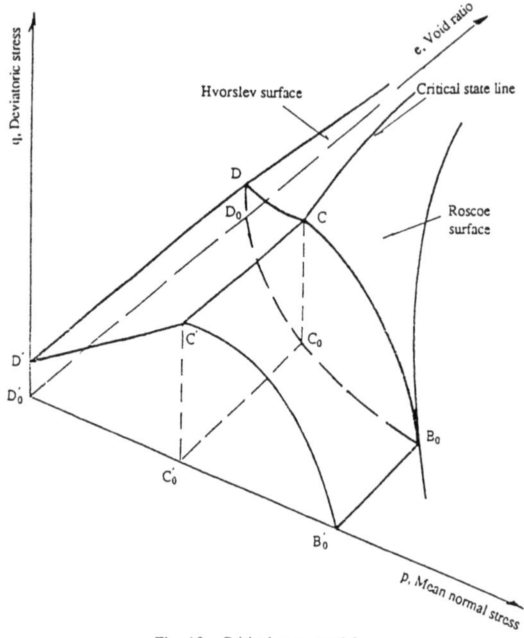

Fig. 18. Critical state model.

material. Then, Roscoe and his co-workers (1963, 1968) extended this idea and developed the so-called *Cam-clay model* and *modified Cam-clay model.* These models were based mainly on triaxial tests on remolded Kaolin clay under normally and lightly over-consolidated condition. In both models, associated flow rule and isotropic behavior were assumed. Elastic shear strain was assumed to be identically zero. For certain stress histories, strain-softening was allowed. In the *cap model,* the *state boundary surface* (including Roscoe and Hvorslev surfaces), and critical state line in the mean normal stress-deviatoric stress-void ratio are shown in Fig. 18.

State points below the state boundary surface are admissible, while those above the surface are not permitted. The *Roscoe and Hvorslev surfaces* which meet the critical state line determine respectively the behaviors of normally consolidated and overconsolidated soils. The state paths which lie below the state boundary surface are associated with elastic behavior, while those which lie on the state boundary surface are associated with strain hardening or softening.

The concept of the cap models has been further modified and extended by Dimaggio and Sandler and Baladi et al.(DiMaggio and Sandler, 1971; Sandler et al., 1976; Nelson and Baladi, 1977; Baladi and Rohani, 1979; Baladi and Sandler, 1980; Chen and Baladi, 1985) to analyze successfully many geotechnical problems, among others.

This so-called *generalized cap model* would allow for the fitting of a wide range of material properties and address to a number of soil behaviors through the concept of the generation of the yield surface and moving cap. These modifications involve anisotropy, strain-softening, cyclic loading behavior, etc. Using pseudo stress invariants (Baladi and Sandler, 1980), the transversely isotropic model can be constructed. Introducing plastic work into the equation of failure surface(Chen and Baladi, 1985), the softening behavior can be described. The hysteretic behavior under cyclic loading can be simulated by allowing the cap to contract in some way and permitting only elastic strain to occur during a reversed loading path(Baladi and Rohani, 1979).

In what follows, an *elliptic cap model* without softening is described. Other more advanced cap models can be found in the books by Chen and Baladi (1985), and Chen and Mizuno (1990).

The failure function may be taken of the Drucker-Prager type:

$$f_1 = \alpha \bar{I}_1 - J_2^{\frac{1}{2}} + k = 0 \tag{3.1}$$

or the function of the form (Dimaggio, et al., 1971) can be assumed as:

$$f_1 = J_2^{\frac{1}{2}} - [A - C \exp(B\bar{I}_1)] \tag{3.2}$$

where A, B, and C are material constants. $\bar{I}_1$ is the first invariant of the effective stress tensor, $J_2^{\frac{1}{2}}$ is the second invariant of deviatoric stress tensor, and α and k are material constants relating friction and cohesion of the soil.

The strain-hardening elliptic cap function has the usual form:

where X is the hardening function that depends on

$$f_2 = (\bar{I}_1 - L)^2 + R^2 J_2 - (X - L)^2 = 0 \tag{3.3}$$

the plastic volumetric strain, L is the $\bar{I}_1$ -value at the center of the elliptic cap, and R is the ratio of the major to minor axis of the elliptic cap (see Fig.19).

3.2 *The NFAP Program Development*

The *NFAP program* was originated by Chang(1980) for nonlinear large deformation finite element analysis of structures. The program was expanded and refined later by Mizuno (1981), and by McCarron (1985) to include the cap model for geotechnical engineering applications. Humphrey(1986a,b) condensed the early version of McCarron's program into a small program intended for use on personal computers.

NFAP program uses an incremental load-displacement analysis procedure. Iterative procedure with the Newton-Raphson or the modified Newton-Raphson method is adopted for solving nonlinear problems. Convergence is based on the difference between two successive displacement norms, and 1% is usually recommended in the numerical studies of geotechnical problems.

Two types of elements are used in the modified NFAP program for geotechnical problems. The first is a two- or three-node truss element which can carry axial load only and is used to simulate geotextile reinforcement. One- or multiple-layer geotextiles can be placed in an embankment above the original surface during construction. The material behavior of geotextile is treated linearly elastic or nonlinearly elastic. The second type of element is a two-dimensional, 4- to 8-node, isoparametric plane strain solid element. The cap model with a practical procedure(will be introduced later) for the determination of model parameters is used to represent the stress-strain relations of soils. The sandy soil is simulated as an elastic-plastic material with Drucker-Prager yield surface, which can be approximated by imposing a relative cap in a cap model.

The large deformation analysis in NFAP program is based on the assumption of a small strain and large rotation, in which the structural stiffness in global equilibrium equation is symmetric. In addition, the node-renumbering system is introduced to minimize the band width of the stiffness matrix for improving solution efficiency. Therefore, the alternative designs for a considered problem need only minor change in the original inputs and the analysis can still proceed. This is more useful for the design and analysis of engineering problems.

Moreover, the modification and extension of the NFAP program are further made for practical engineering problems. These include:

(1) Use double precision for accuracy of plasticity analysis.

(2) Determine initial stress state through an effective method.

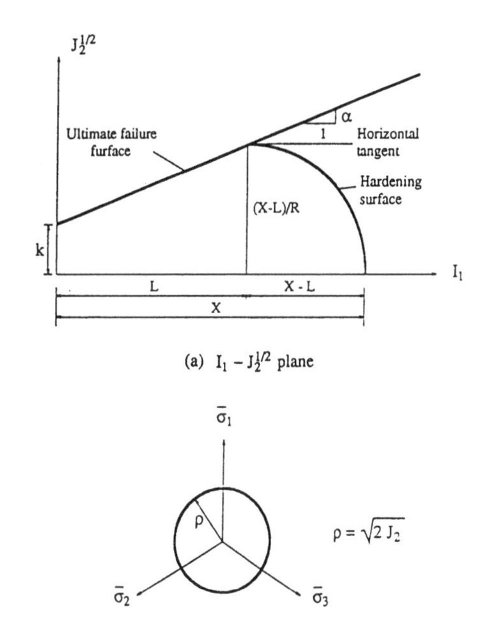

Fig. 19. Generalized cap model.

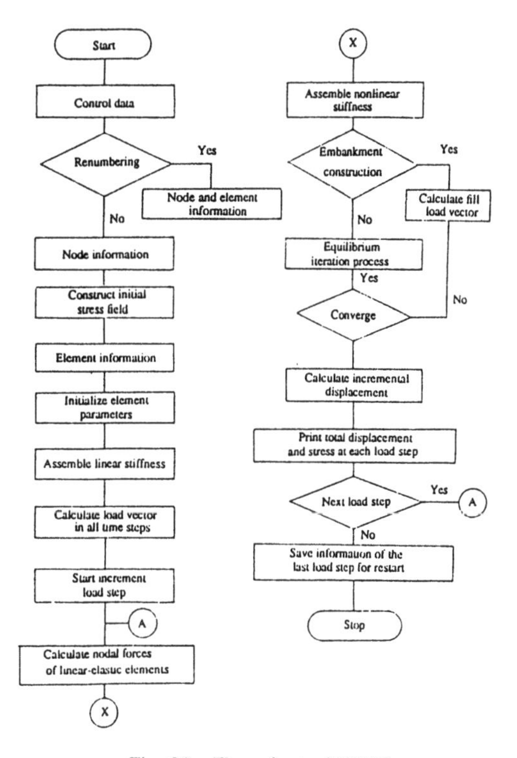

Fig. 20. Flow chart of NFAP

TABLE 1. Properties of Material Constants Required for Determining Cap Model Parameters

Constants	Explanation
K_{min}	minimum value of elastic bulk
ν	Poisson's ratio
e_0	initial void ratio
ϕ	frictional angle
c	cohesion
C_c	compression index
C_r	recompression index
S_u/σ'_{vo}	undrained shear strength to effective overburden pressure
OCR	overconsolidated ratio
T_c	tension cut-off
γ	unit weight of soil
K_0	coefficient of initial lateral earth pressure
β	pore water response factor

where K_f and K_e are the apparent bulk modulus of fluids, and the elastic bulk modulus, respectively.

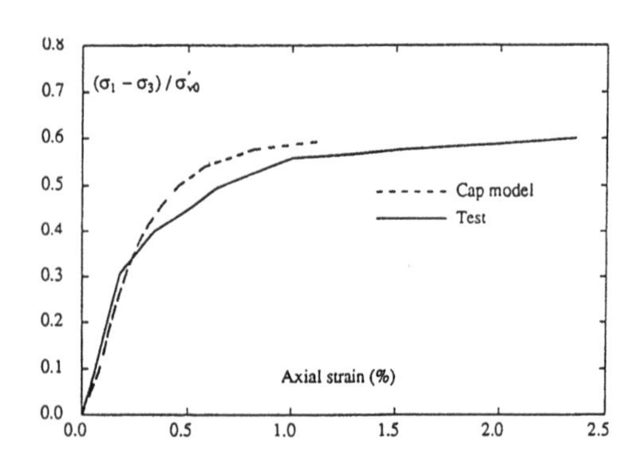

Fig. 21. Undrained triaxial compression tests for Boston blue clay under initial hydrostatic condition.

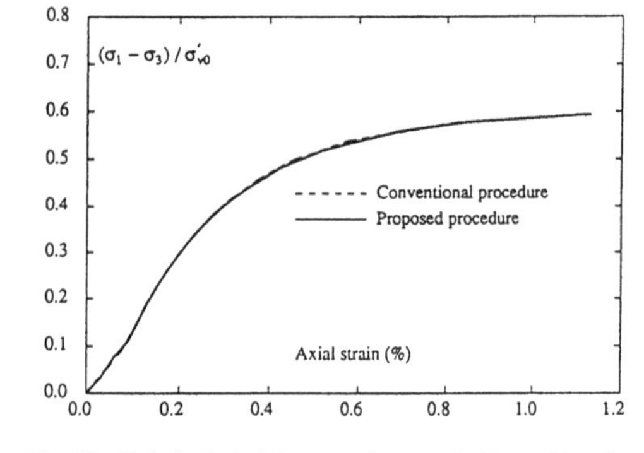

Fig. 22. Undrained triaxial compression tests for Boston blue clay under initial hydrostatic condition

(3) Make the restart procedure available in the analysis.
(4) Handle practical foundation with sloping boundary and varying water level.
(5) Make the analysis of considered problems more realistic.

The flow chart of NFAP program for the example of embankment construction is shown in Fig. 20.

3.3 *Some Recent Developments*

In the finite element analysis of geotechnical engineering problems, the cap plasticity model has received a wide acceptance(e.g. Nelson and Baladi, 1977; Baladi and Rohani, 1979; Chen and McCarron, 1984; Mizuno and Chen, 1984; Daddazio et al., 1987; McCarron and Chen, 1987; Huang and Chen, 1990). The cap plasticity model has been found appropriate for describing soil behavior, including the treatment of stress history, stress path

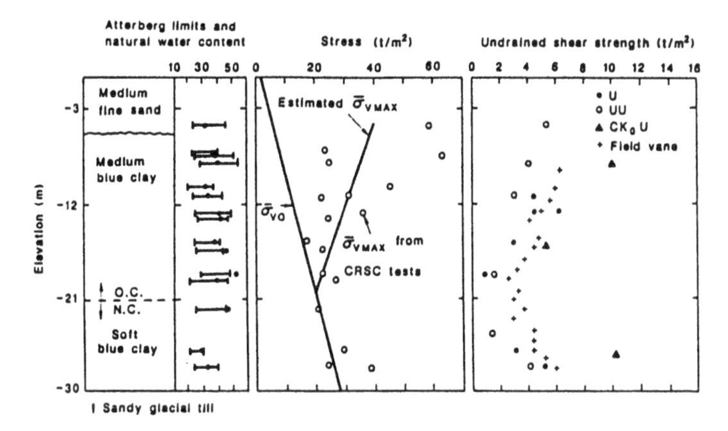

Fig. 23. Subsoil condition for MIT symposium test embankment (MIT, 1975).

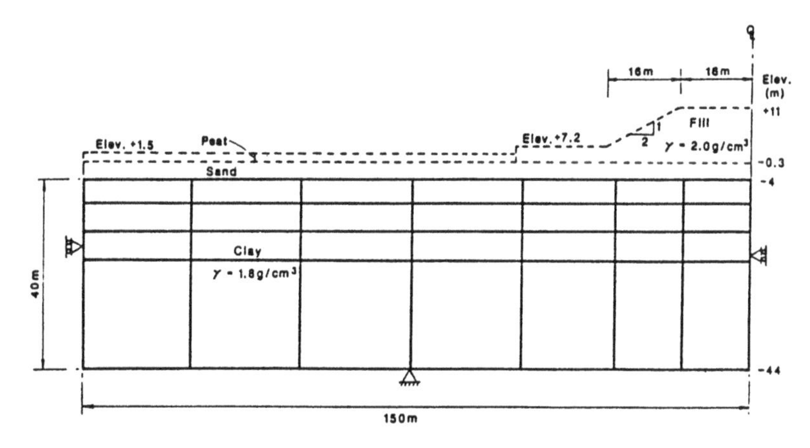

Fig. 24. Foundation model for MIT test embankment.

dependency, dilatancy, and the effect of the intermediate principal stress. However, practicing geotechnical engineers are generally not familiar with the important but difficult task of determining the model parameters. Often, they employ procedures, that were developed mostly by mechanicians from the engineering mechanics viewpoint, rather than from the viewpoint of commonly performed soil tests. To make the cap plasticity model friendly to users in the practical applications of geotechnical engineering problems, a simple and reliable procedure relating the cap model parameters to data available from conventional soil tests has been developed by Huang and Chen(1990) at Purdue University.

The properties of material required for the determination of the cap model parameters using the procedure proposed by Huang and Chen(1990) are summarized in Table 1. These soil constants are commonly used by geotechnical engineers. The unit soil weight γ, the coefficient of at-rest lateral earth pressure K_0, and the location of the ground water level are used to construct the initial stress field, i.e. the initial stress states of the Gaussian points of each finite elements. The pore pressure response factor, β, is used to simulate the effect of pressure(Hermann, ct al., 1982):

$$\beta = \frac{K_j}{K_e} \tag{3.4}$$

where K_f and K_e are the apparent bulk modulus of fluids, and the elastic bulk modulus, respectively.

To verify the effectiveness of the proposed procedure, comparisons were made with conventional procedure (Humphrey, 1986; Chen and McCarron, 1986) for determining the cap model parameters in an undrained triaxial test for Boston blue clay.

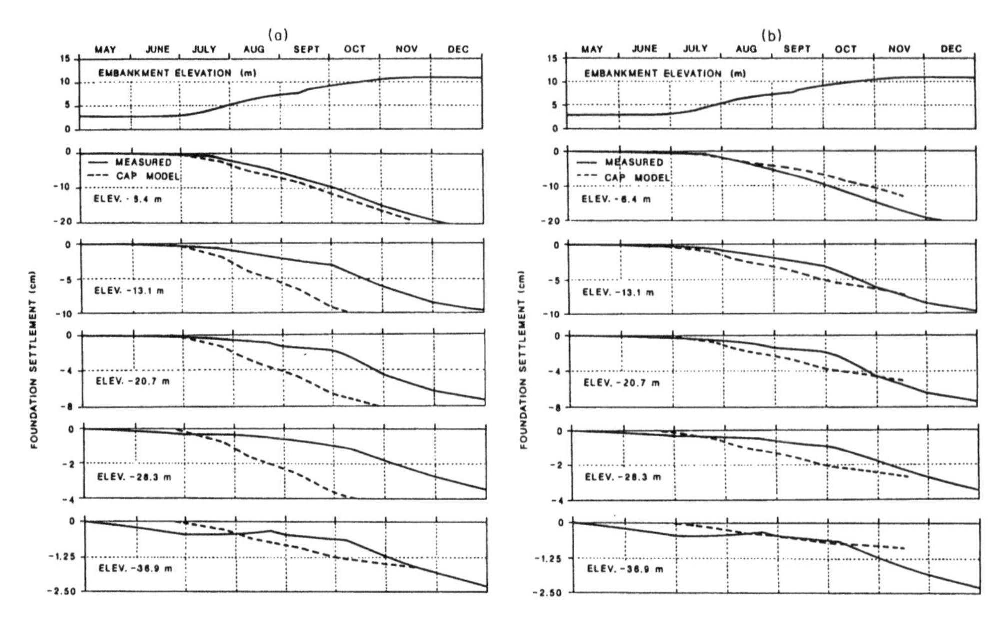

Fig. 25. Settlement below centerline of MIT test embankment:
(a) G = 9.6 MPa; (b) v = 0.1.

Figure 21 illustrates the good matching between the typical undrained response of Boston blue clay (Ladd, 1964) and the predicted response by the cap model through the conventional procedure. The complete stress-strain curve predicted by the two procedures(conventional and the proposed by Huang and Chen) are almost identical(Fig.22). The Huang and Chen's procedure has the following two main advantages over the conventional approach:

(1) The required constants(Table 1) are familiar to geotechnical engineers, while this familiarity is not found in the conventional procedure.

(2) The cap model parameters in the proposed procedure are determined uniquely from the commonly used soil tests. In the case when experimental work is not readily available, they can also be estimated from existing references about soil properties. In the conventional procedure, some parameters are interdependent, and different sets of parameters can lead to similar results. This makes the trial-and-error curve-fitting process difficult and time-consuming.

A more detailed description about the proposed procedure to determine the cap model parameters can be found in the paper by Haung and Chen (1990).

3.4 *Some Engineering Applications Using NFAP*

3.4.1 Boston Blue Clay

An embankment constructed on Boston blue clay using finite element program NFAP is presented. Predictions are compared with actual measured response. Figure 23 describes the initial subsoil conditions at the location of the MIT Symposium test embankment(MIT, 1975). The embankment geometry and finite element description of the foundation for MIT test embankment are shown in Fig. 24. In the analysis, the first layer is assumed to be drained and a constant Poisson's ratio (0.1) is adopted. The initial cap locations and K_0 values for the MIT test embankment are given below:

for layer 1 L = 0.40 MPa, K_0 = 1.20
for layer 2 L = 0.42 MPa, K_0 = 0.80
for layer 3 L = 0.37 MPa, K_0 = 0.55
for layer 4 L = normally consolidated condition, K_0 = 0.50

Figures 25 through 27 compare the predicted and measured deformations and pore pressure within the foundation. Examination of the displacement along the embankment(Fig.25) reveals that the recorded response of the normally consolidated layer (below El. -21.3 m) is initially stiffer than the predicted range. However, as loading continues, the response falls within this range. The predicted pore pressures(Fig.26) at the end of construction compares well with the measured values. However, beneath the centerline of the embankment, where the pore pressures are greatest, the predicted values are too high. Figure 27 compares the predicted and measured lateral displacements within the foundation at the end of construction. A good prediction can be observed from these results.

3.4.2 Safety Evaluation of a Dam

A suitable procedure for evaluating the dam safety after the fill and drawdown of a reservoir is proposed here. Firstly, the piezometric heads at different points in a dam after the fill and drawdown of a reservoir are obtained with a trial-and-error procedure. Then, the numerical analysis of a dam is taken using the finite element method in which the cap model is used for representing soil behaviors and a specific technique to handle the effects of seepage and drawdown is introduced. An example of reservoir completed recently in Taiwan is illustrated. From the results, it shows that the dam for the reservoir is good from both the safety and stability standpoints.

For seepage effect, a finite element procedure proposed by Lacy and Prevost(1986) to locate the free surface of a flow through a dam is used for evaluating the pressure within a dam. Then the seepage force of each element in the finite element mesh can be computed for further analysis. This method is based on the penalized formulation, stemming from a weak form of an equality in terms of the pressure variable. It involves equations and meshes which are consistent with a general formulation of pore pressure and soil displacement for geotechnical engineering problems.

Based on the finite element numerical procedure proposed by Lacy and Prevost(1986), the Galerkin approximation of the weak formulation for solving seepage problem in a dam can be expressed in the following form:

$$K\,P = f \tag{3.5}$$

where P is the vector of nodal pressure, K is the vector of global stiffness assembled from the contributed elements as:

$$K = \sum_1^N (K_{ab})^e \tag{3.6}$$

where N is the number of elements in the finite element mesh, and

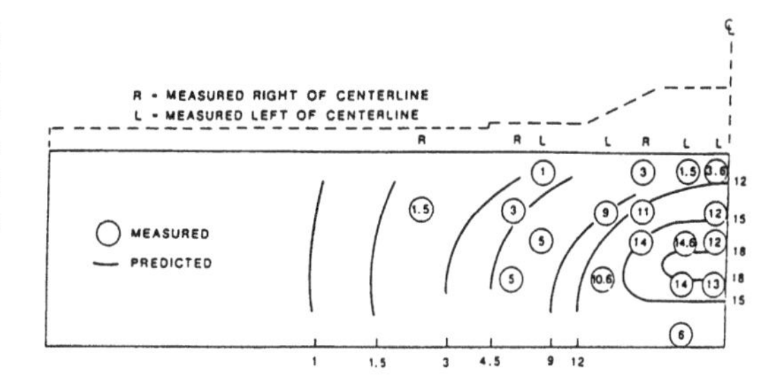

Fig. 26. Measured and predicted excess pore pressure (meters of water) at the MIT test embankment.

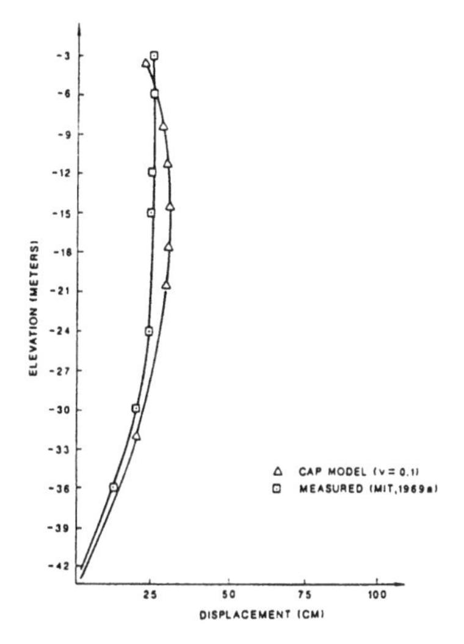

Fig. 27. Lateral displacement within MIT embankment foundation

$$(K_{ab})^e = \int_{\Omega^e} \frac{1}{\gamma}\, \Delta N_a\, k_p\, \Delta N_b\, d\Omega_e \tag{3.7}$$

is the local stiffness contributed from element e in the entire domain of a dam, Ω. N_a and N_b are the shape functions respectively with nodes a and b in element e. k_p is the Darcy permeability, γ is the unit weight of the fluid. Δ is the gradient operator of a scalar.

In Eq.(3.5), $\underline{f}$ is the vector of nodal forces in a form as:

$$f = \sum_1^N (f_a)^e \tag{3.8}$$

where

$$(f_a)^e = \int_{\Omega^e} H_\varepsilon(P)\, \Delta N_a\, k_p\, \Delta y\, d\Omega_e \tag{3.9}$$

is the local nodal force of element e. y is the vertical

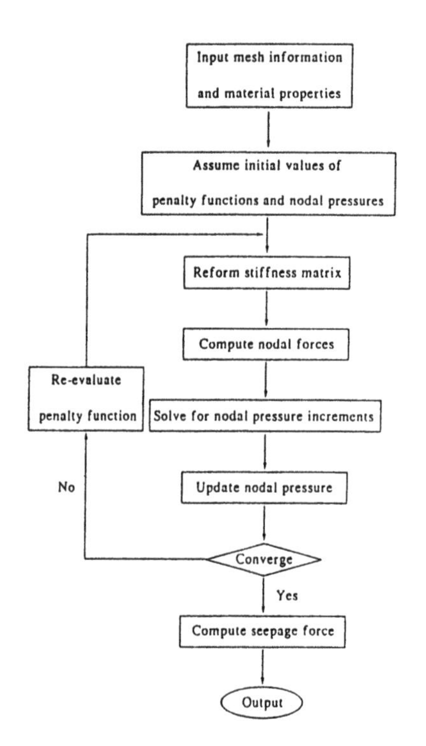

Fig. 28. Flowchart of solution procedure for seepage analysis.

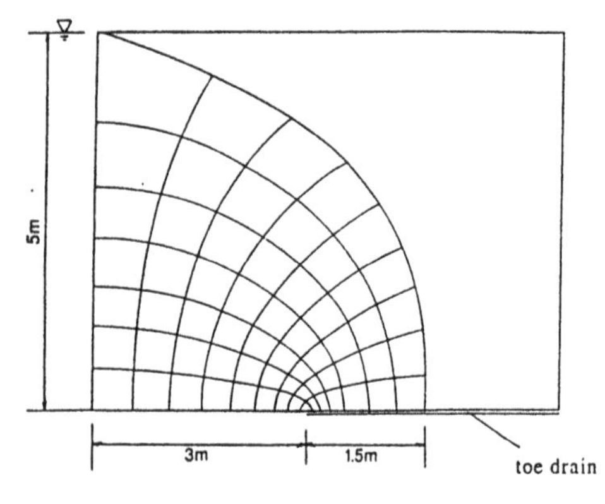

Fig. 29. Rectangular dam with toe drain.

coordinate of the node a in element e. N_a and k_p are defined similarly as before. The $H_\varepsilon(P)$ is the penalty function, namely:

$$H_\varepsilon(P) = \begin{cases} 1 & \text{if } P > \varepsilon \\ (1/\varepsilon)P & \text{if } 0 < P \leq \varepsilon \\ 0 & \text{if } P \leq 0 \end{cases} \tag{3.10}$$

Brezis and his co-workers (1978) show the existence and uniqueness of the pressure P to the penalized problem, as ε reaches a smaller value. The solution procedure for the analysis of seepage through a dam is shown in the flowchart (see Fig.28). The nonlinear systematic equation is solved iteratively by a series of successive substitution. At each iteration, the vectors of stiffness matrix and forces are reformed. In the first iteration, the penalty function $H_\varepsilon(p)$ can be assumed as one throughout the entire domain. In the subsequent iteration, the vectors of stiffness matrix and forces are pressure-dependent. The progress is terminated when the incremental solution satisfies convergence requirements.

Figure 29 shows the example solution using the above procedure to determine the pressure field in a rectangular dam with a toe drain. As for the drawdown of a reservoir, a similar procedure can be taken by imposing the pressure to remain at zero on the assumed free surface. Then the pressure field can be computed for further analysis.

The analysis of the dam after the fill and drawdown of the reservoir was performed with modified NFAP finite element computer program. The procedure to analyze the effect of a dam after the fill and drawdown of the reservoir can be summarized as follows:

(1) Evaluate the initial stress state of a dam prior to the fill of the reservoir. This can be achieved by modeling the dam construction with a gravity buildup to the final stage.
(2) For a given fill level of the reservoir, apply the numerical procedure proposed by Lacy and Prevost (1986) to compute the pressure field within a dam. Then, impose the seepage force and buoyant force at the nodes of finite element analysis mesh to study the behavior of a dam after the fill of the reservoir.
(3) In the analysis of a dam after drawdown of the reservoir, a specified free surface within the dam is selected. Use the same procedure as that in (2) for computing the pressure field during drawdown. It should be noted that the undrained condition is assumed, and the final stress state during the fill is used as the initial stress state in the analysis of drawdown.

The computed results using the NFAP computer program will be incorporated with computer graphic technique. The graphic results include deformed shape, the principal stresses and the local safety factor in each element. The principal stress in each element is evaluated based on the average of the Gaussian point values. The local safety factor in

TABLE 2. Material Constants of Divided Zones for Li-Yue-Tan Dam

Zone	1	2	3	4	5	6	7	8	9	10	11	12
$K_{min}(T/m^2)$	23000	19000	11000	20000	10000	10000	9000	8800	12000	6500	2100	2500
ν	0.35	0.35	0.35	0.35	0.35	0.35	0.35	0.35	0.35	0.35	0.35	0.35
e_0	0.606	0.60	0.335	0.606	0.361	0.361	0.308	0.492	0.452	0.308	0.422	0.283
$\varphi(°)$	23.1	44.5	38.2	38.6	30.7	37.3	38.3	23.9	33.5	35.5	26.4	36.0
$C(T/m^2)$	8.30	8.30	0.80	3.40	3.10	5.30	6.00	7.00	6.80	1.90	7.30	1.30
C_c	0.169	0.025	0.076	0.076	0.113	0.113	0.069	0.103	0.105	0.069	0.098	0.058
C_r	0.015	0.005	0.013	0.013	0.020	0.020	0.010	0.015	0.011	0.010	0.020	0.015
S_u/σ'_{v0}	0.35	0.42	0.38	0.38	0.35	0.38	0.38	0.35	0.38	0.38	0.35	0.38
OCR	10.0	10.0	10.0	10.0	10.0	10.0	10.0	10.0	10.0	10.0	10.0	10.0
$T_c(T/m^2)$	0.01	0.01	0.01	0.01	0.01	0.01	0.01	0.01	0.01	0.01	0.01	0.01
$\gamma\ (T/m^3)$	2.04	2.18	2.16	2.16	2.16	2.25	2.17	2.19	2.20	2.17	2.14	2.21
K_0	0.8	0.8	0.8	0.8	0.8	0.8	0.8	0.8	0.8	0.8	0.8	0.08
β	0 or 10	0	0	0 or 10	0 or 10	0 or 10	0 or 10	0 or 10	0 or 10	0 or 10	0 or 10	0 or 10

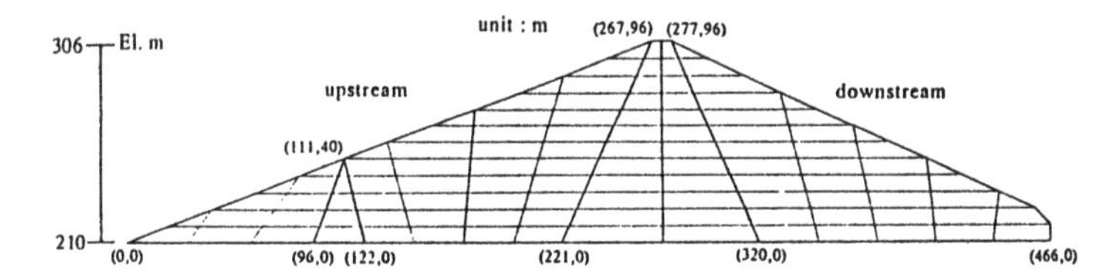

Fig. 30. Finite element mesh of Li-Yue-Tan dam.

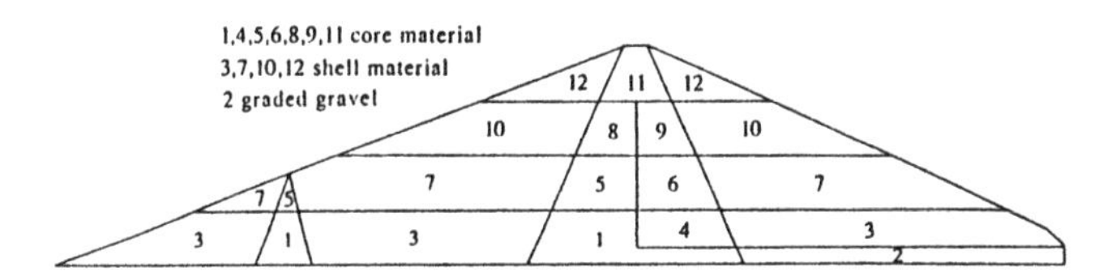

Fig. 31. Divided zones of Li-Yue-Tan dam.

each element is defined as the ratio between the strength and the applied deviatoric stress component under the same hydrostatic pressure in the $\bar{I}_1 - J_2^{\frac{1}{2}}$ plane.

The studied example is the dam of the Li-Yue-Tan reservoir, located in the central part of Taiwan for drinking water and irrigation. The height of the dam is 96 m (from El. 210 m to El. 306m) and the length along the top of the dam is 235 m. The maximum cross section near the middle of the dam is selected for study and its finite element mesh is shown in Fig.30. In the mesh, 365 nodes and 112 eight-node isoparametric plane-strain solid elements are used in the finite element discretization. The properties of the shell and core materials are obtained through the in-situ sampling and testing in the laboratory during different construction stages. Based on the testing results, the dam can be divided into 12 zones as shown in Fig. 31, and the related soil constants for determining the cap model parameters in each zone are summed up in Table 2.

Firstly, the mechanical behavior of a dam after a period of time after final construction is evaluated by simulating the construction of a dam with a gravity buildup of each layer and an assumed drained condition. The final deformation, stress state and

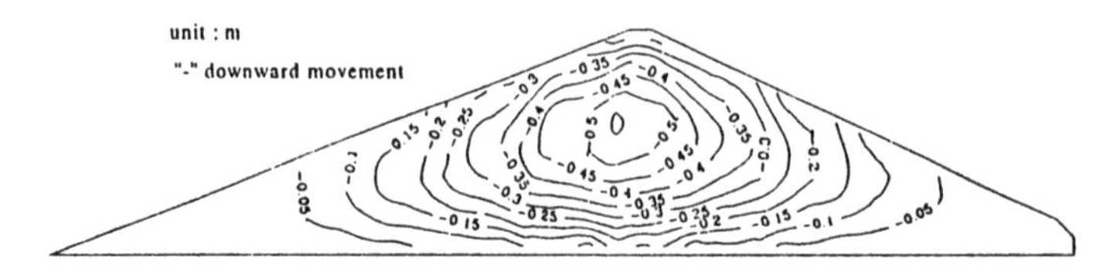

Fig. 32. Contour lines of vertical displacement after completion of the dam.

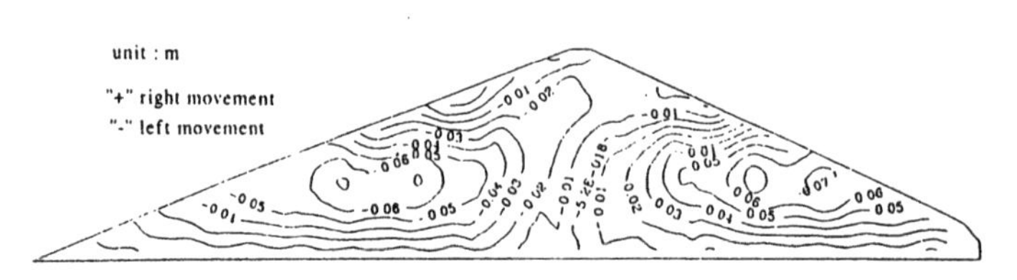

Fig. 33. Contour lines of horizontal displacement after completion of the dam.

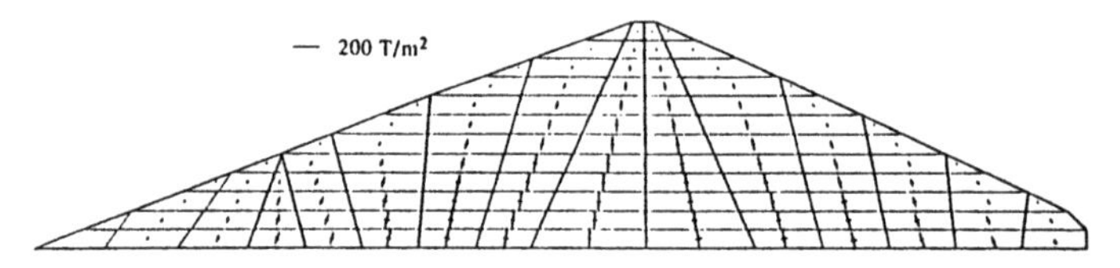

Fig. 34. Principal stresses after completion of the dam.

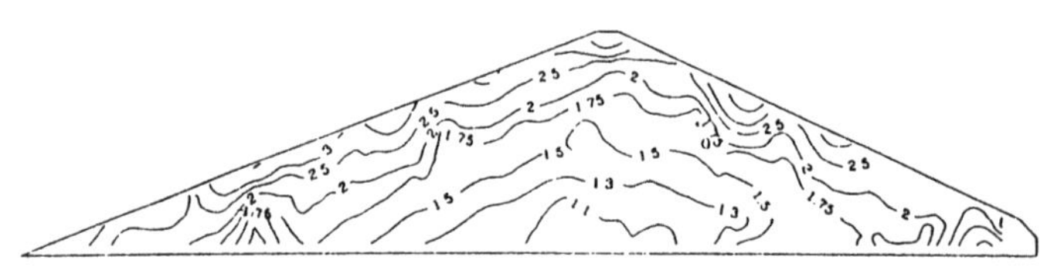

Fig. 35. Contour lines of local safety factor after completion of the dam.

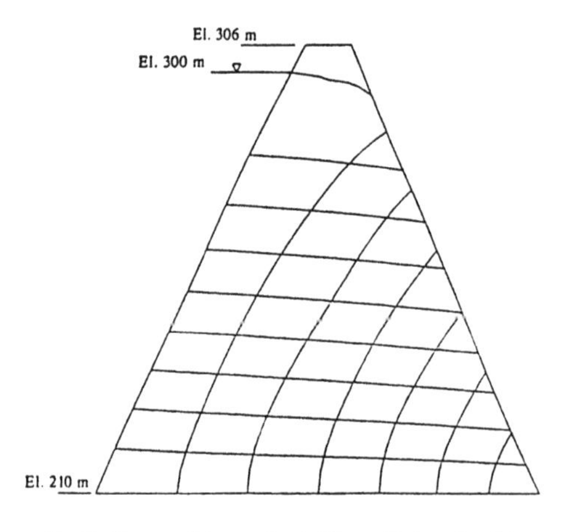

Fig. 36. Flow net of the core after the fill up to El. 300m (k_x - $4k_y$).

local safety factor within a dam are shown in Figs.32-35, and summarized as follows:

(1) The maximum settlement is found to be close to the central part of the core with a value of about 0.55 m.

(2) The spreading of the core is limited to both sides of the shell. The maximum spreading occurs at the surface of downstream shell with a value of about 0.08m.

(3) The principal stresses are very much following dam shape.

The low safety factor for stress concentration is found at the lower part of the core. However, the core is strongly confined by wide and extensive area of safe shells with high safety factor. Therefore, a stability of the dam is expected.

In the following, an analysis of the fill is studied for the design with a water level up to El. 300 m (6 m below the top of the dam). As mentioned before, the seepage effect is evaluated by the numerical

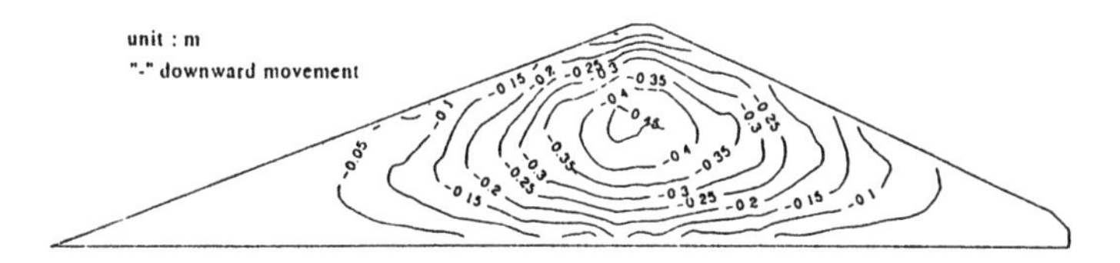

Fig. 37. Contour lines of vertical displacements after fill up to El. 300m.

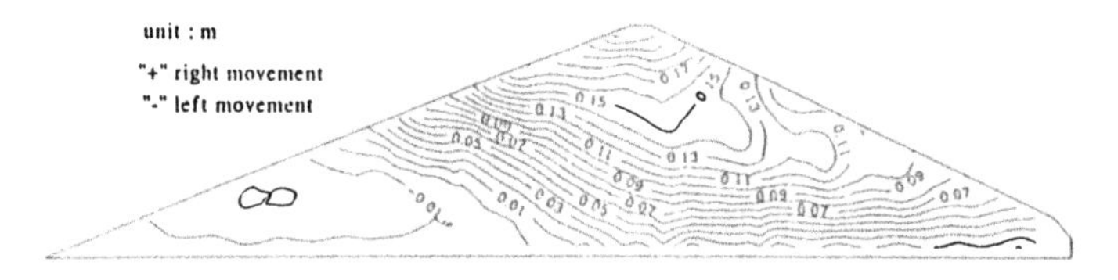

Fig. 38. Contour lines of horizontal displacement after the fill up to El. 300 m.

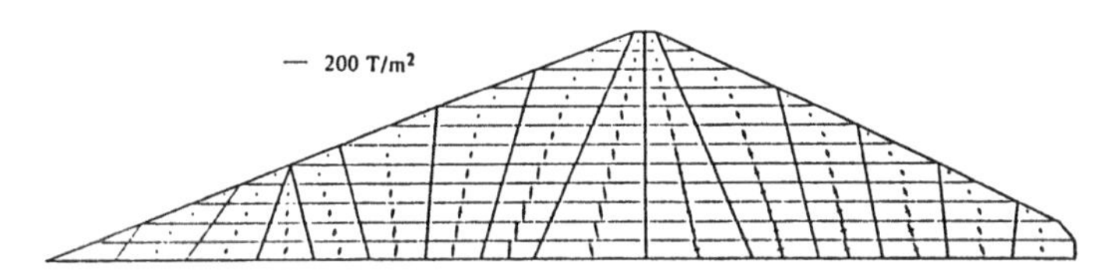

Fig. 39. Principal stresses after the fill up to El. 300 m.

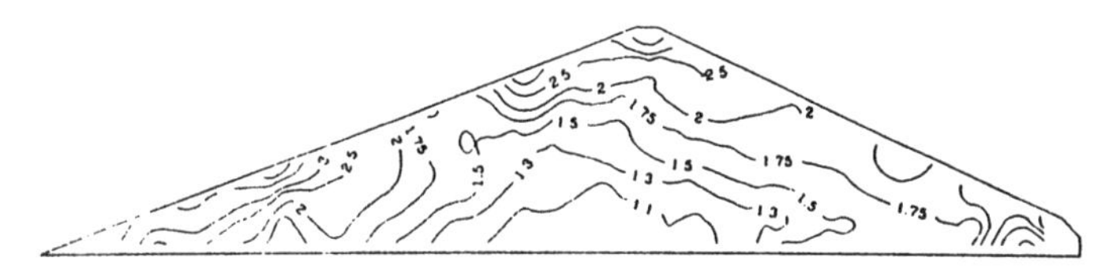

Fig. 40. Contour lines of local safety factor after the fill up to El. 300 m.

scheme (Lacy and Prevost, 1986) and the flow net through the dam is shown in Fig.36. In this figure, the seepage effect is considered only in the core because the permeability is much higher in the shell than in the core. The anisotropy of permeability in the core along the horizontal to vertical direction is assumed to be at the ratio of 4:1.

From the results of seepage analysis, the nodal buoyant force and/or seepage force of the finite element mesh in the upstream shell and core are computed. Considering the initial stress state induced in the previous stage and the computed nodal forces due to seepage effect, the final deformation, stress state and local safety factor are shown in Figs.37-40. Discussions on various results are summarized as follows:

(1) The maximum settlement in the core after the fill becomes 0.45 m. When compared

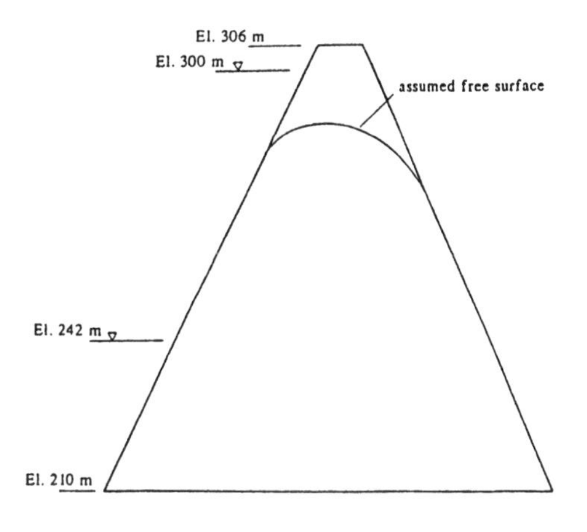

Fig. 41. Location of free surface after sudden drawdown to El. 242 m.

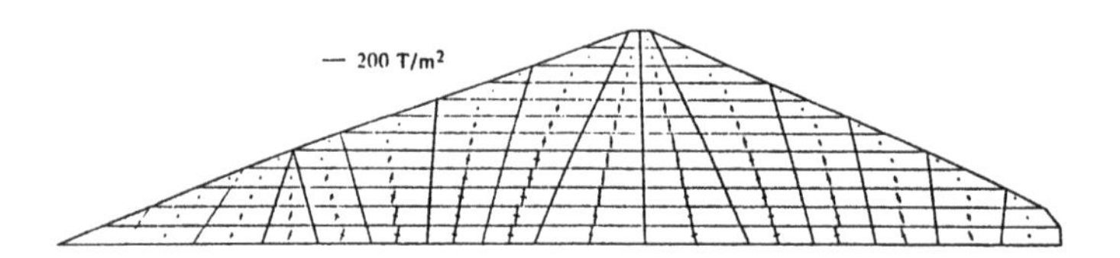

Fig. 42. Principal stresses after sudden drawdown to El. 242 m.

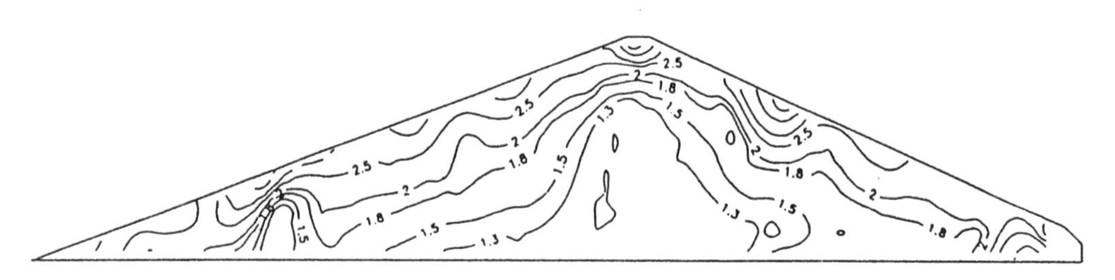

Fig. 43. Contour lines of local safety factor after sudden drawdown to El. 242 m.

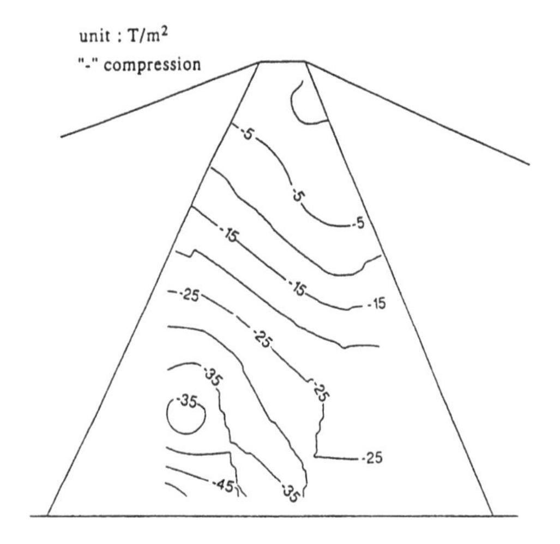

Fig. 44. Pore pressure in the core after sudden drawdown to El. 242 m.

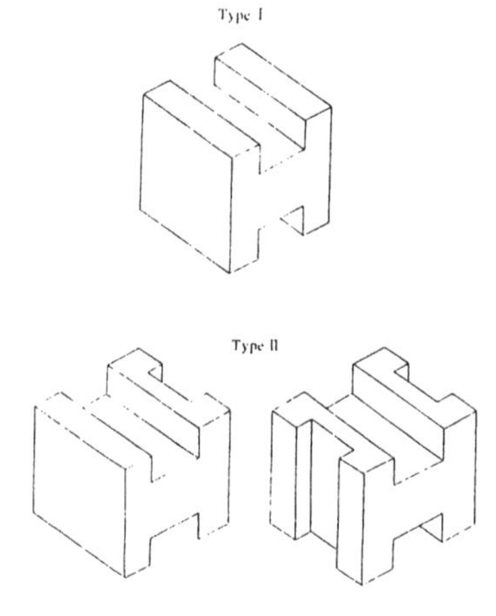

Fig. 45. Concrete blocks.

this settlement with that of previous stage (after the completion of the dam), the total rebound is approximately 10 cm.

(2) The maximum spreading with a value of 0.22 m is found at the top of the dam.

(3) The stress state and the direction of principal stress are changed in a good compliance with the buoyant and seepage effects occurring in the areas of upstream shell and core. The zone of lower safety factor at the lower part of the core extends and propagates wider and upper when compared with that of previous stage. However, the thick shell with high safety factor can still provide a strong resistance against the core movement.

Finally, the safety evaluation of the dam after sudden drawdown from El.300 m to El.252 m is analyzed. The critical free surface is specified as shown in Fig.41. Then, in a similar manner, the nodal forces due to seepage and buoyant effects are computed by using a finite element analysis. It should be noted that seepage force and buoyant force act, respectively along the upstream direction and downward direction at this stage. This has a great effect on the stress state within the dam after sudden drawdown occurs.

With the initial stress state obtained from the analyzed results of the fill and undrained condition assumed in the core, the final stress state and local safety factor are computed and are shown in

Figs.42-43. The effective stress is reduced because of undrained condition. The zone of lower safety factor propagates up and widely in the core area. However, the dam still remains in a good condition with a proper confinement by the safe shell.

As for the induced pore water pressure, the distribution of pore water pressure in the core is shown in Fig.44. It can be seen that a higher value is induced in the upstream side of the lower part of the core. For more safety consideration, the drawdown from El. 300 m to El. 242 m is recommended to operate in a slow and progressive way to let pore water pressure dissipate to some extent during drawdown.

4 GENERALIZED LIMIT ANALYSIS - THE FINITE BLOCK METHOD

4.1 *Basic Concepts*

The finite block method discussed here is originally developed by Shi(1988). Some of the basic concepts of the *finite block method* can be found in earlier works, such as the *discrete element method* by Cundall(1971) and the *discrete structure models* by Kawai (1977).

The whole discontinuous block system generated through the existing separate surface or assumed failure mechanism is considered in the finite block method. A system of equilibrium equations for the block assemblage is derived through the minimization of the total potential energy. These equations are then solved iteratively in time increment under a specific loading condition until the constraining requirements of no tension and no penetration between blocks are fulfilled. A complete kinematic theory is developed by Shi for determining a large assemblage of blocks to move and deform without tension and penetration between blocks.

The path of large displacements and large deformations can be accurately traced through a series of time increments in which a relatively small block deformation is limited in each step. Practical applications of the finite block method can be used to perform the stability analysis and support designs for tunnels, slope, retaining walls, dam abutments and foundations, etc. More detailed descriptions about the finite block method can be found in the Shi's thesis (1988).

4.2 *Historical Sketch*

Much work has been conducted in applying the finite element method for geotechnical problems. Although

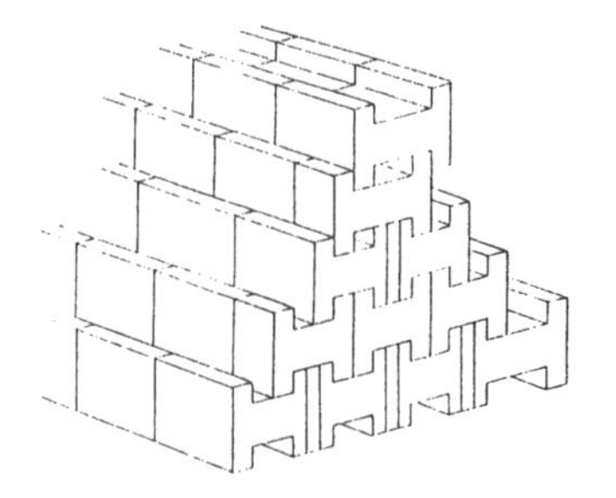

Fig. 46. Typical retaining walls by concrete blocks.

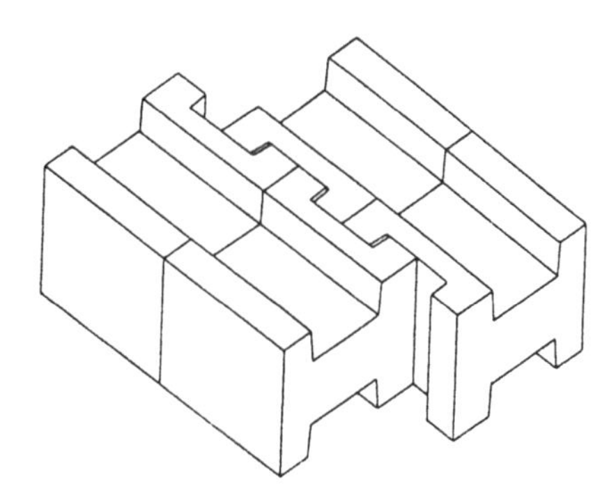

Fig. 47. Interconnection between Type II concrete blocks.

much success can be credited to the finite element method, most researchers would agree that the method has many limitations and difficulties. For example, finite element applications typically require the idealization of soils as elastic-plastic materials and a relatively fine mesh size to capture limit loads. The finite element method is more suited to elastic, *continuous* materials that exhibit equally high tensile and compressive strength. Soils are in the most part nonlinear discontinuous materials typically strong in compression, but weak in tension. Many of the basic assumptions implicit in the finite element method, such as compatibility and continuity, do not apply to soils. Only through significant modifications, the finite block method has been adapted for geotechnical problems.

Limit analysis is recognized as the most powerful procedure for estimating the collapse load of numerous geotechnical problems in a direct manner. In existing techniques of limit analysis, a lower-bound solution is obtained when the load is determined from a distribution of stress alone. The upper-bound technique is concerned mainly with kinematics and consider only the failure mode and the energy equilibrium, while the stress distribution need not be in equilibrium. The proof of the limit theorems is based on the fundamental assumption of perfect

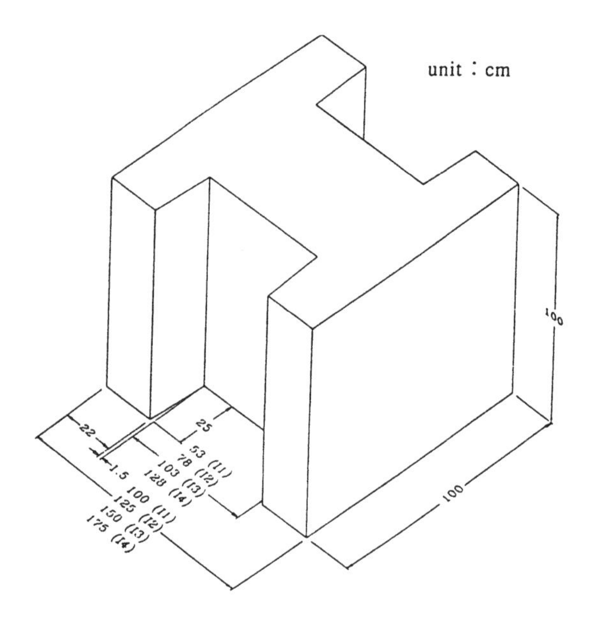

Fig. 48. Detailed dimensions of Type I concrete blocks (I1-I4).

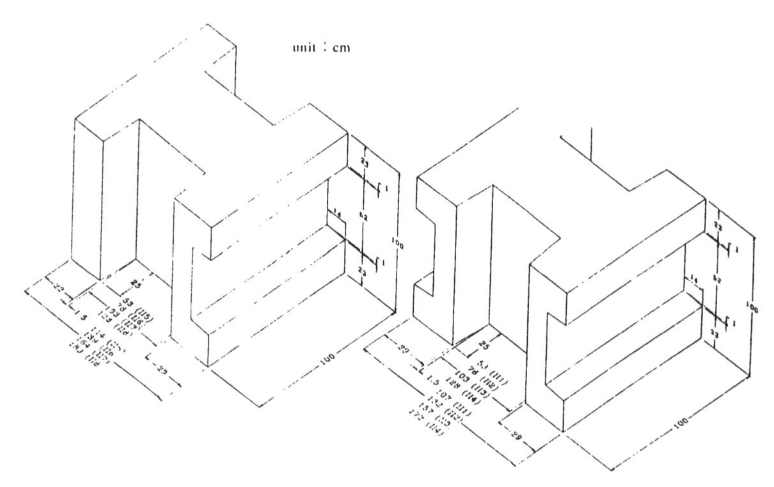

Fig. 49. Detailed dimensions of type II concrete blocks (II1-II4, II5-II8).

plasticity with associated flow rule for the material. If plasticity is not appropriate for the soil, then the answer obtained will not be very significant. Proper alternative theorem and method need be developed. The finite block method may provide such a reasonable alternative for the soil idealized as a frictional material.

The finite block method contains characteristics from both the upper-bound and lower-bound techniques. The method is similar to the upper-bound technique in that a failure mode must be predetermined, block boundaries must be defined by the user, and the solution is based on energy equilibrium. The frictional dissipation requires the normal force on the plane of sliding. Meanwhile, the finite block method satisfies force equilibrium and provide the stress states for all blocks. The capability of the finite block method to model blocks as frictional materials is very significant. Coulomb's law is applied to the contact surfaces when blocks are in contact. Existing upper-bound technique is largely based on materials assuming perfect plasticity. The

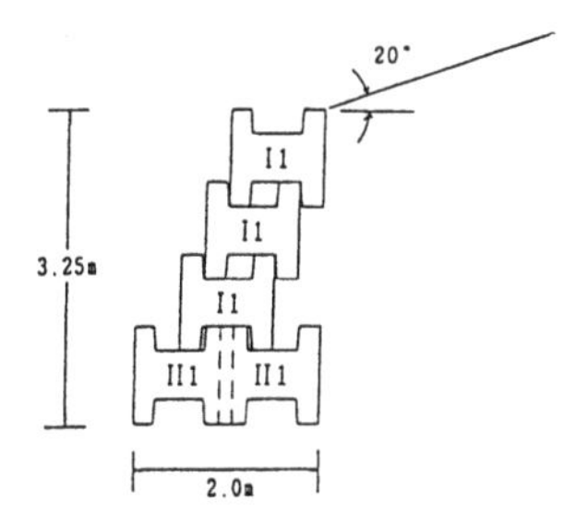

Fig. 50. Example one of proposed retaining wall.

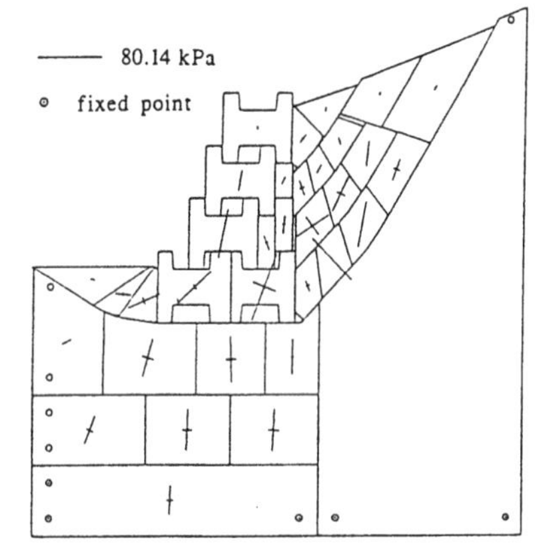

Fig. 51. Deformation and principal stresses of example one.

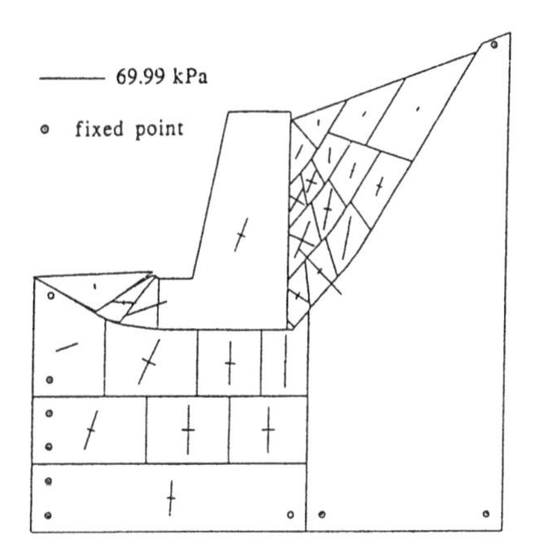

Fig. 52. Deformation and principal stresses of example one (with massive wall).

difficulty of determining normal contact forces between blocks has previously impeded the modelling of frictional materials. Real soils are quite complex and are neither truly frictional in behavior, nor are they plastic. They are probably somewhere inbetween. While upper-bound techniques have

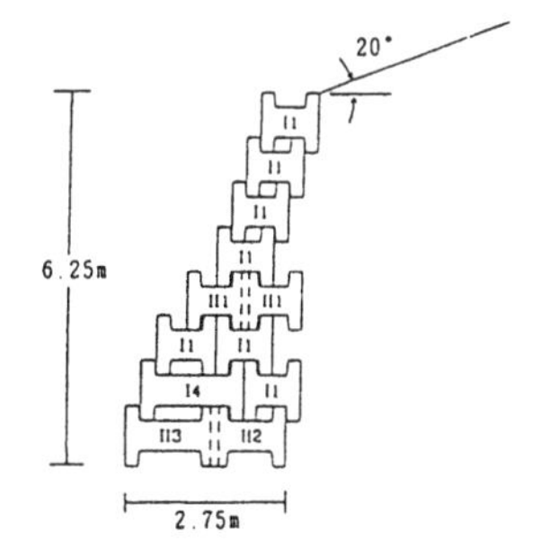

Fig. 53. Example two of proposed retaining wall.

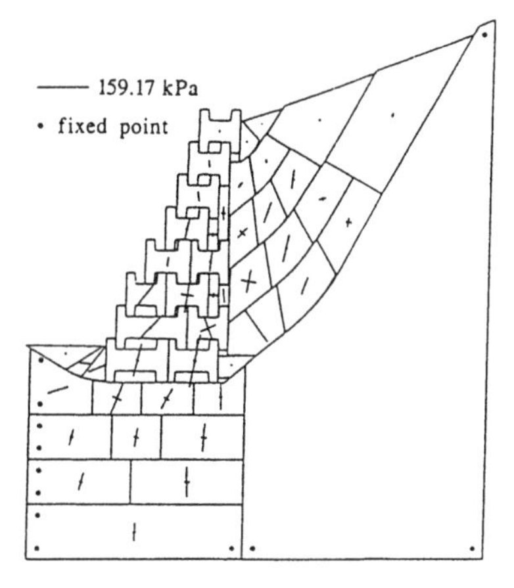

Fig. 54. Deformation and principal stresses of example two.

always provided the ideal plastic solution on the one extreme, the finite block method can now give us the friction solution on the other extreme.

In short, the finite block method extends the existing limit analysis techniques to frictional materials and reduces the range in which the true solution is bracketed by the upper and lower bounds. Overall, the finite block method more resembles the upper-bound techniques since a failure mechanism must be initially defined by the user. As the method is based on the assumption of frictional material and stress equilibrium is satisfied, the limit load would thus be lower than the solution obtained by an upper-bound technique based on the assumption of perfect plasticity. By trying out different configurations for the failure mechanism, the solution that provides the lowest limit load would be much close to the exact solution.

4.3 *Some Recent Development*

The original version of finite block method is

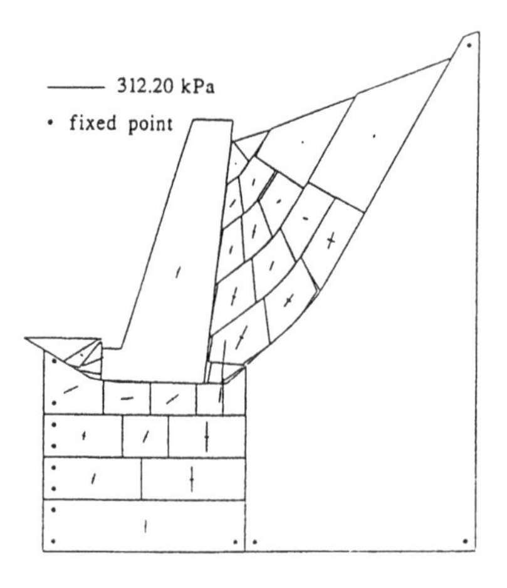

Fig. 55. Deformation and principal stresses of example two (with massive wall).

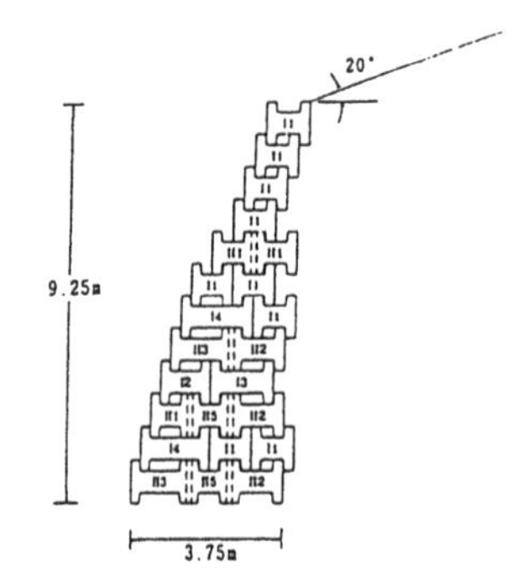

Fig. 56. Example three of proposed retaining wall.

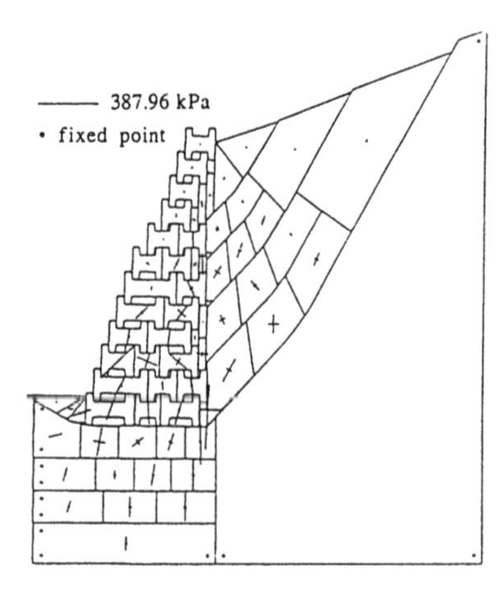

Fig 57. Deformation and principal stresses of example three.

limited to constant-strain state in each block for simplification in the construction of equilibrium. This may not be suitable for problems with large blocks or stress concentration. Therefore, a higher order displacement function was assumed and combined with finite element technique for solving more complex geotechnical problems. On the other hand, the blocks may induce cracks subjected to higher load. How to trace the post-crack behavior after crack occurs is under researching among a number of universities. Recently, the finite block method is extended to the study on the micromechanic behavior for granular soils.

4.4 *Some Engineering Applications*

4.4.1 Retaining Walls

Since most of the current retaining walls can not provide an effective resistance against the action of backfill, a new type of retaining wall constructed by interconnected H-type concrete blocks is proposed. The advantages of this type of retaining wall include stability, proper drainage, durability, allowance of large displacements or differential settlements, and cost effective, etc. Because of the separation among the H-type concrete blocks, the finite block method can be used to perform the numerical analysis for the proposed retaining wall. Some typical examples of the proposed retaining walls will be studied for the mechanical behavior of the H-type concrete blocks and adjacent soils. It indicates that this type of retaining wall provides a good mechanical behavior and stability. It is useful for practical engineering applications.

Two H-type of concrete blocks (see Fig.45) are used to construct the present retaining walls. A typical one is illustrated in Fig.46, in which each layer of the proposed retaining wall is composed of concrete blocks of type I or type II. For the type II layers, two-way interconnection as shown in Fig.47, is developed in the adjacent concrete blocks, while the type I layers only connect adjacent concrete blocks along the wall section. Much more space between the concrete blocks in the layers with type I can be adjusted for the drainage of a large quantity of water. It is very effective for slope stability in the rainy area.

The proposed retaining structure is of flexible type, allowing large displacements and differential settlements. The stress is transferred by distinct concrete blocks in a type of compression or shear. In this way, the tensile weakness commonly happened in the reinforced concrete retaining walls does

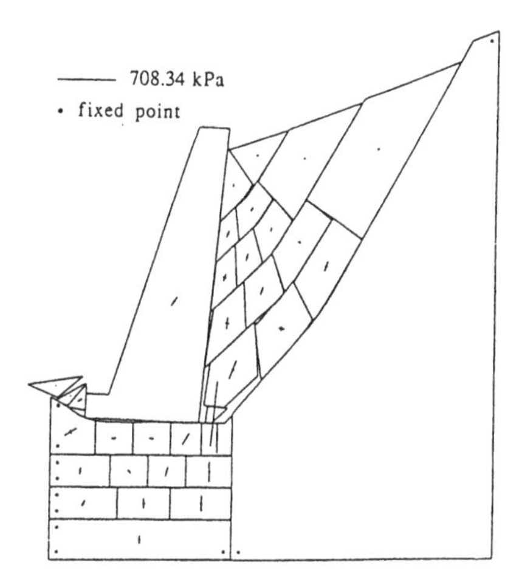

Fig. 58. Deformation and principal stresses of example three (with massive wall).

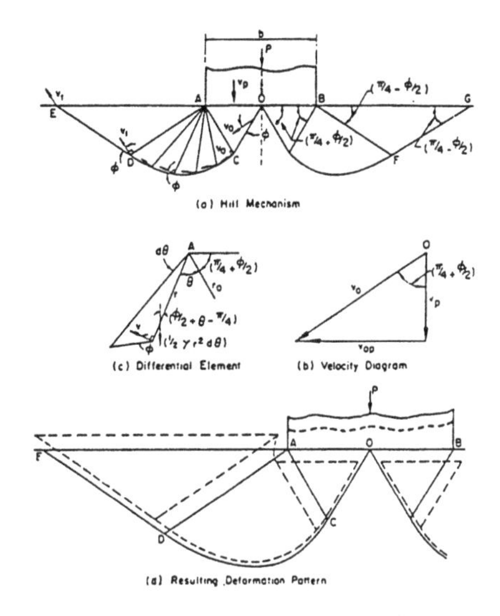

Fig. 59. Bearing capacity calculation based on Hill mechanism.

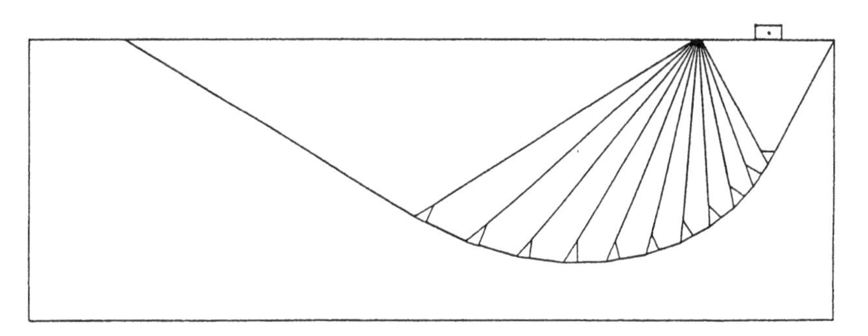

Fig. 60. Finite block method.

not occur in this type of retaining structures and the reinforcing steels are not needed in the concrete blocks. As for the consideration of appearance, the indented portion of concrete blocks in the front part can be used for planting.

Concrete blocks with one meter long in three directions are considered in the later discussions. The size along the horizontal direction of the wall section can vary to fit the retaining wall requirements. Detailed dimensions of the concrete blocks are schematically shown in Figs.48 and 49. For general engineering projects, the size of the concrete blocks can be proportionally changed to meet the practical requirements.

For a better understanding of the mechanical behavior of the proposed retaining walls, three typical examples with height of about 3m, 6m and 9m were studied. The material properties used in the three examples are summarized as:

For concrete blocks

unit weight 23 kN/m^3
Young's modulus $2.1x10^7$ kN/m^3
Poisson's ratio 0.15
Compressive strength 210 kg/cm^2

For soils

unit weight 18 kN/m^3
Young's modulus $2.1x10^4$ kN/m^3
Poisson's ratio 0.25
friction angle 30°
cohesion 0.0

The proposed retaining wall is composed of two H-type (I and II) of concrete blocks. At the top few layers, only one unit of concrete block (type I) is used in the wall section for the economic consider-

Table 3. Summary of the extreme values of three examples (kPa)

	Compression	Tension	Shear
Example One	80.14	0.88	28.86
Example Two	159.17	0.97	74.44
Example Three	304.36	26.05	152.01
Design Value	9270.45	927.05	412.27

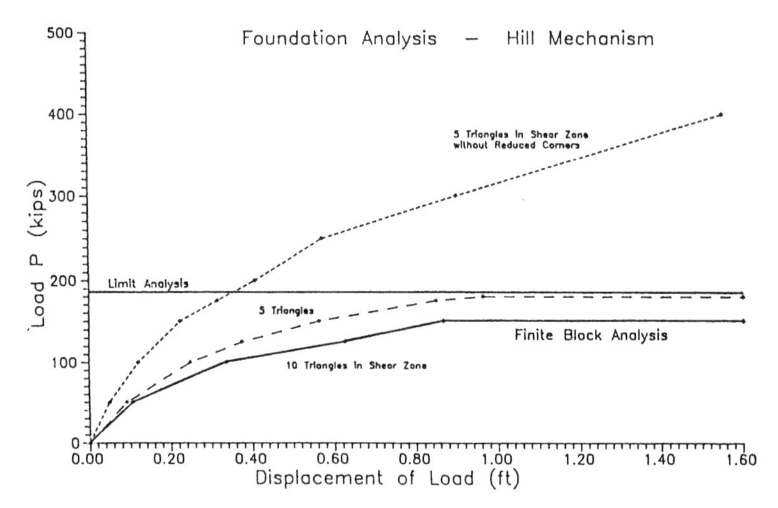

Fig. 61. Analysis of bearing capacity based on hill mechanism.

ation. While in the bottom layers, two or more units of concrete blocks (type I or II) are needed to provide more resistance for the action of deeper and wider backfills. As for the placement of concrete blocks, they should be arranged in such a way to develop more effectively the shear resistance between block connections.

The generation of finite block system in the proposed retaining wall can be divided into two parts. The concrete block system is obtained by directly introducing the vertex coordinates of each block. On the other hand, the block system in the backsoil is generated by assuming logspiral-sandwich failure mechanism (Chen, 1975) and sectioning of the sliding mass along the base of failure mechanism. A more detailed descriptions for generating the critical logspiral-sandwich failure surface in the backsoil of retaining wall can be found in the book by Chen (1975).

Figure 50 shows the cross section of Example one with a full height of 3.25 m. The placement of concrete blocks are also included in the figure. Imposing the constraints on the soil boundary and assuming a plane-strain condition, the final deformed shape and stress state under gravity load of concrete blocks and soil are calculated with results shown in Fig.51. Three thousand iterations using 4 hours in the workstation of HP 9000-720 computer series are experienced for reaching the final stage. The analyzed results show a good stability in the example of proposed retaining wall with 3.25 m high.

For studying the effectiveness of distinct concrete blocks, a rigid and massive wall is used to replace the proposed retaining wall. The reanalyzed results in Fig.52 when compared with those in Fig.51, show some instability in the front of wall and the increase of earth pressure of soil behind the wall.

The retaining wall considered in Example two, as shown in Fig.53, has a full height of 6.25 m. Following the same procedure as that in example one, the final deformed shape and stress state are shown in Fig.54. It took about 6 hours to make three thousand iterations to reach the final result. Figure 54 shows that the structure is stable.

When the proposed retaining wall is replaced by a rigid and massive wall, results shown in Fig.55, indicate instability and stress concentration in the soil near the heel of the wall.

This example shown in Fig.56, has a high wall (9.25 m). After making three thousand iterations with

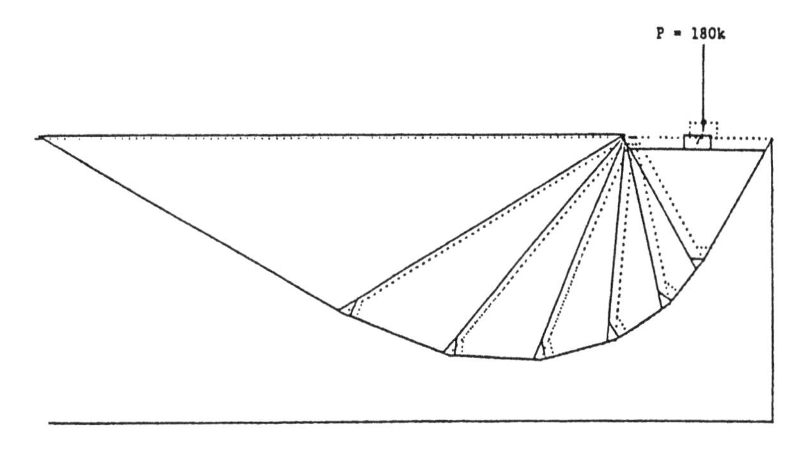

Fig. 62. Deformation pattern resulting from finite block analysis.

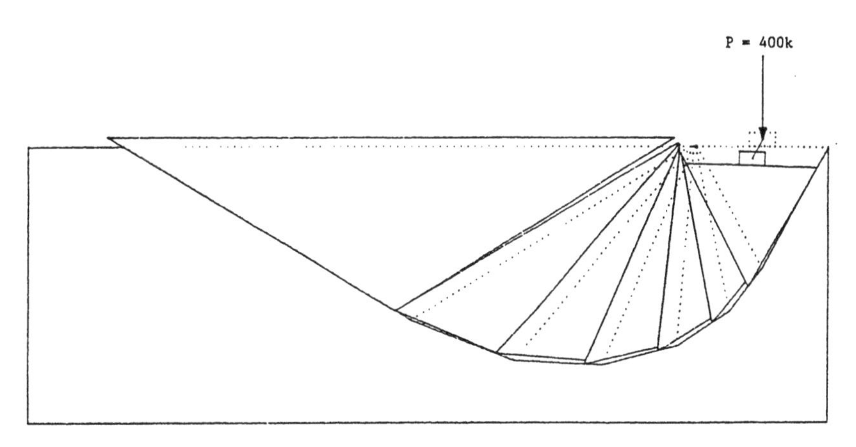

Fig. 63. Deformation pattern of finite block model without reduced corners.

48 hours of computation, the final deformed shape and stress state are shown in Fig.57. From the figure, it can be seen that an apparent settlement is induced in the system due to high backfill and high wall. But the wall system are still in a good condition.

When replacing the proposed retaining wall with a massive wall, instability and stress concentration near the heel of the wall become much more significant (see Fig.58).

As for the stress state in the concrete blocks, the extreme values for all concrete blocks in three examples of the proposed retaining walls are computed and summarized in Table 3. With recommended design values of concrete, it can be seen that the induced extreme values are much less than the design values for concrete blocks. The examples show that the high bearing pressure is developed in the soil beneath the wall, especially for higher walls. Compaction or other soil improvement, and placing plain concrete on the base of the wall prior to the construction of retaining wall are needed in practice.

4.4.2 Bearing Capacity

The bearing capacity of a strip footing based on the Hill mechanism is shown in Fig.59 and is used to illustrate the finite block method. The results will also be compared with the upper-bound limit analysis solution.

Since the Hill mechanism is symmetric about the axis of the footing it is only necessary to consider the left half of the failure mechanism as shown in

Fig.59d. The Hill mechanism consists of three elements : a triangular wedge under the footing(AOC), a triangular wedge under the free surface(ADE), and a logspiral shear zone(ACD). The geometry of the mechanism is dependent on the angle of internal friction ϕ of the soil medium. Complete details of the upper-bound solution for the bearing capacity of the footing can be found in the book by Chen(1975). Briefly, the solution considers the weight of the soil and assumes a smooth surface footing resting on a cohesionless soil(c=0). With the aid of the velocity diagram(Fig.59b) and the differential element of the logspiral of Fig.59c, the work done by the load P and the work done by the soil elements are calculated. The collapse load P_u, for the limit analysis is then given as:

$$P_u = \gamma \frac{1}{2} b^2 N_\gamma \tag{4.1}$$

where b is the width of the footing, γ is the unit weight of the soil, and the dimensionless bearing capacity coefficient N_γ, is defined as:

$$N_\gamma = \frac{1}{4} \tan\left(\frac{1}{4}\pi + \frac{1}{2}\phi\right)\left[\tan\left(\frac{1}{4}\pi + \frac{1}{2}\phi\right) e^{(\frac{1}{2}3\pi)\tan\phi} - 1\right] + \frac{3 \sin\phi}{1+8 \sin^2\phi} \left\{\left[\tan\left(\frac{1}{4}\pi + \frac{1}{2}\phi\right) - \frac{\cot\phi}{3}\right] e^{(\frac{1}{2}3\pi)\tan\phi} + \tan\left(\frac{1}{4}\pi + \frac{1}{2}\phi\right)\frac{\cot\phi}{3} + 1\right\} \tag{4.2}$$

Taking the same example of Fig.59 and applying the finite block method, a block model is developed as shown in Fig.60. In the block model, thelogspiral shear zone is modelled by a series of ten triangular blocks with reduced corners. In the limit analysis, the logspiral shear zone assumes an infinite number of triangular blocks. The reason for reducing the corners of the blocks is important and will be discussed later.

The analyzed results of the finite block method is plotted in Fig.61. The limit analysis by Eqs.(4.1) and (4.2) is also shown in Fig.61 as a horizontal line at the limit load. The results shown in Fig.61 are based on: friction angle $\phi = 30°$, unit weight of soil $\gamma = 120$ pcf, and b = 20 ft. An example of the block displacement produced by a model with five triangular blocks modelling the shear zone is shown in Fig.62.

Comparing the two solid lines in Fig.61, the finite block method results in a lower collapse load than the upper-bound limit load. Since the finite block method is based on a frictional material, the collapse load is expected to be lower than the limit analysis solution which is based on a perfectly plastic material. The true solution would lie between these two solutions as the soil medium in neither a perfectly plastic nor frictional material.

It should be noted that the finite block solution is sensitive to how well the logspiral shear zone is approximated by a finite number of triangular blocks. If only five instead of ten blocks were used as shown in Fig.62, the model becomes excessively stiff and approaches the limit analysis solution as shown by the dashed line in Fig.61.

As mentioned previously, it is important that the finite block model incorporates reduced corners for some of the blocks(see Fig.60). Reduced corners are typically necessary for corner-to-corner contacts such as those in the logspiral shear zone of Fig.60. If corners are not reduced as shown in Fig.63, the block model becomes excessively stiff and 'locking' of the block corners. Extra energy is then required to lift and rotate these blocks, resulting in a much higher collapse load as shown by the dotted dashed line plotted in Fig.61. Reducing the corners is not without theoretical basis. In reality, blocks with perfect sharp corners do not exist and on contact high stress concentrations will be developed to reduce these corners.

REFERENCES

Baladi, G.Y. & B. Rohani 1979a. An elastic-plastic constitutive model for saturated sands subjected to monotonic and/or cyclic loading. *Third Int. Conf. Numer. Methods in Geomech:* 389-404. Aachen: West Germany.

Baladi, G.Y. & B. Rohani 1979b. Elastic-plastic model for saturated sand. *J. Geotechnical Engineering Division* 105: 465-480.

Baladi, G.Y. & I.S. Sandler 1981. Examples of the use of the cap model for simulating the stress-strain behavior of soils. In: R.N. Yong and H-Y. Ko (editors), *Limit Equilibrium, Plasticity and Generalized Stress-Strain in Geotechnical Engineering*, ASCE: 649-710. New York: New York.

Bazant, Z.P. 1985. *Mechanics of geomaterials*. London: John Wiley.

Brezis, H., D. Kinderlehrer & G. Stampacchia 1979. Sur une nouvelle formulation due problem de l'ecoulement a travers une digue. *C.R. Acad. Sic. Paris*: 287A: 711-714.

Chan, S.W. 1980. *Perfect plasticity upper bound limit analysis of the stability of a seismic-infirmed earthslope*. MS Thesis, School of

Civil Engineering, Purdue University: West Lafayette, IN.
Chang, M.F. 1981. *Static and seismic lateral earth pressures on rigid retaining structures.* PhD Thesis, School of Civil Engineering, Purdue University: West Lafayette, IN.
Chang, C.J. 1981. *Seismic safety analysis of slopes.* PhD Thesis, School of Civil Engineering, Purdue University, West Lafayette, IN.
Chang, T.Y. 1980. A nonlinear finite element analysis program. *NFAP*:Vols. 1 & 2. Akron, University of Akron.
Chen, W.F. 1975. *Limit analysis and soil plasticity*, Amsterdam: Elsevier.
Chen, W.F. & W.O. McCarron 1986. Modelling of Soils and Rocks Based on Concepts of Plasitcity. *AIT Symposium and Course of Laboratory & Field Tests and Analysis of Geotechnical Problems*: 467-510. Rotterdam: Balkema.
Chen, W.F. 1982. *Plasticity in reinforced concrete.* New York: McGraw-Hill.
Chen, W.F. & A. Saleeb 1982. *Constitutive equations for engineering materials, Vol 1 - Elasticity and Modeling.* New York: Wiley Interscience.
Chen, W.F. & X.L. Liu 1990. *Limit analysis in soil mechanics.* Amsterdam: Elsevier.
Chen, W.F. 1994. *Constitutive equations for engineering materials, Vol 2 - Plasticity and Modeling.* Amsterdam: Elsevier.
Chen, W.F. & E. Mizuno 1990. *Nonlinear analysis in soil mechanics.* Amsterdam: Elsevier.
Chen, W.F. & M.F. Chang 1981. Limit analysis in soil mechanics and its applications to lateral earth pressure problems. *Solid Mechanics Archive* 6:331-399.
Chen, W.F. 1984. *Soil mechanics, plasticity and landslides*, Mechanics of Material Behavior, Amsterdam: Elsevier: 31-58.
Chen, W.F. & G.Y. Baladi 1985. *Soil plasticity: theory and implementation.* Amsterdam: Elsevier.
Chen, W.F. 1980. Plasticity in soil mechanics and landslides. *J. Eng. Mech. Div.* ASCE 106: 443-464.
Cundall, P.A. 1971. A computer model for blocky rock systems. *Proc. of the Int. Symposium on Rock Fracture.* France.
Daddazio, R.P., M.M. Ettouney & I.S. Sandler 1987. Nonlinear dynamic slope stability analysis. *J. Geotech. Engrg.* 113:285-298.
Dafalias, Y.F. & E.P. Popov 1975. A model for nonlinearly hardening materials for complex loading. *Acta Mechanics* 21: 173-192.
Dafalias, Y.F. & L.R. Herrmann 1980. A bounding surface soil plasticity model. *International Symposium on Soils under Cyclic and Transient Loading*: 335-345.
Dafalias, Y.K. & L.R. Herrmann 1982. *Bounding surface formulation of soil plasticity, Chapter 10 - Soil mechanics-transient and cyclic loads.* New York: Wiley.
Dafalias, Y.K. & L.R. Herrmann 1986. Bounding surface plasticity II: application to isotropic cohesive soils, *Journal Engineering Mechanics Division* 112: 1263-1291.
Desai, C.S. & R.H. Gallagher 1983. *Constitutive laws for engineering materials: theory and application.* London: Wiley.
Desai, C.S. & H.J. Siriwardane 1983. *Constitutive laws for engineering materials.* Englewood Cliffs, NJ: Prentice Hall.
DiMaggio, F.L. & I.S. Sandler 1971. Material models for granular soils. *J. Eng. Mech. Div.* 97: 935-950.
Drucker, D.C. & W. Prager 1952. Soil mechanics and plastic analysis or limit design. *Q. Appl. Math.* 10(2): 157-165.
Drucker, D.C., R.E. Gibson & D.J. Henkel 1957. Soil mechanics and work hardening theories of plasticity. *Trans. 122*: 338-346.
Dvorak, G.J. & R.T. Shield 1984. *Mechanics of material behavior.* Amsterdam: Elsevier.
Fellenius, W.O. 1926. *Mechanics of soils.* Gosstr ollzdat: Statika Gruntov.
Herrmann, L.R., Y.F. Dafalias & J.S. DeNatale 1982. Numerical implementation of a bounding surface soil plasticity model. *Proc. Int. Symp. on Numerical Models in Geomechanics*: 334-343. Switzerland.
Herrmann, L.R., V. Kaliakin, C.K. Shen, K.D. Mish & Z.Y. Zhu 1987. Numerical implementation of plasticity model for cohesive soils. *Journal Engineering Mechanics Division* 113(4):482-499.
Hill, R. 1950. *The mathematical theory of plasticity.* Oxford: Clarendon Press.
Huang, T.K. & W.F. Chen 1990. Simple procedure for determining cap-plasticity-model parameters. *J. Geotech. Engrg.*, 116(3):492-513.
Humphrey, D.N. & R.D. Holtz 1986a. Finite element analysis of plane strain problems with PS-NFAP and the cap model. *Report No. JHRP-86*, School of Civil Engrg., Purdue University, W. Lafayette, IN. 175 pp.

Humphrey, D.N. & R.D. Holtz 1986b. Design of reinforced embankments. *Report No. JHRP-86*, School of Civil Engineering, Purdue University, W. Lafayette, IN. 423 pp.
IABSE 1979. *Proc. colloq. on plasticity in reinforced concrete*. Zurich:IABSE Publ.
Kawai 1977. New discrete structural models and generalization of the method of limit analysis. *Finite Elements in Nonlinear Mechanics*: 885-906. Tronheim: Norweign Institute of Technology.
Lacey, S.J. 1986. Numerical procedures for nonlinear transient analysis of two-phase soil systems. *Ph.D Thesis*, Princeton University.
Lacey, S.J. & Prevost, J.H. 1987. Constitutive model for geomaterials. *Proc. of the Second Inter. Conf. on Constitutive Laws for Engrg. Materials*: 149-160. New York: Elsevier.
Lacy, S.J. & J.H. Prevost 1987. Flow through porous media: a procedure for locating the free surface. *Int. Journal Numer. Anal. Meth. Geomech.* 11:585-601.
Lade, P.V. 1979. Stress-strain theory for Normally consolidated clay. *Proc. 3rd Int. Conf. on Numerical Methods in Geomechanics*:(4) 1325-1337. Germany.
McCarron, W.O. 1985. Soil plasticity and finite element applications. *PhD Thesis*, School of Civil Engrg., Purdue University, W. Lafayette, IN, 266 pp.
McCarron, W.O. & W.F. Chen 1987. A capped plasticity model applied to boston blue clay. *Can. Geotech. Journal* 24(4):630-644.
MIT, 1969a. Performance of an embankment on clay, Interstate-95. *Dept. of Civil Engrg. MIT* R69-67. Cambridge: MIT.
MIT, 1969b. Instrumentation for interstate 95 embankments. *Dept. of Civil Engrg.*. MIT R69-10. Cambridge: MIT.
MIT, 1975. *Proc. of the Foundation Deformation Predictions Symposium*: R75-32.
Mizuno, E. & W.F. Chen 1982. Analysis of soil response with different plasticity models. *Proc. of the Symp. Appl. of Plasticity and Generalized Stress-Strain in Geotechnical Engrg.*: 115-138. New York: ASCE.
Mizuno, E. & W.F. Chen 1981a. Plasticity models for soils - comparison and discussion. *Proc. Workshop on Limit Equilibrium, Plasticity and Generalized Stress-Strain in Geotechnical Engrg.*:871 pp. New York:ASCE.
Mizuno, E. & W.F. Chen 1981b. Plasticity models and finite element implementation. *Proc. Symp. Implementation of Computer Procedures and Stress-Strain Laws in Geotechnical Engineering*: 519-534. Durhan:Acorn Press.
Mizuno, E. 1981. Plasticity modeling of soils and finite element applications. *PhD Thesis*. Purdue University: West Lafayette IN.
Mizuno, E. & W.F. Chen 1984. Plasticity modeling and its applications. *Proc. Symp. on Recent Developments in Laboratory and Field Tests and Analysis of Geotechnical Problems*: 391-426. Rotterdam.
Nelson, I. & G.Y. Baladi 1977. Outrunning ground shock computed with different models. *J. Eng. Mech. Div.* 103:377-393.
Newmark, N.W. 1965. Effects of earthquake on dams and embankments. *The Fifth Rankine Lecture of the British Geotechnical Society*: 137-160. Geotechnique:15(2).
Palmer, A.C. 1973. *Proc. Symp. on Plasticity in Soil Mechanics*: 314 pp. England: Cambridge University Press.
Parry, R.H.G. 1972. *Roscoe Memorial Symp.: Stress-Strain Behavior of Soils*: 752 pp. England: Cambridge University Press.
Prevost, J.H. 1977. Mathematical modeling of monotonic and cyclic undrained clay behavior. *Inter. J. Numer. Anal. Methods Geomech.* 1:195-216.
Prevost, J.H. 1979. Plasticity theory for soil stress-strain behavior. *J. Eng. Mech. Div.* 104:1177-1194.
Prevost, J.H. 1982. Two-surface versus multi-surface plasticity theories: a critical assessment. *Int. J. for Num. and Analy. Methods in Geomechanics* 6:323-338.
Roscoe, K.H., A.N. Schofield & A. Thurairajah 1963. Yielding of clays in state wetter than critical geotechnique, 13(3): 211-240.
Saleeb, A.F. 1981. Constitutive models for soils in landslides. *PhD Thesis*, School of Civil Engineering, Purdue University, West Lafayette, IN.
Sandler, I.S., F.L. DiMaggio & G.Y. Baladi 1976. Generalized cap model for geological materials. *J. of the Geotechnical Engineering Div.* 102:683-699.
Schofield, M.A. & C.P. Wroth 1968. Critical state soil mechanics. London: McGraw Hill.
Shi, G.H. 1988. Discontinuous deformation analysis: a new numerical model for the static and dynamics of block systems. *PhD Thesis*. Purdue University, West Lafayette, IN.
Sokolovskii, V.V. 1965. *Statics of granular media*, New York:Pergamon.

Terzaghi, K. 1965. *Theoretical soil mechanics.* New York: John Wiley.

Wang, Q.Y. & X.L. Liu 1988. Limit analysis based on variational method. *Proc. Int. Conf. of Engineering Problems of Regional Soils*: 469-472. Beijing, China.

Yong, R.N. & H.Y. Ko 1981. Limit equilibrium, plasticity and generalized stress-strain in geotechnical engineering. ASCE: 365 pp.

Yong, R.N. & E.T. Selig 1982. Applications of plasticity and generalized stress-strain in geotechnical engineering. New York: ASCE: 356 pp.

Developments in Geotechnical Engineering, Balasubramaniam et al. (eds) © 1994 Balkema, Rotterdam, ISBN 90 5410 522 4

Recent developments in *in-situ* testing and performance monitoring of soil and rock structures

Helmut Bock
INTERFELS GmbH, Bad Bentheim, Germany

ABSTRACT: Almost any major geotechnical project requires *in-situ* testing and performance monitoring to be carried out as part of the geotechnical design. With the increase in size and complexity of our soil and rock structures both *in-situ* testing and performance monitoring are of over-proportionally increased importance.

Over the last few years there has been a recognizeable revival in the demand of the market for *in-situ* testing. This includes tests carried out in boreholes as well as large-scale tests at accessible surfaces. Within this paper a thorough review is given on the various borehole testing methods and equipment such as pressuremeters, dilatometers and borehole jacks. In particular, it is shown that the common borehole jacks (e.g. Goodman Jack) can influence the test results considerably. Against this background a new borehole jack has been developed which is particularly designed for use in soft rock and soil providing in-depth information on the relevant deformation moduli of the ground.

Mention is made on a number of new large *in-situ* testing equipment such as an 8 MN Plate Load testing equipment and Extra Large Flat Jacks which have been employed in some major dam projects.

In the area of *in-situ* rock stress measurements the trend is clearly away from the overcoring methods (too delicate; too costly; poor cost/benefit ratio). On the increase are hydraulic fracturing and borehole slotting, both of the methods which allow high-definition measurements to be carried out at reasonable costs. Examples are provided with most recent test results.

The trend for high-definition measurements is also recognizeable in performance monitoring. Of particular importance is linewise observation of boreholes or pipes. A combination of mobile probes (mobile inclinometer and mobile extensometer) can provide the full 3-D deformation stage of the ground. Examples are presented for an underground mine and a foundation in collapsing ground.

Finally, an example from inner-city tunnelling is given which highlights the trend for real-time monitoring and computer processing of the measurement data.

1 INTRODUCTION

In-situ testing of geotechnical materials ahead of construction as well as performance monitoring of geo-engineered structures during and after construction are integral parts of any geotechnical design procedure (Fig. 1). The volume of work involved and also the degree of intensity with which *in-situ* testing and performance monitoring are carried out are subject to gradual changes, thus reflecting an ongoing process in the evolution of the geotechnical design procedure. Overall, it can be expected that *in-situ* testing and, in particular, performance monitoring will be of increased importance in geotechnical engineering in the years to come. This expectation is based on market observations and new Professional Standards. The new fundamental European Geotechnical Code EC 7, for instance, makes performance monitoring compulsory for a certain class of geotechnical structures. So far, carrying out monitoring was widely left to the discretion of the design engineer.

Looking now specifically onto *in-situ* testing and performance monitoring, again an ongoing shift of the relative levels of activity is recognizeable. Based on the operating revenues of INTERFELS - a company which since 1961 is specialised in both *in-situ* testing and performance monitoring - a continuously increased relevance of performance monitoring was experienced throughout the Seventies and Eighties. At around the year 1990 approximately 80% of the revenue was in performance monitoring while in *in-situ* testing was at a mere 15 to 20%. In 1992 and 93, however, these figures have changed substantially (Fig. 2). *In-situ* testing now is almost as important as performance monitoring and this trend is expected to continue in the years ahead.

Borehole Deformation Testing

Directional loading of borehole wall by means of load platens	Isotropic loading of borehole wall by means of packer
Borehole Jacking	**Dilatometer / Pressiometer**
for rocks: Goodman Jack ϕ = 76 mm	*measurement of diametral changes:*
for soft rock and soil: "Stuttgarter" Probe	Dilatometer; e.g. Type Rocha ϕ = 76mm
INTERFELS Jack ϕ = 146mm	or INTERFELS IF 096 ϕ = 96 to 101 mm
	volume measuring principle; e.g. Ménard

Tab. 1. Types of borehole deformation principles and instrumentation

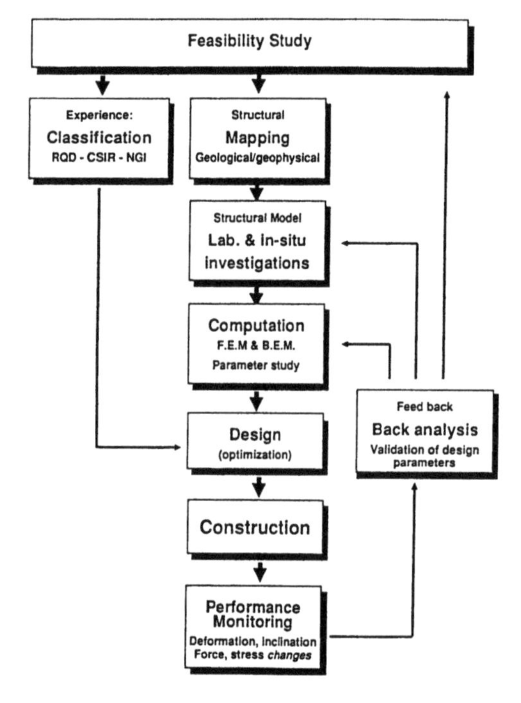

Fig. 1. *In-situ* testing and performance monitoring as components of a typical geo-engineering design

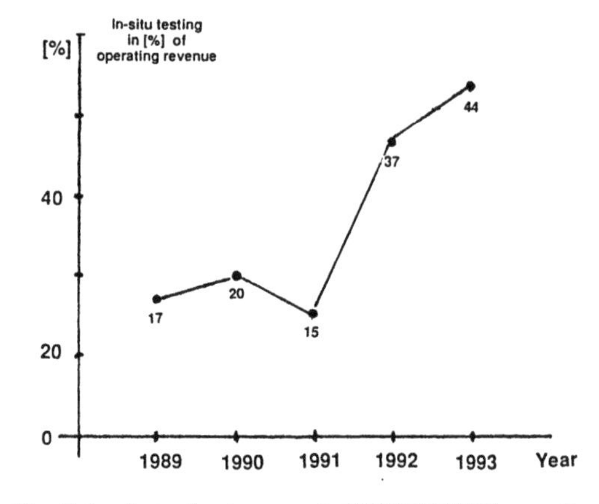

Fig. 2. *In-situ* testing in percent of INTERFELS' operating revenue

The underlying reason for the new emergence of *in-situ* testing is, in our opinion, closely connected with the significant improvement and degree of sophistication in numerical modelling with highly specialised FEM and BEM computer codes commercially available in recent years. Sophisticated models need reliable input data both with regard to material properties and boundary conditions, otherwise all their sophistication would be pretty useless.

It is the scope of this contribution to present a state-of-the-art review on *in-situ* testing (Section 2) and performance monitoring (Section 3) of geotechnical structures.

2. *IN-SITU* TESTING

2.1 *In-Situ Testing in Boreholes*

The bulk of *in-situ* testing work is carried out in boreholes. Within this category borehole deformation is by far the most important type of testing. Table 1 categorizes the common borehole deformation instrumentation and Fig. 3 indicates the range of application of the various techniques.

2.1.1 *Borehole Dilatometer and Pressuremeter*

In our assessment these are the following technical and market trends recognizeable with regard to borehole

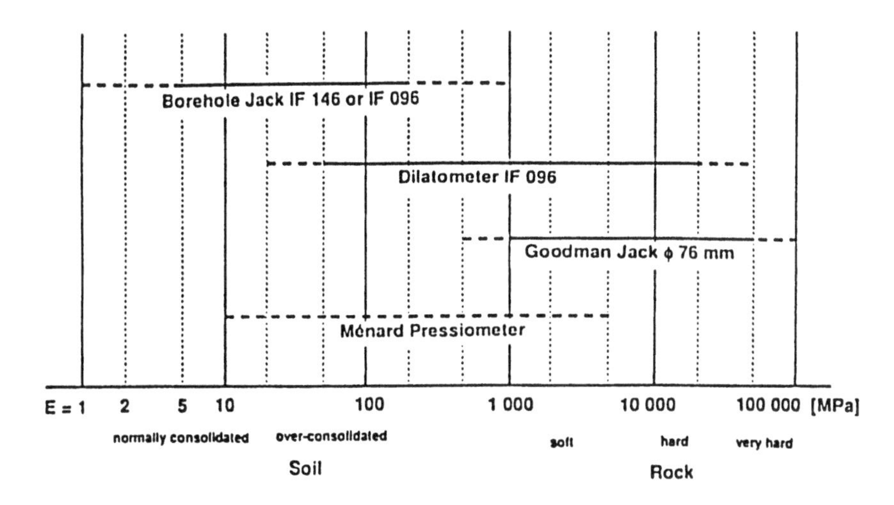

Fig. 3. Range of potential application of different types of borehole deformation instruments

deformation testing:

☐ Increased demands for dilatometer testing at a relative decline in pressuremeter testing.
Main reasons:
> Dilatometer tests with its direct measurement of diametric changes of the borehole wall (Fig. 4) are usually providing more accurate results than pressuremeter tests which are based on the volume measuring principle (System Ménard).
> Dilatometer which is usually equipped with at least three independently oriented displacement sensors can delineate the deformation anisotropy of the rock or soil.
> Dilatometers are providing relatively reliable pressure-displacement curves, even in cyclic loading (Fig. 5), whereas pressuremeter tests tend to yield results which are generally less reliable in cyclic loading (Fig. 6).

☐ Dilatometer testing in boreholes with medium-sized diameters.
Looking specifically into dilatometer testing, the current international trend clearly is for tests to be carried out in medium-sized boreholes, typically of 96 to 101 mm diameter. Smaller borehole diameters (e.g. 76.2 mm) are often considered by geotechnical engineers as providing unacceptably small test volumes, whereas bigger boreholes (e.g. 146 mm diameter) are often seen as too costly in drilling and testing.

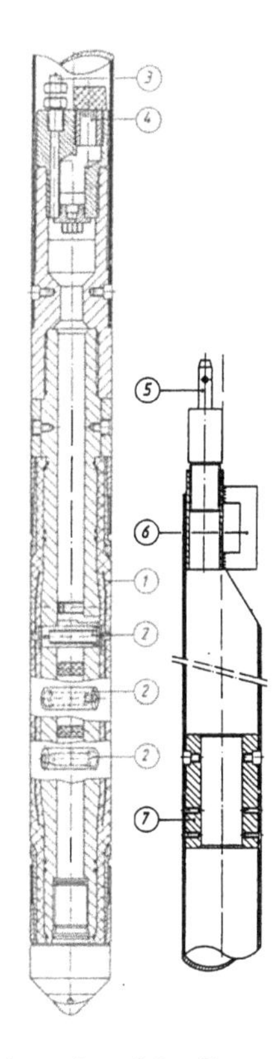

Fig. 4. Schematic section of the dilatometer IF 096 with 3 sensors for measurement of diametric changes of borehole
1 = re-inforced rubber sleeve 2 = displacement transducers; 3 = pressure line; 4 = electric cable plug 5 = protection tube with connection to setting rods 6 = hook for drilling rig rope 7 = connection between dilatometer probe and protection tube

2.1.2 *Experience with the Goodman Jack*

In testing rocks, the widely known Goodman Jack for 76 mm boreholes has much been criticized because of its relatively small size and, more importantly, because of complex interactions between the borehole instru-

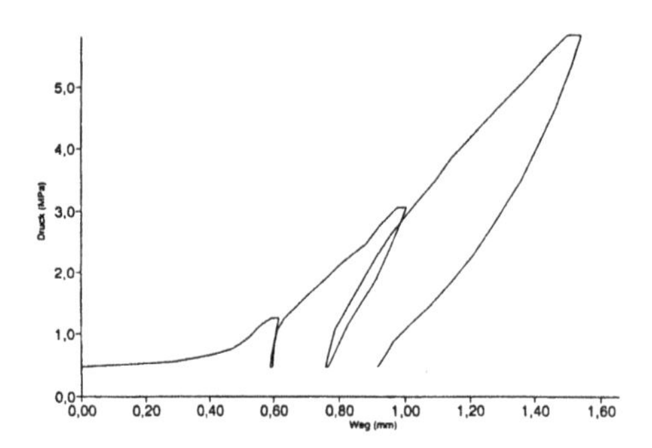

Fig. 5. Example of a dilatometer test record carried out in a Jurassic sandstone

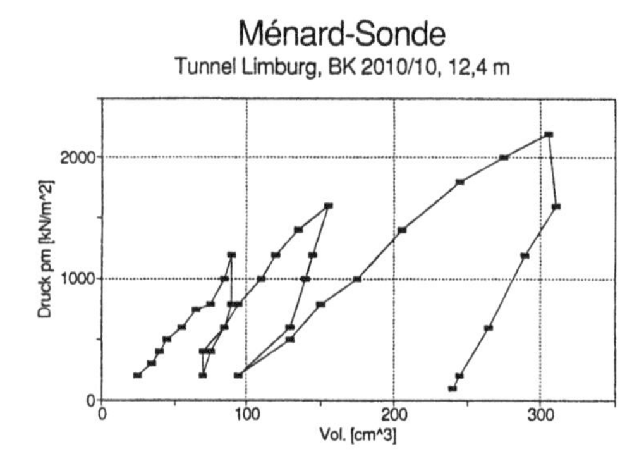

Fig. 6. Example of a Ménard test with 3 load cycles carried out in a weathered shale

ment and the rock to be tested [1]. For these reasons, Goodman jacking should only be carried out and the test results evaluated by an institution which is very experienced in this kind of testing. When avoiding the practical and theoretical traps of this method, Goodman jacking can provide quite useful information, particularly in intensively jointed rock such as limestone, dolomite and slate in which other types of borehole deformation testing might yield erratic results.

Over the last two years, INTERFELS had the opportunity to widen its experience in Goodman testing quite considerably. One of the most important findings is that the non-parallel opening (= tilting) of the jack's platens is much more critical than originally anticipated. From the mechanical design of the jack, tilting is only prevented if the reaction forces of the tested rock are constant over the full length of the load platens. This requires perfectly homogeneous rock conditions at the test location. In the real world, however, such homogeneity is never realized. Consequently, tilting of the load platens will regularly occur in almost any Goodman test. Such tilting, in turn, will put

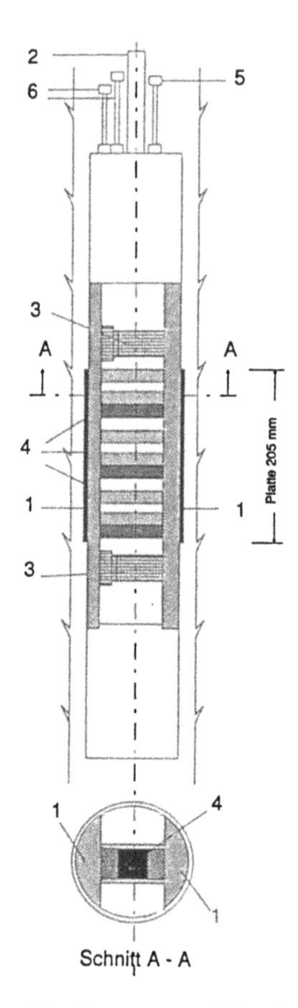

Fig. 7. Section of the "Low-pressure" Goodman Jack, Type 52 102

1 = Load platens 2 = connection to setting rods 3 = Two LVDTs 4 = Three hydraulic pistons (double-acting) 5 = electric cable connection 6 = connections to hydraulic hoses

some lateral forces onto the O-rings of the hydraulic jacks with the consequence that an *uncontrolled* amount of energy is lost in friction of the hydraulic cylinders. The result is that in the pressure/displacement curves of a Goodman test there will always be a certain amount of instrument-induced hysteresis. In the example of Fig. 8 it can be seen that two displacement transducers (LVDTs), built into the borehole jack (Fig. 7), are showing a different degree of hysteresis. Those parts of the jack which have undergone a higher degree of opening also show a comparatively high degree of hysteresis.

Goodman Jacks are commercially available in two versions: A high- and a low-pressure version (Types 52101 and 52102, resp.). The difference in the construction of the two jacks is predominantly related to the numbers of hydraulic elements (12 in the hard rock version and 3 in the soft rock version) and also in the shape of the hydraulic pistons (racecourse section in the hard rock version and circular section in soft rock version). From this construction it is immediately evident that the hard rock Goodman Jack must develop a higher degree of internal friction than the soft rock jack. Fig. 9 is an *in-situ* prove for this with two tests carried out at identical borehole depth, one test with the soft Goodman jack and one with the hard version. Actually, the degree of hysteresis in the hard Goodman jack is so high that this version yields practically useless test curves (right of Fig. 9). Our strong recommendation therefore is to restrict the use of the Goodman jack to the soft rock version.

2.1.3 *New Borehole Jack IF 146*

In response to these new physical insights into the mechanics of hydraulically-operated borehole jacks, INTERFELS has developed a new type of borehole jack which avoids these difficulties. In its new **Jack IF 146** the load platens will always be opened parallel to each other, irrespective of the resistance of the tested soil or rock. This is technically achieved by a patented device in which the jack is constructed similar to a uniaxial testing frame with a load originally applied ***in*** direction of the borehole axis. A system of levers, hinges and abutments is then re-directing the axial load laterally onto the borehole wall (Fig. 10), thus securing that the load platens are *always parallel* to each other, irrespective of the mobilized resistance of the ground.

The Borehole Jack IF 146 is designed for soil and soft rocks with a range of Young's Moduli between 5 MPa and about 500 MPa. The jack is for use in 6-inch boreholes (146 mm diameter); a 4-inch version (96 to 101 mm diameter) is in development. As can be seen from Fig. 11 this jack yields most detailed information on the deformability characteristics of the ground. Usually, one is in a position to clearly distinguish between different types of moduli such as the first-loading, reloading and unloading modulus. So far, such detailed moduli determinations have not been possible with other types of borehole deformation probes. This confirms that the construction of the borehole jack has achieved its purpose in controlling (i.e. minimizing) the degree of internal friction of the borehole instrument, irrespectively of all possible soil reaction behaviours, including the occurence of irregular stratification and of voids within the tested section of the borehole.

2.2 *In-situ Tests at Accessible Surfaces*

In most recent years, INTERFELS gaged an increased

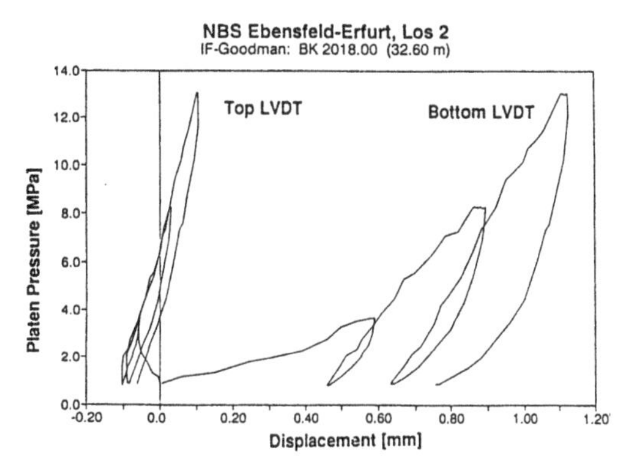

Fig. 8. Example of a "low-pressure" Goodman jacking test indicating substantial tilting of load platens

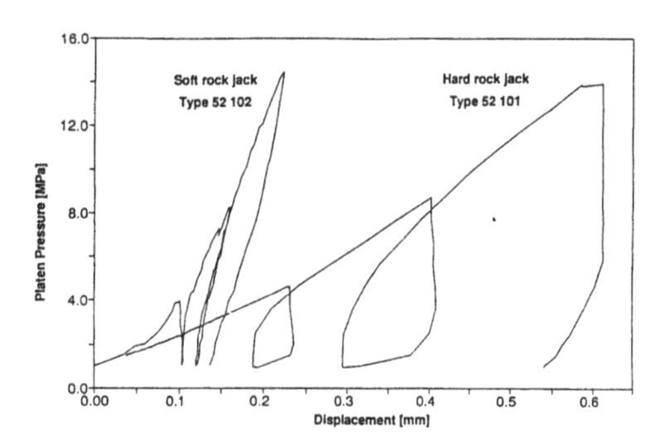

Fig. 9. Comparison of Goodman Jacks designed for soft rock (left) and hard rock (right).
Note the amount of hysteresis of the test curves yielded by the hard-rock jack

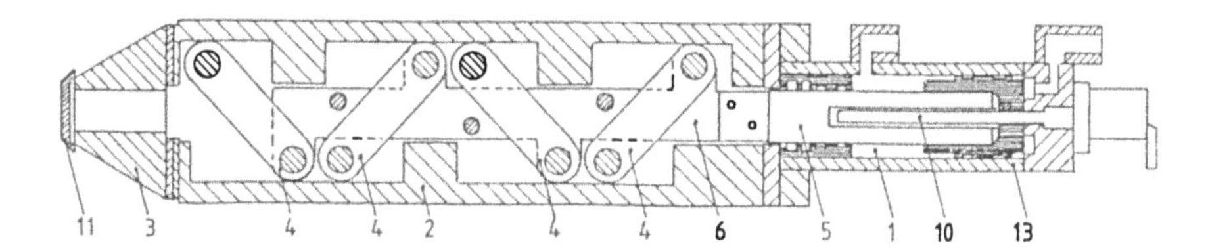

Fig. 10. Schematic section of the new patented Jack IF 146
1 = hydr. cylinder 2 = Load platen 3 = end piece 4 = lever 5 = piston of (1) 6 = pushing rod 10 = displacement transducer 11 = sieve plate 13 = housing of cylinder

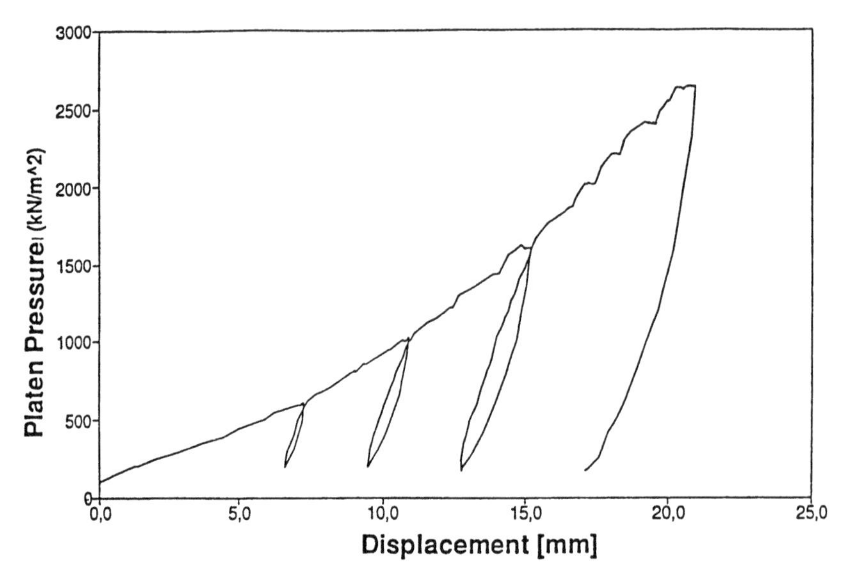

Fig. 11. Example of a borehole deformation test carried out in a glacial till with the INTERFELS Jack IF 146.

Fig. 12. Setting up a 8 MN (800 T) Plate Load Test at Karun 3 (Iran) with central "tree" of extensometer rods for measurement of the rock displacements immediately beneath the load platen

demand for large-scale in-situ tests to be carried out for environmentally sensitive projects such as large dams. Actually, the dimension of these tests are at an almost unprecedented scale as is demonstrated by the following two examples from major dam projects in the Islamic Republic of Iran.

2.2.1 *Plate Load Testing*

INTERFELS has been involved in the supply and in the first phase of testing of a plate load testing equipment with a capacity of 8 MN. The tests were carried out in an underground gallery, specially excavated in the abutment of the Karun-3 dam. The tests were in both horizontal and vertical directions. When loading in horizontal direction (Fig. 12) the impact onto the surrounding rock was sometimes so high, that occasionally some rocks in the roof of the gallery started to fall down indicating a degree of loading and unloading, respectively which has not been experienced in smaller-scale PLT. The bearing pressures approached in this type of testing were close to 10 MPa, which by all technical standards, must be considered as very high.

2.2.2 *Extra Large Flat Jacks*

The biggest type of *in-situ* testing, which is feasible within the budgetary constraints of a normal dam construction project, is Large Flat Jack (LFJ) testing pioneered by LNEC in Portugal. This test is designed for the *in-situ* determination of the rock deformability and also for evaluating some of the *in-situ* stress components in the rock. INTERFELS has developed its own system which, in its dimension, supersedes the LNEC jacking system thus underlining the general trend of increasing the dimension of both equipment and volume of tested rock. The equipment manufactured by INTERFELS is for 1200 mm wide jacks providing a tested area of 1.65 m^2; in comparison: The LNEC jacks are 1000 mm wide providing a test area of 1.14 m^2 (Fig. 13). For this reason the INTERFELS system is named Extra Large Flat Jack (ELFJ 1200).

Fig. 14 gives an example of an ELFJ test carried out at the Godar-e-Landar dam site (Iran). Note that the pressure/slot widening curves are of unusual high degree of linearity, showing a very low degree of hysteresis, particularly when compared with typical Goodman test results (Fig.9). Such very high-quality *in-situ* testing results are quite typical for the ELFJ system. The quality of the results and the more representative scale of these tests are more than justifying the relatively high costs of these tests which, at the moment, are typcially in the order of 25 000 DM per test.

2.3 *In-Situ Stress Measurements*

It is generally accepted that knowledge of the *in-situ* stress state is a precondition for a proper design of large-scale rock structures. However, stress measurements are not very popular with the project engineers and are commonly not carried out to a degree which is desirable from a geotechnical point of view. This situation is caused mainly by the awkwardness of the overcoring stress measuring methods which are currently dominating the market.

Against this background alternative stress measuring methods are gaining increased popularity, in particular hydraulic fracturing and borehole slotting. These alternative methods allow stress measurements to be carried out within a single borehole, not relying on the costly, time-consuming and technically delicate overcoring procedures. This is particularly relevant for *in-situ* testing in remote areas.

By this technical advantage, hydraulic fracturing and borehole slotting are in line with the general trend recognizable in *in-situ* testing and performance monitoring which is a ***trend towards high-definition measurements***. Be it dilatometer testing, deformation measurements along a borehole or *in-situ* stress measurements, there clearly is a market demand for numerous tests and/or measurements to be carried out at reasonable cost. The background of this philosophy is to have a data base which is statistically relevant for the rock which is usually of a relatively low degree of homogeneity.

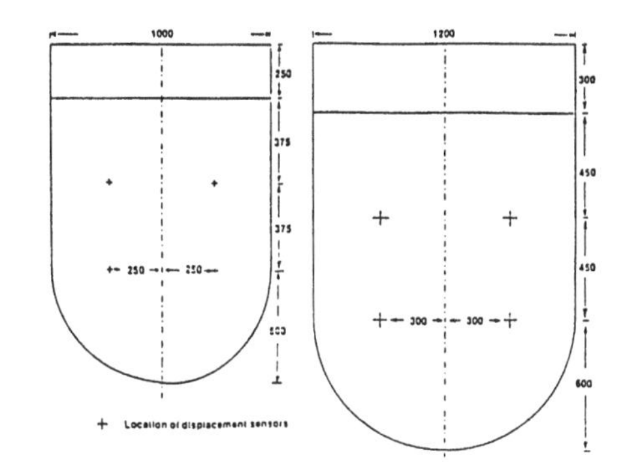

Fig. 13. Dimensions of conventional LFJ (left) and Extra Large Flat Jack (ELFJ) systems (right)

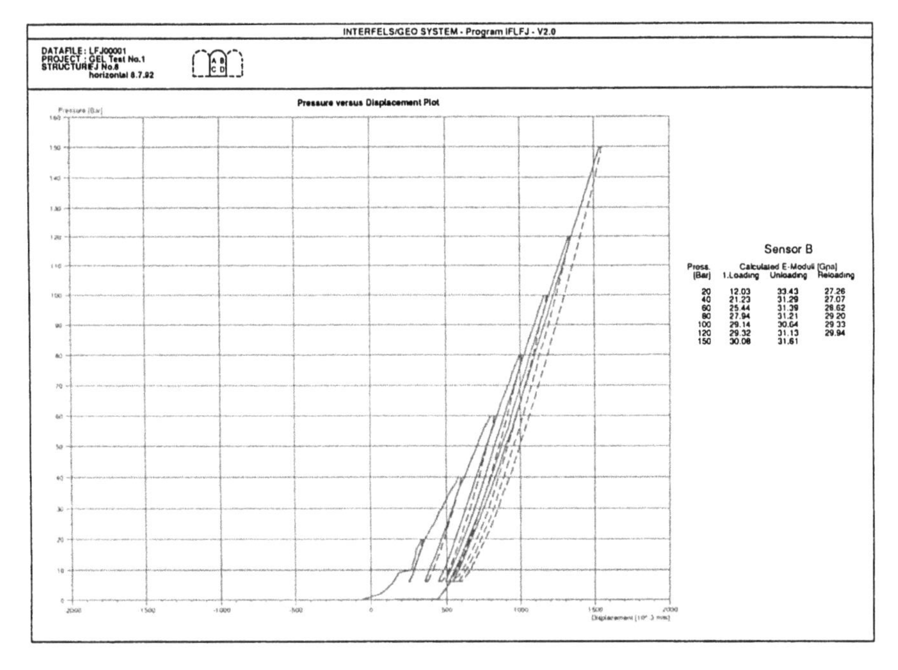

Fig. 14. Example of an ELFJ Test carried out at Godar-e-Landar (Iran). Note the very high degree of linearity and low degree of hysteresis of the pressure-displacement plot.

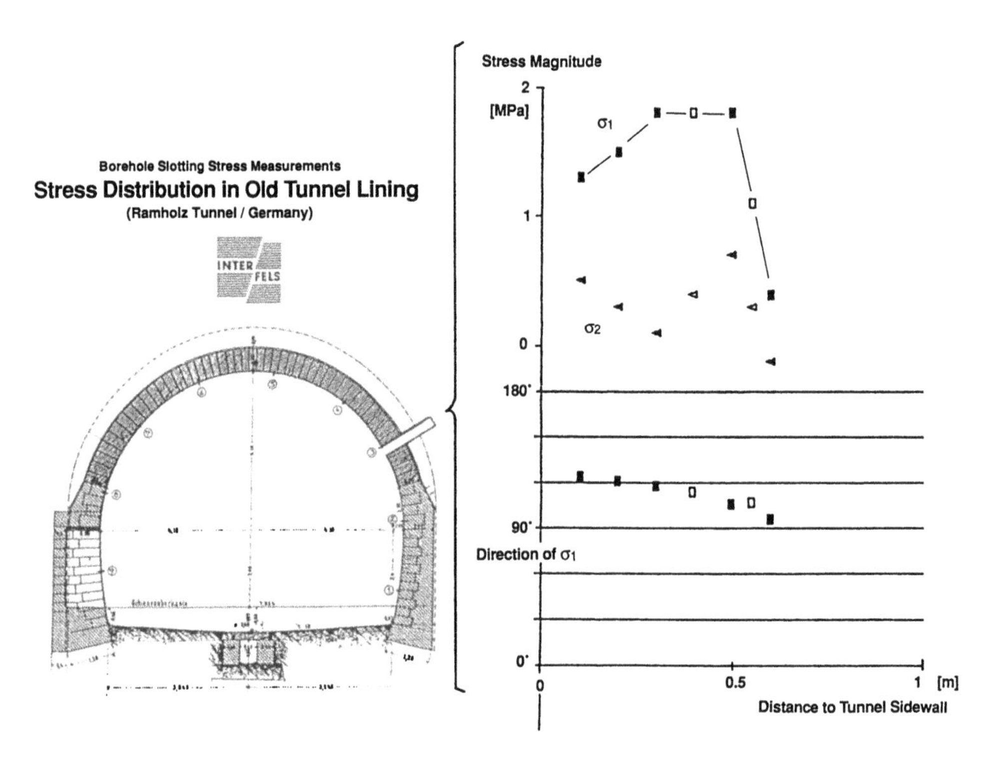

Fig. 15. Borehole Slotter Stress measurements in the masonary brick lining of an old railway tunnel. Note: High-definition stress measurements (1 measurement every 10 cm along a borehole) delineating a stress *profile* within lining

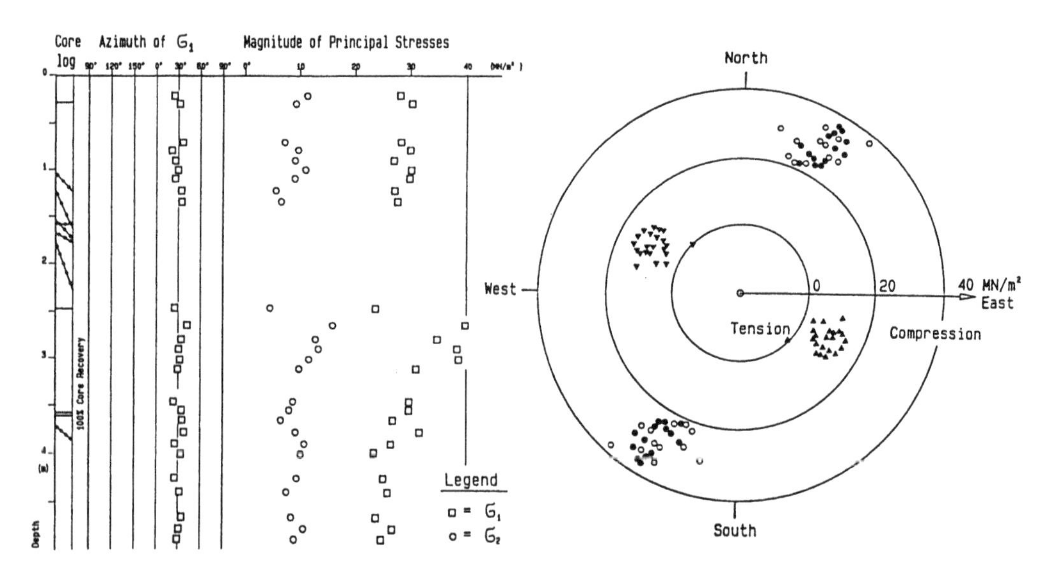

Fig. 16. High-definition borehole slotting stress measurements in an abutment of a gravity dam (Burdekin Falls Dam; Australia). Results are presented in stress distribution plot over depth of borehole (left) and summary polar plot (right). Note: stress disconformity in about 2.4 m depth.

Figures 15 to 17 give examples on most recent tests carried out with the **borehole slotter**. In the first example (Fig. 15) testing was carried out in a 0.8 m thick masonary tunnel lining which was to be rehabilitated after more than 100 years of operation. Note that almost every 10 cm a stress measurement was made, thus delineating a stress profile within the tunnel lining.

Figure 16 presents an example from a gravity dam project showing high-definition stress measurements. In this particular case the *in-situ* stress state turned out to be distinctively directed with regard to the stress orientations as can be depicted on the summary plot (right of Fig. 16).

In July 1993 borehole slotting tests were carried out in a major pump storage scheme in East-Germany. In this particular project conventional 3-D overcoring stress measurements were not successful due to the highly fractured rock with spacing of joints in the order of 5 to 10 cm. Despite of this situation, borehole slotting was capable of delineating the general *in-situ* stress state as is depicted in Figure 17. In total three independently inclined boreholes were drilled for borehole slotting. This allowed the delineation of the general 3-D stress state at the test site. Similar experiences were made recently in Australia [2].

In our opinion it can now be justified for most geotechnical projects (such as underground caverns, mines, dams) to do without dedicated 3-D overcoring tests which, in the past, often proved to be technically over-ambitious and geotechnically problematic. Instead, borehole slotting should be carried out in at least three independently inclined boreholes. This strategy will reward the geotechnical design engineer with all the intrinsic advantages of borehole slotting, such as:

> high measurement density with delineation of stress profiles (e.g. extent of the plastified zone in underground openings).

> possibility of a statistical approach to stress data processing. This approach is highly desirable in materials as inhomogenous as rock.

> computation of the general 3-D state of stress at the test site.

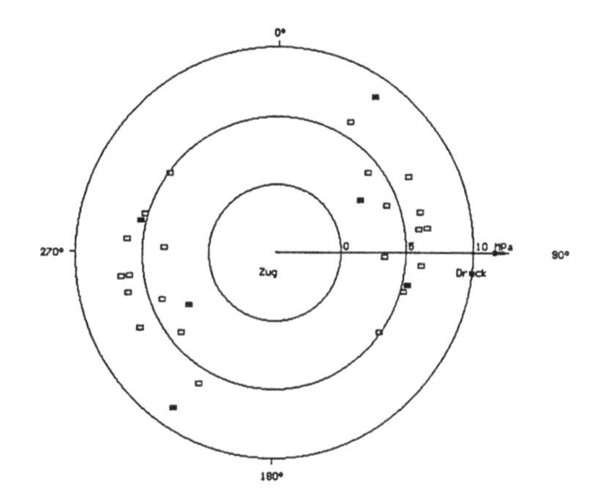

Fig. 17. Example of Borehole Slotting test results for σ_1 of one borehole for hydro-electric scheme "Goldisthal"; E-Germany

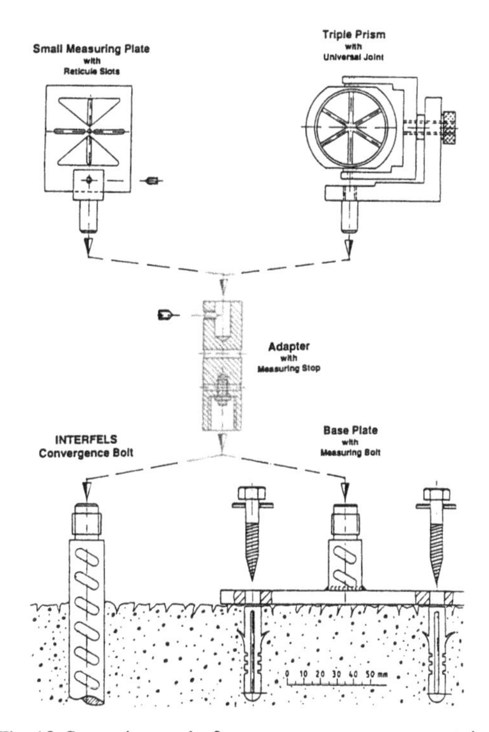

Fig. 18. Surveying marks for convergence measurements in tunnelling (System intermetric/INTERFELS)

3. PERFORMANCE MONITORING

One of the most commonly known example of performance monitoring of complete structures is convergence tape measurement in underground construction. For the New Austrian Tunnelling Method (NATM) such measurements are an integral part of the method.

Conventional convergence measurements by using tapes (e. g. INTERFELS *KM 15 or KM 30*) are more and more replaced by geodetic measurements. However, conventional tapes have to be kept at the tunnelling site for control purposes and also as a back-up for the case that geodetic measurements are not available in critical situations. INTERFELS has developed a set of surveying marks (Fig. 18) which is designed for deformation (including convergence) measurements in underground construction. Note that the system accuracy of the geodetic measurements is

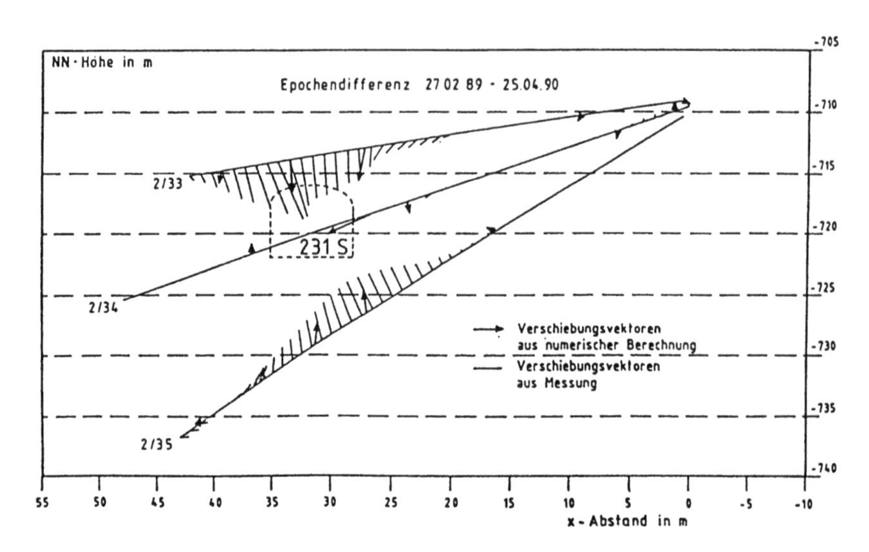

Fig. 19. High-definition rock deformation measurements by linewise observation of three inclined boreholes using mobile extensometer and inclinometer probes. Project: Excavation of a drift in an underground mine at Level -720 m. Big arrows: displacement vectors computed in a FEM model

in the range of +/- 1 mm and by this an order of magnitude lower than with the conventional convergence tapes.

The replacement of the **manual** convergence tape by semi-automatic geodetic measurements is indicative of a **general trend in geotechnical measurements**. There are three points which can be identified in this connection:

- ☐ Significantly **increased measurement density**; in boreholes employment of the **"linewise observation method"**;
- ☐ **automatic data stream**, real-time data collection and processing of data; and closely connected with this
- ☐ **user-friendly software** for IBM compatible PCs for immediate evaluation of the measuring data right at the construction site.

3.1 *Increased Measuring Density - "Linewise Observation" in Boreholes*

A very much increased measuring density has already been identified as one of the significant advantages of the borehole slotting system. In contrast to the conventional convergence measurements using a tape, geodetic deformation measurement in underground construction does not interfere with the tunnelling construction work. This makes an increase in the number of measuring points and measuring sections almost irresistible to the tunnelling engineer, particularly as the additional cost and efforts are quite minimal. A small renaissance of the freezing method in tunnelling (for environmental reasons) in association with increased demands on quality control of the frozen ground lead to an increase in the demand for **temperature chains**. INTERFELS expects a similar increase in the demand for its stationary strain sensor **"INDEX"** for testing of the deformation of in-place concrete piles. INDEX can be assembled into long chains of up to 32 elements for evaluating the skin friction/end bearing relationship of piles.

In the area of deformation measurement of boreholes, the concept of **"Linewise Observation"** [3] is of rapidly increased importance. In this regard INTERFELS offers two borehole probes: the **mobile extensometer INCREX** and the mobile **digital inclinometer** with servo-accelerometers of **particular long-term stability**. The two probes can be employed in a single borehole with a standard ABS inclinometer casing. The evaluation of the complete 3-D state of deformation will then be possible (Fig. 19). With the "linewise observation" technique, large amounts of field data are to be recorded and processed. In the project shown in Fig. 19 more than 16 000 INCREX and 32 000 inclinometer data were recorded within a time span of 2 years.

On the INCREX mobile extensometer a number of detailed case studies were presented in [4]. Mention should be made that INCREX (with or without additional inclinometer measurements) can be successfully employed not only in underground construction but equally meaningfully in foundation and rock slope engineering. So far, the deepest INCREX borehole is 130 m and was sunk in the foundation of a nuclear power plant (Fig. 20). The 130 m long borehole was

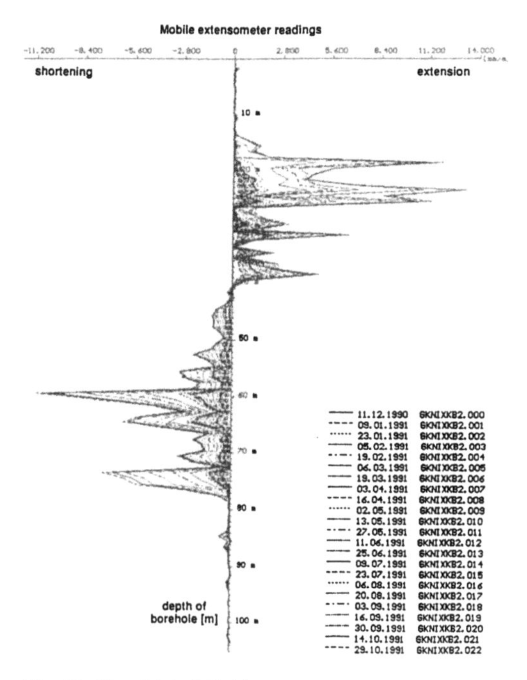

Fig. 20. Ultra high-definition extensometer measurements using the mobile probe "INCREX" in a 130 m deep borehole

evaluated metre for metre on its extension or shortening. Normally however, the depth of an INCREX borehole is in the range of 10 to 30 m.

3.2 *Automatic data stream, real-time monitoring and data transmission*

There is a clear trend towards an automatic data stream. Background of this development is not only the move towards a higher economy but also the sorry experience with "beautified", i.e. manipulated data. The INCREX measurements e. g. can be directly recorded by a battery-operated data-logging / laptop unit and then be transferred onto a PC for detailed evaluation.

Another trend in today's geo-instrumentation is real-time monitoring and evaluation of the measuring data. In this regard all mobile borehole probes are experiencing their inherent limits. In case that real-time monitoring of some loosening processes in underground construction is required, use must be made of conventional multiple-point extensometers instead of INCREX. The extensometers have to be equipped with electrical displacement transducers for automatic monitoring (Fig. 21).

Multiple-point extensometers, one of the oldest geotechnical instruments at all and of straight-forward design, have not escaped technical progress either. For more than three years, INTERFELS is offering single- and multiple-point **extensometers with packer anchors**. These anchors have proven to be most reliable and have been enthusiastically accepted by our customers.

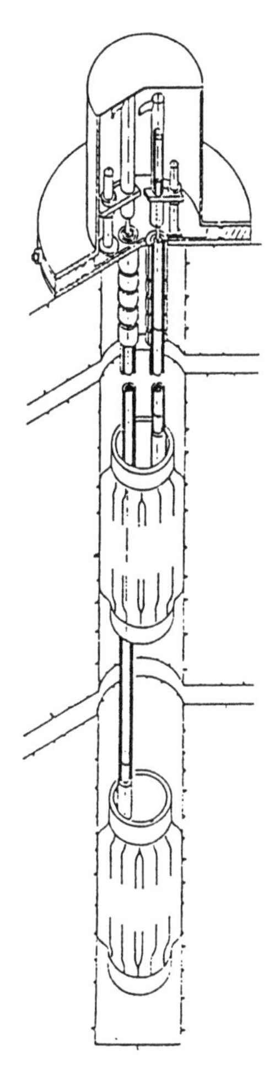

Fig. 21. Multiple-point extensometer with electrical displacement transducers. Maximum reliability by use of packer anchors thus avoiding interactions between extensometer and the ground

Background of this development is an **increased sensitivity towards possible interactions between measuring instrument and soil or rock** which might yield wrong measuring results. In case of conventional rod extensometers, the borehole is normally completely filled with a mortar-based grout. In rock with open joints the grout can run into the surrounding rock and can lead to changes of the rock properties. Moreover, the column of mortar grout in the borehole ("grout nail") might interact with the ground, particularly if the

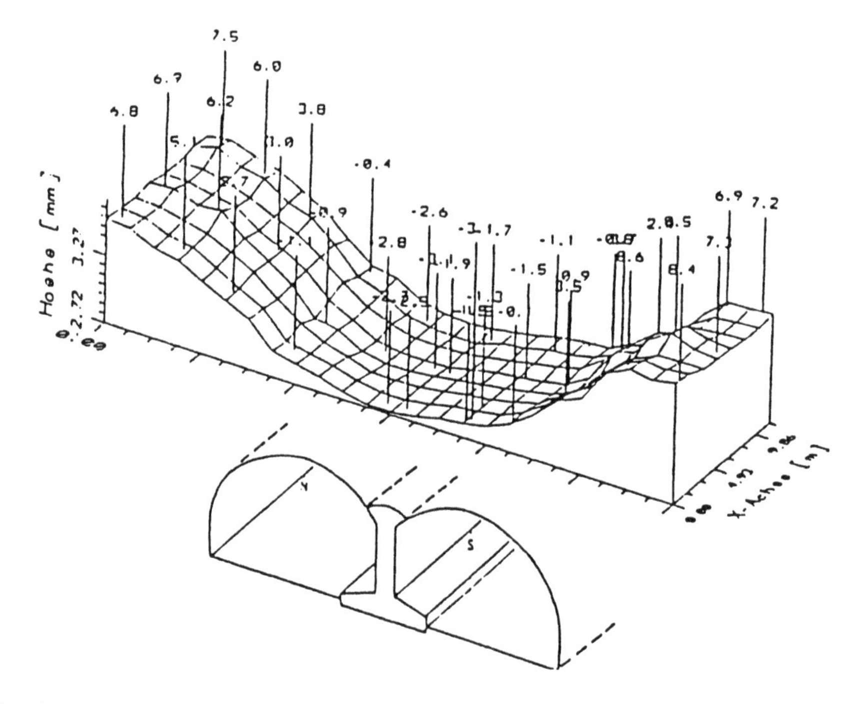

Fig. 22. Real-time monitoring of settlements of a 6-storey building due to inner-city tunnelling [out of 5].

stiffness of the column is higher than that of the surrounding ground. Extensometers with packer anchors avoid all these potential errors.

Back to the subject of real-time monitoring and evaluation of measuring data:- An interesting example was presented in [5], referring to the inner-city tunnelling project B61n in the township of Bielefeld/Germany. In this project, a 25 m wide road tunnel was driven a mere 4.2 m beneath the foundation of settlement-sensitive buildings. By means of a sophisticated electronic water levelling system the foundation settlements were monitored in detail and transmitted to a control centre in which the state of settlement was analysed and displayed on the monitor and in plots in real time (Fig. 22). In response to the actual stage of settlement soil frac measures were undertaken to compensate, if not over-compensate these settlements. For the soil frac procedure the electronic levelling system was also crucially important to prevent excessive uplifting.

This example demonstrates that real-time monitoring and instantaneous processing of geotechnical measurements can be a precondition for the employment of new construction methods, in this case the soil frac method. Moreover, it is a most impressive example for a successful application of the Observational Design Approach in foundation engineering. From the geo-instrumentation point of view it is evident that automatic data transmission is common standard these days.

4. CONCLUSIONS

With reference to some actual projects, which also include a number of Iranian projects, the particular relevance of *in-situ* testing and performance monitoring in geotechnical engineering has been demonstrated. In the years to come, these two parts of the geotechnical design will further increase their importance.

A number of trends has been identified. They include increased efforts with regard to *in-situ* testing both in terms of number of tests carried out and size of the testing equipment used. A further trend is the demand for high-definition measurements which include *in-situ* stress measurements (e.g. borehole slotting) and linewise observations by mobile probes (extensometer and inclinometer) in boreholes. Such observations provide an ideal reference base for a back analysis with numerical modelling codes.

Marrying *in-situ* measurements with numerical modelling is one of the most obvious requirements for a geotechnical design of any major project. This approach, however, is not commonly applied in the day-to-day practice of geotechnical engineering and further efforts from all sides, Engineers, Clients and Testing Institutions, are required to move further ahead towards this goal. With this regard reference is made to most recent efforts undertaken jointly by ITASCA Inc. and INTERFELS [6].

The final objective would be to set up a system which would work very much like the global weather forecast is working; numerical modelling of the global system with continuous upgrading of the input parameters (equivalent to *in-situ* testing) and boundary conditions (performance monitoring) as more information is becoming available. Many more developments are required from the manufacturer, scientists and engineers to make this system work with regard to efficiency, relevance and speed of information transfer.

REFERENCES

[1]Azzam, R. and Bock, H., 1987. A new modified borehole jack for stiff rock. - Rock Mech. Rock Eng., **20**, 191-211, Vienna.

[2]Dugan, K.J.; Hulls, I.H. and Miller, D.R., 1993. Recent experiences with the borehole slotter for measuring in-situ stress. - Proceed. Australian Conf. & Workshop on Geotechn. Instrum. & Monitoring in Open Pit & Underground Mining, June 21-22, 1993, Kalgoorlie.

[3]Kovari, K. and Amstad, Ch., 1981. Das Konzept der "Linien-Beobachtung bei Deformationsmessungen". - Mitt. schweiz Ges. Boden- u. Felsmech., **102**.

[4]+ + interfels news + + , **4**, April 1991.

[5]Otterbein, R. and Raabe, E.W., 1990. Tunnel Bielefeld B61n - Real-time settlement monitoring as the key for safe tunnelling under buildings. - + + interfels news + + , **2**, 1 - 6, May 1990

[6]Konietzky, H., 1993. Numerical back analysis of near-surface tunnelling in soft ground of the Ruhr district / Germany. - + + interfels news + +, **9**, 1-10, 1993 (in print)

Developments in Geotechnical Engineering, Balasubramaniam et al. (eds) © 1994 Balkema, Rotterdam, ISBN 90 5410 522 4

Some unconventional field testing methods for earth materials

B. Ladanyi
Ecole Polytechnique, Montreal, Ont., Canada

ABSTRACT: A new in-situ soil testing method, called "The Sharp Cone Test" (SCT) is described. The method consists of pushing a truncated low-angle cone into a pilot hole, causing a continuous enlargement of that hole as the cone descends, which can be translated into a relationship between radial pressure and radial (or shear) strain. Both load- and penetration-rate-controlled tests can be performed, using either instrumented or plain truncated cones. The results make it possible to deduce from the test the mechanical (short and long-term deformability and strength) properties of a soil or a weak rock.

1. INTRODUCTION

The knowledge of rheological properties of earth materials is an essential condition for the design of structural elements in contact with soils or rocks, to which they transfer the applied loads. Typical rheological properties are the strength of the material and its deformability, both affected by the time or the rate of strain. The earth materials falling into this category are soils, both frozen and unfrozen, ice, and weak rocks, such as rocksalt and potash. Practical problems requiring the knowledge of such rheological properties are, e.g., the design of foundations in frozen and unfrozen soils, the bearing capacity of ice covers, and the design of tunnel and shaft linings.

2. PRESENTLY USED SOIL SOUNDING METHODS

For determining the above mentioned rheological properties, both laboratory and in-situ methods are presently being used. In the former, undisturbed soil samples are taken from borings at selected levels, and are subjected to certain tests pertinent to the purpose at hand. The latter, in-situ methods do not require soil sampling, but they are able to measure only a limited number and scope of rheological properties. Their main advantage over the former is their rapidity and the ability to furnish a continuous picture of the geotechnical profile of the site.

Not considering the dynamic sounding methods, such as the Standard Penetration Test (SPT), or geophysical methods, which measure only certain physical properties of the ground, principal geotechnical in- situ methods presently in use are the following ones:

(a) Cone Penetration Test (CPT)

This is a standardized method in which a pressure-sensitive cone of 3.56 cm (1.4 inch) diameter and a 60° angle at the tip, fixed to the end of drill rods of the same diameter, is pushed into the soil at a rate of 2 cm/sec. From the recorded cone resistance (both total and piezometric pressure), certain mechanical properties of penetrated soils can be deduced, using theoretical models and statistical correlations. Although electrical cone tests have been in geotechnical use since 1950's, such tests have been adapted to measuring also the creep properties of frozen soils and ice only since the 1970's (Ladanyi, 1976, 1982a, 1985).

(b) Pressuremeter Test (PMT)

This test, introduced to geotechnical practice by Menard in the 1950's, consists of placing an inflatable probe into a prebored (or self-bored) borehole of the same diameter. The test is performed by inflating the probe in steps and by recording at each step the relationship between the applied pressure, the hole enlargement and the time. In unfrozen soils, this method has been mostly used for determining the short-term mechanical properties of soils. The theoretical interpretation of the tests in ordinary soils and rocks is presently well developed. Its principles can be found, e.g., in Ladanyi (1963, 1972), Palmer (1972) and Baguelin et al. (1978). In frozen soils and ice, the method has been used also for creep properties determination (Ladanyi and Johnston, 1973; Ladanyi, 1982b; Ladanyi and Eckardt, 1983).

(c) Flat Dilatometer Test (DMT)

The "Flat Dilatometer" is a soil testing tool, introducd by Marchetti (1980). It ressembles a thick sharp wedge, which is pushed into the soil at the end of drill rods. The measurement is made by slightly inflating a metallic diaphragm located at one side of the wedge. The test interpretation is based nearly exclusively on statistical correlations with soil properties deduced from some other field and laboratory tests.
Drawbacks of the currently used methods:

-The Cone Penetration Test (CPT) is certainly a very useful and practical method for continuous soil profiling. However, it furnishes only information on soil strength properties, with few direct data on soil deformability or on the shape of its stress-strain curve.

-The Pressuremeter Test (PMT) furnishes a complete stress-strain and strength information of the soil, but it is not a continuous soil profiling method. In addition, it requires a relatively expensive testing apparatus and a highly skilled operator.

-The Dilatometer Test (DMT) has shown to be a useful practical method for general soil profiling, but unfortunately, it furnishes information which is not clear, because it lacks theoretical background.

3. DESCRIPTION OF THE SHARP CONE TEST

The Sharp Cone Test (SCT) which is the subject of this paper, englobes the advantages of the test methods CPT and PMT, described above, but it is based on a principle which is essentially different from these two methods. It consists in pushing a sharp (low angle) cone (either ordinary or having several pressure transducers along its lateral surface) into a predrilled hole of the same shape as the cone, which ends with a predrilled cylindrical pilot hole of a smaller diameter (Fig. 1). The mechanical properties of the material are then deduced from the recorded relationship between the load applied to the cone, the time (or the rate of penetration), and the amount of vertical displacement of the cone, which is directly related to the enlargement of the pilot hole. This test can be carried out in two different ways:

(a) Testing in a Creep Mode is made by holding constant the load applied to the cone and by recording the relationship between the cone penetration and the time. Such a test makes it possible to determine the creep properties of a geological material, such as frozen soil, ice, rocksalt and other materials having distinct creep properties. A complete

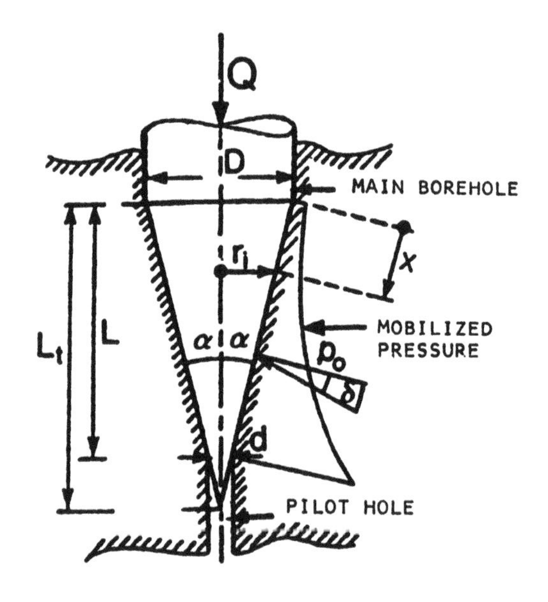

Figure 1. Principle of Sharp Cone Test.

theory of such a test was presented in three previous papers, (Ladanyi and Talabard, 1989; Ladanyi and Sgaoula, 1992; Leite et al., 1993). The papers contain both theoretical and experimental evidence on the validity and practical applicability of this method.

(b) Testing in a Continuous Penetration Mode is made by holding constant either the load on the cone or the rate of penetration into the pilot hole, and by recording the relationship between the penetration or the penetration rate and the resistance of the material against the enlargement of the pilot hole, recorded by a system of pressure transducers (PTD-s), installed on the lateral surface of the cone.

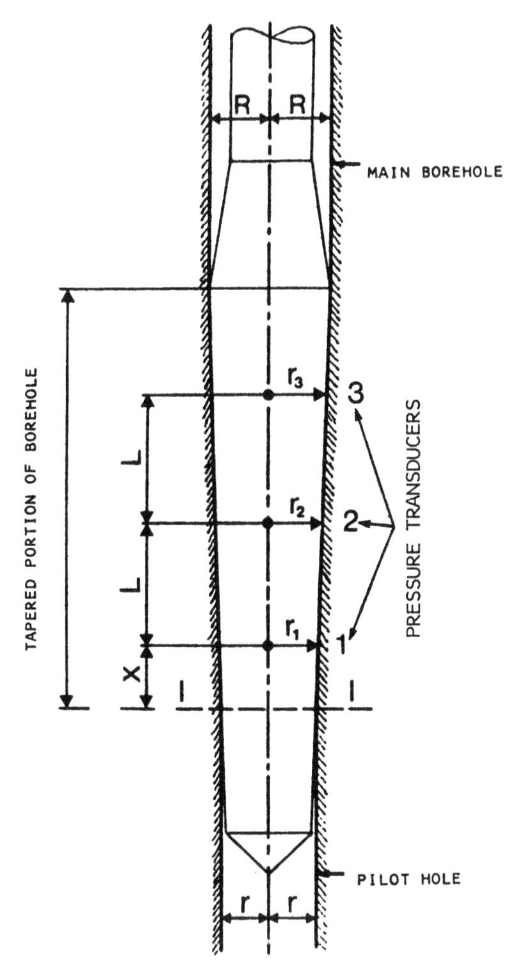

Figure 2. Instrumented Sharp Cone (schematic).

If, for example, the PTD-s are placed at 3 different levels of the lateral surface of the cone, as shown in Fig. 2, each of them will record, at a given level of the soil, a different value of the soil resistance, because at each given level, the total amount of hole enlargement will be different when each successive transducer passes through that level. In other words, for each selected level, this method makes it possible to determine several points of a "pressuremeter curve" (i.e., the relationship between lateral pressures and radial displacements, shown schematically in Fig. 3), similarly as in a pressuremeter test, the interpretation of which in terms of rheological properties is well known for different types of soils. The principle of determination of a "pressuremeter curve" from such a test, is shown in Fig 2, and is described in the following:

Figure 2 shows a low angle cone, with three total-pressure transducers (PTD-s) installed at its lateral surface. The half angle at the cone tip is α. When the cone is pushed downwards, the pilot hole of diameter 2r will be gradually enlarged to the diameter 2R of the main drill hole. Assuming that the PTD-s are placed at a distance L from one another, and that the lowest PTD is at a distance x from the upper end of the pilot hole (Fig.2), then, when the cone is gradually pushed into the pilot hole, traversing successively the distances x, (x+L) and (x+2L), total radial strains, $\ln(r_i/r)$, are, for a constant volume assumption, approximately equal to the circumferential strains ϵ_θ, or to one half of shear strains, γ. At any fixed level "i" (where $r = r_i$) they are equal to:

$$\ln(r_i/r) \approx \Delta r/r_i \approx \epsilon_{\theta i} \approx \gamma_i/2$$

For example, for the level I-I in Fig. 2, one would get:

Penetration	Radial Displacement
x	r_1-r
x+ L	r_2-r
x+2L	r_3-r

Radial Strain

$\ln(r_1/r)$
$\ln(r_2/r)$
$\ln(r_3/r)$

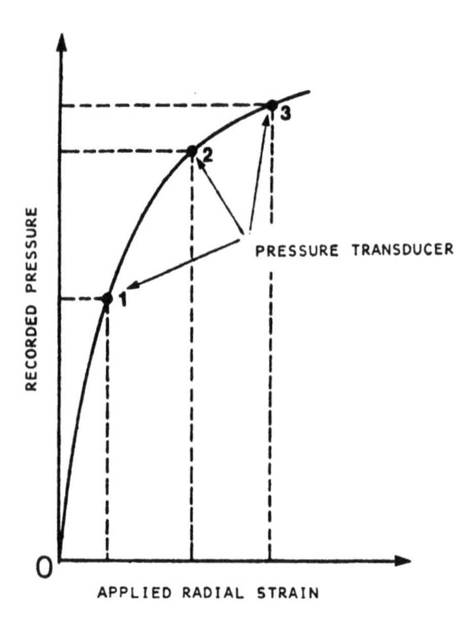

Figure 3. A typical "Pressuremeter Curve" obtained from a Sharp Cone Test (schematic).

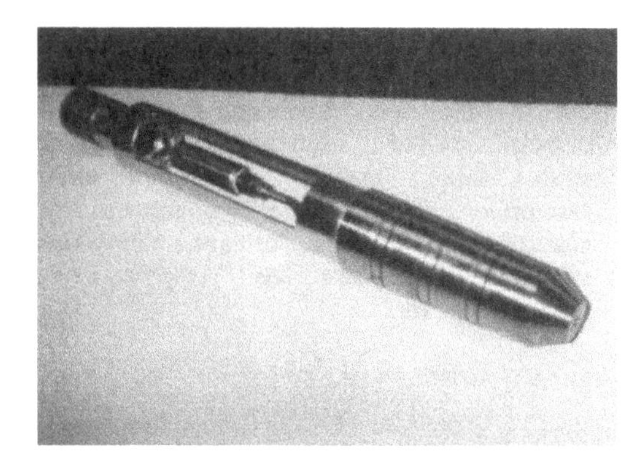

Figure 4. Sharp cone with three levels of lateral pressure transducers.

where

$r_1 = r + x \tan\alpha$
$r_2 = r + (x+L)\tan\alpha$
$r_3 = r + (x+2L)\tan\alpha$

Take, for example, a cone with $\alpha = 1°$, intended to enlarge a pilot hole from r = 3.0 cm to R = 3.5 cm. The tapered portion of the hole, formed by a special tool, would have in this case a length of about 28.6 cm. (Note: The tapered portion of the hole is useful for the beginning of the test, but it is not essential in practice). Assume further that the PTD-s are placed at distances of 5 cm, 15 cm, and 25 cm, respectively, from the upper end of the pilot hole. For a penetration of x = 5 cm, one would get at the level I-I (Fig.2) a radial strain, equal to: ln(1 + 5x0.01746/3) = 0.0287, and the corresponding pressure recorded by the PTD 1 will be p_1. A penetration of 15 cm gives the strain: ln(1 + 15x0.01746/3) = 0.0837, and the pressure recorded by PTD 2 will be p_2. Finally, a penetration of 25 cm leads to: ln(1 + 25x0.01746/3) = 0.1358, and the pressure recorded by PTD 3 will be p_3.

Had, for example, an angle of 2° been selected for the cone instead of 1°, the corresponding radial strains would have been: 0.057, 0.161, and 0.255, respectively.

At any given lateral pressure measurement level, the strains will remain the same as long as the pilot hole precedes the cone, but the recorded pressures will vary according to the soil properties.

Obviously, by selecting an appropriate taper of the cone, it is possible to adapt the system to any expected deformability of the tested material. The experience acquired to date indicates that a practical range of the semi-angle of the cone should preferably be between 1° and 10°, with a most frequent range of 1° and 5°. In particular, in the creep testing mode, an angle of 5° was found convenient for testing ice and frozen sand, while 2° was found preferable for testing a much stronger rocksalt.

In the continuous penetration mode, angles of 1° to 2° are found convenient for testing saturated clays, because they are able to produce the most important portion of the stress-strain curve. On the other hand, larger angles of up to and above 5° may be found more appropriate when testing very compressible materials, such as peat or loose sands.

By relating the radial (or shear) strains with the corresponding pressures recorded by the PTD-s at different levels of the pilot hole, one can get an unlimited number of "pressuremeter curves", such as the one shown schematically in

Fig.3. The curves can be treated in a conventional manner, described, e.g., in Ref. (1, 2, 3, 6, 8, 9, 14).

4. STATE OF DEVELOPMENT OF SCT METHOD

The non-instrumented sharp cone constant-load method for creep properties determination has up to now been the subject of three experimental studies, namely, in ice (Ladanyi and Talabard, 1989), in frozen sand (Ladanyi and Sgaoula, 1992), and in rock salt (Leite et al., 1993), the last one including also a FEM verification.

An instrumented cone for continuous sounding with three lateral pressure measuring levels has recently been developed by ROCTEST LTD., Montreal (Fig. 4). It uses a lateral pressure measuring system developed originally by Huntsman (1985) and Tseng (1989) at the University of California, Berkeley.

Figure 5 shows a first result obtained at a sensitive clay site near Montreal. The test produces an infinity of pressuremeter curves, from which only 7 were shown in a $\Delta V/V$ versus pressure plot for 7 selected levels at 0.5 m distance. As well known, (e.g., Ladanyi 1972) such a plot makes it possible to determine a portion of the stress-strain curve for each selected depth.

5. CONCLUSION

Compared to the Pressuremeter Test (PMT), the proposed SCT method:
(a) Uses a simpler and less expensive equipment, by furnishing at the same time the data comparable to those measured by a PMT test;
(b) It can be used for a continuous sounding, which is not possible with the PMT;
(c) Its capacity is not limited by the loading system, as in a PMT, which makes it applicable also to very strong and stiff materials, like rocksalt.

Compared to the Cone Penetration Test (CPT), it furnishes more complete rheological information, while CPT is limited only to the extreme conditions at failure.

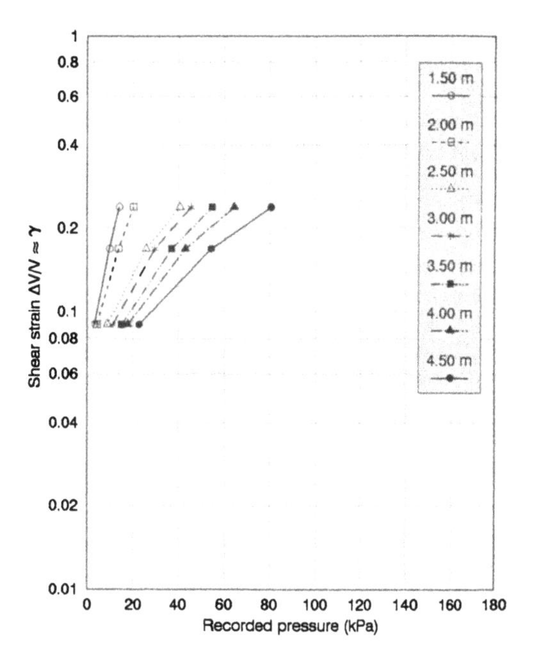

Figure 5. Typical "pressuremeter curves" obtained by a Sharp Cone Test in a sensitive clay.

Compared to the Flat Dilatometer Test (DMT), it is based on a clear and well defined interpretation theory, and it can deform the soil much more extensively than DMT, making it possible to determine a portion of the stress-strain curve of the soil.

On the other hand, the proposed method requires a predrilled pilot hole, which must remain stable before and during the cone penetration. This is easy to realize in strong materials, but it may present certain difficulties in weak soils, such as water-bearing sands and weak clays, in which a self- boring tool with the use of drilling mud will be necessary. In that case, a free passage of drilling mud across the cone will be required. On the other hand, in very stiff materials, such as frozen sands and the rocksalt, the method requires an efficient greasing of the cone during penetration.

Besides the PTD-s for measuring the total lateral pressure, some piezometric transducers can also be installed on the cone, for measuring generation and dissipation of pore pressure around the advancing or static cone.

6. REFERENCES

Baguelin, F. Jezequel, J.F. and Shields, D.H. (1978). The pressuremeter and foundation engineering. Trans Tech Publ., Clausthal, Germany.

Huntsman, S.T. (1985). Determination of in-situ lateral pressure of cohesionless soils by static cone penetrometer. Ph.D. Thesis University of California, Berkeley.

Ladanyi, B. (1963). Evaluation of pressuremeter test in granular soils.Proc. 2nd Panamerican Conf. on Soil Mech & Found. Engrg., Sao Paulo, 1, 3-20.

Ladanyi, B. (1972). In-situ determination of undrained stress-strain behavior of sensitive clays with the pressuremeter. Canad. Geotech. J., 9, 313-319.

Ladanyi, B. (1976). Use of the static penetration test in frozen soils. Canad. geotech. J., 13, 95-110.

Ladanyi, B. (1982 a). Determiation of geotechnical parameters of frozensoils by means of the cone penetration test. Proc. 2nd Europ. Symp. on Penetration Testing, Amsterdam, 1, 671-678.

Ladanyi, B. (1982 b). Borehole creep and relaxation tests in ice-rich permafrost. Proc. 4th Canad. Permafrost Conf., Calgary, 1981, R.J.E. Brown Memorial Volume, CNRC, Ottawa, 406-415.

Ladanyi, B. (1985). Use of the cone penetration test for the design of piles in permafrost. ASME J. of Energy Resources Technology, 107, 183-187.

Ladanyi, B. and Eckardt, H. (1983). Dilatometer testing in thick cylinders of frozen sand. Proc. 4th Int. Conf. on Permafrost, Fairbanks, Alaska, Nat. Acad. Press, Washington, D.C., 677-682.

Ladanyi, B. and Johnston, G.H. (1973). Evaluation of in situ creep properties of frozen soils with the pressuremeter. Proc. 2nd Int. Conf. on Permafrost, Yakutsk, North Amer. Contr. Vol., NAS, Washington, D.C., 310-318.

Ladanyi, B. and Sgaoula, J. (1992). Sharp cone testing of creep properties of frozen sand. Canad. Geotech. J., 29, 757-764.

Ladanyi, B. and Talabard, Ph. (1989). Sharp-cone testing of frozen soils and ice. Proc. 5th Int.Conf. on Cold Regions Engrg., St.Paul, Minn., 282-296.

Leite, M.H., Ladanyi, B. and Gill, D.E. (1993). Determination of creep parameters of rocksalt by means of an in situ sharp cone test. Int. J. of Rock Mech. and Mining Sci., 30, 3, 219-232.

Marchetti, S. (1980). In situ tests by flat dilatometer. J. of Geotech. Engrg. Div., ASCE, 106, GT 3, 299-321.

Palmer, A.C. (1972). Undrained plane strain expansion of a cylindrical cavity in clay.: a simple interpretation of the pressuremeter test. Geotechnique, 22, 451-457.

Tseng, D. (1989). Prediction of cone penetration resistance and its application to liquefaction assessment. Ph.D. Thesis, University of California, Berkeley.

Developments in Geotechnical Engineering, Balasubramaniam et al. (eds) © 1994 Balkema, Rotterdam, ISBN 90 5410 522 4

It took 2000 years of penetration to arrive at the Amap'Sols static-dynamic penetrometer

G. Sanglerat
Ponts et Chaussées, Lyon, France

ABSTRACT : Penetration tests to evaluate soil resistances have been used for centuries. Their modern versions were developed in Sweden, The Netherlands and in France. Lyons took a leading position in the development of the static-dynamic penetration, first in 1959 and then again in 1967. A new static-dynamic penetrometer, called Amap'Sols, was constructed in 1993. This paper presents the unique features and first results obtained with this new device which represents an important improvement in the methods for soil investigations.

1 INTRODUCTION

For ever, man has been faced with the problem to evaluate the bearing capacity of soils, to either support structures foundations or to construct roads and railways. Over 2,000 years ago, before Jesus-Christ, bamboo sections were driven into the soil in China for that purpose and also to obtain water. During the Middle Ages in France, prototype wooden piles would be driven before starting construction of pile foundations for some large bridges or cathedrals.

In 1846, Collin developed a static penetrometer of the Vicat needle type in order to evaluate cohesion of soils through which canals had to be excavated.

In 1917, the Swedish Railways standardized the use of a dynamic penetrometer and used it for the construction of railway lines and bridges.

In 1929, Terzaghi was one of the first investigators to measure in a continuous manner, the variation of static penetration resistance of sand with the use of a cone pushed vertically into the soil (wash-point hydraulic probe).

In the Netherlands in 1932, the static cone penetrometer test (CPT), now a classic tool, was invented by Barentsen (ref.1). This device found widespread use due to the work of the Delft Soil Mechanics Laboratory and the Goudsche Machinefabriek of Gouda, and later, by Van Den Berg.

Both the purely static and purely dynamic penetrometer types have advantages and drawbacks (ref.2). The French developed the static-dynamic penetrometer in order to combine the advantages and eliminate the short comings.

2 THE FIRST STATIC-DYNAMIC PENETROMETERS

It was in Lyons, in 1950, that the Jangot-Bonneton static-dynamic penetrometer was constructed. This device had a fixed cone with a 60 mm diameter. It was driven mechanically from inside a casing. The casing had the same length as the inner rods and pushed down on a collar through a special connection. The load was applied to either the point of the casing independently or to both simultaneously. This enabled measurements of point resistance and total side friction separately by dynamometers. In 1955, Meurisse in Northeast France developed a similar device.

In 1967, in Lyons, an important improvement to the method was introduced by Andina by adding a friction sleeve to measure skin friction during the static mode of testing (ref.1).

The dynamic method was also improved by providing a hammer weighing 7,500 kN, falling 0.16 m.

Until 1992, this penetrometer was the most

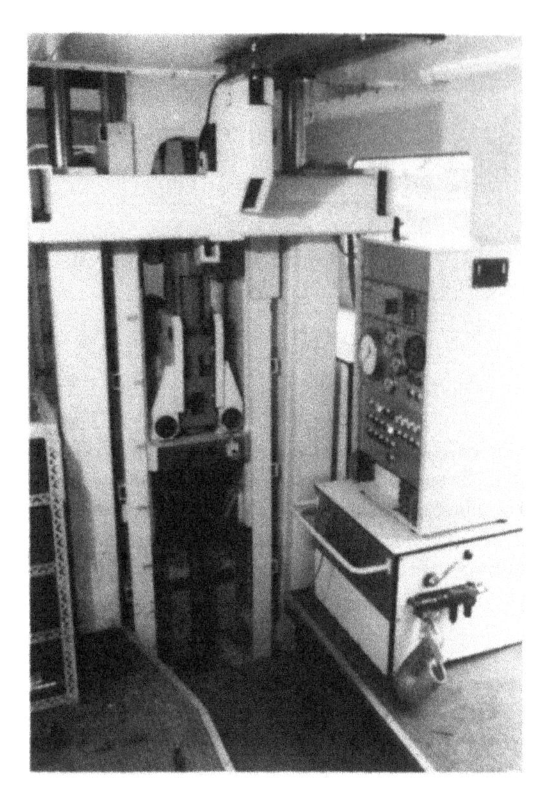

Fig.1 - View of the hydraulic hammer and measuring equipment for the static-dynamic penetration of Amap'Sols

Fig.2 - General view of the Amap'Sols truck

powerful soil testing tool in use anywhere and was widely used in France and Lebanon. It performed very well near Grenoble where record depths of 77 meters of penetration were reached under favorable conditions. However, all of its operations were manual.

3 THE AMAP'SOLS STATIC-DYNAMIC PENETROMETER

3.1 Principle of operation

In 1992, geotechnical engineers in Lyons and Saint-Etienne decided to develop a new static-dynamic penetrometer to improve this type of soil exploration.

The important improvements brought about consisted of :

Static mode : totally automatic operations, penetration at 2 cm/sec with continuous numeric recording every 2 cm on memory board with simultaneous drawing of the pressure diagram in real time. This allows for instantaneous control of the penetration.

The records and evaluations are transmitted by modem to the office which permits rapid engineering interpretation of the test results to evaluate the soil parameters needed for the determination of designs and settlements (ref.3,4 and 5).

Dynamic mode : the old fashion way of driving a probe by a free falling hammer has been replaced by a powerful, fast-action hydraulic hammer (see fig.1), with adjustable energy, capable of going through extremely dense layers and penetrate into altered bedrock.

Installation : the penetrometer is mounted on 6x6 Mercedes truck, of 260 kN.
The leveling of the truck to insure true vertical penetration is done automatically by five hydraulic jacks controlled by electronic sensors. Four of the jacks bear on retractable track part of the truck auxiliary propulsion mechanism (see fig.2).

The catlike tracks insure fast and easy mobility on soft soil terrains where conventional trucks would bog down. They are a patented device of Van Den Berg.

The system of penetration in the static-dynamic modes was conceived and built in Lyons and after, mounted on a Mercedes truck in the Netherlands, by Van Den Berg who has a great experience with

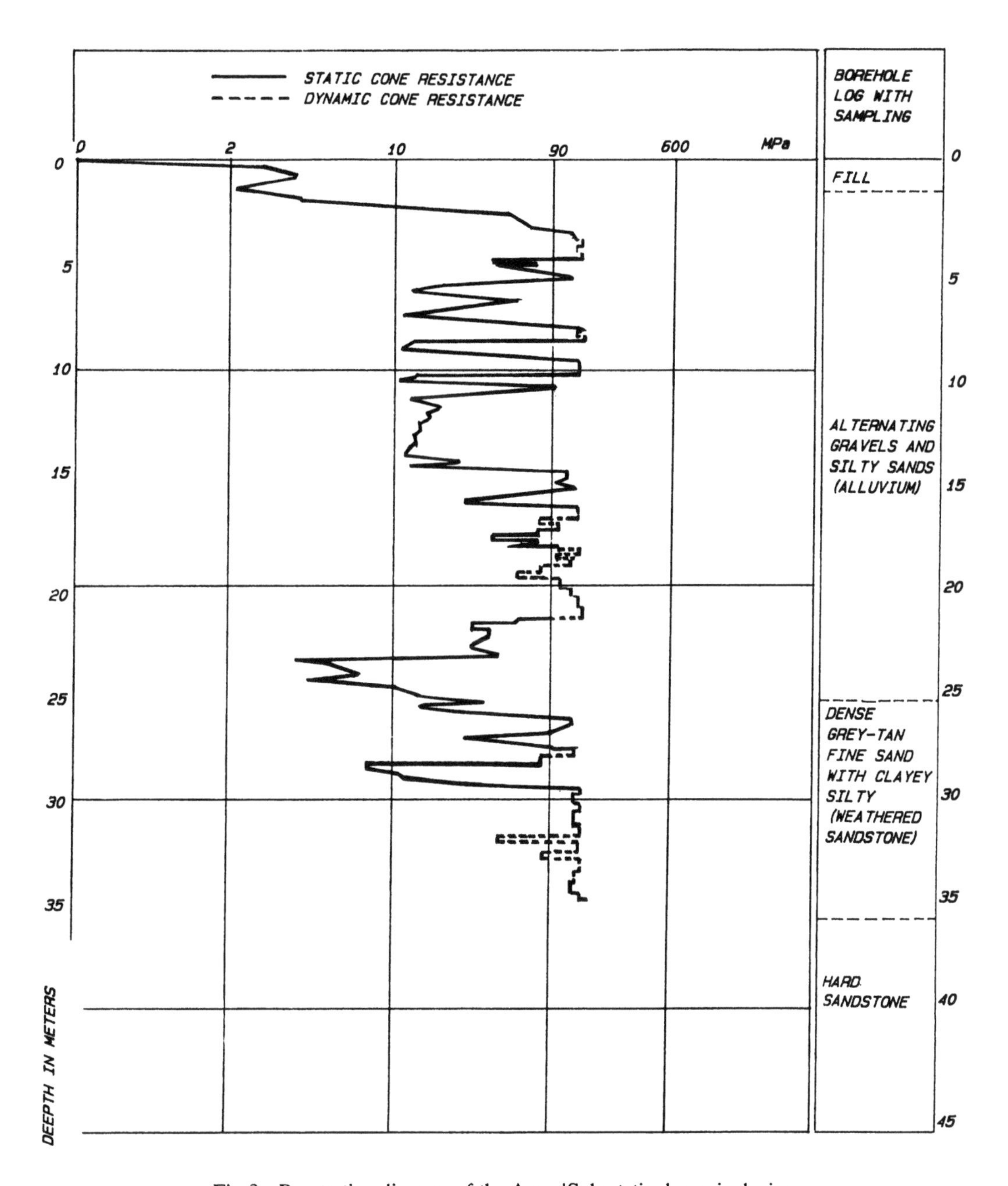

Fig.3 - Penetration diagram of the Amap'Sols static-dynamic device and log of boring obtained from an Infrasoil boring made 2 m away

the static penetrometer over many years. His high degree of technical expertise in hydraulic systems and the recording of data has greatly contributed to the success of the device.

The Amap'Sols can, therefore, be considered to be an European penetrometer with the idea coming from Lyons and its construction from both France and The Netherlands.

The name Amap'Sols means *"Ateliers Mobiles d'Auscultation par Pénétration des Sols"* (Mobile Soil Testing Unit by Penetration).

3.2 Characteristics

This penetrometer offers the possibility of all manners of penetration into soils for specific purposes. It can push all the known measuring tips varying from 10 to 100 cm^2, be they of the mechanical or the electronic types such as piezocones and others.

Usually, static penetration is done either with an 80 mm diameter tip equipped with a skin friction sleeve of 250 mm in length, or with a 75 mm tip and a 200 mm sleeve.

In the static mode, the following measurements are made :

q_c: cone resistance (up to 30 MPa)

f_s : skin friction, which permits calculation of the friction ratio, FR

Q_{st}:total resistance to penetration (up to 220 kN)

When refusal is met with the 75 or 80 mm tip in a hard soil layer, static penetration may be continued with a 39 mm tip not equiped with a friction sleeve. The cone resistance of the 39 mm tip can reach 125 MPa. When refusal is met, dynamic penetration is used (see fig.3).

It is obvious that at this level of stress, the evaluation of the ultimate shear strength of soils no longer translates into significant physical meanings.

So, dynamic penetration is used only to get through dense soil layers.

For additional information, every 25 cm, a static test is performed up to 125 MPa (which is very high and well above the capacity of any other penetrometer). Each time that q_c < 125 MPa, the static penetration mode is employed. A sound alarm is triggered every time any of the load limits are reached for each of the tubing configurations.

In the event of a sudden drop of tip resistance, the dynamic driving mechanism automatically stops instantaneously at the upper boundary of the less resistant layer. This prevents the penetration to occur without any measurements made of the softer layer.

3.3 Other uses

The penetrometer accepts different cone tips, such as the piezocone and the envirocone of Van Den Berg (ref.6).

In each of these two cases, the computer programs used are those defined by Van Den Berg.

Depending on its manner of use, the piezocone can measure q_c, f_s as well as pore water pressures.

With the envirocone, the following measurements, besides q_c, are made possible:

soil conductivity
H^- and O^{++} concentration
redox potential
temperatures
pH and porewater pressure

This range of utilization brings about a considerable improvement of the resources available to study environmental and waste management problems, thanks to the quality of the data obtained.

4 - SAMPLE PENETRATION DIAGRAMS

Penetration tests with the Amap'Sols penetrometer were made at sites which had previously been studied and whose geology was well known.

Usually, in the Lyons area, the static penetration tests of the Gouda type are totally useless because refusal is met between 3 or 4 meters depth. With classic static-dynamic penetrometer refusal is too often encountered either in dense upper alluvial layers or at best, in the marine sandstone encountered around 20 m.

The first tests performed with the static-dynamic Amap'Sols device gave a depth of 35 m, which represents a considerable performance improvement.

Over the upper 20 m the results obtained are identical to those coming from other static-dynamic penetrometers. For the penetration through the following 15 m, no comparison with previous penetrometer test data was possible since none could reach these depths. Comparison therefore was made with borehole information with parameters recordings (ref.7).

Fig.3 shows static-dynamic penetration test results made 2 m away from the location of a drill hole in which rate of penetration, vertical pressure and torque were recorded down to 26.90 m. From that depth to 34.50 m, disturbed soil samples were recovered and the drill hole was continued without sampling but with the measurements of the three previously mentioned parameters to a depth of 44.50 m. The log of the drill hole is shown on Fig.3 adjacent to the static-dynamic penetration record obtained with the Amap'sols.

The Amap'sols penetrated to 35 m, a feat never

before achieved since in 1989 the previously most powerful static-dynamic penetrometer had met refusal between 17.20 and 19.80 m at the same site.

5 CONCLUSIONS

It is obvious that because of the choice of different tips, the Amap'sols static-dynamic penetrometer is a fast and thorough way of studying soil properties. Its mobility and set up are made easy thanks to its self levelling jacks resting on tracks.

It provides the means of using a piezocone or an envirocone, thereby yielding additional important data.

It offers a significant improvement over other methods and is expected to gain wide acceptance, both in France and many other countries. It is now the state of the art for penetration techniques since no other device can be compared to it.

REFERENCES

1-Sanglerat, G. 1979. *The penetrometer and soil exploration*. Second enlarged edition, 488 p. Amsterdam, New York : Elsevier.

2-Stefanoff, G., Sanglerat, G., Bergdal, U.Melzer, K.J. 1988. *Dynamic Probing (DP) International Reference Test Procedure.* ISOPT-1. Orlando, Rotterdam : Balkema, pp 53,70.

3-Sanglerat, G., Girousse, L., Bardot, F. 1977. *Settlements predictions buildings based on the static penetrometer data. Fifth Southeast Asian Conference on Soil Engineering*: 27-40. Thai Watana Panich Press.

4-Sanglerat, G., Olivari, G., Cambou, B. 1984. *Pratical Problems in Soil Mechanics and Foundations Engineering.*
Vol.I *Physical Characteristics Soils, Plasticity, Settlement Calculation, Interpretation of in situ Test*. 283 p.
VOl.II *Retaining Walls, Sheetpile Walls, Shallow Footings, Deep Foundations, Slope and Dams*. 253 p. Amsterdam : Elsevier.

5-Gielly, J., Lareal, P., Sanglerat, G. 1969. *Correlation between in situ penetrometer test and the compressibility characteristics of soils. Conference on in situ Investigation in Soils and Rocks*. 167-172 and 189-191. London.

6-Lunne, T., Eidsmoen, T.E., Powell, J.J.M., Quaterman, R.S.T. 1986. *Piezocone testing in overconsolidated clays. 39th Canadian Geotechnical Conference on in situ Testing and Field Behaviour*. Ottawa.

7-Darricau, C., Deletie, P. 1990. *"Quinze années d'utilisation des enregistreurs de paramètres en forages. Evolution de la méthode et du traitement des données. Application aux reconnaissances et aux travaux*. Paris : Société Géologique de France. NS 1190, n° 167.

Developments in Geotechnical Engineering, Balasubramaniam et al. (eds) © 1994 Balkema, Rotterdam, ISBN 90 5410 522 4

Measurement of ground vibrations

P.J. Moore & J.R. Styles
The University of Melbourne, Vic., Australia

ABSTRACT: This paper explores, theoretically and experimentally, the use of surface and embedded mountings with transducer attached for the measurement of free field vibrations on the ground surface. A theoretical study for both vertical and horizontal vibrations has been carried out. Field measurements of vibrations generated by an electromagnetic vibrator and by a falling weight have been made at sandy and clayey sites. Comparisons between vibration observations and predictions indicate an approximate level of agreement.

1. INTRODUCTION

The measurement of amplitudes of vibration has been the subject of much comment in the literature. Major attention has centered on the transducers themselves and on their characteristics (e.g. Bradley and Eller (1961), Richart et al. (1970) and Hanna (1985)). Relatively less attention has been given to the mounting of transducers. Brown (1971), for example, describes measurements of ground vibrations caused by construction equipment and by blasting operations. The transducers were mounted on a block of aluminium but no further details were provided. More recently Grant (1983) described an investigation in which he concluded that the measured ground motion can be considerably influenced by the performance of the mounting. He found that mounting shape, size, material type and method of installation all played prominent roles.

Dowding (1985), in relation to the measurement of ground vibrations from blasting, has mentioned that the type of mounting is least critical when the maximum particle accelerations are less that 0.3g. For greater accelerations he suggests that the mounting should be partially or completely buried if the ground surface consists of soil. These recommendations are quite inadequate.

Regarding earthquake vibrations, Crouse et al. (1984) have commented on the desirability of satisfying two criteria:

(a) the base dimensions of the mounting are much smaller than the wavelengths of the seismic waves and

(b) the natural frequencies of the mounting are much greater than the seismic wave frequencies.

While these criteria are necessary for the accurate measurement of ground vibrations, they must be quantified to be useful in practice.

A theoretical examination into the performance of mountings for measurement of free field vertical and horizontal vibrations has been carried out by Moore (1986, 1988). This work showed how the magnitude of the error in the measurement of vibration amplitude depends upon the vibration frequency, the mounting geometry, the ground rigidity and the natural frequency of the mounting.

2. VERTICAL VIBRATIONS - THEORETICAL CONSIDERATIONS

The precision with which the vibration transducer correctly measures the ground vibration amplitude may be indicated by means of the displacement transmissibility. It is clear that the transducer will provide increasingly accurate measures of the amplitude of ground motion as the displacement transmissibility more closely approaches unity. As illustrated in Fig. 1, this is best achieved at low values of the frequency ratio. For example Fig. 2 shows that for zero damping and provided the frequency of ground motion is no greater than 30% of the natural frequency (f_n) the maximum error in the measurement of ground motion amplitude is 10%. Fig. 2 also shows that the errors in amplitude measurement may become quite large when the frequency of ground motion is greater than 50% of the natural frequency of the mounting unless considerable damping is present in the system.

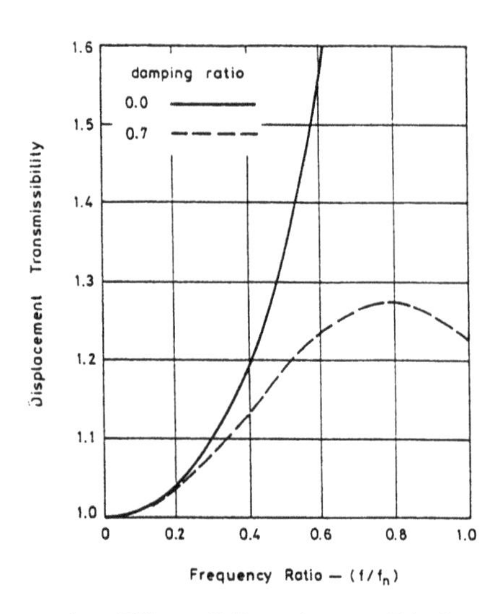

Figure 1. Effect of Damping on Displacement Transmissibility

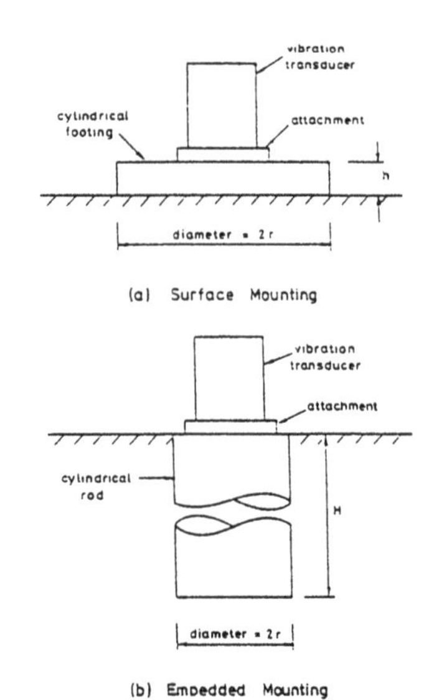

Figure 3. Mounting Types for Vibration Transducers

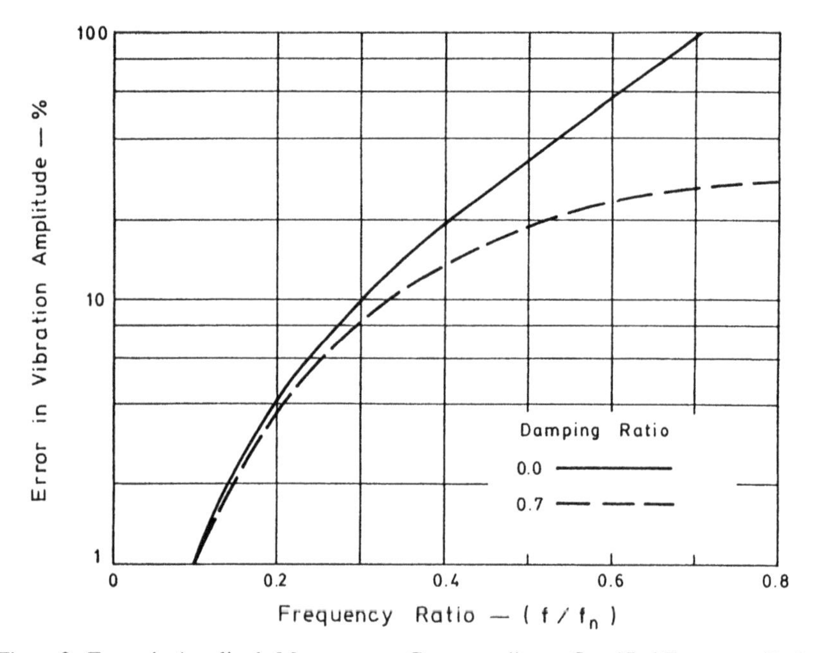

Figure 2. Errors in Amplitude Measurement Corresponding to Specified Frequency Ratios

The damping ratio for many transducers is either in the vicinity of zero or around 0.7. The types of transducers mostly used by the authors are Brüel & Kjaer piezoelectric accelerometers. As stated by Serridge and Licht (1986) the damping ratios for these instruments are very low. In the following discussion the damping ratio will be taken as zero.

2.1 *Surface Mountings*

The upper limit of the ground motion frequency can easily be evaluated for certain stated magnitudes of the maximum error in the amplitude measurement after the natural frequency (f_n) of the mounting has been evaluated. This has been done for surface mountings which cylindrical footings, usually made of steel and to which an accelerometer is attached. The unit is simply placed on the ground surface (Fig. 3).

The upper limits of the ground vibration frequency corresponding to various maximum errors in the measured amplitude of ground motion are illustrated, for example, in Figure 4 for soft ground (represented by shear modulus (G) of 10MPa). These calculations confirm that there are serious limitations to the accurate measurement of high frequency vibrations. Consider for example, a mounting with a radius of 20cm and a mass ratio of 10. If the maximum acceptable error in the amplitude measurement is, say 10%, then the greatest frequency that can be measured is about 13 hertz on soft ground and about 59 hertz on hard ground. These frequency limits could be raised, however, by accepting a less accurate amplitude measurement or by altering the geometry of the surface mounting.

In the dynamic analyses described above it has been assumed that the surface mounting behaved as a rigid block. A rigidity requirement has been developed from the work of Brown (1969) and for a steel mounting this can be expressed as:

$$(h/r)^3 = 13.3 \times 10^{-5} G \qquad (1)$$

where G is the shear modulus of the ground in MPa.

The measuring range for a transducer on a surface mounting can be represented in a diagram such as that of Figure 5. For this diagram a particular surface mounting size (b = 10, r = 5cm) a particular transducer (Brüel & Kjaer accelerometer # 4371) and a particular ground rigidity (G = 200MPa) have been selected. Note that the mass ratio (b) incorporates the surface mounting, attachment and accelerometer, all acting as a single rigid unit. To avoid the surface mounting losing contact with the ground an upper acceleration limit of one gravity has been shown. The lower acceleration limit for the accelerometer has been obtained from manufacturers literature. The upper frequency limits have been derived from the dynamic analysis. The area remaining within the boundaries in Fig. 5 indicates the measuring range.

2.2 *Embedded Mountings*

The embedded mounting considered here consists of a cylindrical rod which may be pushed or excavated into the ground and to the top end of which is attached the transducer (see Fig. 3). As in the case of the surface mounting, the embedded mounting is considered as an undamped mass-spring system for purposes of analysis of vibration response.

Such analyses yielded the following observations:

(a) In terms of frequency, the upper limit of the measuring range (for a given maximum error in amplitude) with an embedded mounting increases as the ground becomes more rigid.

(b) For a specified rod length (H) and maximum error in measured amplitude, the upper limit of frequency increases as the rod radius is decreased.

(c) For rod radii less than about 2 cm, the upper limit of frequency increases as the rod length increases.

(d) For rod radii greater than about 2 cm, the upper limit of frequency increases as the rod length decreases.

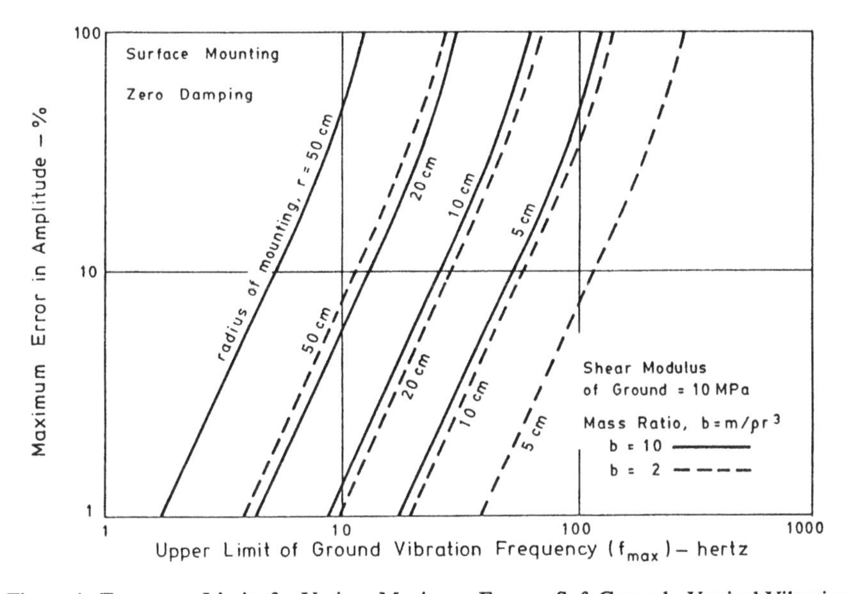

Figure 4. Frequency Limits for Various Maximum Errors - Soft Ground - Vertical Vibration

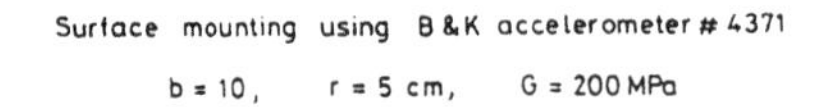

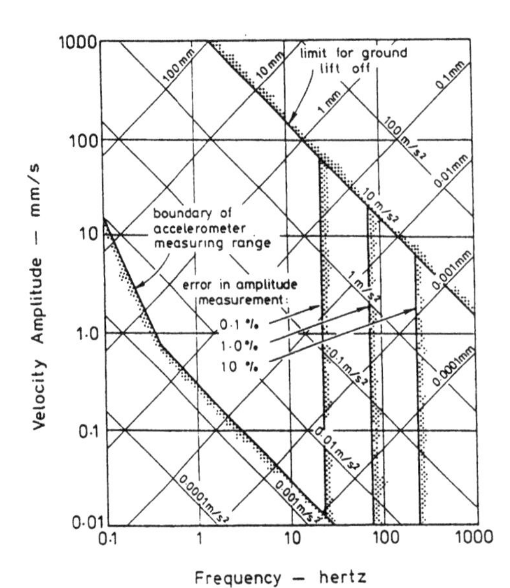

Figure 5. Measuring Range for Surface Mounting with B & K Accelerometer

In the analyses described above it was assumed that the embedded rod behaved as a rigid body. This assumption has been checked by adapting the work of Mattes and Poulos (1969) who examined the top and bottom displacement of a loaded compressible pile. The required relative rigidity of the rod (made of steel) for the rod to behave as a rigid body depends upon the ratio of length to radius of the rod. The requirement, which is approximate only, is illustrated in Figure 6 in terms of the maximum ground rigidity corresponding to rigid mounting behaviour. It is seen that for slender rods (large (H/r) ratios) this requirement places serious limitations on the rigidity of the ground in which vertical vibrations may be accurately measured.

3. HORIZONTAL VIBRATION - SURFACE MOUNTINGS

In a similar way to the vertical vibration case, the maximum ground motion frequency (f_{max}) corresponding to specified magnitudes of the maximum acceptable error in amplitude measurement, can be evaluated from the displacement transmissibility. These analyses demonstrate the limitations placed on accurate measurement of high frequency vibrations. For example, for a 10% maximum acceptable error with a 10 cm radius mounting with mass ratio of 10, the maximum frequency is about 25 hertz on soft ground and about

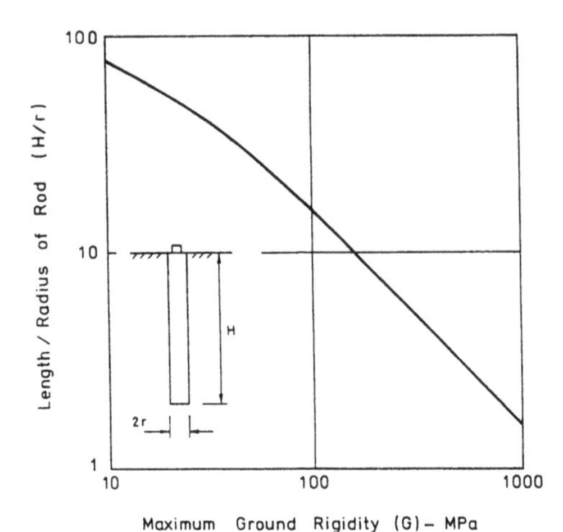

Figure 6. Rigidity Requirement for Embedded Rod

113 hertz on hard ground. The analyses also show that the maximum frequency can be increased if the mounting mass ratio is decreased provided the mounting radius is not increased, a condition that may be difficult to satisfy in practice.

For a particular transducer, mounting size and ground rigidity, the measuring range for a surface mounting can be represented by diagrams such as Fig. 7. The acceptable measuring range is represented by the area enclosed within a number of lines which are located on the figure in accordance with the following criteria:

(a) a boundary of the transducer measuring range and provided by the manufacturer,

(b) a specified acceptable maximum error in the amplitude measurement, and

(c) a boundary beyond which the surface mounting is likely to slide horizontally.

In the dynamic analyses previously described it was assumed that the surface mounting behaved as a rigid block. This assumption was examined and it was found that the following limitation needs to be applied

$$(h/r) \nless 0.05\, G^{1/3} \qquad (2)$$

where G is in units of MPa.

4. COUPLED VIBRATIONS

In the preceding analysis and discussion it was assumed that the mounting would undergo purely horizontal vibrations. In reality the mounting would experience coupled rocking and horizontal sliding vibrations. An analysis for this,coupled motion has been examined to identify the differences in results from the simpler single degree of freedom analysis above.

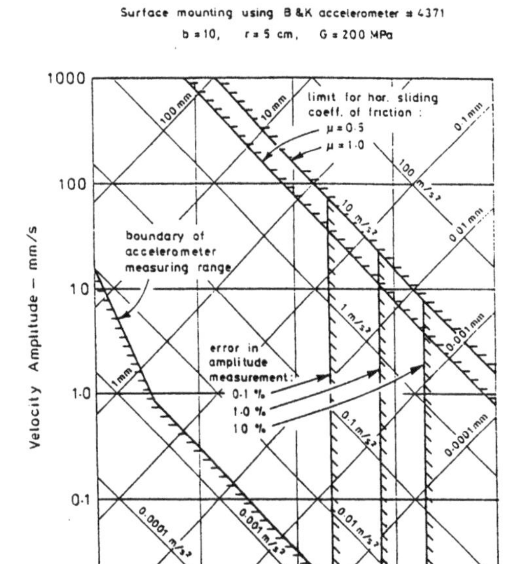

Figure 7. Measuring Range for Surface Mounting with B & K Accelerometer

For surface mountings, it was found that the relationships between maximum frequency and maximum acceptable error were almost identical with those for the single degree of freedom provided that mass ratio did not exceed 10. It is most unlikely that surface mountings with mass ratios greater than about 10 would be used in practice. Consequently, the uncoupled one degree of freedom analysis would be generally acceptable for the evaluation of the maximum permissible frequency corresponding to specified maximum errors in the amplitude measurement.

For embedded mountings, the analyses showed that:

(a) the maximum frequency corresponding to a particular maximum error in the amplitude measurement, increases as the ground rigidity increases,

(b) this maximum frequency increases or decreases with increasing rod length, depending upon the rod radius,

(c) for rod lengths greater than about 10cm, the maximum frequency increases as the rod radius decreases.

It was found that the effects of mounting geometry, ground rigidity and maximum allowable error in the amplitude measurement, on the maximum permissible frequency can be represented in a single diagram as given in Fig. 8. The parameter plotted in this figure is $(f_{max}/(Ge)^{1/2})$ where:

f_{max} = maximum permissible frequency of ground motion in hertz

G = shear modulus of the ground in MPa

e = maximum acceptable error in amplitude measurement in %.

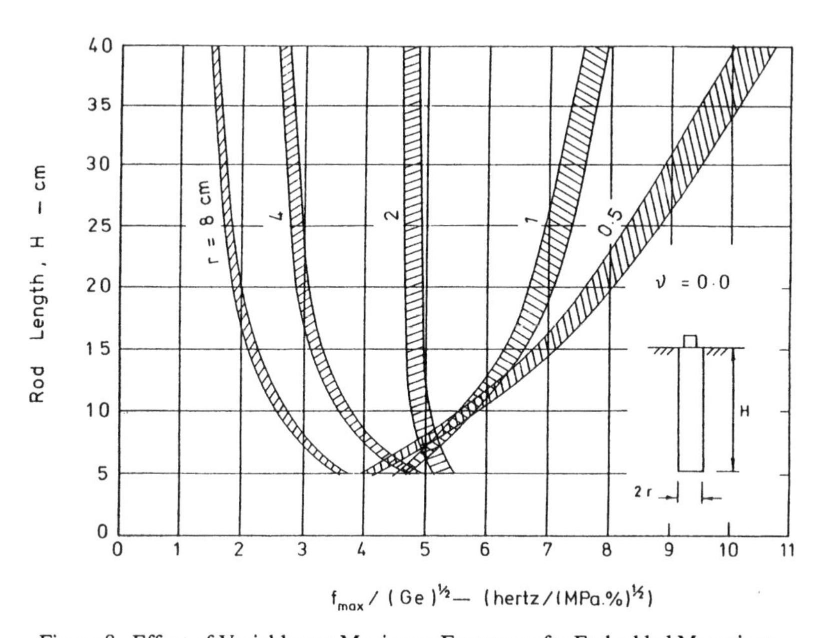

Figure 8. Effect of Variables on Maximum Frequency for Embedded Mountings

In the analysis described above it was assumed that the embedded rod behaved as a rigid body. An examination of this assumption showed that the rigidity requirement may be satisfied by the following approximate expression

$$(H/r) \ngtr 16/G^{1/4} \quad (3)$$

where the shear modulus (G) is in units of MPa.

5. FIELD TESTING EQUIPMENT

In order to determine experimentally the magnitudes of errors in free field vibration amplitude measurements at different frequencies of ground motion, and in order to check on some of the conclusions derived from the theoretical considerations above, a program of field testing was commenced. This involved using vertically mounted transducers and various plate and rod mountings covering a range of sizes. The plate mountings were made of timber, aluminium or steel and varied in diameter from 100mm to 400mm and in mass from 140g to 126kg. The rods were made of steel and varied from 10mm to 80mm in diameter and from 50mm to 200mm in length. The transducer used was a Bruel & Kjaer piezoelectric accelerometer type 4370. This was used in conjunction with a Bruel & Kjaer Charge Amplifier type 2635 which produces an output voltage proportional to the charge coming from the accelerometer.

One source of vibrations was provided by a Ling dynamic System Model 409 electromagnetic shaker. This shaker has an operating frequency range from 5Hz to 9kHz, and a maximum thrust of 196N. The shaker was attached to a steel footing placed on the ground surface. A transducer was attached to this footing to monitor the vibrations transmitted into the ground. The mass-spring resonance within the shaker occurred at 30Hz so this frequency was avoided in field testing. For most of the field testing the displacement amplitude of the shaker footing was kept at 2μm.

The signal generating system for the shaker was a Hewlett Packard Model 2000 wide range oscillator. For the field testing all of the electrical equipment was run off a portable petrol driven generator.

After passing through the charge amplifier, the signal generated by the transducer was fed into a Hewlett Packard (Model HP 35660A) dynamic signal analyser for provision of the final output.

A second source of vibrations was provided by means of a falling weight which was dropped from a known height on to a footing consisting of one or more steel plates. This generated an impact vibration with a range of frequencies present.

6. SITE CHARACTERISTICS

Testing was carried out at two sites near Melbourne. The first site at Lyndhurst consisted of a deposit of siliceous dune sand of fine to medium grain size. The second site, Maribyrnong River, consisted largely of a saturated silty clay deposit of more than 20m thickness.

The dynamic moduli at the testing sites were evaluated by means of a number of seismic refraction surveys. The geophones were placed collinearly in the ground and observations of first arrival times were made with forward and reverse shots. The results of one such test at Lyndhurst are shown in Fig. 9. The results indicated that the P wave velocities varied slightly from test to test and that there was a small increase in velocity with depth. Overall it was concluded that there existed a surface layer of sand at least 5m in thickness with a shear modulus that was within the range of 27MPa to 33MPa. In calculating the shear modulus the Poisson's Ratio was assumed to be 0.3 and the soil density was measured at 1.9t/m^3. For the Maribyrnong River site the density measured was 1.9t/m^3 and the shear modulus was within the range of 13MPa to 17MPa with an assumed Poisson's Ratio of 0.5.

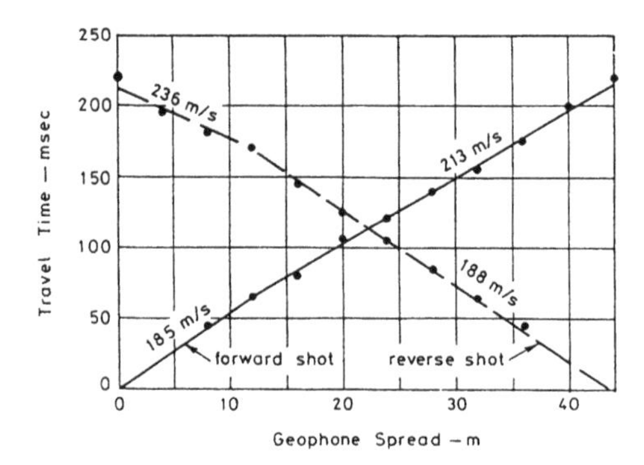

Figure 9. Seismic Refraction - Lyndhurst Site

7. SINUSOIDAL VIBRATION TESTING

Vibration amplitude measurements were carried out with the different mountings over a frequency range of 40 to 260 hertz (Chang (1992)). In most cases the mountings were located 1m from the shaker. Results of measurements taken with aluminium surface mounts at the Lyndhurst site are shown in Fig. 10 and the calculated natural frequencies are given in Table 1.

At low frequencies, the measurements show some scatter with the ground motion being generally within the range of 5 to 10 x 10^{-8}m. At the high frequency end of the observation range there appears to be a noticeable difference between the amplitudes measured with different size mountings. The larger mountings with the lower natural frequencies generate the higher measured amplitudes. This observation is consistent with theoretical findings discussed above that greater over-registration of the true ground motion will occur as the ground vibration frequency more closely approaches the natural frequency of the mounting being used for the measurement.

Table 1. Natural Frequencies for Aluminium Mountings (Lyndhurst Site)

Diameter (mm)	Mass (kg)	Natural Frequency (hertz)	
		G = 27MPa	G = 33MPa
100	0.50	630	690
200	2.56	390	430
400	10.00	280	310

Table 2. Natural Frequencies for Steel Mountings (Lyndhurst Site)

Diameter (mm)	Mass (kg)	Natural Frequency (hertz)	
		G = 27MPa	G = 33MPa
200	5.06	280	310
200	9.75	200	220
200	19.55	140	160
400	47.0	130	140
400	66.5	110	120
400	105.9	90	100

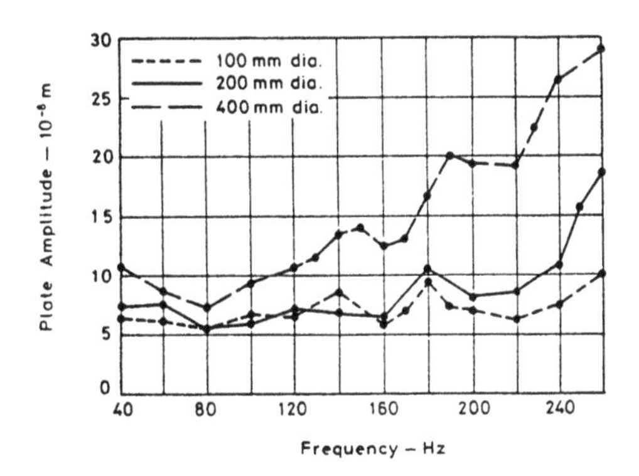

Figure 10. Vibration Measurement with Aluminium Mountings

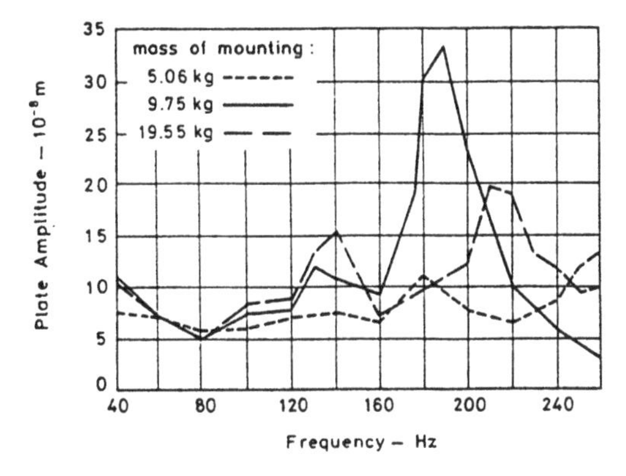

Figure 11. Vibration Measurements with Steel Mountings - 200mm dia.

Figures 11 and 12 show results of amplitude observations at the Lyndhurst site with steel mountings of different diameters and different total mass. With these observations actual peaks occur with the measured amplitudes. These peaks should occur at or slightly below the natural frequency of the mounting. From the tabulated natural frequencies in Table 2, it is seen that this is only approximately correct.

The data in Figures 11 and 12 again support the theoretical conclusion that over-registration of the ground vibration occurs when the ground frequency is in the vicinity of the natural frequency of the mounting. For the rod mountings the calculated natural frequencies were quite high compared with the ground frequency measuring range of 40 to 260 hertz (see Table 3).

There did not appear to be a significant over registration of the ground motion for the rod mountings, although this was not obvious from the measured data because of the scatter of results.

Because of the wide differences between natural frequencies of the surface mountings and rod mountings it was considered that the average amplitude measurement from the rod mountings should provide a reasonable estimate of the true ground motion. Figure 13 shows comparisons between calculated and observed amplitude ratios for one particular mounting. The observed amplitude ratio is the ratio between the observed amplitude of the plate mounting and the observed average amplitude of the rod mountings. The calculated amplitude ratios are based on the two estimates of natural frequency, the curves plotted corresponding to zero damping. If

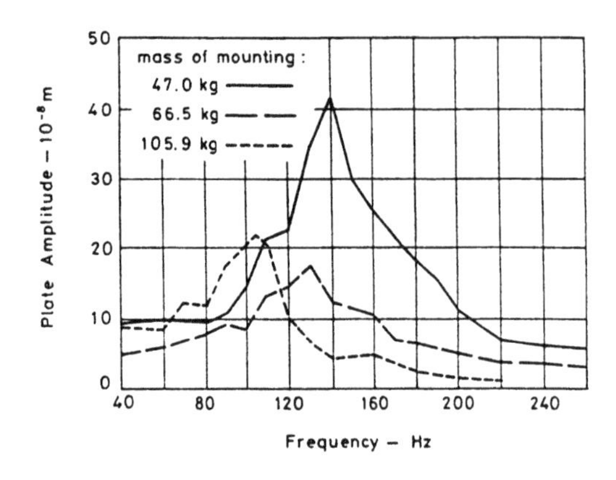

Figure 12. Vibration Measurements with Steel Mountings - 400mm dia.

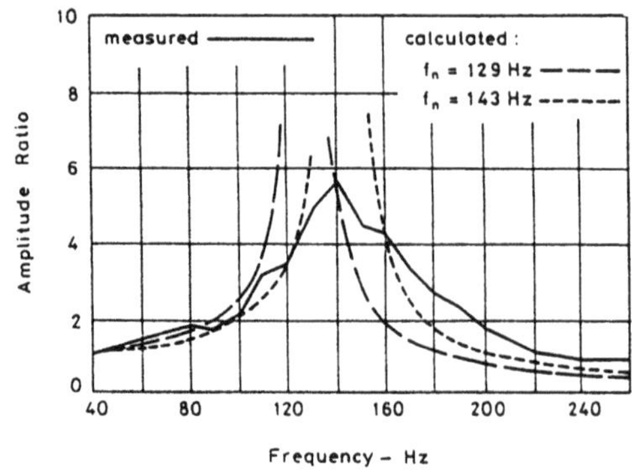

Figure 13. Calculated and Observed Amplitude Ratios for 400mm dia. 47kg Mounting

Table 3. Natural Frequencies for 10mm Dia. Rod Mountings (Lyndhurst Site)

Length (mm)	Mass (g)	Natural Frequency (hertz)
50	25	2100
100	52	2000
200	108	1900

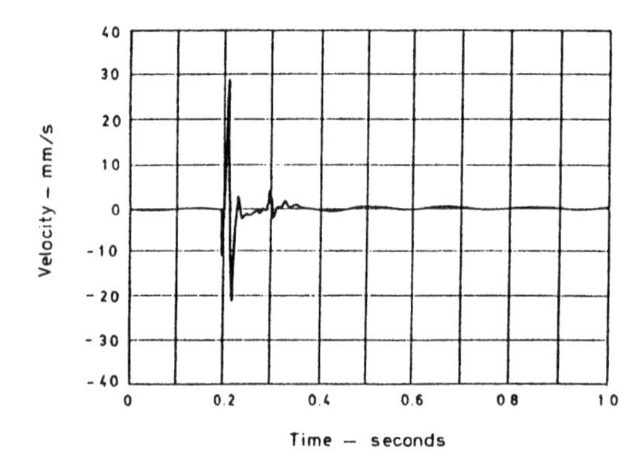

Figure 14. Typical Time Trace Obtained from Impact Tests

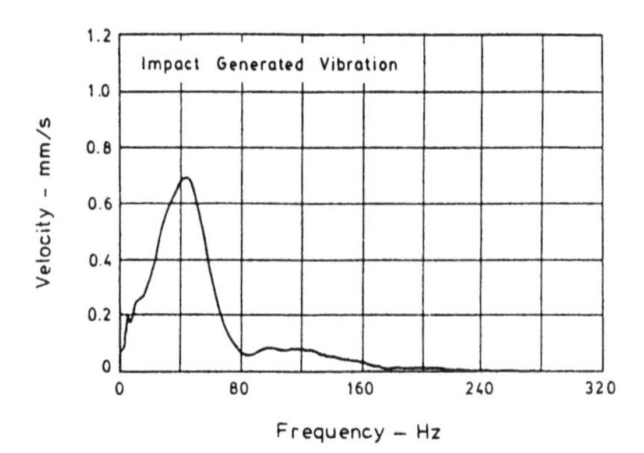

Figure 15. Typical Frequency Trace from FFT Analysis

allowance is made for damping which is clearly present in the observations then approximate agreement between calculated and observed amplitude ratios is evident. In some other cases with surface mountings it was found that the observed peak amplitude ratio occurred at frequencies quite different from the calculated natural frequency for reasons that are not fully understood.

8. IMPACT TESTING

Much of the impact testing was carried out at the Maribyrnong River site. The bulk of the testing involved an impact generated by means of a 50kg steel ball falling heights of 100mm and 250mm onto a 400mm diameter by 20mm thick steel plate located on the ground surface. A typical time trace of the impact is shown in Figure 14, from which the peak velocity amplitude could be read. To obtain the frequency trace, a Fourier Transform procedure was applied using the Dynamic Signal Analyser. Figure 15 shows

Table 4. Natural Frequencies for Surface Mountings (Maribyrnong River Site)

Diameter (mm)	Thickness (mm)	Material	Natural Frequency (hertz)
100	20	timber	1100
200	30	timber	800
400	30	timber	590
100	20	aluminium	560
100	10	steel	475
200	30	aluminium	345
100	20	steel	345
400	30	aluminium	245
100	40	steel	245
400	10	steel	245
400	20	steel	175
200	40	steel	175
400	60	steel	96
400	100	steel	76
400	120	steel	70

Table 5. Natural Frequencies for Steel Rod Mountings (Maribyrnong River Site)

Diameter (mm)	Length (mm)	Natural Frequency (hertz)
10	100	1200
20	100	845
20	200	760
40	100	500
80	100	320

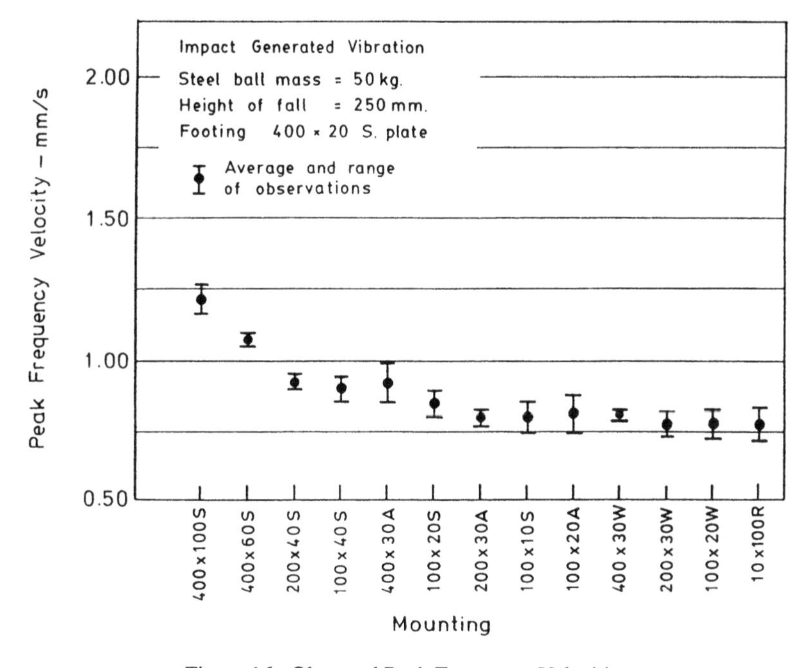

Figure 16. Observed Peak Frequency Velocities

a typical result. The peak velocity corresponding to the dominant frequency in Figure 15 was defined as the "peak frequency velocity" to avoid confusion with the peak velocity on the time-trace. Both of these peak velocities were used for comparisons of test results. Results obtained with many of the mountings are summarised in Figure 16.

The mountings are arranged in order of increasing natural frequency. The natural frequencies for the mountings used are listed in Tables 4 and 5. The calculations were based on an average shear modulus of 15MPa. The method used to identify the mountings in Figure 16 is diameter x thickness (or length) x material type (steel, aluminium, wood or rod). Figure 16 confirms that over registration of peak frequency velocity occurs as the dominant frequency more closely approaches the natural frequency of the mounting used. Results in terms of peak velocities showed no significant trend and were widely scattered. The dominant frequencies for the data in Figure 16 varied over a small range of 30 to 50 hertz.

9. CONCLUSIONS

Surface or embedded mountings may be used for the measurement of ground vibrations but the most important parameter governing the magnitude of errors in measurements is the natural frequency of the mounting. The rod mountings with their higher natural frequencies compared with surface mountings would be expected to provide more accurate measurements of vibration amplitude. At both the Lyndhurst and Maribyrnong River sites with sinusoidal and impact generated vibrations, the observed results confirmed the trend that vibration amplitudes were amplified as the natural frequency of the mounting more closely approached the dominant frequency of ground vibrations. For the majority of the vibration amplitudes observed, there was approximate agreement with theoretical predictions.

REFERENCES

Bradley, W. & Eller, E.E.1961. *Introduction to Shock and Vibration Measurements,* Ch. 12 of Shock and Vibration Handbook, C.M. Harris and C.E. Crede, editors, Vol. 1, pp. 12.1 - 12.24, McGraw Hill Book Co., New York.

Brown, L.M. 1971. *Measurements of Vibrations caused by Construction Equipment and Blasting*, Dept. of Highways, Ontario, Report No. RR 172.

Brown, P.T., 1969. *Numerical Analyses of Uniformly loaded Circular Rafts on Elastic Layers of Finite Depth*, Geotechnique, Vol. 19, No. 2, pp 301-306.

Chang, A. 1992. *The Influence of Mounting Systems on the Measurement of Ground Vibrations*, M.Eng.Sc. thesis, University of Melbourne.

Crouse, C.B., Liang, G.C. & Martin, G.R. 1984, *Amplification of Earthquake Motions Recorded at an Accelerograph Station*, Proc. 8th World Conf. on Earthquake Engineering, Vol. 2, pp 55-62.

Dowding, C.H. 1985. *Blast vibration Monitoring and Control*, Prentice-Hall Inc. New Jersey.

Grant, J.R.T. 1983. *Investigation into Transducer - Ground Coupling Techniques for Surface Blast Vibration Measurement*, M.Eng.Sc. thesis, Faculty of Engineering, James Cook University of North Queensland.

Hanna, T.H. 1985. *Field Instrumentation in Geotechnical Engineering*, Trans Tech Publications, Federal Republic of Germany.

Mattes, N.S. & Poulos, H.G.1969. *Settlement of Single Compressible Pile*, Jnl. Soil Mech. and Found. Div. ASCE, Vol. 95, No. SM1 pp 189-207.

Moore, P.J., 1986. *Measurement of Vertical Ground Vibrations*, Proc. of In-Situ '86, Use of In Situ Tests in Geotechnical Engineering, Blacksburg, Virginia, ASCE Geotechnical Special Publication No. 6, p 840-853.

Moore, P.J. 1988. *Performance of Mountings for Horizontal Ground Vibration Measurements*, Fifth Aust.-N.Z. Conf. on Geomechanics, Sydney, pp 427-431.

Richart, F.E., Hall, J.R. & Woods, R.D., 1970, *Vibration of Soils and Foundations*, Prentice-Hall Inc. New Jersey.

Serridge, M. & Licht, T.R. 1986. *Piezoelectric Accelerometer and Vibration Preamplifier Handbook.* Brüel and Kjaer, Denmark.

Developments in Geotechnical Engineering, Balasubramaniam et al. (eds) © 1994 Balkema, Rotterdam, ISBN 90 5410 522 4

Evaluation of in-situ liquefaction resistance of sands based on high-quality undisturbed samples

Yoshiaki Yoshimi
Shimizu Corporation, Tokyo, Japan

ABSTRACT: The quality of samples of saturated sands obtained by an in-situ freezing method is examined on the basis of laboratory and field tests, and case history of soil liquefaction during earthquakes. It is concluded that the method can yield high-quality undisturbed samples that retain the in situ liquefaction resistance as well as density and elastic shear modulus. The liquefaction resistance of the in-situ frozen samples is well correlated with corrected SPT *N*-values over a wide range in *N*-values. The significance of the correlation for earthquake, wind and wave loadings is discussed. For dense sands, the in-situ frozen samples are denser and much stronger than the samples obtained with conventional tube samplers. A method is suggested by which to evaluate the quality of an in-situ frozen sample of silty sands by thawing and refreezing a specimen prepared from the sample.

1 INTRODUCTION

In seismically active regions, soil liquefaction has become a standard entry in our checklist for seismic design of foundations on sandy soil deposits. For dense sands that are expected to support important structures such as high-rise buildings, high dams and nuclear power plants, we must carefully examine the stability and deformation of the bearing strata during unusually strong earthquakes or winds. The same can be said about wave loading on offshore structures during severe storms. The dense sands include both natural deposits and artificially densified sands.

For an actual design situation, the engineer needs site-specific information on in-situ soil properties rather than general information on the mechanism of liquefaction based on reconstituted samples. In this respect, liquefaction criteria based on the field performance data of liquefied and non-liquefied sites during past earthquakes give us guidelines by which to estimate liquefaction resistance of sands from Standard Penetration Test (SPT) *N*-value (e.g., Tokimatsu and Yoshimi, 1983; Seed et al., 1985).

However, such means has limitations in two respects: (1) Field data for unusually strong earthquakes are sparse and the failure of dense sands is not as conspicuous as the liquefaction of loose sands, and (2) the method does not provide stress-strain relationship of sands for detailed analyses. The only alternative to overcome these limitations would be to run laboratory tests on truly undisturbed samples. It is well known, however, that the acquisition of high-quality undisturbed samples of cohesionless soils from below groundwater table has been one of the most challenging tasks faced by geotechnical engineer.

The object of this paper is to show that we can obtain high-quality undisturbed samples of clean sands over a wide density range by in-situ freezing, and to show that the liquefaction resistance of these samples determined in the laboratory can be correlated with SPT *N*-values with appropriate correction.

2 TERMINOLOGY

Because the term "liquefaction" was once a topic of intense debate in the past (Castro, 1975; Casagrande, 1976; Seed, 1979), it is perhaps worthwhile to devote a few paragraphs to define the terminology related to soil liquefaction problems.

2.1 *Flow slides and limited cyclic strains*

Regardless of density, sand skeleton is contractive when it is subjected to small shear strain. Thus, the application of cyclic shear stresses under undrained conditions causes a buildup of pore pressures in saturated sand in both loose and dense states. The pore pressure buildup occurs while shear strains are small, and thus, the small strain behavior of sand is important in evaluating the strength and deformation characteristics of saturated sands under cyclic loading conditions.

Depending on the combination of density and confining stress, and on whether or not "driving" shear stresses are present, the behavior of saturated sand involving pore pressure buildup may be classified into flow slides and limited cyclic shear strains as shown in Table 1. According to Castro (1987),

Table 1. Factors influencing field behavior of saturated sand under cyclic loading conditions

Density	Confining stress	Dilatancy at large shear strain	Field behavior of saturated sand	
			Level ground (without driving shear stresses)	With driving shear stresses
Very low	Any	Contractive	Ground settlement, cracks	Flow slides
Low	High			
	Low	Dilatant	Limited cyclic strains	Limited lateral spreading, settlement of heavy structures, floating of buried structures
Medium to high	Any			

driving shear stresses are those required to maintain the soil mass in equilibrium, and exist, e.g., within an earth embankment and below a structure. For flow slides to occur the sand mass must be contractive even at large shear strain. Thus, under a moderate confining stress, this is possible only if the sand is so extremely loose that it forms a "meta-stable" structure that is likely to undergo "spontaneous" liquefaction by slight disturbance including an increased static load, as described by Terzaghi and Peck (1967). Such extremely low densities can be obtained in the laboratory only by gently packing bulked sand. In the field, extremely loose deposits of clean sand with *N*-values of 1 or less are reported to exist in backfills of sand from which iron sand has been extracted as shown in Fig.1. The sand deposits at the Hachinohe, Aomori, site and the Oshamambe, Hokkaido, site underwent severe liquefaction during the Tokachi-oki earthquake of 1968 (Ohsaki, 1970) and the Hokkaido Nansei-Oki earthquake of 1993 (Tokimatsu et al., 1994), respectively.

When a sand is loose but not extremely so, on the other hand, the confining stress must be very high, e.g., on the order of 1 MPa, for flow slides to occur (Ishihara, 1993). Although such high level of stress may exist below unusually high dams or buildings, it is not likely that natural deposits of loose sand exist under such high overburden stresses that correspond to depths of the order of 100 m.

This paper concerns only the behavior of level ground of saturated sand involving limited cyclic shear strains as shown in the shaded box in Table 1, although this is called "liquefaction" in its broad sense of the word.

2.2 *Liquefaction resistance*

In this paper, the term "liquefaction resistance" is analogous to undrained cyclic strength for saturated soil that is defined as the shear stress amplitude required to cause a certain shear strain amplitude (or zero effective stress) in a given number of stress cycles when a saturated specimen is subjected to completely reversed shear stresses under undrained conditions. A typical test result is shown in Fig. 2. Different curves can be drawn for different criteria that define liquefaction. When a higher strain amplitude is selected to define liquefaction, the curve shifts towards the right as shown in broken line. In a statistical sense, the number of cycles tends to increase with an increase in earthquake magnitude (Seed et al., 1983). Also shown in the figure is a probable range of the number of cycles for wind and wave loadings which may be considerably more numerous than earthquake shaking. As a result, the

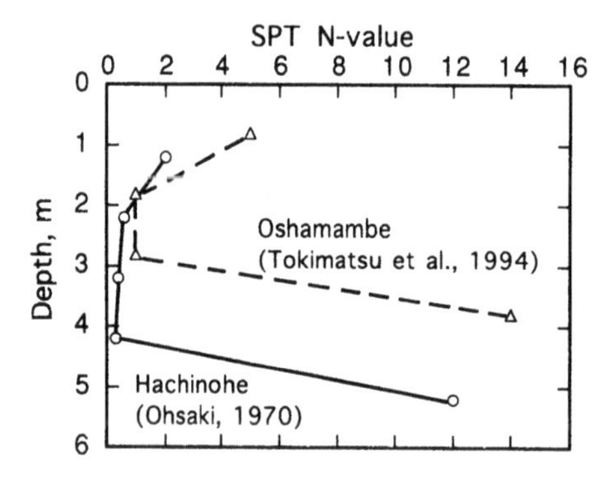

Fig. 1 Examples of extremely loose deposits of clean sands from which iron sand has been extracted

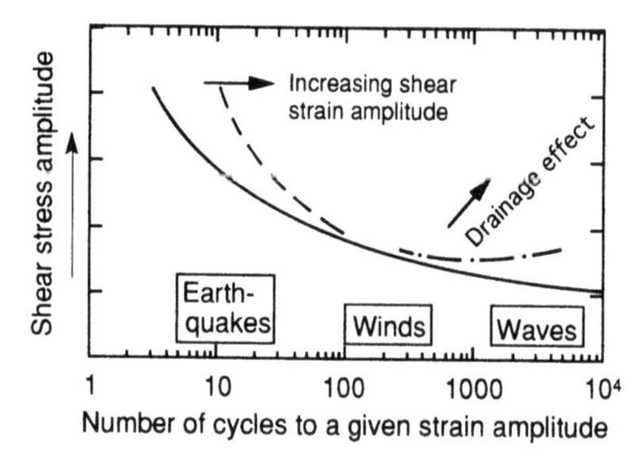

Fig. 2 Typical results of undrained cyclic tests on saturated sand

liquefaction resistance for the wind or wave problem may be lower than that for the earthquake problem unless compensated by drainage effects as shown in chain-dotted line in Fig. 2.

In the triaxial test, the shear stress amplitude is equal to $\sigma_d/2$ in which σ_d is deviator stress, and the shear strain amplitude is usually expressed as double amplitude axial strain, ε_{da}. In most cases, liquefaction resistance is expressed as a ratio, $\sigma_d/2\sigma_o'$, in which σ_o' is initial confining stress.

3 ESTIMATION OF LIQUEFACTION RESISTANCE OF LEVEL GROUND BASED ON FIELD PERFORMANCE DATA AND *N*-VALUES

By comparing SPT *N*-values that were obtained in Niigata, Japan, before and after the Niigata earthquake of 1964, Koizumi (1966) proposed "critical *N*-value" as a function of depth below ground surface that would separate liquefiable and non-liquefiable conditions. This criterion was in fairly good agreement with another criterion by Kishida (1966) based on the degree of damage to building foundations.

In an attempt to seek a more general criterion that would be applicable to any soil and to any earthquake, Seed and Idriss (1971) proposed a procedure in which the amplitude of cyclic shear stress was compared with the liquefaction resistance expressed as a function of relative density which, in turn, was estimated from SPT *N*-value and effective overburden stress. The procedure has since been modified to replace relative density with "corrected *N*-value" (Castro, 1975; Seed, 1979; Seed et al., 1983; Tokimatsu and Yoshimi, 1983; Seed et al., 1985). The *N*-value is preferred to relative density because the liquefaction resistance and *N*-value are influenced by common factors in nearly the same way, while relative density is only one of these factors (Schmertmann, 1978; Tokimatsu, 1988). Another limitation of relative density is that high-quality undisturbed samples are needed to determine its in-situ values (Yoshimi et al., 1994).

The simplified procedure in its current form is illustrated in Fig. 3. After plotting shear stress ratios against corrected *N*-values for both liquefied and non-liquefied sites, a line can be drawn to separate a group of (solid) points for liquefied conditions and another group of (open) points for non-liquefied conditions. The ordinate of such line can then be interpreted as liquefaction resistance as a function of corrected *N*-values. An early example by Castro (1975) is shown in Fig. 4 in which the ordinate is the "average" shear stress ratio that is 0.7 times the maximum shear stress ratio caused by peak horizontal acceleration. The original *N*-values that were normalized to an overburden stress of 245 kPa have been converted to *N*-values normalized to 98 kPa (1 kgf/cm^2) to make it comparable to more recent data.

According to Tokimatsu and Yoshimi (1983), the equivalent shear stress ratio for 15 uniform stress cycles may be given by

$$\left(\frac{\tau_d}{\sigma_v'}\right) = 0.1(M-1)\frac{\alpha_{max}}{g}\frac{\sigma_v}{\sigma_v'}(1-0.015z) \tag{3.1}$$

in which τ_d = amplitude of uniform shear stress cycles equivalent to actual seismic shear stress time history, M= earthquake magnitude, α_{max}= maximum horizontal acceleration at ground surface, g= gravitational acceleration, z = depth from ground surface in m, σ_v= total overburden stress at depth z, and σ_v' = initial effective overburden stress at depth z.

Concerted efforts by many investigators to enhance the reliability of SPT *N*-value have been summarized by Seed et al. (1985) and Skempton (1986). In this paper, the *N*-value for clean sand is corrected for rod energy, rod length, and confining stress. No correction seems necessary for the borehole diameter in the current Japanese practice (Oh-oka and Tatsui, 1987; Yoshimi and Tokimatsu, 1983).

Surface acceleration
Earthquake magnitude
Number of load cycles
Equivalent shear stress ratio τ/σ'
Unit weight of soil
Overburden pressure σ'
Groundwater table
Liquefied
τ/σ'
Not liquefied
Fines content
SPT N-value
Normalized N-value N1
Corrected N-value (N1)c
(N1)c
SPT hammer correction
SPT rod length correction

Fig. 3 Simplified procedure to correlate liquefaction resistance with SPT *N*-value based on field performance data

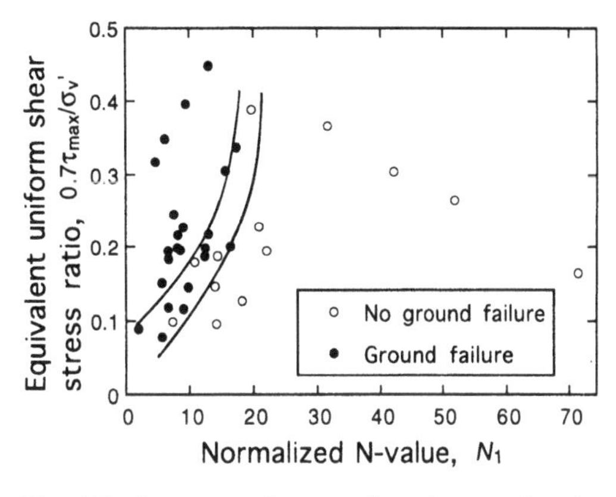

Fig. 4 Performance of saturated sands at earthquake sites (adapted from Castro, 1975)

3.1 *Correction of N*-value for rod energy

The percentage of the energy of the SPT hammer that is transmitted to the rod depends primarily on the type of hammer and anvil, and the method to release the hammer. For the donut hammer that is exclusively used in Japan, the rod energy ratio for a free-fall method using a trigger mechanism locally called "tombi" is estimated 78% whereas the ratio for a rope and pulley method with two turns of rope is estimated 65% (Skempton, 1986). The *N*-value is considered inversely proportional to the rod energy ratio (Schmertmann and Paracios, 1979). Thus, the *N*-values with rod energy ratios of 78% and 65% may be converted to the *N*-value with a rod energy ratio of 60%, $(N)_{60}$, as follows:

$$(N)_{60} = (78/60)(N)_{78} = 1.3(N)_{78}$$
$$(N)_{60} = (65/60)(N)_{65} = 1.1(N)_{78} \quad (3.2)$$

The use of $(N)_{60}$ in this paper is intended to facilitate comparison with recent papers including those by Seed et al. (1985) and Skempton (1986).

To minimize variability in *N*-values, it is desirable to automate the free-fall mechanism for releasing the hammer. The Japan Geotechnical Consultants Association (1993) has recently developed two types of such equipment with a mechanism to grab and release the hammer automatically. One type is fully automatic in that the hammer is raised with a hydraulic jack. The other is called semiautomatic because the hammer is raised with a rope and pulley. The Association has also developed an automatic recorder-plotter that can be used with either type of equipment.

3.2 *Correction of N*-value for rod length

Schmertmann and Palacios (1979) reported that the rod energy ratio for a safety hammer released by a rope and pulley method decreased with a decrease in rod length when the rod lengths were less than about 10 m, because reflected stress waves in the rod returns to the anvil while the hammer is still descending. The correction factor, C_r, based on their average curve may be formulated as follows:

$$C_r = 1, \qquad L \geq 10\text{ m}$$
$$C_r = 0.562\sqrt[4]{L}, \qquad L < 10\text{ m} \quad (3.3)$$

in which L= rod length in m.

It will be assumed tentatively that Eq. (3.3) is applicable to the Japanese donut hammer released by either the tombi method or the rope and pulley method. According to Robertson et al. (1992), however, the correction factor for rod lengths less than 5 m may be considerably less than the above.

3.3 *Correction of N*-value for confining stress

The *N*-value measured at a given overburden stress, σ_v', is normalized with respect to σ_v' =1 kgf/cm2 (=98 kPa) as follows:

$$N_1 = \frac{1.7N}{\sigma_v'(\text{kgf/cm}^2) + 0.7} = \frac{167N}{\sigma_v'(\text{kPa}) + 69} \quad (3.4)$$

in which N=measured *N*-value, and N_1=normalized *N*-value.

The data points of Fig. 5 show the relationship between the equivalent uniform shear stress ratios of earthquakes given by Eq. (3.1) and the *N*-values corrected by Eqs. (3.2) through (3.4) for the well-documented body of data given by Tokimatsu and Yoshimi (1983). The solid points show liquefied conditions and open points non-liquefied or marginal conditions. Note that the liquefied and non-liquefied conditions are clearly separated by a curve for *N*-values from 7 to 25. For higher *N*-values, the reliability of the field performance data suffers from scarcity of data as mentioned in Introduction.

For corrected *N*-values greater than about 25, therefore, the only alternative seems to be laboratory tests on high-quality undisturbed samples. This subject will be discussed below for block samples, conventional tube samples, and for in situ frozen samples.

4 EVALUATION OF LIQUEFACTION RESISTANCE BASED ON CONVENTIONAL "UNDISTURBED" SAMPLES

4.1 *Block samples*

The block sampling method which can be effective

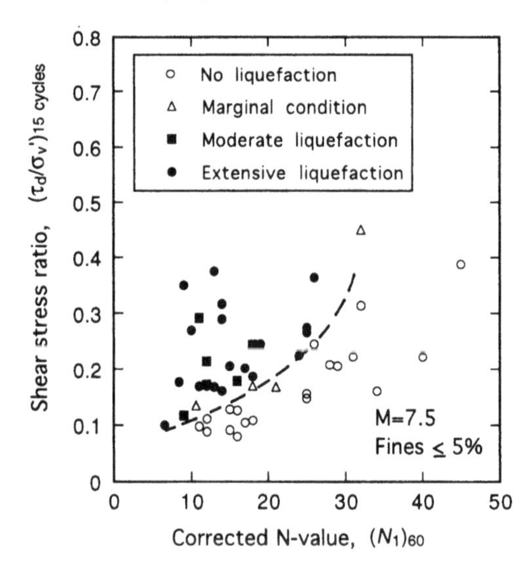

Fig. 5 Field correlation between shear stress ratio and corrected *N*-value (adapted from Tokimatsu and Yoshimi, 1983)

for soils with cohesion or cementation is not readily applicable to saturated, uncemented sand at considerable depths because the method requires dewatering and excavation. Not only costly and time-consuming, dewatering and excavation tend to change the state of stress in the sand which, in turn, may affect the density and fabric of the soil. On the basis of very careful block sampling in the laboratory, Seed et al. (1982) showed that the liquefaction resistance of block samples taken from a large reconstituted sample of clean sand that had been subjected to cyclic strain history lost some of the additional strength gained by shear strain history.

4.2 *Tube samples*

Tube samplers of various types have been used to obtain "undisturbed" samples of saturated sands (e.g., Osterberg and Varaksin, 1973; Tatsuoka et al., 1978; Ishihara et al., 1979). On the basis of sampling from reconstituted sands and subsequent laboratory tests on the samples, Marcuson et al. (1977) and Seed et al. (1982) showed that sampling with a push-type sampler tends to densify loose sand and loosen dense sand.

The penetration of a sampling tube is likely to affect soil fabric as well as soil density. Thus, liquefaction resistance that depends on both soil fabric and soil density is bound to be affected by tube sampling. The loosening and possible deterioration of soil fabric of tube samples for higher N-values is reflected in Fig. 6 in which the liquefaction resistance of tube samples increases only slightly with a significant increase in N_1. Such negative curvature (concave downward) is contrary to the field performance data shown in Figs. 4 and 5 in which the curves show distinctly positive curvature. Concerning Fig. 6, a fixed piston sampler was used for looser sands and a Pitcher sampler for denser sands.

The circular and triangular points in Fig. 7 show other results of cyclic triaxial tests on tube samples of clean sands obtained with a hydraulic piston sampler (Ishihara et al., 1979) and a double tube core barrel (Yoshimi et al, 1989; Tokimatsu et al., 1990; Iai and Kurata, 1991). The diamond shaped points in the figure show the data of Fig. 6 whose ordinates have been multiplied by a factor of 0.94 so that the number of cycles to define liquefaction is changed from 10 to 15. The factor of 0.94 is based on Castro's typical test results. Although the rod energy ratios for Castro's N-values are unknown, 60% is assumed. The assumption is not expected to cause excessive errors because the curve of Fig. 6 is relatively flat. Although different types of samples are involved, Fig. 7 shows a strange negative correlation for $(N_1)_{60} < 20$ in that liquefaction resistance decreases as $(N_1)_{60}$ increases.

Another indication of disturbance of tube samples of sandy soils can be seen in Fig. 8 which compares elastic shear moduli of tube samples measured in the laboratory with those computed from shear wave velocities measured in the field (Yasuda and Yamaguchi, 1984). The data points for higher shear moduli

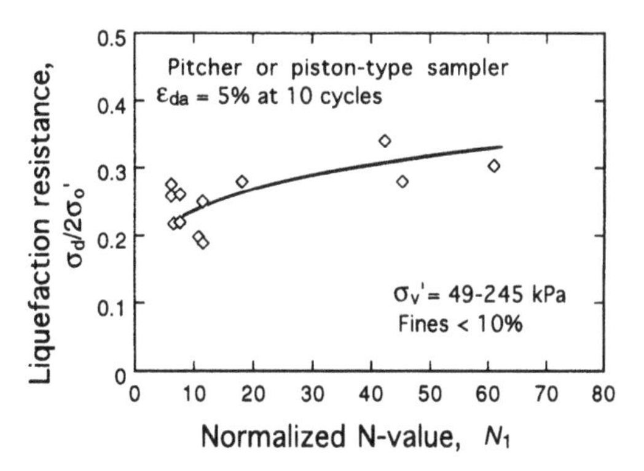

Fig. 6 Relationship of liquefaction resistance and normalized N-value for tube samples (adapted from Castro, 1975)

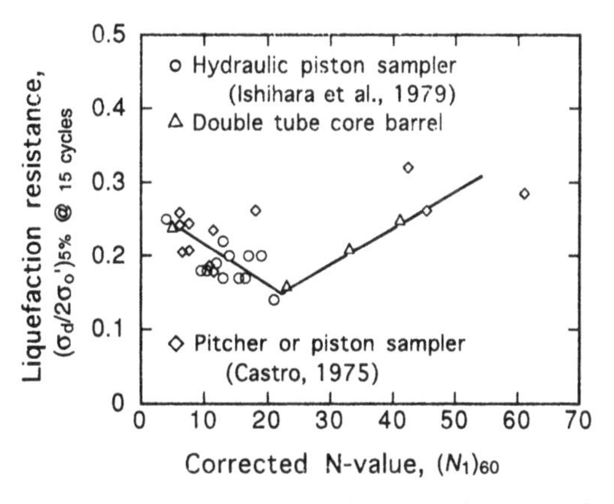

Fig. 7 Relationship of liquefaction resistance and corrected N-value for tube samples (Yoshimi, 1994)

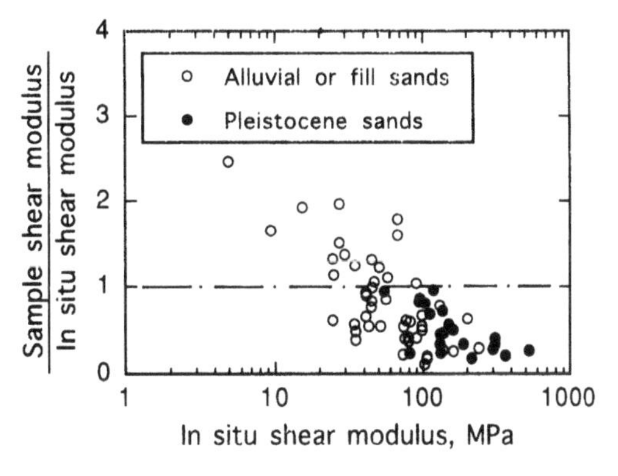

Fig. 8 Comparison between shear moduli of tube samples and in situ shear moduli (Yasuda and Yamaguchi, 1984)

including all Pleistocene sand samples lie below the chain-dotted line where the sample moduli are smaller than the in-situ moduli, presumably because the soil fabric and density were affected by sample disturbance. The reverse is true for very low shear moduli presumably because the effect of densification outweighed possible deterioration of soil fabric by sample disturbance.

Figs. 6 through 8 show clearly that "undisturbed" samples of clean sands obtained with conventional tube samplers suffer from significant disturbance that may cause their liquefaction resistance and elastic shear modulus to deviate considerably from the in-situ values. For a dense sand with an $(N_1)_{60}$ of 40, Tokimatsu and Hosaka (1986) deduced on the basis of laboratory tests that the decrease in its liquefaction resistance and shear modulus was caused by static shear strain history of the order of 1%.

The failure of the tube samples to retain in-situ liquefaction resistance and shear modulus can be attributed to the fact that these properties reflect small strain characteristics and therefore are governed by soil fabric that is more sensitive to sample disturbance than large strain property such as static shear strength. Thus, the need for high-quality undisturbed samples cannot be overemphasized for the evaluation of in-situ liquefaction resistance of sand deposits, particularly for dense ones.

5 QUALITY OF UNDISTURBED SAMPLES OBTAINED BY IN-SITU FREEZING

Since 1977, the author and his colleagues have conducted laboratory and field tests on the sampling of sands by in-situ freezing with a single freezing pipe (Yoshimi et al., 1977, 1978, 1984, 1989). In addition, Singh et al. (1982) and Goto (1993) conducted extensive laboratory tests to study the effect of a freeze and thaw cycle on the liquefaction resistance of sands. The results of these studies will be reviewed below to show that we can obtain high-quality undisturbed samples of saturated sands that retain not only their in-situ density but also their in-situ liquefaction resistance, provided that: (1) Freezing progresses without impeding drainage at the freezing front, e.g., radially outward from a single freezing pipe; (2) the sand does not contain too much fines; (3) the effective confining stress is adequate; and (4) only the part of the frozen sample far enough away from a possible zone of disturbance due to boring and insertion of the freezing pipe is used for tests.

5.1 *One-dimensional freezing tests in the laboratory*

Early laboratory studies showed that the unidirectional freezing and thawing of clean saturated sand under moderately high confining stresses caused no change in the density and static stress-strain-dilatancy characteristics under drained conditions (Yoshimi et al., 1978).

Singh et al. (1982) sought more direct confirmation that the liquefaction resistance of saturated sand was not affected by a history of freezing and thawing. They froze one-dimensionally reconstituted samples of a loose, clean sand that had been "tempered" by applying cyclic strain history, and ran cyclic triaxial tests on the samples after they had been thawed and saturated. By comparing the results with those on the samples that had not been frozen, it was shown that the history of freezing and thawing did not affect their liquefaction resistance, as shown in the lower curve in Fig. 9. Goto (1993) conducted similar tests on a dense, clean sand, and reached the same conclusion that a freeze-thaw cycle had no effect on its liquefaction resistance as shown in the upper curve in Fig. 9.

Singh et al. (1982) further showed that a sample 71 mm in diameter cored from a frozen sample 305 mm in diameter retained the liquefaction resistance of the larger sample before it had been frozen.

5.2 *Large-scale radial freezing tests in the laboratory*

A steel bin 1.0 m in diameter and 2.9 m deep was used to study the effect of radial freezing on the density of saturated sand (Yoshimi et al., 1977). One sample of sand was prepared for each of three relative densities of 27, 69 and 91%. A freezing pipe 73 mm in diameter was placed in the center of the sand in the bin, and the sand around the pipe was frozen to a diameter of about 400 mm by circulating coolant in the pipe in the same way as in the field. Small specimens were cut out of the frozen sand and their densities were determined by submerging them in chilled mercury to measure their volume. The dry

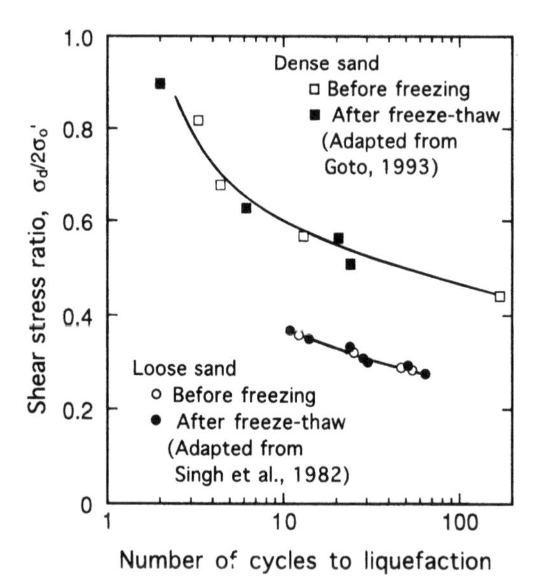

Fig. 9 Effect of a freeze-thaw cycle on the liquefaction resistance of clean sands

densities of the specimens at seven depths were averaged and plotted against the distance from the surface of the freezing pipe as shown in Fig. 10. The arrows in the figure show the overall average density before freezing. It can be seen that near the freezing pipe the loose sand became denser while the dense sand became looser. On the contrary, the outside part of the sand, more than about 70 mm away from the freezing pipe, successfully retained the original density before freezing.

5.3 *Density measurements of a sample frozen in the field*

Measurements similar to those described in the preceding section were made on a frozen sample obtained in the field, and the result is shown in Fig. 11. The density is nearly constant in the outside part of the sample, more than 70 mm away from the freezing pipe. We can only deduce, however, that the outside part retained the in-situ density because no independent measurements of the in-situ density were made.

Fig. 12 shows the relative density of in-situ frozen samples plotted against $(N_1)_{60}$. The coefficient of correlation is 0.80 which may be considered fairly high in light of the variability in both relative density and N-value.

5.4 *Coordinated field and laboratory tests on elastic shear modulus*

Fig. 13 compares the elastic (small strain) shear modulus of in-situ frozen samples and that computed from the shear wave velocity measured in the field (Tokimatsu and Ohara, 1990). Unlike Fig. 8 in which similar comparison is made for tube samples, the in-situ frozen samples retained the in-situ shear moduli fairly well.

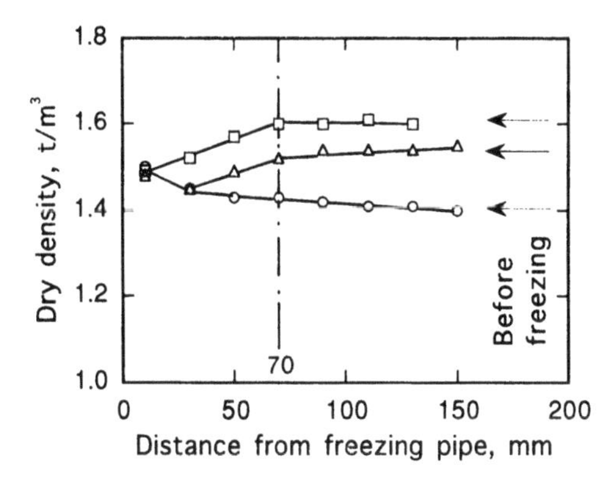

Fig. 10 Density distribution in radially frozen columns of clean sand: laboratory tests (adapted from Yoshimi et al., 1977)

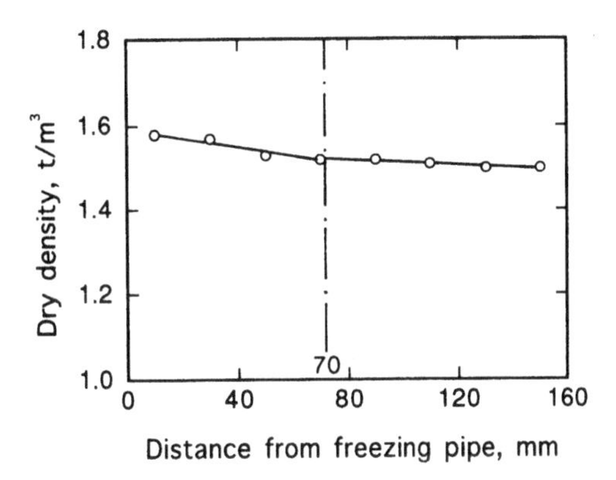

Fig. 11 Density distribution in a radially frozen column of clean sand: field test (adapted from Yoshimi et al., 1977)

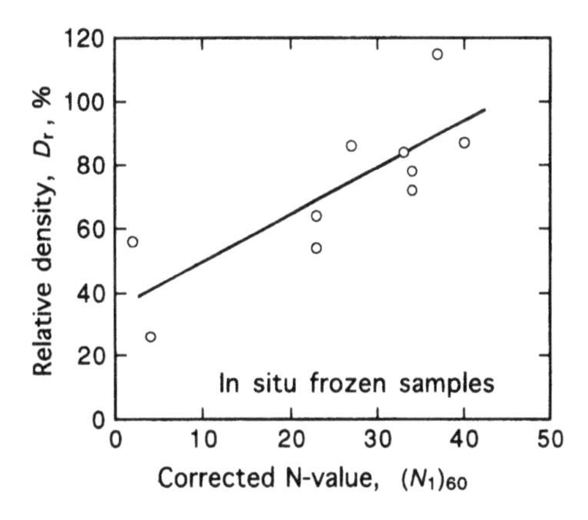

Fig. 12 Relationship of relative density and corrected N-value for in situ frozen samples (Yoshimi, 1993)

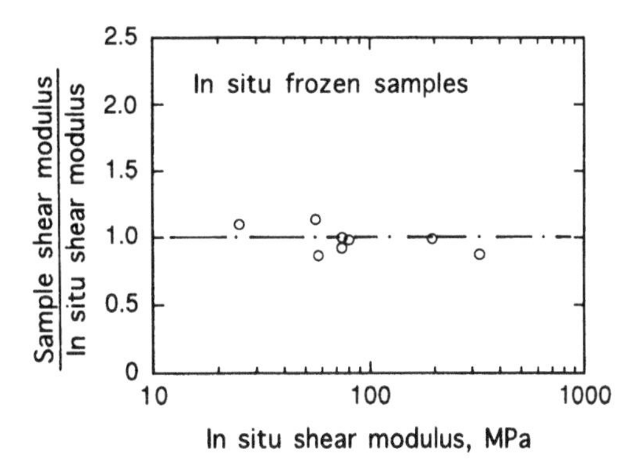

Fig. 13 Comparison between shear moduli of in situ frozen samples and in situ shear moduli (Tokimatsu and Ohara, 1990)

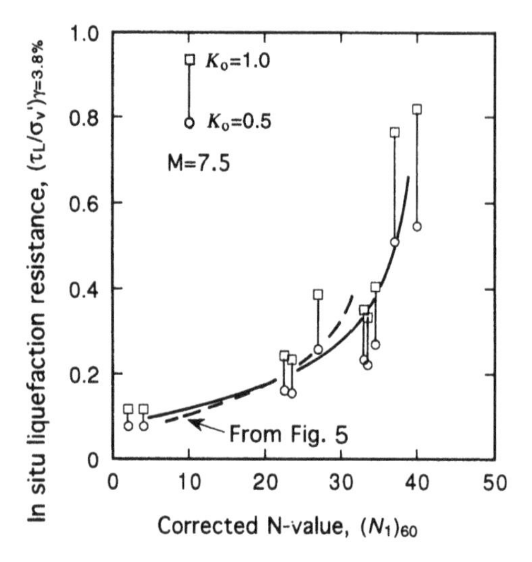

Fig. 14 In-situ liquefaction resistance of clean sands for magnitude 7.5 earthquakes estimated from laboratory tests on in situ frozen samples: comparison with field performance data of Fig. 5 (Yoshimi, 1993)

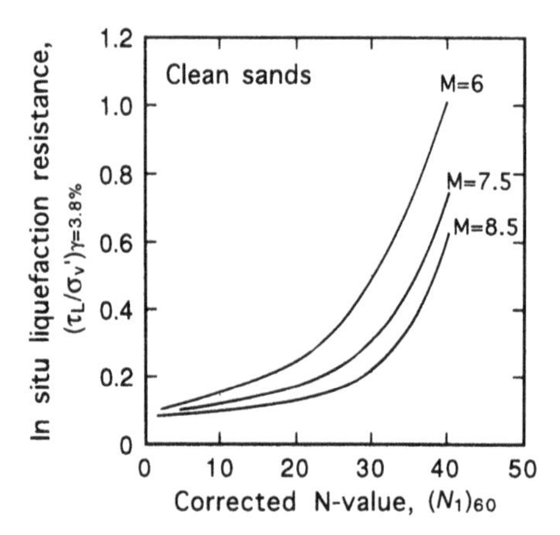

Fig. 15 In-situ liquefaction resistance of clean sands during earthquakes estimated from laboratory tests on in situ frozen samples (adapted from Yoshimi et al., 1994)

5.5 *Comparison with field performance data of liquefaction*

The data points in Fig. 14 show the in-situ liquefaction resistance for 15 cycles estimated from the cyclic triaxial test results on specimens prepared from in-situ frozen samples (Yoshimi et al., 1994). The number of stress cycles of 15 is assumed to correspond to an earthquake magnitude of 7.5 (Seed et al., 1983). The ordinate has been computed from the results of cyclic triaxial tests as follows:

$$\left(\frac{\tau_L}{\sigma_v'}\right)_{\text{in situ}} = 0.9\frac{1+2K_o}{3}\left(\frac{\sigma_d}{2\sigma_o'}\right)_{\text{lab.}} \quad (5.1)$$

in which 0.9 is a factor to correct for multidirectional shear (Seed et al., 1978), and K_0 denotes the coefficient of earth pressure at rest. Because the values of K_0 are unknown, they are assumed to lie between 0.5 and 1.0. On the other hand, the double amplitude axial strain of 5% in the triaxial test corresponds to a shear strain amplitude of 3.8% because the Poisson's ratio is 0.5 for the undrained conditions.

The solid curve in Fig. 14 is drawn through the data points, and the broken curve is the line that separates liquefiable and non-liquefiable conditions based on the field performance data shown in Fig. 5. For a range of $(N_1)_{60}$ between 7 and 25, the agreement between the two curves is good. For higher N-values not covered by the field performance data, the solid curve based on the laboratory tests is the only correlation available at this time.

6 EVALUATION OF LIQUEFACTION RESISTANCE OF CLEAN SANDS BASED ON IN-SITU FROZEN SAMPLES

Figs. 10 through 14 show that the outside part of the samples of clean sands obtained by the in-situ radial freezing method can be considered to retain the in situ density, shear modulus and liquefaction resistance, and that the liquefaction resistance of clean sand deposits can be correlated with the corrected N-value. The earthquake problem will be discussed separately from the wind and wave problems.

6.1 *Liquefaction resistance of clean sand during earthquakes estimated from corrected N-value*

The middle curve in Fig. 15 represents the correlation given in Fig. 14. The other curves show similar correlations for 5 and 26 load cycles which correspond to earthquake magnitudes of 6 and 8.5, respectively. These curves may be used to obtain rough estimates of liquefaction resistance of clean sands on the basis of corrected N-values. Note that the ordinates of these curves for higher N-values are much greater than those in Figs. 6 and 7 based on cyclic triaxial tests on tube samples.

Fig. 15 shows that the liquefaction resistance begins to rise at a steep slope when $(N_1)_{60}$ exceeds about 30 and becomes very large when $(N_1)_{60}$ reaches about 40. Thus, sands whose $(N_1)_{60}$ is greater than 40 may be considered stable even during unusually strong earthquakes, provided that a shear strain amplitude of a few percent is allowed. The fact that liquefaction resistance is sensitive to a small change in N_1 at large values of N_1 seems to justify the thorough studies on the correction of N-values described earlier in this paper.

The above discussion concerns level ground subjected to horizontal shaking. Soil-structure interaction problems involving driving shear stresses as shown in the right column of Table 1 are outside the scope of this paper.

6.2 *Liquefaction resistance of clean sands against wind or wave loading*

Cyclic loading by winds or waves differ from seismic loading as follows:

1. Unlike the earthquake problem discussed in the preceding section in which the liquefaction resistance of level ground is dealt with, the behavior of level ground itself is irrelevant for wind or wave loading, unless the sand is very loose. Instead, what concerns us is the stability and deformation of sand deposits below structures. Below the edge of a mat foundation, for example, the soil is subjected to cyclic loading conditions somewhat resembling the cyclic triaxial test, except for lateral constraint.
2. Unlike seismic loading that is of short duration and is highly irregular, the time histories of wind loading on tall buildings and wave loading on offshore structures consist of much more cycles of less erratic amplitudes.

Thus, for preliminary evaluation of the stability of sand deposits below a structure subjected to wind or wave loading, we may be able to use the results of cyclic triaxial tests carried out to an adequate number of cycles, e.g., 5000 cycles for a 100-year storm for the wave problem (Lee and Focht, 1975). An example for 100 cycles which may correspond to a wind loading problem is shown in Fig. 16.

If neither the Standard Penetration test nor sampling by in-situ freezing can be done offshore, the following procedure proposed by Tokimatsu (1988) may be adopted to evaluate the liquefaction resistance of sand deposits: (1) Determine in-situ shear wave velocity and in-situ density (e.g. by radioactivity methods); (2) obtain tube samples and modify their densities and shear moduli to match the in-situ values by applying cyclic shear strain history at small strain level; and (3) run cyclic tests on the modified samples.

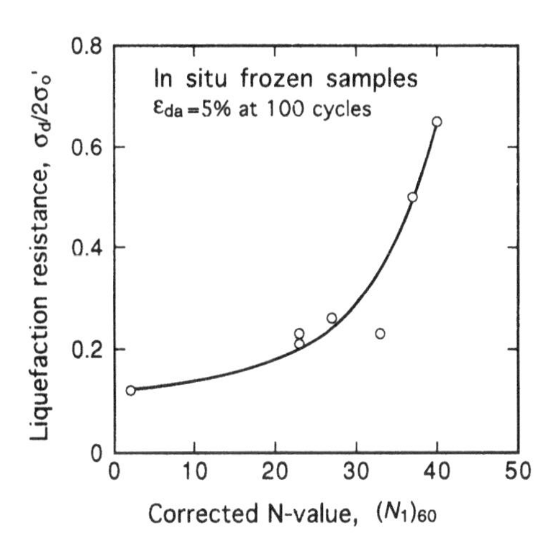

Fig. 16 Liquefaction resistance of clean sands for 100 load cycles (Yoshimi, 1993)

7 DISCUSSION

7.1 *Comparison between in-situ frozen samples and tube samples*

The solid symbols in Fig.17 show the liquefaction resistance of in-situ frozen samples defined at a double amplitude axial strain of 5% at 15 cycles. The open symbols show the liquefaction resistance of tube samples reproduced from Fig. 7. The two curves seem to cross each other at an N-value of about 13, and diverge as the N-values either increase or decrease, the divergence being more pronounced at higher N-values. Thus, by relying on the undrained cyclic tests on tube samples, we may greatly underestimate the liquefaction resistance of dense sands or overestimate that of very loose sands.

7.2 *Effect of confining stress*

In this paper liquefaction resistance has been expressed as a ratio between shear strength and initial effective stress, $\sigma_d/2\sigma_o'$. This normalization does not present any problem if the ratio is independent of the initial effective stress as shown in Fig. 18 in which the results of undrained cyclic triaxial tests on reconstituted samples of clean sands are plotted. For these samples the liquefaction resistance as defined

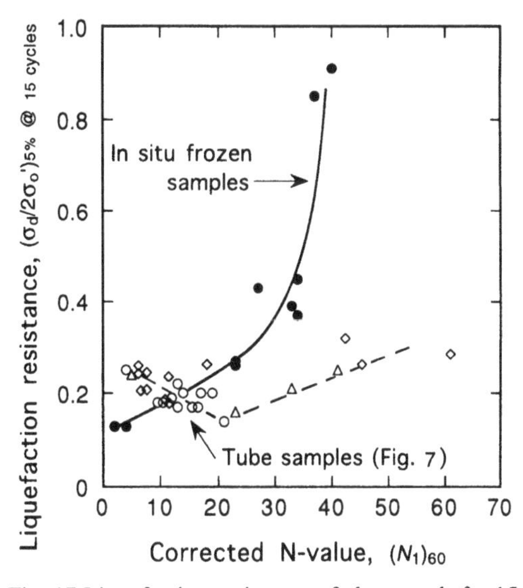

Fig. 17 Liquefaction resistance of clean sands for 15 load cycles: Comparison of in situ frozen samples and tube samples (Yoshimi, 1994)

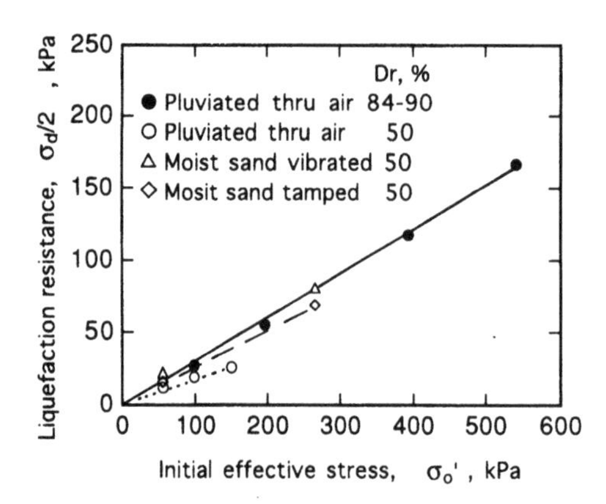

Fig. 18 Effect of initial confining stress on the liquefaction resistance of reconstituted samples of clean sands

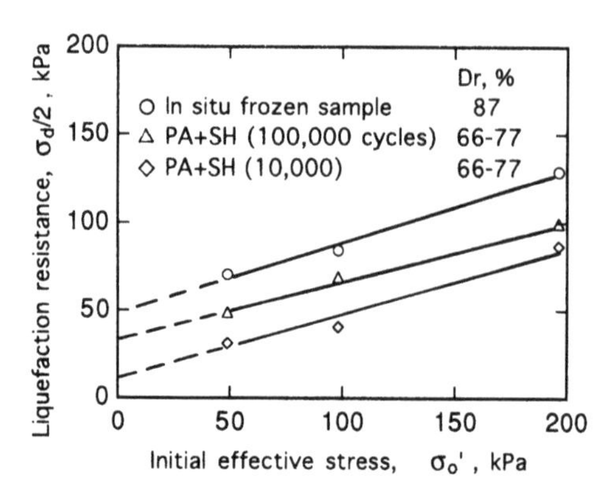

Fig. 19 Effect of initial confining stress on the liquefaction resistance of in situ frozen samples and reconstituted samples with significant cyclic strain history (PA: pluviated through air; SH: cyclic shear strain history at small amplitude)

by the ratio, $\sigma_d/2\sigma_o'$, is independent of the initial effective stress in that the data points lie on straight lines passing through the origin.

As shown in Fig. 19, on the contrary, the liquefaction resistance is not independent of the initial confining stress for a dense sand obtained by in-situ freezing and for reconstituted samples of a medium dense sand with significant shear strain history. It can be seen in the figure that the stress ratios that are equal to the secant moduli decrease with an increase in confining stresses. Thus, we must be careful when we estimate liquefaction resistance in terms of stress ratios from Fig. 15 or 16 for a higher confining pressures than those used to obtain the test data, i.e., between 36 and 106 kPa in the present case. Two possible cases may be considered.

Case 1: The overburden stress at which the N-value is measured is significantly higher than 100 kPa.

Case 2: The effective stress is significantly increased by, e.g., the weight of a structure beyond the overburden stress at which the N-value is measured.

An example of Case 2 is shown in Fig. 20 which covers a much wider range in initial effective stresses than Fig. 19. The sample was obtained from depths of 25m to 29m in Edogawa stratum in Tokyo that consists of Pleistocene sands. The sampler used was a double tube core barrel with protruding inner barrel with a PVC liner. The shear wave velocity of the laboratory specimens computed from the Young's modulus at very small axial strain is 350 m/s, which is smaller than the in-situ shear wave velocity of 450 m/s measured by the down-hole method, indicating that the tube sample was somewhat disturbed and that the ordinates of Fig. 20 may be lower than the in-situ values. Qualitatively, however, the shape of the curve in the figure is considered a reasonably good representation of the in situ properties of the sand. The curve resembles a static undrained strength curve for an overconsolidated clay. The preconsolidation stress of about 1.1 MPa shown in the figure was estimated on the basis of consolidation tests on samples

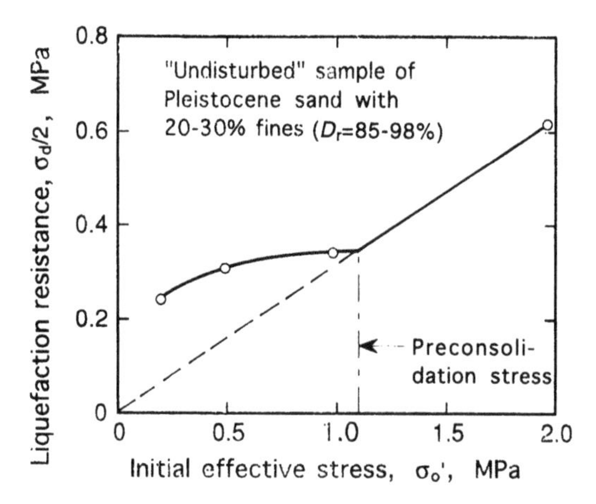

Fig. 20 Effect of initial confining stress on the liquefaction resistance of tube samples of a Pleistocene sand (adapted from Mori et al., 1993)

obtained from a clay stratum overlying the sand. It is interesting to note that the preconsolidation stress seems to divide the overconsolidated condition on its left and normally consolidated condition on its right. The curve is drawn, however, on the assumptions that the data points are free of scatter and that the right half of the curve for initial effective stresses exceeding the preconsolidation stress lies along a straight line passing through the origin. Fig. 20 seems to show that even a dense sand that has more than adequate liquefaction resistance under moderate confining stresses may not necessarily possess very high liquefaction resistance in terms of shear stress ratio under a high confining stress of the order of 1 MPa, e.g. below a 100-meter class dam or below the foundation of a 100-story class high-rise.

7.3 *Effect of fines*

Sand deposits containing some fines (silty sands) are not uncommon. It has been recognized that the *N*-values of silty sands tend to be lower than those of clean sands having comparable liquefaction resistance. Thus, the correlation for clean sand as shown in Fig. 15 should be displaced toward the left as fines contents increase (Seed et al., 1985).

Eventually we need high-quality undisturbed samples of silty sands to enhance the reliability of in-situ frozen samples and of correlations involving *N*-values. Goto (1993) studied under various confining stresses the effect of a freeze-thaw cycle on the liquefaction resistance of sands containing some fines (5 to 10% kaolin or 5 to 20% fines extracted from alluvial soil). As shown in Fig. 21, the freeze-thaw cycle has no effect on the liquefaction resistance if the volumetric expansion during freezing stays below 0.5%. The greater the volumetric expansion beyond 0.5%, the smaller the liquefaction resistance. Thus, the results of undrained cyclic tests on in-situ frozen samples of silty sands may be used as conservative estimates of the in-situ liquefaction resistance.

If less conservative estimates are needed, on the other hand,the liquefaction resistance of in-situ frozen samples of a silty sand may be corrected by measuring the volumetric expansion during the process of refreezing a thawed specimen in the laboratory, provided that the relationship as shown in Fig. 21 is unique. Further research will be needed to confirm that the volumetric expansion during in-situ freezing can be estimated from that during refreezing. It will also be required to simulate field conditions that may affect the volumetric expansion during freezing, e.g., the rate of cooling and confining stress.

8 CONCLUSIONS

The following conclusions may be drawn on the basis of a series of field and laboratory tests on clean sands of Japan:

1. The in-situ freezing method of sampling with a single freezing pipe can yield high-quality undisturbed samples of clean sands that retain the in-situ liquefaction resistance as well as density and elastic shear modulus.
2. The liquefaction resistance of the in-situ frozen samples of clean sands is well correlated over a wide density range with SPT *N*-values corrected for overburden stresses, rod energy ratios and rod lengths.
3. The in-situ liquefaction resistance estimated from the results of undrained cyclic tests on the frozen samples is consistent with the correlation based on field performance data during earthquakes.
4. The liquefaction resistance of the in-situ frozen samples of clean sands begins to increase at a steep rate when corrected *N*-values, $(N_1)_{60}$, exceed about 30.
5. For dense sands,the samples obtained with conventional tube samplers are looser and much weaker than those obtained by the in-situ freezing method. The reverse is true for very loose sands. Thus, by

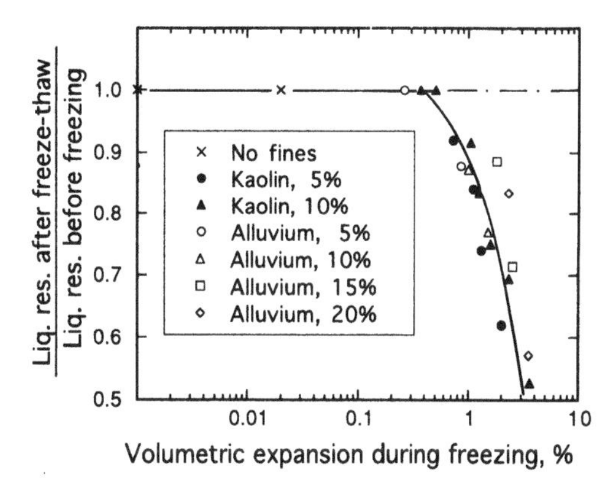

Fig. 21 Effect of volume increase during unidirectional freezing under isotropic stresses on liquefaction resistance of a clean sand and silty sands (adapted from Goto, 1993)

relying on the undrained cyclic test on tube samples, we may greatly underestimate the liquefaction resistance of dense sands or overestimate that of very loose sands.

6. A method is suggested whereby the quality of an in-situ frozen sample of silty sand can be evaluated by thawing and refreezing a specimen prepared from the sample, then by measuring the volume expansion of the specimen during the refreezing process.

ACKNOWLEDGMENTS

The author is grateful to Professor Kohji Tokimatsu of Tokyo Institute of Technology and Dr. Hiroshi Oh-oka of the Building Research Institute, the Ministry of Construction, for their valuable comments on the draft of this paper. Sincere thanks are also due to Dr. Gonzalo Castro for kindly providing supplementary information regarding Fig. 6.

REFERENCES

Casagrande, A. 1976. Liquefaction and cyclic deformation of sands — a critical review, *Harvard Soil Mechanics Series* No. 88, Harvard Univer-sity.

Castro, G. 1975. Liquefaction and cyclic mobility of saturated sands. *Jour. Geotech. Eng. Div., ASCE*, 101, GT6, 551-569

Castro, G. 1987. On the behavior of soils during earthquakes — liquefaction. *Soil Dynamics and Liquefaction*, Elsevier & Computational Mechanics Publications, 169-204

Gibbs, H. J. and W. G. Holtz 1957. Research on determining the density of sands by spoon penetration testing. *Proc. 4th ICSMFE*, London, 1: 35-39

Goto, S. 1993. Influence of a freeze and thaw cycle on liquefaction resistance of sandy soil. *Soils and Foundations*, 33: 4, 148-158

Iai, S. and E. Kurata 1991. Pore water pressures and

ground motions measured during the 1987 Chiba-Toho-Oki Earthquake (in Japanese). Technical Note 718, Port and Harbour Research Institute, Ministry of Transport, Yokosuka, Japan, 1-18

Ishihara, K. 1993. Liquefaction and flow failure during earthquakes. *Géotechnique*, 43: 3, 351-415

Ishihara, K., M. L. Silver and H. Kitagawa 1979. Cyclic strength of undisturbed sands obtained by a piston sampler. *Soils and Foundations*, 19: 3, 61-76

Japan Geotechnical Consultants Association 1993. Development of automatic equipment for the standard penetration test (in Japanese). *Chishitsu-to-Chosa*, 1, 20-26

Kishida, H. 1966. Damage to reinforced concrete buildings in Niigata City with special reference to foundation engineering. *Soils and Foundations*, 6: 1, 103-111

Koizumi, Y. 1966. Changes in density of sand subsoil caused by the Niigata earthquake. *Soils and Foundations*, 6: 2, 38-44

Lee, K. L. and J. A. Focht, Jr. 1975. Liquefaction potential at Ekofisk tank in North Sea. *Jour. Geotech. Eng. Div.*, ASCE, 101: GT1, 1-18

Marcuson, W. F. III, S. S. Cooper and W. A. Bieganusky 1977. Laboratory sampling study conducted on fine sands. *Proc. Specialty Session on Soil Sampling*, 9th ICSMFE, Tokyo, 15-22

Meyerhof, G. G. 1958. Discussion, *Proc. 4th ICSMFE*, London, 3: 110

Mori, N. et al. 1993. Effect of confining pressures on liquefaction resistance of a Pleistocene sand (in Japanese). *Proc. 28th Japan National Conf. on SMFE*, 919-920

Oh-oka, H. and T. Tatsui 1987. Effects of borehole diameter on *N*-values and energy efficiency measurements just before hammer impact (in Japanese). *Proc. 22nd Japan National Conf. on SMFE*, 87-88

Ohsaki, Y. 1970. Effects of sand compaction on liquefaction during the Tokachioki earthquake. *Soils and Foundations*, 10: 2, 112-128

Osterberg, J. O. and S. Varaksin 1973. Determination of relative density of sand below ground water table. *Special Technical Publication 523*, ASTM, 364-378

Robertson, P. K., D. J. Woeller and K. O. Addo 1992. Standard penetration test energy measurements using a system based on the personal computer. *Canadian Geotechnical Journal*, 29: 4, 551-557

Schmertmann, J.H. 1978. Use the SPT to measure dynamic soil properties? -- Yes, But..! *Dynamic Geotechnical Testing*, ASTM STP 654, 341-355

Schmertmann, J. H. and A. Paracios 1979. Energy dynamics of SPT. *Jour. Geotech. Eng. Div.*, ASCE, 105: GT8, 909-926

Seed, H. B. 1979. Soil liquefaction and cyclic mobility evaluation for level ground during earthquakes. *Jour. Geotech. Eng. Div.*, ASCE, 105: GT2, 201-255

Seed, H. B. and I. M. Idriss 1971. Simplified procedure for evaluating soil liquefaction potential. *Jour. Soil Mech. and Found. Div.*, ASCE, 97: SM9, 1249-1273

Seed, H. B., C. K. Chan and T. F. Viela 1982. Considerations in undisturbed sampling of sands. *Jour., Geotech. Eng. Div.*, ASCE, 108: GT2, 265-283

Seed, H. B., I. M. Idriss and I. Arango 1983. Evaluation of liquefaction potential using field performance data. *Jour. of Geotech. Eng.*, ASCE, 109: 3, 458-482

Seed, H. B., R. M. Pyke and G. R. Martin 1978. Effects of multidirectional shaking on pore pressure development in sands. *Jour., Geotech. Eng. Div.*, ASCE, 104: GT2, 265-283

Seed, H. B., K. Tokimatsu, L. F. Harder and R. M. Chung 1985. Influence of SPT procedures in soil liquefaction resistance evaluations. *Jour. of Geotech. Eng.*, ASCE, 111: 12, 1425-1445

Singh, S., H. B. Seed and C. K. Chan 1982. Undisturbed sampling of saturated sands by freezing. *Jour., Geotech. Eng. Div.*, ASCE, 108: GT2, 247-264

Skempton, A. W. 1986. Standard penetration test procedures and the effects in sands of overburden pressure, relative density, particle size, ageing and overconsolidation. *Géotechnique*, 36: 3, 425-447

Tatsuoka et al. (1978): A method for evaluating undrained cyclic strength of sandy soils using standard penetration resistances, *Soils and Foundations*, 18: 3, 43-58

Terzaghi, K. and R. B. Peck 1967. *Soil Mechanics in Engineering Practice, 2nd ed.*, John Wiley & Sons, 108-109

Tokimatsu, K. 1988. Penetration tests for dynamic problems. *Penetration Testing 1988*, Proc. Int. Symp. Penetration Testing, Balkema, 117-136

Tokimatsu, K. and Y. Hosaka 1986. Effects of sample disturbance on dynamic properties of sand. *Soils and Foundations*, 26: 1, 53-64

Tokimatsu, K. and J. Ohara 1990. Soil sampling by freezing (in Japanese). *Tsuchi-to-Kiso*, JSSMFE, 38: 11, 61-68

Tokimatsu, K., Y. Suzuki and S. Tamura 1994. Preliminary report on the geotechnical aspects of the Hokkaido-Nansei-Oki earthquake of July 12, 1993. *Performance of Ground and Soil Structures During Earthquakes*, Special Volume for 13th ICSMFE, New Delhi, 75-86

Tokimatsu K. and Y. Yoshimi 1983. Empirical correlation of soil liquefaction based on SPT *N*-values and fines content. *Soils and Foundations*, 23: 4, 56-74

Yasuda, S. and I. Yamaguchi 1984. Dynamic shear moduli measured in the laboratory and the field (in Japanese). *Proc. Symp. Evaluations of Deformation and Strength Characteristics of Sandy Soils and Sand Deposits*, JSSMFE, 115-118

Yoshimi, Y. 1994. Relationship among liquefaction resistance, SPT *N*-value and relative density for undisturbed samples of clean sands (in Japanese). *Tsuchi-to-Kiso*, JSSMFE, 42: 4, 63-67

Yoshimi, Y., M. Hatanaka and H. Oh-oka 1977. A simple method for undisturbed sand sampling by freezing. Proc. Specialty Session on Soil Sampl-

ing, *Proc., 9th ICSMFE*, Tokyo, 23-28
Yoshimi, Y., M. Hatanaka and H. Oh-oka 1978. Undisturbed sampling of saturated sands by freezing. *Soils and Foundations*, 18: 3, 59-73
Yoshimi, Y. and K. Tokimatsu 1983. SPT practice survey and comparative tests. *Soils and Foundations*, 23: 3, 105-111
Yoshimi, Y., K. Tokimatsu and Y. Hosaka 1989. Evaluation of liquefaction resistance of clean sands based on high-quality undisturbed samples. *Soils and Foundations*, 29: 1, 93-104
Yoshimi, Y., K. Tokimatsu, O. Kaneko and Y. Makihara. 1984. Undrained cyclic shear strength of a dense Niigata sand. *Soils and Foundations*, 24: 4, 131-145
Yoshimi, Y., K. Tokimatsu and J. Ohara 1994. In situ liquefaction resistance of clean sands over a wide density range. *Géotechnique*, 44: 1

Developments in Geotechnical Engineering, Balasubramaniam et al. (eds) © 1994 Balkema, Rotterdam, ISBN 90 5410 522 4

Mechanism in the formation of kink bands

P. Habib
Laboratoire de Mécanique des Solides, Ecole Polytechnique, Palaiseau, France

ABSTRACT : When a schisty material is tectonically compressed in a direction parallel to the schistosity a kind of buckling of the layers occurs, so-called kink band. A model is proposed for this phenomena which shows that the angle of the kink band with the normal to the layers is equal to the angle of dilatancy between the layers. It seems that the elements in the kink, between the hinges, may be in any length.

INTRODUCTION

When certain anisotropic materials, such as shales or schists, some frozen soils, some crystals, or wood, are compressed in a direction parallel to the layers, cleavage or fibres, failure occurs by grouped bucklings of the constitutive elements. In the case of schisty materials, it is accepted that so-called kink bands have formed under the effect of tectonic compression parallel or almost parallel to the direction of the layering. The discontinuous formation of these bands generates an overall shortening of the material. This phenomenon can be reproduced in the laboratory in a triaxial cell, for example by subjecting to compression a test piece cut in a lance-head gypsum crystal parallel to the cleavage direction.

The explanation for the occurrence of these discontinuities falls into the field of rock mechanics, but the dilatancy phenomena which accompany the formation of kink bands are of a nature to cast light on those occurring in shear bands in soil mechanics.

Fig. 1 represents a set of kink bands in a micashale : the width of deformed bands here is around some milimetres, but it can quite often reach some decimetres. The orientation of the direction taken by these discontinuities with respect to the direction of compression of layers is clearly above 45°, which excludes that this phenomenon be assimilated to the formation of a slip plane where the slope would always be less than 45°.

Fig. 2 represents a well-developed kink band in a shale sample taken from the Col de Seigne located between France and Italy in the Mont Blanc mass. The width of bands deformed in bayonet shape is around 10-15 mm. For this test piece, the thickness e′ of the deformed stacks is almost equal to the initial thickness e of the shale stack (*fig. 3*).

Figure 1 : Kink band in a schist

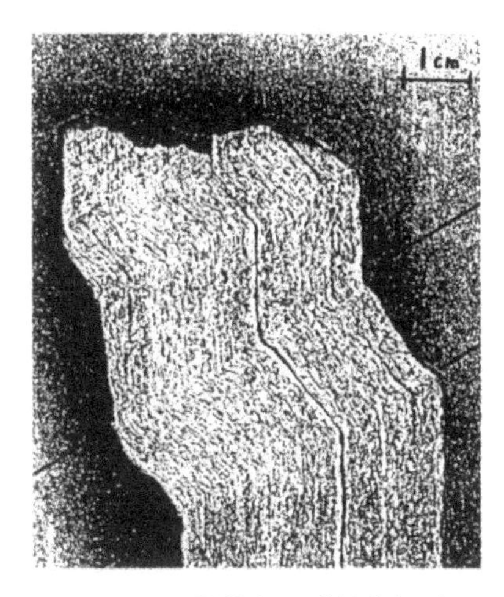

Figure 2 : Well formed kink band

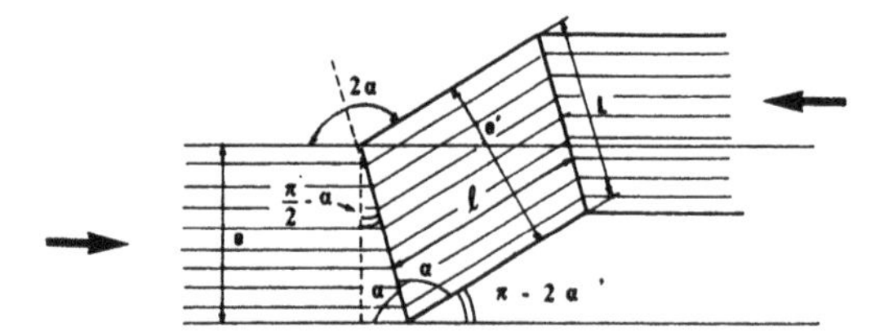

Figure 3 : Notations

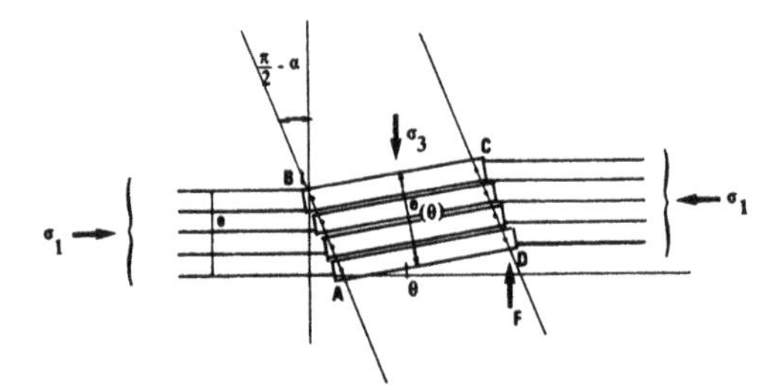

Figure 4 : Modelling

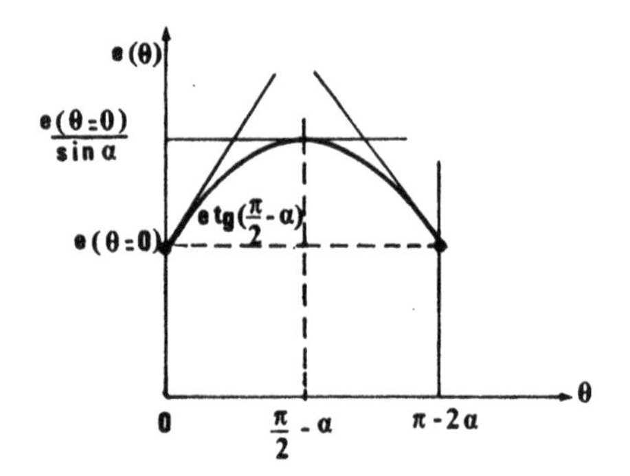

Figure 5 : Thickness variation as a function of θ

If it is accepted that the length of layers in the band is equal to the inital length or again that the volume of material in the deformed band has remained equal to the initial volume after the transformation which generated the double fold, it follows that the hinge in the kink is a bisector of the angle 2α of the kink. This is quite well verified on Fig. 2 where the rock is as compact in the fold as outside the fold.

Generally speaking, the orientation of the kink band hinges is not so regular and sometimes even the kink bands are not perfectly plane. Because of the interlocking of kinklines and perhaps also later tectonic deformations, there is a certain scattering in the orientation of kink bands around ± 5° and the kink hinge is rarely exactly bisectrix of the angle 2α. According to Stubley (1990), the deviation occurring in one direction or the other but on an average, the angle α in the band is rather greater than the external angle of the band.

MODELLING

The formation of a kink band appears as a local instability, very much like buckling. Let us examine the mechanical model made up of a pile of plates hinged at each end, the articulation points being lined up on two parallel straight lines inclined at angle $\frac{\pi}{2}-\alpha$ perpendicular to the layers. The model is subjected to a stress σ_1 in the plane of the plates and a stress σ_3 perpendicular to the plates (*fig 4*). Under the effect of the deviatoric stress $\sigma_1 - \sigma_3$ the system is unstable on only one side, the right side of the figure can only displace upwards. At the beginning of the movement when the plates turn following angle θ they separate from each other. The thickness of the articulated band passes through a maximum for $\theta = \frac{\pi}{2}-\alpha$ then returns to initial thickness e_0 when $\theta = \pi - 2\alpha$. At this moment, the mechanism blocks and cannot go further.

a) The variation in the thickness of the articulated band can easily be calculated, that is to say the variation in the volume of the model as a function of θ (assuming thant the length of the plates remains constant). The length L of the hinge is :

$$L = e/\sin\alpha$$

and the thickness $e(\theta)$ of the stack :

$$e(\theta) = e\,\sin(\alpha+\theta)/\sin\alpha$$

The variation in the thickness is :

$$\frac{de(\theta)}{d\theta} = e\,\cos(\alpha+\theta)/\sin\alpha$$

It is zero for $\theta = \frac{\pi}{2} - \alpha$ and maximum for $\theta = 0$ where it is expressed :

$$\frac{de\,(\theta=0)}{d\theta} = e/\tan\alpha = e\,\tan\left(\frac{\pi}{2} - \alpha\right)$$

a value which is obviously identical, almost to a sign, to $\frac{de\,(\theta=\pi-2\alpha)}{d\theta}$

Fig. 5 indicates the variation in the thickness of the stack of hinged plates as a function of θ. It is seen that the maximum perpendicular expansion occurs at the beginning of the instability (that is to say for $\theta = 0$) and it is proportional to $\tan\left(\frac{\pi}{2} - \alpha\right)$.

Now, the work of perpendicular expansion will oppose instability, this leads to instability occurring all the more easily as the angle of the hinge perpendicular to the layers will be small.

b) Now let us consider the relative movement of two plates (fig 6).

A plate being considered as fixed, the neighbouring one is displaced by a circular sideways movement and the angle of the tangent to the circle with the layer planes at the beginning of the movement is $\frac{\pi}{2} - \alpha$.

But the mineral layers are not perfect planes and if an attempt is made to slide them one on the other at the beginning of the movement, dilatancy of angle δ will occur linked to the irregularities of the surfaces in contact. If δ is greater than $\frac{\pi}{2}-\alpha$ the movement of the plates is blocked and the instability cannot begin ; the set of plates forms a monolithic structure. This means that the angle $\frac{\pi}{2}-\alpha$ formed by the hinge with the perpendicular to the layer must be greater than the angle of dilatancy δ to allow the mechanism to occur. Now, according to the previous paragraph (a), the angle $\frac{\pi}{2}-\alpha$ must be the smallest possible. Hence we have $\frac{\pi}{2}-\alpha = \delta$. On figure 2 we read $\frac{\pi}{2}-\alpha = 25°$.

c) Let us now consider the beginning of instability. Fig. 4 shows the beginning of the kink band deformation ; under the effect of external forces, the parallelogram is subjected to a distorsion $\Delta\theta$. The force F is due, for example, to the fact that the principal direction of compression is not strictly in the plane of the layers, which causes a disturbance. On the facet CD the stress deviator exerts force $e(\sigma_1 - \sigma_3)$ which is offset by $\ell\Delta\theta$ in respect to the force which is exerted on the facet AB, hence a couple $e(\sigma_1 - \sigma_3)\,\ell\Delta\theta$. On the facets BC and AD the stress σ_3 resists the perpendicular expansion, which sets up an opposing couple. The expression for the variation in thickness of stacking of plates is given in paragraph a). The variation in volume as a function of θ is :

$$\frac{dV(\theta)}{d\theta} = e\ell\,\frac{\cos\alpha\cos\theta - \sin\alpha\sin\theta}{\sin\alpha}$$

and for small θ we can set down :

$$\frac{dV(\theta)}{d\theta} = e\ell\,(\tan^{-1}\alpha - \Delta\theta)$$

The corresponding deformation work is :

$$\Delta W = \sigma_3\, e\ell\,(\tan^{-1}\alpha - \Delta\theta)$$

Finally, the return torque of the elastic deformation $Ge\ell\sin\alpha\Delta\theta$ must be brought in. Hence we obtain the equation of equilibrium :

$$e(\sigma_1 - \sigma_3)\,\ell\Delta\theta + F\ell = \sigma_3 e\ell\tan^{-1}\alpha - \sigma_3 e\ell\Delta\theta + Ge\ell\sin\alpha\Delta\theta \qquad (1)$$

with which we can calculate the distorsion $\Delta\theta$:

$$\Delta\theta = \frac{\frac{F}{e} - \sigma_3\tan^{-1}\alpha}{G\sin\alpha - \sigma_1}$$

We see that $\Delta\theta$ becomes infinite, that is to say there is instability, for $\sigma_1 = G\sin\alpha$.

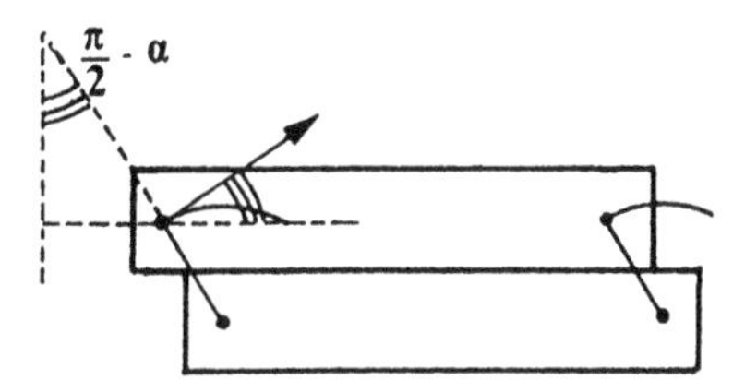

Figure 6 : Relative movement of two small plates

Figure 7 : A very thick kink band : Hamersley Range fold in Australia (photo J.P. Ferrero/Jacana)

Unfortunately, it is difficult to estimate the shear modulus G since this modulus can be very high if there is little cohesion between the layers, or very small if this is just a bending of the material at the hinges in the layers. This result simply shows the existence of a critical compression stress σ_1.

In equation (1) we can notice that the length ℓ is present in the two parts and consequently disappears. This means that the length of the elements in the kink does not intervene. Indeed, in fig. 7 we can see a kink band of metric thickness.

d) Fig. 8 shows a particularly interesting detail. The top of the sample is a break which occurred on the bend of a kink, inclined as in figure 2. Below we can see two kink bands 2-4 mm thickness which are crossed towards the left by what obviously becomes a slip line. Above and below this zone of discontinuities, we observe the displacement 6 mm leftwards of two rigid blocks. The kink bands are compact but the shear band, which has not the correct inclination for the layers to block after the movement, is strongly dilated ; the black spots which appear on the photograph are the vacuoles dotted over all the shear plane.

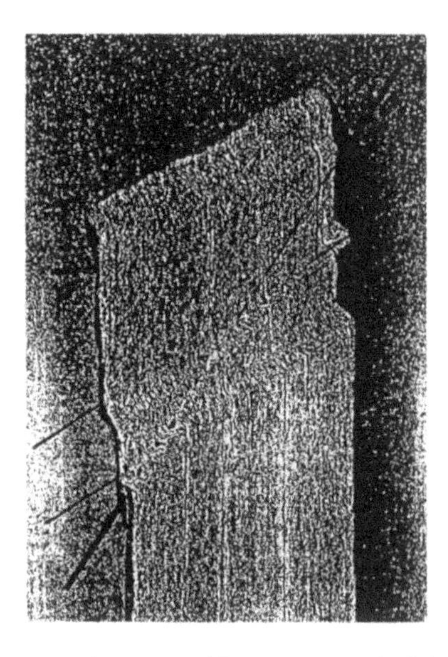

Figure 8 : Slip surface dilatancy and two kink bands

CONCLUSIONS

With the hypotheses adopted, the modelling proposed to interpret the formation of a kink band leads to the following conclusions :

- the direction of the kink band is close to the bisectrix of angle 2α of the kink. If these is any shortening (or pleating) of layers in the kink, the angle external to the band is smaller than the angle internal to the band ;
- the angle between the direction of the band and the perpendicular to the layers of schistosity is equal to the dilantancy angle of slip of the layers over each other ;
- the thickness of the kink band is not a parameter of the model. It can therefore be of any value within reasonable bounds ;
- the kink band can only occur if the stress in the direction of the layers exceeds a certain threshold, but a confinement σ_3 is necessary around the layered medium ;
- finally, if the stress σ_3 increases greatly, the contact forces between layers restores the mechanical continuity. It is well known then that anisotropy disappears and if failure occurs, it is materialized by a slip surface cutting the layers and no longer kink bands. The orientations are then quite different.

BIBLIOGRAPHIE

APARICO M., SIRIEYS P. (1978), *Déformation naturelle des roches par kink bands.* Colloque Int. "Sciences de la Terre et Mesures". Orléans 1977. Mémoires du BRGM, n⁰ 91, (1978) pp. 187-198.

ARCHAMBAULT G. et LADANYI B. (1993), *Failure of jointed rock masses by kink zone of instability : an important and dangerous mode of rupture.* Eurock'93. Balkema. Volume I pp. 761-769.

DONATH F.A. (1968), *Experimental study of kink band development in Martinsburg slate in "Kink Bands and Brittle Deformation".* Géol. Surv. Pap. Can. 68-52, pp. 255-288.

HABIB P. (1991), *Orientation de la charnière d'un Kink band.* Revue Française de Géotechnique, n⁰ 56, juillet 1991, pp. 51-55.

SIRIEYS P. (1985), *Mécanismes de glissement plastique des roches à structure planaire.* Coll. Int. du CNRS n⁰ 319 - "Comportement plastique des solides anisotropes", pp. 475-488.

STUBLEY M.P. (1990), *The geometry and kinematics of a suite of conjugate kink bands, southeastern Australia.* Journal of Structural Geology, vol. 12, n⁰ 8, 1990, pp. 1019-1031.

Developments in Geotechnical Engineering, Balasubramaniam et al. (eds) © 1994 Balkema, Rotterdam, ISBN 90 5410 522 4

State parameter of granular soils

N. Moroto
Hachinohe Institute of Technology, Japan

ABSTRACT: Mogami's theory of granular materials were evaluated in terms of the author's state parameter. It was said that Mogami's entropy can be related consistently to the author's state parameter.This relationship yielded a reasonable parameter transformation from (e,s) to (e,γ). The coefficient k which has appeared in the state functions was found to be dependent on grain shape and grading. This nature was used to a material classification.

INTRODUCTION

In this paper, let us pay a particular interest to the two state parameters which were obtained by Mogami(1965) and Moroto(1976).
Employing the approach of statistical mechanics, Mogami first introduced the idea of the entropy to the mechanics of granular materials.
The entropy H was written from the average void ratio,e, and the deviation of void ratio,s, as:

$$H=H(e,s)$$

$$= e \log e - (1+e)\log(1+e) + \frac{s}{2(1+e)e} \qquad (1)$$

Moroto introduced the state function S_s from phenomenological point of view, which was defined by

$$S_s = \int dW_s^p/p$$

$$= \int dv_d + \int (q/p)d\gamma \qquad (2)$$

where

W_s^p:plastic work done due to shear force
p;confining pressue
v_d:volumetric strain due to dilatancy
γ:shear strain
q:shear stress

The author recognizes that a reasonable (e,s) and (e,γ) transformation can be performed to compare the incremental form of the two functions, dH and dS_s, while Mogami assumed that a function F(e-kγ) was equivalent to H(e,s), where k is a coefficient.
Here, the author examines the relationship between Mogami's entropy and the Moroto's state parameter,then he studies the coefficient k which is characterized by the grain properties.

SOME MODIFICATIONS OF MOGAMI'S THEORY

Mogami assumed the following relation:

$$dW_s^p = K' \, dH \qquad (3)$$

where K' is a constant
In the above relation, the plastic work done W_s^p depends greatly on the

stress path as indicated by the author (Moroto,1976). It is unreasonable to relate H to the plastic work done. It is, therefore, recommended that the function H must be connected with the state parameter S_s.

The increment of S_s can be written as

$$dS_s = \frac{-1}{1+e}(de - kd\gamma) \qquad (4)$$

and

$$k = (1+e)(q/p) \qquad (5)$$

Now we define a function

$$d\psi = (1+e)dS_s$$
$$= (-de + kd\gamma) \qquad (6)$$

The increment of H can be written as

$$dH = A(e,s)(de - \frac{1}{A(e,s)\ B(e)} ds \qquad (7)$$

where

$$A(e,s) = \log\frac{1+e}{e} + \frac{(1+2e)s}{2e^2(1+e)^2} \qquad (8)$$

$$B(e) = \frac{1}{2e(1+e)} \qquad (9)$$

In Eqs.(6) and (7), the incremental forms at the right hand side of the equations are assumed to be equivalent as follows:

$$de - \frac{1}{A(e,s)B(e)} ds$$
$$\equiv de - kd\gamma \qquad (10)$$

This leads to

$$ds = 2kd\gamma e(1+e)A(e,s) \qquad (11)$$

When s is small, Eq.(11) becomes

$$ds = 2kd\gamma\, e(1+e) \log\frac{1+e}{e} \qquad (12)$$

Eqs.(11) and (12) are none other than (e,s) to (e,γ) transformation in Mogami's theory.

Substituting Eq.(11) into Eq.(7) and using Eq.(4) gives:

$$dH = -(1+e)A(e,s)dS_s \qquad (13)$$

Rewriting Eq.(13), we have the following expression:

$$dS_s = K\ dH \qquad (14)$$

where $K = -1/(A(e,s)(1+e))$

From Eq.(14), one can understand that Mogami's entropy and the author's parameter are simply related to each other. In the mechanics of granular materials, we must use Eq.(14) in place of Eq.(3). Sign of dH and dS_s is opposite. This problem remains to be solved.

STRENGTH COEFFICIENT, k

From Eq.(5), the coefficient k can be expressed by

$$k = (1+e) \sin \phi \qquad (15)$$

where ϕ is the angle of internal friction.

The strength coefficient,k, can be determined from drained shear tests (mostly triaxial compression tests) on granular soil specimens which have different initial void ratios e_o. Thus,

$$k = (1+e_o)\sin\phi \qquad (16)$$

From Eq.(16),the consistency of k and its average value $\bar{k}$ can be examined and calculated for each material. Moroto(1982) showed that the

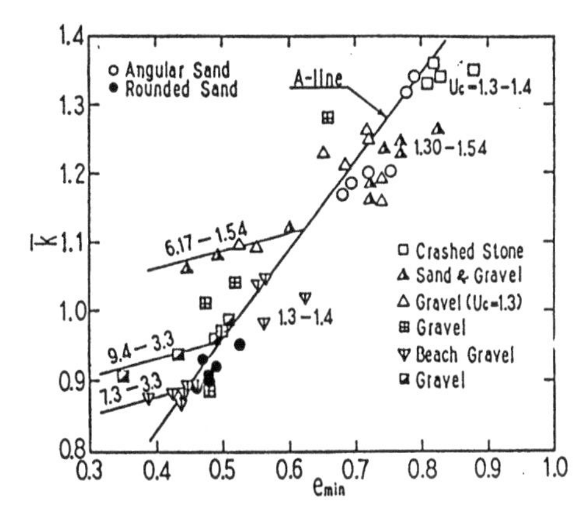

Fig. 1 $\bar{k}$-e_{min} relationship of sands and gravels

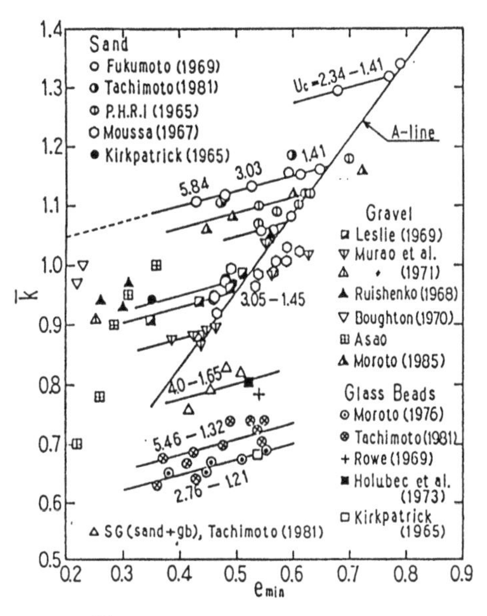

Fig. 2 $\bar{k}$-e_{min} relationship of coarse grained materials

consistency of the coefficient k was reasonable for different granular soils and that the dependence of k on the particle properties of material was notable. Further, he concluded that k was a material constant closely related to the minimum void ratio e_{min}. This could be represented as

$$k = a\, e_{min} + b \tag{17}$$

where a and b are constants depending on particle shape and grading. This feature of Eq.(17) can be observed for a confining pressure range of (σ_3 = 50 - 500 or 700 kN/m^2).

Shimobe and Moroto(Moroto and Shimobe 1993), applying the $\bar{k}$-e_{min} relationship of Eq.(17) to data obtained by many researchers from tests on sands and gravels, investigated the effects of particle characteristics such as roundness, R, and coefficient of uniformity, Uc, on the constants a and b. It was found that:

(a) For many sands with a uniform grading (Uc ≒ 2) and different particle shape, the following equation was obtained:

$$\bar{k} = 1.290\, e_{min} + 0.314 \tag{18}$$

The 'A-line' given by Eq.(18) for uniform samples represents a particular characteristics line and the position of each material on A-line depends on the degree of roundness(also from Fig. 1).

(b) For granular materials with similar roundness but with a variation in grading, the $\bar{k}$-e_{min} relationship for each soil is linear. The lines are at an angle and to the left of the A-line and parallel to one another as shown in Figs. 1 and 2. These lines can be expressed by

$$\bar{k} = 0.260\, e_{min} + b_2 \tag{19}$$

where b_2 is constant depending on the roundness.

(c) In order to relate the constant b_2 as defined by Eq.(19) to roundness, R, the $\bar{k}$-e_{min} relationship for uniform quarts sand with varying roundness and supplementary data from tests on sands with same roundness and varying grading was established by

$$b_2 = 1.114 - 0.544R \tag{20}$$

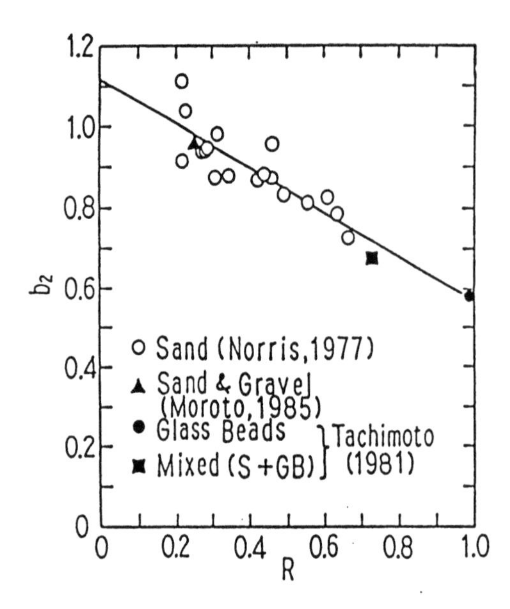

Fig. 3 b_2 and R relationship

This relationship is shown in Fig. 3.

CONCLUDING REMARKS

(a) Mogami's entropy can be closely related to Moroto's state parameter.
(b) The strength constant k at failure is dependent on the particle properties and is related linearly to the minimum void ratio.

REFERENCES

1) Mogami,T.(1965):A statistical approach to the mechanics of granular materials, Soils and Foundations, Vol. 5, No. 2,pp. 26-36
2)Moroto,N.(1976):Anew parameter to measure degree of shear deformation of granular material in triaxial compression tests, Soils and Foundations, Vol. 16, No. 4,pp.1-10
3)Moroto,N.(1982):An application of Mogami's strength formula to the classification of granular soils, Soils and Foundations, Vol.22, No.1, pp.82-90
4)Moroto,N. and Shimobe,S.(1993): Compactive measure of cohesionless soils, ENGINEERED FILLS, edited by B.G.Clarke et al, Thomas Telford pp.205-213

Developments in Geotechnical Engineering, Balasubramaniam et al. (eds) © *1994 Balkema, Rotterdam, ISBN 90 5410 522 4*

Multi-stage triaxial compression test under a constant cell pressure

Sang-Kyu Kim
Department of Civil Engineering, Dongguk University, Seoul, Korea

Hyun-Tae Kim
Rural Development Corporation, Kyungki-do, Korea

ABSTRACT: In this paper, a new procedure for carrying out a series of consolidated-undrained triaxial compression tests with a single specimen is proposed. In the proposed procedure, a high cell pressure is applied to the specimen, and then kept constant until the completion of the test. In the first stage of test, a specified effective cell pressure can be applied by allowing some drainage from the constant pressure, and then it is sheared until failure commences. The same process is repeated for the following stages. With this procedure, the overall testing time is significantly reduced by saving consolidation time of each stage, while performing a series of triaxial tests with a single specimen. In order to verify the applicability of the proposed test, comparisons are made with the results obtained from the standard triaxial compression test and the conventional multi-stage triaxial test. It is shown that the strength parameters determined by using the proposed procedure agree quite well with those obtained from other tests.

1 INTRODUCTION

The standard triaxial compression test has been widely used to determine strength parameters for design purposes. In order to obtain strength parameters from this test, at least three homogeneous specimens are required from a thin-walled tube, and thus difficulties in determining the parameters may be encountered if homogeneous samples are not enough, as in the case when the soil contains some sea shells, sand seam or silt seam. Furthermore, much efforts are needed in preparing several homogeneous specimens, in particular, for testing granular materials.

In order to overcome those problems, the so-called multi-stage triaxial compression test has been proposed (Kenney and Watson, 1961; Lumb, 1964; Anderson, 1974). The major advantage of the former method is that the strength parameters may be obtained by repeating consolidation and shearing for a single specimen within a cell.

Using this procedure, Kenney and Watson (1961) carried out the consolidated undrained triaxial test successfully for determining shear strength parameters on both undisturbed and remoulded soils. Lumb (1964) applied this procedure on undisturbed unsaturated residual soils of Hong Kong and reported that there was excellent agreement between multi-stage and standard tests in terms of deviator stresses at failure and drained shear strength parameters. Anderson (1974) determined undrained strength parameters on stony boulder clay with this procedure. Ho and Fredlund (1982) applied this procedure for measuring the increase in shear strength resulting from soil suction in an unsaturated soil. Kawabara et al. (1986) performed the test on granular materials with maximum size of 50mm under consolidated undrained (CU) and consolidated drained (CD) conditions.

It has been demonstrated that the conventional multi-stage test results are in good agreement with those obtained using the standard triaxial test, in particular, for the shear strength parameters. However, the testing time for a specimen is still long in CU test because the consolidation time to achieve over 95% degree of consolidation would take usually 24 hours for clayey soils. Therefore, at least 4 days are required even with this test procedure using a single test specimen.

In this experimental study, a high cell pressure is applied to a specimen and remained constant until the completion of the test. The first stage of confining pressure can be applied by allowing some drainage from the initial cell pressure. By doing this, the first stage of consolidation is achieved very quickly by means of an increase of effective stress. After applying the deviator stress until failure commences, it is released and then the second effective cell pressure is applied in the same way. This process is repeated until the completion of the test. By using this technique, a series of CU triaxial test on a saturated clay can be completed within 1.5 day.

2 BASIC CONCEPT AND TEST PROCEDURES OF THE MODIFIED MULTI-STAGE COMPRESSION TEST

2.1 *Basic Concept*

When a saturated clay is consolidated isotropically or under K_0 condition, volume changes occur with time, and a relationship between the change of void ratio and square root of elapsed time is drawn as shown in Fig. 1. It is noted that a linear relationship is obtained at the first part of the curve until around 60% degree of consolidation has been reached. The curve,

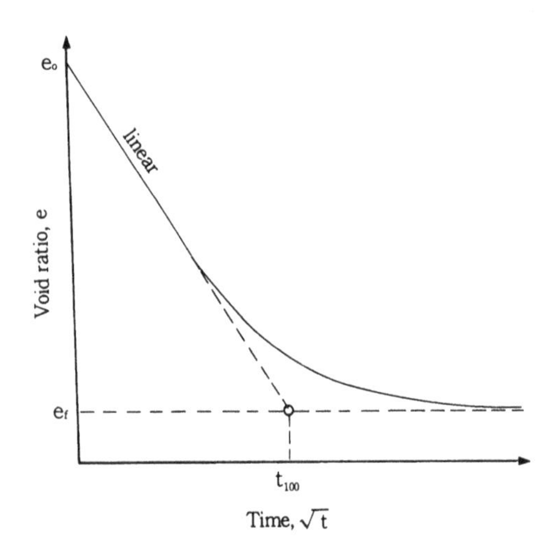

Fig. 1. Relationship between void ratio decrese and elapsed time during consolidation.

thereafter, becomes concave because consolidation rate decreases. In the standard triaxial compression or the multi-stage triaxial compression tests (CU test), the volume change of a specimen in consolidation stage takes place along this curve. Consolidation over 95% requires usually 24 hours for clayey soils.

If we apply a high cell pressure and then let some pore pressure allow to dissipate, a void ratio capable of obtaining by consolidating a specimen during 24 hours with smaller cell pressure can be also obtained within the linear portion of the curve in Fig. 1. Fig. 2 shows two different methods applicable in obtaining a certain void ratio. In Fig. 2(a), the first cell pressure is applied to a specimen during 24 hours and then sheared to failure under a constant volume. The same procedure is followed for the second and the third stages. This gives a typical void ratio-elapsed time relationship for the conventional multi-stage CU test. In the lower figure (Fig. 2(b)), a high cell pressure (in this figure, σ_{33}) is applied to a specimen. Then the void ratio, e_1, corresponding to the first consolidation stage as obtained in Fig. 2(a) can be obtained by means of pore pressure dissipation. After the shearing of the specimen under constant volume, e_2 can be obtained in the same way as before. However, it should be noted that the final stage of consolidation in Fig. 2(b) does not follow a straight line.

It can be seen clearly from Fig. 2 that the void ratio capable of obtaining from each consolidation stage in the conventional multi-stage test can be also obtained by pore pressure dissipation after applying a high cell pressure. In this case, the consolidation time attainable to a given void ratio or an effective cell pressure can be estimated easily by examining the consolidation behaviour of soils. It is assumed herein

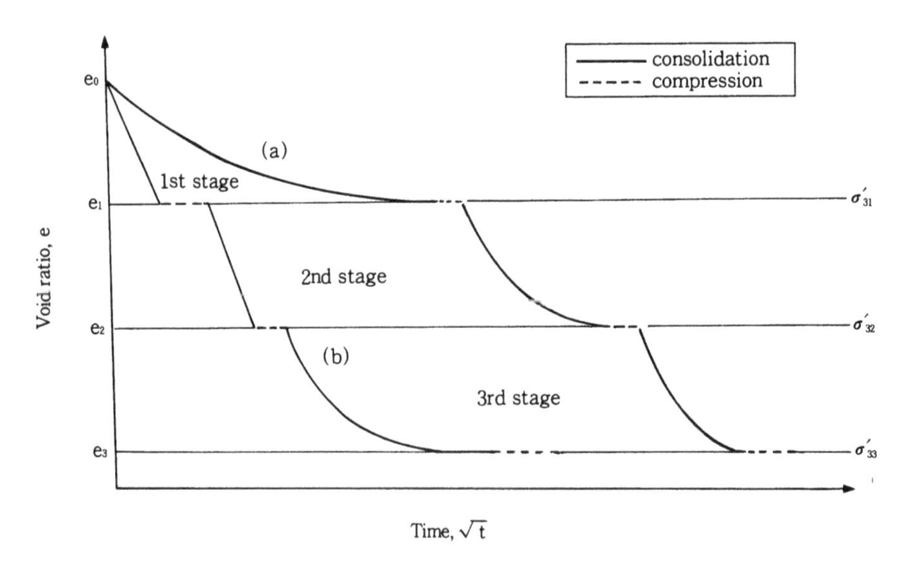

Fig. 2. Comparison of consolidation methods for obtaining a given void ratio.

that the pore pressure distribution within a specimen is uniform during the process of the consolidation.

From Fig. 1, the following linear relationship is established:

$$e = a - b\sqrt{t} \tag{1}$$

where a, b = constants, t = elapsed time and e = void ratio.

From the consolidation characteristics of soils, it is known that there exsists a linear relationship between the void ratio decrease and the consolidation pressure increase in log scale in the normally consolidated range. Thus, the following relationship can be written:

$$e = a' - b' \log \sigma_c' \tag{2}$$

From Eq. (1) and (2),

$$a' - b' \log \sigma_c' = a - b\sqrt{t}$$

$$\log \sigma_c' = \frac{a'}{b'} - \frac{a}{b'} + \frac{b}{b'}\sqrt{t}$$

$$\log \sigma_c' = a'' + b''\sqrt{t} \tag{3}$$

where a'', b'' = constants and σ_c' = effective confining pressure.

It is clear from Eq. (3) that there exists a linear relationship between $\log \sigma_c'$ and $\sqrt{t}$. If the constants are determined by measuring pore pressure changes with elapsed time, the time required to obtain a given effective cell pressure can be estimated from the following expression:

$$\sqrt{t} = \frac{1}{b''}\log \sigma_c' - \frac{a''}{b''}$$

$$t = \left(\frac{1}{b''}\log \sigma_c' - \frac{a''}{b''}\right)^2 \tag{4}$$

It should be noted, however, that Eq. 4 is only valid within the range in which the void ratio decrease is proportional to the square root of time.

By using Eq. 4, the consolidation time can be determined experimentally by pore pressure measurement. Points *a* and *b* in Fig. 3(b) are obtained at a consolidation stage and they should be on a straight line as indicated in Eq. (3). If a cell pressure σ_3' is given, the intersection between the extended straight line and the cell pressure becomes the time required for the first stage consolidation. When the time is reached, drainage from the specimen can be interrupted by closing the drainage valve of the triaxial compression apparatus. A shear test can then be performed under constant volume change.

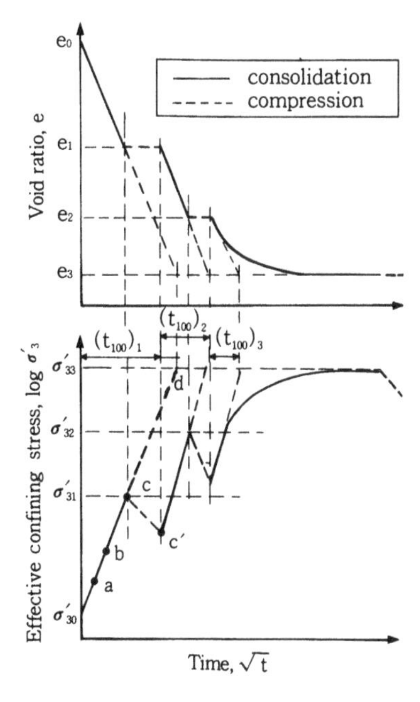

Fig. 3. Determination of consolidation time from a linear relationship between σ_3' and $\sqrt{t}$

While the specimen is being sheared, pore water pressure is generated. This causes the decrease of the effective cell pressure as indicated by point *c'* in the figure. The second stage follows the same procedure, however, in the third stage of consolidation there exists no linear relationship unless the initial cell pressure applied is high enough.

If a straight line is drawn connecting points *a* and *b*, this line will intersect σ_{33} line which will give the time coordinate t_{100} , as shown in Fig. 3. This is the time of 100% degree of consolidation obtained from a linear relationship of σ' and $\sqrt{t}$, which may be used for the estimation of the shear strain rate on the test specimen (Henkel, 1962).

2.2 *Stress Path*

Fig. 4 gives a comparison of the stress paths of the proposed test (also refered herein as modified test) in terms of total stress and effective stress for consolidated undrained condition. When a high cell pressure (for example, σ_{33}) is applied to the specimen, the total stress paths go from *a* to *h*, and when the pore pressure dissipation corresponding to σ_{31}' occurs, the effective stress path goes from *a* to *b*.

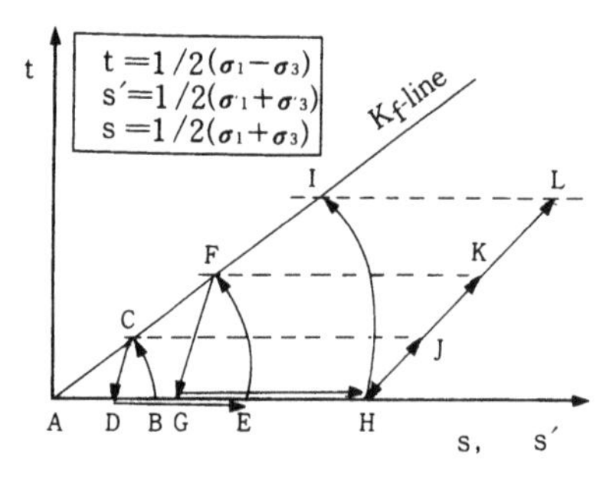

Fig. 4. Stress paths of the modified multi-stage compression test for consolidated undrained condition.

Table 1. Physical properties of soils tested.

Soils	Natural water content (%)	Liquid limit (%)	Plasticity index (%)	Unified soil classification
Undisturbed marine clay	60.1	57.8	36.0	CH
Compacted silty clay	24.8	36.8	13.8	CL
Compacted silty sand	21.0		NP	SM

By the application of deviator stress in the first stage, the total stress path moves up from *h* to *j* while the effective stress path moves up from *b* to *c* and arrives at K_f line. Since the increased deviator stress have to be released for the application of the second effective cell pressure, the total stress path goes back from *j* to *h* along the first path, while the effective stress path moves down from *c* to *d*. This is plotted lower than *b* because pore pressure generated during the application of the deviator stress is not reduced immediately with the release of the deviator stress. In the second stage of consolidation, the total stress path remains constant at *h*, while the effective stress moves from *d* to *e*. During the second shearing process, the total stress path goes up from *h* to *k*, while the effective stress path moves up from *e* to *f*. Since the deviator stress increased is released after the shearing of the specimen, the total and effective stress paths come down from *j* to *h* and *f* to g, respectively. The stress paths of the third stage are drawn in the same manner as shown in Fig. 4.

3 TESTING APPARATUS, SPECIMEN, AND TESTING PROCEDURE

3.1 *Testing Apparatus*

The triaxial compression apparatus used herein was commercially available which was manufactured in ELE in England. It is equiped with strain control type capable of measuring axial strain from 0.0001 to 5.0mm/min. Pressure is applied by means of motorised oil-water system and pore pressure can be measured and automatically recorded using a transducer. Volume change is measured by a rolling diaphragm recording system.

3.2 *Specimen*

Three kinds of specimens were used, namely: undisturbed marine clay; compacted silty clay; and compacted silty sand. The undisturbed marine clay is typical of the Korean marine clay, and is found along the coast of the Korean Peninsula. Specimens for disturbed soils were prepared by compacting with 97% degree of compaction of standard compaction test. Physical properties of the soils tested are given in Table 1.

3.3 *Testing Procedure*

Three kinds of tests were performed for the above soils, namely: standard triaxial compression test (standard test); conventional multi-stage triaxial test (multi-stage test); and modified multi-stage test (modified test), as proposed in this paper. The standard test was performed with three different homogeneous specimens, while each one of the other two tests was done using a single specimen. For the standard and multi-stage tests, test specimens were consolidated for 24 hours at each consolidation stage and for the modified test, the specimen was consolidated by letting pore pressure dissipate. All tests were performed under consolidated undrained condition.

The specimens used were 35mm in diameter and 70mm in height. The specimens for both compacted clay and silty sand were subjected to 97% degree of compaction by using the Proctor method.

Back pressures were applied to saturate the specimens. It was assumed that they could be completely saturated if their pore pressure coefficient *B* had reached over 97%. On the basis of this assumption, back pressures applied were 2 to 2.5kgf/cm^2 for the undisturbed marine clay, and 2.5 to 3.5 kgf/cm^2 for the compacted clay and silty sand.

Much effort was required to carry out the modified test. During the consolidation process, pore pressure measurements were made more than two times by closing the drainage valve. Time lag in response of pore pressure was observed. This means that measured values of pore pressure are not immediately stabilised after closing the drainage valves so as to interrupt the pore pressure dissipation.

In order to investigate the response of pore pressure, a simple experiment was conducted. Fig. 5 shows the test results, in which 4.5kgf/cm^2 of cell pressure inclusive of 1.5kgf/cm^2 of back pressure was applied and then, by closing the drainage valve at different time, pore pressures were measured. The general trend of the pore pressure response is that a rapid increase occurs up to 4 min. of elapsed time for the marine clay, and 1 min. for the silty sand, and thereafter the response is more or less constant. After around 15 min. the pore pressure response keeps almost constant regardless the property of soils. From the results obtained, the pore pressure measurements have been taken at 15 min. after closing the drainage valve.

With the pore pressure measured as described above, the effective cell pressure can be calculated, and then the consolidation time for the first and the second stages corresponding to given cell pressures can be estimated according to Eq. (4). After the completion of consolidation up to a given effective cell pressure, a shear test was done under constant volume with the strain rate recommended by Head (1985). The strain rates used are as follows: 0.041 to 0.044% for the undisturbed marine clay; and 0.083% for the compacted silty clay and silty sand.

The important consideration in the the multi-stage triaxial test is the determination of the failure criteria. Kenney and Watson (1961) have discussed this in detail. The general criterion used herein is to shear until the effective stress path arrives at K_f line. For the undisturbed marine clay, once the stress path had reached the line, both the deviator stress and the pore pressure did not show any increase. On the other hand, for the compacted soils the effective stress path went up along K_f line even after reaching it, showing an increase of deviator stress and a slight decrease of pore pressure. Thus, the criterion applied herein is to shear until the value of M as defined below remains constant:

$$M = \frac{\Delta u}{\Delta(\sigma_1 - \sigma_3)} \tag{5}$$

On the basis of the above criterion, the strain to commence failure was 3 to 5% in the first and second stage for all soils tested. In the final stage of shear test more than 10% of strain were applied so that enough deformation occurs.

4. TEST RESULTS AND ANALYSIS

4.1 *Effective Pressure Increase with Pore Pressure Dissipation*

By measuring pore pressure during the consolidation process, the effective stress can be estimated according to effective stress principle. Shown in Fig. 6(a) is the relationship between the effective stress and square root of elapsed time obtained after applying a cell pressure of 2kgf/cm^2 for the undisturbed marine clay. It is noted that the all experimental points in the first and second consolidation stages lie on a straight line. The consolidation time corresponding to the first stage effective cell pressure of 0.5kgf/cm^2 is possible to be estimated by obtaining constants, a'' and b'', in Fig. 6(b) and using Eq. (4).

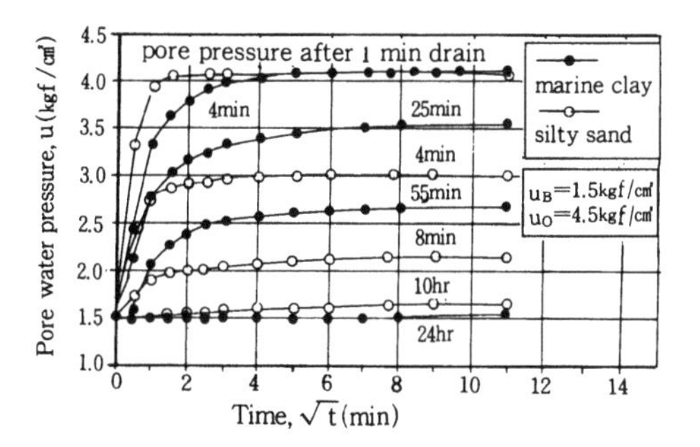

Fig. 5. Pore pressure response after different dissipation time of pore water pressure.

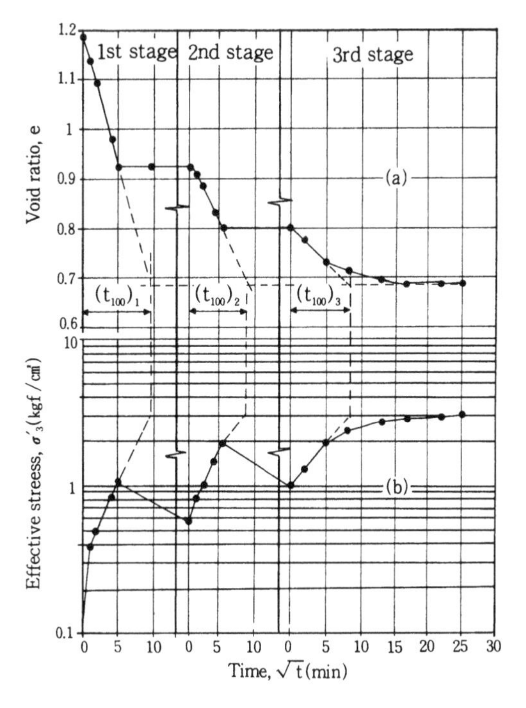

Fig. 6. (a) Relationship between void ratio and elapsed time and (b) relationship between effective cell pressure and elapsed time for the undisturbed marine clay.

The estimated consolidation time in the first and

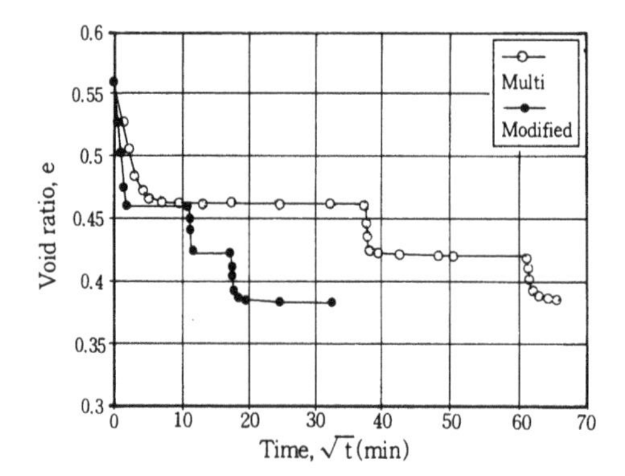

Fig. 7. Relationship between voil ratio and elapsed time for the compacted silty sand.

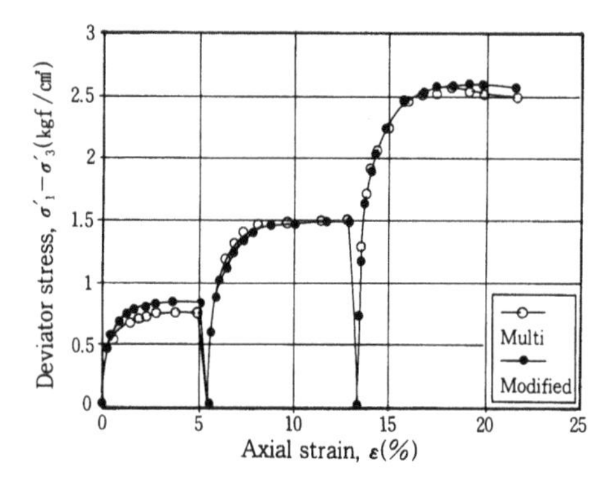

Fig. 8. Comparison of stress-strain curves for the undisturbed marine clay.

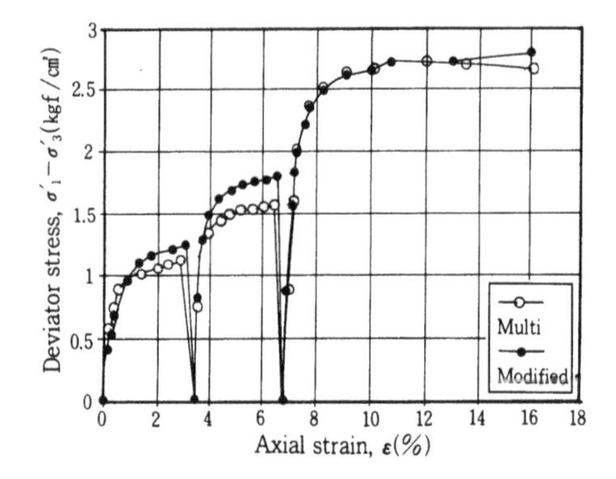

Fig. 9. Comparison of stress-strain curves for the compacted silty cly.

second stages for the undisturbed marine clay were 16 and 49 min., respectively. This indicates significant saving of consolidation time for the proposed test procedure, comparing with usually 24 hours for the conventional multi-stage test. Similar results were obtained for the compacted silty clay, in which the consolidation time in the first and second stages were 6 and 12 min., respectively.

As shown in Fig. 6(b), it is noted that the effective cell pressure at the beginning of the second consolidation stage was less than the effective cell pressure of the first stage. This is because the pore pressure had developed during shearing of the specimen. The decrease of effective cell pressure occurred as well at the third stage of consolidation.

Fig. 7 give a comparison of the consolidation characteristics obtained from the modified multi-stage test and the conventional multi-stage triaxial test for the compacted silty sand. It is apparent from this figure that significant reduction of consolidation time has been achieved for the modified test.

4.2 *Characteristics of Stress and Strain Relationship*

Fig. 8 shows a comparison between stress-strain curves performed by the modified test and the multi-stage test for the undisturbed marine clay. It can be seen from the figure that two curves agree very well. Shown in Fig. 9 are test results for the compacted silty clay, in which the curves of the first and second stages for the modified test are plotted slightly above those of the multi-stage test. This is because longer pore pressure dissipation in the former test caused less void ratio. In other words, σ'_{31} of 0.6kgf/cm^2 was applied for 0.5kgf/cm^2 and σ'_{32} of 1.1kgf/cm^2 for 1.0kgf/cm^2. It must be noted, however, that the determination of soil strength parameters is practically not affected by the application of the pressures which are slightly different from the specified values.

4.3 *Pore Pressure Dissipation and Development During Testing*

A comparison of pore pressures developed during the shearing of the specimens for the undisturbed marine clay is given in Fig. 10. In the modified test, the pore pressure is dissipated until a required cell pressure is obtained, and then the pressure is developed when sheared by axial loading. Because of different test procedures, different curves are drawn, but it is observed that the shapes of the two curves are similar for the first two stages while the curves of the final stage are practically identical. The same trend was shown for both the compacted silty clay and silty sand.

4.4 *Pore Pressure Coefficient A*

According to Skempton (1957), an increase of pore pressure is written as follows:

$$\Delta u = B[\Delta\sigma_3 + A(\Delta\sigma_1 - \Delta\sigma_3)] \quad (6)$$

where A, B = pore pressure parameters.

If a triaxial compression test is carried out under a constant cell pressure with completely saturated specimens, there is no change of the cell pressure, and thereby the pore pressure parameter A becomes:

$$A = \frac{\Delta u}{\Delta \sigma_1} \tag{7}$$

Figs. 11, 12, and 13 show a comparison of pore pressure parameter A obtained from three different test methods for the three soils. If test specimens are isotropic and homogeneous, the change of A with an increase of strain would show similar curves for every cell pressure. This has been proved to be valid, as shown in Fig. 11, in the standard triaxial compression test of the second and third stages for the undisturbed marine clay. However, its lower values obtained from the first stage are likely to be due to slightly overconsolidated behaviour of the specimen because the sample has been taken at shallow depth.

The values of parameter A obtained from other two tests were generally lower than those of the former. In the modified and multi-stage tests, the specimens which were subjected to axial loading would have experienced anisotropic consolidation. This may cause the specimens to be stiffer, which give lower values of parameters. Inspite of the different values for diferent consolidation stages, it is shown in Fig. 11 that the pore pressure parameters obtained from both the multi-stage and the modified tests are in good agreement.

For the compacted silty clay, the values of A are different for different cell pressures and different testing procedures, as shown in Figs. 12 and 13. This observation is to be expected since the compacted clay behaves like overconsolidated clay. The overconsolidation behaviour is apparent from the effective stress path and will be explained below.

4.5 *Effective Stress Path*

Shown in Fig. 14 are effective stress paths for the undisturbed marine clay when tested by three different test methods. The stress paths for the standard triaxial test are typical of normally consolidated clay, while the stress paths of final stage obtained by both the modified and multi-stage tests shows slightly overconsolidated behaviour. This may be due to anisotropic consolidation occurring during shearing in the previous stages. Although the paths go along slightly different ways, it is observed that all paths reach a common K_f line.

Fig. 15 shows the effective stress paths for the compacted silty clay. As expected, the stress paths for the first stages follow a pattern of overconsolidated clays. This is because the residual

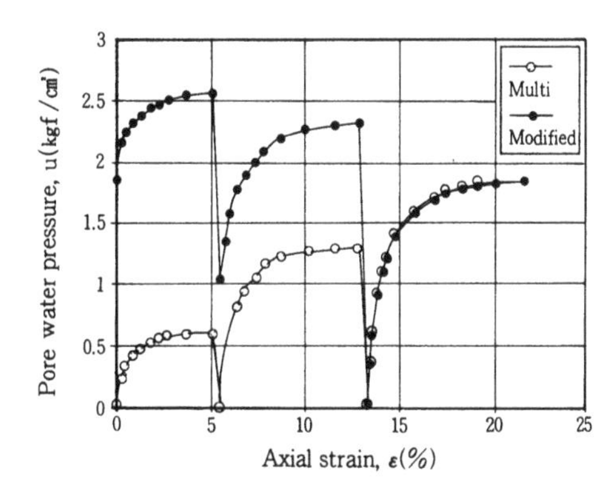

Fig. 10. Comparison of pore pressure development during shearing between the modified and multi-stage tests for the undisturbed clay.

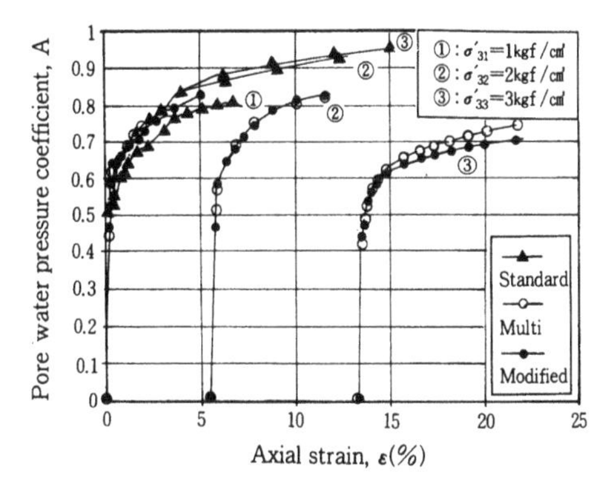

Fig. 11. Pore pressure parameter A for the undisturbed marine clay.

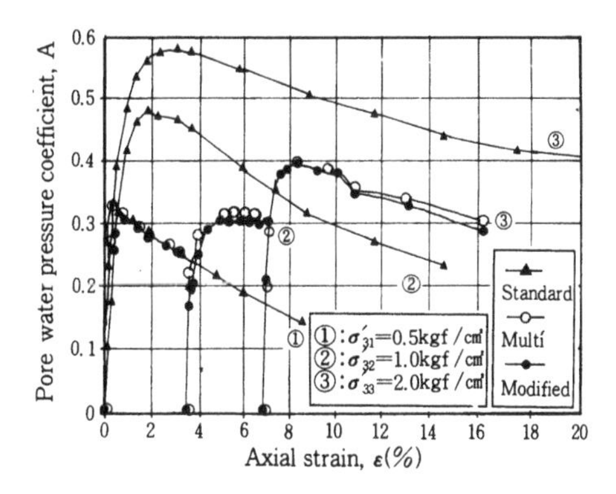

Fig. 12. Pore pressure parameter A for the compacted silty clay.

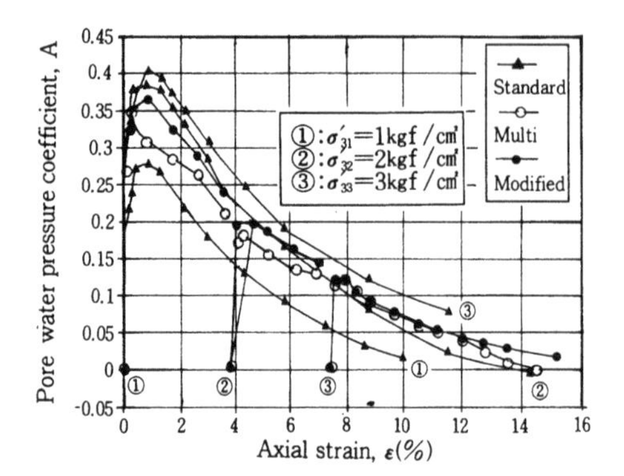

Fig. 13. Pore pressure parameter A for the compacted silty sand.

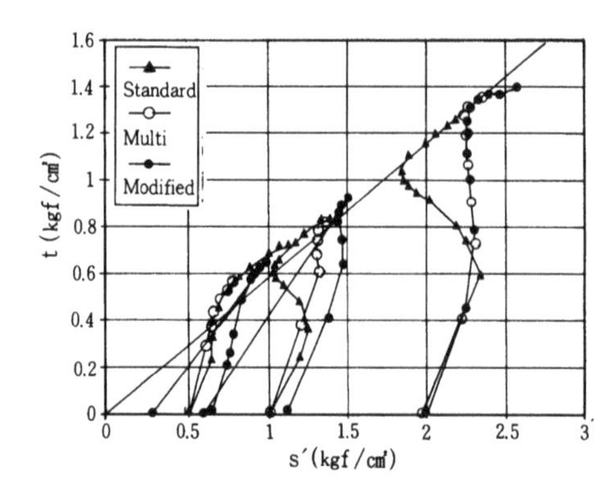

Fig. 15. Effective stress paths for the compacted silty clay.

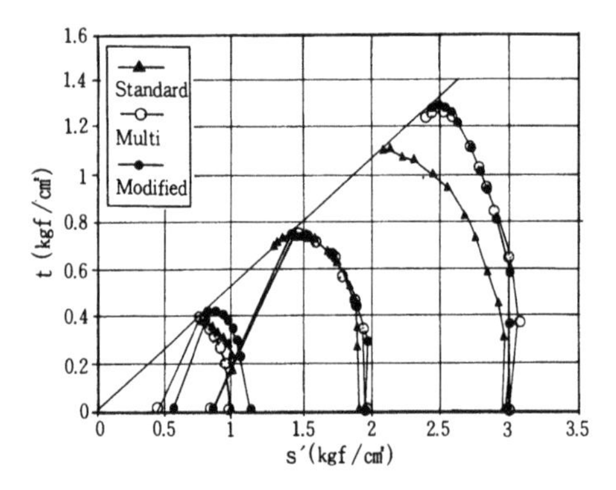

Fig. 14. Effective stress paths for the undisturbed marine clay.

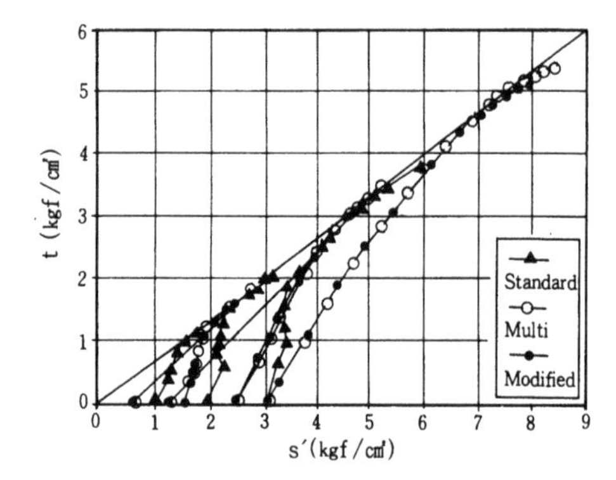

Fig. 16. Effective stress paths for the compacted silty sand.

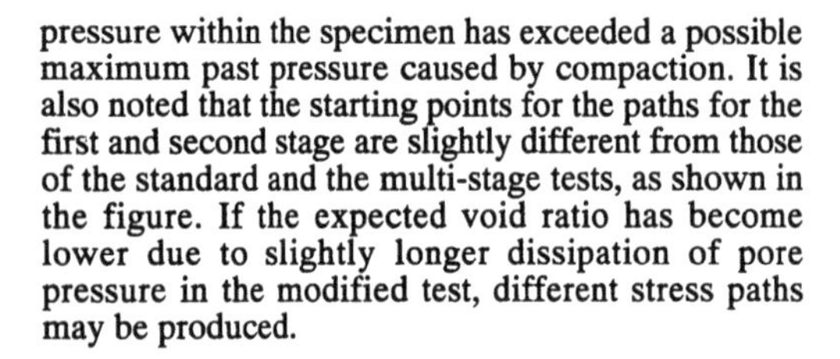
pressure within the specimen has exceeded a possible maximum past pressure caused by compaction. It is also noted that the starting points for the paths for the first and second stage are slightly different from those of the standard and the multi-stage tests, as shown in the figure. If the expected void ratio has become lower due to slightly longer dissipation of pore pressure in the modified test, different stress paths may be produced.

Fig. 16 gives the effective stress paths for the compacted silty sand which shows similar trend of the compacted silty clay but showing much more overconsolidated behaviour.

As shown from Figs. 14 to 16, all effective stress paths lie on a common K_f line regardless of the test method used. This means that there is no difference of the method in determining peak shear strength parameters. The test results show that the differences of cohesion in terms of both total stress and effective stress were only 0.02kgf/cm^2 and that of friction angle only 1.2 degree. This difference of the order of 1 degree is negligible for engineering purposes. Thus, the strength parameters determined from the modified test are useful in practice and may bring a great benefit because of significant reduction of time in performing the test.

CONCLUSION

The modified multi-stage test proposed herein is to carry out a series of triaxial compression test with a single specimen in which a high cell pressure is applied, and each stage of consolidation is accomplished by means of pore pressure dissipation.

A major advantage of this test is the significant reduction of consolidation time of a specimen. The dissipation time of pore pressure for obtaining a given effective cell pressure can be predicted by measuring pore pressure during consolidation process.

To examine the applicability of this test, the standard triaxial compression test, the conventional multi-test, and the modified multi-stage test have been carried out under consolidated undrained condition for different soils. Comparisons for the test results were made for pore pressure development, deviator stress, pore pressure parameter A, stress paths, and strength parameters.

It was shown that the pore pressure parameters A obtained from the conventional multi-stage test and the modified test show slightly lower values in the second and third stages compared with those of obtained from the standard test. This is because the specimens in multi-stage tests have been anisotropically consolidated during the shearing process of the previous stages, which may be suitable to in-situ stress condition. The effective stress paths of the former two methods show the behaviour of overconsolidated soils because of the same reason described above.

A comparison of strength parameters obtained from the standard test with those of the modified test shows a good agreement with the difference in the order of 1 degree in the angle of shearing resistance, in terms of both the total stress and the effective stress. It is concluded that this proposed method may be used with sufficient degree of accuracy in evaluating shear strength parameters for engineering purposes.

REFERENCES

Anderson, W. F. 1974. The use of multi-stage triaxial tests to find the undrained strength parameters of stony boulder clay, *Proc. Inst. Civ. Eng.*, Technical Note No. TN 89.

Kawabara, H., Yamaguchi, S., Iwamoto, T. 1986. The study on application of multiple-stage triaxial compression test for coarse grain (Japanese), *Strength Characteristics on Coarse Materials and Their Testing Methods*, The Japanese Society of Soils and Foundations: 23-26.

Head, K. H. 1985. *Manual of Soil Laboratory Testing, Vol. 3 : Effective Stress Tests,* New York: John Wiley and Sons.

Ho, D. Y. F. and Fredlund, D. G. 1982. A multi-stage triaxial test for unsaturated soils, *Geotechnical Testing Journal*, Vol. 5, No. 12: 309-364.

Kenney, T. C. and Watson, G. H. 1961. Multiple-stage triaxial test for determining c' and ϕ' of saturated soils, *Proc., 5th Int. Conf. SMFE*, Paris, Vol. 1: 191-195.

Lumb, P. 1964. Multi-stage triaxial tests on undisturbed soils, *Civil Engineering. and Public Works Review,* May, 1964: 592-595.

Skempton, A. W. 1954. The pore pressure parameters *A* and *B*, *Geotechnique*, 4: 143.

Developments in Geotechnical Engineering, Balasubramaniam et al. (eds) © 1994 Balkema, Rotterdam, ISBN 90 5410 522 4

The correlation between some Dutch-cone penetrometer results and soil properties in some areas of Indonesia

D.S. Soelarno
Department of Geotechnical and Civil Engineering, Parahyangan Catholic University, Bandung, Indonesia

M. Irwan
Department of Civil Engineering, ITB, Bandung, Indonesia

ABSTRACT: The main purpose of this study is to examine the applicability of the correlation between the Dutch-cone penetrometer results with soil properties, such as types, consistency, plasticity index, and D50, which were obtained by Soelarno - Ferrynaga, 1990 for Jakarta area, and using soil data from outside Jakarta, South Sumatera, Lampung, Riau, Central Java, East Java and from other places in Aceh, Medan, Bengkulu, Jambi, Jakarta, Bandung, West and East Kalimantan, and Sulawesi.

Results of the study indicate that the correlation of Dutch-cone penetrometer to soil properties obtained by Soelarno and Ferrynaga, 1990 can be concluded as applicable for areas outside Jakarta, and that, there was also no influence of groundwater in the correlation mentioned above.

For the next changes, the author will examine the results of this study using data from the east part of Indonesia.

INTRODUCTION

From the past and until now, Begemann type of Dutch-cone penetrometer apparatus has been familiarly used by many engineers in Indonesia, because it is very easy to use and fast for getting cone-bearing, friction ratio, raft groundwater table and also hard layer (qc larger than 150 kg/cm2).

However for direct use of the Dutch-cone penetrometer results, for example in case of foundation design, slope stability, etc., we need to correlate it with soil types (Begemann, Schmertmann, 1978, Searle, 1979, Olsen, 1981, Robertson and Campanella, 1983, Burland and Burbidge, 1985, Soelarno and Ferrynaga, 1990), consistency (Schmertmann, 1978, Searle, 1979, Soelarno and Ferrynaga, 1990), and plasticity index (Soelarno and Ferrynaga, 1990).

The main purpose of this paper is to study mainly the applicability of the correlation between Dutch-cone penetrometer results with soil properties which have been done by Soelarno and Ferrynaga, 1990 for only in Jakarta area, is also applicable for the other areas as shown in Table 1 and 2. Especially, using the data from Table 2, we discuss also the influence of water table and at last we get additional correlation between cone bearing value and unconfined compressive strength.

For the next changes, the authors still anticipate to discuss the results of this study by using soil data from the east part of the Indonesia area.

Soil data

Soelarno and Ferrynaga, 1990, made a study according to the research results of Robertson and Campanella, 1983, especially in Jakarta area. In this study, the authors used Soil Data as shown in Table 1. This table shows the location of the projects, amount of soil test data, capacity of Dutch cone apparatus and bore. The soil properties related to our study can be obtained in Table 3 for South Sumatera, Table 4 for Lampung, Table 5 for Riau, Table 6 for Central Java, and Table 7 for East Java. In these tables, we show soil parameters related to this study as follows; natural water content, liquid limit, plastic limit, plasticity index, unit weight, cone bearing value, friction ratio, and unconfined compressive strength. Soil data shown in Table 2, are used to evaluate, especially, the influence of ground water table and also can be used for evaluating the results obtained by Soelarno and Ferrynaga, 1990, if applicable or not for this additional data.

Purpose and Results of Study

The purpose of this study is to examine the applicability, the correlation of Dutch cone penetrometer results and soil properties obtained by Soelarno and Ferrynaga, 1990. And so the results of study can be classified as follows : the correlation of Dutch Penetrometer results and soil types, consistency, plasticity index, D50 and unconfined compressive strength.

1. The correlation between Dutch cone penetrometer results and soil types are shown in Fig. 15.1. Fig. 15.2, Fig. 15.3, Fig. 15.4, Fig. 15.5, respectively, for South Sumatera, Lampung, Riau, Central Java, and East Java. The general plot of these correlations can be obtained in Fig. 15.6.

2. The correlation between Dutch cone penetrometer results and plasticity index. The correlation are shown in Fig. 16.1, Fig. 16.2, Fig. 16.3, Fig. 16.4 and Fig. 15.5 respectively for South Sumatera, Lampung, Riau, Central Java, and East Java. The general plot of the results can be obtained in Fig. 16.6.

3. The correlation between Dutch cone penetrometer results and consistency. The correlations are shown in Fig. 17.1, Fig. 17.2, Fig. 17.3, Fig. 17.4, Fig. 17.5, respectively, for South Sumatera, Lampung, Riau, Central Java and East Java. The general plot of the results can be obtained in Fig. 17.6.

4. The correlation between qc/N and D50. The correlation can be obtained in Fig. 18.1, Fig. 18.2, Fig. 18.3, Fig. 18.4 and Fig. 18.5 for South Sumatera, Lampung, Riau, Central Java and East Java respectively. The general plot of the results is shown in Fig. 18.6.

5. The influence of Ground Water Table in the correlation obtained in this study. The influence of Ground Water Table in the correlation in between Dutch Cone Penetrometer results and soil types, plasticity index, consistency, qc/N, D50 are shown respectively in Fig. 19.1, Fig. 19.2, Fig. 19.3 and Fig. 19.4. From these data, it seems that there is no influence of Ground Water Table for the correlations mention above.

CONCLUSIONS

1. The correlation between Dutch cone penetrometer results and soil type obtained by Soelarno and Ferrynaga 1990, Robertson and Campanella 1983, and of this study are plotted in Fig.20 and Fig. 22. The results of this study can be said almost the same with Soelarno and Ferrynaga 1990, and have different results from Robertson and Campanella, 1983.

2. The correlation between Dutch cone penetrometer results and plasticity index in this study is almost the same for PI smaller than 12% and for PI larger than 12%. The value is smaller than the data obtained by Soelarno and Ferrynaga, 1990. This fact is shown in Fig. 23.

3. The correlation between Dutch cone penetrometer results and consistency of this study and Soelarno and Ferrynaga, 1990 is shown in Fig. 24. There are differences in the limit zone for medium to stiff and very soft to soft.

4. The correlation of qc/N and D50 of this study, Robertson and Campanella, 1983, Soelarno and Ferrynaga, 1990 is plotted in Fig. 25. For D50 larger than 0.03 mm the results of this study is near the results of Robertson and Campanella, 1983, for D50 smaller than 0.03 mm, the results of this study is in between the data of Robertson and Campanella, 1983, and Soelarno and Ferrynaga, 1990.

5. The authors anticipate that these results of study can be applied roughly in the Indonesian area although it still need to check the applicability of the results by soil data from the east part of Indonesian area.

Acknowledgement

The authors deliver their thankfulness and deep appreciation to all soil test companies, Geotechnical Division of Hydraulic Research Institute as well as the Road Research Institute in Bandung which support many valuable soil data and also to Civil Engineering Department of ITB, Faculty of Civil Engineering, Parahyangan Catholic University that gave many facilities for completing this study.

REFERENCES

1. AASHTO (1982), Standard Specification for Transportation Materials and Methods of Sampling and Testing, Part II, Washington.

2. Baligh, M.M., et.al., April, 1980, Cone Penetration in Soil Profiling, Journal of the Geotechnical Engineering Division, ASCE, Vol.106, No. GT4, p.447.

3. Bowless, J.E. (1984), Foundations Analysis and Design, McGraw Hill International Book Company, Singapore.

4. Bowless, J.E. (1984), Physical and Geotechnical Properties for Soil, McGraw Hill Book Company.

5. Cernica, J.N. (1982), Geotechnical Engineering, CBS College Publishing, New York.

6. Das, Braja, M. (1989), Advanced Soil Mechanics, McGraw Hill Book Co., Inc., New York, 1989.

7. Das, Braja, M. (1983), Fundamental of Soil Dynamics, Elsevier Science Publishing Co., Inc., New York.

8. De Ruiter, J. (1988), Penetration Testing 1988, A.A. Balkema, Rotterdam, 1988.

9. De Ruiter, J. (Feb., 1971), Electric Penetrometer for Site Investigation, Journal of the Soil Mechanics and Foundation Division, ASCE, Vol.97, No. SM2, Proc.paper 1907, p.457.

10. Harr, M.E., (1966). Foundation of Theoritical Soil Mechanics, McGraw Hill Book Company, Tokyo.

11. Hunt, R.E. (1984), Geotechnical Engineering Investigation Manual, McGraw Hill Book Company Taiwan.

12. Ingles, O.G., Metcalf, J.B. (1972), Soil Stabilization : Principles and Practice, Butterworth & Co., London.

13. Lee, I.K., et.al. (1983), Geotechnical Engineering, Pitman Publishing Pty Ltd., Melbourne.

14. Lambe, T.W., Whitman, R.V. (1969), Soil Mechanics, John Wiley & Sons, Inc., London.

15. Meigh, A.C., Cone Penetration Testing methods and interpretation, CIRIA, Ground Engineering Report, 1987.

16. Mitchell, J.K. (1976), Fundamentals of Soil Behavior, John Wiley & Sons, Inc., New York.

17. Proceeding Konperensi Geoteknik Indonesia IV, Bandung 26-27 Februari 1990.

18. Robertson, P.K., Campanella, R.G. (1984), Guidlines for Use and Interpretation of the Electronic Cone Penetration Test, Hogentogler & Co., Gaithersburg.

19. Sanglerat, G. (1972), The Penetration and Soil Exploration, Elsevier Publishing Company, Amsterdam.

20. Schmertmann, J.H. (May, 1970), Static Cone to Compute Static Settlement Over Sand, Journal of the Soil Mechanics and Foundation Division, ASCE, Vol.96, No. SM3, Proc. paper 7302, p.1011.

21. Schmertmann, J.H. (1977), Guidlines for Cone Penetration Test, U.S. Department of Transportation.

22. Smith, G.N. (1974), Element of Soil Mechanics for Civil and Mining Engineers, Crosby Lockwood Staples, London.

23. Soelarno, D.S. (1988), Pengenalan Alat-Alat Geoteknik dan Aplikasinya, Seminar in Departemen Pekerjaan Umum, Jakarta.

24. Sudjana (1992), Teknik Analisis Regresi dan Korelasi, Tarsito, Bandung.

25. Suharsini Arikunto (1990), Manajemen Penelitian, Rineka Cipta, Jakarta.

26. Verhoef, P.N.W., Drs., Geologi Untuk Teknik Sipil, Penerbit Erlangga, Jakarta 1989.

27. Wesley L.D., Mekanika Tanah, Badan Penerbit Pekerjaan Umum, Jakarta , 1977.

28. Yong, R.N. (1975), Soil Properties and Behaviour, Elsevier Scientific Publishing Company, Amsterdam.

Table 1. Location of Projects and Soil Test Data

No.	Location	Soil Data	Dutch cone R	Dutch cone B	Bore D	Bore DL
1.	South Sumatera	81	226	-	158	25
2.	Lampung	12	36	-	21	4
3.	Riau	8	10	24	22	5
4.	Central Java	10	70	16	22	30
5.	East Java	40	98	68	81	50

Note : R = Dutch-cone 2.5 ton
B = Dutch-cone 10 ton
D = Hand Boring
DL= Machine Boring

Table 2. Additional Location of Projects and Soil Test Data for Evaluating the Influences of Water Table

No.	Location	Projects
1.	Aceh	-Access Road & Bridge District of Singkil
2.	Medan	-Jl. Guru Patimpus Gedung OPMC
3.	Bengkulu	-Jb. Air Sano
4.	Jambi	-PLTD Muara Bungo
5.	Jakarta	-PRJ area, Kemayoran
6.	Bandung	-Guest House BTN
7.	West Kalimantan	-PLTG Siantan
8.	East Kalimantan	-PT Pupuk Kaltim Jb. S. Petai Balikpapan Hospital Samarinda Port Balikpapan Port
9.	North Sulawesi	-PLTG Toli-Toli
10.	Southeast Sulawesi	-Jb. Sabilambo
11.	Central Sulawesi	-Jb. Salumbia Jb. Marantale

Table 3. Ranges of Soil Data, South Sumatera

Parameters	Ranges	Unit
Water content (Wn)	8.70 - 112.00	%
Liquid limit (LL)	38.00 - 151.00	%
Plasticity index (PI)	12.00 - 105.00	%
Unit weight (TW)	1.40 - 1.80	t/m3
Cone bearing (qc)	1.00 - 150.00	kg/cm2
Friction ratio (FR)	0.30 - 6.00	%
Unconfined strength (qu)	0.01 - 3.58	kg/cm2

Table 4. Ranges of Soil Data, Lampung

Parameters	Ranges	Unit
Water content (Wn)	21.29 - 128.00	%
Liquid limit (LL)	34.50 - 142.30	%
Plasticity index (PI)	12.80 - 99.70	%
Unit weight (TW)	1.55 - 1.78	t/m3
Cone bearing (qc)	1.00 - 150.00	kg/cm2
Friction ratio (FR)	0.80 - 5.80	%
Unconfined strength (qu)	0.12 - 3.60	kg/cm2

Table 5. Ranges of Soil Data, Riau

Parameters	Ranges	Unit
Water content (Wn)	5.00 - 105.00	%
Liquid limit (LL)	44.00 - 114.00	%
Plasticity index (PI)	13.00 - 66.43	%
Unit weight (TW)	1.35 - 1.92	t/m3
Cone bearing (qc)	2.00 - 135.00	kg/cm2
Friction ratio (FR)	0.80 - 5.90	%
Unconfined strength (qu)	0.13 - 2.34	kg/cm2

Table 6. Ranges of Soil Data, Central Java

Parameters	Ranges	Unit
Water content (Wn)	15.68 - 168.80	%
Liquid limit (LL)	31.25 - 203.00	%
Plasticity index (PI)	6.59 - 161.60	%
Unit weight (TW)	1.43 - 1.87	t/m3
Cone bearing (qc)	2.00 - 125.00	kg/cm2
Friction ratio (FR)	0.60 - 6.80	%
Unconfined strength (qu)	0.31 - 2.86	kg/cm2

Table 7. Ranges of Soil Data, East Java

Parameters	Ranges	Unit
Water content (Wn)	15.02 - 113.70	%
Liquid limit (LL)	29.00 - 140.00	%
Plasticity index (PI)	12.70 - 92.20	%
Unit weight (TW)	1.44 - 1.96	t/m3
Cone bearing (qc)	2.00 - 120.00	kg/cm2
Friction ratio (FR)	0.89 - 5.85	%
Unconfined strength (qu)	0.08 - 3.13	kg/cm2

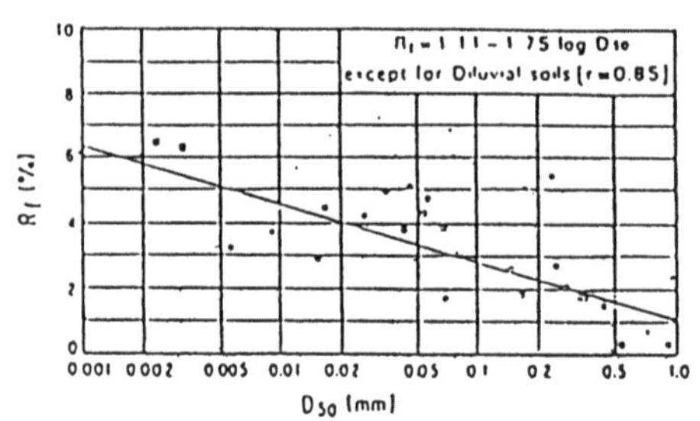

N.B. :

o Reclaimed soils

. Alluvial soils

. Diluvial soils

Fig. 1. Correlation of FR and D50 (Muromachi, 1981)

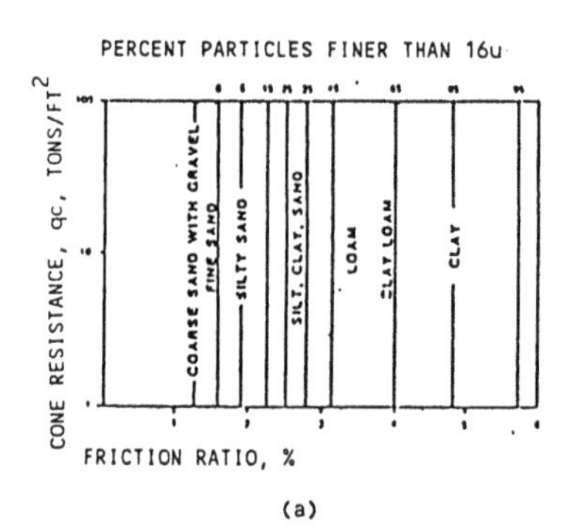

(a)

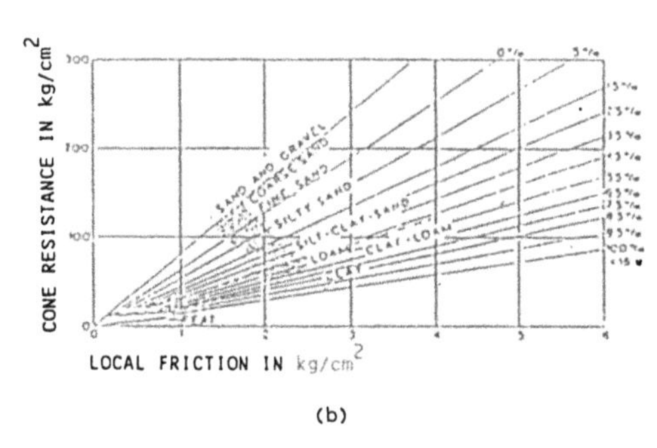

(b)

Fig. 2. Soil Classification Diagram from Dutch Cone Penetrometer Results (Begemann)

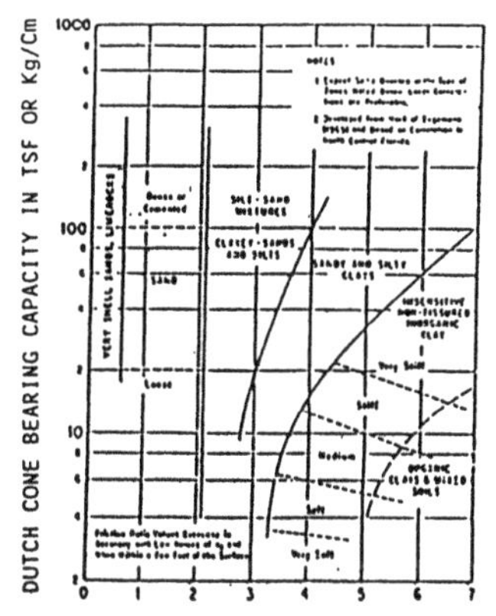

Fig. 3. Soil Classification Diagram from Dutch Cone Penetrometer Results (Schmertmann, 1978)

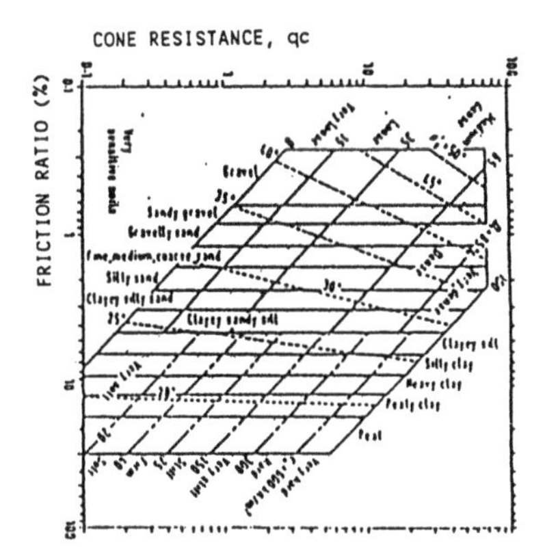

Fig. 4. Soil Classification Diagram from Dutch Cone Penetrometer Results (Searle, 1979)

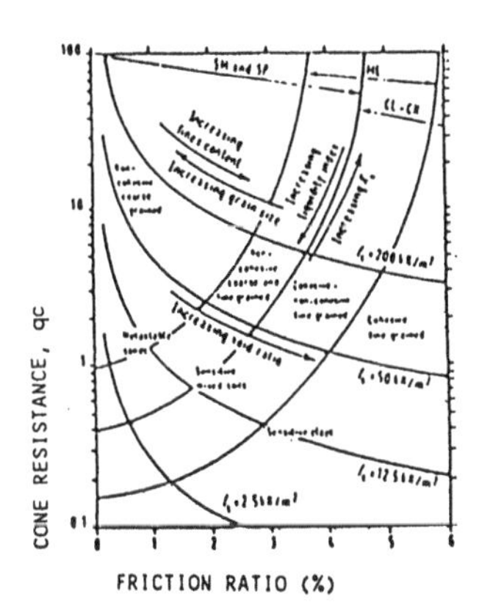

Fig. 5. Soil Classification Diagram from Dutch Cone Penetrometer Results (Olsen, 1981)

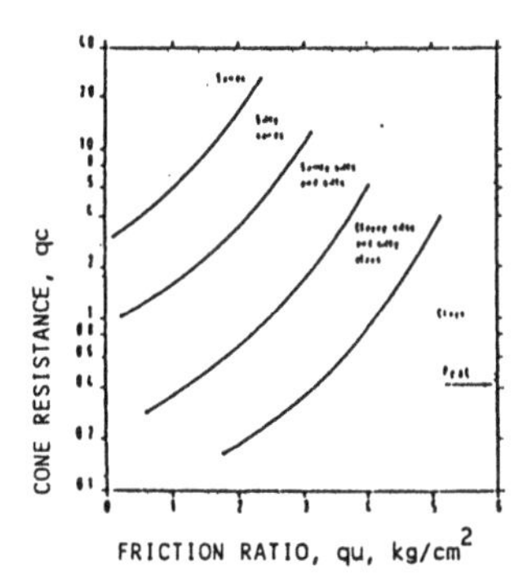

Fig. 6. Soil Classification Diagram from Dutch Cone Penetrometer Results (Robertson & Campanella)

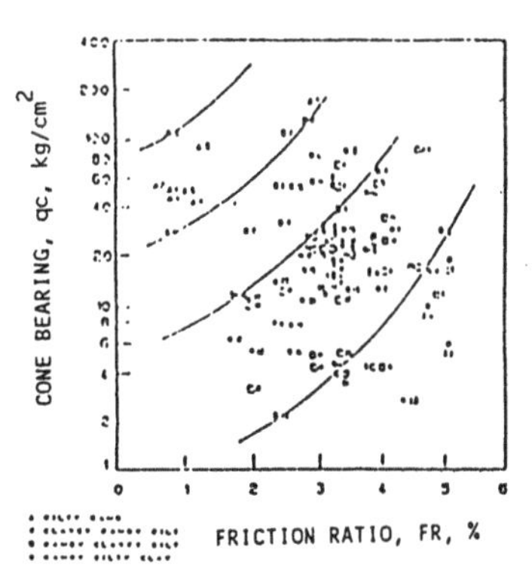

Fig. 7. Soil Classification Diagram from Dutch Cone Penetrometer Results in Jakarta area (Soelarno and Ferrynaga, 1990)

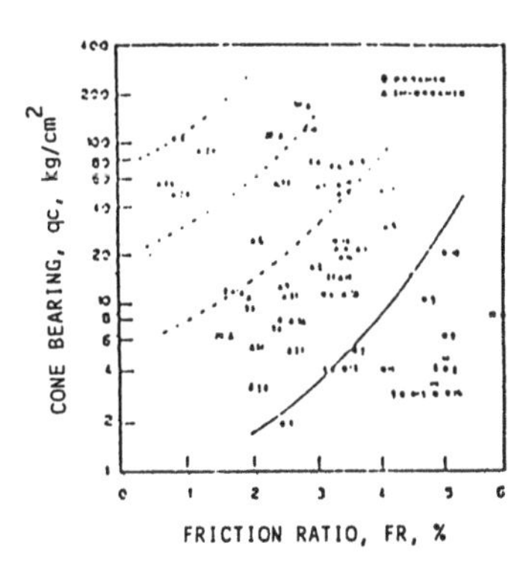

Fig. 8. Correlation of qc and FR for organic and inorganic soils (Soelarno and Ferrynaga, 1990)

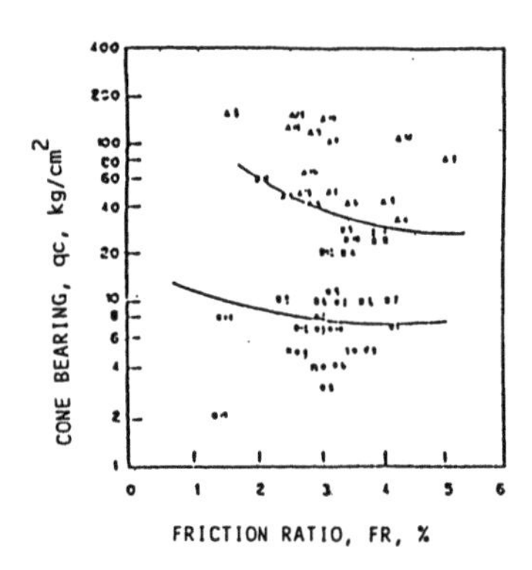

Fig. 9. Correlation of qc, FR and consistency of soils in Jakarta (Soelarno and Ferrynaga, 1990)

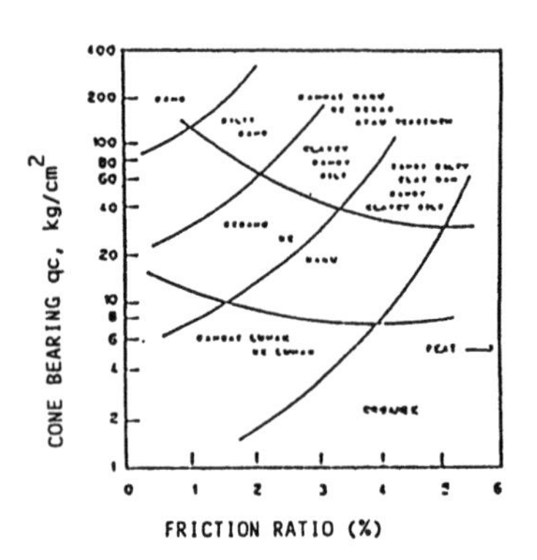

Fig. 10. Correlation of qc, FR, soil type and consistency in Jakarta area (Soelarno and Ferrynaga, 1990)

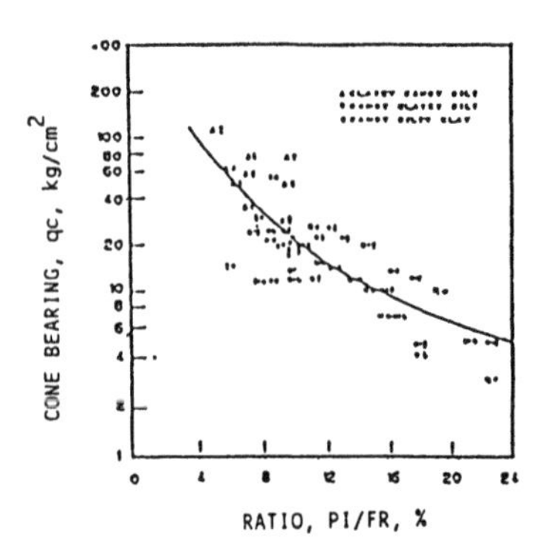

Fig. 11. Correlation of qc, FR, and Plasticity Index in Jakarta (Soelarno and Ferrynaga, 1990)

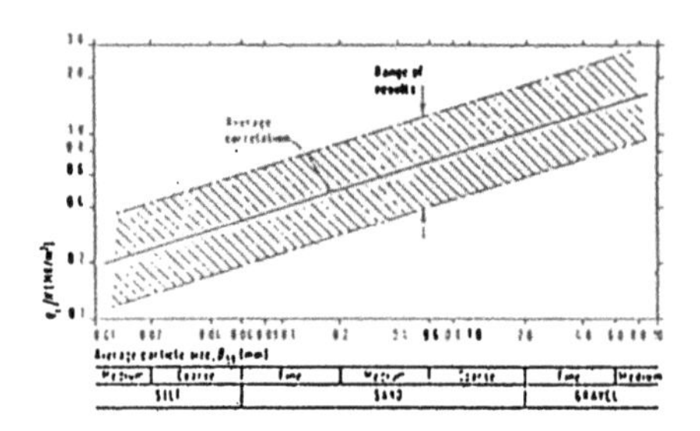

Fig. 12. Correlation of qc/N and D50 (Burland and Burbidge, 1985)

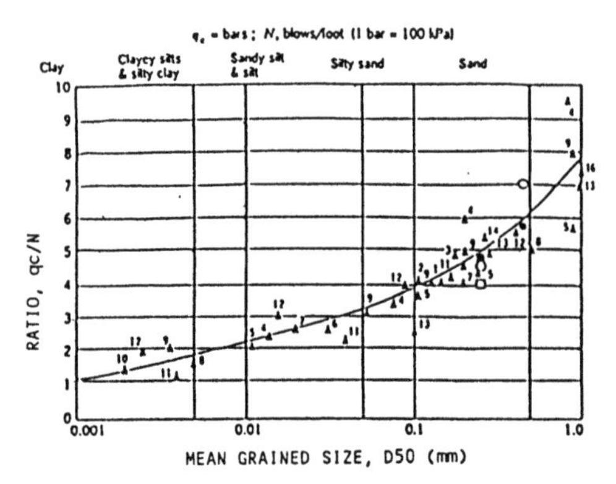

Fig. 13. Correlation of qc/N and D50 (Robertson and Campanella, 1983)

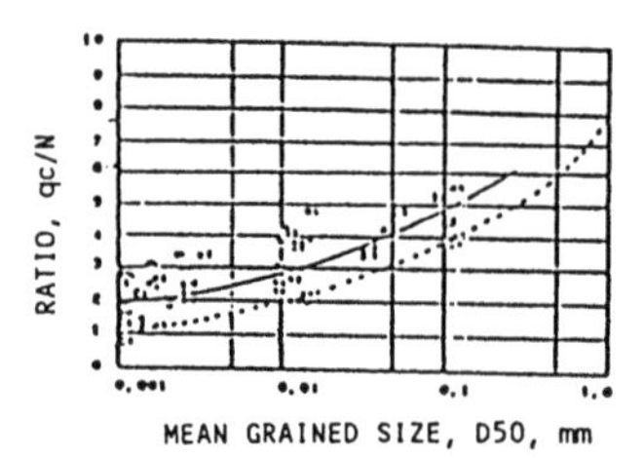

Fig. 14. Correlation of qc/N and D50 in Jakarta area (Soelarno and Ferrynaga, 1990)

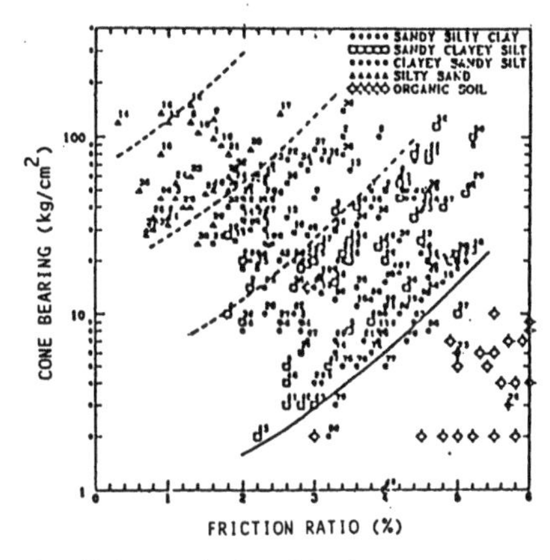

Fig. 15.1. Correlation of Dutch cone penetrometer results and soil types in South Sumatera area

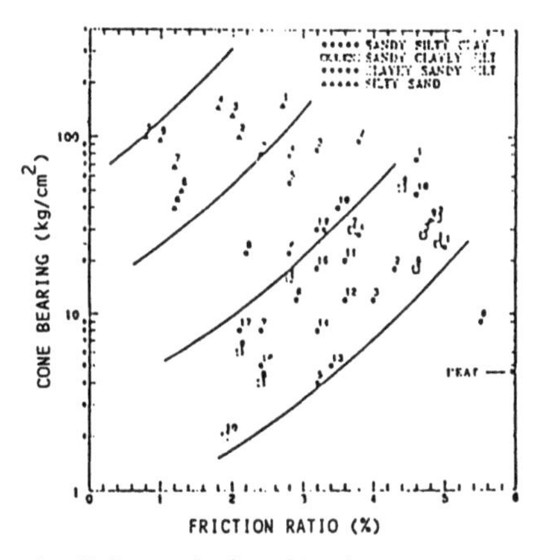

Fig. 15.2. Correlation of Dutch cone penetrometer results and soil types in Lampung area

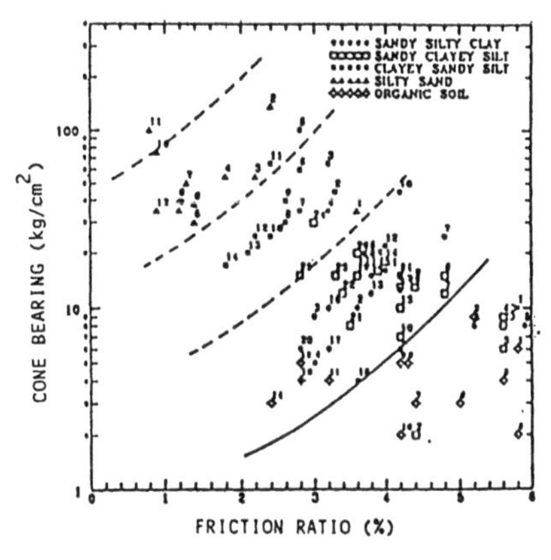

Fig. 15.3. Correlation of Dutch cone penetrometer results and soil types in Riau area

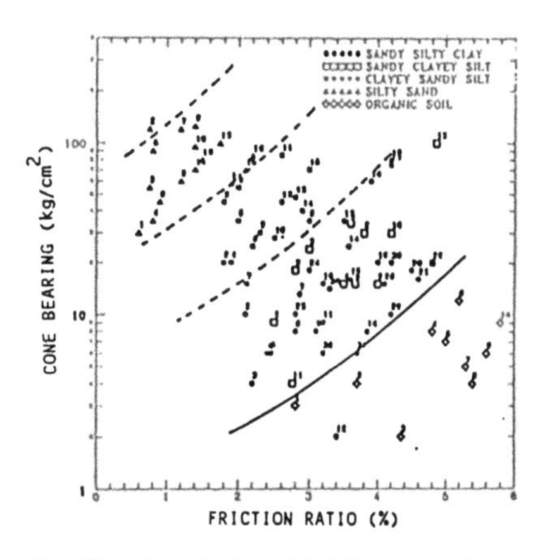

Fig. 15.4. Correlation of Dutch cone penetrometer results and soil types in Central Java area

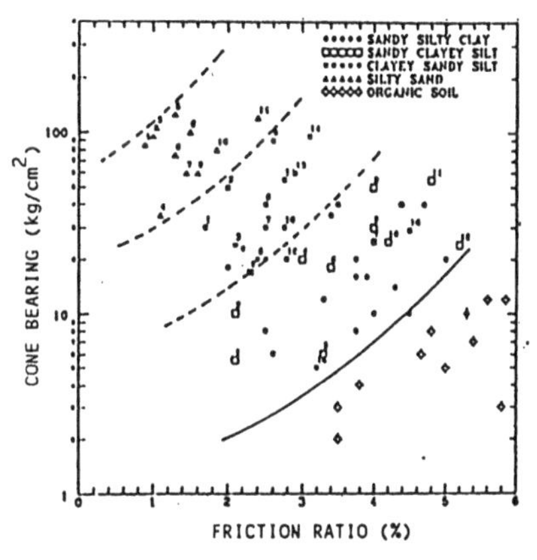

Fig. 15.5. Correlation of Dutch cone penetrometer results and soil types in Java area

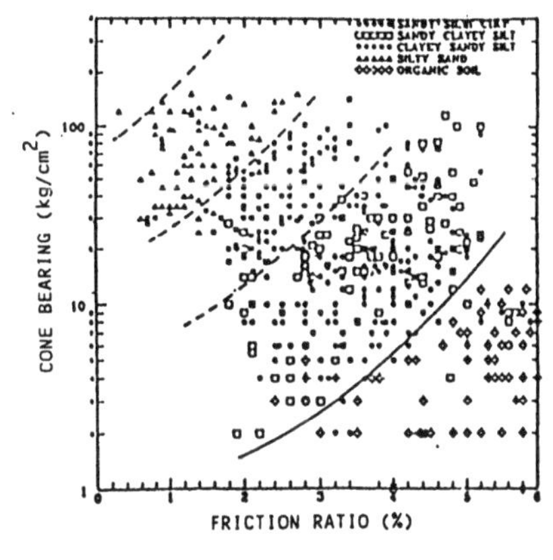

Fig. 15.6. Correlation of Dutch cone penetrometer results and soil types in general

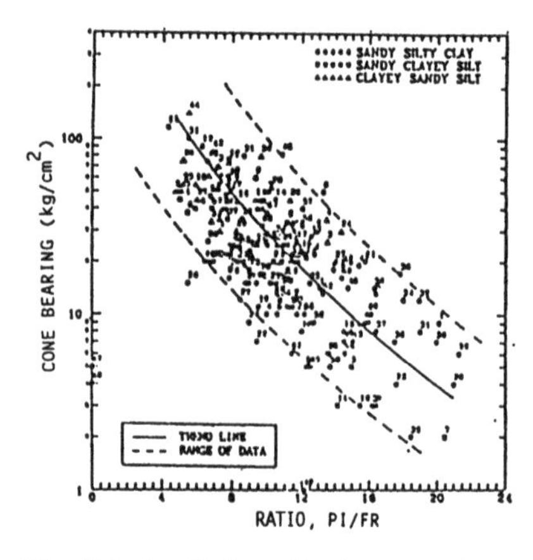

Fig. 16.1. Correlation of Dutch cone penetrometer results and plasticity index in South Sumatera area

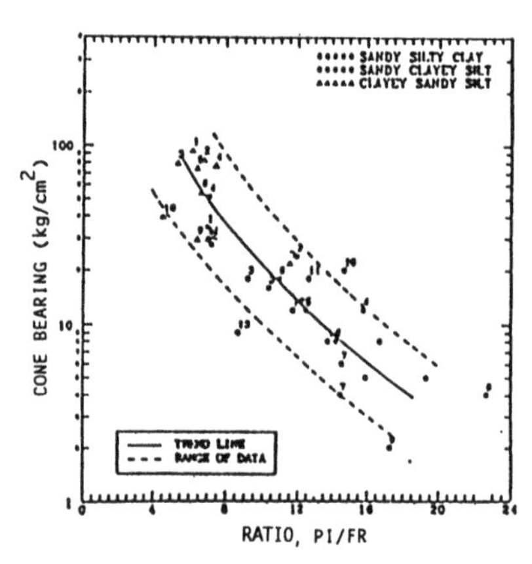

Fig. 16.2. Correlation of Dutch cone penetrometer results and plasticity index in Lampung area

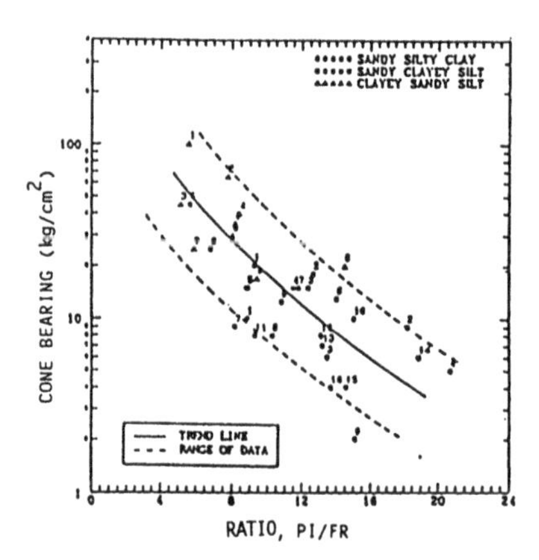

Fig. 16.3. Correlation of Dutch cone penetrometer results and plasticity index in Riau area

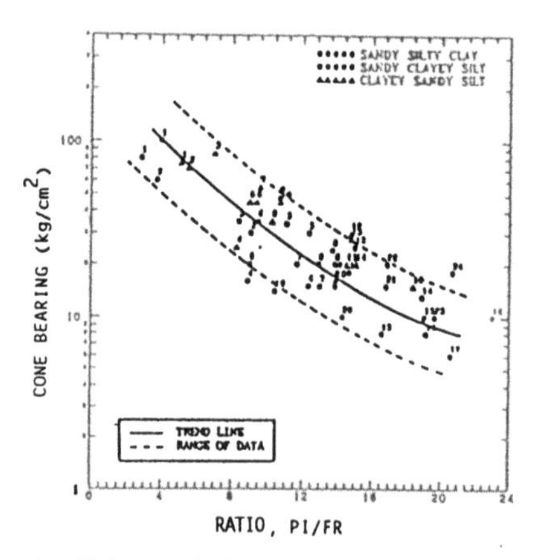

Fig. 16.4. Correlation of Dutch cone penetrometer results and plasticity index in Central Java area

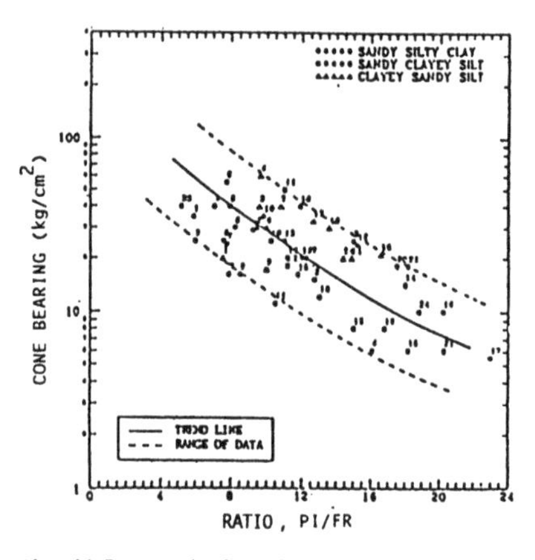

Fig. 16.5. Correlation of Dutch cone penetrometer results and plasticity index in East-Java area

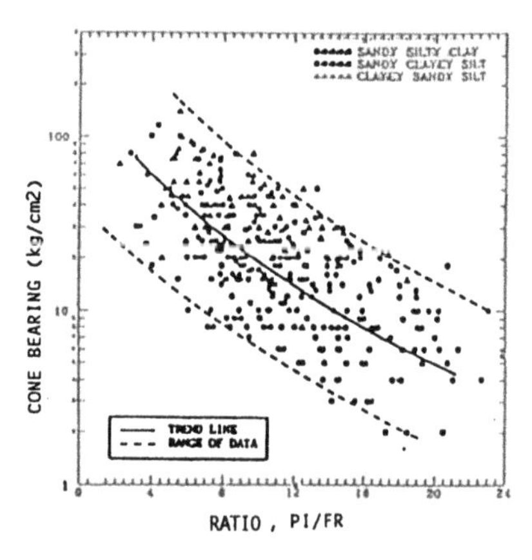

Fig. 16.6. Correlation of Dutch cone penetrometer results and plasticity index in general

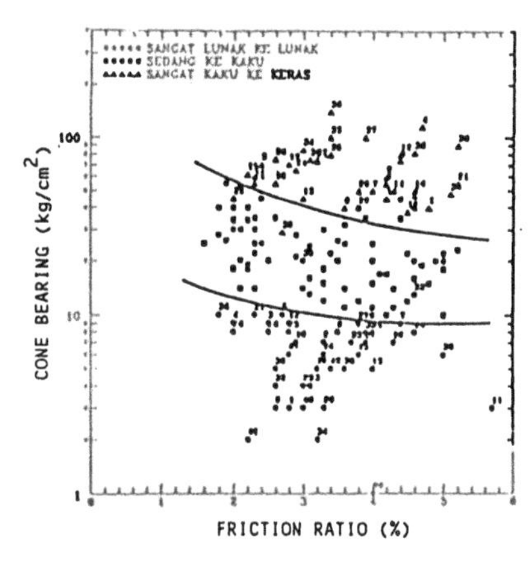

Fig. 17.1. Correlation of Dutch cone penetrometer results and consistency in South Sumatera area

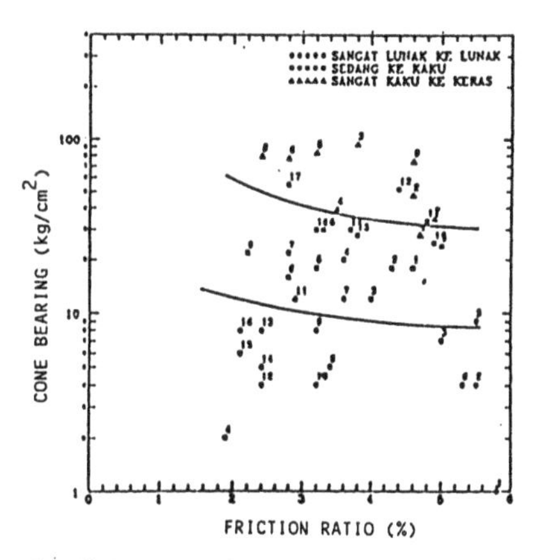

Fig. 17.2. Correlation of Dutch cone penetrometer results and consistency in Lampung area

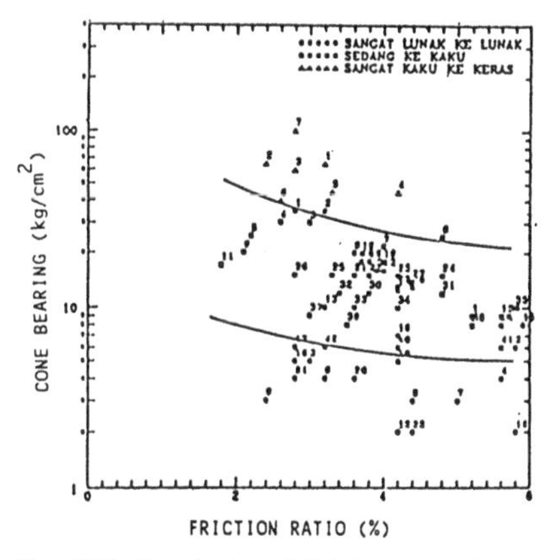

Fig. 17.3. Correlation of Dutch cone penetrometer results and consistency in Riau area

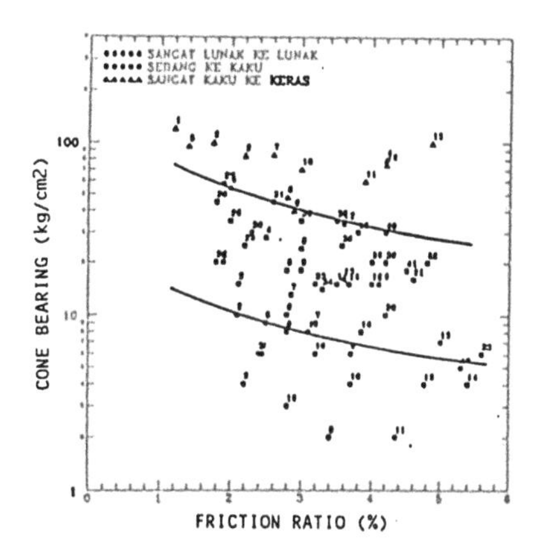

Fig. 17.4. Correlation of Dutch cone penetrometer results and consistency in Central Java area

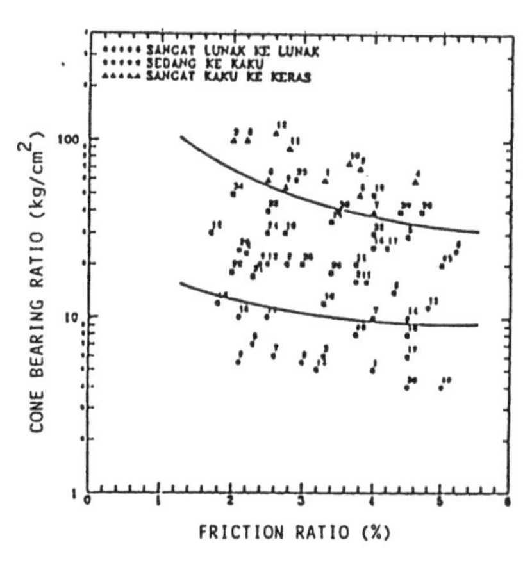

Fig. 17.5. Correlation of Dutch cone penetrometer results and consistency in East Java area

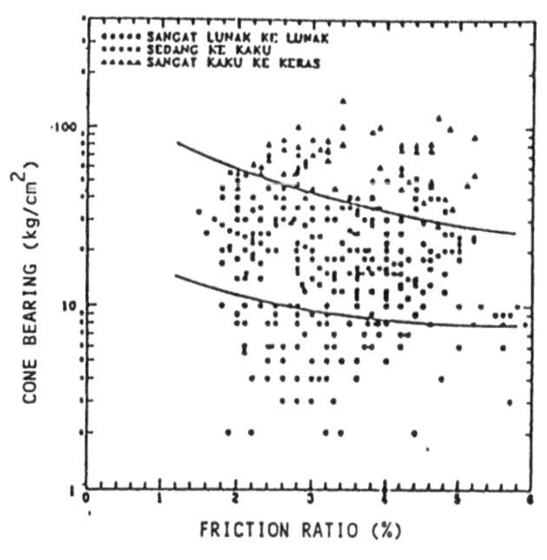

Fig. 17.6. Correlation of Dutch cone penetrometer results and consistency in general

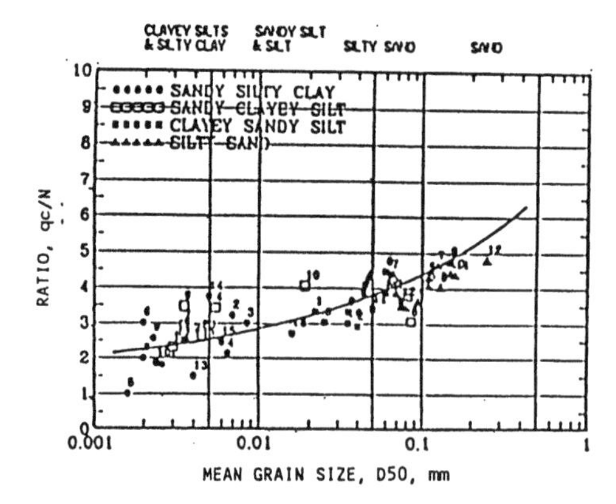

Fig. 18.1. Correlation of qc/N and D50 in South Sumatera area

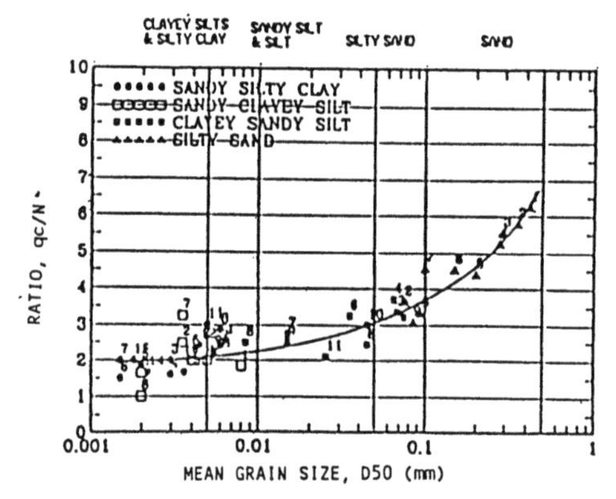

Fig. 18.2. Correlation of qc/N and D50 in Lampung area

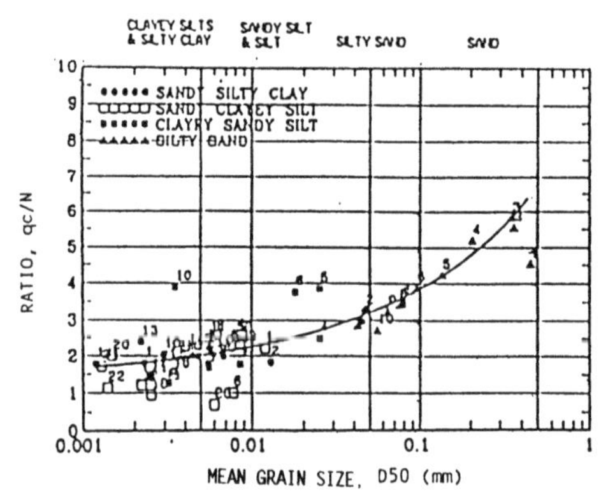

Fig. 18.3. Correlation of qc/N and D50 in Riau area

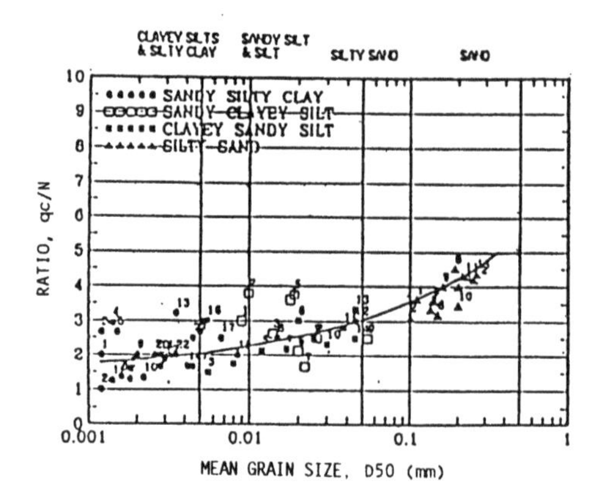

Fig. 18.4. Correlation of qc/N and D50 in Central Java area

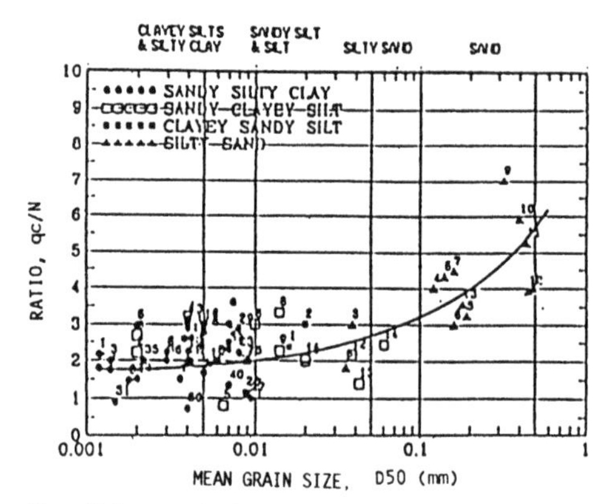

Fig. 18.5. Correlation of qc/N and D50 in East Java area

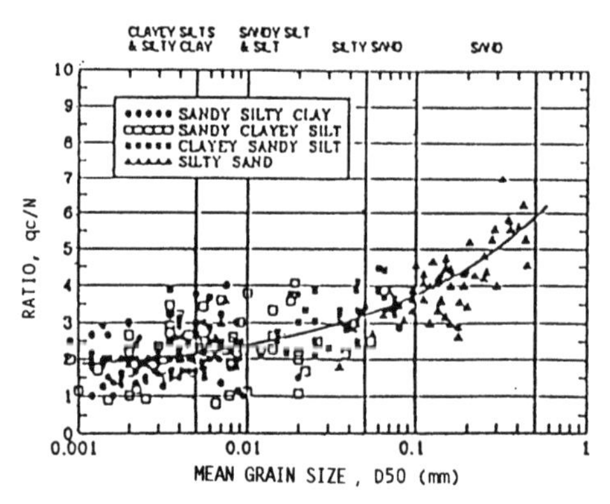

Fig. 18.6. Correlation of qc/N and D150 in general

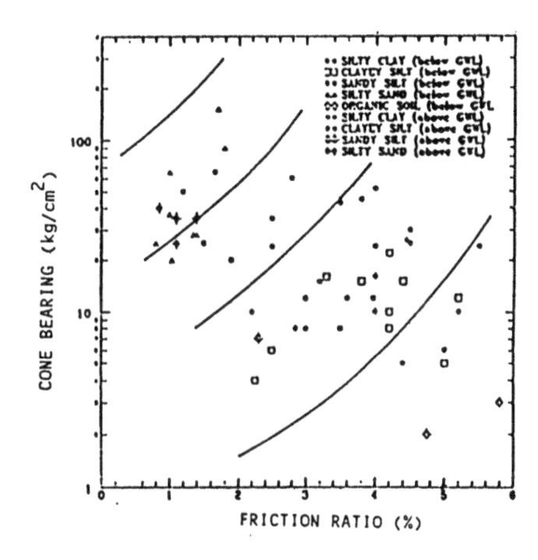

Fig. 19.1. Correlation of Dutch cone penetrometer results and soil types from area pointed out in Tabel 2

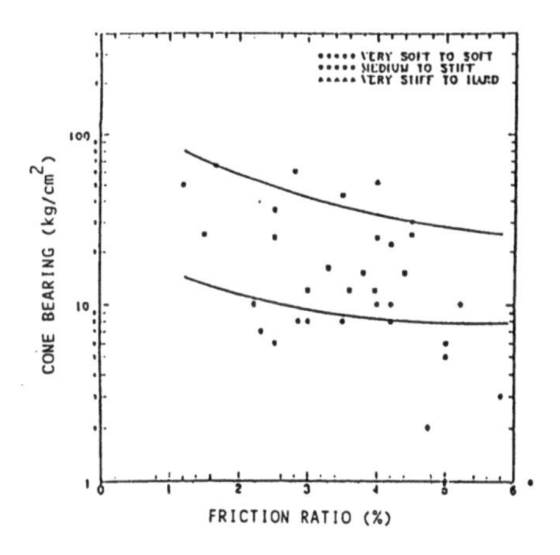

Fig. 19.3. Correlation of Dutch cone penetrometer results and consistency from area pointed out in Tabel 2

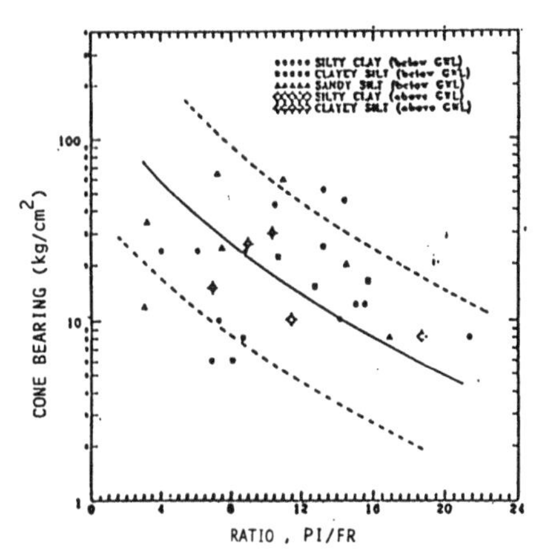

Fig. 19.2. Correlation of Dutch cone penetrometer results and plasticity index from area pointed out in Tabel 2

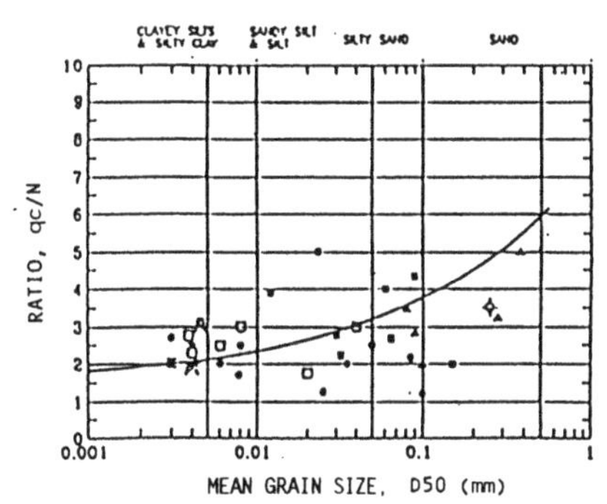

Fig. 19.4. Correlation of qc/N and D50 from area pointed out in Tabel 2

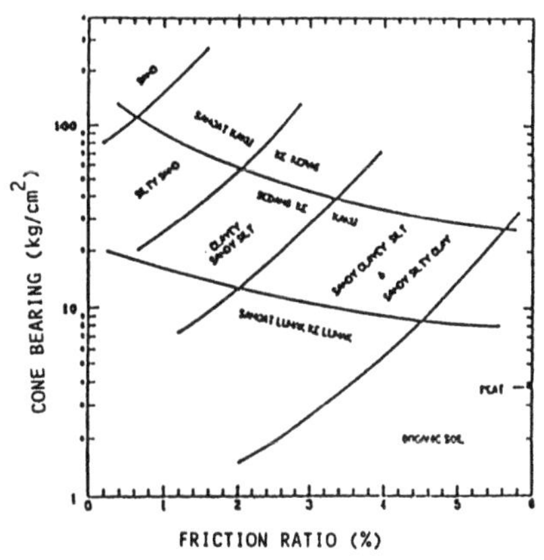

Fig. 20. Correlation of Dutch cone penetrometer results, soil types and consistency

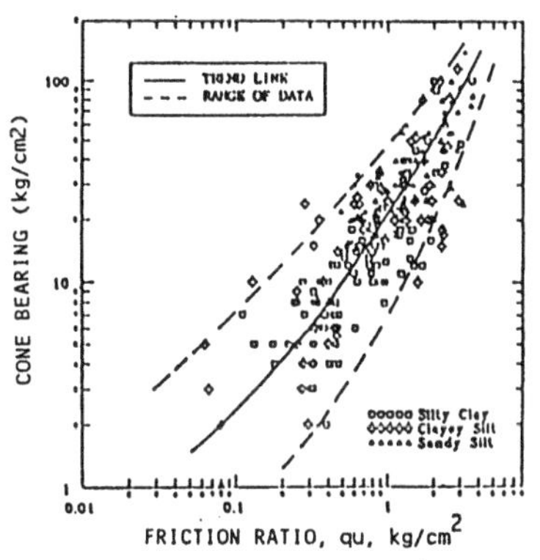

Fig. 21. Correlation of cone bearing value (qc) and unconfined compressive strength (qu)

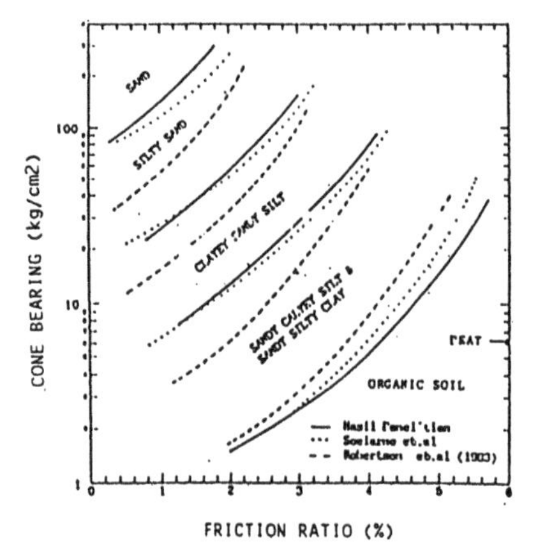

Fig. 22. Correlation of Dutch cone penetrometer results and soil types from this study, Soelarno-Ferrynaga 1990, Robertson and Campanella 1983

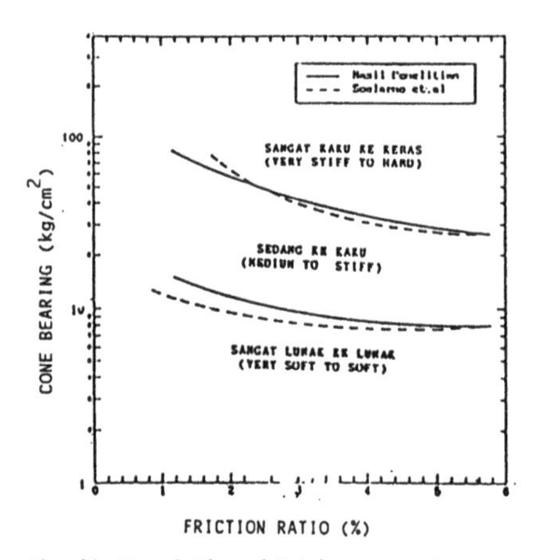

Fig. 24. Correlation of Dutch cone penetrometer results and consistency from this study and Soelarno-Ferrynaga 1990

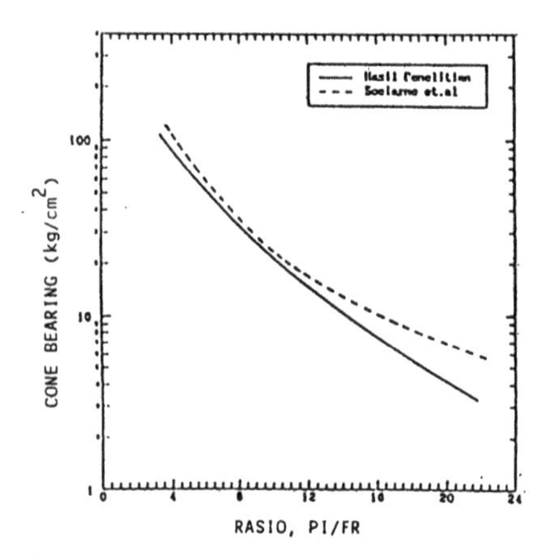

Fig. 23. Correlation of Dutch cone penetrometer results and plasticity index from this study and Soelarno - Ferrynaga 1990

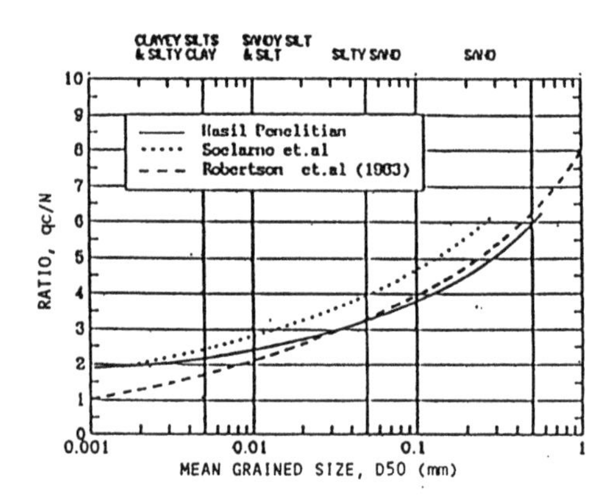

Fig. 25. Correlation of qc/N and D50 from this study, Soelarno-Ferrynaga 1990 and Robertson - Campanella 1983

Developments in Geotechnical Engineering, Balasubramaniam et al. (eds) © 1994 Balkema, Rotterdam, ISBN 90 5410 522 4

Horizontal mass permeability of a clay stratum

T. Akagi & T. Ishida
Toyo University, Saitama, Japan

ABSTRACT: A new simple laboratory permeability test is proposed; an undisturbed soil sample encased in a thin wall (Shelby) tube may be tested directly for horizontal mass permeability. Through a row of small holes drilled on each side of the tube, water with a constant head is supplied from one end to the other across the diameter of the soil sample until a steady state seepage is established through the soil encased in the sample tube. From the rate of discharge measured, the coefficient of permeability may readily be computed over each divided section or over the entire length of the thin wall sample, thus providing an important information on horizontal mass permeability of a thick compressible stratum to evaluate more logically and reliably the rate of settlement of a structure on a soft clay foundation.

1 INTRODUCTION

It has repeatedly been pointed out that the observed rates of settlement of earth structures founded on soft clay are more often than not much greater than those calculated on the basis of the one-dimensional consolidation theory using the results of oedometer tests carried out on small specimens.

It is apparent that efforts should be directed to more careful evaluation of drainage characteristics of soft clay deposits which often contain such deficiencies as cracks, fissures, silt seams, sand pockets, decayed vegetation, etc., some of which may serve effectively as drainage paths principally in the horizontal direction. It is essential that more reliable field and laboratory techniques should be developed to determine more accurately soil profiles and mass permeability of soft clay strata.

With the present-day level of routine subsurface investigations, it is still not possible to pinpoint which seams or layers have horizontal continuity and drainability capable of shortening drastically the time required for consolidation. It appears in many cases the difference between the predicted and observed time-settlement relationships amounts to 1 to 2 orders of magnitude in terms of the coefficient of consolidation c_v or in terms of the time required to attain a certain degree of consolidation. It goes without saying that such a state of the art is totally inadequate particularly when one has to determine the necessity of vertical drains to accelerate the expected consolidation.

Generally the coefficient of permeability of a saturated clay is determined by an oedometer test in the laboratory. When permeability in the horizontal direction is required, an oedometer specimen is cut out of the sample so that the axis of compression and the direction of flow of porewater during the test may coincide with the horizontal of the soil sample.

Oedometer tests make it possible to evaluate the coefficient of permeability of a specimen under various effective stresses. However, it has been demonstrated time and again that small oedometer specimens often fail to indicate the correct order of magnitude of the permeability of the actual compressible stratum with considerable thickness, particularly in the horizontal direction.

It is often said that the horizontal coefficient of consolidation, c_h, is generally equal to or only a few times greater than the vertical coefficient of consolidation, c_v, in many alluvial clays. It should be remembered, however, that such an empirical rule is based mainly on oedometer test results on small specimens.

It is interesting to note that in accumulating such experience we have the habit of selecting "good looking" portions of clay for oedometer specimens, that is, the most homogeneous and perhaps the most impervious parts of the clay sample, excluding carefully the portions containing deficiencies such as cracks, fissures, silt and sand seams, veins of decayed vegetation, etc., frequently encountered in soft clay deposits. These defects may have a profound influence on the horizontal permeability of the clay mass if they serve as drainage paths within a thick compressible stratum.

While an in-situ permeability test appears to be an attractive means to evaluate the mass per-

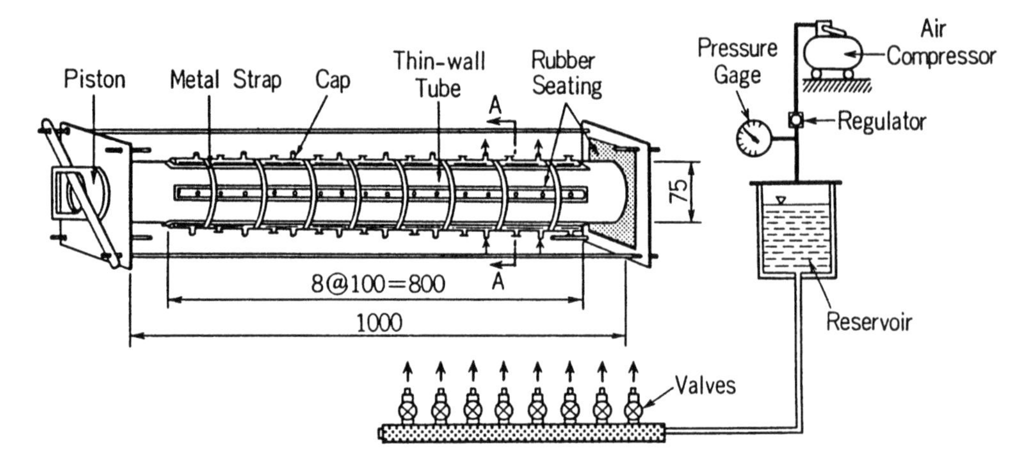

Fig. 1 New permeability test setup

meability, it may not always be regarded as a low-cost, easy-to-run, reliable method. It is obviously difficult to interpret field test results unless the subsurface conditions are well defined and the soil layers involved are very uniform and isotropic.

Since an idea of a simple laboratory permeability test was first presented (Akagi 1983), considerable developments and improvements have been made both analytically and experimentally (Akagi and Ishida 1988 and 1989), but very little on this subject has been published in the English language (Akagi 1989).

2 FLOW NETS AND THE COEFFICIENT OF PERMEABILITY

Characteristically this test is conducted directly on an undisturbed soil sample which is still encased in a thin wall (Shelby) tube to measure horizontal permeabilities over divided sections (usually each 100 mm in length) or over the entire length of the encased sample, Fig. 1.

It is well recognized that thin wall tube samples of cohesive soils are of good quality in general if taken with due care and right equipment. The proposed test requires drilling a row of closely-spaced small holes or opening a line of narrow slits on each side, diametrically opposed, of the tube, but requires neither pushing the soil sample out of the tube nor trimming it.

When water with a constant head flows from one side to the other across the diameter of a cylindrical soil sample encased in a tube, the flow soon reaches a steady state which may be illustrated by a two-dimensional flow net such as shown in Fig. 2. This flow net is based on the assumption that the soil is homogeneous and isotropic two-dimensionally on any cross-section and is in good contact with the inner surface of the thin wall tube.

When water with a constant head H is supplied and discharge Q is obtained within time t and over distance d along the axis of the tube, the rate of discharge q is given by:

$$q = Q/(t\ d) = k\ H\ (N_f/N_d) \tag{1}$$

where k = the coefficient of permeability, N_f = the number of flow channels and N_d = the number of potential drops. Then,

$$k = \alpha\ (q/H) \tag{2}$$

where $\alpha = N_d/N_f$.

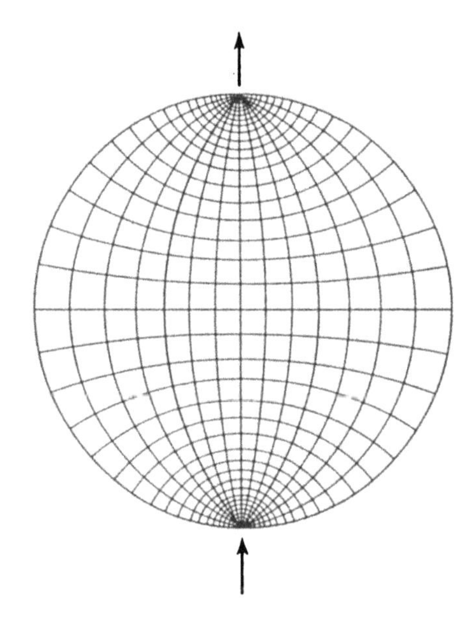

Fig. 2 Two-dimensional flow net, Section A-A

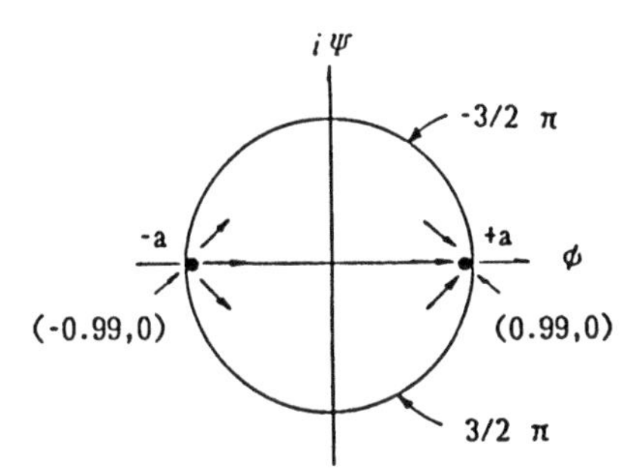

Fig. 3 Inlet and outlet on the w-plane

On the basis of the theoretical and experimental analyses which are briefly described in the following, it has been concluded that the value of α is approximately 4 for the purpose of calculating the horizontal coefficient of permeability k.

The average coefficient of horizontal permeability, k_{av}, over the entire length of the thin wall sample, D, or the mass permeability of the tube length may be computed readily from the following formula:

$$k_{av} = 1/D \ \Sigma \ (k_i \ d_i) \tag{3}$$

where k_i = the coefficient of horizontal permeability of the i-th test section and d_i = the length of the i-th section. Equation 3 makes it possible also to evaluate the mass permeability in the horizontal direction of a thick clay stratum.

The following types of analyses were conducted on the seepage flow across the section of a circular cross-section: i) mathematical, ii) numerical, iii) of electrical analogy and iv) experimental. Details of the mathematical and numerical treatments have been given by Akagi and Ishida (1988).

2.1 *Mathematical and Numerical Analyses of Flow Nets*

If two-dimensional isotropy is assumed over a cross section of a soil sample encased in a tube, a flow net may be constructed on the basis of a mathematical analysis making the use of the theory of complex variables.

A two dimensional flow from one end to the other across the diameter within a circular boundary may be transformed into a formula:

$$z = \zeta + a^2/\zeta \tag{4}$$

where a = the radius and $\zeta = a e^{i\theta}$.
Then the following expression may be derived:

$$\frac{1 + r e^{i\theta}}{1 - r e^{i\theta}} = e^{\phi + i\phi} \tag{5}$$

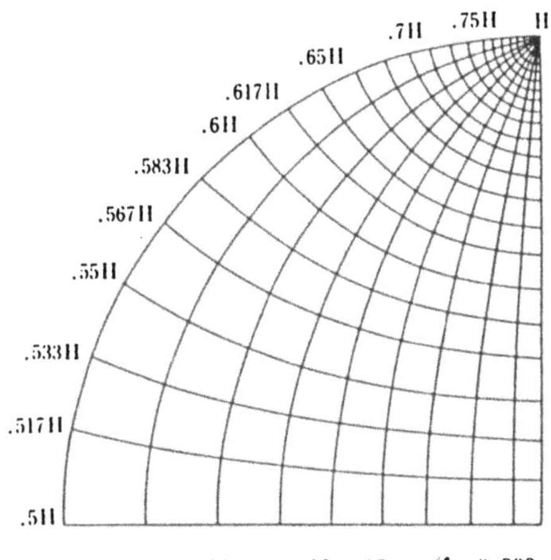

$N_d=60$, $N_f=17$, $re^{i\theta}=0.999$

(a) Mathematical solution

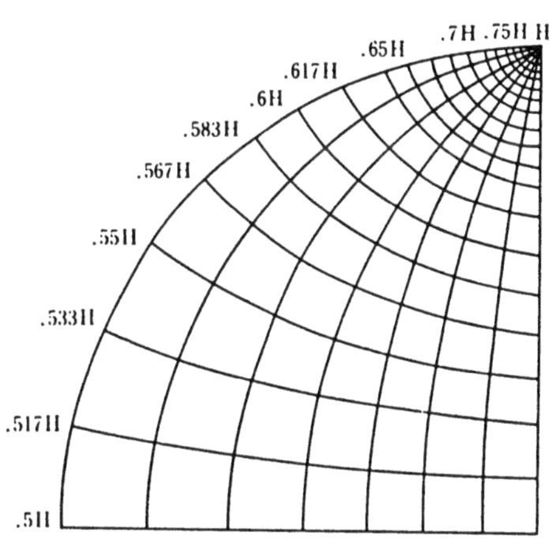

$N_d=60$, $N_f=14$, (Case 1)

(b) Boundary element solution

Fig. 4 Flow nets, isotropic cases

Giving a potential of unity to the inlet and a zero potential to the outlet as shown in Fig. 3, the exponent of the right side exprcssion of Equation 5 is given by:

$$\phi + i\phi = 2 \times 5.29(\phi - 1/2) + i\ 3/2\pi\ \phi \tag{6}$$

in which a value of 5.29 is obtained when the value of $r e^{i\theta}$ is assumed to be 0.99. Thus Equation 5 may be divided into the real and the imaginary parts, making it possible to construct two sets of curves for construction of a flow net;

$$\phi(x, y) = \text{constant and } \phi(x, y) = \text{constant} \tag{7}$$

Since the pattern is symmetrical with respect

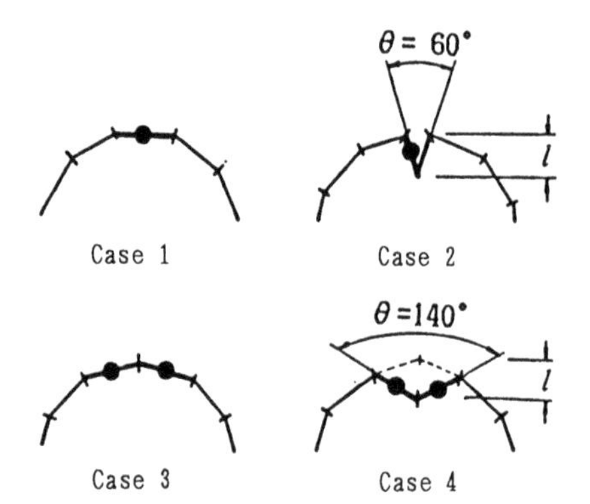

Fig. 5 Sizes and shapes of the inlet opening for the boundary element analysis

Table 1 Shape factor α determined by mathematical, numerical and analogical methods

Method	Conditions	α value
Mathematical	$r e^{i\theta} = 0.99$	3.33
	$r e^{i\theta} = 0.999$	3.53
Finite difference	$O_r = 0.005$	4.00
	$O_r = 0.055$	2.75
	$O_r = 0.105$	2.20
Boundary element	Case 1	4.28
	Case 2	3.75
	Case 3	3.75
	Case 4	3.12
Electric analogy	$O_r = 0.03$	2.86

Notes:
1) For the expression $r e^{i\theta}$, see Equation 5. 2) O_r means the opening ratio which is the ratio of the width of an inlet opening to the diameter of the circular cross section. 3) For Cases 1 through 4, see Fig. 5.

to both the vertical and horizontal axes, a quarter of a flow net constructed on the basis of a mathematical solution ($r e^{i\theta} = 0.999$) is illustrated in Fig. 4(a) indicating N_d = 60 and N_f = 17 and hence α = 3.53 where H is the input potential.

Numerical analyses of the Laplace Equation with a circular boundary and the input and output openings were conducted by means of the finite difference method and the boundary element method on several different sizes and shapes of openings. If the ratio of an opening width to the diameter of the tube sample is defined as the opening ratio, O_r, a 1.0 mm wide slit corresponds to an opening ratio of 0.013. A quarter of the flow net based on the solution for the case of an isotropic sample with O_r = 0.01 determined by means of the boundary element method is given as an example on Fig. 4(b), indicating N_d = 60 and N_f = 14 and hence α = 4.28.

The finite difference analysis also gives solutions which make it possible to construct flow nets for different opening ratios. The boundary element analysis gives with relative ease solutions not only for various opening ratios and opening shapes such as shown in Fig. 5, but also for anisotropic cases. Figs. 6(a) and (b) illustrates two such cases; (a) the permeability in the x-direction is 10 times as large as that in the y-direction, or k_x = 10 k_y and (b) the x-direction permeability is one tenth of the y-direction, or k_x = 0.1 k_y. The former indicates α = 7.2, while the latter α = 3.0. Fig. 7 shows the variation of the α values with the change in the permeability ratio, k_x/k_y, when the ratio O_r is 0.01. It is to be noted that the α value is relatively insensitive to a change in anisotropic permeability and does not change the order of magnitude of the k values obtained.

Table 1 summarizes some of the typical results obtained from the foregoing analyses. The α values appear to fall within a relatively narrow range and may be considered to be roughly of the order of 4 for the purpose of computing a coefficient of permeability by means of Equation 2 when the two-dimensional cross-section is assumed to be isotropic. The k value may be refined by means of Fig. 7 when the permeability tests have been conducted in two right-angled directions (separately) and found a significantly anisotropic nature of the soil tested.

2.2 *Experimental Analyses on Seepage across a Soil Specimen*

Electric analogy tests were conducted simulating the seepage condition of the proposed permeability test. A thin "conductive" plastic sheet, 0.3 mm thick, was cut into a 500 mm diameter circle and a pair of input and output terminals of different sizes and shapes. 10 volt DC was applied to the terminals and voltage was measured at each grid intersection of the 10 mm grid system marked on the entire plastic sheet circle.

The following experimental studies were also conducted and supported the foregoing results of computations and the electric analogy;

(i) Comparison between the flows along the axis of the tube and across the tube filled with uniform, isotropic fine sand gave α values ranging from 2.0 to 4.5 except an isolated value of 12 when water was supplied through 1 mm diameter holes on 5 mm spacings (Case F3), as summarized in Table 2. These data indicate a close relationship between Cases F1 and F2 when water was fed through 1 mm wide side slits (Case F1) and 1 mm diameter holes on 2.5 mm spacings (Case F2), respectively. This suggests that a row of

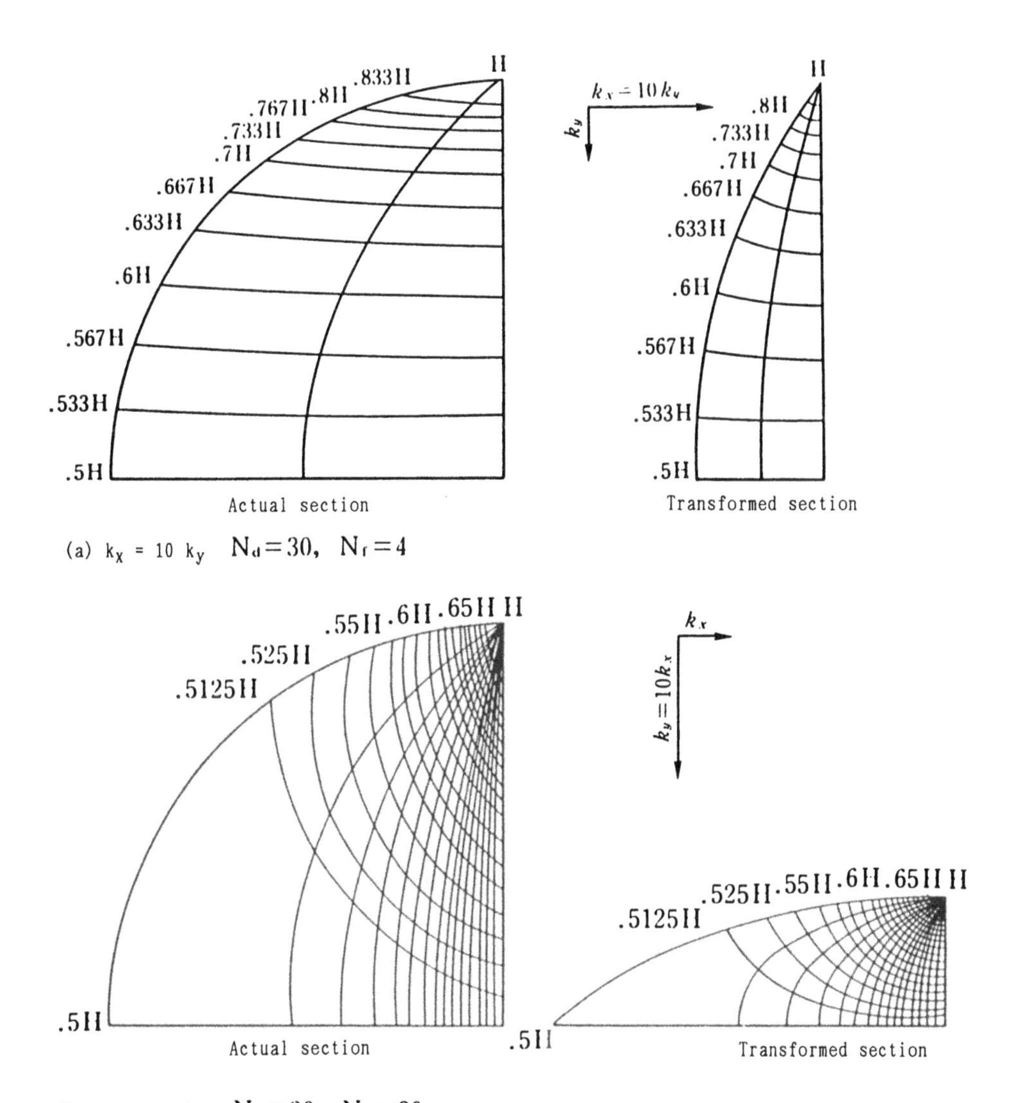

(a) $k_x = 10\ k_y$ $N_d = 30$, $N_f = 4$

(b) $k_x = 0.1\ k_y$ $N_d = 80$, $N_f = 30$

Fig. 6 Flow nets, anisotropic cases

closely spaced small holes functions equally well as a narrow slit,

(ii) Porewater pressure measurements of the seepage across the circular section of a 300 mm diameter tube filled with sand showed a distribution of potential which indicated a similar pattern as that of theoretical analyses, and

(iii) 3% sodium chloride by weight was added to and thoroughly mixed with salt-free Kanto loam which was then filled in a 300 mm diameter tube. Water was fed from one side to the other across the tube. The sodium chloride content was measured at various parts of a soil sample after water had been kept seeping through it for 24 and 48 hours. The results appear to indicate that water percolates through the soil quite uniformly throughout the circular section.

3 TESTING PROCEDURE

An undisturbed soil sample is taken in accordance with standard sampling procedure. Normally thin wall (Shelby) tubes of brass or stainless steel, 75 mm in diameter, are pushed into a soil stratum using a piston sampler. 300 mm diameter thin wall samples may also be taken for this test.

As shown schematically in Fig. 8, a row of small holes, 1.0 mm in diameter, are drilled at close spacings of 2.5 mm on each diametrically

Table 2 Shape factor α determined by comparison between one- and two-dimensional flows

Gradation of Sand	Void Ratio	k ($x10^{-3}$cm/s) (a) 1-D	(b) 2-D F1	F2	F3	Shape factor α F1	F2	F3
250-420 μm	0.74	5.8	2.5	2.4	1.9	2.3	2.4	3.1
105-250 μm	0.71	2.4	0.7	0.6	0.2	3.4	4.0	12.0
Standard sand	0.66	4.5	1.7	2.2	1.0	2.6	2.0	4.5

Notes:
1. (a)1-D is the case of a one-dimensional flow along the axis of the tube and (b)2-D the case of a two-dimensional flow across a cross-section of the tube.
2. For the two-dimensional flow across the circular cross section, the inlet and outlet openings, located diametrically opposed, consist of 3 cases; (F1) 1 mm wide slits, (F2) 1 mm diameter holes on 2.5 mm spacings and (F3) 1 mm diameter holes on 5 mm spacings.

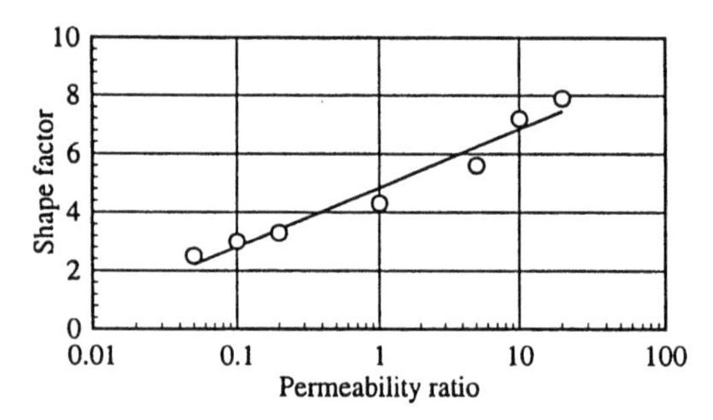

Fig. 7 Permeability ratio k_x/k_y versus shape factor α

opposed side of the thin wall tube. It has so far been found much easier to drill small holes rather than cutting manually narrow slits on the side of the tube. Should an automatic cutting device be developed, however, no difficulty should be anticipated either to cut a 1 mm wide slit or to drill small holes on close spacings on each side of the tube.

Holes are covered by special caps with an inlet or outlet holes. An inlet cap has an air vent also. Hard rubber seating is placed between the thin wall tube and each cap, the whole of which is then put together and tightened up by metal straps, Fig. 9(a) and (b).

After the inside of both the ends are thoroughly cleaned up and some bentonite paste is placed to assure the water-tightness, the top end of the thin wall tube is sealed by a steel piston with an O ring and the bottom end by a steel plate with hard rubber seal, Fig. 1. These attachments are so designed that they may readily be put together and removed as may be seen in Fig. 8.

Water with a constant head may be supplied to the thin wall sample from a reservoir positioned at a predetermined height or from a pressure tank connected to an air compressor system capable of maintaining a constant pressure in it. Discharge is measured at appropriate time intervals together with measurement of temperature.

When the amount of water coming out of the outlet is nominal which is usually the case when clay is tested, it may generally be regarded that the discharge has no potential. Otherwise, the water should be led into a reservoir maintaining a constant head tail water and the amount of overflow should be measured as seepage discharge.

4 TYPICAL TEST RESULTS

The permeability tests were conducted on undisturbed sample of soft alluvial clay taken from a depth of 1 to 2 m at Oppe River, Saitama. The clay is relatively homogeneous having a liquid limit of 66.5% and a plastic limit of 31.7% with a natural water content of 50.0%. It is slightly overconsolidated with an apparent preconsolidation pressure of the order of 40 to 60 kPa.

Fig. 10 shows the relationship between discharge Q and elapsed time t when the water fed was under the pressure of 36.7 kPa. It is basically linear and the slope of a straight line is directly proportional to the horizontal coefficient of permeability, k_h. The number on each straight line indicates that of each test section which is 100 mm in length. The numbers in parentheses indicate the k_h value in cm/sec to be multiplied by 10^{-7}.

It is important to continue the test until a linear relationship Q vs. t is well established on each test section. It takes time varying from a few minutes to several hours to assure the the linearity of this relationship, depending mainly on the permeability and the degree of saturation of the soil tested. Measurements of Q are taken at time intervals ranging from a few minutes to an hour depending on the permeability of the soil.

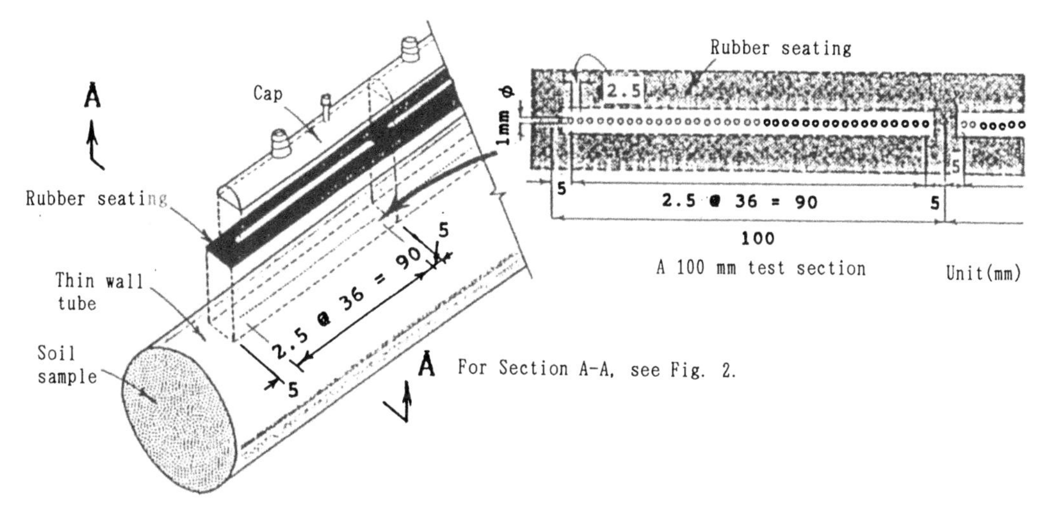

Fig. 8 Details of the openings for inlet flow

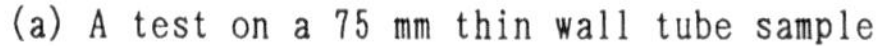

(a) A test on a 75 mm thin wall tube sample

(b) A test on a 300 mm thin wall tube sample

Fig. 9 Typical test setup

The discharge should be proportional to the head given to the water at the input if the coefficient of permeability is constant. It is often found, however, that the value of k_h either increases or decreases with an increase in the pressure applied to water. Fig. 11 shows a relationship between the coefficient of horizontal permeability k_h and the water pressure given at the inlet. This diagram happens to show a downward trend of permeability with increasing pressures due probably to silting of some water passages in the soil sample with increase in water pressure. Some samples, however, have shown an upward trend with increasing pressure of seepage.

In most of the tests discharge was clean in which no soil particles were visible throughout the test. During some of the tests, however, when the water pressure was increased to 150 to 200 kPa, a few of the outlets suddenly started yielding so much muddy water, the test had to be discontinued. Hydraulic fracturing was suspected as a cause of this phenomenon. The same was observed during the test on other soft clay samples at a pressure as low as 80 kPa.

A series of oedometer tests was conducted on undisturbed clay samples taken from the same clay stratum in Oppe River. Both the vertical and horizontal permeabilities, k_v and k_h, were determined from oedometer tests run on the specimens

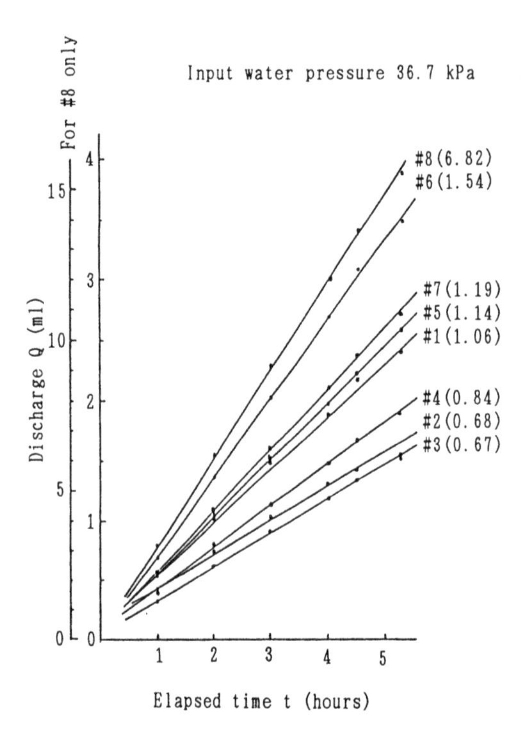

Note: Numbers with # indicate the numbers of test sections, each 100 mm in length. Numbers in parentheses indicate the coefficient of permeability in cm/s to be multiplied by 10^{-7}.

Fig. 10 Discharge vs. time in each test section

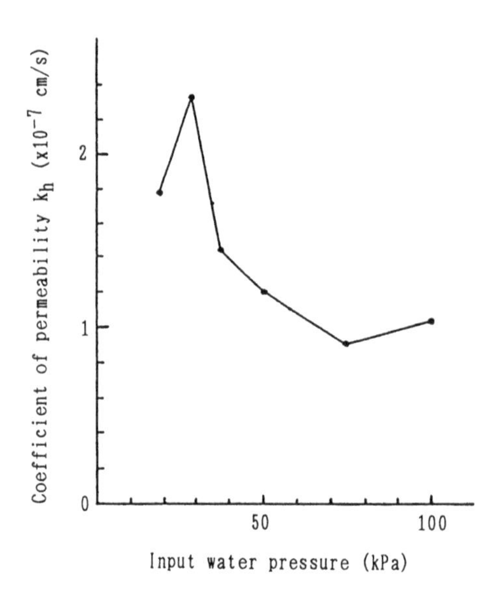

Fig. 11 Coefficient of permeability versus water pressure

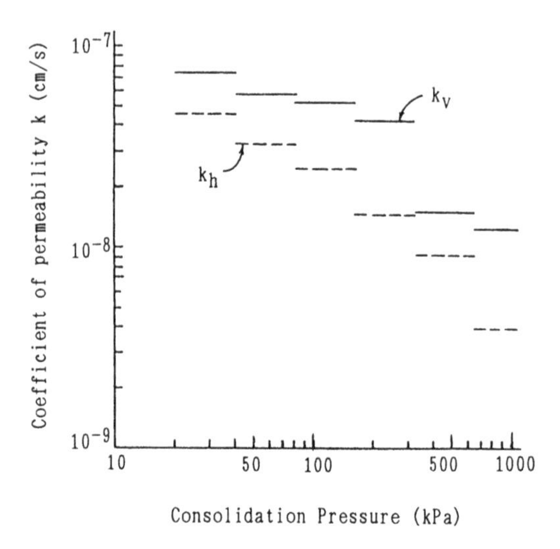

Fig. 12 Oedometer test Results

cut out horizontally and vertically, respectively, with respect to the axis of the thin wall tube. Fig. 12 shows a relationship between consolidation pressures and the coefficients of permeability computed from oedometer test results.

In spite of the fact that the clay tested appears relatively homogeneous and that there is no physical relationship between the abscissas of the diagrams on Figs. 11 and 12, one obtains an impression that the k values as determined by the proposed permeability test are several times greater than those obtained from the oedometer test.

5 AN APPLICATION TO A FIELD CASE

Undisturbed samples were taken from two borings DH-1 and DH-2, 17.5 m and 20.2 m deep, respectively, at a low lying site where a test embankment was to be constructed near Isehara City, Kanagawa. Fig. 13 shows a geologic section, starting from the ground level downward, consisting of highly compressible peat shown as P_t, a thin layer of alluvial sand as A_s, and soft alluvial clay as A_c, underlain by stiffer clay, D_c, a dense gravel layer, D_g, and a dense sand layer, D_s, of diluvial origin.

These two borings were made specifically for taking thin wall tube samples for the proposed permeability tests near Boring No. 2 which had been drilled earlier for exploratory purposes. In the first hole DH-1, a total of 22 thin wall samples were taken almost continuously with depth, but the soils were so soft only 12 of them were usable for the test and the rest 10 were either of poor recovery or of very poor quality.

It is important to make sure the soil sample is in good contact with the thin wall tube. This

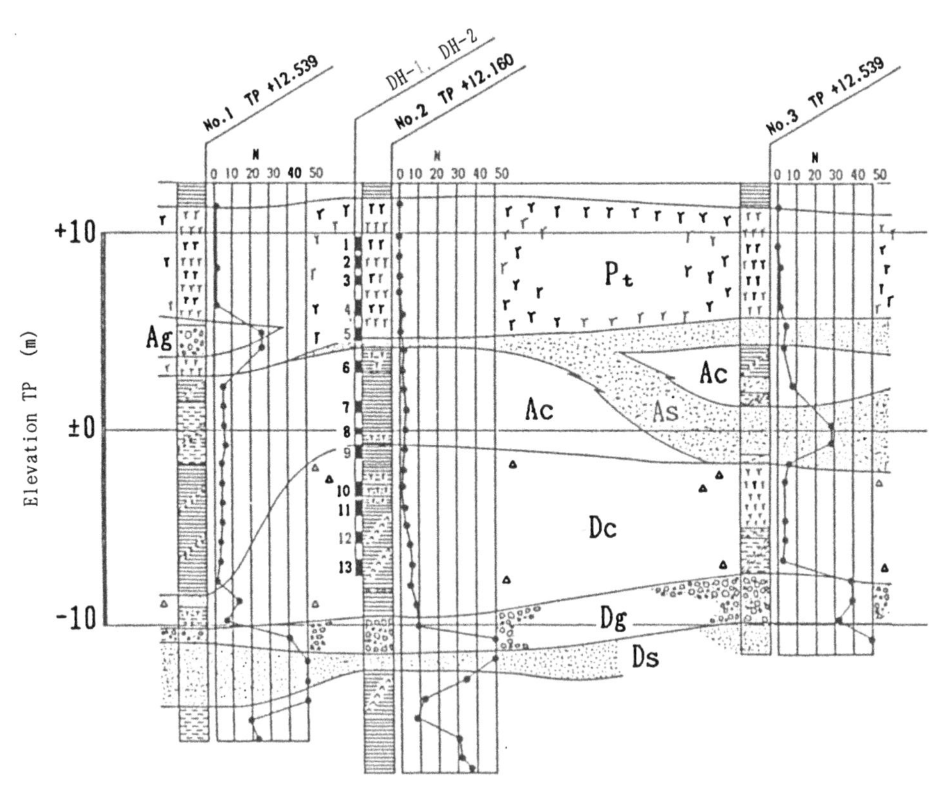

Fig. 13 Geologic section at the Isehara site

should carefully be checked not only before the permeability test but also after the test when the sample is being extruded by measuring the force required to push it out.

In the second hole DH-2 a total of 13 thin wall samples were obtained all in a satisfactory condition and the depths of sampling are indicated alongside Boring No. 2 in Fig. 13. Fig. 14 summarizes the soil properties determined of these samples after the new permeability tests were performed on them.

The permeability tests were conducted on each thin wall sample after drilling two rows of 1 mm diameter holes on 2.5 mm spacings on its sides located diametrically opposing each other. Water with a constant head was supplied through the inlet caps, each of which covered a 100 mm long section of the tube. The seepage discharge was measured at each section when it reached a steady state. The coefficient of horizontal permeability was then computed by Equation 2 for each 100 mm long section of the sample. The average coefficient of horizontal permeability, k_{av}, over the entire length of the thin wall sample, or the mass permeability was computed readily from Equation 3.

Since another set of two rows of small holes had been drilled but closed temporarily during the first testing, the direction of seepage flow was rotated by 90 degrees after closing the first set of holes. Fig. 15 shows typical test results of the thin wall samples Nos. 9 and 13, showing variation of the k_h value under a constant head of 3 m over each 100 mm test section numbered #1 through #7 on the abscissa. The directions of the flow were 90° away from each other as shown in an illustration "Flow Dir." given in the upper right-hand corner of Fig. 15. The samples tested appeared to be fairly isotropic in terms of the permeability on the horizontal plane.

After the permeability test was completed, the samples were pushed out of thin wall tubes and visually examined. Where relatively high permeability was indicated during the test, almost always revealed was presence of organic matters, fine sand seams, hard clay pebbles, etc., which had apparently contributed to a higher permeability.

Fig. 16 indicates the coefficients of horizontal permeability as determined by the new permeability test with depth in DH-1 and DH-2. The "max. and min." points on Fig. 16 correspond to the maximum and minimum values of the horizontal mass permeability, k_{av}, over the entire length of each thin wall tube sample while changing the direction of seepage and the head of input flow.

The k_h values of the peat layer appear to be in a range of 10^{-3} to 10^{-4} cm/s, whereas those of

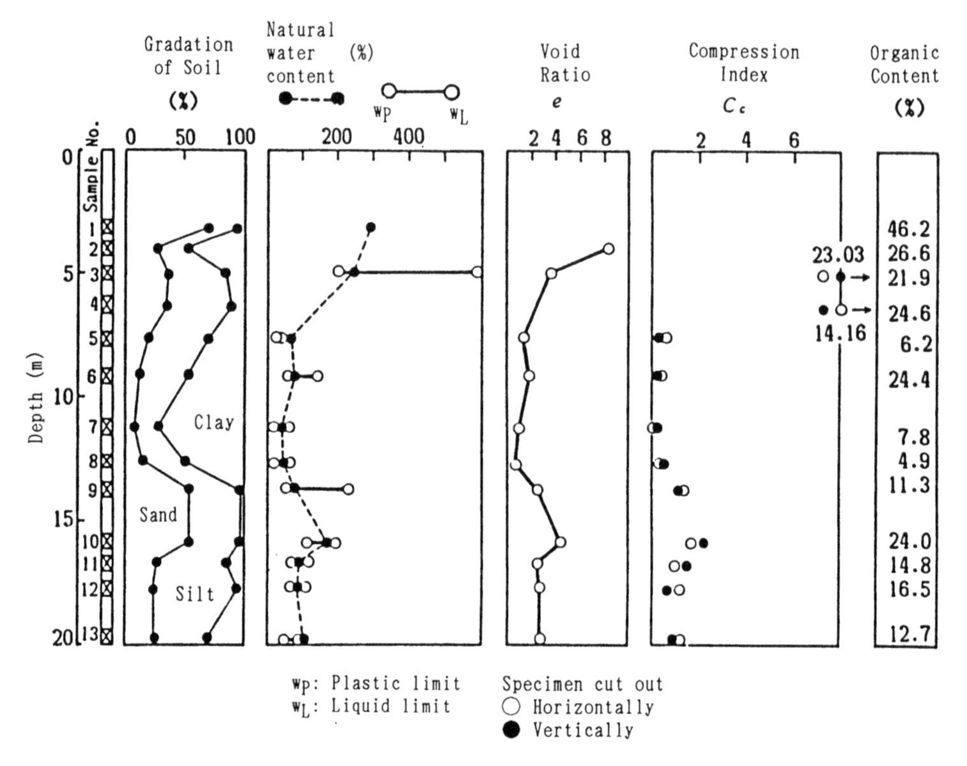

Fig. 14 Soil properties with depth at the location of DH-2

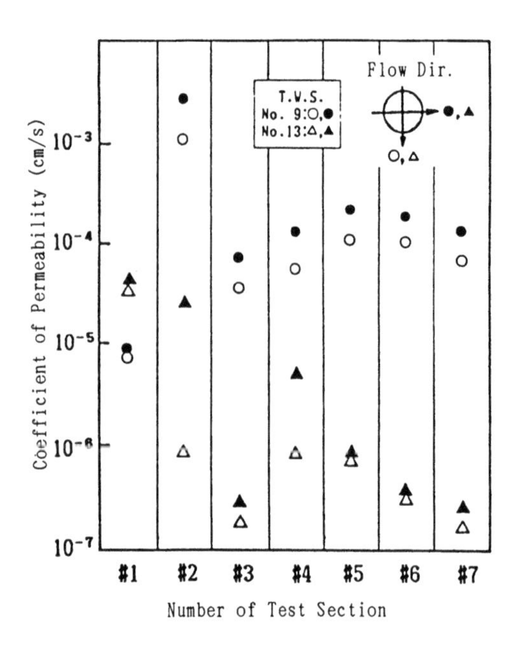

Fig. 15 Coefficients of horizontal permeability in different directions in each test section

the organic clay roughly in a range between 10^{-4} and 10^{-6} cm/s. In Fig. 16(a) in particular, a great variation in k_h values at depths below about 10 m indicates in fact the presence of alternating thin layers of relatively pervious peat and impervious clay.

Fig. 17 shows the coefficients of permeability in the vertical and horizontal directions which were determined by oedometer tests on the thin wall tube samples after the permeability test had been conducted on them. These data were obtained when the consolidation pressure, p, was 39.2 kPa. Unlike the k_h values in Fig. 16, it appears that the values of k_v and k_h fall in a low range between 10^{-6} and 10^{-8} cm/s and the values for the peat at shallow depths and the clay at greater depths exhibit no significant difference.

Presence of a thin smear zone having a much lower permeability is suspected around the soil sample just inside the thin wall tube, which may affect the measured values of k_h to a significant degree. However, the foregoing test results clearly demonstrate that the new permeability test gives far greater values of permeability with a difference as much as a few orders of magnitude from the coefficients of permeability evaluated by other conventional methods such as oedometer tests. The smear zone problem certainly requires an elucidation, but the difference in the test results is so great it ap-

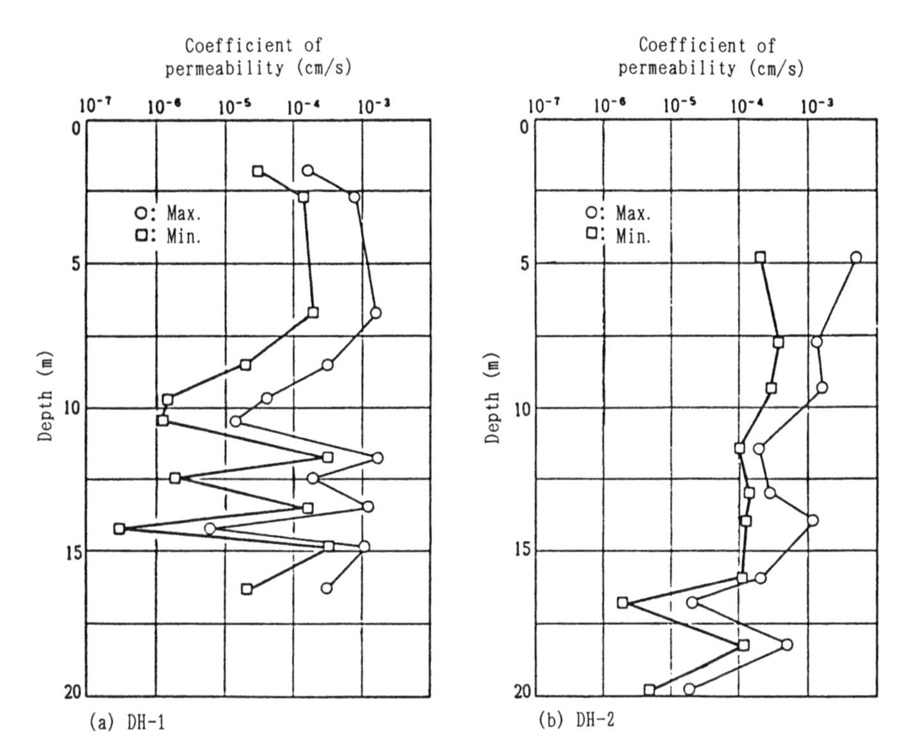

Fig. 16 Coefficient of horizontal permeability determined by the proposed permeability test versus depth

pears to be relatively a minor factor.

Fig. 18 shows preconsolidation pressures and unconfined compressive strengths (shown by blank circles) determined by the conventional laboratory tests which were conducted on the specimens trimmed out of the thin wall tube samples which had been subjected to the new permeability tests. The solid circles are the data obtained from a previous subsurface study made at the same site. It is to be noted that the blank and solid circles fall in the same range, demonstrating clearly that the new permeability test did not disturb the samples which therefore may be tested for strength and compressibility as "undisturbed samples."

In the vicinity of DH-1 at this site, a 6 m high test embankment was constructed and field measurements were taken of porewater pressures and settlements. The excess porewater pressure measured reached a maximum of 64 kPa at a depth of 6 m at the completion of the fill in about 20 days after the initiation of filling operations and dissipated almost completely in about 40 days after the peak was reached.

Fig. 19 shows the time-settlement relationships; blank circles indicate the field measurements of the settlement plate placed on the ground surface, a dotted curve is a prediction based on oedometer test results and a one-dimensional consolidation theory, a solid curve is a prediction using parameters derived from the results of the new permeability test and a two-dimensional consolidation theory. On the basis of the e-log p curve, the total ultimate settlement was estimated to be 3.03 m.

The oedometer test prediction assumes soil parameters given in a two layer profile in Fig. 20. A finite difference analysis was made on the differential equation for consolidation given by Terzaghi. The new permeability test run under a constant head of 5 m gave an average coefficient of horizontal permeability, k_{av}, of 1.6×10^{-4} cm/s for the peat and 5.3×10^{-6} cm/s for the underlying organic clay.

By means of the use of the average coefficient of volume compressibility, m_v, these values may be converted into the coefficient of consolidation, c_h, since the m_v values are found in general to be of the same order of magnitude as those in the horizontal directions, although the c_h values are often a few orders of magnitude greater than the c_v values on the basis of the field observations. Thus the above k_{av} values may be converted into c_{h-av} values of 2.7×10^4 cm^2/d for the peat and 1.3×10^4 cm^2/d for the organic clay, which are roughly two orders of magnitude larger than the c_v values given in Fig. 20.

By means of the two-dimensional consolidation analysis method by Davis and Poulos (1972), a

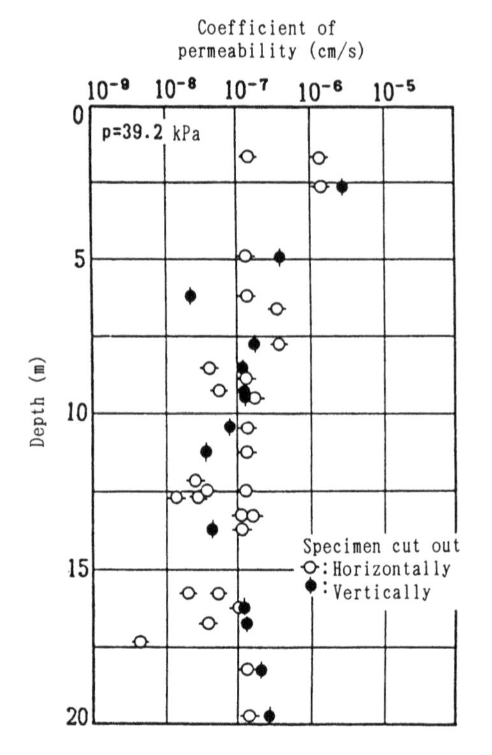

Fig. 17 Coefficient of permeability determined by oedometer tests

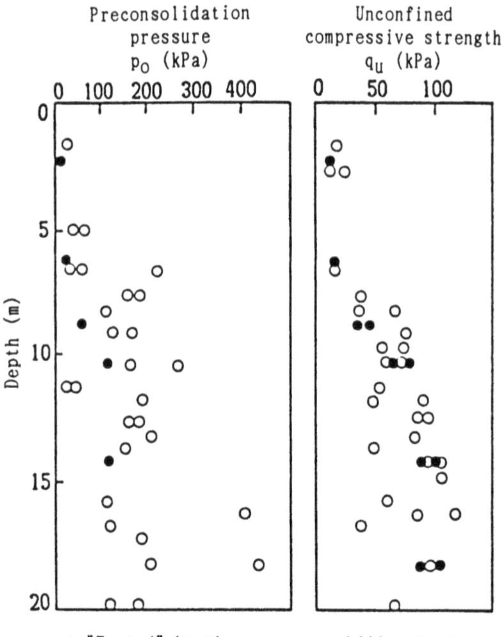

Fig. 18 Preconsolidation pressures and unconfined compressive strengths of samples "tested" and "not tested" by the proposed permeability test

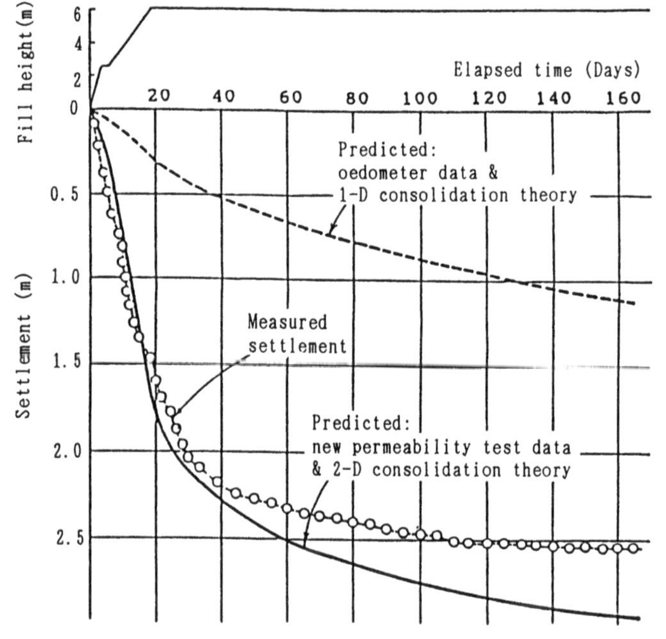

Fig. 19 Time-settlement relationship of a test embankment near DH-1

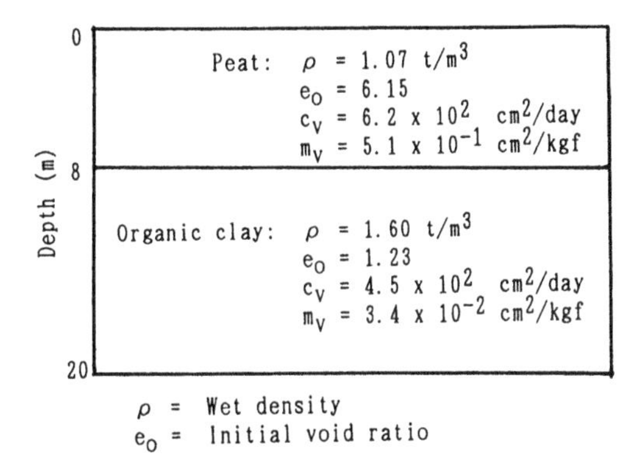

Fig. 20 Soil profile and parameters used for settlement analyses

combination of the c_v values in Fig. 20 and the c_h values obtained in the above makes it possible to construct the solid curve in Fig. 19, showing a close fit to the actual measured settlements.

6 SUMMARY AND CONCLUSIONS

1. It is felt essential to improve our capability, which is often totally inadequate, to predict the time-settlement relationship of a structure founded on a natural soft clay foundation. Without ameliorating it, it is not only costly and embarrassing but also not even possible to determine whether an improvement such as installation of vertical drains are indeed required. In this regard it is important to be able to define clearly the soil profile and to evaluate the mass permeability of each clay stratum.

2. The proposed permeability test is conducted directly on a thin wall tube sample after drilling a row of small holes on close spacings or opening a slit on each side of the tube. Water with a constant head, H, is supplied from one end to the other across the diameter of the soil sample until a steady state of flow is established through the soil still encased in the tube. From the rate of flow discharge measured, q, the coefficient of horizontal permeability, k_h, may readily be computed over each divided section or over the entire length of the soil sample by Equation 2, i. e., $k_h = 4(q/H)$. Thus an important information can be obtained on horizontal mass permeability of a thick compressible stratum.

3. This test can be conducted on a wide range of soils as long as a good quality undisturbed sample is obtained by means of a thin wall tube.

4. In addition to the fact that this permeability test is simple and easy to run, the soil sample may be extruded out of the tube after the test and be examined visually in detail. Furthermore the extruded sample may be tested for various soil properties including strength and compressibility, for the test does not disturb the sample but merely saturates it with water. It is hoped that this permeability test will constitute part of a routine laboratory testing program for a geotechnical investigation for soft clay deposits, for which horizontal mass permeability has a decisive role on their consolidation behavior.

REFERENCES

Akagi, T. 1983. A simple laboratory permeability test. Proc. Symposium on Recent Developments in Laboratory and Field Tests and Analysis of Geotechnical Problems. 82-95. Bangkok.

Akagi, T. and Ishida, T. 1988. A horizontal permeability test for a cohesive soil stratum. Proc. Japan Society of Civil Engineers. 394/III-9, 123-130 (in Japanese).

Akagi, T. and Ishida, T. 1989. A study on mass permeability and settlement of soft ground. Tsuchi-to-Kiso(J. JSSMFE), 37/6:41-47 (in Japanese).

Akagi, T. 1989. Settlement and strength of soft ground stabilized by driven vertical drains. 596-612. The Art and Science of Geotechnical Engineering. New Jersey:Prentice Hall.

Davis, E. H. and Poulos, H. G. 1972. Rate of settlement under two- and three-dimensional conditions. Geotechnique. 22/1:95-114.

2 Engineering behaviour of soils (including collapsible and expansive as well as other problematic soils)

Developments in Geotechnical Engineering, Balasubramaniam et al. (eds) © 1994 Balkema, Rotterdam, ISBN 90 5410 522 4

Effect of coarse inclusions on strength of residual/colluvial soils

Noppadol Phienwej & Bhupatindra Gopal Vaidya
Asian Institute of Technology, Bangkok, Thailand

ABSTRACT: Effects of coarse inclusions on strength behavior of a compacted residual soil were investigated through a laboratory direct shear test program. Factors investigated included inclusion content, size, shape, orientation and spacing, and saturation of matrix materials. It was found that critical inclusion content , below which inclusions have no effect, for this clayey matrix material was between 10% and 20%. The important factors controlling a strength increase of soils are inclusion surface angularity and roughness, inclusion spacing, and height of inclusion perpendicular to shear plane.

1 INTRODUCTION

Engineering behavior of residual soils is complex and is much different from that of transported soils. It reflects the nature of parent rock and weathering environment that produced the soil. In a weathering profile, particle sizes of the soil may vary from clay size particles to very coarse corestones. The corestones act as coarse inclusions in the finer grained soil matrix and is one of the principal factors causing complex and unusual behavior of the soils. However, its effect is generally neglected in usual laboratory tests for strength determination of residual soils as the coarse materials are usually excluded (Brand, 1992). Consequently, laboratory strength values do not correlate well with observed field behavior. The performance of the soils tends to be better than that estimated, resulting in a conservative design (De Mello, 1972). For instance, it is an observed fact that slopes in residual soils stand at slope angles that has theoretical factors of safety much less than one (Brand, 1992).

Strength of a residual soil is much variable as it is influenced by a number of factors, namely,

1. matrix strength and coarse inclusions,
2. bond strength, void ratio and relict structure,
3. soil suction and degree of saturation.

While many research works have dealt with the last two groups of factor, there are only few attempting to investigate the effect of coarse inclusion owing to difficulties in sample preparation. Due to the lack of knowledge on this subject, a comprehensive laboratory strength test program was conducted by the authors to make a detailed investigation and obtain a better understanding on the subject. The research was inspired by a similar research work on colluvial soils, which has to deal with coarse inclusion, conducted at the Hong Kong Geotechnical Control Office (GCO) (Brand, 1992). The purpose of this paper is to present the author's findings and compare them with those reported in the literature, so that the influencing role of the coarse inclusion can be properly identified and evaluated. From the work on colluvial soils by Irfan & Tang (1992), factors related to coarse inclusions that govern shear strength of residual soils are: a) inclusion content; b) size; c) shape; d) elongation ratio; c) spacing; f) density; and, g) matrix strength. These factors were investigated to a certain extent in the test program and the results are presented herein.

2 TEST PROGRAM

The main purpose of this study is to investigate the effect of coarse inclusions on strength behavior of residual soils. Therefore, direct shear tests on a reconstituted residual soil with various conditions of coarse inclusion were performed.

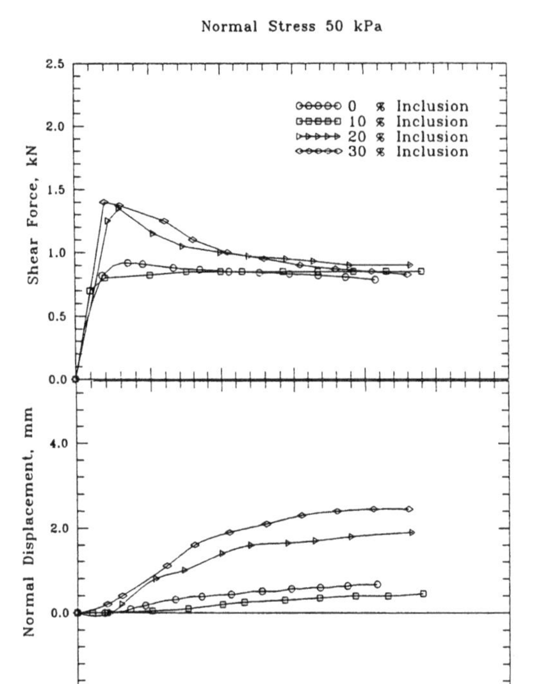

Fig. 1 Influence of Inclusion on Shear Force Displacement Behavior

Due to the nature of the test specimens, the effect of other influencing factors, namely: *in-situ* density, bond strength, and relic structure, were not investigated and, therefore, excluded from the scope of the study. Direct shear test, which has become a preferred test method for strength determination of residual soil (GCO, 1984), was adopted because it allowed less difficulties and time for specimen preparation than the triaxial test, thus, a larger number of tests could be made for this comparative study. Altogether, more than 80 tests were made during the six-month testing period of the master thesis work of the second author (Bhupatindra, 1993).

Granitic residual soils (Grade V to VI) from the eastern seaboard of Thailand were chosen as the matrix material since it is the type of residual soil that has posed problems in foundation and slope works owing its vulnerability to weakening when wetted. The soil is classified as clayey to silty sand (SC-SM) with 20% gravel, 55% sand, 14% silt and 11% clay size particles. The plasticity index is 30-35%. It has maximum dry density of 18.3 kN/m^3 at optimum water content of 12.3% in the Standard Proctor Compaction.

Three different types of inclusions were used for comparison, i.e.

- crushed rock (sizes 6.3-8.0 mm and 12.7-16.0 mm)
- river gravel (size 6.3-8.0 mm)
- cement blocks (7.5 mm and 10 mm cubes, and 7.5 mm x 7.5 mm x 15 mm prismatic blocks)

The limestone crushed rocks represent angular shaped inclusions. They were sieved and manually sorted to obtain the specified size ranges. Flaky and elongated pieces were discarded. Only pieces with an elongation ratio (maximum length to minimum length) smaller than two were used for the test. The bulk density of the limestone is 26.7 kN/m^3. The rounded shape river gravels, bulk density of 27.8 kN/m^3, were sorted and only near spherical pieces were used. High strength cement mortar (compressive strength of 45 MPa, bulk density of 24.0 kN/m^3) was used to make cement cubes and rectangular prismatic blocks of uniform sizes to investigate the effect of inclusion elongation ratio and orientation. Three orientations were simulated, i.e. HAA - long dimension along the direction of shear, HPA -long dimension perpendicular to the direction of shear, and VERT - long dimension in vertical direction.

The inclusions were hand placed in each compaction lift of matrix material in a 154 mm by 154 mm direct shear box. The thickness of each lift and the pattern of inclusion placement were made in such a way that the inclusions were uniformly distributed and their positions on the shear plane, and the matrix-material was uniformly compacted to the specified density. The density was targeted at 16.5 kN/m^3 for all specimens in order to eliminate the effect of the variation in the matrix properties on the test results. A higher density closer to the compaction maximum density could not be achieved in the preparation of specimens with high inclusion content. Therefore it was necessary to adopt a smaller density. Inclusion contents were varied up to 30% by volume. A higher content could not be prepared to obtain the targeted matrix density.

After compaction with hand tamping on every lift, the specimens were soaked in a bucket for 12 hours as a referance condition. However, soaking time was also varied for some specimens in order to investigate the effect of the degree of saturation of the matrix material on the effect of coarse inclusion. After soaking, the specimens were set in the direct shear machine and allowed to consolidate for 12 hours at two main normal stresses of

50 and 100 kPa before shearing. Other normal stresses were also used for some specimens. A shear rate of 0.084 mm/min as recommended for granitic residual soil (GCO, 1984) was used.

3 TEST RESULTS AND DISCUSSIONS

3.1 *Inclusion Content*

Fig. 1 shows the effect of inclusion content on shear force and displacement behavior. As it can be seen, the presence of adequate inclusion content resulted in an increase in dilation and thus strength of the soil. Fig. 2 is a summary plots of the variation in strength of the composite material with percent inclusion content for various inclusion types at two normal stresses (50 and 100 kPa). Fig. 3 shows the variation in Mohr-Coulomb shear strength parameters, ϕ and c with inclusion content for different shapes and sizes.

For coarse inclusion of 10% or below the strength of the composite soil is mainly governed by the matrix material for all groups of inclusion. The uniformly distributed inclusions have no influence at this level of inclusion content. Inclusions started to show an effect as the content increased from 10 to 30%, resulting in strength increase for all normal stresses tested. The amount of strength increase varied with inclusion shape, orientation, size and spacing which will be explained in the subsequent sections. The increase was very significant (60% increase) for some types of inclusion but very small (10% increase) for the others depending on the above factors.

It can be stated that the critical inclusion content (below which the strength is predominantly controlled by the matrix material) of this type of residual soil is somewhere between 10 and 20%. Similar finding was reported by Irfang and Tang (1992) from their triaxial tests on a colluvial sandy granitic soil in Hong Kong. Slightly higher ranges of the critical inclusion content (30 to 55% by weight equivalent to 20 to 40% by volume) were reported for clayey soils with gravel inclusions based on triaxial and direct shear tests (Holtz and Ellis, 1961; Dodiah et al. 1969; Padwardhane et al. 1970; Hender and Martin, 1982). However, the values were much lower than the critical content of clay sand mixes, ranging from 60 to 80% by weight (Miller and Sower, 1957; Kurata and Fujishita, 1960; Toh and Ting, 1983; Nishioke et al. 1988). This indicated that the critical inclusion content is smaller for larger sized inclusions.

The increase in the coarse inclusion content to

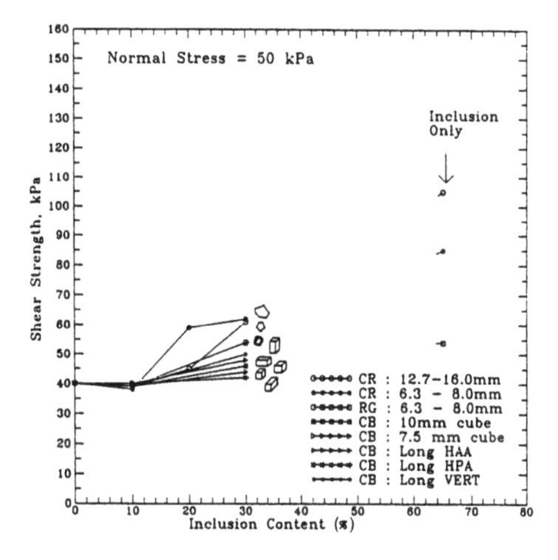

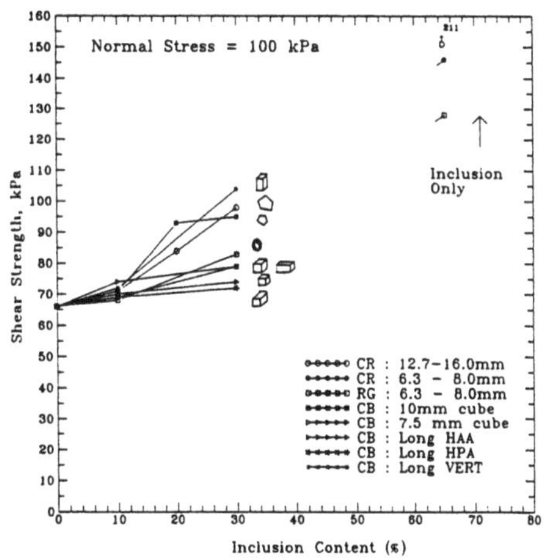

Fig. 2 Variation in Strength of Residual Soil with Inclusion Content & Inclusion Properties

30% resulted in increases in both Φ and c for all sizes and shapes of inclusion (Fig.3). However, the cohesion c will decrease and approach zero as the inclusion content increases.

3.2 *Inclusion Size*

The effect of inclusion size was investigated using the crushed rock and cement cubical inclusions. For the two sizes of the inclusions under investigation (6.3-8.0 mm and 12.7-16.0 mm for crushed rock and 7.5 mm and 10 mm for cement cubes) the smaller sized inclusions yielded a significantly higher strength increase (48% vs. 15% strength increase at 50 kPa normal stress) at 20% inclusion content. But as the inclusion content increases to

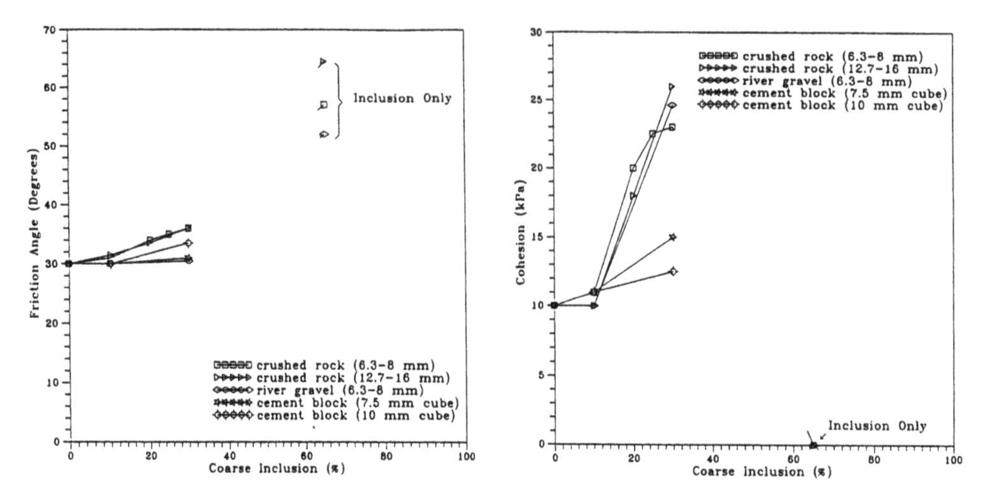

Fig. 3 Variation in Shear Strength Parameters with Inclusion Content for Various Inclusion Properties

30%, the difference reduced and became insignificant (both around 55% strength increase at 50 kPa normal stress), although the larger sized inclusions consistently showed a very slightly higher strength increase for both shapes and both normal stresses. It suggests that at a higher content, the effect of larger sized inclusions become more significant. The amount of strength increase was higher at smaller normal stress. In addition, there was a distinct difference in shear displacement in that the soil with smaller sized inclusions reached peak strength at a smaller shear displacement. This behavioral response was mainly related to the spacing between the inclusions in the composite soil. This factor controls the interaction condition between the stiff and hard inclusions in the much softer matrix material.

3.3 *Inclusion Shape*

The behavior of residual soils with three inclusion shapes of similar size, namely angular (crushed rock), rounded (river gravel) and cubical (cement cube) shapes, were compared. It was clearly shown that angular shaped inclusion (crushed rock) with rough surfaces has a much more influence in increasing the strength of the composite soil than the rounded and cubical inclusions (strength increase was 55% for the former compared to 35% and 12% for the latter two). It yielded the highest dilation during shear. This was attributed to its promotion for a better frictional contact with the matrix material and more interlocking interaction between adjacent inclusions. Round shaped inclusions show a higher effect in increasing strength than the cubical shaped inclusion. A likely explanation is the difference in interaction between inclusions and the matrix soil due to difference in surface conditions of the inclusions, i.e. curved surface of the former versus flat and smooth surfaces of the latter. Very little strength increase was provided by the cubical shaped inclusions even at 30% content (12 % increase). Little dilation was induced during shear of both types of inclusion. This is in agreement with Irfang and Tang (1992) who showed that steel ball inclusions (with round and slikensided surface) showed no influence on strength of colluvial soils at all contents. Very poor friction existed in such a case. It can thus be stated that the contact between the surface of inclusions and surrounding matrix has a profounded effect on the role of the inclusions.

3.4 *Inclusion Orientation*

The rectangular prismatic cement blocks (7.5 mm × 7.5 mm × 15 mm) were used to investigate the strength variation of the composite soil with inclusion orientation. At 10% inclusion content, there was practically no difference in strength among the three orientations investigated (HAA, HPA and VERT). At a higher inclusion content (30%), the strength increases were not so significant for HAA and HPA oriented inclusions at both normal stresses (only 10-20% increase) while the VERT oriented inclusions showed a large increase at 100 kPa normal stress (56% increase). Among the three orientations, the VERT oriented inclusions always gave the largest strength increase while the HPA oriented inclusions yielded the lowest. This observed behavior may be attributed to two main factors, i.e. inclusion height perpendicular to the shear plane and spacing between neighboring inclusions along the shear direction. The former showed a much more effect than the latter especially at a high normal stress. The specimen with VERT oriented inclusion showed the highest dilation which greatly promoted strength increase. The strength of specimens with HAA oriented inclusions developed faster than that of the HPA specimens which clearly reflects the effect of inclusion spacing. The HAA inclusions were situated closer and yielded a slightly higher strength (20% vs. 10% strength increase). The effect of inclusion spacing is also clearly shown when comparing strengths of HPA and HAA oriented inclusions with the cubical inclusions. All three had the same height perpendicular to the shear plane but different spacings. At 30% inclusion content, the amounts of strength increase at both normal stresses were in a reverse order of the inclusion spacing although the amount of increase was not high (10-20%).

The effect of inclusion content, shape and orientation found in this experiment program seemed to be comparable with the result of theoretical analysis on soil slope stability containing various inclusion properties reported by Irfan and Tang (1992) (Fig. 4). In their analysis, a trial for factor of safety were made to determine critical surface which was assumed not to pass through the inclusion.

3.5 *Inclusion Spacing*

It was very clear when comparing the effect of inclusion size and orientation as discussed above that the spacing between adjacent inclusions along the direction of shear has a profounded influence on the strength of the composite soil especially at a low inclusion content. The spacing between the inclusions controls the degree of interaction between the adjacent inclusions and the deformation behavior of matrix material located in between. The closer the spacing, the higher the confinement effect during shear displacement on the matrix material. Consequently, the closer the inclusion, the higher the strength increase.

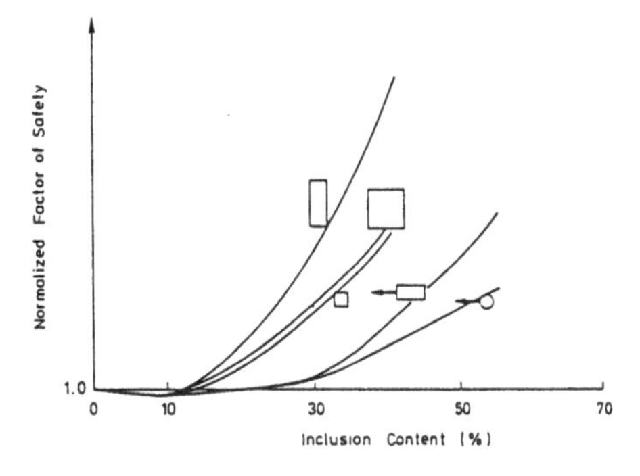

Fig. 4 Theoretical Results on Effect of Various Inclusion Properties on Stability of a Slope (After Irfan & Tang,992)

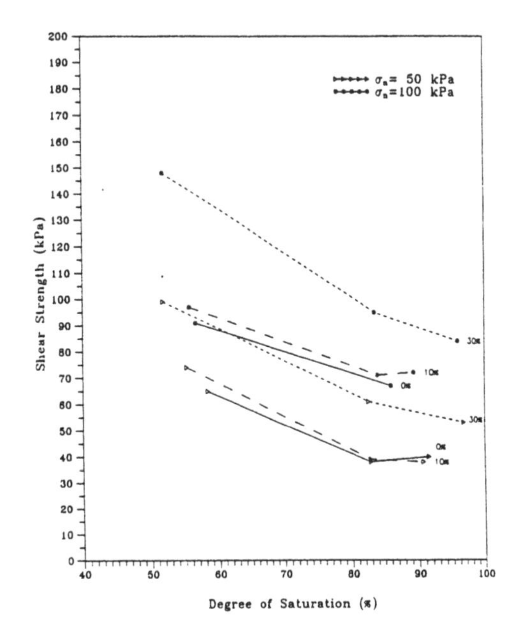

Fig. 5 Effect of Degree of Saturation on Strength at Various Inclusion Contents

3.6 *Matrix Material Saturation*

The influence of saturation of the matrix material on the role of coarse inclusions on the strength of the composite soil were investigated at three saturation conditions (i.e. 0, 12 and 60 hours of soaking). After 60 hour soaking, the specimens almost reached 100% degree of saturation while the specimens with 12 hours soaking already reached very high degrees of saturation. The variations in peak strength of the composite soil with various contents of crushed rock inclusions with degree of saturation are summarized in Fig. 5. The strength of the 12-hour soaking is nearly the same as that of the 60-hour soaking. The test results showed that the effect of inclusion in increasing the strength of the material was enhanced by a decrease in the degree of saturation for both the low and high normal stresses. In other words the role of inclusion is more pronounced in relatively dry soil.

4 CONCLUSIONS

The following can be concluded from the direct shear test program on compacted reconstituted granitic residual soil with various conditions of coarse inclusions.

1. The critical inclusion content below which the strength of the soil is mainly controlled by the matrix material is between 10 and 20% by volume. This is lower than those of the clayey soil with sand inclusion as reported in the literature.
2. The significant factors related to inclusion conditions that control the strength increase of residual soil with the presence of inclusions are shape and surface angularity of inclusions, spacing between inclusions along a shear plane, and height of inclusion perpendicular to a shear plane. Inclusion size seems to have little effect once an inclusion content is large enough.
3. Presence of inclusions in a soil promotes dilation during shear and an increase in confinement of the matrix material along shear plane resulting in a strength increase.
4. The effect of inclusion on the amount of strength increase for a given inclusion condition and normal stress is more pronounced in a dry soil than in a wet soil.

5 REFERENCES

Brand, E.W. (1992), Slope Instability in Tropical Areas. Proceedings of the Sixth International Symposium on Landslides, Christchurch, New Zealand, Vol 3, 1992.

Bhupatindra, G. V. (1993), Effect of Coarse Inclusion on Shear Strength of Granitic Residual Soil. MSc. Thesis, Asian Institute of Technology.

De Mello, V.F.B. (1972), Thoughts of Soil Engineering Applicable to Residual Soils, Proceedings of the Third Southeast Asian Conference on Soil Engineering, Hong Kong, 5-34.

Dodiah, D., Bhat, H.S., Somasekhar, P.V., Sosalegowda, H.B., and Ranganath, K.N. (1969), Shear Characteristics of Soil-Gravel Mixtures, The Journal of the Indian National Society of Soil Mechanics and Foundation Engineering Vol. 8, No. 1, January 1963, 57-66.

GCO (1984), Geotechnical Manual For Slopes, Second Edition, Geotechnical Control Office, Hong Kong.

Hencher, S.R. and Martin, R.P. (1982), Applica tion of Back Analysis to Some Hong Kong Land slides, Proceedings of the Seventh Southeast-Asian Geotechnical Conference, Hong Kong, Vol. 1, 125-142.

Holtz, W.B. and Ellis, W. (1961), Triaxial Shear Characteristics of Clayey Gravel Soils, Proceedings of the Fifth International Conference of Soil Mechanics and Foundation Engineering, Paris, Vol. 1, 143-149.

Irfan, T. Y. and Tang K. Y. (1992), Effect of Coarse Fraction on the Shear Strength on Colluvium, a Draft Report, Geotechnical Engineering Office, Hong Kong.

Kurata, S. and Fujishita, T. (1960), Research on the Engineering Properties of Sand-Clay Mixtures, Proceedings: List of Papers, First Regional Conference of the Asian Region of the International Society of Soil Mechanics and Foundation Engineering, Delhi, India, 1-12.

Miller, E.A. and Sowers, G.F. (1957), The Strength Characteristics of Soil Aggregates Mixtures, Highway Research Board Bulletin, No. 183, 16-23.

Nishioke, T., Sonobe, Y. and Yoshida, N. (1988), A New Approach to Determine the Shear Strength of Clay Sand Mixed Soil, Engineering Problems of Regional Soils, Proceedings of the International Conference on Engineering Problems of Regional Soils, Beijing, China, 659-664.

Patwardhana, A.S., Rao, J.S. and Gaidhane, R.B. (1970), Interlocking Effects and Shearing Resistance of Boulders and Large Size Particles in a Matrix of Fines on the Basis of Large Scale Direct Shear Tests, Proceedings of the Second Southeast Asian Conference on Soil Mechanics, Singapore, 265-273.

Toh, C.T. and Ting, W.H. (1983), Characteristics of a Composite Residual Granite Soil, Recent Developments in Laboratory and Field Tests and Analysis of Getechnical Problems, Bangkok.

Developments in Geotechnical Engineering, Balasubramaniam et al. (eds) © 1994 Balkema, Rotterdam, ISBN 90 5410 522 4

Experience with swelling and collapsing soils

G.Wiseman
Department of Civil Engineering, Technion-Israel Institute of Technology, Haifa, Israel

ABSTRACT: The paper concentrates primarily on the problems and methods of site exploration, field and laboratory testing of swelling and collapsing soils. The paper is based on 40 years of experience with swelling soils in the semi-arid north, and with loess soils in the arid south of Israel. A method of predicting heave based on one dimensional swelling tests is discussed. Heave measurements on 3 different airfield pavements on a swelling soil are analyzed. Prototype load testing of footings and cast insitu piles at a site suspected of collapse potential are presented.

1 PREFACE

The subtitle given this symposium - From Harvard to New Delhi, is possibly the clue to my having been invited to participate. I was a graduate student, studying Soil Mechanics at Harvard University in 1952 during the days of Critical Void Ratio, (before the days of Critical State Soil Mechanics), and at a time when the development of pore pressure during undrained shearing was presented as the shifting of effective stress circles as described by Professor Casagrande in his Working Hypothesis in the 3rd Progress Report in 1941 to the U.S. Waterways Experiment Station, (and not A and B coefficients).

I returned to Israel in 1955 knowing something about the swelling clay shales in the spillway of the Fort Peck dam, and after having done some testing of the properties of Mississipi Gumbo (a highly plastic clay containing a lot of Montmorillonite). However, having studied at Harvard, I knew most about Boston Blue Clay and was not knowledgeable about the problems I was going to be so concerned with over the years - swelling and collapsing soils in a semi-arid region, which are the subject of this paper.

Before proceeding further it might be useful to review the contributions on the subject of foundations on swelling and collapsing soils in the proceedings of the International Conference on Soil Mechanics and Foundation Engineering held at Harvard University in 1936.

2 HARVARD - 1936

There are two contributions on experience with swelling soils in the proceedings of the conference held at Harvard in 1936, which relate to serious damage to pavements and buildings.

The first, by Porter (1936) reported on movement observations on a 500 meter long concrete pavement laid in Texas in 1931. Cracks 15 cm wide, 3.5 meters deep were observed in the pavement shoulders and seasonal fluctuations in elevation of 12 cm at the pavement edge and 6 cm at the pavement centerline were measured. The subgrade soils at the site were reported to have a Liquid Limit of 80 to 100, a Plasticity Index of 52 to 74 and a Shrinkage Limit of 8 to 10. It is worth noting that based on the Atterberg Limits alone, the subgrade soils would be identified today as a highly plastic clay (CH) probably exhibiting large volume change if large variations in subgrade moisture content were to occur.

The second, a lengthy paper by Wooltorton (1936) contains detailed descriptions of damage to numerous buildings in the Mandalay District of Burma due to vertical and horizontal ground movements. The area has high rainfall, but also extended dry seasons. Seasonal moisture variations were measured at depths of 3.5 meters and cracks in the soil to even greater depths. About 100 damaged buildings were examined in detail and the conclusion arrived at, that the building damage was not due to poor workmanship, or to a low bearing capacity of the soil, as was presumed at the time, but to seasonal ground movement. Computations are presented in the paper showing that the cracked grade beams, whose movement had been restrained, probably had swelling pressures of 100 kPa acting upward on the bottom of the beams. Wooltorton made various recommendations how to build structures on swelling clay subgrades so as to minimize potential damage. In a discussion of Wooltorton's paper, Simpson (1936), already referring to 20 years experience, expressed the opinion that the only solution was to use deep piers and suspended floors.

There was also a contribution by Abeleff (1936) who in referring to experience in the U.S.S.R. with collapsable soils wrote "if the soil has a large voided structure .... deformations of the soil can be produced through accidental moistening of the soil e.g. from a water supply, etc".

3 INTRODUCTION

The International Soil Mechanics community has continued to show an interest in swelling and collapsing soils. There have been numerous international conferences, devoted primarilly to swelling soils. The technical literature contains recommendations for what to avoid, and what to incorporate into designs so as to minimize potential damage. Nevertheless there are still numerous reported cases of damage. Futhermore our ability to predict behaviour of structures on potentially swelling or collapsing soils is still poor, except for limited circumstances. We know how to identify, potentially troublesome soil conditions, using simple index tests. We shall describe these below. For swelling soils we can also do a fair job of predicting movements and/or pressures developed, for those limited cases where we can predict moisture changes during the design life of the structure. However we shall report below a case where even predicting the direction of movement (whether swelling or collapse) proved to be difficult.

4 DENSITY CRITERIA

Swelling and/or collapse of foundation soils is dependant on soil type, insitu density and moisture content, the surcharge stress, and the availability of water. Simple density criteria, as related to the Liquid Limit have been found to be surprisingly useful in identifying the existence of a potential foundation problem with respect to both swelling and collapse of unsaturated soils upon wetting. Some early criteria with respect to collapse of loess refer to a critical density (approx. 14 kN/cu m) as being the density below which soil is most likely to collapse upon wetting. A more logical and generally used criteria is to compute the Density Ratio, defined as the insitu dry density of the soil divided by the dry density the soil would have if completely saturated at a void ratio corresponding to the moisture content at the Liquid Limit.

It has been reported (Zur & Wiseman 1973) that if the Density Ratio as defined above is less than 1.1, the soil should be considered prone to collapse. Whereas if the Density Ratio is greater than 1.3, the soil should be considered prone to swelling. Examination of data from many laboratory studies on undisturbed samples of unsaturated soils tested in rigid ring consolidometers, and a more limited amount of field data, have shown the above criteria to be very useful in identifying the existence of a swelling or collapse problem.

Table 1.a. Collapse @ 100 kPa (%) (ranked by density)

Sample No	Soil Class	Dry Density kN/cu m	Collapse (%)
1	CL	13.2	1
2	ML	13.2	11
3	CL	13.5	4
4	ML	13.9	8
5	ML	14.5	7
6	SM	15.1	12
7	CL	15.2	0
8	CL	15.5	-0.5
9	SM	17.3	0

Shown in Table 1a are laboratory test results on undisturbed samples from a large site in the arid south of Israel that was being investigated for suspected collapse phenomena. The samples were tested in rigid ring consolidometers, loaded to 100 kPa at their natural moisture, and the collapse settlement on inundation measured. Test results are listed in order of increasing dry density. The same test results are listed in Table 1b in order of increasing Density Ratio. It may be clearly observed that density alone when used as a single criteria for a range of soil types is a poor predictor, whereas the Density Ratio would appear to be useful. Furthermore the results are consistent with the criteria given above.

Table 1.b. Collapse @ 100 kPa (%) (ranked by density ratio)

Sample No	Soil Class	Density Ratio	Collapse (%)
5	ML	0.69	7
4	ML	0.79	8
6	SM	0.79	12
2	ML	0.83	11
3	CL	0.98	4
9	SM	1.00	0
1	CL	1.10	1
8	CL	1.15	-0.5
7	CL	1.25	0

Komornik and David (1969) published a data base of laboratory swelling studies that they used for developing an expression for predicting the swelling pressure of a soil from a knowledge of certain index properties. Shown in Table 2a are the percentage of test results below 40 kPa and 100 kPa, respectively, grouped by Dry Density and listed in order of increasing density. Shown in Table 2b are the same test data grouped by Density Ratio. It can be seen that Density alone is also a poor predictor for swelling phenomena whereas the Density Ratio is more useful and also consistent with the criteria given above. (Fig. 1)

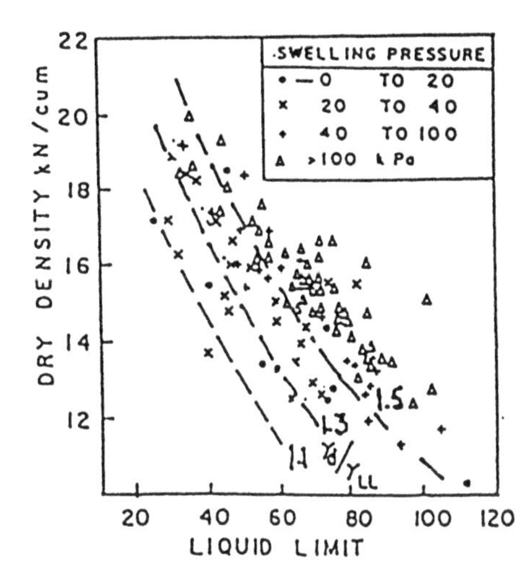

Fig. 1. Swelling pressure studies

Table 2.a. Swelling Pressure versus Density.

Dry Density kN/cu m	Tests <40 kPa (%)	Tests <100 kPa (%)
<12	33	100
12 - 14	42	67
14 - 16	24	46
16 - 18	21	38
18 - 20	33	56

Table 2.b. Swelling Pressure versus Density Ratio.

Density Ratio	Tests <40 kPa (%)	Tests <100 kPa (%)
<1.3	100	100
1.3 - 1.5	50	80
>1.5	7	30

5 SWELLING PRESSURE

Komornik and David (1969) used their data base to develop a prediction equation for the numerical estimation of the swelling pressure as related to Liquid Limit, Dry Density and Moisture Content. (It should be noted that the Density Ratio does not include any consideration of moisture content). Their equation is widely referenced and fills a need for identifying swelling potential at a stage when only limited test data is available.

Shown in Fig. 2 are contours of equal swelling pressure, on a Dry Density - Moisture Content plot, for two different Liquid Limits, as computed using the Komornik - David equation. Even for the data base from which it was developed, there is considerable scatter when plotting measured versus predicted swelling pressure. It is therefore important, that when using the equation, their coefficients be adjusted using some site specific data. A procedure for doing this has been described elsewhere (Wiseman et al., 1985).

6 PREDICTION OF HEAVE

We use a simple procedure for quantifying expected heave based on the data obtained from one-dimensional swelling tests on undisturbed, or when applicable on

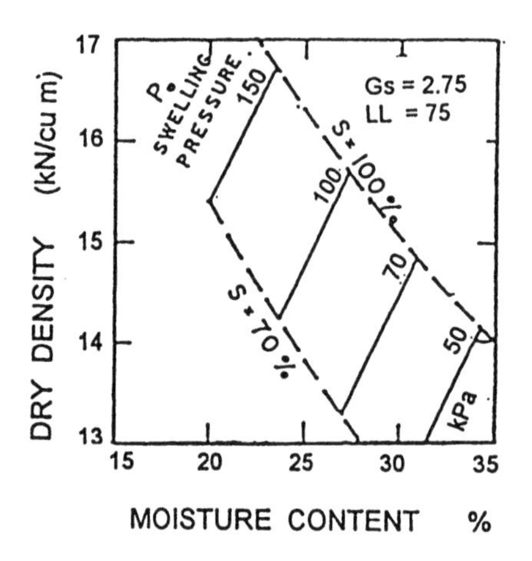

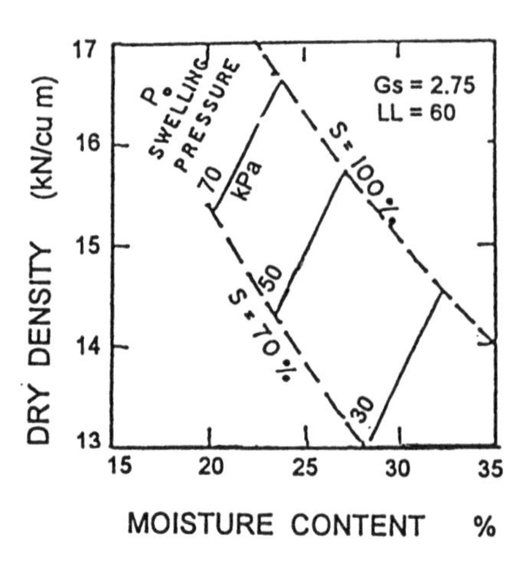

Fig. 2. Contours of equal swelling pressure on moisture-density plots

compacted samples, suitably surcharged and then submerged. The expected equilibrium suction conditions are accounted for in the computation of heave by assuming, that in addition to the total stress acting due to overburden and applied load, an additional surcharge stress can be estimated that would restrain swelling to the same degree that the equilibrium pore water suction would. We have called this additional surcharge stress the "suction stress equivalent".

This procedure is therefore not based directly on considerations of suction but rather on our ability to estimate the "suction stress equivalent" as defined above. At high degrees of saturation the "suction stress equivalent" would be expected to be close to the actual suction, whereas at low degrees of saturation it would be expected to be considerably less than the actual suction.

The "Active Depth" for predicting heave then becomes that depth at which the total overburden pressure plus the "suction stress equivalent" is equal to the Swelling Pressure determined for zero vertical movement under submerged conditions. In effect the boundary conditions for the heave prediction are then this total vertical stress.

We have subsequently included the "lateral restraint factor" to account for the difference between the lateral boundary conditions in the field and those prevailing in the one-dimensional consolidometer. This factor could be as high as 1.0 if there are no vertical cracks and as low as 0.3 for a badly cracked soil profile.

Prior to submergence the specimens are always incrementally loaded to at least the existing overburden pressure. After loading to existing overburden pressure, the specimen is submerged and the "swelling pressure" determined for zero vertical deformation followed by rebound at various reduced vertical pressures so as to develop a complete percent swell versus vertical pressure curve from one specimen. If the budget and time available for testing permits, the use of separate specimens for each vertical pressure is a preferred procedure.

The soil profile must be divided into layers that may be assumed to be characterized by a single percent swell versus pressure curve such as shown in the upper right hand part of Fig. 3. The total heave is the sum of the heaves computed for all such layers within the "Active Depth". An example of such a computation where one swelling curve is considered to be applicable over the "active depth" is demonstrated in Fig. 3. The total heave is the result of integrating the area under the curve in the lower left hand part of the figure. The 1:1 slope in the upper left is for a "lateral restraint factor", (f) = 1.0.

The applicability of the procedure to field conditions is therefore predicted on the assumption that we have a basis for estimating the additional surcharge stress that can be assumed to restrain swelling to the same degree that the equilibrium pore water suction would. There are cases where this may be done with a fair degree of reliability. The more common situation is that sources of water are unpredictable.

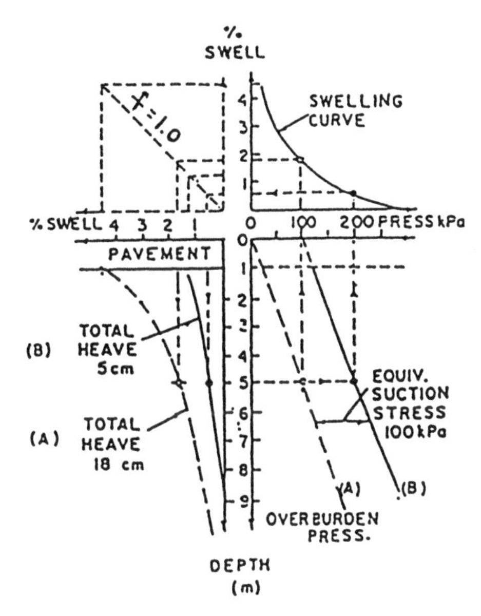

Fig. 3. Computation of heave

Though we have been using the above procedure for many years, we do not have direct confirmation as to its ability to predict heave under field conditions. We have primarily been using this procedure as a tool for evaluating sites having swelling clay soil profiles and deciding on appropriate foundation solutions for both pavements and structures.

7 SITE INVESTIGATION - SWELLING SOILS

Deciding on the potential severity of a swelling soil problem for a specific structure at a specific site, and on an appropriate field and laboratory site investigation program, is a very critical matter for any project. This subject is treated at considerable detail in a microcomputer based Expert System for Foundations on Expansive Soils (Wiseman et al. 1992) in the section dealing with Site Investigation and Classification.

The purpose of the above section in the Expert System is to guide the user as to the sort of information that is generally required in order to classify a site as to swelling potential. Based on the answers to a set of questions, recommendations are made as to the type and extent of field sampling, and laboratory testing, that would be required to provide data for designing foundations on swelling soils.

The user is asked to provide answers chosen from NO, YES, UNKNOWN, to a series of questions in five categories as described in the following:

SOIL: - Is the % colloidal material > 15? Is the Plasticity Index > 18? Is the Shrinkage Limit < 15? Is the Free Swell > 50%? For any question to which the user knows the answer, he is then asked to provide numerical values.

PROFILE: - Is the depth to the water table more than 4.0 m? Is the moisture content / plastic limit ratio less than 1.2? Are shrinkage cracks observable at the ground surface? In the driest years are there more than 3 continuous months without rainfall?

BUILDING ENVIRONMENT: - Have there been recent changes in drainage patterns? Have trees been recently planted or felled? Are there sources of water such as from car washing? Are there sources of heat such as kilns that might dry out the soil? Are there septic tanks?

If there are existing buildings located nearby the following questions are asked:

EXISTING BUILDINGS: - Is there evidence of tilting or horizontal movement? Are there distorted window or door frames? Is there diagonal cracking of walls? Is there excessive movement in joints? Is there evidence of distress in foundation columns that might be due to differential loading?

PROPOSED BUILDINGS: - Are there retaining or basement walls? Are there slabs on grade for which 1.0 cm of movement would be considered excessive? Are there elements in the structure that are sensitive to even small differential movements (large swinging doors, crane rails on columns)? Are there plastered block walls or other features that would show distress for even small differential movement of the foundations?

The system then offers comments such as pointing out inconsistencies in the responses. When necessary it draws attention to the fact that site conditions are very difficult and that therefore the client should be warned of all possible dangers.

Based on the user's responses, four alternate programs of field sampling and laboratory testing are recommended (ranging from simple index tests on disturbed samples to complete swelling studies on undisturbed samples). The extent of the program depends on the evaluation by the system of the severity of the expansive soil problem. The proposed investigations are exclusive of any program for developing strength and settlement parameters, or for the depth of boring required to assess bearing capacity.

8 CASE HISTORY - PAVEMENTS

At an airfield site in the north of Israel having a semi-arid climate and a deep, highly swelling clay profile, heave measurements were carried out for many years on various pavement structures having varying degrees of access to surface water. These included concrete aprons, conventional flexible pavements and a full-depth asphalt pavement (Wiseman et al, 1981). Measured heaves varied from as little as 1 cm to as much as 18 cm. An example of such measurements over an 8 year period for a typical cross-section of an 80 cm conventional flexible airfield pavement is shown in Fig. 4. The edge of the pavement with less heave had been provided with sub-drains at a depth of 1.4 m.

Using the procedure described in the previous section the back-calculated associated "suction stress equivalent" values ranged from 160 kPa to zero. (Table 3) These values were consistent with the drainage provisions and/or the measures taken for restricting access to surface water at each of the pavement structures, and have since been used as a guide for estimating "suction stress equivalent" values for use in heave prediction calculations.

Several case studies on airfield

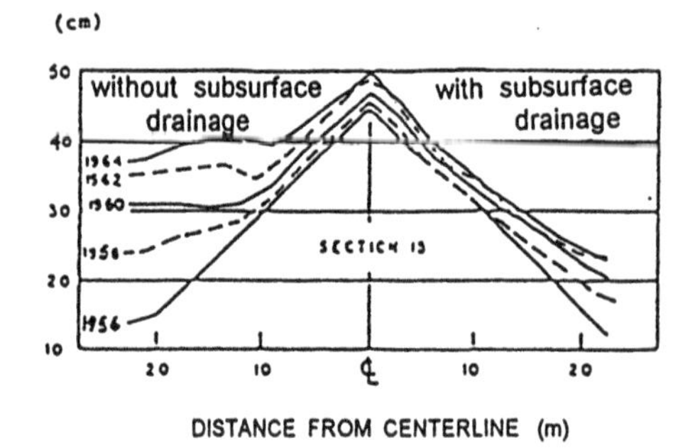

Fig. 4. Typical cross section showing pavement heave

Table 3. Measured heave and back calculated suction equivalent.

Pavement Type	Location	Meas. Heave (cm)	Eqiv. Suction (kPa)
Flexible (80 cm)	Center	5	95
	Edge (Drains)	9	50
	Edge (No Drains)	18	0
Full Depth (50 cm) AC		2	160
Concrete (30 cm)		12	25

pavements on expansive clay subgrades in Israel indicate that pavement roughness due to differential subgrade heave starts to become troublesome 5 years after construction and often becomes intolerable after about 10 years. It is frequently the major factor in deciding to overlay a pavement. (Uzan et al, 1984).

9 SITE INVESTIGATION - COLLAPSE

One of the main difficulties facing the geotechnical consultant in arid zones is in estimating the expected moisture regime during the design life of the structure. In Israel, experience has been that collapse of low density soils has always been due to "accidental" wetting of the foundation soils. (see also Abeleff (1936)). Since this possibility always exists, the site investigation should be aimed at providing information on the expected performance of the foundation soils under conditions of at least partial wetting.

Described below is the investigation for a large project which included both residential and industrial buildings in the Southern Negev of Israel (Wiseman & Lavie 1983). The area has a desert climate with only occasional rainfall. The soils are coarse and fine grained desert aluvium and loess.

To evaluate the collapse potential of the soils at the site, it was necessary to obtain good quality undisturbed samples at their low natural moisture content. A Denison triple tube core barrel was found to work well using compressed air as the

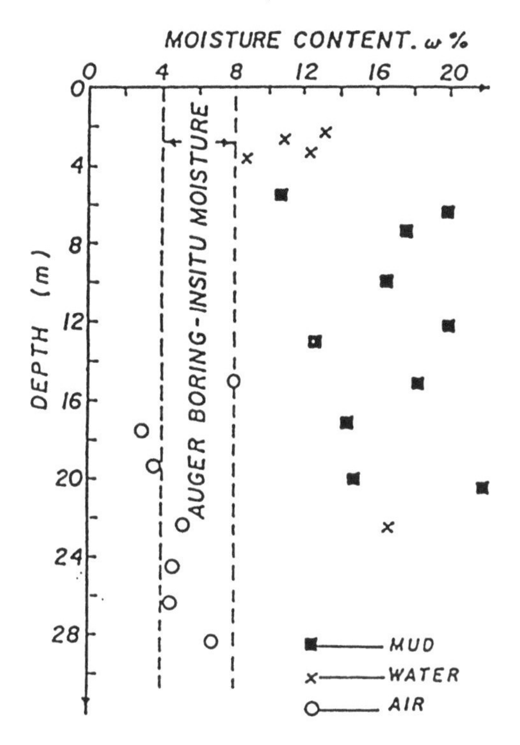

Fig. 5. Influence of drilling fluid on moisture content of undisturbed samples.

drilling "fluid". The use of either water or mud as a drilling fluid caused a considerable increase in moisture content of the samples. (Fig. 5)

A study of the influence of wetting and loading on the volume change in a consolidation test on an undisturbed loess sample is shown in Fig. 6. The natural moisture content was 5%, the dry density 14.8 kN/cu m and the Liquid Limit was 24. The Density Ratio was 0.90 and therefore the loess would be expected to have a high collapse potential. Samples were tested at various moisture contents, achieved by artificial wetting up to a moisture content of 30% corresponding to a degree of saturation of 95%. Tests of this sort are very useful in identifying potential collapse. It can be seen from examining Fig. 6, that without some estimate of the increase in moisture content that can be reasonably expected during the lifetime of a structure, there is no possibility of making quantitative predictions of foundation behaviour.

The subsurface investigation included SPT tests in most of the boreholes. Since the natural moisture contents were all very low, an attempt was made to pre-wet the soil in the boreholes prior to each SPT corresponding to some possible future condition. This procedure was abandoned almost immediately as being impractical, both because of cost consideration and time scheduling. It was hoped that it would be possible to extract some useful information from the SPT values at natural moisture and this proved to be the case. It became apparent that the site could be divided up into 3 areas based on SPT values (as will be demonstrated below) that also coincided with the different depositional environments. The coarser soil profiles (silty gravels) were at the southern end of the site - Area C; whereas the finer grained soil profiles (silty clays of very low plasticity - loess) were at the northern end of the site. Area B was intermediate in location and soil characteristics.

SPT blow counts from more than 14 borings in Area B are plotted in Fig 7. Though many of the blow counts (N) are more than 100 blows/30 cm, there is a clear pattern of increasing blow count with depth if particular attention is paid to the lower N values at each depth. Listed in Table 4 are the same test data where the number of test results with N values less than N = 10 to N = 90 are recorded for each depth. This type of analysis of the blow counts in the SPT test was done for each of the 3 Areas and 20 percentile values (low end) plotted

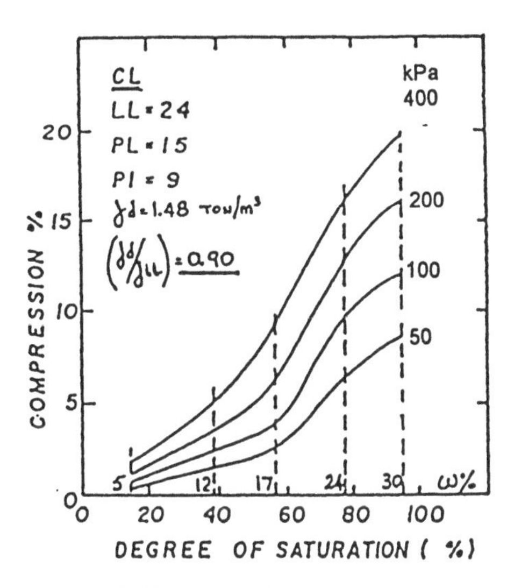

Fig. 6. Collapse studies on undisturbed loess specimens taken from a single block sample.

Table 4. Analysis of the SPT blow counts for Area B.

DEPTH (m)	NUMBER OF TEST RESULTS N=< 20	30	40	50	60	70	80	90		Σ
1.5	0	3	8	11	14	16	17			17
3.0		0	2	6	9	11	13	15		17
4.5		0	1	3	5	7	9	12		16
6.0			0	2	5	8	8	11		17
7.5			0	1	3	4	8	11		16
9.0				0	1	2	5	10		14
10.5				0	3	3	5	11		14

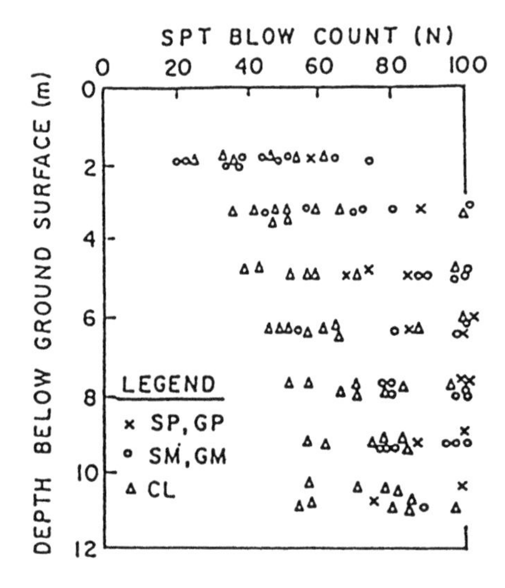

Fig. 7. SPT blow count for Area B

versus depth (Fig 8). The difference between the 3 areas, even at the low natural moisture at the site, is clearly demonstrated in this figure.

The danger of collapse of loess soils is primarily dependent on density. Based on the insitu density of the undisturbed samples taken from the exploratory borings in which SPT tests were performed, it was possible to develope a reasonably substantiated relationship between dry density, depth and SPT (Wiseman and Lavie 1983). For example, at a dry density of 16 kN/cu m, the SPT was about 40 at a depth of 4 meters and increased to about 100 at a depth of 24 meters. The above relationship was found to be site specific. However, the overburden effect was also observed at another site, where the loess soils had a higher plasticity.

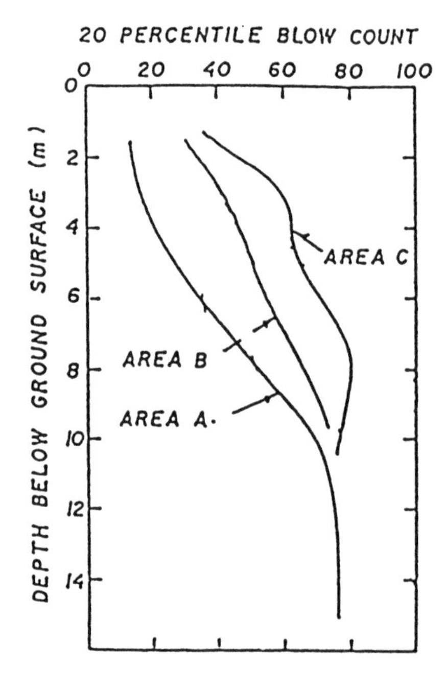

Fig. 8. 20 Percentile blow count Areas A, B, and C

10 CASE HISTORY - FOUNDATIONS

Before developing foundation designs for the site described above, full scale foundation elements (footings and drilled cast insitu piling) were load tested under conditions of controlled wetting. Two locations were chosen, one at the southern end of the site where the soil profile was primarily silty sands and gravels (GM), the other at the northern end of the site where the soil profile was primarily silty clay loess (CL). At each location both a prototype footing and a cast insitu bored pile were tested (Wiseman and Lavie, 1983).

10.1 Footing loading tests

The concrete footings were 2.0m by 2.0m and 1.0 m deep. The load was applied in small increments by jacking 6 soil anchors sequentially against the footings. First, loading to about 500 kPa was performed at natural moisture. The total settlement at this stage was less than 5 mm, at both the locations tested. Subsequently the load was reduced, and at a contact stress of of 230 kPa the subgrade soils were wetted, by dripping a total of 100 cu m of water, during a 2 week period into drill holes backfilled with fine gravel. At both sites this caused the subgrade soils to swell which increased the load on the anchors. The rigidity of the anchoring system was made such that an upward movement of the footing of 1.0 mm increased the contact stress by about 20 kPa and vice versa.

During the two week wetting period at the southern loading test (GM), after an upward movement of about 1.0 mm, the footing then settled about 1.0 mm, therefore returning to the 230 kPa contact stress. Increasing the load by jacking to 300 kPa caused a total settlement of about 25 mm. The load was then increased to 400 kPa with negligible settlement, but again, with a slight additional load the settlement increased to 55 mm. There were indications that the settlement collapse with volume change, and not bearing capacity failure.

During the two week wetting period at the northern loading test (CL) the footing heaved about 5 mm, and therefore the contact stress increased to 330 kPa. The load was then further increased in steps by jacking. There was negligible settlement, until at 450 kPa the footing suddenly settled 25 mm. At a slight increase in load it then settled another 40 mm. There were indications that in this case, we had a bearing capacity failure at 450 kPa because of the reduced subgrade strength upon wetting.

10.2 Pile loading tests

At each of the above two locations loading tests to 4 MN were conducted on 60 cm diameter, 20 m long, bored, cast insitu piles. Telltales were installed in pairs at mid-height and at the bottom of the piles. These made it possible to determine the average load in the top 10 m and the bottom 10 m, respectively, for each increment of load. The piles were loaded while the soil profile was at natural moisture and then the load was removed. Subsequently the soil in the vicinity was wetted by dripping about 100 cu m of water into 8 m deep boreholes. At both sites there was a marked decrease in SPT values which at the northern site (CL) extended to 12 m depth and at the southern site to 19 m depth (Wiseman and Lavie 1983).

At the southern site in the silty granular soil (GM), the settlement of the pile head was less than 10 mm at the maximum load of 4 MN even after wetting. Measurements on the telltales clearly indicated that all the load was taken by friction and that none of the load reached the bottom of the pile. The pile at the

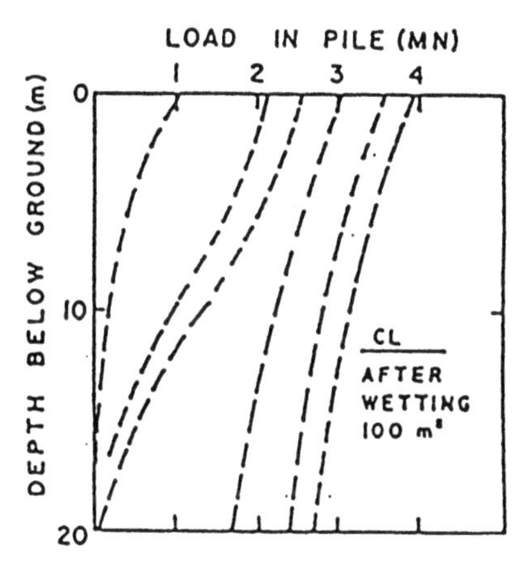

Fig. 9. Pile load test (CL)
Analysis of teltales

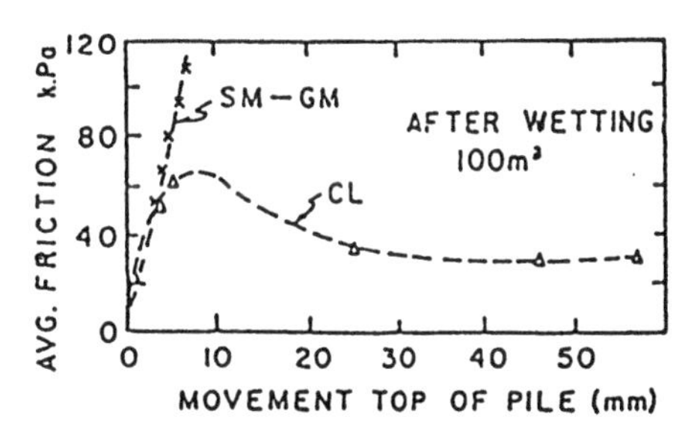

Fig. 10. Pile load tests
Avg. friction vs movement

northern site in the silty clay loess (CL) after wetting failed at a load of 2.5 MN. Interpretation of the telltales (Fig 9) showed that at this load, (at 5 mm of movement) the average peak side friction of about 60 kPa was reached. The mobilized friction reduced to about 40 kPa with additional movement. The average mobilized friction versus movement for both piles is shown in Fig 10.
These loading tests provided invaluable information as to the possible response of footing and pile foundations to accidental wetting. This sort of information would have been almost impossible to obtain any other way. The comparatively poor performance of the footing on the silty granular subgrade was to a certain extent unexpected. In the final designs, measures were adopted to minimize the possibility of wetting of the foundation soils. Buildings that were deemed to be particularly sensitive to possible foundation movements were founded on drilled cast insitu piles.

11 CONCLUDING REMARKS

Distress associated with swelling and collapsing soils run the full gamut, from minor problems associated with differential movement over the years between floor slabs on grade and other parts of the structure that are on more stable foundations, to complete structural failure caused by large movements of foundations. A tendency for swelling or collapse can also be the cause of excessive loads on columns due to restraint of the movement provided by the rigidity of the structure. What is generally common to these cases is that the distress first manifests itself only several years after the completion of construction. This of course complicates the apportioning of professional and financial responsibilities, and hence, creates employment for the legal profession.

After more than 50 years Simpson's (1936) advice for buildings is still valid - use deep foundations and suspend floors. Though mentioned in connection with swelling soils, it remains good advice for collapsing soils as well. The extent to which this can be economically justified depends on many variables beyond our present scope. For pavements and many other facilities, no equivalent simple rules apply.

To further complicate matters, the satisfactory performance of a pavement or a building in terms of acceptable distortions or cracking is not judged only by the engineer. The final judgement is in the hands of the public and the courts. Under these circumstances the geotechnical engineer should see to it that he can at least claim to have executed an adequate study of site conditions, when called upon to explain what went wrong.

REFERENCES

Abeleff, J.M. 1936. Principles governing interpretation of results obtained through the exploration of soils for foundation purposes. Proc. 1st ICSMFE, Harvard Univ., Cambridge, Mass. Vol. 2. p 280.

Komornik, A. & David, D. 1969. Prediction of swelling pressures of clays", Jour. of Soil Mechanics and Foundation Division ASCE, Vol 95, SM1.

Porter, H.C. 1936. Observations of the Texas State Highway Department on the subsequent effects of the uniformity and non-uniformity of foundation soil types on pavements; and also the effects of uniformity of moisture content fluctuations in soil foundations of high volumetric change". Proc. 1st ICSMFE, Harvard Univ., Cambridge, Mass. Vol. 2, p.256.

Simpson, W.E. 1936. Discusssion on paper No. Z-27 (Wooltorton), Proc. 1st ICSMFE, Harvard Univ., Cambridge, Mass. Vol. 3, p 257.

Uzan, J., Frydman, S. & Wiseman, G. 1984. Roughness of airfield pavements on expansive clay". 5th Int. Conf. on Expansive Soils, Adelaide, Australia.

Wiseman, G., Livneh, M. & Uzan, J. 1981. Performance of a full depth asphalt pavement on an expansive clay subgrade. Proc. 3rd Conf. of Road Eng. Assoc. of Asia & Australasia, Taiwan.

Wiseman, G. & Lavie, Y. 1983. Arid zone subsurface exploration and load testing. Proc. 7th Asian Regional Conf. on SM&FE. Haifa, Israel. Vol. 1, p 98-103.

Wiseman, G., Komornik, A. & Greenstein, J. 1985. Experience with roads and buildings on expansive clays. Trans. Research Record, No.1032, TRB, U.S.A. p 60-67.

Wiseman, G., Zeitlen J.G., & Komornik, A. 1992. An expert system for foundations on expansive soils. Proc. 7th Intl. Conf. on Expansive Soils, Dallas, Texas. Vol. 1, p 495-499.

Wooltorton, D. 1936. A Preliminary investigation into the subject of foundations in the black cotton and Kyatti soils of Mandalay district, Burma. Proc. 1st ICSMFE, Harvard Univ., Cambridge, Mass., U.S.A. Vol.3, p 242.

Zur, A. & Wiseman, G. 1973. A study of collapse phenomena of an undisturbed loess. Proc. 8th ICSMFE, Moscow. Vol 2, Part 2, p 265.

Developments in Geotechnical Engineering, Balasubramaniam et al. (eds) © *1994 Balkema, Rotterdam, ISBN 90 5410 522 4*

Swelling and collapsible behaviour of arid soils

Mohamed A. El-Sohby
Faculty of Engineering, Al-Azhar University, Cairo, Egypt

ABSTRACT: Classification of dry soils into collapsible and swelling soils, and identifying the undesirable deformation will help in avoiding possible damage for the structure. In this study charts were constructed using data from 72 prepared soil samples to relate sand-clay and silt-clay unsaturated soils to collapsible and swelling behaviour upon wetting, and to evaluate the swelling pressure if any. Ten undisturbed soil samples taken from five different geographic locations were chosen to investigate the validity of the proposed charts for predicting the soil behaviour upon wetting. The study emphasises the importance of soil fabric on the swelling and collapsible behaviour of arid soils.

1 INTRODUCTION

In arid and semi arid regions, vast areas are covered with unsaturated clayey and sandy soils. These soils are generally sensitive to variation in water content, and show excessive volume changes upon wetting. The problem arises from the unpredictable volume changes which occur when such soils are flooded. Swelling and collapsible soils are typical examples of such soils. In some regions the occurrence of these types of soils may exist with different degree of sensitivity to volume change within limited areas.
It seems that there is no border line which distinguishes between these two types of undisturbed soils, and that there is a common area which it is difficult to classify a certain soil as swelling or collapsible. For instance, Jennings 1967, Komornik 1967, Dudly 1970, and El- Sohby and Rabba 1984 have mentioned the possibility of swelling occurring under low pressure in soils which might be expected to collapse. Depending on the combined effect of certain factors, an unsaturated soil upon wetting may swell, collapse . or behaves normally. The most important of these factors are the initial water content, dry unit weight, the type of clay mineral, the coarse material fraction and the applied pressure.

Most studies in the literature concentrate on either swelling or collapsible behaviour of soils. However, it is useful from both theoretical and practical points of view for work on unsaturated soils not to be restricted to either swelling or collapsibility. Combining the work on both types of soils will improve our understanding, and will lead to more significant advances in the collective work on unsaturated soils.The purpose of this study is to improve and extend previous works on prediction of swelling and collapsible behaviour of natural clayey and sandy deposits.

Experimental data from swelling tests carried out on 72 specimens of prepared soils to present a wide range of soil properties were used. Test results were combined to build up charts to relate the swell or collapse percent and the swelling pressure to the soil properties.

In order to check the applicability of these charts, ten different undisturbed samples chosen from five different geographic locations were tested. The swelling and collapsible properties of these undisturbed samples were measured in the laboratory and deduced from the charts and the results were discussed.

2 PARAMETERS AFFECTING DEFORMATION

The geological setting up of a site dictates the nature of the soil fabric developed within its boundaries (El-Sohby et al. 1993).
The fabric of a soil mass and its properties govern the behaviour of the soil. The natural fabric is quite complicated (Collins and McGown 1974), however it can be simplified to a structure of coarse grains of

sand and/or silt and a matrix of clay packets (e.g. El-Sohby et al. 1987).

The sand or silt grains are bonded together with clay packets at their contact points by higher curvature menisci acting at their contact points. Volume change of unsaturated soils having such a structure occur when water is taken up as a result of one or more of the following effects (Torllope 1960):

1. slippage of large grains (sand or silt) due to loss of shear strength by submergence.
2. distortion of the clay packets under the applied loads.
3. swelling of the clay packets due to the water take-up.

At low dry density and clay content, the structure of soil will be mainly composed of an open structure of sand or silt. Such a structure will be subjected to collapse under wetting. On the other hand, at high dry density and clay content, the structure of soil will be mainly composed of a matrix of clay particles containing floating chains of sand or silt grains. Such a structure will swell upon wetting. The swelling of the matrix will depend on the type of clay mineral. The extent to which either effect will predominate depends on many factors, the chief ones being the initial structure, which will be represented in this investigation by dry unit weight and the type of coarse grained fraction, the clay content and the type of clay mineral and the initial water content.

3 EXPERIMENTAL DATA

Prepared soils

In order to simulate natural soils of wide range of deformational characteristics in the laboratory, two groups of soil were prepared. One group is composed of sand and clay and the other is of silt and clay. Both sand and silt were mixed with different percentages of clay 0, 10, 20, 30, 40, 50, 60, 80, 90 percent. Thus, the two main groups of sand-clay and silt-clay mixtures were divided into 18 subgroups, giving a total of eighteen different soils. For each soil, four specimens with initial dry densities 14.0, 15.5, 17.0 and 18.0 kN/m^3 were tested. These will give us data for 72 soil specimens. All specimens were unsaturated and of 8% initial water content.

The specimens were tested in the oedometer apparatus. After the specimen reached equilibrium under 100 kN/m^2, it was flooded with water and allowed to swell. When the swelling ceased, the specimen was loaded in stages of 200, 400, 800 and 1,600 kN/m^2.

For each load increment, the specimen was allowed to attain its equilibrium. The specimen was then unloaded at the same loading increments and the test was finished. The deformational characteristics of the tested specimens, namely swelling or collapsible percent and swelling pressure were determined from the results of the loaded tests. The swelling or collapsible percent were determined under a load of 100 kN/m^2, and the swelling pressure was determined by the preswelled method.

Advantages and disadvantages of this method have been discussed by El-Sohby and Mazen 1980.

Natural soils

The natural soils used in this investigation were taken from five locations, which are as follow:

1. New Ameriyah, which is located in the Mediterranean coast, about 20 km to the south-west of Alexandria (soil C1).
2. El- Saff, an existing town lies on east bank of the river Nile, about 65 km to the south of Cairo (soil C2).
3. El- Maadi, which is located about 10 km to the south of Cairo (soil C3).
4. Six of October, which is located on upland where the elevation varies from 180 to 190 m above sea level. The site is at a distance of about 32 km to the south-west of Cairo (Soil C4).
5. Nasr City, where the elevation ranges from 45 to 135 m above sea level. It lies at a distance of about 10 km to the north-east of Cairo (soils S1, S2, S3, S4, S5 and S6).

Properties of the tested soils are summarized in Table 1.

4 DEVELOPMENT OF CORRELATION

From previous research works, many relationships have been established by which swelling pressure and swelling potential can be estimated based on index properties and the physical state of the soil. Most of these equations predict the swelling properties of the soil with different degree of accuracy and within a limited range of the behaviour (Oloo et al. 1987).

The main problem when applying these relationships arises from the fact that both the swelling pressure and swelling percent are dependent to initial soil fabric, the stress path and the boundary conditions.

Consequently, undisturbed samples have to be used. The swelling pressure and swelling percent have also to be measured under standardized conditions. In this investigation, 72 soil specimens of prepared soils were used in order to construct the proposed

Table 1 Properties of tested undisturbed soils

Soil / Properties	C1	C2	C3	C4	S1	S2	S3	S4	S5	S6
Depth m	1.5	2	4	2.7	3.5	7	3	3.3	3.5	7
Water cont.%	2	2	5.5	2	6.1	13.8	18.4	21.7	22.2	8.3
Dry density kN/m^3	12.6	15	14.6	16.5	19.6	19.8	14.8	14.9	16.2	19
Specific gravity	2.67	2.68	2.65	2.66	2.72	2.69	2.7	2.71	2.75	2.7
Liquid limit	26	31	41	n.p.	81.5	51.5	85.5	85	89	59
Plastic limit	15	14	24	n.p.	27.8	35.5	42.5	46	44	23
Plasticity index	11	17	17	n.p.	53.7	16	43	39	45	36
Shrinkage limit	-	-	-	n.p.	13.8	24.5	14	18.5	12	13.7
Clay %	17	15	18	8	59	20	46	44	62	50
Silt %	56	45	64	5	37	31	44	45	24	33
Sand %	27	40	18	87	4	49	10	11	14	17

classification charts, and develop the recommended equations.

The prepared samples were divided into two main groups, each one was composed of 36 soil specimens. The first main group was composed of sand-clay mixes, and the other was of silty-clay mixes. This was planned in itself to investigate the importance of soil fabric.

The swelling percent was measured under 100 kN/m^2 after inundation. The swelling pressure was measured by the preswelled method for simplicity, and fortunately for this type of soil, there is an experimental evidence that the swelling pressure by the preswelled and different pressure methods gave similar results (Rabba 1975).

The relationship between swelling pressure, unit weight and clay content was described by El- Sohby and Rabba 1981 as follows:

$$\text{Log } P_s = k(\gamma_d + k_2 C - k_1) \quad \text{....(1)}$$

Measured values of k, k_1 and k_2 for the tested soils were as follows:

	sand-clay soils	silt-clay soils
k	2.17	2.50
k_1	1.80	1.60
k_2	0.0084	0.0060

It is evident that the constants in this equation will depend on the internal fabric of the soil, however it is left at this stage to personal judgement, and these constants can be determined experimentally for any particular coarse-grained fraction.

In this investigation,the dry unit weight and the type of coarse-grained fraction were used to allow for the internal structure of the soil. The clay content was used to represent the matrix of the internal structure. As the coarse-grained fraction of soil is represented in the equation by specific constants, the effectiveness of the clay content can also be allowed for by the corresponding liquid limit or activity.

El- Sohby and Mazen 1987 applied equation (1) to some natural soils and compared the results with data obtained from equations recommended by some investigators. They proposed a development to equation (1) by introducing the liquid limit in the equation as follow:

$$\text{Log } P_s = k(\gamma_d + k'_2 w_L + k'_1) \quad \text{...(2)}$$

where,

	sand-clay soils	silt-clay soils
k	2.17	2.50
k'_1	2.00	1.83
k'_2	0.10	0.007

Table 2 Activities and free swell tested soils

Soil No.	w_L	A	A_m	A_m/A_m*	F.S.	$F.S._m$	$F.S._m/F.s._m$*
C1	26	0.64	-	-	25	-	-
C2	31	1.13	-	-	40	-	-
C3	41	0.88	-	-	18	-	-
C4	n.p	n.p	-	-	18	-	-
S1	81.5	0.9	0.75	1	100	150	1
S2	51.5	0.8	1.53	2.04	150	260	1.73
S3	85.5	0.93	0.78	1.04	105	155	1.03
S4	85	0.89	0.71	0.95	40	135	0.9
S5	89	0.73	0.84	1.12	140	170	1.13
S6	59	0.72	0.57	0.76	65	95	0.63

* Reference sample
A_m Activity of El- Sohby, 1981
A Activity of Skempton, 1953

Table 3 Measured and deduced axial swell and swelling pressure

Soil	Axial swell at $10kN/m^2$	Axial swell at $100kN/m^2$ measured	Axial swell at $100kN/m^2$ deduced	P_s (kN/m^2) measured*	P_s (kN/m^2) deduced
C1	-1.2	-12	-12	-	-
C2	2	-4	-1	30	-
C3	6.8	2	-2	180	-
C4	-0.6	-5.5	-0.5	-	-
S1	15.8	12	14	1,000	2,000
S2	14	13	5	2,000	800
S3	21.6	8	5	1,000	300
S4	19	12.5	4	530	250
S5	25.5	18	11	1,250	1,000
S6	10.4	6.3	12	900	1,800

P_s= Swelling pressure

* Measured by different pressures method

However, it is believed in this investigation to keep the clay content in the equation as a distinct property of the internal fabric, and to allow for the effectiveness of the clay fraction by the activity. It is also preferable to use the activity of clay minerals (A_m) as recommended by El- Sohby (1981), and as shown in Table 2 that A_m is a direct assessment of the effectiveness of the clay fraction, compared with the free swell.

The test results obtained from the 72 prepared soils, in developing equations (1) and (2), were used to construct the charts relating the dry unit weight and clay content to the swelling pressure and swelling percent under 100 kN/m^2. It is left to the user to allow for the difference in the activity of clay or the coarse grained fraction.

The only parameter not mentioned in the above discussion is the initial water content. Previous studies proved that it has a small influence if its value is less than the shrinkage limit of the soil (El- Sohby and Rabaa, 1981).

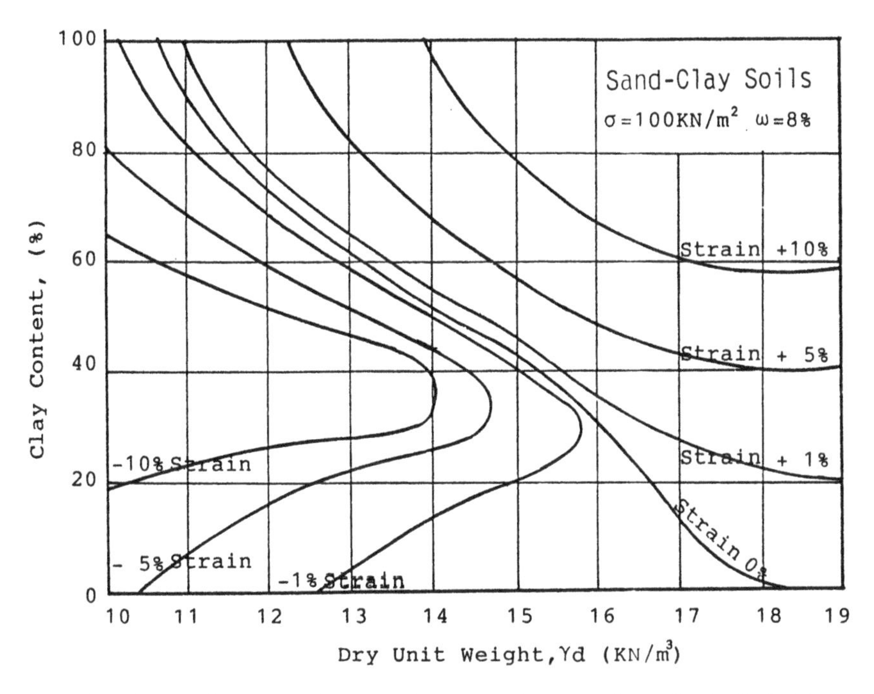

Fig.1 Swelling-Collapse Percent Versus Dry Unit Weight and Clay Content for Sand-Clay Soils.

5 SOIL DEFORMATION CHARTS

Figures 1 and 2 present the charts relating the percentages of swell and collapse under a load of 100 kN/m² to dry unit weight and clay content of sand-clay and silt-clay soils respectively.

In order to investigate the validity of these charts, ten different undisturbed soil samples chosen from five different geographic locations where arid and semi arid problematic soils predominate were tested.

Samples were selected to contain a wide range of soil properties. Details of these properties are shown in Table 1.

Summary of measured and deduced axial strains at 100 kN/m² are shown in Table (3). It can be noted that the differentiation between soils which exhibit collapse and those which show swell was clearly achieved, comparing the results of soils C1 to C4 and S1 to S6.

Soils C2 and C3 show slight swell although both usually classified as collapsible soils.

On the other hand for soils S1 to S6 reasonable agreement between measured and deduced axial swell were approved for soils S1,S3,S5,S6. Whereas appreciable differences were noted for soils S2 and S4. These might be attributed for difference in fabric and activity.

6 SWELLING PRESSURE CHARTS

The charts showing the relationships between the swelling pressure to the dry unit weight and clay content for sand-clay and silt-clay soils are shown in Figs 3 and 4 respectively. The calculated relationships from equation (1) are also presented. In order to check the validity of these relationships, the measured values of swelling pressure for tests carried out on the chosen 10 natural samples mentioned before were used.

Table 3 compares the results of the measured and deduced swelling pressure. The soils can be divided into three categories:

1. soils C1 and C4 are collapsible soils.
2. soils C2 and C3 normally classified as collapsible but measured swell pressure were 30 and 180 kN/m² respectively. Deduced swelling pressure for both samples were less than 100 kN/m².
3. soils S1 to S6 were proven to be swelling, however the relationship between the measured and deduced values of swelling pressure varied from

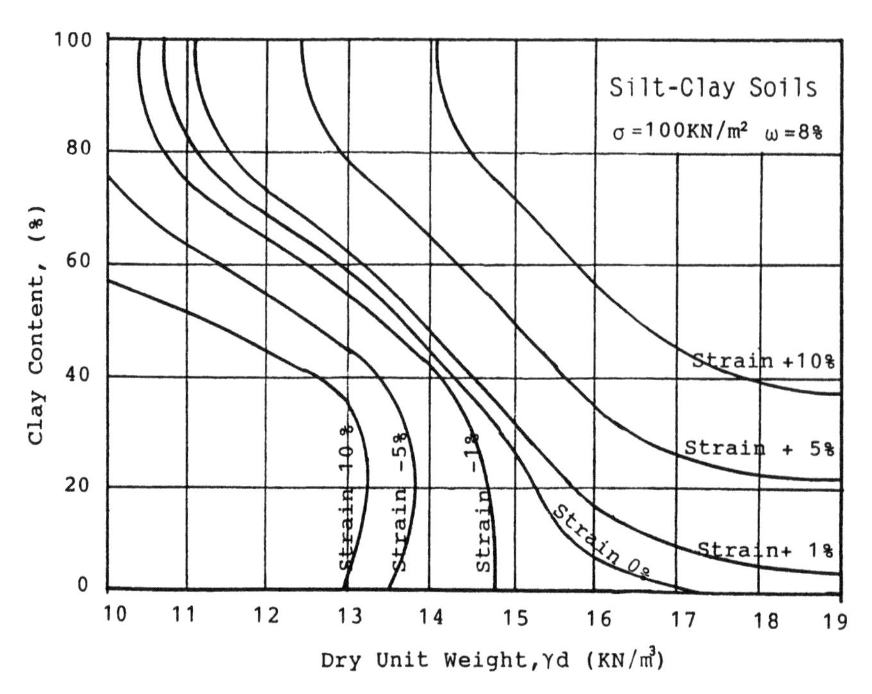

Fig.2 Swelling-Collapse Percent Versus Dry Unit Weight and Clay Content for Silt-Clay Soils.

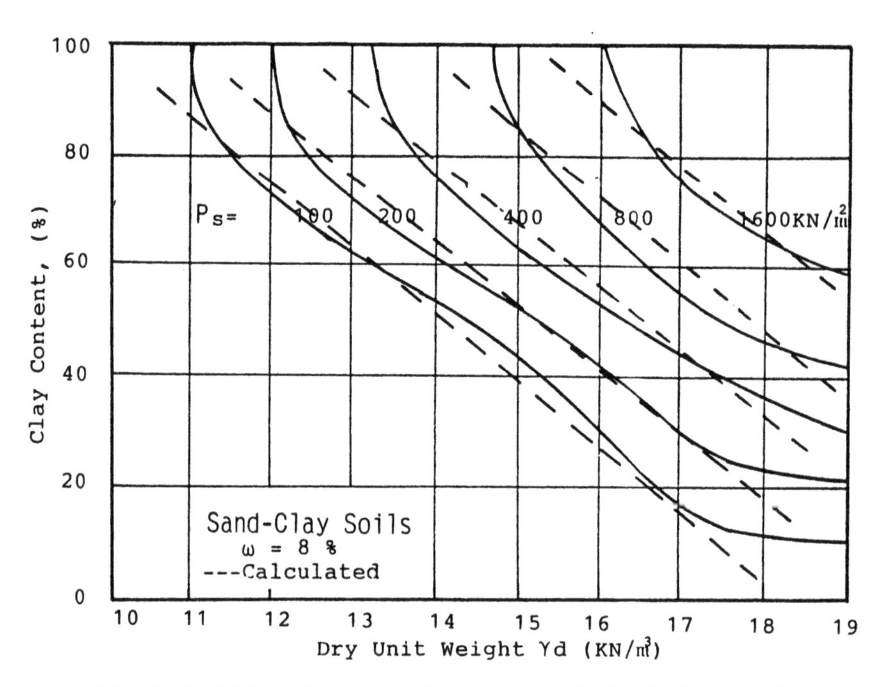

Fig.3 Swelling Pressure Versus Dry Unit Weight and Clay Content for Sand-Clay Soils.

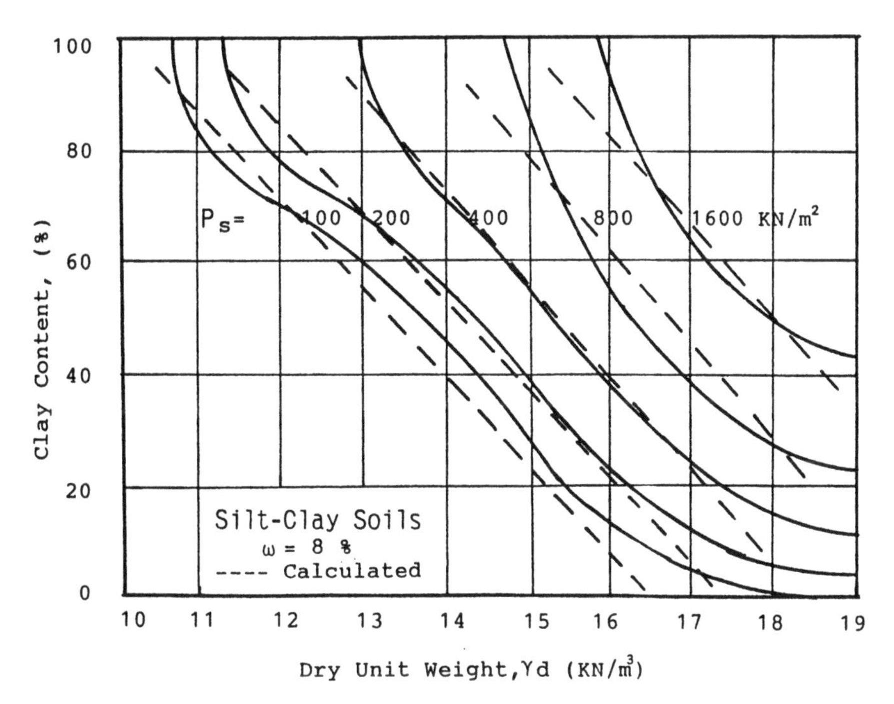

Fig.4 Swelling Pressure Versus Dry Unit Weight and Clay Content for Silt-Clay Soils.

agreement to a considerable differences. Allowance for clay activity improved the results for soils S2 and S5 and S6. This also emphasized the importance of natural soil fabric of swelling and collapsible soil behaviours.

7 IDENTIFICATION OF SWELL AND COLLAPSE POTENTIAL

Both swelling and collapsible soils can be identified by several means. Qualitative evaluation of the expansibility of soils using the index properties have been adapted by many investigators e.g. Holtz and Gibbs, 1956 and Seed et al., 1962. Another method utilizing the ratio of the natural water content to the liquid limit was mentioned by Vijayvergiya and Ghazzaly 1975. For collapsible soils measuring the collapsible potential proposed by Jennings and knignt 1975 has been widely used.

Based on the correlation developed in this investigation a simple way of identifying both the swell and collapsible potential can be qualitatively evaluated from the dry unit weight and the percentage of clay.

8 CONCLUSIONS

Based on data collected from swelling and collapsible percent tests under 100 kN/m^2 and swelling pressure measurements of 72 prepared soils composed of silt-clay and sand-clay mixes an effort has been made to develop relationships and charts to relate the main soil properties and collapse or swell percent and swelling pressure.

Results from tests carried out on ten different undisturbed soil samples chosen from five different geographic locations were used to investigate the validity of the proposed charts.

According to the proposed charts the soils can be classified into:

1. collapsible
2. collapsible- swelling
3. swelling

Deduced values of collapse or swell percent and swelling pressure showed reasonable agreement with the measured values for the different soils. However, more effort is still needed in presenting the fabric of soil and the activity of clay matrix in the fabric.

Efforts are also needed to standardize the measurement of swelling percent and swelling pressure.

These results can be used to classify the unsaturated soils to collapsible, collapsible- swelling and swelling upon wetting. The degree of swell and collapse can be evaluated reasonably to most soil used.

REFERENCES

Collins, K. & A. McGown. 1974. The form and function of microfabric. Geotechnique 24, No.2, pp. 223- 254.

Dudley, J.H. 1970. Review of collapsing soils. Proc. Am. Soc. Civ. Engrs, Vol. 96, No. SM3, PP. 925- 948.

El-Sohby, M. A. & S.O. Mazen. 1980. On measuring swelling pressure by two methods. 7th Regional for Africa on Soil Mech. & Found. Engng., Acra, Vol. 2, pp. 773- 783.

El-Sohby, M. A. 1981. Activity of soils. Proc. of 10th Int. Conf. on S.M.F.E., Stockholm, pp. 587-591.

El-Sohby, M. A. & S.A. Rabba. 1981. Some factors affecting swelling of clayey soils.Geotechnical Engineering. 12, pp. 19- 38.

El-Sohby, M. A. & S.A. Rabba. 1984. Deformational behaviour of unsaturated soils upon wetting. Proc. of 8th Regional Conf. for Africa on S.M.F.E., Harare, pp.129- 137.

El-Sohby, M. A. & S.O. Mazen. 1987. On the prediction of swelling pressure and deformational behaviour of expansive soils. Proc. 9th Conf. for Africa on S.M.F.E., Lagos, pp. 129-133.

El-Sohby, M. A.; El- Kharaibi, M.C.; Rabaa, S. A. & M.A. El- Saadany. 1987. Role of soil fabric in collapsible soils. Proc. of 8th Pan Amrica Conf. on S.M.F.E., Cartagena.

El-Sohby, M. A.; Mazen, S. O.& M.I Aboushook 1993. Geotechnical aspect on behaviour of weakly cemented and overconsolidated soil formations. Proc. of Int. Symp. on Geothecnical Engineering of Hard Soils-Soft Rocks, Vol. 1, pp. 91- 97.

Holtz, W. G. & H.J. Gibbs.1956. Engineering properties of expansive clays. Transactions, American Soc. of Civ. Engineers, Vol. 121, pp. 641- 663.

Jennings, J.E. 1967. Discussion on aeolian soils properties and engineering problems. 3rd Asian Regional Conf. on S.M.F.E., Vol. 2, pp. 102-104.

Jennings, J. E. & K. Knignt.1975. A guide to construction on or with material exhibiting addition settlement due to "collapse" of grain structure. Proc. 6th Reg. Conf. for Africa on S.M.F.E., Durban.

Komornik, A. 1967. Written contribution, discussion on aeolian soils properties and engineering problems. 3rd Asian Regional Conf. on S.M.F.E., Vol. 2, No. 1, pp.106-109.

Oloo, S.; Schreiner & Burland 1987. Identification and classification of expansive soils. Proc. of 6th Int. Conf. on Expansive Soils. PP. 23- 29.

Rabaa, S. A. 1975. Factors affecting engineering properties of expansive soils. M. Sc.Thesis, Al- Azhar University, Cairo, Egypt.

Seed, H. B.; Woodward, R.J. and &R. Lundgren. 1962. Prediction of swelling potential of compacted clays. Journal ASCE, SMFE Division, Vol. 88, pp. 53- 87.

Vijayvergiya, V. N. & O.I. Ghazzaly. 1975. Prediction of swelling potential for natural clays. Proc. of 3rd Int. Conf. on Expansive Soils, pp. 227- 234.

Developments in Geotechnical Engineering, Balasubramaniam et al. (eds) © 1994 Balkema, Rotterdam, ISBN 90 5410 522 4

Geomorphology as the prime determinant of current engineering problems in the Tamar River basin at Launceston, Tasmania

Owen G. Ingles
Owen Ingles Pty, Ltd, Soil Engineering & Risk Management Consultants, Swan Point, Tas., Australia

ABSTRACT: The city of Launceston is situated at the confluence of two major rivers, where they enter a drowned river valley. It is traversed by major fault lines (thought to be inactive), and encompasses both swamp plains and steep slopes of more than 40°. It has grown to a city of 60,000 inhabitants in less than 200 years since it was founded, and currently faces an unusually wide range of major engineering problems which were not recognised at the time of settlement.

These problems include: highly expansive soils, major and extensive landslip, flooding and river sedimentation, deep soft soils (nearly quicksilt), variable layered and lensed alluvials, and the largely unknown risks associated with seismic action. Almost no part of the city is exempted from one or more of these problems requiring engineering solution.

Not all the problems are natural hazards. Landslip and sedimentation, for example, are substantially influenced by development practices. Thus the geomorphology of the Tamar river basin has greatly accelerated since first settlement. This report draws on historical records as well as recent investigations to show the importance of recognising how dynamic landform change influences the choice of engineering solutions for the various soil stability problems. In effect, there is a need to work with nature rather than against it.

1. Geology

The basement rock in the Launceston region is Jurassic dolerite in the form of a sill about 300m thick, faulted along a line NNW-SSE and tilted about 10° towards the west to produce a series of parallel asymmetric fault troughs. These fault troughs have been filled by clay, silt, sand and gravel lacustrine sediments in the Tertiary period. These Tertiary sediments have subsequently suffered erosion and in places been intruded by basalt, so that both dolerite gravel and basalt talus are found in soils in and near the city. The downcuttting of the Tamar valley has been followed by infilling with Quaternary sediments, primarily a fine silt-clay mix, to depths of around 20m in the Tamar basin at Launceston.

2. Geomorphology

The climate is cool temperate with an annual rainfall of some 700mm average. Weathering of the dolerite has been and is extremely slow, and soil depths on the dolerite seldom exceed 1m. Intruded basalt has weathered more rapidly, but in situ soil seldom exceeds 1.5m. The deeper soils of the Tamar basin are all either Tertiary or Quaternary alluvials. Their depths, which can sometimes be very considerable (tens of metres) can vary over quite short distances due to extensive faulting which occurred in late Tertiary and Quaternary times.

Earth tremors still occur, and caused some damage to buildings in the 1880's, but the frequency and magnitude of seismic risk is largely unknown. The peak ground intensity (Modified Mercalli) with a 10% chance of exceedance in 50 years is V, and the peak ground acceleration for the same return risk is $0.3 ms^{-2}$, according to current earthquake mapping from a very sparse data base. The present author believes that a major earthquake occurred on the south-east edge of Launceston between 5 and 15 thousand years ago, judging from laterisation found in an old landslip, but there is no confirmation. This would coincide roughly with the 60m rise in sea level known to have occurred about 11,000 years ago and which drowned the Tamar Valley as far as Launceston (1).

The deep Tertiary alluvials - mostly clays, but with

waterworn gravels interbedded or lensed; and in one case (at South Norwood) comprising pockets of buried ultrafine quartz silt (2-10 micron size) some 2 or 3m in diameter - have been partly eroded in Quaternary times so that the clays are often overconsolidated. Landslips have been common in these soils for at least the past 10,000 years (2) and have been accentuated since European settlement led to clearing of the native forests.

The Quaternary alluvials accumulated principally at the confluence of the North and South Esk rivers, where they join to form the Tamar, at the present day city of Launceston. They formed a broad flood plain which, at first settlement in 1806, was mapped as a marsh. These alluvial sediments accumulated rapidly after settlement because of clearing and erosion in the upper reaches of the North and South Esk rivers; and by 1860 maps show the flats north of Launceston by the name of "Swamp Town". The swamps were gradually reclaimed by filling, but continued to be inundated by floods until 1930 (though protective levees of low height had been pushed up in earlier days). The alluvial soil on these flood plains has attained a depth of up to 20m, equal to an average rate of 2mm/year assuming 10,000 years. The soil deposited is an almost equal proportion of clay and silt.

In summary, therefore, the present dolerite and basalt exposures are very stable, apart from rockfalls mentioned later. The Tertiary beds are a source of serious and frequent topographic change due to landslides triggered both by climate and by human activity, as discussed presently. The Quaternary beds are composed of soft, deep clays of very low bearing capacity and vulnerable to flooding as well as to changes of course by the meanders of the North Esk river. The whole area is subject to uncertain seismic risk and traversed by major fault lines. On this complex a city of 60,000 inhabitants is now established, and growing. The resultant problems and solutions are now summarised.

3. Rockfall

Rock (dolerite) is at or near surface on the western side of the City; the principal exposures being along the course of the South Esk river, which flows through a steep rock gorge.

Periodic flooding has carried fallen rocks from this gorge and deposited them in the Tamar River Basin, about 250m from the Gorge mouth (a deep scour basin lies between the mouth and the first rockpile). In 1833, the map shows "rapids" at this point. The deposited rock sizes become progressively smaller with distance from the gorge mouth, becoming sand at a distance of some 500m. The discharge rockfield remained more or less undisturbed until 1992, when it was partly dredged to deepen it for small boats (this may lead to future flood flows impacting more seriously on the defensive levees downstream).

From the dredging, the size and shape of rocks falling from the dolerite rockfaces could be assessed. The largest were estimated to have a volume of $1.55m^3$ and a weight of 4.1 tonnes. They had a shape factor of 2.0 and a Zingg classification of "short rods". Later random sample estimates of mean rock size/volume on the cliff faces (i.e. from jointing and bedding discontinuities, which are, respectively, at about 1m and 2m spacings) indicated some $2.5m^3$ and 6.6 tonnes, with a shape factor of 1.9 and a "short rod" classification. The rocks swept out by flood have presumably been somewhat abraded before reaching their final point of deposition.

Dip and strike of the dolerite now largely determine soil and rock conditions and hazards for the western side of the City. Wherever the slope is under 25°, the soil is typically shallow and stable (apart from expansive properties). Slip is possible, but usually only from human intervention sources (e.g. a broken water pipe). On slopes above 30° the "soil" is largely a fine rock "flour" from which all clay has been leached. It has a friction angle when dry of 49° and when saturated of 33°. As a consequence, the sides of the Gorge have equilibrated at approximately 35°.

Rock release from slopes between 30° and 50° is thus governed by friction, and is greatest in times of accentuated or sustained rainfall. Such rocks will roll or slide as they fall and because of their weight (see above) are best controlled by cable stays or counterforts or both. Rock release from slopes greater than 50° would lead to bouncing and high impact forces, so again active defences are needed, and cable staying or overmeshing is appropriate. Due to the very slow weathering of this dolerite, the rate of rock release is low (one or two a year), but a seismic shock would have serious consequences. Zones of high and low risk need to be identified, perhaps in the manner used for the French Alps (3).

To what extent rockfalls correlate with tree cover on high angle slopes also deserves future investigation.

4. Tertiary sediments

These pose two major problems in the Launceston region. Firstly, the clay in these alluvials is highly

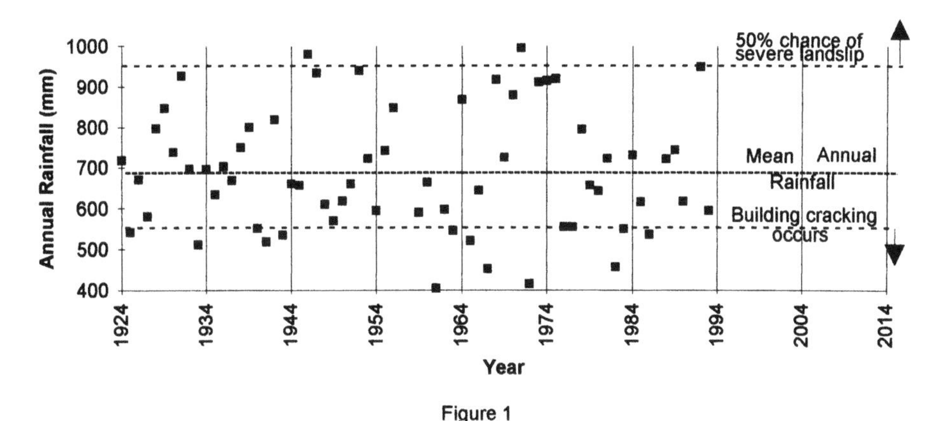

Figure 1

Annual Rainfall Scattergram showing suggested Slip & Cracking Limits (Launceston, Tasmania)

expansive; and secondly, they are prone to severe landslip. It is best to address these two problems separately.

4.1 Expansive clays

It is common to find seasonal moisture movements of more than 100mm in the Launceston area, and the present author has observed movement as much as 180mm. An Australian standard (AS2870) has been produced as a guide to designing foundations on expansive soils, but its highest swell class is just 70mm.

Several legal judgements have been given against builders for house cracking on expansive soils in the Launceston region. Most of this seems to have risen from drying shrinkage during the January to March low rainfall period in northern Tasmania. The depth of seasonal cracking is poorly defined or even unknown on these Tertiary sediments, which are sometimes gravelly, sometimes lateritic, sometimes just stiff clay. In the circumstances, it is extremely difficult to make accurate prediction of future foundation movement, and better foundation design systems seem necessary.

Linear Shrinkage is considered as unreliable (and even in some cases unconservative) measure for expansive soils since shrinkage stops long before all moisture is lost, and hence the moisture content-shrinkage gradient is seriously in error. A similar criticism applies to the initial moisture content used in the linear shrinkage test. This author prefers and uses psychrometric measurements over a defined moisture range, but supplemented by filter paper suction measurements for that part of the year when, in Tasmania, the soils are too wet for a reliable psychrometer reading (pF 3.2 or less).

For the expansive Tertiary clays, the best solution may well be to control the "geomorphology" by controlling the moisture regime under footings - which means to inhibit drying out during drought periods. This can be done by the use of moisture barriers as "aprons" around house footings, designed so as to lengthen the path for moisture loss sufficient to prevent any substantial drying during some defined period of drought. In Launceston, the return period for substantial damage to domestic housing due to drought conditions appears to be currently as little as about 5 years, climatic changes not accounted. Figure 1 shows the rainfall pattern.

4.2 Landslides

There have been, and continue to be, major losses in and around Launceston from landslides. Although triggered principally by high rainfall events with a return period of about 20 - 25 years at Launceston, there are many important contributory factors. Most of these are well recognised, such as the clearing of trees (which reduces evapotranspiration), trenches or drainage lines laid on a contour (increasing the infiltration), loading the head of a marginal slope, etc.

Most of the slips do not conform with classical slip circles, but are cycloidal in the manner first analysed by Collin in 1846 (4). This is due to the presence of planes of weakness in these alluvial soils, which are often tilted (at Lawrence Vale, in the southern suburbs, the tilt is 10.5°). This inclination of the strata has led to slips being frequent on west-facing slopes and rare on east-facing slopes of the same soil.

Figure 2: Launceston, Tasmania, showing principal soil problem areas

1. *Problems with Consolidation and Low Bearing Capacity*
2. *Problems with Extreme Shrink/Swell and Landslips*
3. *Problems with Landslip and Rockfalls from steep scarps*

Moreover, slips (strictly, slides) occur chiefly on one easily recognised layer of gray "clay", which is actually a very fine silt, hence permeable. The presence of gray "clay" in Launceston region soils is an excellent warning of ultralow bearing capacity, potential landslide, poor water containment, and many other engineering problems.

There are remedies and/or defensive measures which reduce landslide risk. For example, with house design; service pipes and trenches should never be located on the upslope wall, but on either of the side walls from where any leakages can rapidly discharge downslope. Drainage or absorption trenches must always be set at an angle to the contours, the higher the better; and with falls never less than 3% (cut-off drains must have a good fall, else their risk become "tension cracks"). On slopes of 12° or more, either soil should be excavated sufficient to compensate largely for the weight of the proposed structure (a "status quo" solution) or else piling installed to a depth exceeding that of any weak layers, and rigidly locked to resist superficial soil movements. Perhaps a "Noah's Ark" house could be designed, able to remain intact and upright whilst floating gently downhill on the moving soil mass? Such a concept would be the only one to survive massive losses in the event of a serious seismic shock during the wet winter months.

In summary, slopes of 8° or less are considered stable in and near Launceston; slopes of 12° or more on Tertiary sediments are landslide prone (but sometimes mitigated by favourable bedding dip); and slopes between 8° and 12° must be treated on their merits as potential risks, very sensitive to human intervention (good or bad).

5. Quaternary sediments

These are associated solely with flood plain and river bank, hence their engineering problems are principally problems of bearing capacity and consolidation (because of their great depth). A substantial part of northern Launceston has been built over the flood plain of the north Esk, and important roads and port facilities, as well as residential and industrial areas require protection from recurrent floods.

The Tamar silt properties may be summarised as in the following table (taken from the author's files). The depth of silt varies between about 10 and 20m, and its stable angle underwater is approximately 14°. Whilst very variable, typical values of liquid limit would be 177 and plasticity index of 136. The intertidal zone barely supports pressures of 16 - 32 kPa and the river flats support approximately 52 kPa only. The intertidal equilibrium slope is 6° and major slips have occurred at banking heights as low as 2.45m. The latter value illustrates the major problem of providing Launceston with safe flood levees, since the tidal range alone is 3.3m at Launceston; and in historic times - the last 190 years - river banks have moved as much as 80m to and fro.

Table 1. Guide to Tamar silt properties

Composition:	silt 60%, clay 40%
Moisture Content:	100 - 225%, average 120%
Wet Density:	1.3 - 1.45 t/m^3, average 1.35 t/m^3
c'	1.4 - 14.4 kPa, average 6 kPa
ø'	14° - 23.5°, average 22°
Residual Shear Strength:	2 - 20 kPa
c_v	0.08 - 0.15 m^2/yr, average 0.12 at 50kPa
m_v	average 3×10^{-3} m^2/kN for 30-120kPa

The early surveyors had great difficulty in defining the river banks on the flood plain not only because it varied with every flood, but because it was a reedy swamp over a large area. When much of this swamp was reclaimed in the second half of the century, fill was placed on top of the reeds so that today a layer of well preserved reeds lies about 2m below ground surface, acting as a drainage layer or an aquifer. Often the fill used for reclamation was obtained from dredging the river (wholly so at Royal Park), so that the surface soil is also Tamar silt and liable to large consolidation settlements (up to 1m) over long periods of time (more than 20 years) for comparatively light loading.

For roading, the problem of deep soft subgrades has been addressed most successfully by lime stabilisation. The initial CBR on the Tamar silts ranges between 0.2 and 0.4, and N-values between 0.6 and 0.8. This can be increased tenfold by lime stabilisation, using circa 4 - 5% lime. The reconstruction of Goderich Street, one of the most heavily trafficked of the flood plain roads, used 300mm of lime-stabilised in-situ soil as a subgrade, with notable success for the past 20 years.

A useful conversion for the Tamar silts is:-
CBR 2 = N 7 = 70 kPa (shear strength)

The Tamar silt contains enough clay and water to react quite spectacularly with quicklime, hence in recent years quicklime piles have been used to strengthen levee bank foundations and thus allow the levees to be raised safely, free from the risk of collapse in times of major flood. It is the writer's contention that the quicklime pile method is applicable only for soils with a moisture content in excess of 50% (because of soil/lime suction considerations) and with a range of clay contents sufficient to ensure some clay, but not so much as to reduce the permeability too far. Thus a range of clay contents between 10 and 50% is considered ideal, and results of trials on the Tamar silts have certainly supported this.

Success in raising the levee banks to safer flood defence heights will bring new engineering problems however, which must be solved in the future. The earliest maps of the Tamar basin at Launceston show the North and South Esk rivers impacting head on (at the stonefield "rapids" referred to earlier). Land reclamation works and river dredging have moved the mouth of the North Esk some 200m northwards, and at the same time its meandering course across the Invermay flood plain has been checked by filling and constrained within levee banks. In times of peak flood, the effect is likely to be similar to trying to hold a part-coiled garden hose against jet flow - there will be extreme pressure, causing collapse. Moreover, serious consideration must be given to whether the expected volume of flood water can be accommodated within the restrictive walls until discharged, especially if the South Esk is also in flood and high tides have raised the level in the Tamar River itself. It may be that some kind of "spillway" will ultimately be needed to a lower reach of the Tamar to cater for major flood events.

6. Conclusions

This paper provides only an overview of the engineering problems for Tamar Basin development planning. It has sought to collect into one source hitherto fragmentary and sparse information on engineering properties of its soils and rocks. In so doing, it has become clear that geomorphological changes arising both from climatic factors and from ill-considered human interventions are critically important for the on-going sound development of the Launceston region.

In particular, more work is desirable on flood modelling, and various lime-stabilisation techniques for the river silts (Yamanouchi's (5) lime-banking comes especially to mind). Landslip dating techniques would be helpful, in particular for defining the return period of major damage events. Seismic records are sparse and, because of Launceston's special vulnerability to seismic-triggered landslide and rockfall, need to be monitored closely. Curiously, initial analysis suggests that the silts are not likely to present liquefaction problems during an earthquake; but a quake during winter months when the soils are saturated could cause major landslip losses.

In the past, clearing of trees and reclamation of river flats, as well as dredging, has been effected without much thought for the longer term. Consequently, the engineering solutions become more costly today than would have otherwise been the case. It is better to work with nature than against it. Had the north Esk been encouraged to shorten its meander path to Stephenson's Bend, for instance, much levee bank cost would have been saved and greater security achieved. Likewise with landslip, better zoning and planning between 1850 and 1950 could have avoided much of the present day hazard.

Little can be done to ameliorate highly expansive soils, but measures to restrict moisture content change are likely to prove much more economical than massive foundations; and patience (not building whilst the soil is saturated) is likely to be better than costly repairs to a cracked structure.

REFERENCES

Davies, J.L. 1965. Landforms. Atlas of Tasmania, p.22. Lands & Surveys Dept., Hobart

Gill, E.D. 1961. The geological background to problems of landslip in Launceston. Tasmania.

Groupe d' études des Falaises. 1978. éboulements et chutes de pierres sur les routes. LCPC, Rapport de recherche LPC No. 81.

Collin, A. 1846. Recherches expérimentales sur les glissements spontanés des terrains argileux, accompagnées de considerations sur quelques principes de la mécanique terrestre. Paris.

Yamanouchi, T. & Miura, M. 1967. Multiple sandwich method of soft clay banking using cardboard wicks and quicklime. Proc. 3rd Asian Reg. Conf. S.M. & F.E. Haifa, pp. 256 - 260.

Developments in Geotechnical Engineering, Balasubramaniam et al. (eds) © 1994 Balkema, Rotterdam, ISBN 90 5410 522 4

The static and dynamic properties of schistous sands

Jing Zhou & Can-wen Yang
China Academy of Railway Sciences, People's Republic of China

ABSTRACT: In this paper, the laboratory experimental results of the static and dynamic properties of schistous sands are presented: e. g. compressibility, stress-strain relationship, strength, influence of anisotropy, dynamic modulus and potential of liquefaction et al. These results have been compared with those of siliceous sands. Practical applications are discussed.

Introduction

Schistous sand deposit is widely distributed along the Lower Yang-zi River Valley, China. It consists predominantly of the weathering particles of mica and other heavy minerals with schistous shape. Their engineering properties have not yet been thoroughly studied as compared with the siliceous sands. Whether these concepts or experiences based on the studies on siliceous sands can be applied to them or not are very important in practice. A series of research work on the static and dynamic properties of schistous sands have been carried on in China Academy of Railway Sciences. It is supported by National Natural Science Foundation of China.

Two kinds of sand samples, Nanjing schistous Sand(NS) and Fujian siliceous Sand(FS), are used for comparative tests, of which the fundamental physical properties are listed in Table 1.

Table 1. Physical properties of sand sample used

Sand	Gs	d_{10} (mm)	d_{60} (mm)	d_{50} (mm)	e_{max}	e_{min}
NS	2.72	0.034	0.10	0.09	1.17	0.612
FS	2.68	0.26	0.42	0.38	0.853	0.555

1 STATIC PROPERTIES

1.1 Compressibility

Samples were prepared by raining sands

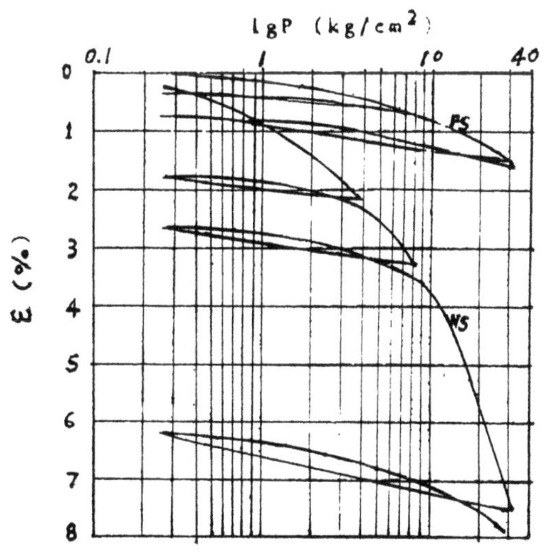

Fig. 1 Compression Curves of Both Sands

through water to simulate the process of natural deposition. Under one dimensional compression, the compressibility of NS is greater than that of FS, about 5-6 times. In Figure 1, the compression curves of both sands with D_r=60% are shown, in which the difference of both sands in unrecoverable deformation is also very clear. Schistous sand in natural deposit has shown significant anisotropic compressibility. A sample of 60×44×60 mm was used for isotropic compression test. In Figure 2, it is shown that the compressibility in horizontal direction is greater than in vertical direction and that after the pressure greater than a certain value (1.5 kg/cm²

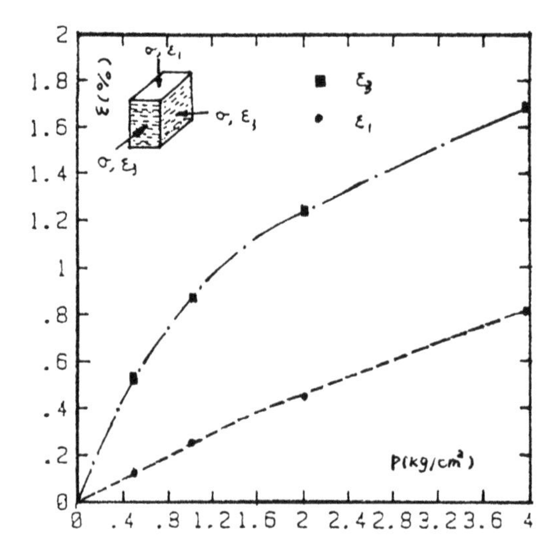

Fig. 2 $P-\varepsilon_1$ & $P-\varepsilon_3$ Curves of NS of D_r=50%

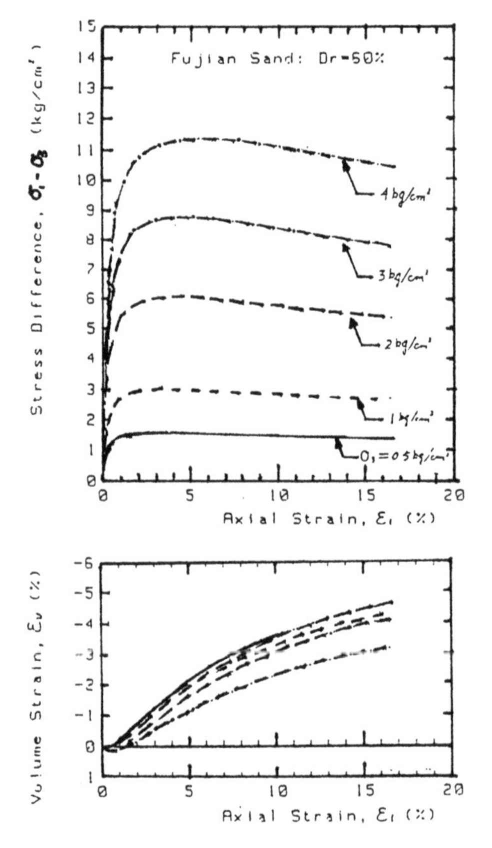

Fig. 3 $(\sigma_1-\sigma_3)\sim\varepsilon_1$, $\sim\varepsilon_v$ Curves

as in Figure. 2) . the sample behaves an isotropic compressibility.

1.2 Triaxial compression test

Both sands of various relative density were tested in triaxial cell and the typical test results of both sands are shown in Figures 3 & 4. In the Figures. it is shown that the axial strain at peak strength of NS is much larger than that of FS. and that the volume strain at peak strength of FS is in dilation but that of NS is in volume decreasing when the confining pressure is higher than 3 kg/cm^2. The peak strength envelope of FS is a straight line but the envelope of NS has a slight curvature. If a straight line envelope were assumed for both sands, their internal friction angles at various relative density as well as the residual friction angle are listed in Table 2.

Table 2. Φ and Φ_r of both sands

Sand	D_r=80%	=60%	=40%	Residual
FS	39.6	37.2	34.7	33.0
NS	39.8	37.0	36.0	35.8

It is well known that it is hard to run the test for the residual strength. A graphic method to find the residual strength on the ordinary triaxial test result is suggested as shown in Figure 5. Analysis of test data has shown that the normalized work (Moroto. 1976) can describe the dilation behaviour of both sands very well. So the residual strength of both sands might be calculated by:

$$(\sigma_1-\sigma_3)/\frac{1}{3}(\sigma_1+2\sigma_3)=\frac{3}{2}\lambda \qquad (1.1)$$

where λ is called normalized factor.

Although the NS has greater compressibility than FS. the strength parameter Φ of NS is almost the same or even greater than that of FS.

In conventional triaxial test. the Φ of NS at compression (b=0) and at extension (b=1) are very close. If tests are run under a stress path different from that of conventional triaxial test (e.g. σ_1=const. and σ_r decreased for compression; σ_1= const. and σ_r increased for extension) . the Φ at compression in both cases are very close but at extension in latter case is smaller. More works have to be done.

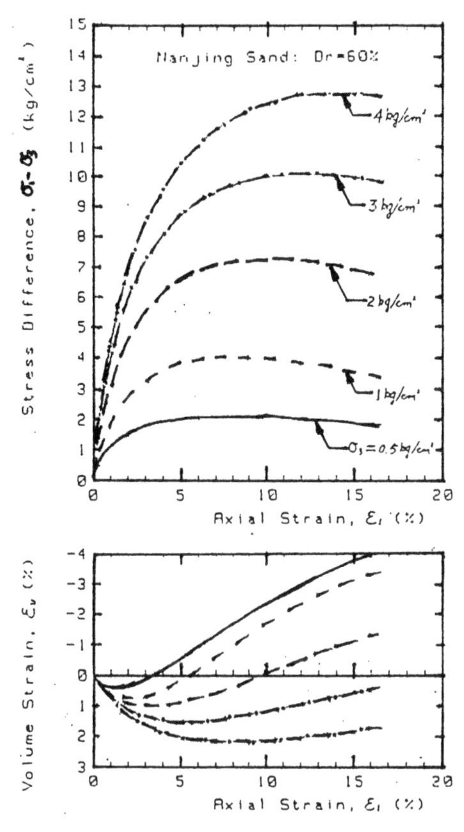

Fig. 4 $(\sigma_1-\sigma_3)\sim\varepsilon_1\sim\varepsilon_v$ Curves

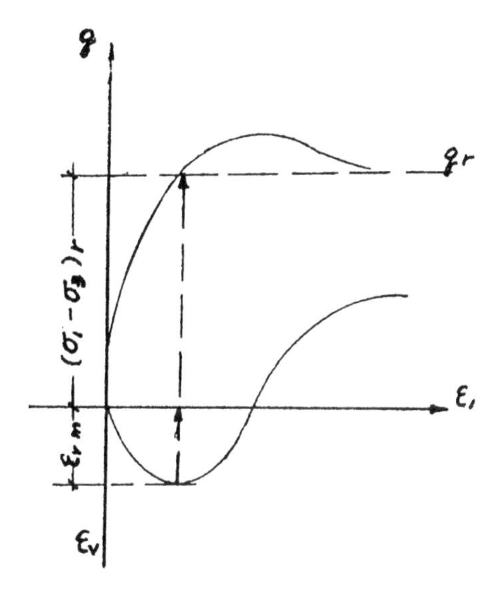

Fig. 5 Sketch for Determining Φ_r

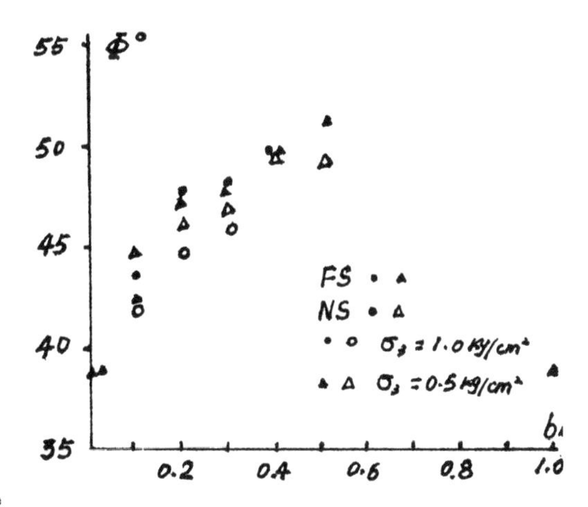

Fig. 6 Influence of b on Φ

1.3 Three stress and plane strain test

The size of specimen is 60×44×60 mm. Due to the limitation of the function of the apparatus, tests could be run for b=0 to b =0.5, but not for the higher values of b. A typical test result of the influence of the intermediate stress on Φ is shown in Figure 6. The influence of b on Φ is greater for FS than that for NS. The Φ of b=1 in Figure 6 is measured in triaxial extension test. The shape of the stress-strain- volume strain curve is greatly influenced by the change of b value. In plane strain tests, the variation of intermediate principal stress has been monitored so that the variation of the value of b can be calculated. The value of b is increasing at the beginning and finally approaches a steady value of 0.4 for NS and about 0.35 for FS. Comparing the results of plane strain test with that of three stress test (b = 0.4 for NS; b = 0.35 for FS), the stress-strain-volume strain curves are very close. If the internal friction angle is defined as $\Phi=\arcsin((\sigma_1-\sigma_3)/(\sigma_1+\sigma_3))$, the Φ obtained by plane strain test is 10% greater than that by triaxial compression test for NS and 15% greater for FS. The influence is greater for FS than that for NS.

1.4 P=constant test

On the π -plane of p=200 kpa and p=300 kpa, a series of test with different Lode angle θ have been done. The failure loci of both sands on π -plane have been calculated and plotted in Figure 7 together with various failure criteria. It seems that the test data of both sands fell between SMP and Lade-Duncan criteria. The test data of θ =60 (b=1) have not been plotted on because

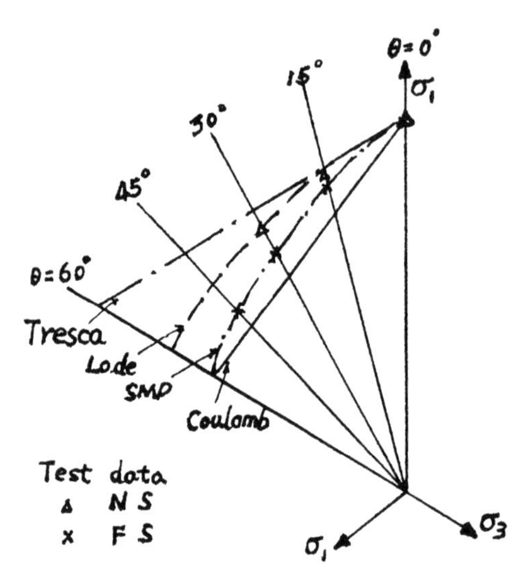

Fig. 7 Failure Loci of FS and NS

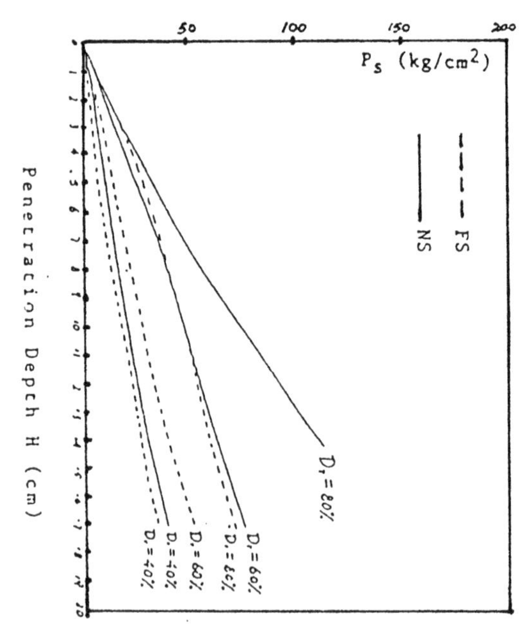

Fig. 8 Penetration Curves of Simple Cone

there are still some contradiction among the results obtained by different stress path as mentioned above.

1.5 Model test of cone penetration.

A series of model test on both sands with small cone penetrometer have been carried on in order to find out the effect of special mechanical properties of schistous sands on the cone resistance.

In the first series of tests, the sand container used is Φ 17.7×25.64 cm. The diameter of the cone is 2.5 cm with 60° angle and the diameter of the connecting push rod is only 7 mm which is smaller than that of cone. This model cone simulates the early Dutch simple cone that left enough space behind the cone for the sand to squeeze in when the cone is pushed into the sand. Three density states of sand, D_r=0.4, 0.6 and 0.8 are made for testing. The rate of penetration is 3.45 mm/min. The typical test results are shown in Fig. 8.

It is seen in the Figure that the cone resistance is continuously increased with the penetration depth and that the cone resistance in NS is higher than that in FS. When the surface of the sand is loaded with surcharge weight, the cone resistance will be increased correspondingly and increased with penetration depth in decreasing rate.

In the second series of tests, the container is Φ 35×43cm. The test cones are with two different base area, A=10 and 4.9cm ^{2}. The connecting push rod used has the same diameter as that of the cone, that means no gaps left behind the cone. One of the test results are shown in Figure 9.

It is seen in the Figure that the cone resistance is increased with the penetration depth up to an ultimate value and that the ultimate cone resistance in NS is lower than that in FS.

The second series of tests simulate the deep foundation and also the type of static penetration cone used in China. In such case, the cone penetration resistance in NS will be lower than that in FS if both sands have the same relative density.

The results of the model tests have shown that the failure mechanism of two types of cone are different. The failure in simple cone is caused by shear squeezing, but for the second type of cone, the failure phenomena varied gradually from shear squeezing to compression with the increase of penetration depth.

2 DYNAMIC PROPERTIES

2.1 Dynamic strength parameter

The effective angle of internal friction Φ_d of both sands were obtained by the effective stress path from strain-controlled undrained cyclic triaxial tests on saturated specimens. The test results are listed in the Table 3. As can be seen that the Φ'_d of FS is approximately equal to the Φ'_s under the same relative density D_r= 0.6, but the Φ'_d of NS is much lower than the Φ'_s. It indicates that the strength of NS would

Table 3. Test results of Φ'_d and Φ'_s

Sand	Φ'_d	Φ'_s
FS	36.00	37.20
NS	29.00	37.00

be significantly decreased during cyclic loading.

2.2 Dynamic shear modulus at low strain level

For clean sand, it has been found that the maximum shear modulus at low shear strains ($r=10^{-6}-10^{-4}$) G_{max} is mainly dependent on effective confining pressure $\bar{\sigma}_o$ and void ratio e. Analytical expressions have been developed by Hardin et al. as follows:

$$G_{max}=700 \cdot \frac{(2.17-e)^2}{1+e} \cdot \bar{\sigma}_o^{0.5} \quad (2.1)$$

for round-grained sands and

$$G_{max}=326 \cdot \frac{(2.97-e)^2}{1+e} \cdot \bar{\sigma}_o^{0.5} \quad (2.2)$$

for angular-grained sands,
where G_{max} and $\bar{\sigma}_o$ are in kg/cm^2.

The results form a series of resonant column tests at confining pressures of 1, 2, 3, and 4 kg/cm^2 and several void ratios and G_{max} calculated are listed in Table 4. It shows that the tested G_{max} and calculated G_{max} (according to Hardin) are in excellent agreement for FS but much different for NS.

Based on the test results, the analytical expression by Hardin should be modified for Nanjing Schistous Sand as follows:

$$G_{max}=210 \cdot \frac{(2.97-e)^2}{1+e} \cdot \bar{\sigma}_o^{0.58} \quad (2.3)$$

The values of G_{max} computed with this expression are also given in Table 4.

Fig. 10 shows the relationships between G_{max} and $\bar{\sigma}_o$ for two different sample preparation methods (wet-tamping and water-pluviation) at the angles of 0°, 45°, and 90° between the shear plane and the deposit surface. It is seen from this figure that the effects of sample preparation method on the G_{max} of NS are insignificant and may be neglected in using the emperial expression (2.3).

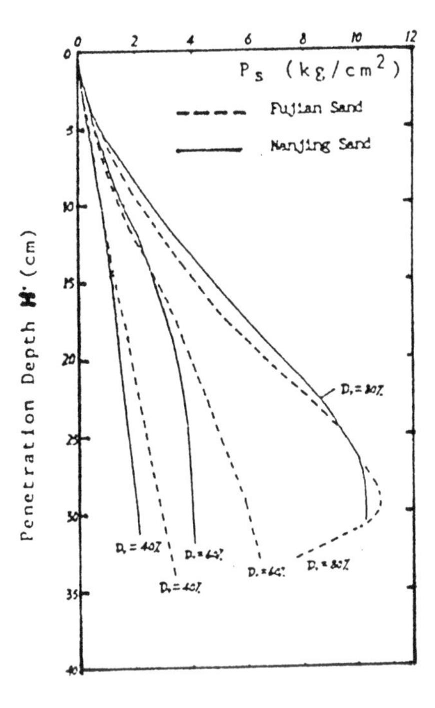

Fig. 9 Penetration Curves of Cone A=10cm²

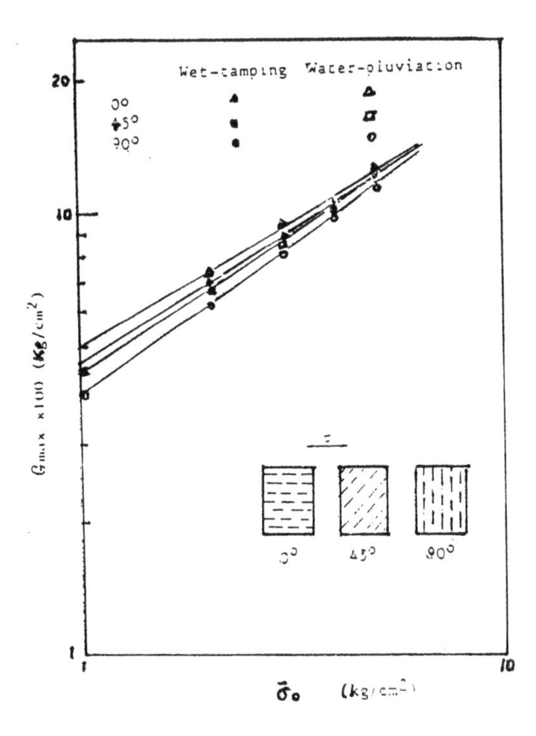

Fig. 10 Effects of Sample Preparation Methods on G_{max} of NS

Table 4. G_{max} of both sands

	e	σ_o / G_{max}	1	2	3	4	Maxium absolut error(Kg/cm^2)	Maxium relative error(%)
Fujian sand	0.71	Measured	800	1163	1451	1710	73	9
		Equ. (1)	873	1234	1511	1745		
	0.62	Measured	976	1412	1758	2058	63	6
		Equ. (1)	1039	1468	1798	2076		
Nanjing sand	0.88	Measured	430	679	844	1014	500	76
		Equ. (2)	757	1071	1311	1514		
		Equ. (3)	488	729	923	1090	79	13
	0.84	Measured	451	720	931	1070	538	78
		Equ. (2)	804	1137	1393	1608		
		Equ. (3)	518	774	979	1157	87	15
	0.80	Measured	499	791	1006	1100	606	71
		Equ. (2)	853	1206	1477	1706		
		Equ. (3)	549	821	1039	1228	128	12
	0.77	Measured	530	838	1020	1240	542	68
		Equ. (2)	891	1260	1543	1782		
		Equ. (3)	574	858	1086	1283	65	8

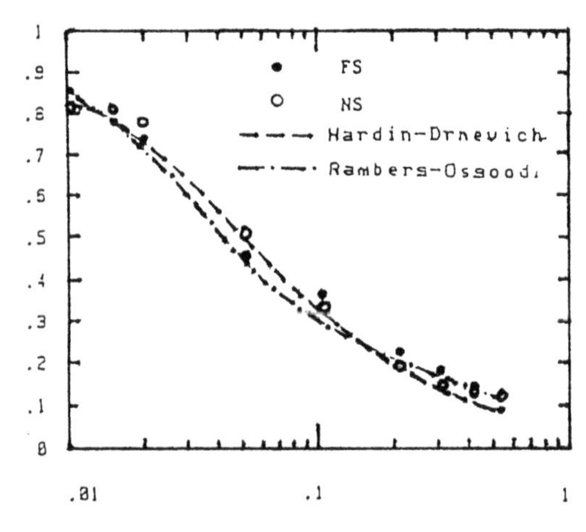

Fig. 11 $G1/G_{max}$ versus γ_o

2.3 Cyclic stress-strain behavior

The results of undrained-controlled cyclic triaxial tests on saturated specimens of both sands show that the relationship between G1, the secant shear modulus in the first cycle of shear loading, and the cyclic shear strain γ_o can be characterized by Hardin-Drvenich hyperbolic equation and Ramberg-Osgood equation (Fig. 11). After several cycles of loading, the excess pore water pressures Δu of saturated sands buildup and the shear moduli decline. The relations between G/G1 and normalized pore water pressure $\Delta u/\sigma'_{3o}$ are plotted in Fig. 12. It is found that G/G1 may be directly related to (1-U) and can be fitted to the equation:

$$G/G1=\begin{cases} 1 & U<0.3 \\ \lg[1+13(1-U)] & U\geq 0.3 \end{cases} \qquad (2.4)$$

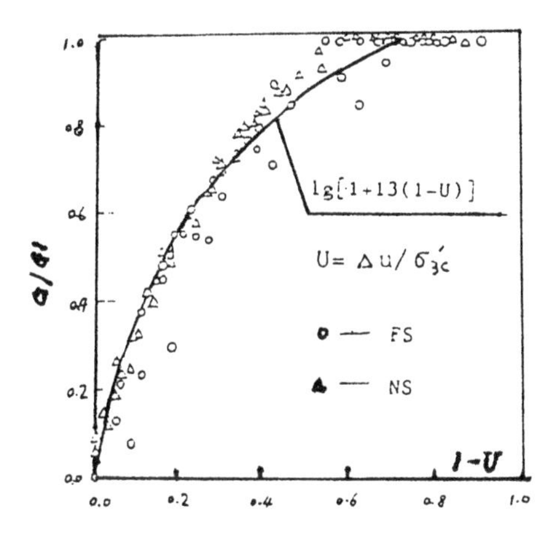

Fig. 12 G/G1 versus (1-U)

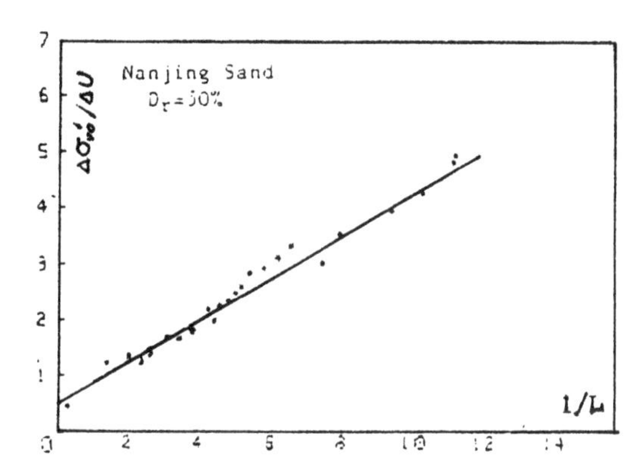

Fig. 14 Relationship Between $\Delta u/\sigma'_{vo}$ and L

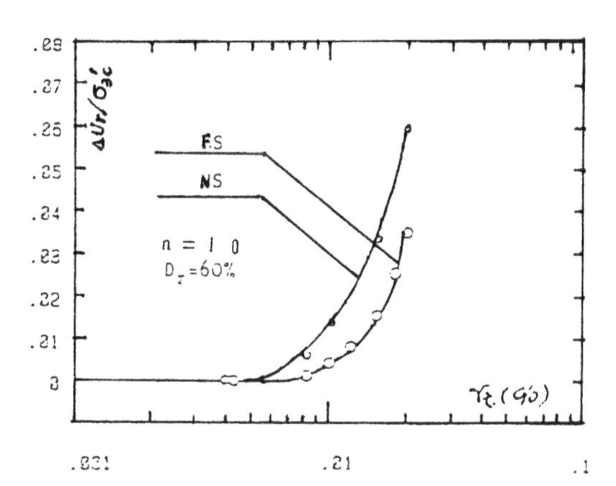

Fig. 13 Threshold Shear Strain

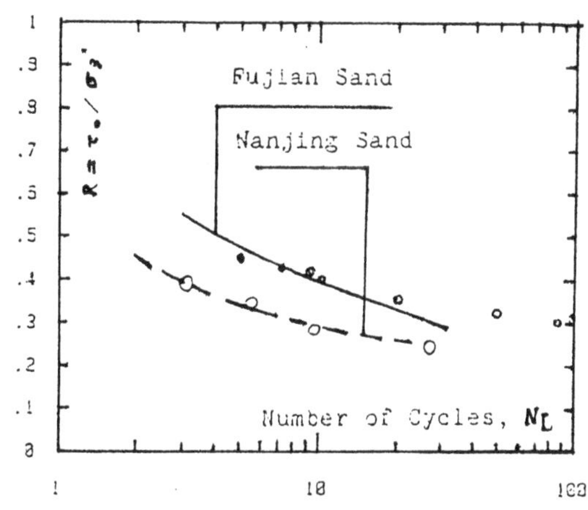

Fig. 15 Relationship between Cyclic Stress Ratio and Number of Cycles Required to Cause Liquefaction under Triaxial Compression Test

2.4 Buildup of excess pore water pressure during cyclic loading

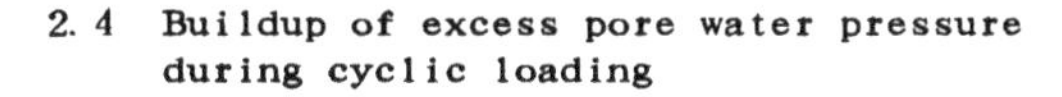

More recent research has conclusively demonstrated that the buildup of excess pore water pressure of sand during cyclic loading is controlled mainly by the magnitude of the cyclic shear strain. There is a threshold strain γ_t existed. below which no pore water pressure developed under cyclic loading. Fig. 13 illustrates the relevant results on γ_t of both sands. $\gamma_t \doteq 0.8\times10^{-4}$ for FS and $\gamma_t \doteq 0.5\times10^{-4}$ for NS. which is somewhat lower than that of FS. It implies that the excess pore water pressure in NS is easier to buildup than in FS under the same cyclic loading conditions.

The test results have shown that the excess pore water pressure developed during cyclic loading may be expressed a function of a single parameter L called damage parameter. incorporating the influence of current cyclic shear strain γ_c and past strain accumulation $\zeta_c = \sum\gamma_c$. The damage parameter L is determined from the relation:

$$L = \gamma_c \cdot \lg(1+\zeta_c) \qquad (2.5)$$

The curve of $\Delta u/\sigma'_{vo}$ versus L is shown in Fig. 14 and may be fitted to an equation as follows:

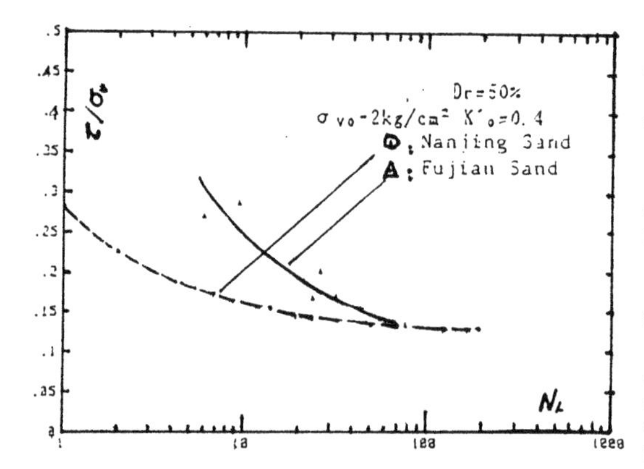

Fig. 16 Relationship Between Cyclc Stress Ratio and Number of Cycles Required to Cause Liquefaction under Simple Shear Test

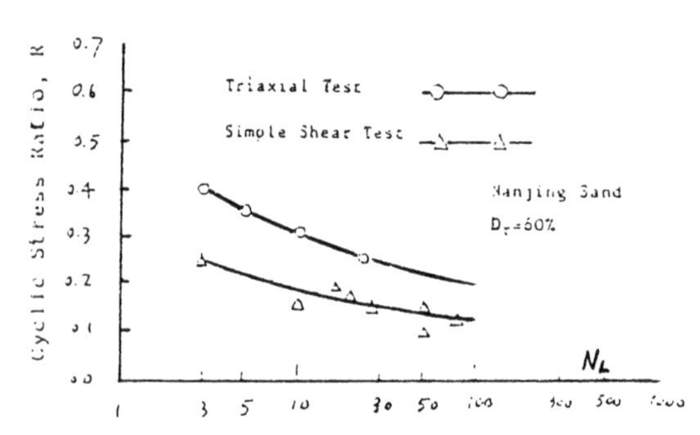

Fig. 17 The Cyclic Stress Ratio Required to Cause Liquefaction under Triaxial Compression and Simple Shear Conditions

$$\frac{\Delta u}{\sigma'_{vo}} = \frac{L}{a+b \cdot L} \qquad (2.6)$$

where the constants a and b are determined by nonlinear least-squares curve fitting.

The results of liquefaction test under dynamic triaxial compression test and simple shear test for both sands are shown in Fig. 15 and Fig. 16 respectively.

To compare the relative magnitudes of the cyclic stress ratio required to cause liquefaction under triaxial compression and simple shear conditions, samples with a relative density of 60 percent were tested at confining pressure of 1 kg/cm² in both types of test. The results are plotted in Fig. 17. It illustrates that the cyclic stress ratio required to cause initial liquefaction under simple shear condition (τ/σ'_{vo}) is considerably less than the cyclic stress ratio required to cause initial liquefaction under triaxial compression condition ($\sigma_d/2 \cdot \sigma'_{3o}$). The ratio between ($\tau/\sigma'_{vo}$) and ($\sigma_d/2 \cdot \sigma'_{3o}$) agrees with C_r proposed by Seed. C_r is a correction factor to be applied to laboratory triaxial test data to obtain the stress conditions causing liquefaction in the field.

CONCLUSIONS AND DISCUSSION

1. Schistous sand has much higher compressibility than siliceous sand, especially under shearing stress. This feature should be considered in practical works.
2. High volume reduction under shear stress of schistous sand is the main factor that causes the low dynamic shear strength, high potential of liquefaction and faster rising of excess pore water pressure under cyclic loading.
3. The results of the model cone penetration tests in both schistous and siliceous sands might suggest:
   a. End bearing capacity of pile in schistous sand may be smaller than that in siliceous sand at same relative density.
   b. End bearing capacity of pile in schistous sand will be influenced by the geometry of pile(straight side or enlarged end).
4. In some code, the relative density is used to evaluate the potential of liquefaction of saturated sand, which is based on the experiments on siliceous sand, and therefore, it has to be modified if used in schistous sand.
5. The empirical metheds SPT and CPT for evaluating the static and dynamic properties of sands have to be modified if used in schistous sand for the same reason.

ACKNOWLEDGEMENTS

This paper is only a part of the results of our research works. The authors gratefully offer their thanks to the Research Group members who have worked for this project.

REFERENCES

Li, S. 1985. Experimental study on the stress-stain relationship of schistous sand. Master's Thesis, CARS

Liu, G. N. 1985. One dimensional compressibility of schistious sand. Master's Thesis, CARS.

Luo, M. Y. & Zheng, W. B. & Han, Z. L. 1987. Model cone penetration test in sand.

Shan, C. 1987. Three dimensional stress-strain relationship and normalized work

for sand. Master's thesis. CARS.
Shen. W. 1987. Experimental study on the dynamic property of schistous sand. Master's thesis. CARS.
Tang. Z. H. 1987. Experimental studyon the shear modulus and damping properties of schistous sand. Master's thesis. CARS.
Wu. Y. H. 1986. Study on the dynamic properties of schistous sand by simple shear test. Master's thesis. CARS.
Yao. C. W. 1986. Experimental study on anisotropy of schistous sand. Master's thesis. CARS.
Zen. M. L. 1988. Effect of stress state and stress path on the engineering properties of schistous sand. Master's thesis. CARS.
Zheng. W. B. 1986. Stress-strain charactristics of schistous sand under three dimensional stresses. Master's thesis. CARS.
Zhou. J. & Zheng. W. B. & Li. S. 1987. The strength properties of schistous sands. Proc. 8th ARC on SMFE. Vol. 1

Developments in Geotechnical Engineering, Balasubramaniam et al. (eds) © 1994 Balkema, Rotterdam, ISBN 90 5410 522 4

Black cotton soils – Highly expansive clays of India

Dinesh Mohan
Central Building Research Institute, Roorkee, India

ABSTRACT: This paper examines the characteristics of highly expansive black cotton soils of India and presents some experiences with foundation design and construction. Foundation failures in these soils are attributed mainly to the differential movement of the structure as a result of uneven ground movements due to alternate swelling and shrinkage of the soil. The use of under-reamed piles in black cotton areas has resulted in economy of as high as 30-60% when compared to strip footings. It is also quick and needs no exta backfilling thus providing for better and more uniform conditions for floor finishes adjacent to the walls.

1 INTRODUCTION

India has large tracts of expansive soils known as black cotton soils. This name derives from their black colour and cotton being grown in many regions on these soils. The major area of their occurrence is Central India, South of Vindhyachal range, covering an area of about 0.8 million sq. km., thus forming about 20 per cent of the total land area of the country[1].

The Indian black cotton soils are generally heavy clays containing predominantly of clay mineral montmorillonite, exhibiting high shrinkage and swelling characteristic. During shrinkage due to drying, there is a formation of hexagonal columnar structure with vertical cracks up to 8 cm wide at the ground level, extending up to 2.5m depth[2]. Shrinkage in horizontal direction is nearly two thirds of the total volumetric shrinkage[3]. Similar soils occur in other countries also e.g. 'Cheronozems' of U.S.S.R., 'Badole' of Japan, 'Pampus' of Argentina, 'Tirs' of Morocco, 'Margilatic' of Indonesia and black earths of Australia, Java, & Sumatra[4]. Vast areas near Irbid (Jordan) are also full of expansive clays. Geologically, their formation is associated with basalts but their occurrence on granite, shale, sand-stone and slate is also recognized. They occur both as residual and transported. In the former case, where the parent rock lies underneath, their depth is shallow, averaging about 1m, but in low lying and flat areas when they develop over alluvium, after being transported by the wind and water, they go deep and usually average 5m.

2 BASIC SOIL PROPERTIES: CORRELATION WITH ENGINEERING PROPERTIES.

The author procured samples of black cotton soil from twenty different places in India (Fig.1) covering practically all the regions where such soils exist. The depth of the samples varied from 1-3m. The liquid limit ranged between 46 and 97, plasticity index between 22 and 49 and the shrinkage limit between 11 and 14. The plot on the plasticity chart (Fig.2) showed that practically all the soils fall above the Casagrande 'A' line. The organic content varied from 0.4 to 2.4 per cent and specific gravity had an average value[5] of 2.7.

A straight line relationship was observed between shear strength (plotted to log scale) and liquidity index (Fig.3). For this experiment the soil was compacted at the plastic limit in a Proctor's mould by the standard Proctor rammer. Cylindrical test specimen of diameter 5cm and height 10cm were obtained from the compacted soil mass in the mould by pushing in a thin walled steel sampling tube. These were subjected to gradual air drying and compressive strength was determined at four different moisture conditions in a decreasing order in a compression testing machine working at a uniform rate of loading. The curve has been plotted only upto liquidity index range of 0.3 which gives the minimum moisture content at which black cotton soils usually exist at a depth of 1m. The liquidity index has a negative value in all cases as the moisture content was below the plastic limit. The concentration of points is along a straight line and at the plastic limit, the strength of the soil is very low.

No appreciable difference was found in strength between undisturbed and remoulded samples (Table I). This was checked by taking a portable unconfined compression test equipment to the site and testing undisturbed and remoulded samples from various depths. The sensitivity of Indian black cotton soil is therefore close to unity.

Differential free swell test was used to get a comparative idea of the swelling potential of the

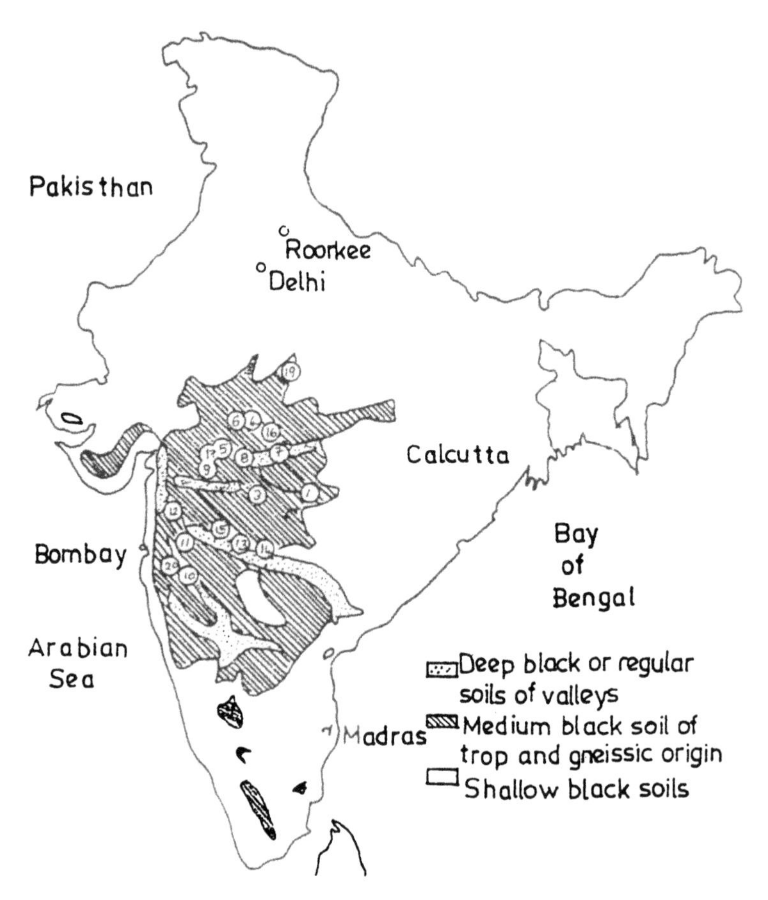

Fig. 1 Soil map of India showing location of black soils.

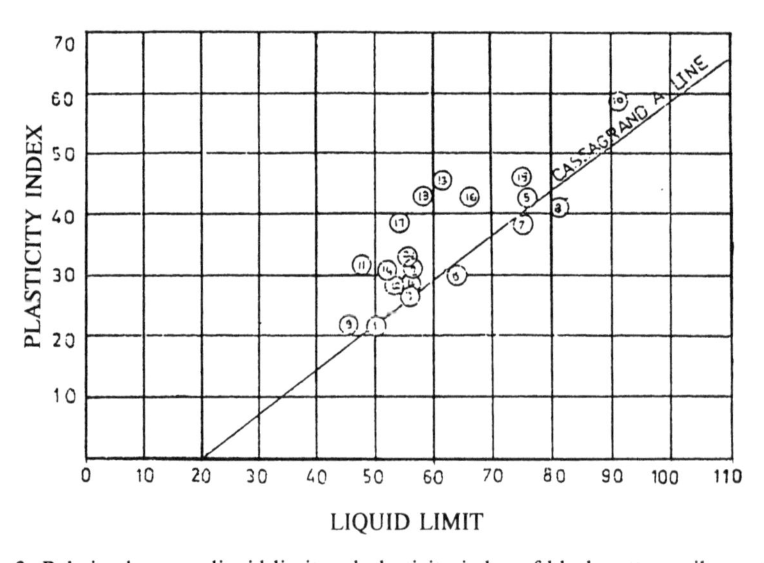

Fig. 2 Relation between liquid limit and plasticity index of black cotton soil samples.

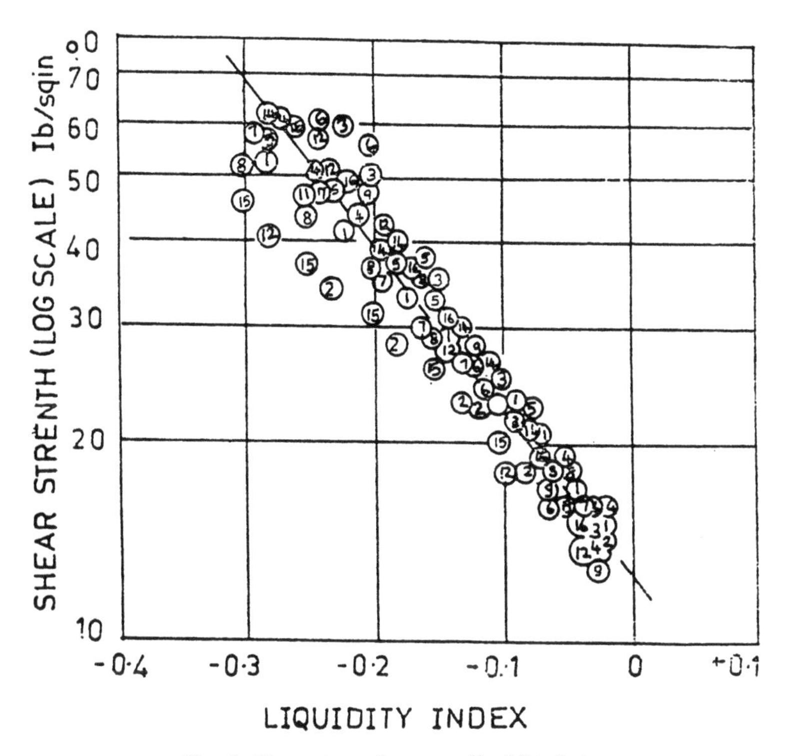

Fig. 3 Shear strength versus liquidity index.

Table 1. Shear strength (kPa) of undisturbed and remoulded samples.

Location	Depth (m)	Undisturbed samples	Remoulded samples
Powerkhera	1.22	131	145
	2.44	145	117
	3.66	138	124
Indore	1.22	117	110
	2.44	151	158
	3.66	200	200

soil. The normal free swell test suggested by Holtz and Gibbs[7] consists of gently pouring 10cc of oven dried soil, passing 36 BS sieve (Particle size less than 0.42mm), into distilled water and noting the increase in volume of the soil. The main drawback of the method is lack of uniformity in packing and long time required for the soil to come to constant volume in a soaked state. The differential free swell test developed by RDSO[8] overcomes this drawback. In this method, 10g of two oven-dried samples of soil are soaked, one in distilled water and the other in kerosine oil or some other non-polar liquid, their respective volumes being measured. The difference in volume expressed as a percentage of the volume in kerosine oil gives the differential free swell. The values were found to range from 28 to 122. A curve was plotted between the differential free swell on one hand and volume expansion and swelling pressure on the other. A straight line relationship was observed (Fig.4).

3 FOUNDATION DESIGN AND CONSTRUCTION

Foundation of buildings and other structures on Indian Black cotton soils have been a matter of great concern to the engineers and builders. It is heart-rending to see some of the structures cracking badly within one year after their construction.

Until 1955, the normal methods for constructions of foundations on black soils were:

(1) Provision of reinforced concrete bands at plinth and lintel level.

(2) Provision of reinforced concrete raft or beam to support the superstructure.

(3) Removing the black soil entirely or to a considerable depth and backfilling the trench with cohesionless soils.

Method (1) was not very effective and method (2) was costly. Method (3) is uneconomical when depth of black soil exceeds 1m.

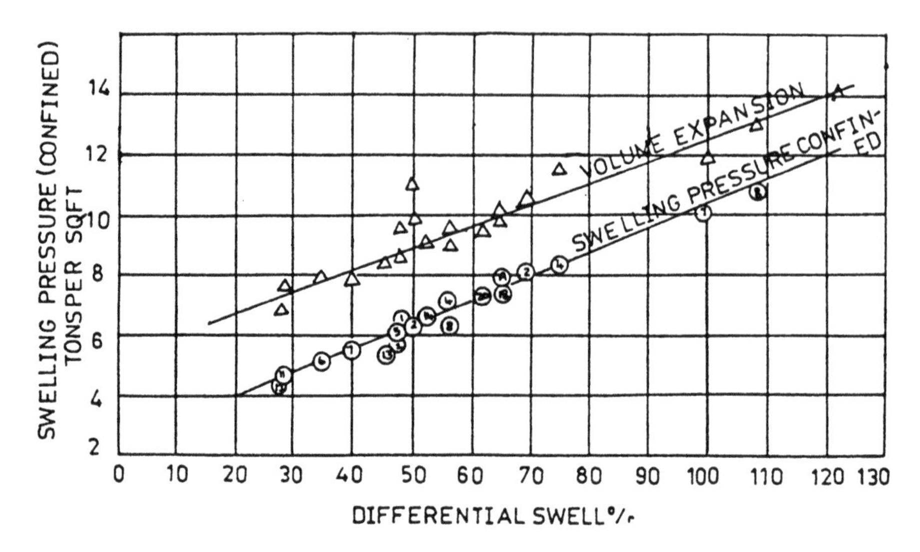

Fig. 4 Relation between differential swell, and swelling pressure and volume expansion of black cotton soil samples.

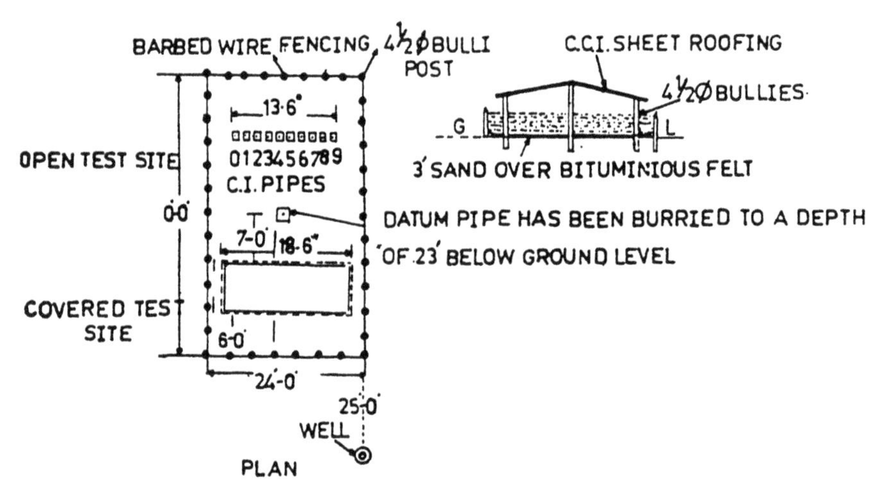

Fig. 5 Plan of the test site.

The main reason leading to the failure of foundations in black soils is the differential movement of the structure with an uneven ground movements due to alternate swelling and shrinkage of the soil. The vertical ground movements decrease with depth and are negligible at a particular depth. Investigations were carried out by CBRI to locate this depth in two different regions with deep layers of black soils. Test sites were established at both places and each of the sites had a row of G.I. pipes anchored at different depths from 15cm to 4.5m deep. There were two such rows at each site, one in the open and the other under a shed and protected by bituminous felt to simulate the cover produced by a building (Fig.5). Fortnightly, levels of each pipe were recorded over a period of 13 to 18 months and these levels have been plotted in Figs. 6 and 7. From a study of the curves it would be seen that 3.5 m is the depth at which the vertical ground movements are inappreciable.

Another test site 25m x 35m was set up after a lapse of about 15 years in a black cotton soil area where the top 2.7m was black silty clay overlying yellow clay which extended to 5m and beyond[11]. A number of surface movement indicators, depth gauges and *in situ* swelling pressure measuring instruments were installed and measurements were carried out over a period of 3 years (Fig.8).

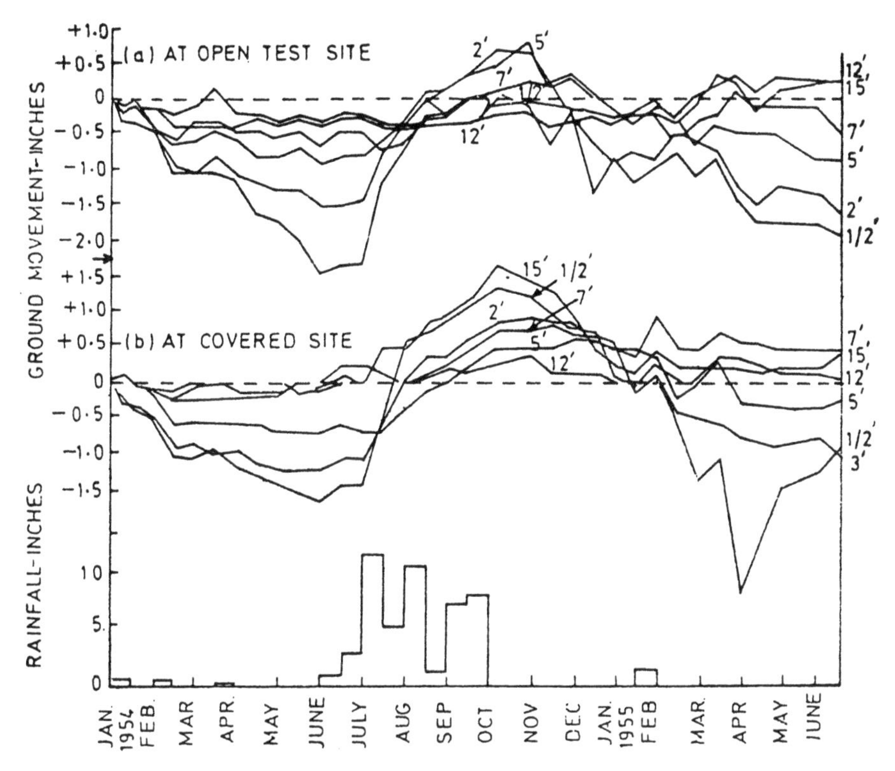

Fig. 6 Record of vertical ground movement at Hoshangabad.

Maximum value of ground movement at the surface was found to be 65mm which decayed to a negligible value at about 5m depth. Bulk of the movement was found to occur within the top 2m and 10 per cent of maximum occurred at nearly 3/4 of the depth of negligible heave. A depth of 5 x 3/4 = 3.75m could therefore be taken as a safe depth for foundations. *In situ* swelling pressure studies were carried out at depths varying from 0.5m to 2.5m using individual beam set up (Fig.9). In order to accelerate swelling, the area was continuously flooded for a period of five months. Out of the total six swelling pressure set ups, two were measured *in situ* swelling pressures by accelerated tests. Swelling pressure values (σ_s) as high as 5.5kg/cm^2 was observed with depth (D) which could be expressed by the following relationship $\sigma_s = 8 - \delta\ 1.6D$. Having found 3.50 - 3.75m as depth of inappreciable ground movement, a design was developed of under-reamed pile foundations with the belled out portion anchored at a depth of 3.5m or earlier if water table or a stable strata was encountered. The boring for the pile was carried out by a spiral auger and its bottom was under-reamed to about 2½ times the shaft diameter. For this purpose a portable hand-operated under-reaming tool was designed (Fig.10). The tool is simple to fabricate in a workshop. It consists of an assembly of four steel blades fixed around a central shaft and a bucket to hold the cut soil. The blades widen out as the shaft is pressed downwards and by rotating them the bore hole is widened. The shaft has a number of holes and a pit[12] inserted in a particular hole controls the maximum diameter of the bell.

The pile diameter varies from 20 to 50cm depending on the load it is expected to carry. For small works, in out of the way places, manual operation is usually preferred but for larger diameter and deeper piles, a mechanized boring rig is normally necessary.

The piles are provided at appropriate locations keeping in view the layout of the building and the load carrying capacity of the piles. As much as possible, all piles are uniformly loaded and the spacing adjusted so as to keep the door and window openings mid-way between two piles. A typical layout of a residential building and details of piles is given in Fig. 11(a) and a section through the pile foundation is given in Fig. 11(b). It would be seen that the grade beam is kept 8cm clear of the ground so that the soil does not heave against it. These beams, carrying the masonry superstructure, are designed for a bending moment of WL/50 to allow for panel action in the masonry. The shuttering supporting the beams is therefore not removed for about a week.

The bearing capacity (Q_u) of under-reamed piles can be determined by the following expression normally used for bored piles:

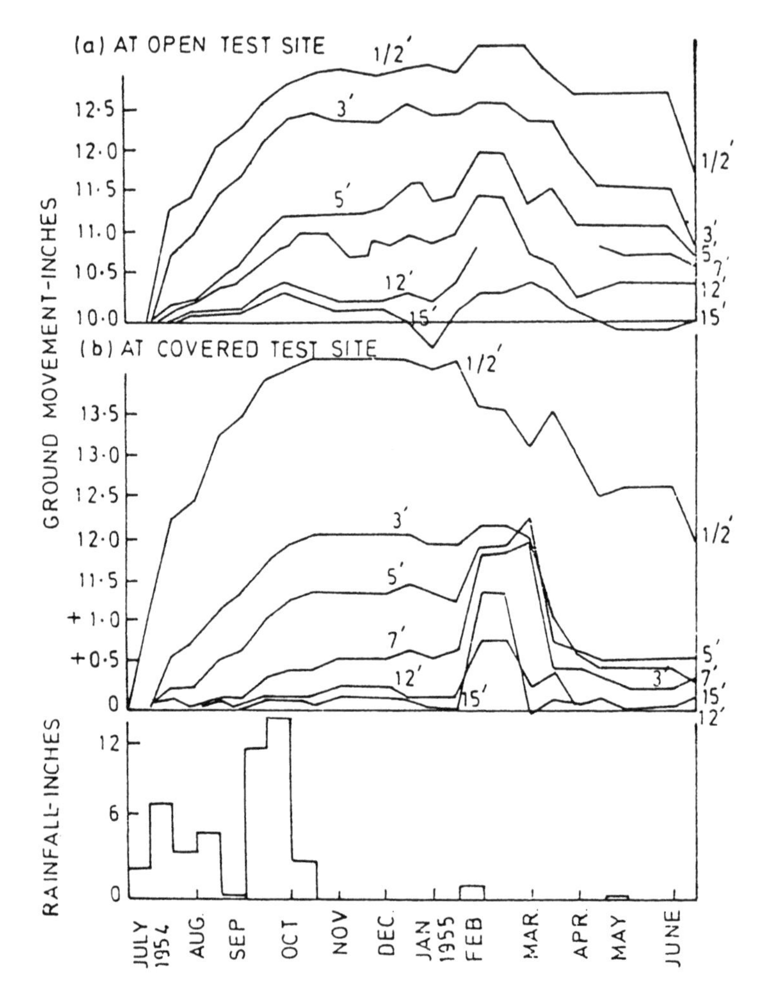

Fig. 7 Record of vertical ground movement at Indore.

$$Q_u = A_p N C_p + \alpha C A_s \qquad (1)$$

where,

A_p	=	area of the pile tip (under-reamed base of the pile
N	=	bearing capacity factor (may be taken as 9)
C_p	=	undisturbed shearing strength of the soil at the bearing level
α	=	reduction factor for bored piles
C	=	average undisturbed shearing strength of the soil along the pile length
A_s	=	surface area of the pile shaft

The value of reduction factor (α) may be taken as 0.5. This has been determined by carrying out a series of loading and pull out tests on *cast-in-situ* bored concrete piles in black cotton soils at four different sites in India[10].

The adoption of under-reamed pile in expansive soils has resulted in economy to the extent of 30 - 60 per cent when compared to the traditional strip footings. An additional advantage is that the process is quick and no extra excavation or backfilling required. This provides better and more uniform conditions for floor finishes adjacent to the walls.

A full scale load test[12] was carried out on a 7.6 m long and 45 cm stem diameter multi-under-reamed pile, having four bulbs, each 112.5 cm in diameter, in a deep layer of black cotton soils, having undrained cohesion varying from 0.9 to 1.45 kg/cm^2, liquid limit from 65 to 70 and plasticity index from 35 to 50. Twelve under-reamed piles, arranged in a circle, were used as anchors for the loading frame designed to carry a load of 400 tonnes. A sectional view of the test set up is shown in Fig.12. Load carrying capacity of the test pile as worked out from table *Appendix*-A was found to be 290 tonnes. The computed value, using bearing

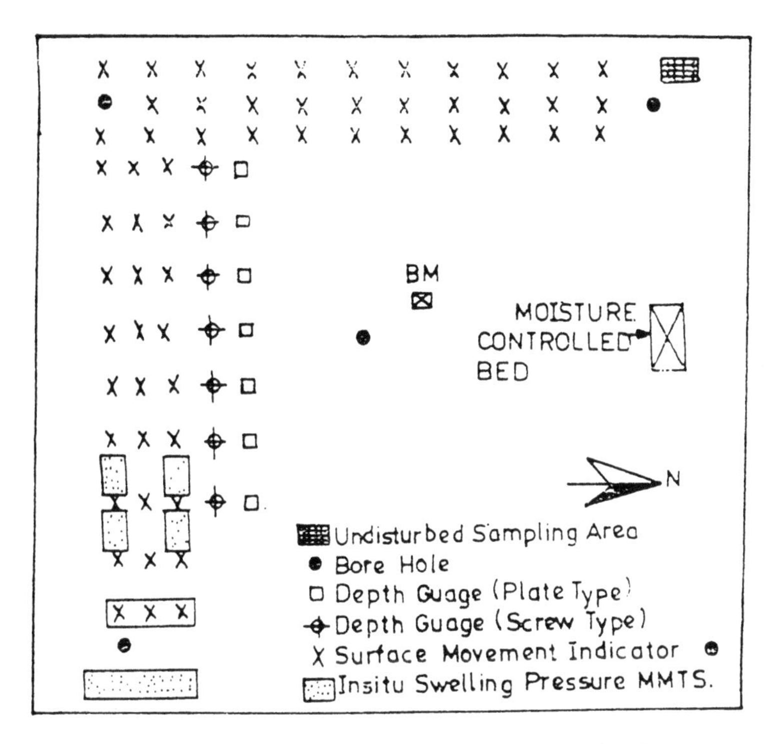

Fig. 8 Site of field experiment showing layout of measuring instruments.

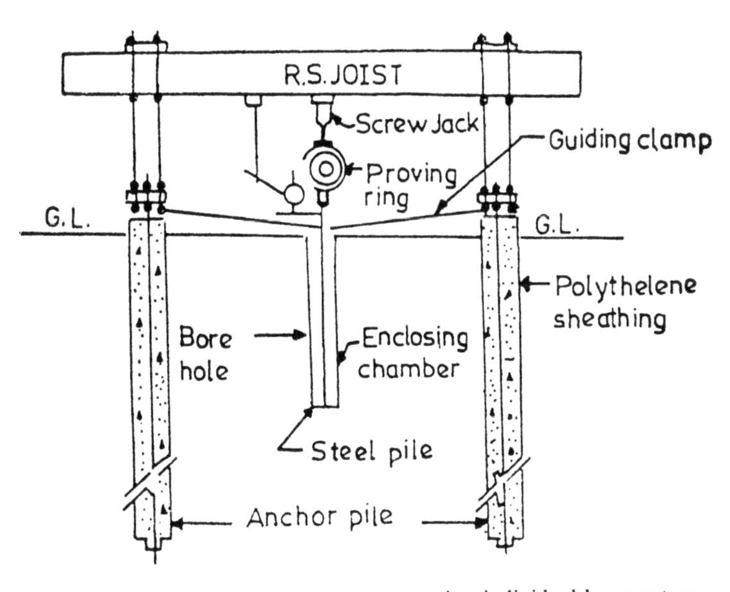

Fig. 9 Swelling pressure measurement using individual beam set up.

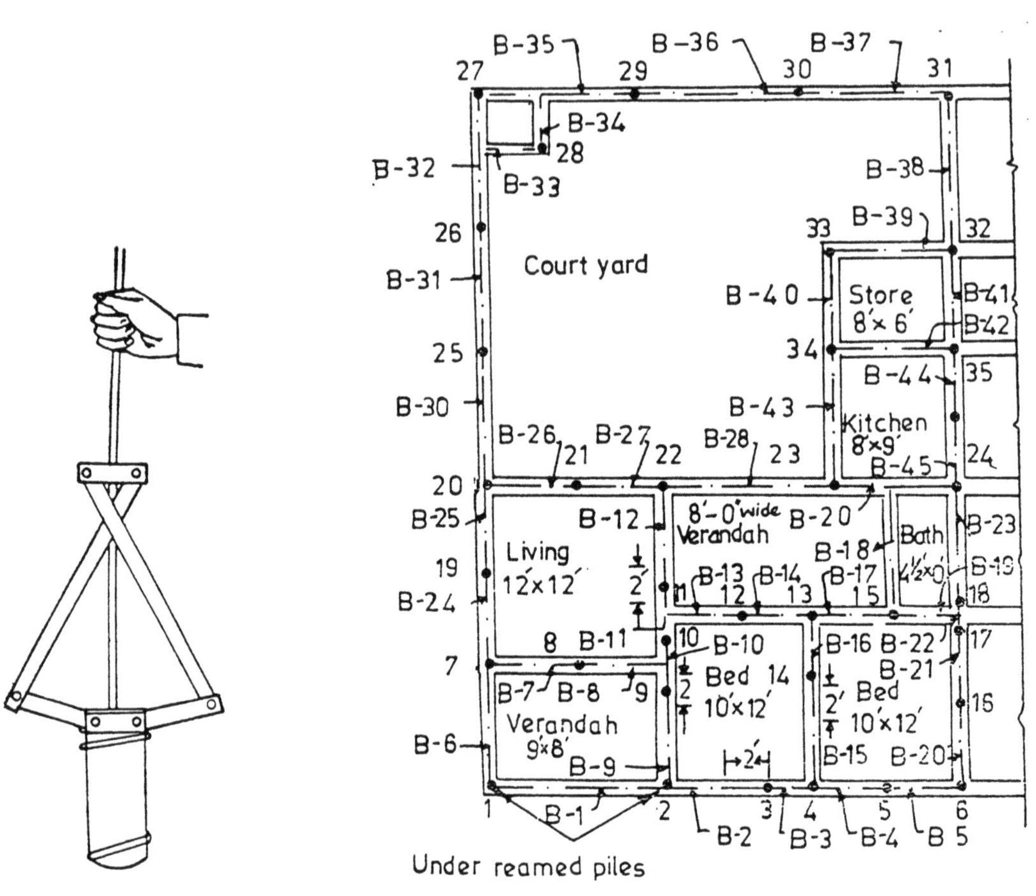

Fig. 10 Under-reaming tool.

FIG 11 (a)

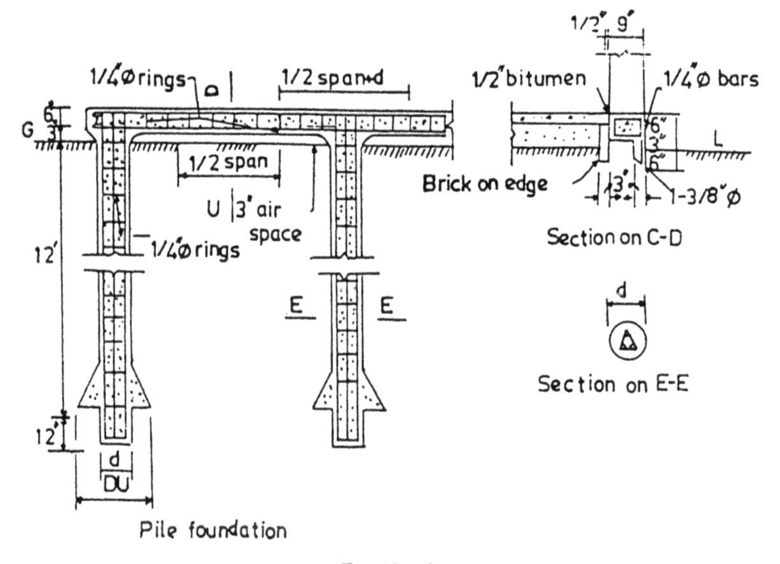

FIG 11 (b)

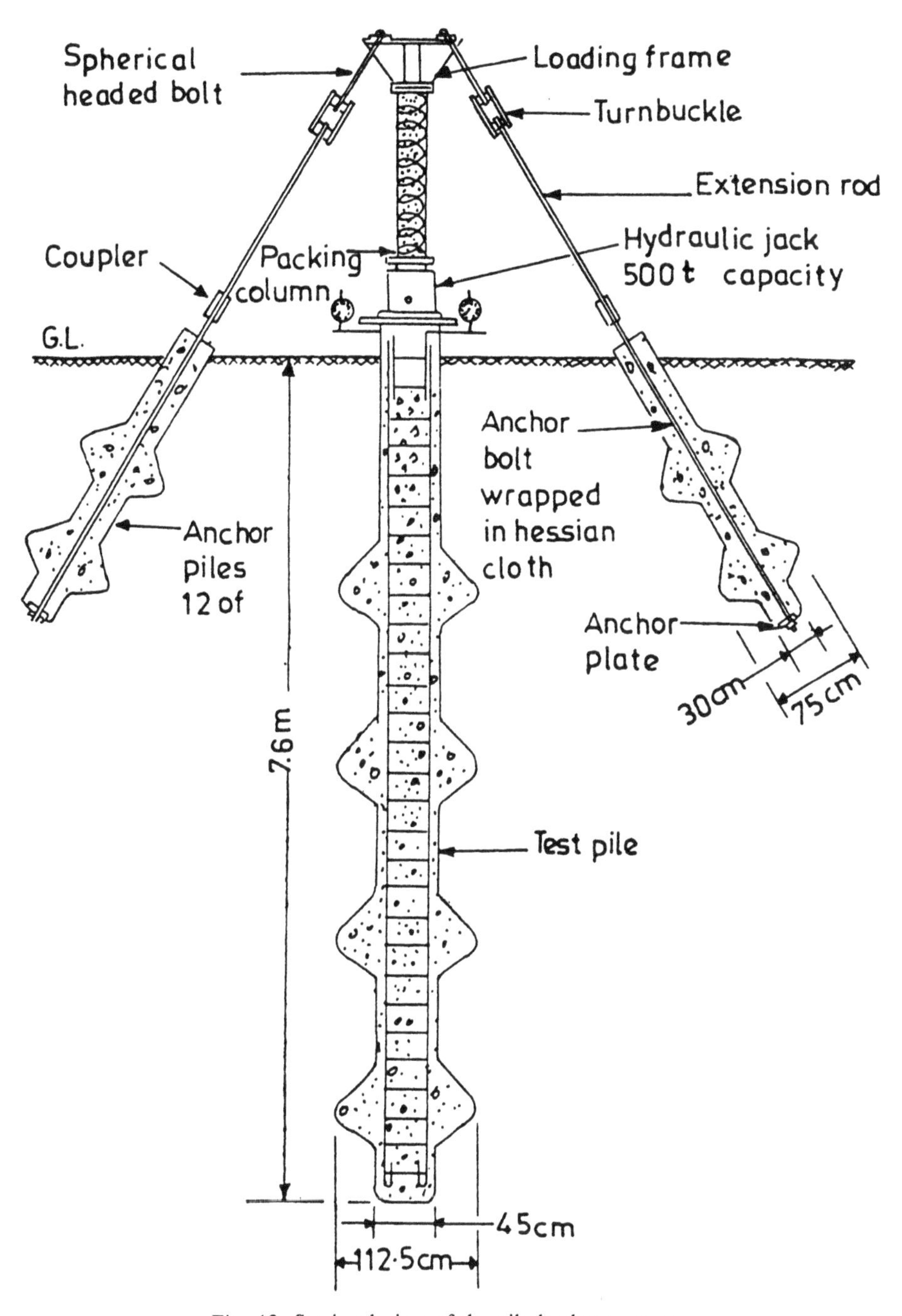

Fig. 12 Sectional view of the pile load test set-up.

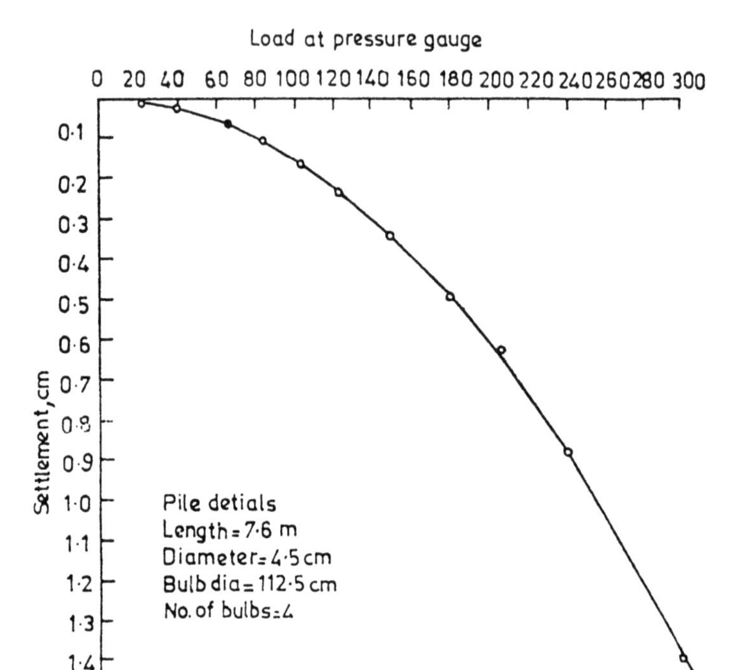

Fig. 13 Load Settlement curve.

capacity formula (1), gave a value of 328 tonnes. The load settlement curve of the test pile is shown in Fig. 13. The pile could only be tested upto 300 tonnes at which the settlement was only 14 mm. Further testing was not possible due to yielding of the anchor piles. It is clear from the load test that the ultimate capacity of the pile is more than 300 tonnes and could be close to the computed value of 328 tonnes.

CONCLUSION

The use of under-reamed piles in India is now an accepted practice in black cotton soil areas. It is no longer restricted to expansive soils or buildings alone but is used in other types of soil conditions such as filled up grounds and poor soils overlying firm strata, foundations subjected to uplifts and thrusts such as transmission line towers, underground tanks etc. A private developer has even adopted them for normal grounds in view of their speedy construction. An Indian Standard Code of Practice [IS: 2911 (Part III) 1980] on under-reamed piles has been brought out to help the designer and a safe load table provided in the code has been given in *Appendix*-A.

It can therefore be safely claimed that the under-reamed pile foundations have provided a foolproof, economical and quick to construct solution for foundation in expansive clays in India.

REFERENCES

Anon (1958). Some classification tests for shrinkage and swelling soils, *Indian Railway Tech Bull* 15 (128), pp. 31.

Chaudhri, R.M. Sulaiman and A B Bhuiyn (1943) Physico-chemical and minerological studies of black and red soil profile near Coimbatore, *Indian Journal of Agricultural Science,* pp. 13.

Gupta, S.P. and R K Bhandari (1983). *In-situ* engineering behaviour of Indian black cotton soil deposit, *Proc VII Regional Conference ISSMFE Haifa (Israel)*, pp. 19-23.

Holtz, W.G. and H J Gibbs (1956). Engineering properties of expansive clays, *Transactions of ASCE Paper* No 2814, pp. 641.

Wooltorton, F.L.D. (1954). *The Scientific Basis of Road Design*, Edward Arnold Ltd, London.

Vogel, K.J. (1954). Foundations in shrinkable soils, *Ministry of Railways Government of India Tech paper* No. 329.

Shanmugam, N. (1962). Building Foundations in black cotton soil, *Journal of Indian Natan Society SMFE* 1(3), pp. 60.

Mohan, D. (1957). Consolidation and characteristics of Indian black cotton soils, *Proceedings of the International Conference on Soil Mechanics and Foundation Engineering, London,* 74-76

Mohan, Dinesh and R.K. Goel, (1959). Swelling pressure and volume expansions of Indian black cotton soils, *Journal of the Institution of Engineers (India)*, XL(2), pp. 58-62.

Mohan, Dinesh and Giri Raj Singh Jain (1955). Under-reamed pile foundation on black cotton soils, *Indian Conf News*, Dec 25-28.

Mohan, Dinesh and G S Jain (1979). Pile testing using raked under-reamed pile, *Gr Engn*, April, pp. 47-52

Mohan, Dinesh and Subhas Chandra (1961). Frictional resistance of bored piles in expansive clays, *Geotechnique*, pp. 294-301.

Appendix A

Extract from Indian Standard Code of Practice for Under-reamed Piles–IS : 2911 (Part III)–1980

TABLE

Safe load for vertical bored cast in situ *under-reamed piles in sandy and clayey soils including black cotton soils*

Size		Length		Mild Steel Reinforcement			Compression				Safe Loads in Uplift Resistance					Lateral Thrust	
Diameter of pile	Under-reamed diameter	Single under-reamed	Double under-reamed	Longitudinal reinforcement		Rings spacing of 6mm dia rings	Sinlge under ream-ed	Double under-reamed	Increase per 30cm length	Decrease per 30cm length	Single under-ream-ed	Double under-reamed	Increase per 30cm length	Decrease per 30cm length	Single under-reamed	Double under-reamed
cm	cm	m	m	No.	Dia mm	cm	t	t	t	t	t				t	t
(1)	(2)	(3)	(4)	(5)	(6)	(7)	(8)	(9)	(10)	(11)	(12)	(13)	(14)	(15)	(16)	(17)
20	50	3.5	3.5	3	10	18	8	12	0.9	0.7	4	6	0.65	0.55	1.0	1.2
25	62.5	3.5	3.5	4	10	22	12	18	1.15	0.9	6	9	0.85	0.70	1.5	1.8
30	75	3.5	3.5	4	12	25	16	24	1.4	1.1	8	12	1.05	0.85	2.0	2.4
37.5	94	3.5	3.75	5	12	30	24	36	1.8	1.4	12	18	1.35	1.10	3.0	3.4
40	100	3.5	4.0	6	12	30	28	42	1.9	1.5	14	21	1.45	1.15	3.4	4.0
45	112.5	3.5	4.5	7	12	30	35	52.5	2.15	1.7	17.5	25.75	1.60	1.30	4.0	4.8
50	125	3.5	5.0	9	12	30	42	42	63	1.9	21	31.5	1.80	1.45	4.5	5.4

1. The safe bearing, uplift and lateral loads for under-reamed piles given in the Table on p. 2-65 apply to both medium compact ($10 < N < 30$) sandy soils and clayey soils of medium ($4 < N < 8$) consistency including expansive soils. The values are for piles with bulb diameter equal to two-and-a-half times the shaft diameter.
   The columns (3) and (4) of Table provide the minimum pile lengths for single and double under-reamed piles, respectively, in deep deposit of expansive soils. Also the length given for 375 mm diameter double under-reamed piles and more in other soils are minimum. The values given for double under-reamed piles in columns (9) and (13) are only applicable in expansive soils. The reinforcement shown is mild steel and it is adequate for loads in compression and lateral thrusts (Columns (8), (9), (16) and (17)). For up lift (Columns (12) and (13)), requisite amount of steel should be provided. In expansive soils, the reinforcement shown in Table is adequate to take upward drag due to heaving up of the soil. The concrete considered is M 15.
2. Safe loads of piles of lengths different from those shown in Table can be obtained considering the decrease or increase as from Columns 10, 11, 14 and 15 for the specific case.
3. Safe loads for piles with more than two bulbs in expansive soils and more than one bulb in all other soils (including non-expansive clayey soils) can be worked out from Table by adding 50 per cent of the loads shown in columns (8) or (12) for each additional bulb to the values given in these columns. The additional capacity for increased length required to accommodate bulbs should be obtained from columns (10) and (14).
4. Values given in columns (16) and (17) for lateral thrusts shall not be increased or decreased for change in pile lengths. Also for multi-under-reamed piles the values shall not increase than those given in column (17). For longer and/or mult-under reamed piles higher lateral thrusts may be adopted after establishing from field load tests.
5. For dense sandy ($N > 30$) and stiff clayey ($N > 8$) soils, the safe loads in compression and uplift obtained from Table may be increased by 25 per cent. The lateral thrust values should not be increased unless the stability and strength of top soil (strata upto a depth of about three times the pile shaft diameter) is ascertained and found adequate. For piles in loose ($4 < N < 10$) sandy and soft ($2 < N \leq 4$) clayey soils, the safe loads should be taken as 0.75 times the values shown in the Table. For every loose ($N \leq 4$) sandy and very soft ($N < 2$) clayey soils the values given in the Table should be reduced by 50 per cent.
6. The safe loads obtained from Table, should be reduced by 25 per cent if the pile bore holes are full of subsoil water or drilling mud during concreting.
7. The safe loads in uplift and compression given in Table or obtained in accordance with 2 to 6 should be reduced by 15 per cent for piles with bulb of twice the stem diameter. But no such reduction is required for lateral loads shown in Table.
8. The safe loads in Table and the recommendations made to obtain safe load in different cases (2 to 8) are based on extensive pile load tests obtained may be taken equal to two-thirds the loads corresponding to deflection of 12mm for loads in compression and uplift. The deflections corresponding to respective safe loads will be about 6mm and 4mm. The deflection at safe lateral load will be about 4mm. The values given in Table will be normally on conservative side. For working out ultimate compressive and uplift loads, if defined as loads corresponding to 25mm deflection on load-deflection curve, the value obtained from Table can be doubled. In case of lateral thrust twice the values in Table should be considered corresponding to deflection of 12mm only.
9. For piles/subjected to external moments and/or larger lateral loads than those given in Table the pile should be designed properly and requisite amount of steel should be provided.

3 Natural hazards and environmental geotechniques

Developments in Geotechnical Engineering, Balasubramaniam et al. (eds) © 1994 Balkema, Rotterdam, ISBN 90 5410 522 4

Seismic deformations in embankments and slopes

W.D. Liam Finn
Department of Civil Engineering, University of British Columbia, Vancouver, B.C., Canada

R.H. Ledbetter & W.F. Marcuson III
Geotechnical Laboratory, US Army Engineer Waterways Experiment Station, Vicksburg, Miss., USA

ABSTRACT: This paper traces the history of seismic response analysis of level and sloping ground and embankment dams from 1960 to 1993 and evaluates the current state of the art. The main emphasis is on estimating permanent seismic deformations and designing to satisfy deformation criteria. The necessary technology is now available. The U.S. Army Engineer Waterways Experiment Station (WES) has experience using the dynamic effective stress analysis methods of Finn and they are used extensively in this paper to illustrate the major role that seismic deformation analyses can contribute to safety evaluations and design of remediations.

The most difficult problems are posed by structures which contain potentially liquefiable soils in the embankments and/or the foundation. Problem areas are: the reliable determination of the residual strength of liquefied soils for post-liquefaction stability analyses, the assessment of post-liquefaction deformations, the criteria used for planning cost effective remediation and the evaluation of remedial measures given seismically induced liquefaction occurs. Suggestions for dealing with these problems are proposed based on critical assessment of practice and research findings.

1 INTRODUCTION

Prior to 1960, practice was concentrated solely on stability which was evaluated using limit equilibrium methods. The stability of embankment dams during earthquake loading was assessed in practice pseudostatically using the seismic coefficient method. This involved conducting conventional slope stability analyses while incorporating a force on the sliding mass to represent the effects of earthquake shaking. This force, F, was usually applied horizontally and expressed as a fraction of the weight of the sliding mass, W, by the equation $F = KW$. This implies that K is an equivalent horizontal acceleration expressed in units of gravity. Values of K ranged from 0.05 to 0.15 but no scientific basis was provided for the selection of an appropriate K value.

The rudimentary state of geotechnical earthquake engineering in the 1960's is not surprising given the state of development of geotechnical engineering itself at that time. Soil mechanics as practiced today had its formal beginnings in 1925 (Terzaghi, 1925) and by the early 1960's had developed good static methods for investigating slope stability (Bishop, 1960; Morgenstern and Price, 1965) and was examining critically for the first time proper constitutive modelling of soils. The state of the art in 1960 is well exemplified by the proceedings of the important conference on shear strength held at Boulder, Colorado (ASCE, 1960). This conference showed the beginnings of collaboration between proponents of applied mechanics and geotechnical engineers as illustrated by the keynote lecture by N.M. Newmark (1960) who became one of the leading figures in both structural and geotechnical earthquake engineering.

A gradual shift to more innovative approaches to seismic design is illustrated by experimental studies of the 90-m-high Kenny Dam in British Columbia by Clough and Pirtz (1958). They conducted shake table tests on models of the dam to develop an appreciation of modes of failure and the pattern of deformations. This study also drew attention to the problems associated with retaining similitude in the testing of earth structures on shake tables.

Two major developments provided a focus for research in geotechnical earthquake engineering: the development of civilian nuclear power and the technical challenges raised by the widespread damage caused by liquefaction during the major earthquakes in Alaska and Japan in 1964. These two earthquakes focused the attention of the geotechnical engineering profession on the problem of seismically induced liquefaction for and the need of fundamental study of how earth structures and foundations responded to earthquake loading.

The historical development of geotechnical earthquake engineering, as it relates to the estimation of seismic deformations, will be traced in broad outline from 1960 to 1993. This review will provide a framework for understanding the evolution of practice to its current state.

2 HISTORICAL DEVELOPMENT

Major analytical contributions leading to a more fundamental understanding of the seismic response of embankment dams were made by Ambraseys (1960a, b, c). He assumed that soil was a visco-elastic material and treated the dams as one dimensional (1-D) and two dimensional (2-D) shear beams in his analyses. He demonstrated how the incoming motions were amplified throughout the dam, the contribution of the different modes of vibration of the dam to the global response, and how the seismic coefficient varied along the height of the dam. Ambraseys (1960c) studied the elastic response of dams in both wide and narrow rectangular valleys and showed that if the ratio of the width of the valley to the height of the dam was less than 3, the seismic response changed significantly. This confirmed earlier results by Hatanaka (1952, 1955). Despite the limitations of the visco-elastic model of soil behavior, this analysis captured many of the important characteristics of seismic response and provided the starting point for subsequent developments. Seed and Martin (1966) carried out similar analyses for a variety of dam sizes and material properties and provided a comprehensive database for selecting appropriate values of seismic coefficients. They also drew attention to the deficiencies in the seismic coefficient method should the materials in the dam lose strength during an earthquake.

Newmark (1965) clarified many aspects of the problem of seismic stability. He pointed out that, although the factor of safety in an equilibrium analysis incorporating the seismic coefficient might show a factor of safety less than 1, this need not imply that the performance of an embankment dam would be unsatisfactory or its stability compromised. The factor of safety was less than 1 only for short intervals during which the dam underwent some deformation. Newmark stressed that what counted was whether the deformations that the dam suffered during the earthquake were tolerable or not. Obviously, large deformations that resulted in loss of freeboard and extensive cracking of the dam were not acceptable. The level of tolerable deformation should be based on the particular characteristics of the dam under study, judgment of experienced dam designers, and an appreciation of the reliability with which the deformations can be estimated.

A fundamental requirement for implementing the criteria based on limited deformation is to have a method for reliably estimating the deformations. Newmark (1965) presented a simplified method for estimating permanent deformations based on a sliding block model.

One of the most significant events which contributed to the rapid development of geotechnical earthquake engineering and the estimation of seismic displacement was the application of finite element methods to the analysis of embankment dams for the first time by Clough and Chopra (1966). This was followed by the seismic response analysis of slopes by Finn (1966a, b) and the analysis of central and sloping core dams by Finn and Khanna (1966). The latter study demonstrated the effects of the stress transfer between core and shell. All these analyses were conducted using a visco-elastic constitutive model of the soil and therefore were not capable of modelling the porewater pressure development or permanent deformations. To overcome this problem Finn (1967) outlined a procedure for interpreting the effects of the dynamic stresses computed by the visco-elastic analysis with the help of data on porewater pressures and strains from laboratory cyclic loading tests.

A major improvement in analysis occurred in 1972 when Seed and his colleagues at the University of California at Berkeley developed the equivalent linear method of analysis for approximating nonlinear behavior. This method was incorporated in the 1-D shear wave propagation program SHAKE (Schnabel et al., 1972). The technique was extended to 2-D finite element analysis by Idriss et al. (1973) and Lysmer et al. (1975) in the programs QUAD-4 and FLUSH, respectively. These programs took into account the strain-dependence of damping and shear modulus. However, the analysis was still elastic and, as a result, permanent deformations could not be estimated directly. Despite the limitation of elastic behavior, these programs led to more realistic analyses of embankment dams under earthquake loading, especially under strong shaking, and have remained the backbone of engineering practice to the present day.

While this program development was going on, the capability of testing soils under cyclic loading was also being developed. The cyclic triaxial test was developed by Seed and Lee (1966) and made possible the study of liquefaction potential. The test also made possible estimations of seismically induced deformations from the strains developed in the samples. At around the same time, the resonant column test was developed for measuring dynamic shear modulus and damping at low strains (Hall, 1962; Hardin and Music, 1965; Hardin and Black, 1966,1968; Drnevich, 1967; Hardin, 1970; Hardin and Drnevich, 1970). In the early 1970's, the use of the cyclic simple shear test was pioneered by Seed and Peacock (1971) and Finn et al. (1971).

By 1975, geotechnical engineers had many of the analytical and laboratory capabilities necessary for realistic assessments of the seismic safety and deformation behavior of embankment dams. These methods were put to the test when Seed et al. (1973, 1975a, 1975b), undertook a comprehensive study of the liquefaction-induced slide in the Lower San Fernando Dam which occurred as a result of the San Fernando earthquake of 1971. The analyses predicted that the dam would undergo large deformations upstream during the earthquake. In fact, the dam did not deform significantly until some 20 to 30 seconds after the earthquake (Seed, 1979). This post-earthquake slide was attributed by Seed (1979) to porewater pressure redistribution.

The equivalent linear method of analysis used in the study of the San Fernando Dam is a total stress analysis and therefore does not take into account the effect of porewater pressures on soil properties and dynamic response during the earthquake. Therefore, the analyses tend to predict a stronger response than actually occurs. As a result, the San Fernando case history provided the stimulus for the development of effective stress methods of dynamic analysis which could take the effects of porewater pressures into account directly. The Martin-Finn-Seed (MFS) model for generating porewater pressures during earthquake loading based on the strain response of the soil was developed by Martin et al. (1975) and paved the way for dynamic effective stress analysis.

The first non-linear dynamic effective stress analysis based on the MFS porewater pressure model was developed by Finn et al. (1975, 1976) and was incorporated in the 1-D program DESRA-2 by Lee and Finn (1978). A rudimentary 2-D version of this program was developed by Siddharthan and Finn (1982). An updated comprehensive program TARA-3 was developed by Finn et al. (1986). TARA-3 has the capability to conduct both static and dynamic analysis under total stress or effective stress conditions and can compute permanent deformations directly. The program uses properties that are normally measured in connection with important engineering projects.

Since the mid 1980's, other non-linear effective stress programs have been developed, for the most part based on some version of plasticity theory. Detailed presentations of some of these programs may be found in Pande and Zienkiewicz (1982) and comprehensive critical reviews in Finn (1988a,b) and Marcuson et al. (1992). A number of models were recently used in numerical predictions of centrifuge earthquake-induced experiments (Arulanandan and Scott, 1993). These programs are mathematically and analytically quite powerful but use some properties which are not routinely measured in the laboratory or the field.

The estimation of post-liquefaction deformations is an important part of assessing the consequences of liquefaction in embankment dams. Finn and Yogendrakumar (1989) developed the program TARA-3FL to track large post-liquefaction deformations using an updated Lagrangian technique for coping with the large strains and deformations.

3 EMPIRICAL AND SEMI-EMPIRICAL METHODS FOR ESTIMATION OF DEFORMATIONS

3.1 Ground Displacements

From the late 1970's there has been considerable interest in the global distribution of displacements in level ground and relatively gentle slopes as a result of liquefaction in underlying layers. When foundation soils liquefy, considerable displacements can occur even on slopes of a few percent. Such movements are especially large near an open face such as the banks of a river or a canal. In Niigata, Japan, during the 1964 earthquake, large displacements occurred toward the Shinano River (up to 10 m) and toward the Tsusen River (displacements up to 2 m).

Hamada et al. (1986), Youd (1980), Youd and Perkins (1978), and Bartlett and Youd (1992) have carried out extensive investigations of ground deformations associated with liquefaction and they have developed equations for estimating these movements. These types of deformations have a major impact on lifelines and on building foundations. On the basis of several post-earthquake studies of damaged and undamaged buildings, Youd (1989) concluded that most buildings can withstand 50-100 mm of differential ground displacement with little damage. On the other hand, few buildings can survive more than 1 m of displacement without major or catastrophic

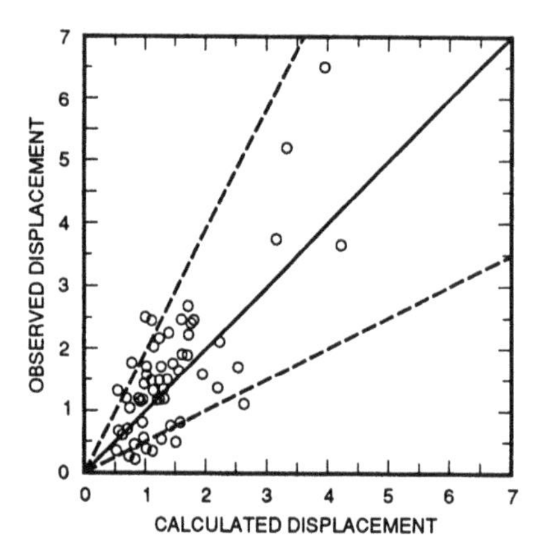

Fig. 1 Comparison between estimated and measured permanent horizontal ground displacements (after Hamada et al., 1986)

damage. Buildings on pile foundations, unless the piles are designed to resist the forces of the moving soil, can be severely damaged also.

3.2 Method of Hamada

Hamada et al. (1986) related permanent ground displacement, D (m), to the thickness of the liquefied layer, H (m), and the steepness of the slopes, Sl, of the ground surface or the bottom of the liquefied layer expressed in percent (%). The relation was developed by a regression analysis using 60 data sets from the 1983 Nihon-kai-Chubu, the 1964 Niigata, and the 1971 San Fernando earthquakes. The resulting equation is

$$D = 0.75\ H^{0.48}\ Sl^{0.33} \qquad (1)$$

Predictions based on this equation are compared with field data in Fig. 1. This equation is heavily weighted by the data from the Japanese earthquakes. It probably performs best when estimating deformations on slopes less than 5 percent and for earthquake magnitudes $M = 7.2$ to 7.5 causing ground accelerations in the range of 0.2 g to 0.3 g. Probably somewhat unreliable outside this range, it nevertheless offers some guidance as to relative deformation potentials that might be expected in different zones of a metropolitan area. Finn (1988a) reviewed these developments in some detail.

3.3 Method of Bartlett and Youd

Bartlett and Youd (1992) have made detailed studies of all readily available data in Japan and the United States and developed two models for estimating displacements during liquefaction. These equations are for displacements near a free face such as the bank of a river and for sloping ground conditions far from a free face. The equations are

Free face:

$$\mathrm{Log}(D_H + 0.01) = -16.366 + 1.178\ M - 0.927\ \mathrm{Log}R - 0.013\ R + 0.657\ \mathrm{Log}W + 0.348\ \mathrm{Log}T_{15} + 4.527(100 - F_{15}) - 0.922\ D50_{15} \qquad (2)$$

and

Sloping ground conditions:

$$\mathrm{Log}(D_H + 0.01) = -15.787 + 1.178\ M - 0.927\ \mathrm{Log}R - 0.013\ R + 0.429\ \mathrm{Log}S + 0.348\ \mathrm{Log}T_{15} + 4.527(100 - F_{15}) - 0.922\ D50_{15} \qquad (3)$$

where D_H (m) is the horizontal ground displacement, M (M_w) is earthquake magnitude, R (km) is the distance from the seismic source, W (%) is the ratio of the height of the free face divided by the distance from the free face, S (%) is the gradient of the ground slope, T_{15} (m) is the cumulative thickness of saturated, granular soils in the profile having a modified standard penetration value (i.e., $(N_1)_{60}$) less than or equal to 15, F_{15} (%) is the average fines content in the T_{15} layer(s), and $D50_{15}$ (mm) is the average mean grain size of the T_{15} layer(s). Because log(0) is undefined, 0.01 m was expediently added to all values of D_H prior to fitting these equations. This expediency allowed log (D_H) to be calculated for all zero displacement values that are included in the database.

Predictions by the Bartlett and Youd (1992) model are compared with Japanese and U.S. field data in Fig. 2.

3.4 Mapping Damage Potential from Ground Displacements

Structural damage resulting from seismically induced permanent ground displacements during earthquakes is a function of the type and amount of differential ground movement. With the exceptions of tectonic fault movements and failures of relatively steep or initially weak slopes, significant permanent ground deformations are usually associated with liquefaction. Therefore, in assessing the severity of expected displacements, Youd and Perkins (1978) developed a parameter S which they called Liquefaction Severity. S is defined as the measured amount of differential ground displacement in millimeters divided by 25.

Based on their study of the damage resulting from ground displacements during the Alaska and Niigata earthquakes in 1964 and the San Fernando earthquake in 1971, Youd and Perkins (1978) suggested that little damage would be expected in ordinary buildings for $S < 5$, but that moderate to severe damage would be likely for $5 \leq S < 20$ and major damage for $S > 30$.

The S parameter can vary widely even over limited regions. Therefore, while estimates of S are useful for site-specific studies, it is not a useful parameter for mapping expected levels of deformations in a given area with a given probability of occurrence.

To simplify this problem, Youd and Perkins (1978) introduced the liquefaction severity index (LSI) which is the general maximum S value in a specific topographic, hydro-logic, and sedimentologic environment. Typically, LSI is considered to be the maximum S for lateral spreads in geological units such as wide flood plains, deltas and other areas of gently sloping late Holocene river deposits. Large S values due to localized untypical soil or topographic conditions are excluded. Since LSI is the maximum recorded S, it gives a conservative estimate of displacements and associated damage potential.

Youd and Perkins (1978) have shown that LSI for a given environment which undergoes liquefaction may be related to the moment magnitude of the earthquake, M_w, and the distance, R (km), from the energy source by

$$\mathrm{Log(LSI)} = -3.49 - 1.86 \log R + 0.98 M_w \tag{4}$$

This equation is based on data obtained in the Western United States. Variations in the attenuation of acceleration from the source varies from seismic region to seismic region and the attenuation of LSI may be expected to do likewise. If equations similar to Eq. (4) are not derived specifically for a seismic region under study, the results of applying Eq. (4) should be modified empirically for any difference in

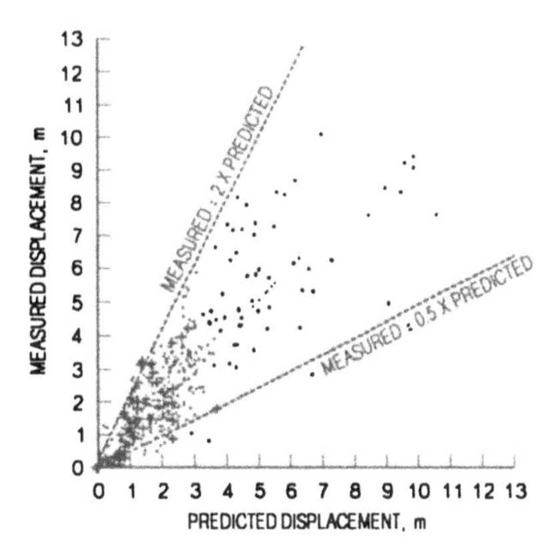

Fig. 2. Comparison between Bartlett and Youd (1992) model and field data from Japan and United States.

the attenuation characteristics of the study area compared with the Western United States.

3.5 The Newmark Method

Newmark (1965) developed and published a method based on a sliding block analogy for estimating earthquake induced relative displacements. It is interesting to note that on a project for the U.S. Army Corps of Engineers, the late D.W. Taylor (letters from D.W. Taylor, 1953) appears to have independently developed a similar model for a similar purpose. The Taylor model was first implemented by R.V. Whitman at M.I.T. in 1953. A remarkable sentence is taken from the 20 May letter; "The procedure therefore can not be expected to have much validity if, as in the writer's opinion, the threat of damage from earthquake action lies not in an increase of activative force but in a progressive decrease in shearing resistance as a result of many cycles of application of the activating force."

Deformation of a dam is modelled as the deformations of a rigid block sliding on an assumed failure surface under the action of the ground motions at the site. Various potential sliding surfaces in the embankment are analyzed statically to find the inertia force $F_I = (W/g)a_y$ required to cause failure (Fig. 3). The average yield acceleration a_y is then deduced from this force. The sliding block is assumed to have the same acceleration time history as the ground. The yield acceleration is deducted from the acceleration time history, and the net acceleration (the shaded area in Fig. 3) is available to generate permanent displacements. The analysis

Table 1: Desirable Downstream Slopes for Concrete Faced Rockfill Dams.

Earthquake Magnitude	Peak Base Acceleration	Crest Peak Acceleration	Average D.S. Slope* for Displacements of 0.6 m or less	Average D.S. Slope* for Displacements of 0.3 m or less
6.5	0.5g	1.0g	1.5	1.55
7.5	0.5g	1.0g	1.6	1.65
8.5	0.5g	1.0g	1.8	1.80

* Change in the horizontal with respect to change in the vertical.

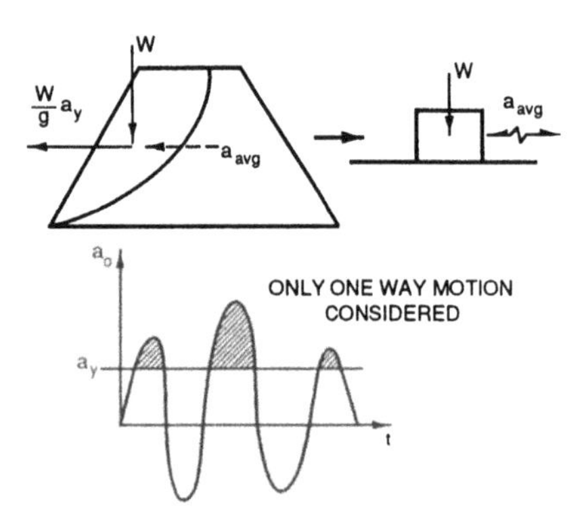

Fig. 3. Elements of Newmark's deformation analysis.

is conducted on the equivalent model of a horizontal sliding block on a plane with only one-way motions allowed (Fig. 3).

Makdisi and Seed (1978) modified the Newmark method by taking the flexibility of the dam into account. The average acceleration time record of the sliding block is obtained usually from a QUAD-4 analysis. The method differs from the Newmark approach in generating relative displacements by the net accelerations above the sliding surface, whereas Newmark used the net accelerations below the sliding surface. The QUAD-4 accelerations in the sliding block are determined without taking the yield accelerations into account. Therefore, in many cases of strong motions, estimates of displacements would probably be conservative.

The Newmark method was introduced at a time when there were no direct methods of computing permanent deformations. It is still widely used despite all the evidence that the sliding block model is not a very good representation of how embankment dams deform, especially embankment dams with low factors of safety. This was shown as early as 1958 by Clough and Pirtz (1958) in their shake table tests on models of Kenny Dam. The method is useful in comparing the deformation potential of alternative design proposals, or in comparing a design with that of an existing dam. However, in many cases, similar results to those of the Newmark method can be achieved by using Makdisi and Seed's (1978) simplified approach. Using current technology and finite element analysis, permanent deformations can be calculated directly without restrictive assumptions about the mode of deformation.

Seed (1983) made a study of concrete-faced rockfill dams in highly seismic environments using the Makdisi and Seed (1978) method. He investigated the stability requirements for rockfill dams for different earthquake magnitudes with base accelerations of the order of 0.5 g and crest accelerations of the order of 1 g to determine desirable slopes. Results of his study are shown in Table 1 with performance categories specified in terms of displacements in the downstream slope. Requirements for downstream slopes (D.S.) would be applicable generally to all rockfill dams.

The Jordanelle Dam is a modern example of a compacted zoned earthfill dam which exemplifies the use of defensive measures (Wilson et al., 1992). A cross-section of the embankment is shown in Fig. 4. It consists primarily of compacted material which consists of well graded sand, gravel, cobbles and boulders. The upstream slope is 2 horizontal to 1 vertical and the downstream slope is 1.5 horizontal to 1 vertical. These slopes are unusually steep for earth dams in seismically active areas. The upstream sloping core is protected by a two stage filter drain to prevent piping and to control seepage. The inclusion of the core and filter drain system in the upstream portion of the dam leaves most of the

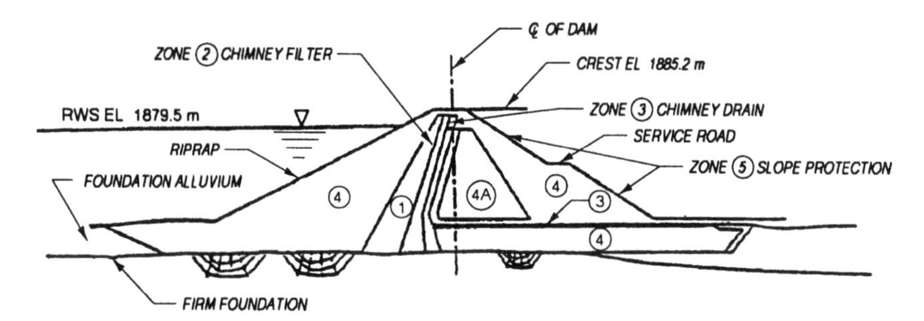

Fig. 4. Section of Jordanelle embankment dam (Wilson et al., 1992).

downstream embankment in an unsaturated condition promoting stability.

The deformations of the Jordanelle Dam were analyzed for three different earthquake scenarios using the U.S. Bureau of Reclamation's version of Newmark's (1965) method of analysis. The largest displacements were generated by a local earthquake of magnitude M=6.5 with a peak ground acceleration of 0.65. This acceleration was amplified to 1.2g at the crest. Analysis indicates that up to 1.4 m of vertical displacement is possible for a potential slip circle passing through the center of the crest and exiting one-fourth of the distance down the slope. This displacement is about twice that predicted by Makdisi and Seed's method and indicates sharp sensitivity to peak acceleration increases once the yield acceleration is exceeded.

Some dams have major deficiencies in either their embankment structure or the foundation when judged by today's understanding of the factors controlling the seismic response and safety of embankment dams. Generally speaking, comprehensive analyses have to be conducted on these dams in conjunction with detailed in situ and laboratory testing to determine their seismic safety and the extent and nature of the remediation that may be necessary. Some examples of this type of study will be discussed later.

4 DYNAMIC ANALYSIS OF SOIL STRUCTURES: OVERVIEW

The state of the art of earthquake analysis procedures for concrete and embankment dams was summarized in Bulletin 52 of the International Commission on Large Dams (Zienkiewicz et al., 1986). The bulletin outlined a general framework for analysis in both total and effective stress modes applicable to embankment dams using equations which coupled the response of soil and water. It recommended three levels of analyses:

1) Simple total stress methods including pseudostatic analysis using seismic coefficients when porewater pressures are negligible and no significant degradation in soil properties occurs.

2) The equivalent linear method of analysis coupled with the use of laboratory data (Seed et al., 1973; Seed, 1979) when substantial porewater pressures are generated.

3) Effective stress analysis conducted in "a direct and fundamental manner."

Pseudostatic analysis with seismic coefficients might be used safely in areas where a long history of use has calibrated the seismic coefficients to reflect experience with dam behavior during earthquakes, such as Japan. It is not recommended where such direct experience is not available. The equivalent linear method is still the most widely used in practice, but "direct and fundamental" methods are finding increasing application. This is especially true in dealing with the complex problems that must be faced when evaluating the safety of existing dams which contain potentially liquefiable soils. However, the optimistic assertion regarding reasonable computational cost has not been borne out in the case of programs based on the general framework outlined in Bulletin 52.

4.1 Equivalent Linear Analysis

The dynamic response of an earth dam is usually computed in engineering practice using an equivalent linear (EQL) method of 2-D analysis such as that incorporated in the computer programs QUAD-4 (Idriss et al., 1973) or FLUSH (Lysmer et al., 1975). The results may be corrected approximately for three dimensional (3-D) effects (Mejia and Seed, 1983). These corrections were used in the back analyses of Oroville Dam for the 1975 earthquake (Vrymoed, 1975). The correction is based on altering the shear modulus in the 2-D analysis so that the fundamental 2-D period matches the equivalent 3-D period.

Dakoulas and Gazetas (1986a,b) studied the problem again and Gazetas (1985) points out that, despite matching the fundamental period, the contributions of higher harmonics may be substantially underestimated. Therefore assessing the seismic response of embankment dams in narrow valleys requires the exercise of engineering judgment, since the higher harmonics are likely to have their greatest effect at the crest of the dam.

The EQL analyses are conducted in terms of total stresses and the effects of seismically induced porewater pressures on elemental shear stiffness are not reflected in the computed strains, stresses, and accelerations. Since the analyses are elastic, they cannot predict the permanent deformations. Therefore, equivalent linear methods are used only to get the distribution of maximum accelerations and maximum shear stresses in the dam. Semi-empirical methods are often used to estimate the permanent deformations using either the acceleration or stress data from the equivalent linear analyses.

4.1.1 Deformations from Acceleration Data

Deformations are often estimated from the acceleration data using the Newmark method as modified by Makdisi and Seed (1978). The resulting deformations do not represent the deformation patterns of embankment dams under strong shaking, but they may provide a useful index of potential deformation. If a sliding wedge can be found which undergoes large deformations, one would expect to estimate large deformations by an appropriate nonlinear finite element analysis. However, the deformations computed by the Newmark approach should not be used for estimating whether the seismic deformations will satisfy displacement criteria.

4.1.2 Deformations from Stress Data

A more detailed picture of potential strains and deformations is obtained using Seed's semi-empirical method (Seed, 1979). The computed dynamic stresses in soil elements in the dam are converted to equivalent uniform stress cycles and are applied to laboratory specimens in consolidated states similar to corresponding elements in the dam. The resulting strains in the laboratory specimens are assigned to the corresponding elements in the dam. This procedure gives an incompatible set of strains which, are an indication of the potential for straining at selected locations within the dam.

These procedures were used to investigate the slide in the Lower San Fernando Dam during the 1971 earthquake (Seed, 1979). Large upstream displacements were predicted to occur during the earthquake. In fact, the failure occurred under static loading conditions shortly after the earthquake shaking had ceased. A major motivation for the development of more general constitutive relations has been the need to model non-linear behavior in terms of effective stresses and to provide reliable estimates of porewater pressures and permanent deformations under seismic loading.

4.2 Nonlinear Methods of Analysis

A hierarchy of constitutive models is available for the direct and fundamental analysis of the dynamic response of embankment dams to earthquake loading. The models range from the relatively simple equivalent linear model to complex elastic-kinematic hardening plasticity models. Detailed critical assessments of these models may be found in Finn (1988a,b) and Marcuson et al. (1992). This review presents the main procedures used in current practice and outlines their advantages and limitations.

4.2.1 Elastic-Plastic Methods

It is generally recognized that elastic-plastic models of soil behavior under cyclic loading should be based on a kinematic hardening theory of plasticity using either multi-yield surfaces or a boundary surface theory with a hardening law giving the evolution of the plastic modulus. These constitutive models are complex and incorporate some parameters not usually measured in field or laboratory testing. Soil is treated as a two-phase material using coupled equations for the soil and water phases. The coupled equations and the more complex constitutive models make heavy demands on computing time (Finn, 1988b).

Validation studies of the elastic-plastic models suggest that, despite their theoretical generality, the quality of response predictions is strongly stress path dependent (Saada and Bianchini, 1987; Finn, 1988b). When loading paths are similar to the stress paths used in calibrating the models, the predictions may be good. As the loading path deviates from the calibration path, the prediction becomes less reliable. In particular, the usual method of calibrating these models, using data from static triaxial compression and extension tests, does not seem adequate to ensure reliable estimates of response for the dynamic cyclic shear loading paths that are important in many kinds of seismic response studies. It is recommended that calibration studies of elastic-plastic models for dynamic response analysis should include appropriate cyclic loading tests, such as triaxial, torsional shear, or simple shear tests. The accuracy of pore pressure prediction in the coupled models is highly dependent on the accurate characterization of the soil properties. It is difficult to characterize the volume change characteristics of loose sands and silts which control porewater pressure development because of the problems obtaining and testing undisturbed samples representative of the field conditions. As a check on the capability of these models to predict porewater pressure adequately, it is helpful to use them to predict the field liquefaction resistance curve as derived from normalized Standard Penetration test data (Seed et al., 1986).

Typical elastic-plastic methods used in current engineering practice to evaluate the seismic response of embankment dams are DYNAFLOW (Prevost, 1981), DIANA (Kawai, 1985), DSAGE (Roth, 1985), DYNARD (Moriwaki et al., 1988), FLAC (Cundall and Board, 1988), DYSAC2 (Muraleetharan et al., 1988,1991), and SWANDYNE 4 (Zienkiewicz et al., 1990a,1990b). There is no published information on the current version of DIANA which is an extensive modification of the earlier program. Programs DSAGE, DYNARD, and FLAC are proprietary to their developers.

The constitutive model of DYNAFLOW is based on the concept of multi-yield surface plasticity. The initial load and unload (skeleton) stress-strain curve obtained from laboratory test data is approximated by linear segments and the curves for loading, unloading and reloading follow the Masing criteria (Masing, 1926). The procedure can include anisotropy. The program allows dissipation and redistribution of porewater pressures during shaking. Validation of the program has been by data from centrifuge tests. The computational requirements of the code are quite intensive.

DSAGE is predecessor of the program FLAC. The latter is a microcomputer implemented code based on the explicit finite difference method for modelling nonlinear static and dynamic problems. The program uses an updated Lagrangian procedure for coping with large deformations.

DYNARD was developed by M. Beikae and Y. Moriwaki, Woodward-Clyde Consultants, in the 1980's. The program uses an explicit finite difference method for Lagrangian nonlinear analysis allowing large strains and displacements. It analyzes the deformation and response of earth structures to the simultaneous effects of gravity and seismic shaking using undrained strength and degradable undrained soil moduli. The cyclic and nonlinear behavior of soils is incorporated in the analysis by 2-D bounding surface model, similar to that of Cundall (1979) and Dafalias and Hermann (1982).

DYSAC2 is a fully coupled non-linear dynamic analysis procedure. The constitutive model is also based on bounding surface plasticity. The program has been validated in a preliminary way using the results of centrifuge model tests (Muraleetharan et al., 1993).

SWANDYNE 4 is a general purpose elastic-plastic computer code which permits a unified treatment of such problems as the static and dynamic nonlinear drained and undrained response analyses of saturated and partially-saturated soils to earthquake loading. The formulations and solution procedures, upon which the computer code is based, are presented in Zienkiewicz et al. (1990a,1990b).

4.2.2 Direct Nonlinear Analysis

The direct nonlinear approach is based on direct modelling of the soil nonlinear hysteretic stress-strain response. The WES has been working with the direct nonlinear dynamic effective stress analysis methods of Finn for more than ten years. This approach is represented here by the program TARA-3 (Finn et al., 1986), which is proprietary to Finn.

WES has extensive experience using this method in practice and a number of studies are available. The TARA-3 programs and studies are mainly used in the remainder of this paper to simply illustrate the use of dynamic effective stress and seismic deformation analyses in evaluating and/or remediating for seismic safety of embankment dams.

The objective during analysis is to follow the stress-strain curve of the soil in shear during both loading and unloading. Checks are built into the program to determine whether or not a calculated stress-strain point is on the stress-strain curve and corrective forces are applied to bring the point back on the curve if necessary. To simplify the computations, the stress-strain curve is assumed to be hyperbolic. This curve is defined by two parameters which are fundamental soil properties, the strength τ_{max} and the in situ small strain shear modulus, G_{max}. The response of the soil to uniform all round pres-

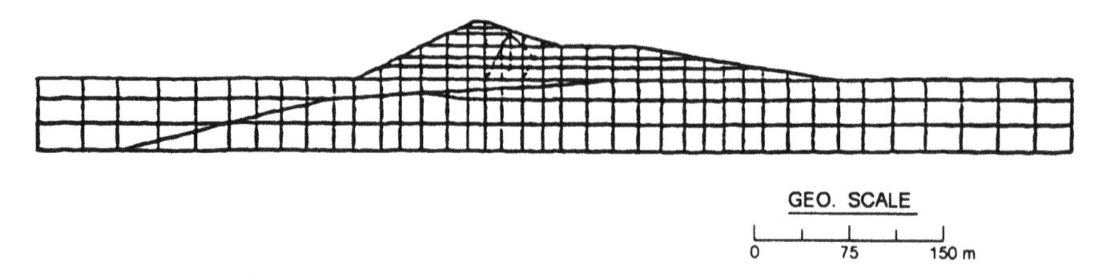

Fig. 5. Section of Lukwi Tailings Dam showing the finite element mesh.

sure is assumed to be nonlinearly elastic and dependent on the mean normal effective stress.

The response of the soil to an increment in load, either static or dynamic, is controlled by the tangent shear and tangent bulk moduli appropriate to the current stress-strain state of the soil. The moduli are functions of the level of effective stress, and therefore, excess porewater pressures must be continually updated during analysis and their effects on the moduli taken progressively into account.

During seismic shaking, two kinds of porewater pressures are generated in saturated soils -- transient and residual. The residual porewater pressures are due to plastic deformations in the sand skeleton. These persist until dissipated by drainage or diffusion and therefore they exert a major influence on the strength and stiffness of the soil skeleton. These pressures are modelled in TARA-3 using the MFS porewater pressure model (Martin et al., 1975).

4.2.3 Structure of the TARA-3 Analysis

TARA-3 can be used to simulate the gradual construction of a dam and conducts both static and dynamic analysis. A static analysis is first carried out to determine the stresses and strains in the dam at the end of construction.

Dynamic analysis of a dam starts from the static stress-strain condition in each element. Methods of dynamic analysis commonly used in practice ignore the static strains in the dam and start from the origin of the stress-strain curve in all elements, even in those which carry high shear stresses.

TARA-3 also allows the analysis to start from the zero stress-strain condition. The program takes into account the effects of the porewater pressures on moduli and strength during dynamic analysis and estimates continuously the additional deformations due to gravity acting on the softening soil and due to consolidation.

5 GLOBAL DEFORMATION PATTERN FROM NONLINEAR ANALYSIS

An example will now be given to illustrate the global picture of dam response that the general nonlinear methods of analysis can provide to an engineer. The example shows that these analyses can provide clear pictures of how the different zones in the dam behave individually and how they interact with each other. This kind of information is extremely useful in assessing the potential mechanisms for failure or large deformations and for the planning of remedial measures.

5.1 Lukwi Tailings Dam, New Guinea

The analysis of Lukwi Tailings Dam by TARA-3 is a good illustration of the kind of global information on dam behavior provided by a dynamic nonlinear deformation analysis. The finite element representation of the proposed dam is shown in Fig. 5. The sloping line in the foundation is a plane between two foundation materials. Upstream to the left is a limestone with shear modulus $G = 6.4 \times 10^6$ kPa and a shear strength defined by $c' = 700$ kPa and $\Phi' = 45°$. The material to the right is a siltstone with a low shearing resistance given by $c' = 0°$ and $\Phi' = 12°$. The shear modulus of the siltstone is approximately $G = 2.7 \times 10^6$ kPa.

The difference in strength between the foundation materials is reflected in the dam construction. The upstream slope on the limestone is steeper whereas the downstream slope on the weaker foundation is much flatter and has a large berm to ensure stability during earthquake shaking. The designers were interested in the foundation and berm interaction during shaking.

A finite element model of the dam was strongly shaken with a peak acceleration of 0.33g (Finn, 1988b, 1990). Shear stress-shear strain response of the limestone foundation is almost elastic (Fig. 6).

The response of the siltstone foundation is strongly nonlinear. The deformations increase progressively in the direction of the initial static shear stresses as shown in Fig. 7. Since the analysis starts from the post-construction stress-strain condition, subsequent large dynamic stress impulses move the response close to the highly nonlinear part of the stress-strain curve. Note that the hysteretic stress-strain loops all reach the flat part of the stress-strain curve. An element in the berm also shows strong nonlinear response with considerable hysteretic damping resulting from large strains in the siltstone (Fig. 8).

Despite the highly nonlinear response of the foundation, the estimated maximum displacements of the dam were limited to about 250 mm due to the effect of the berm. The deformed shape of the central part of the dam is shown in Fig. 9.

6 VALIDATION OF CONSTITUTIVE MODELS

Constitutive models are normally validated by using them to predict response in single element tests such as the static or cyclic triaxial test. However, single element tests may be a necessary but not a sufficient test because they do not provide an adequate validation of the predictive capability of a model. The stresses or the strains are known *a priori* and there is no need to solve the boundary value problem using the constitutive model to predict the response. All practical applications involve the solution of the equilibrium equations and the continuity equations under a prescribed set of boundary conditions and a prescribed input. In other words, adequate model validation requires an inhomogeneous stress field which is assumed not to be the case in the laboratory test.

The centrifuge test offers the best opportunity for validating models by the solution of boundary value problems. Centrifuge models can be extensively instrumented, prepared under controlled conditions and shaken by prescribed input. Constitutive models, numerical procedures, and finite element models can be clearly tested by seeing how well the performance of the centrifuge model can be predicted. Also, numerical models and procedures can be calibrated and improved or modified for phenomena that may not have been adequately accounted for in a model.

Until recently (Arulanandan and Scott, 1993), little centrifuge validation testing had been conducted on constitutive models used in geotechnical earthquake engineering practice. The TARA-3 model has beenn subjected to validation studies on the centrifuge over

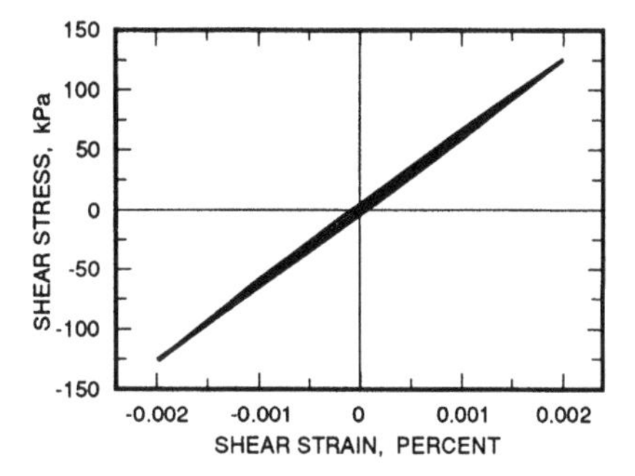

Fig. 6. Elastic response of the foundation limestone.

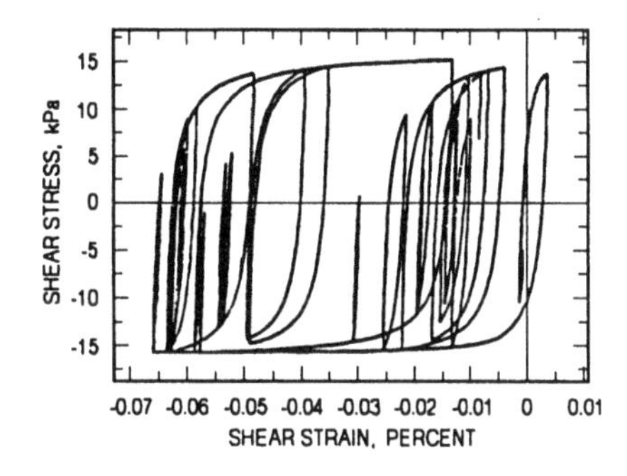

Fig. 7. Strongly nonlinear behavior of foundation siltstone.

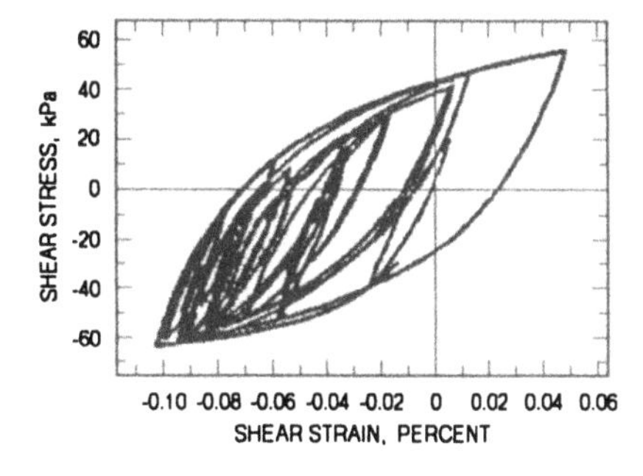

Fig. 8. Strongly hysteretic response of the downstream berm.

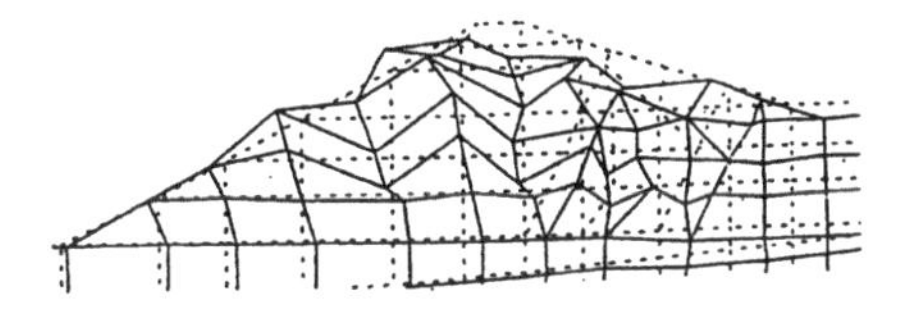

Fig. 9. Deformed post-liquefaction shape of Lukwi Tailings Dam.

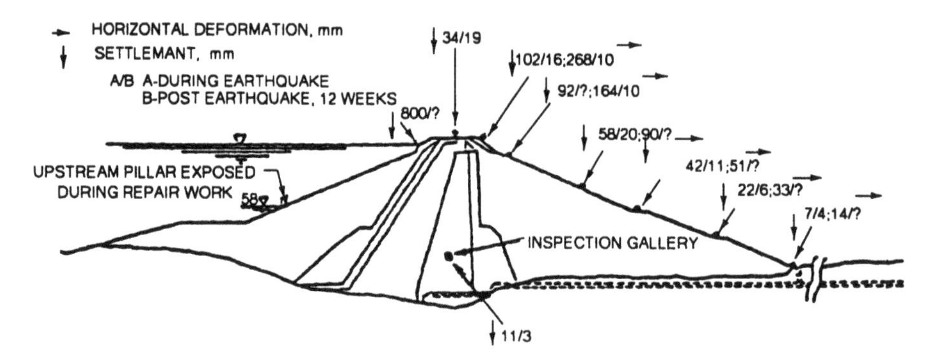

Fig. 10. Cross-section of the Matahina Dam showing the locations of displacement measurements both during the earthquake and twelve weeks after. The vertical and horizontal components are given for displacements during the earthquake.

a three-year period. The tests were conducted at Cambridge University on behalf of the Nuclear Regulatory Commission of the United States. Results from the validation studies have been reported by Finn (1988b, 1991b).

A validation program based on centrifuge tests has recently been conducted in the United States under the auspices of the National Science Foundation called the VELACS program. The acronym arose from the title Verification of Liquefaction Analysis by Centrifuge Studies. A conference on predictions made under this program was held at the University of California at Davis in October 1993 (Arulanandan and Scott, 1993).

6.1 Simulation of Response of Matahina Dam

Opportunities for quantitative validation by case histories in the field are quite limited, primarily because structures are not generally adequately instrumented and earthquakes are rare. The 1987 Edgecumbe Earthquake in New Zealand M=6.7 provided an opportunity to see whether the acceleration response and the permanent deformations could be adequately modeled in Matahina Dam.

The dam is located on the Rangatake River in the eastern Bay of Plenty Region of New Zealand about 23 km from the earthquake epicenter and about 11 km from the main surface rupture.

Founded on rock, the dam is 86 m high and has a crest length of 400 m (Fig. 10). The core is low plasticity weathered greywacke and slopes upstream. Dam shells are compacted rockfill of hard ignimbrite. The transition zones adjacent to the core are the fines and soft ignimbrite stripping from the rockfill quarry and left abutment excavation (Finn et al., 1992).

Matahina Dam was instrumented at three locations along the crest to measure accelerations at the top of the crest, at the mid point between the crest and the base and at the base. Lateral and vertical displacements of the downstream slope have been monitored consistently at many locations since the dam was constructed. Readings were taken shortly before the earthquake and immediately afterward. These data provided a base for checking the capability of predicting permanent deformations (Fig. 10).

A study was conducted to simulate the performance of the dam during the Edgecumbe earthquake using the TARA-3 program preparatory to calculating how the dam might behave under the design earthquake which was substantially larger than the Edgecumbe earthquake (Finn et al., 1992). Analyses assumed that no pore pressure developed in the rockfill and the core performed undrained. The properties of the clay core were obtained by laboratory testing and in situ measurements. Stiffness of the rockfill was estimated by measuring average shear wave velocities at various locations. Strength was conservatively taken from the literature, and as a first step, volume change properties of the rockfill were estimated by inverse analysis from the measured deformations.

The computed and recorded accelerations at the crest for the Edgecumbe earthquake are shown in Fig. 11 and appear satisfactory for engineering purposes. Recorded and computed deformations during the earthquake at points on the downstream slope are shown in Table 2.

There is good agreement except for the node at the crest. The discrepancy here may be due to the fact that appurtenant structures on the crest were not modelled. The model was considered to be satisfactory for estimating the response under the design earthquake.

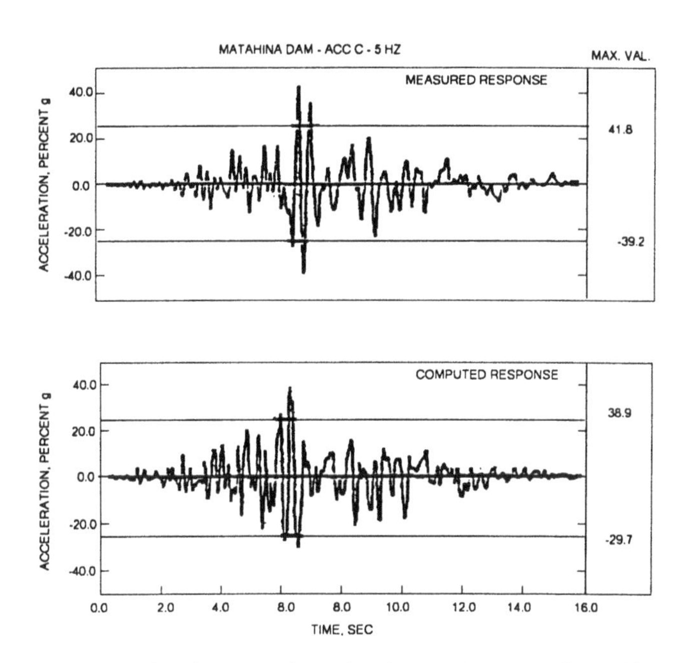

Fig. 11. Measured and computed accelerations at the crest of Matahina Dam.

7 EVALUATION OF POST-LIQUEFACTION BEHAVIOR

A challenging technical problem for geotechnical earthquake engineers involves the post-liquefaction behavior of existing dams with potentially liquefiable zones in the structure or foundation. Two major challenges are: (1) estimating the post-liquefaction behavior of the dam, and (2) planning cost-effective remedial measures.

In the context of this section, liquefaction is synonymous with strain softening of sand in undrained shear as illustrated by curve 1 in Fig.12. When the sand is strained beyond the point of peak strength, the undrained strength drops to a value that is maintained more-or-less constant over a large range in strain. This is called the undrained steady state or residual strength.

If the driving shear stresses due to gravity on a potential slip surface in an embankment are greater than the undrained steady state strength, deformations will occur until the driving stresses are reduced to values compatible with static equilibrium. The more the driving stresses exceed the steady state strength the greater the deformations needed to achieve equilibrium. Clearly, the residual strength is a key parameter controlling the post-liquefaction behavior.

If the strength increases after passing through a minimum value, the phenomenon is often called limited liquefaction and is illustrated by curve 2 in Fig. 12. Limited liquefaction may also result in significant deformations because of the strains necessary to develop the strength to restore stability.

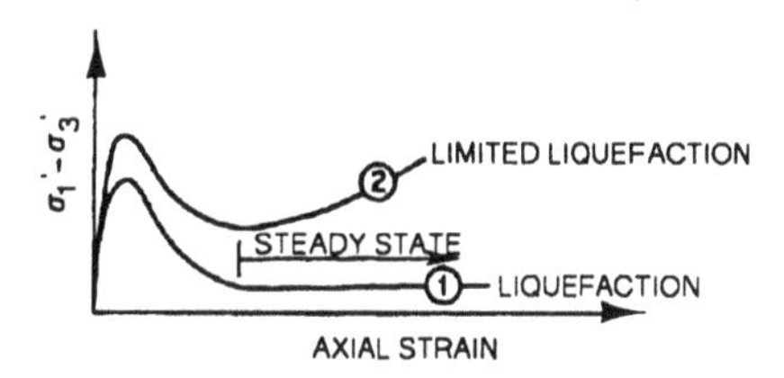

Fig. 12. Types of liquefaction behavior.

Table 2: Measured and Computed Seismic Displacements of Matahina Dam in mm.

Node	X meas	X comp	Y meas	Y comp
215	-	85	-34	-44
235	268	234	-102	-99
271	164	153	-92	-88
323	90	98	-58	-53
340	51	54	-42	-41
366	33	33	-22	-22
202	-	10	-11	-5

7.1 Residual Strength

The controversy surrounding residual strength makes it difficult to assess reliably the consequences of liquefaction. There are two methods generally recognized for determining the residual strength. The first, due to Poulos et al. (1985), consists of testing high quality undisturbed samples in static triaxial tests and correcting the results for the disturbance effects of sampling, transportation and testing. Residual strengths derived in this way appear to be higher than those derived from the back analysis of past flow slides. On the basis of such analyses Seed (1987) developed a correlation between $(N_1)_{60}$ and the undrained residual strength, S_{ur}. An updated version of this correlation was proposed by Seed and Harder (1990) which is shown in Fig. 13.

Lower bound strengths from this correlation are often used in practice although, occasionally, values approaching the 33rd percentile have been used. In either case these values are small at low $(N_1)_{60}$ values and frequently result in the prediction of instability and the need for substantial remediation.

An alternative approach which has recently come into practice consists of relating the residual strength to the effective overburden pressure.

$$S_{ur} = Cp' \tag{5}$$

Where p' is the effective vertical stress that exists before liquefaction and C is an empirical constant.

In the safety evaluation of Sardis Dam, Mississippi, Finn et al. (1991) used C = 0.075. This value gave residual strengths substantially higher than the lower bound values from the Seed correlation at low $(N_1)_{60}$ values. A re-examination of Seed's case histories by Lo et al. (1991) and McLeod et al. (1991) suggests that C may have a value of about 0.1. Values of C ranging from 0.06 to 0.08 have been used in other dams.

Vaid and Thomas (1994) have shown that the residual strength in extension is a function not only of void ratio but of the confining pressure (Fig. 14). In a test sequence on Fraser River sands they showed that irrespective of void ratio, the ratio of residual strength to initial effective overburden pressure is 0.15. Baziar and Dobry (1991) report a ratio of 0.12 for loose silty sand from undrained static compression tests. These test data support the alternative approach discussed above.

The wide divergence in estimates of the residual strength could be resolved by a focused research program. Part of the answer may lie in the effects of stress paths (Vaid et al., 1989; Finn, 1991a).

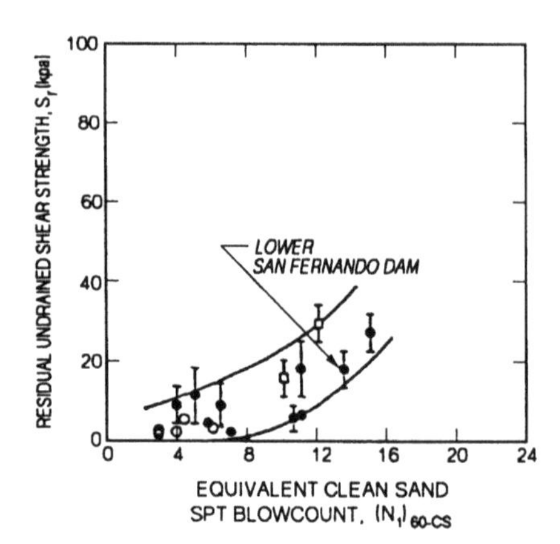

Fig. 13. Correlation of residual strength, S_{ur}, with $(N_1)_{60}$ (after Seed and Harder, 1990).

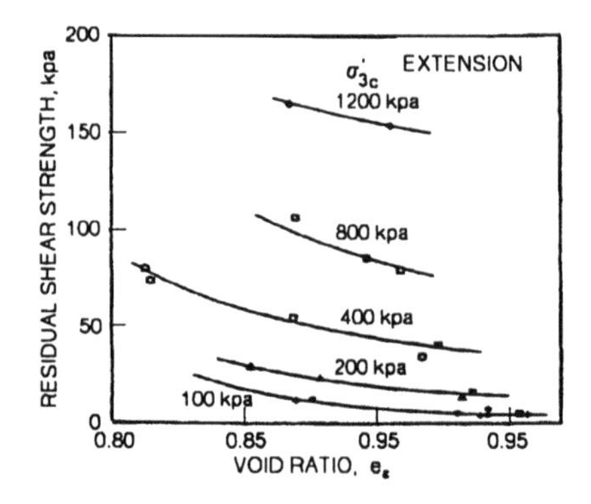

Fig. 14. The effects of overburden pressure on residual strength (Vaid and Thomas, 1994).

The work by Vaid et al. has shown factors of 6 to 10 between residual strength measured in compression and extension testing modes. On a curved failure surface the stress conditions vary roughly over this range so that using compression values, as has been the custom, may lead to a non-conservative estimation of the average mobilized residual strength.

Another factor that confuses the issue of residual strength determination is the method of sample preparation used in fundamental research studies. Some researchers (Poulos et al., 1985) have used moist tamped samples which might be considered representative of compacted fills. Vaid et al. (1989) and Vaid and Thomas (1994) used water pluviated samples which would be more representative of natural deposition and placement by hydraulic fill methods. There is a grave need for a research program to resolve these controversial issues

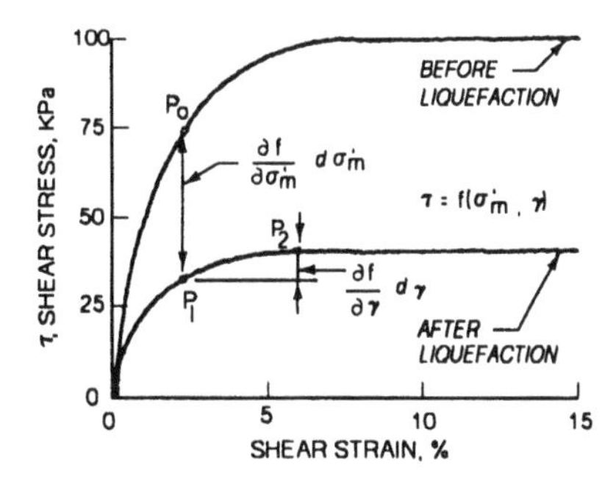

Fig. 15. Adjustment of stress-strain state to post-liquefaction conditions.

because estimates of residual strength have a major impact on the costs of remediation.

Procedures for assessing post-liquefaction behavior will be presented here and illustrated by a number of case histories.

7.2 Post-liquefaction Response

Major difficulties are associated with estimating reliably what will happen after liquefaction in order to plan for cost-effective remediation. The most basic approach to the problem is to investigate the stability of the embankment by limiting equilibrium analysis which incorporates the residual strength of the liquefied soils. The major uncertainties associated with residual strength were noted earlier. Usually a factor of safety 1.1 to 1.2 is considered acceptable. Reliance on acceptable factors of safety alone is not adequate. Test data (Vaid and Thomas, 1994) show that large strains may be necessary to mobilize the residual strength or a significant level of post-liquefaction shearing resistance. The associated deformations can result in unsatisfactory behavior of the dam despite adequate factors of safety. Additionally, as a liquefied soil deposit thickness increases, the assumption of well-defined failure surfaces becomes less reliable, and the dam may significantly deform from bearing capacity failure in the deposit. Therefore, it is necessary to conduct post-liquefaction deformation analyses to investigate the full consequences of liquefaction.

7.3 Deformation Analysis

Analysis of post-liquefaction deformation is an essential adjunct to stability analysis. The global picture of dam behavior provided by such an analysis allows the designer to adopt deformation criteria for evaluating dam performance in addition to factors of safety. Deformation analysis (a) suggests the failure mode that is likely to develop and (b) makes clear where in cross-section to best intervene to remediate the structure and foundation. The factor of safety is not a discriminating tool for deciding on the type or extent of remedial measures. A factor of safety 1.2 can have different connotations depending on the dam geometry and the extent and location of liquefied zones. But the displacement can be interpreted in the light of dam-specific criteria about the allowable potential loss of freeboard or the tolerable extent of potential horizontal deformation. Engineers can make sounder and more cost-effective decisions based on both factor of safety and deformation data than they can using the factor of safety alone.

An independent assessment of the equilibrium of the final position should be conducted using a conventional static stability analysis. The factor of safety determined in this way should be unity or greater depending on whether the deformations occurred relatively slowly after the earthquake or during it when inertia forces were acting.

7.3.1 Method of Analysis

In a particular element in a dam, the shear stress-shear strain state which reflects pre-earthquake conditions may be specified by a point P_0 on the stress-strain curve as shown in Fig. 15 (Finn and Yogendrakumar, 1989).

When liquefaction is triggered, the undrained shear strength will drop to the residual strength. The post-liquefaction stress-strain curve cannot now sustain the pre-earthquake stress-strain condition and the unbalanced shear stresses are redistributed throughout the dam. This process leads to progressive deformation of the dam until equilibrium is reached at the state represented by P_2. The form of the post-liquefaction stress-strain curve in Fig. 15 is only symbolic of behavior. Other shapes can be incorporated in analyses.

A computer program, TARA-3FL, which is a variation of the general computer program TARA-3, has been developed by Finn and Yogendrakumar (1989) for estimating large post-liquefaction deformations based on the above concepts.

The analysis is based on the undrained stress-strain curve of the liquefied materials. In the liquefied elements, the stresses are adjusted according to the following equation (Finn and Yogendrakumar, 1989),

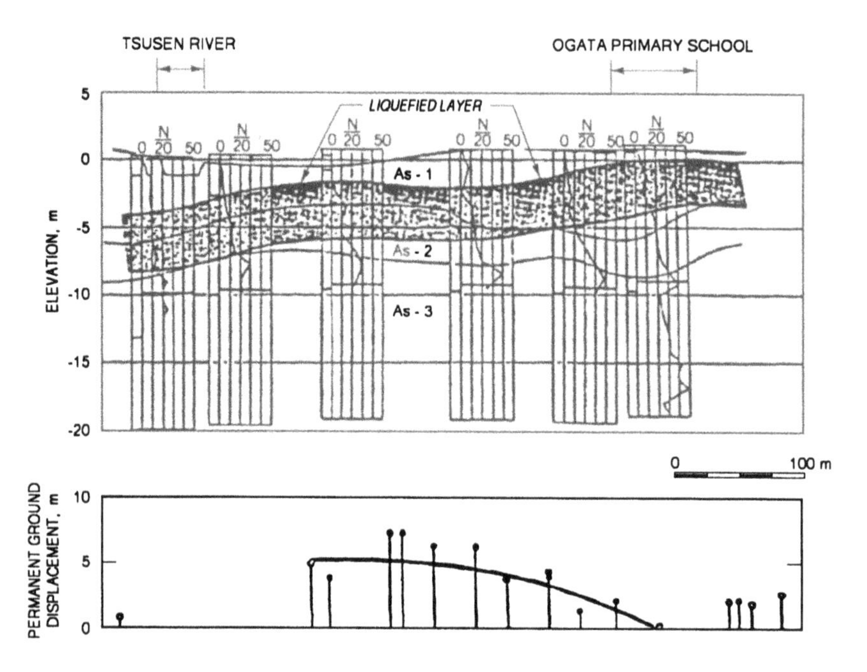

Fig. 16. Section showing liquefied layer with measured and computed displacements (Hamada et al., 1988).

$$\partial\tau = [\partial f/\partial\sigma_m']d\sigma_m' + [\partial f/\partial\gamma]d\gamma \quad (6)$$

where shear stress τ is a function of effective mean confining stress σ_m' and shear strain γ [$\tau=f(\sigma_m',\gamma)$].

Since the deformations may become large, it is necessary to update progressively the finite element mesh. Each calculation of incremental deformation is based on the current shape of the dam, not the initial shape as in conventional finite element analysis.

The application of post-liquefaction analysis in practice will be illustrated by three case histories; Niigata, Sardis Dam, and the Upper San Fernando Dam. A detailed study will be presented in the next section on Mormon Island Auxiliary Dam. This example is especially useful for illustrating the procedures for evaluating proposed remediation measures and evaluating measures which have been completed, and therefore, is treated in detail separately.

7.4 Case Histories

7.4.1 Ground Deformations in Niigata

Large ground deformations occurred in liquefied soils in Niigata, Japan, during the 1964 earthquake. Hamada et al. (1988) conducted detailed studies of these deformations.

A cross-section of the ground between Ogata School and the Tsusen River is shown in Fig. 16, together with the distribution of displacements along the section. Peak displacements are about 7 m. Ogata School is a stationary point with the displacements leading from it in opposite directions.

The Tsusen River provided a free surface which facilitated displacements of the gently sloping ground. Considerable inflow into the river was reported by residents but was quickly eroded away after the earthquake.

A deformation analysis was conducted with TARA-3 using a simplified model of the topography and of the geometry of the liquefied layer. The residual strengths were selected on the basis of the N-values shown in Fig. 16, using the correlation by Seed et al. (1988). As pointed out earlier, at low penetration resistances, there is great uncertainty in selecting an appropriate residual strength. In this case, with low driving shear stresses and large measured deformations, it was clear that the minimum values from the correlation would be appropriate.

The computed deformations are shown by the solid curve in Fig. 16. Predictions are good except at the location of maximum displacements. Part of the

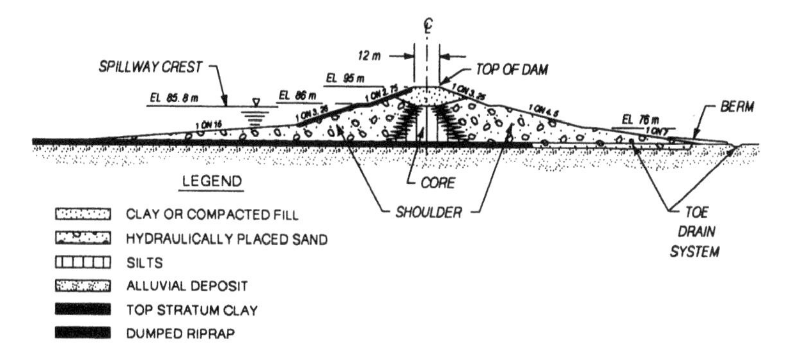

Fig. 17. Cross-section of Sardis Dam.

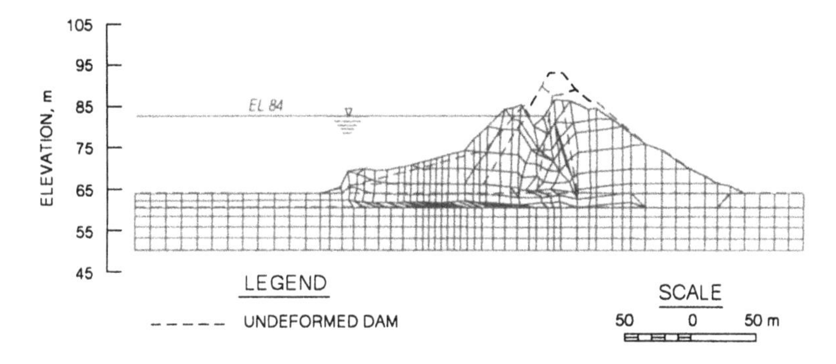

Fig. 18. Deformed cross-section of Sardis Dam.

reason for low computed values in this area may be that an average slope was assigned to the liquefied layer rather than the actual slightly undulating form. Therefore the slope leading into the area of peak displacements in the calculations was somewhat flatter than the original slope.

These results are tentative because of the uncertainties in the residual strength and the simplified geometry of the section used in the analysis. Nevertheless over a section approximately 0.5 km in length, the pattern of computed displacements is representative of the displacements that occurred during the earthquake.

7.4.2 Sardis Dam, Mississippi

Sardis Dam is a U.S. Army Corps of Engineer dam constructed in the late 1930's located in northwestern Mississippi, 16 km southeast of the town of Sardis on the Little Tallahatchie River. The dam is approximately 4,600 m long with a maximum height of 36 m. It was constructed by hydraulic filling and consists of predominantly a silt core surrounded by a sand shell (Fig. 17).

The foundation consists of a 3 m to 6 m thick zone of natural silty clay called the topstratum clay as shown in Fig. 17. The topstratum clay is underlain by pervious alluvial sands (substratum sands) approximately 12 m thick which in turn are underlain by Tertiary silts and clays.

The U.S. Army Engineer District, Vicksburg, evaluated the seismic stability of Sardis Dam for a maximum credible earthquake having a peak horizontal acceleration of 0.20 g. Field and laboratory testing and seismic stability analyses indicated significant strength loss or liquefaction which threatens upstream stability may occur in (1) the hydraulically placed silt core, (2) a discontinuous layer (1.5 m to 4.5 m thick) of clayey silt located in the topstratum clay, and (3) the upper 3 m - 9 m of sand shell along the lower portion of the upstream slope (U.S. Army Engineer District, Vicksburg 1988; Finn et al., 1991; Finn and Ledbetter, 1991).

The liquefaction or strength loss potential of the silty clay was judged on the basis of a modification (Finn et al., 1991) to the Chinese criteria developed by Wang (1979). The residual strength (S_{ur}) of the

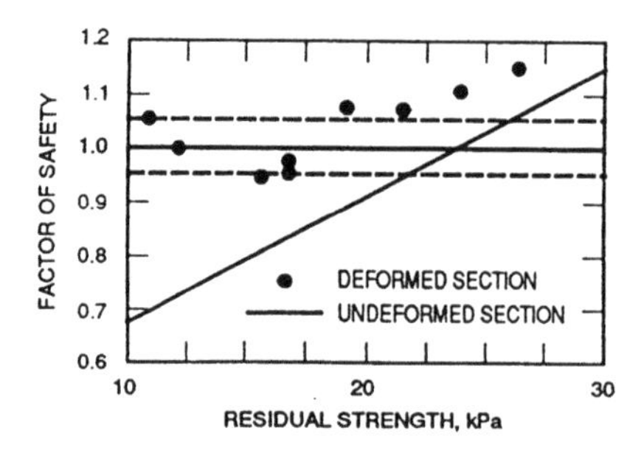

Fig. 19. Factors of safety for Sardis Dam.

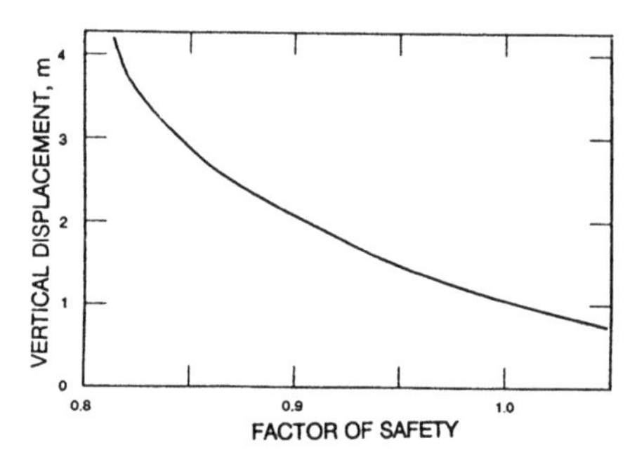

Fig. 20. Variation of loss of freeboard with factor of safety of undeformed dam.

silty clay was estimated from field vane tests and laboratory investigations to be 0.075 times the effective overburden pressure (p'), $S_{ur} = 0.075p'$ (Finn et al., 1991). The following discussions are for a section where the weak clayey silt layer is 1.5 m thick.

Deformation Analysis of Sardis Dam

Deformation analyses by TARA-3FL supplemented by slope stability analyses were used to investigate the post-liquefaction response of the dam and to develop the remediation requirements.

The large differences between the initial and post-liquefaction strengths in Sardis Dam resulted in major load shedding from liquefied elements. This put heavy demands on the ability of the program to track accurately what was happening and on the stability of algorithms used.Therefore, it was imperative to have an independent check from conventional stability analysis that the computed final deformed positions were indeed equilibrium positions. If the major deformations occur during the earthquake, the resulting factor of safety should be greater than unity because some of the deformation field is driven by the inertia forces. If the major deformations occur relatively slowly after the earthquake, the factor of safety could be closer to unity.

The initial and final deformed shapes of the dam are shown in Fig. 18. for a particular distribution of residual strengths. Substantial vertical and horizontal deformations may be noted, together with intense shear straining in the weak thin layer. Different deformed shapes resulted from different assumptions about the distribution of residual strengths. The static stability of each of these deformed shapes was analyzed by Spencer's method (1973) using the program UTEXAS2 (CAGE, 1989). In the clearly unstable region defined by a factor of safety less than one for the undeformed section, computed factors of safety for deformed sections were in the range of 1.0 ± 0.05 (Fig. 19).

The variation of vertical crest displacement with factor of safety of the undeformed dam is shown in Fig. 20. This type of plot gives much more meaning to the factor of safety by associating with each factor an index of overall critical displacement such as loss of freeboard.

Remediation Requirements for Sardis Dam

The deformation analyses supplemented by slope stability analyses were used to investigate various proposals for remediation. Driven reinforced concrete piles were selected to control the post-liquefaction deformations of the dam (Fig. 21). The location of the zone of remediation is controlled by the conservation level of the pool and a desire to avoid driving the piles through riprap on the upper slope above the break.

During shaking by the design earthquake, the saturated portion of the core and the weak foundation clay outside the remediated zone are still expected to liquefy. This will result in increased lateral forces against the piles. Therefore, the piles must fulfill two functions: they must have sufficient strength to prevent shearing along the level of the weak clay layer and also have sufficient stiffness to prevent significant horizontal bending deformations that could lead to unacceptable loss of freeboard.

The static and dynamic loads for design of the piles were estimated by TARA-3FL and TARA-3 analyses. The time history of peak moments in the leading row of piles is shown in Fig. 22 under the assumption that liquefaction occurred at the beginning of the earthquake. The final design of the pile installation was based on limiting embankment vertical deformations of the crest to about 1.5 m.

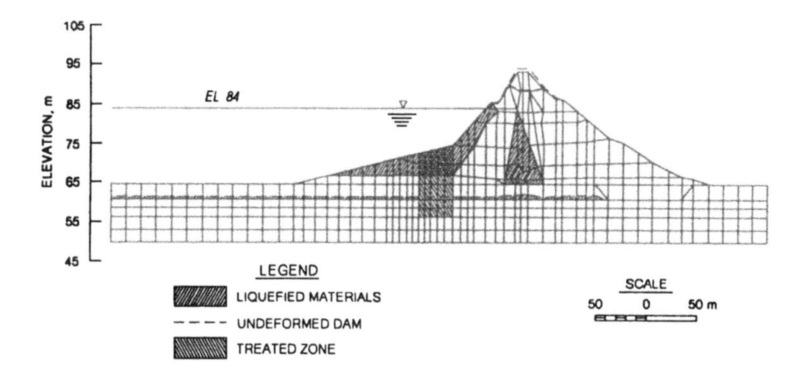

Fig. 21. Cross-section of Sardis Dam showing location of remediation piles and weak thin layer (after Finn and Ledbetter, 1991).

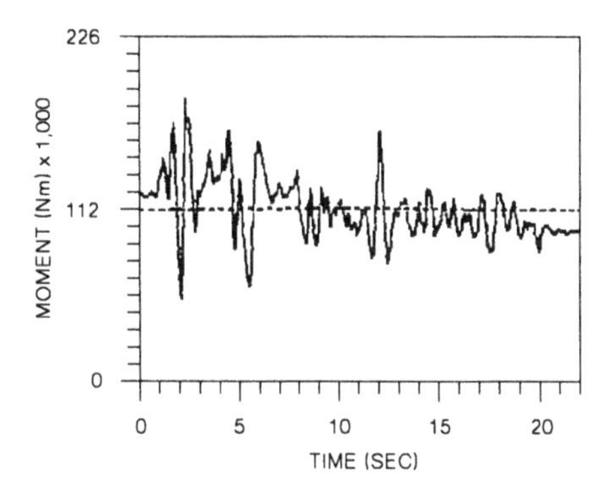

Fig. 22. Variation in total moment during the earthquake.

The pile design is for 0.6 m square steel reinforced concrete piles placed 1.2 m on center perpendicular to the dam and 2.4 m on center parallel to the dam axis for the first three rows closest to the dam center line and 3.7 m on center for the remaining seven rows.

7.4.3 Upper San Fernando Dam

Inel et al. (1993) incorporated a simple soil model in the general purpose program FLAC (Cundall and Board, 1988) to investigate the deformations of the Upper San Fernando Dam during the 1971 San Fernando earthquake. The program uses an updated Lagrangian procedure similar to TARA-3FL for coping with large deformations. The constitutive model incorporated the Mohr-Coulomb failure criterion and elastic shear and bulk moduli dependent on the mean normal effective stresses. Porewater pressures are generated by an incremental scheme (Roth et al., 1991, 1992).

The Upper San Fernando Dam (Fig. 23) is an earth embankment with a maximum height of 24.4 m. During the 1971 San Fernando earthquake, M_l = 6.6, severe longitudinal cracks developed along the crest of the dam near the upstream slope. The crest moved downstream 1.5 m and settled 1 m. A pressure ridge, 0.6 m high, was created at the downstream toe. The pre-earthquake configuration is shown in Fig. 23 as a dashed line.

The cross-section, soil properties and modified Pacoima Dam record used by Seed et al. (1973) in a previous study of the dam were used in FLAC analysis. The cross-section used is shown in Fig. 24.

As shown in Fig. 25, the porewater model satisfactorily predicted the laboratory-based liquefaction resistance curves. The deformed mesh at the end of shaking is shown in Fig. 26. A settlement of the crest of about 1 m was predicted and a bulging of the downstream toe of between 0.3 m and 1 m. These agree well with the measured data. However, the computed overall deformation pattern differs significantly from the field pattern. The field data suggest the entire upper portion of the dam moved downstream. The computed deformation pattern suggested two different deformation modes with the upstream and downstream slopes moving apart creating significant deformations both in the upstream and downstream directions.

8 MORMON ISLAND AUXILIARY DAM: REMEDIATION MEASURES

Deformation analyses using TARA3 and TARA-3FL are being used to guide the remedial treatment of Mormon Island Auxiliary Dam (Ledbetter et al.,

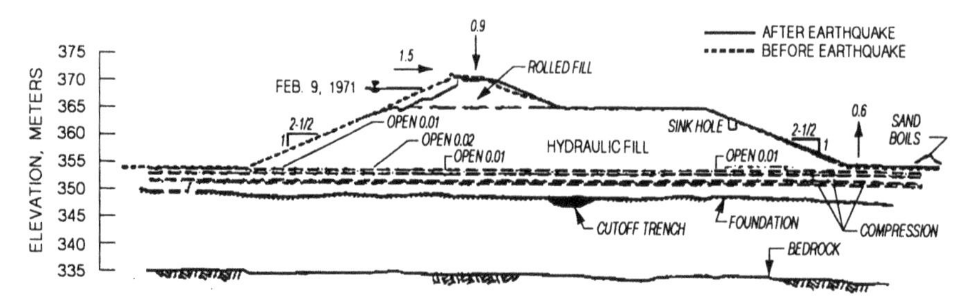

Fig. 23. Upper San Fernando Dam.

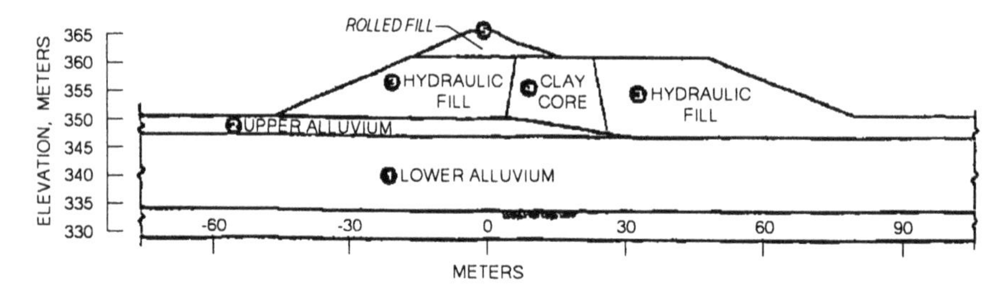

Fig. 24. Upper San Fernando Dam representative cross-section (Seed et al., 1973).

1991, 1993; Ledbetter and Finn, 1993). These analyses continue at this writing. The deformations in the dam were evaluated for earthquake-induced strength degradation and liquefaction in the embankment and foundation soils. Critical regions of high strains were identified which contributed to unacceptable deformations and thereby controlled the performance of the dam. A cost-effective remediation plan was developed using results of the deformation analyses to keep post-earthquake deformations within tolerable limits by optimizing the location and degree of remediation.

8.1 Description of Mormon Island Auxiliary Dam

The Folsom Dam and Reservoir Project is located on the American River about 32 km upstream of Sacramento, California. Seven kilometers of man-made water retaining structures comprise the Folsom project and include the earth-fill Mormon Island Auxiliary Dam. The Folsom project was designed and built by the Corps of Engineers in the period 1948 to 1956, and is now owned and operated by the U.S. Bureau of Reclamation.

Mormon Island Auxiliary Dam was constructed across an ancient channel of the American River which is about 1.6 km wide at the dam site.

The channel gravels are about 20 m in thickness and were dredged for their gold content in the deepest portion with the cobbly-gravelly-tailings placed back into a partially water-filled channel in a loose condition. The tailings are variable with particle size distribution in the ranges: (1) 5-55 % cobbles, (2) 5-75 % gravel, (3) 5-75 % sand, and (4) 5-40 % silt and clay.

Mormon Island Auxiliary Dam is a zoned embankment dam 1,469 m long and 50 m high from core trench to crest at the maximum section. The narrow, central impervious core is a well-compacted clayey mixture founded directly on rock to provide a positive seepage cutoff. Two 3.7 m wide transition zones flank the core. Dam shells were constructed of dredged tailing gravels and are founded on rock, undredged alluvium, and dredged alluvium. The alluvium was excavated to obtain slopes of approximately 1 vertical to 2 horizontal.

The seismic threat to the Folsom project was determined to be an earthquake with peak acceleration of 0.35 g. Material zones and properties for Mormon Island Auxiliary Dam were derived from extensive field, laboratory, and geophysical investigations (Hynes-Griffin et al., 1988; Wahl et al., 1988).

Equivalent linear dynamic stress analyses (Hynes-Griffin et al., 1988; Wahl et al., 1988) resulted in the idealized section with zones of liquefiable and excess pore pressure soils shown in Fig. 27. Extensive liquefaction is expected in the dredged gravel foundation over a 245 m long section of the dam.

8.2 Deformation Analysis of Mormon Island Auxiliary Dam

Large differences between the initial and post-earthquake strengths in Mormon Island Auxiliary Dam result in major load shedding from those elements undergoing strength loss due to excess residual pore pressures and liquefaction. This results in substantial deformations throughout the dam. The deformed zones were used to identify: (1) severely strained regions, (2) locations for remediation which are critical to control dam performance, (3) the geometric extent for remediation treatment areas, and (4) the strength requirements for the remediated zones.

Fig. 28 shows a typical projected deformation pattern for the dam during the process of progressive loss of strength to 50 percent of initial strength in the zones predicted to liquefy in Fig. 27. The pore pressures were progressively increased to 30 to 40 percent of the effective vertical stress in the regions identified by P in Fig. 27. Because a reduction in shear strength of 80 to 95 percent is required to reach the residual shear strength, the dam will continue to significantly deform.

As shown in Fig. 28, the critical areas controlling the dam performance are in the foundation materials beneath the shells of the dam. These materials are losing strength, deforming, and trying to move from beneath the dam. The largest strains are developing in the foundation soils just above bedrock. Remedial measures must contain and control the areas of large strain in order to control the performance of the dam.

8.3 Remediation for Mormon Island Auxiliary Dam

Deformation studies to provide guidance and insight for engineering judgments in the development of remediation requirements involved variations in location, width, depth, and strength of areas to be treated. These led to the remediation plan shown in Fig. 29, which includes foundation treatment and an upstream berm.

8.3.1 Upstream Remediation

Due to the drought in California and the low reservoir, upstream remediation (Fig. 29) was completed in November 1990. Dynamic compaction was chosen as the method for strengthening the foundation material. This would control the deformations on the upstream side of the dam.

A 32,000 kg weight was dropped up to 30 times from a 30 m height in a grid pattern with a minimum spacing of 4 m on center. Upstream post-treatment evaluation is continuing. Results are being incorporated in analyses with expected downstream treatment to obtain the anticipated global behavior of the dam. Significant increases (more than 50) in $(N_1)_{60}$ values resulted in the top 12 m with decreasing improvement from depths of 12 m to 20 m.

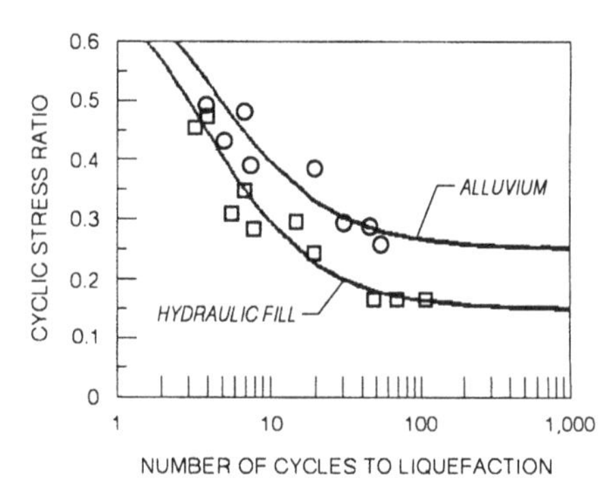

Fig. 25. Upper San Fernando Dam cyclic shear strength curves.

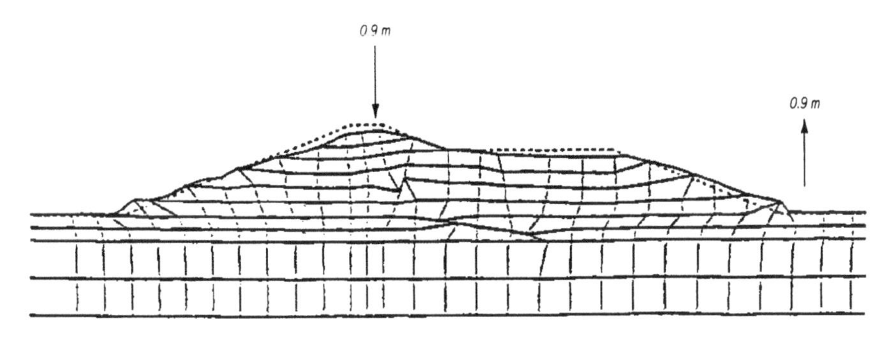

Fig. 26. Upper San Fernando Dam deformed mesh.

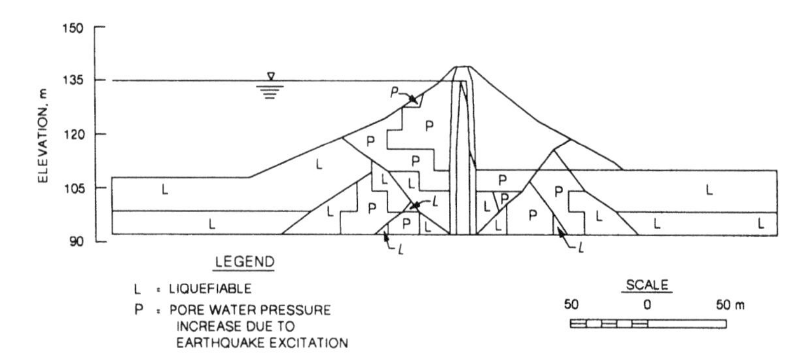

Fig. 27. Mormon Island Auxiliary Dam zones of liquefaction and excess pore pressure.

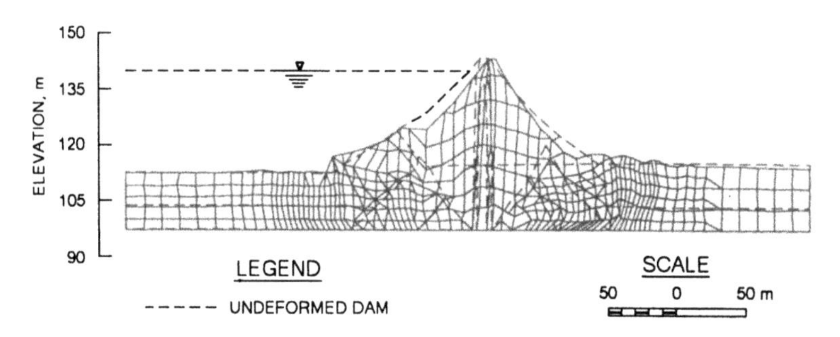

Fig. 28. Snapshot of deformation patterns at 50 percent strength reduction state in liquefying materials.

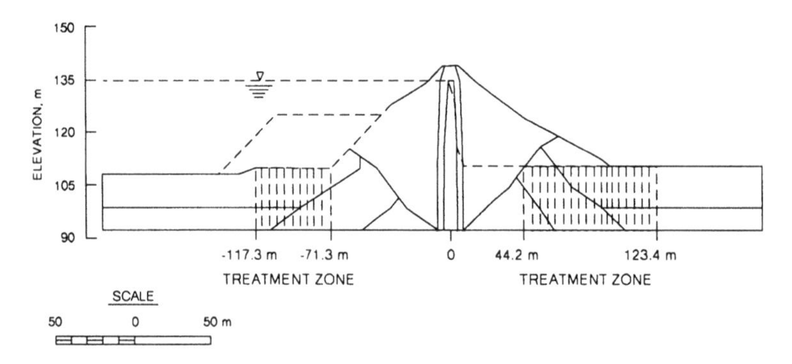

Fig. 29. Cross section of Mormon Island Auxiliary Dam showing zones to be treated.

8.3.2 Downstream Remediation Design Studies

TARA3 and TARA-3FL were used to perform design parameter studies to determine sensitivities and optimize geometry and strength. Typical effects on crest movement of various remediation zone widths and locations in the downstream foundation area are shown in Fig. 30. These data indicate the desirability of the treatment zone extending as far as possible beneath the shell.

Pore pressure migration including the effective stress affects were studied. Results showed the potential for excess pore pressure to rise to significant levels within the treated zone from outside migration leading to strength loss in the zone.

Fig. 31 indicates the sensitivity of the dam and treatment zone width to excess residual pore pressure in the treated zone from both generation within and migration from outside. The deformation of the dam was evaluated for various combinations of strengths and excess pore pressures in the treatment zone. Pore pressures must be controlled in the treated zone to limit crest movement to less than 3 m. Therefore, excess pore pressures in the treated zone during the earthquake should be limited to about 30 percent of the vertical effective stress and should be dissipated quickly after the earthquake.

Excess pore pressure control by stone columns was an option studied. Simultaneous generation and dissipation of pore pressures during the earthquake was simulated.

Results showed that clean-high-permeability stone columns could be used as drains to control earthquake generated and migrated excess pore pressures within the treatment zone.

8.3.3 Downstream Remediation

Based on test section results of achieving significant densification with stone column construction, the current downstream remediation plan is shown in Fig. 32. Small diameter (250 mm) clean stone columns are used to intercept and dissipate the migrating pore pressures. Large diameter (0.8 m) stone columns are used in the interior of the treated zone to densify the material and to assist in pore pressure dissipation. At this writing, downstream remedial treatment is in progress.

Due to static ground water conditions the clean stone columns are anticipated to remain clean until an earthquake occurs that induces excess pore pressures.

Field evaluation of remediation achievements is being made by penetration and shear wave tests. The penetration criteria were based on limiting the pore pressure generation potential to about 30 percent of the vertical effective stress. A factor of safety against liquefaction (FSL) can be calculated as the ratio of cyclic stress ratio causing liquefaction (Seed et al., 1986) to cyclic stress ratio induced by the design earthquake. The FSL for different values of $(N_1)_{60}$ is related to residual excess pore pressure (Seed and Harder, 1990). From these relations, $(N_1)_{60}$ criteria can be developed which limits pore pressure generation potential.

8.3.4 Remediated Dam Response

Analyses of the assumed treated dam indicate that during the design earthquake excess pore pressures in the downstream treated zone do not exceed 30 percent of the vertical effective stress and are dissipated quickly after the earthquake, Fig. 33. (It

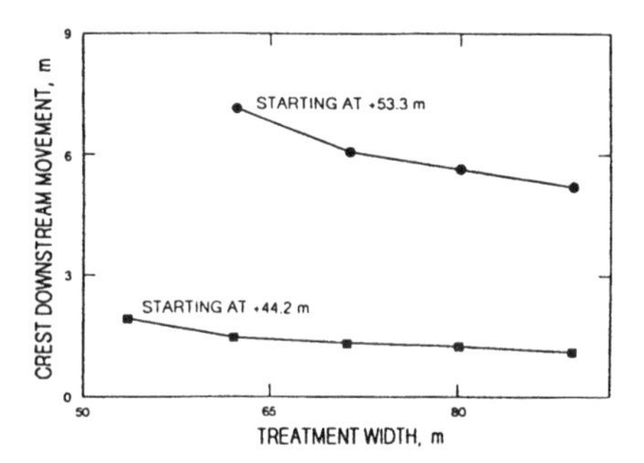

Fig. 30. Typical effects from varying downstream treatment zone width and starting location.

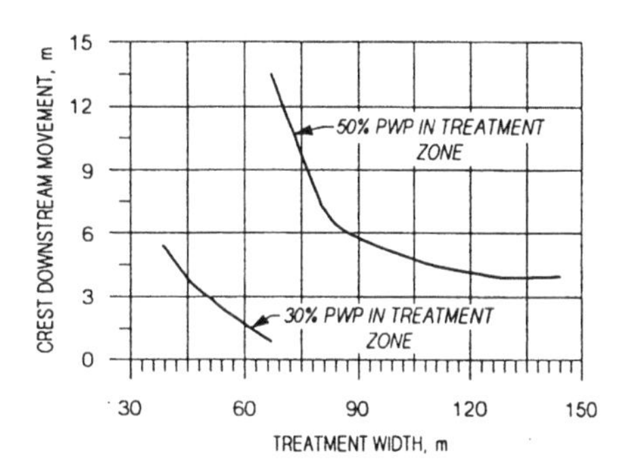

Fig. 31. Effects of pore pressure in the treatment zone.

is realized that the dissipation times and peak pore pressures are dependent on a number of factors and could vary significantly; therefore, conservative assumptions are made.) The four curves in Fig. 33 are for the first five layers from the bottom of a mesh similar to that shown on Fig. 28. The top is for a drainage blanket.

Shown in Fig. 34 are typical results of the estimated deformations for the assumed remediated dam. The behavior results from the progressive loss of strength to the liquefied state and residual strength values in all materials labeled L (Fig. 27) outside the treated zones. The pore pressures were progressively increased to 30 to 40 percent of the vertical effective stress outside the treated zones for the foundation and embankment materials labeled P in Fig. 27.

A comparison of Fig. 34 with Fig. 28 shows that the fully and partially liquefied soils between the dam center line and the treated zones are expected to be contained, and that the performance of the dam is controlled.

Dynamic analyses of the assumed remediated dam show that the deformations are mainly post-earth-

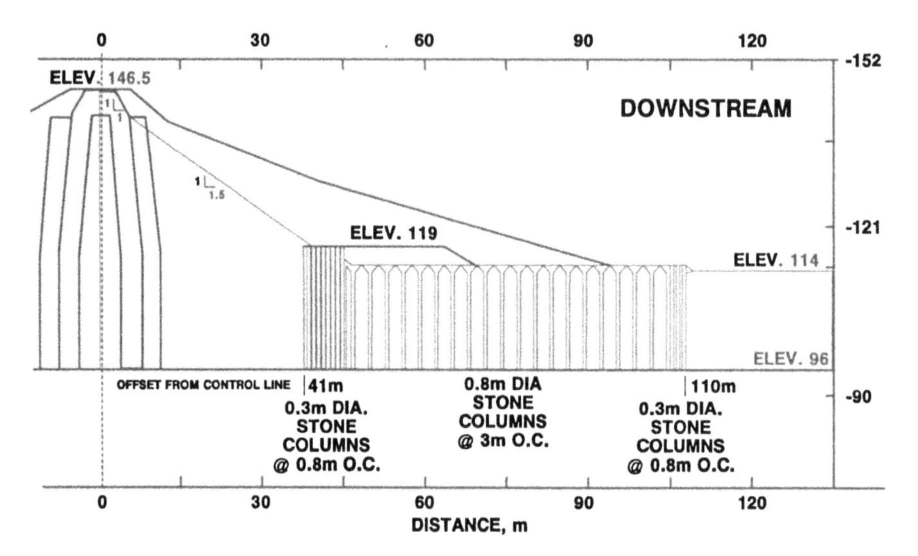

Fig. 32. Downstream remediation plan.

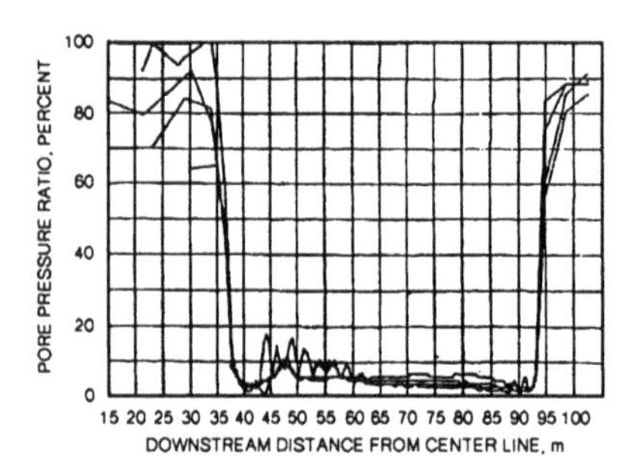

Fig. 33. Typical post-earthquake excess residual pore pressures in the assumed treated downstream zone.

quake gravity driven. A study was made to evaluate the deformed dam to an aftershock equal to the main shock. In order to be conservative, the two earthquakes were assumed to occur back to back with no dissipation of pore pressure assumed. The dam is predicted to respond satisfactorily to a strong aftershock.

Analyses are planned to evaluate the earthquake deformation performance of the remediated dam after completion of remediation.

9 IMPACT OF SOME RECENT DEVELOPMENTS

The above analyses show the key roles played by liquefaction resistance and residual strength. The former controls the triggering of liquefaction, and the latter, the post-liquefaction stability and deformation. Recent researches by Vaid and Thomas (1994) and Thomas (1992) may have considerable impact on the estimation of liquefaction resistance using the procedure described by Seed and Harder (1990) and on the evaluation of residual strength.

It is generally recognized that the rate of increase of resistance to liquefaction decreases with increase in effective confining stress. Liquefaction resistance derived from Seed's liquefaction resistance chart (Seed et al., 1985) has been normalized to an effective overburden pressure of 100 kPa. The liquefaction resistance at higher confining pressures is derived by multiplying the resistance at 100 kPa by a factor K_σ. Seed and Harder (1990) have plotted K_σ for many soil types against confining pressures up to 800 kPa and suggested a conservative curve for use in practice (Fig. 35).

Vaid and Thomas (1994) and Thomas (1992) report detailed studies on the effects of confining pressure on the liquefaction resistance of Fraser River sand. Their study confirms in a general way, the findings of Seed and Harder (1990), but they demonstrate clearly that K_σ is a function of relative density. Their results are shown in Fig. 36 along with data from Fig. 35. At 59% relative density and a confining pressure of 1200 kPa, the value of K_σ is about 0.75, compared with a value of 0.4 for the Seed and Harder (1992) curve at a pressure of only 800 kPa. These results strongly suggest that job specific values of K_σ should be determined. In the case of Fraser River sand, for example, the differ-

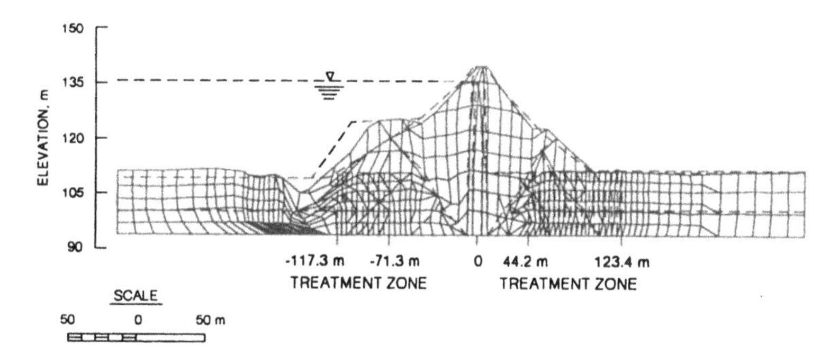

Fig. 34. Estimated permanent deformations after treatment.

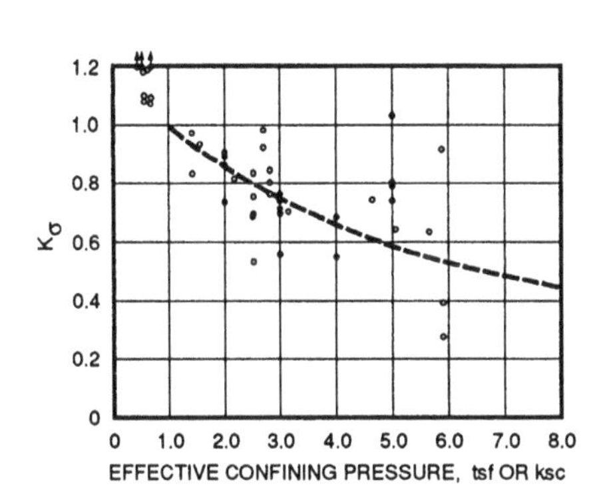

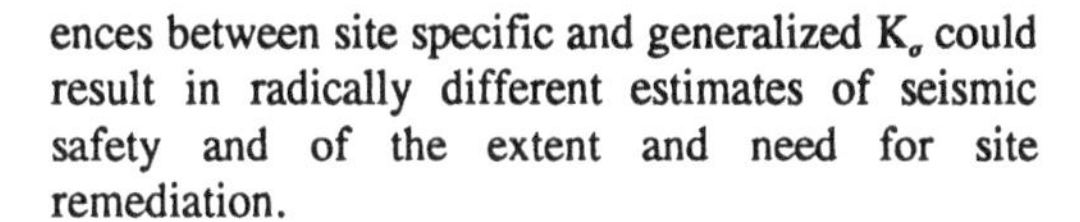

Fig. 35. Relationship between effective vertical stress (σ_v') and K_σ (after Seed and Harder, 1990).

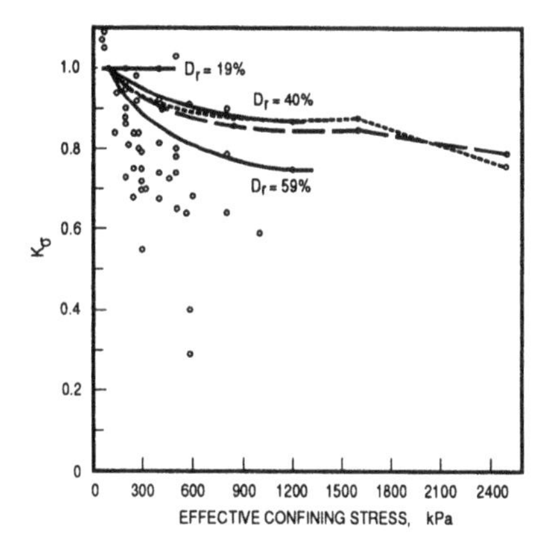

Fig. 36. Relationship between effective confining stress and K_σ (Vaid and Thomas, 1994).

ences between site specific and generalized K_σ could result in radically different estimates of seismic safety and of the extent and need for site remediation.

10 CONCLUSIONS AND RECOMMENDATIONS

The technology is available today for constructing safe embankment dams in any seismic environment. Indeed even in the extreme cases where a major fault passes under the dam it has been confidently asserted that "an embankment dam can be theoretically made safe against any feasible fault displacement", (Sherard et al., 1974).

The major geotechnical problems facing dam designers in a seismic environment arise in the evaluation of the safety of existing dams. The most common factor leading to potential instability is the presence of loose saturated cohesionless soils in the dam itself and/or in the foundation which may liquefy during an earthquake. There are three difficult technical problems associated with potential instability induced by liquefaction. Will liquefaction be triggered? If so what will be the consequences? How can cost-effective remediation measures be designed to mitigate or prevent the consequences?

The scaling parameter K_σ used to extrapolate the existing database on the incidence of liquefaction to the depths and stress conditions associated with embankment dams shows a wide variation with respect to soil type and relative density. Since this parameter has a major impact on the requirements for remediation if liquefaction is allowed to trigger, it should be determined on a project specific basis.

The post-liquefaction behavior of dams should be assessed using both limiting equilibrium analysis and deformation analysis. The extent and location of remediation should be determined primarily on the basis of calculated deformation patterns. For many dams, especially those with substantial freeboard, criteria based on factor of safety alone can result in unnecessary remediation costs.

Whether equilibrium or deformation procedures are used, the post-liquefaction undrained behavior of the liquefied material is the essential factor controlling the cost of remediation. It has two elements which should be well defined, the residual strength and the strain level required to reach it.

The controversy surrounding residual strength carries significant cost penalties for the remediation of embankment dams. A focused research program is required to resolve issues. The residual strength has traditionally been considered a function of void ratio only. The key question that has surfaced lately is, does it also depend on the loading path and the confining pressure?

The dynamic response analyses of embankment dams are still largely based on technology developed in the 1970's and represent our first attempts to carry out nonlinear analyses by equivalent linear procedures. The stresses and accelerations determined in this way are input into other procedures for determining the performance of the dam. These procedures appear to work quite well provided the behavior of the dam is not strongly non-linear and significant pore pressures do not develop. More comprehensive methods are available which can deal with these problems directly especially for evaluating the permanent displacements resulting from strong shaking with or without the presence of liquefaction. These procedures should be used when appropriate.

The Newmark procedure for estimating permanent deformations based on sliding block analysis is widely used despite all the evidence that deformations do not occur in this way. These methods are particularly inappropriate when a large zone has liquefied in the embankment or foundation.

The seismic safety evaluation of dams has evolved from very empirical procedures in 1960 to a mature sophisticated professional practice in 1993. As pointed out above, the evaluation of well designed dams is well within the capabilities of the profession. The assessment of existing dams which may have potentially liquefiable zones can be done safely but the uncertainties associated with the critical elements of the procedure are such that conservative judgments are being made. The challenge for the profession in the immediate future is to reduce these uncertainties and thereby the extent and cost of the remediation of embankments.

11 ACKNOWLEDGEMENTS

The financial support of the National Sciences and Engineering Research Council of Canada under grant 5-81498 for research studies on embankment dams and the support of the U.S. Army Engineer Waterways Experiment Station (WES) during the preparation of the manuscript is appreciated and acknowledged. The authors greatly appreciate the detailed review of the manuscript and helpful suggestions from Paul H. Hadala of the WES. Permission to publish this paper was granted by the Corps of Engineers.

12 BIBLIOGRAPHY

AMBRASEYS N.N., "The Seismic Stability of Earth Dams", Proceedings, 2nd World Conference on Earthquake Engineering, Vol. 2, pp. 1345-1363, July, 1960a.

AMBRASEYS N.N., "On the Seismic Behavior of Earth Dams", Proceedings, 2nd World Conference on Earthquake Engineering, Vol. 1, pp. 331-356, July, 1960b.

AMBRASEYS N.N., "On the Shear Response of a Two-Dimensional Truncated Wedge Subjected to an Arbitrary Disturbance," Bull. Seismic Society of America, Vol. 50, pp. 45-46, 1960c.

AMERICAN SOCIETY OF CIVIL ENGINEERS, "Shear Strength of Cohesive Soils", Proceedings, Research Conference on Shear Strength of Cohesive Soils, University of Colorado, Boulder, 1164 pp., June, 1960.

ARULANANDAN K. AND SCOTT R.F., EDITORS, Proceedings, International Conference on the Verification of Numerical Procedures for the Analysis of Soil Liquefaction Problems, Davis, California, Published by A.A. Balkema, October, 1993.

BARTLETT S.F. and YOUD T.L., "Empirical Analysis of Horizontal Ground Displacement Generated by Liquefaction-Induced Lateral Spreads", Technical Report CEG-92-01, Dept. of Civil Engineering, Brigham Young University, Provo, Utah, 1992.

BAZIAR M.H. and DOBRY R., "Liquefaction Ground Deformation Predicted From Laboratory Tests", Proceedings, Second International Conference on Recent Advances in Geotechnical Earthquake Engineering and Soil Dynamics, St. Louis, Vol. 1, pp. 451-458, March, 1991.

BISHOP A.W., "The Use of the Slip Circle in the Stability Analysis of Slopes", Geotechnique, Vol. 5, No. 1, pp. 7-17, January, 1960.

CAGE, G-CASE TASK GROUP ON SLOPE STABILITY, "Users's Guide: UTEXAS2 Slope Stability Package, Vol. II, Theory", Instruction Report GL-87-1, U.S. Army Engineer Waterways Experiment Station, Vicksburg, Mississippi, 1989.

CLOUGH R.W. and CHOPRA A.K., "Earthquake Stress Analysis in Earth dams", Journal of the Engineering Mechanics Division, ASCE, Vol. 92, No. EM2, pp. 197-211, April, 1966.

CLOUGH R.W. and PIRTZ D., "Earthquake Resistance of Rockfill Dams", Transactions, ASCE, New York, NY, pp. 792-810, 1958.

CUNDALL P.A., "The 'Failure Seeking Model' for Cyclic Behavior in Soil-An Initial Formulation for Two Dimensions," Technical Note PLAN-1, Peter Cundall Associates, July, 1979.

CUNDALL P.A. and BOARD M., "A Micro - Computer Program for Modelling Large-Strain Plasticity Problems", Proc. 6th International Conference on Numerical Methods in Geomechnics, Innsbruck, Austria, pp. 2101-2108, April, 1988.

DAFALIAS Y.F. and HERMANN L.R., "Bounding Surface Formulation of Soil Plasticity," Soil Mechanics - Transient and Cyclic Loads, G. Pande and O.C. Zienkiewicz, Eds., John Wiley & Sons, Inc., London, U.K., pp. 253-282, 1982.

DAKOULAS P. and GAZETAS G., "Seismic Shear Strains and Seismic Coefficients in Dams and Embankments", Soil Dynamics and Earthquake Engineering, Vol. 5, No. 2, pp. 75-83, April, 1986a.

DAKOULAS P. and GAZETAS G., "Seismic Shear Vibration of Embankment Dams in Semi-Cylindrical Valleys", Earthquake Engineering and Structural Dynamics, Vol. 14, No. 1, pp 19-40, January, 1986b.

DRNEVICH V.P., "Effect of Strain on the Dynamic Properties of Sand", Ph.D. Thesis, University of Michigan, Ann Arbor, Mich, 1967.

FINN W.D. Liam, "Static and Dynamic Stresses in Slopes", Proceedings, First International Congress on Rock Mechanics, Lisbon, Vol. II, pp. 167-171, 1966a.

FINN W.D. Liam, "Static and Seismic Analysis of Slopes", Rock Mechanics in Engineering Geology, Vol. IV/3, Springer-Verlag/Wien, New York, pp. 268-277, 1966b.

FINN W.D. Liam, "Behavior of Earth Dams During Earthquake", Proceedings, 9th International Congress on Large Dams, Istanbul, pp. 355-367, September, 1967.

FINN W.D. Liam, "Permanent Deformations in Ground and Earth Structures During Earthquakes", State of the Art Report, Proc., 9th World Conf. on Earthquake Eng., Tokyo-Kyoto, Japan, Vol. VIII, August, 1988a.

FINN W.D. Liam, "Dynamic Analysis in Geotechnical Engineering", Proceedings, Earthquake Engineering and Soil Dynamics II - Recent Advances in Ground Motion Evaluation, ASCE Geotechnical Engineering Division, Park City, Utah, pp 523-591, June, 1988b.

FINN W.D. LIAM, "Seismic Analysis of Embankment Dams," Dam Engineering, Vol. 1, Issue 1, pp. 59-75, January, 1990.

FINN W.D. LIAM, "Assessment of Liquefaction Potential and Post-Liquefaction Behavior of Earth Structures: Developments 1981-1991," Proceedings: Second International Conference on Recent Advances in Geotechnical Earthquake Engineering and Soil Dynamics, St. Louis, Missouri, pp. 1833-1850, March, 1991a.

FINN W.D. Liam, "Estimating How Embankment Dams Behave During Earthquakes", Water Power and Dam Construction, London, pp. 17-22, April, 1991b.

FINN W.D. Liam, "Seismic Safety Evaluation of Embankment Dams", Proceedings, International Workshop on Dam Safety Evaluation, Grindelwald, Switzerland, Vol. 4, pp. 90-136, April, 1993.

FINN W.D. Liam and KHANNA J., "Dynamic Respdnse of Earth Dams", Proceedings, Third Symposium on Earthquake Engineering, Roorkee, pp. 315-324, 1966.

FINN W.D. Liam, PICKERING D.J. and BRANSBY P.L., "Sand Liquefaction in Triaxial and Simple Shear Tests", Journal of the Soil Mechanics and Foundations Division, ASCE, pp. 639-659, April, 1971.

FINN W.D. Liam, LEE K.W. and MARTIN G.R., "Stress-Strain Relations For Sand in Simple Shear", Session No. 58, Seismic Problems, in Geotechnical Engineering, ASCE Annual Meeting, Denver, Colorado, November, 1975.

FINN W.D. Liam, LEE K.W. and MARTIN G.R., "An Effective Stress Model For Liquefaction", Journal of the Geotechnical Engineering Division, ASCE, Vol. 103, No. GT6, Proc. Paper 13008, pp. 517-533, June, 1976.

FINN W.D. Liam, YOGENDRAKUMAR M., YOSHIDA N. and YOSHIDA H., "TARA-3: A Program to Compute the Response of 2-D Embankments and Soil-Structure Interaction Systems to Seismic Loadings", Department of Civil Engineering University of British Columbia, Canada, 1986.

FINN W.D. Liam and YOGENDRAKUMAR M., "TARA-3FL; Program for Analysis of Liquefaction Induced Flow Deformations, Department of Civil Engineering, University of British Columbia, Vancouver, Canada, 1989.

FINN W.D. Liam and LEDBETTER R.H., "Evaluation of Liquefaction Effects and Remediation Strategies by Deformation Analysis", Proceedings, International Conference on Geotechnical Engineering for Coastal Development, GEOCOAST 91, pp. 1-20, 1991.

FINN W.D. Liam, LEDBETTER R.H., FLEMING R.L.M TEMPLETON A.E.M., FORREST T.W. and STACY S.T., "Dam on Liquefiable Foundation - Safety Assessment and Remediation", Proceedings, 17th International Congress on Large Dams, Vienna, pp. 531-553, June, 1991.

FINN W.D. Liam, GILLON M.D., YOGENDRAKUMAR M. and NEWTON C.J., "Stimulating the Seismic Response of a Rockfill Dam", Proceedings, NUMOG-4, Vol. I, A.A. Balkema, Rotterdam, pp. 379-391, 1992.

GAZETAS G., Discussion of "Seismic Analysis of Concrete Face Rockfill Dams", Proceedings, Symposium on Concrete-Face Rockfill Dams - Design, Construction and Performance", ASCE, Detroit, MI, pp. 1247-1251, October, 1985.

HALL J.R. Jr., "Effect of Amplitude on Damping and Wave Propagation in Granular Materials", PhD. Thesis, University of Florida, at Gainesville, Florida, 1962.

HARDIN B.O., "Suggested Methods of Test for Shear Modulus and Damping of Soils by the Resonant Column", STP 479, American Society for Testing and Materials, pp 516-529, 1970.

HARDIN B.O. and BLACK W.L., "Sand Stiffness Under Various Triaxial Stresses", Journal of the Soil Mechanics and Foundation Division, ASCE, Vol. 92, No. SM2, pp. 27-42, March, 1966.

HARDIN B.O. and BLACK W.L., "Vibration Modulus of Normally Consolidated Clay", Journal of the Soil Mechanics and Foundations Division, ASCE, Vol. 94, No. SM2, pp. 353-369, March, 1968.

HARDIN B.O. and DRNEVICH V.P., "Shear Modulus and Damping in Soils, I, Measurement and Paramenter Effects", Technical Report UKY 26-70-CE2, University of Kentucky, 1970.

HARDIN B.O. and MUSIC J., "Apparatus for Vibration During the Triaxial Test", STP 392, Symposium on Instrumentation and Apparatus for Soils and Rocks, American Society for Testing and Materials, 1965.

HAMADA M., YASUDA S., ISOYAMA R., AND EMOTO K., "Study on Liquefaction Induced Permanent Ground Displacements," Published by the Association for the Development of Earthquake Prediction in Japan, p. 87, 1986.

HAMADA M., YASUDA S., AND WAKAMATSA K., "Case Study on Liquefaction-Induced Ground Failures During Earthquakes in Japan", Proceedings, First Japan-U.S. Workshop on Liquefaction, Large Ground Deformations and their Effects on Lifeline Facilities, Tokyo, Japan, National Center for Earthquake Engineering Research, Buffalo, New York, 1988.

HARDER L.F., "Evaluation of Becker Soundings Performed at Morman Island Auxiliary Dam in Conjunction With Upstream Remediation", Report prepared U.S. Army Engineer District Sacramento, October, 1992.

HATANAKA M., "Three Dimensional Consideration on the Vibration of Earth Dams", Transactions of the Japanese Society of Civil Engineers, 37:10, pp. 423-428, 1952.

HATANAKA M., "Fundamental Considerations on the Earthquake Resistant Properties of the Earth Dam", Bull. Disaster Prev. Res. Inst. No. 11, 1955.

HYNES-GRIFFIN M.E., WAHL R.E., DONAGHE R.T. and TSUCHIDA T., "Seismic Stability evaluation of Folsom Dam and Reservoir Project", Report No. 4, Mormon Island Auxiliary Dam - Phase I, T.R. GL-87-14, U.S. Army Engineer Waterways Experiment Station, Vicksburg, Mississippi., 1988.

IDRISS I.M, LYSMER J., HWANG R. and SEED H.B., "QUAD-4: A Computer Program for Evaluating the Seismic Response of Soil Structures by Variable Damping Finite Element Procedures", Report No. EERC 73-16, University of California, Berkeley, 1973.

INEL S., ROTH W.H. and DE RUBERTIS C., "Nonlinear Dynamic Effective-Stress Analysis of Two Case Histories", Proceedings: Third International Conference on Case Histories in Geotechnical Engineering, St. Louis, Missouri, pp. 1604-1610, June, 1993.

KAWAI T., "Summary Report on the Development of the Computer Program DIANA - Dynamic Interaction Approach and Non-linear Analysis", Science University of Tokyo, 1985.

LEDBETTER R.H., FINN W.D. LIAM, NICKELL J.S., WAHL R.E., and HYNES M.E., "Liquefaction Induced Behavior and Remediation For Mormon Island Auxiliary Dam", Proceeding, International Workshop on Remedial Treatment of Liquefiable Soils, Tsukuba, Japan, pp. 215-233, January, 1991.

LEDBETTER R.H. and FINN W.D. LIAM, "Development and Evaluation of Remediation Strategies by Deformation Analysis," ASCE Specialty Conference on Geotechnical Practice In Dam Rehabilitation, Raleigh, North Carolina, pp. 386-401, April, 1993.

LEE M.K. and FINN W.D. Liam, "DESRA-2, Dynamic Effective Stress Response Analysis of Soil Deposits with Energy Transmitting Boundary Including Assessment of Liquefaction Potential", Soil Mechanics Series No.38, Department of Civil Engineering, University of British Columbia, Vancouver, Canada, 1978.

LO R.C., KLOHN E., and FINN W.D. LIAM, "Shear Strength of Cohesionless Materials Under Seismic Loadings", Proceedings, IX Pan-American Conference, Santiago, Chile, pp.1047-1062, August, 1991.

LYSMER J., UDAKA T., TSAI C.F., and SEED H.B., "FLUSH - A Computer Program for Approximate 3-D Analysis of Soil Structure Interaction Problems", Report No. EERC 75-30, Earthquake Engineering Research Center, University of California, Berkeley, November, 1975.

MASING G., "Eigenspannungen und Verfestigung beim Messing", Proceedings, 2nd International Congress of Applied Mechanics, Zurich, Switzerland, 1926.

MAKDISI F.I. and SEED H.B., "Simplified Procedure for Estimating Dam and Embankment Earthquake-Induced Deformations", Journal of the Geotechnical Engineering Division, ASCE, Vol. 104, No.GT7, pp. 849-867, July, 1978.

MARCUSON W.F., HYNES M.E. and FRANKLIN A.G., "Seismic Stability and Permanent Deformation Analyses: The Last Twenty Five Years", Proceedings, ASCE Specialty Conference on Stability and Performance of Slopes and Embankments - II, Geotechnical Special Publication No. 31, Eds. R.B. Seed and R.W. Boulanger, ASCE, New York, NY, Vol. I, pp. 552-592, June, 1992.

MARTIN G.R., FINN W.D. Liam and SEED H.B., "Fundamentals of Liquefaction Under Cyclic Loading", Journal of the Geotechnical Engineering Division, ASCE, Vol. 101, No.GT5, pp. 423-438, May, 1975.

MCLEOD H., CHAMBERS R.W., and DAVIES M.P., "Seismic Design of Hydraulic Fill Tailings Structures", Proceedings, IX Pan-American Conference, Santiago, Chile, pp. 1062-1081, August, 1991

MEJIA L.H. and SEED H.B., "Comparison of 2-D and 3-D Dynamic Analysis of Earth Dams", Journal of the Geotechnical Engineering Division, ASCE, 109 (GT11), pp 1383-1398, September, 1983.

MORGENSTERN N.R. and PRICE V.E., "The Analysis of the Stability of General Slip Surfaces", Geotechnique, Vol. 15, pp. 79-93, 1965.

MORIWAKI Y., BEIKAE M. and IDRISS I.M., "Nonlinear seismic analysis of the upper San Fernando Dam under the 1971 San Fernando Earthquake", Proceedings, 9th World Conference on Earthquake Engineering, Tokyo and Kyoto, Japan, Vol. III, pp. 237-241, 1988.

MURALEETHARAN K.K, MISH K.D., YOGACHANDRAN C., and ARULANANDAN K., "DYSAC2: Dynamic Soil Analysis Code for 2-dimensional Problems", Computer Code, Department of Civil Engineering, University of California, Davis, California, 1988.

MURALEETHARAN K.K, MISH K.D., YOGACHANDRAN C., and ARULANANDAN K., "User's Manual for DYSAC2: Dynamic Soil Analysis Code for 2-dimensional Problems", Report, Department of Civil Engineering, University of California, Davis, California, 1991.

MURALEETHARAN K.K, MISH K.D., and ARULANANDAN K., "A Fully Coupled Nonlinear Dynamic Analysis Procedure And Its Verification Using Centrifuge Test Results", Submitted to the International Journal for Numerical and Analytical Methods in Geomechanics, 1993.

NEWMARK N.M., "Failure Hypotheses for Soils", American Society of Civil Engineers, "Shear strength of cohesive soils", Proceedings, Research Conference on Shear Strength of Cohesive Soils, University of Colorado, Boulder, pp. 17-32, June, 1960.

NEWMARK N.M., "Effects of Earth-quakes on Dams and Embankments", 5th Rankine Lecture, Geotechnique, Vol. 15, No. 2, pp. 139-160, 1965.

PANDE G.N. and ZIENKIEWICZ O.C., Editors, "Soil Mechanics - Transient and Cyclic Loads", John Wiley & Sons, Inc., London, U.K., 627 pg., 1982.

POULOS S.J., CASTRO G. and FRANCE W., "Liquefaction Evaluation Procedure", Journal of the Geotechnical Engineering Division, ASCE, Vol. 111, No. 6, pp. 772-792, June, 1985.

PREVOST J.H., "DYNAFLOW: A Nonlinear Transient Finite Element Analysis Program", Princeton University, Department of Civil Engineering, Princeton, NJ, 1981.

ROTH W.H., "Evaluation of Earthquake Induced Deformations of Pleasant Valley Dam", Report for the City of Los Angeles, Dames & Moore, Los Angeles, 1985.

ROTH W.H., BUREAU G. and BRODT G., "Pleasant Valley Dam: An Approach to Quantifying the Effect of Foundation Liquefaction", Proceedings, 17th International Congress on Large Dams, Vienna, pp. 1199-1223, 1991.

ROTH W.H., FONG H. and DE RUBERTIS C, "Batter Piles and the Seismic Performance of Pile-Supported Wharves", Proceedings, Ports '92 Conference, ASCE, Seattle, Washington, 1992.

SAADA A. and BIANCHINI G.S., Editors, Proceedings, International Workshop on Constitutive Equations for Granular Non-Cohesive Soils, Case Western Reserve University, Cleveland, Ohio, Published by A.A. Balkema, July, 1987.

SCHNABEL P.L., LYSMER J. and SEED H.B., "SHAKE: A Computer Program for Earthquake Response Analysis of Horizontally Layered Sites", Report No. EERC 72-12, University of California, Berkeley, 1972.

SEED H.B., "Considerations in the Earthquake-Resistant Design of Earth and Rockfill Dams", 19 Rankine Lecture, Geotechnique, Vol. 29 No.3, pp. 215-263, 1979.

SEED H.B., "Earthquake Resistant Design of Earth Dams", Seismic Design of Embankments and Caverns, Proceedings of ASCE Symposium, Philadelphia, pp. 41-64, 1983.

SEED H.B., "Design Problems in Soil Liquefaction", Journal of Geotechnical Engineering, ASCE, Vol. 113, No.7, pp. 827-845, August, 1987.

SEED H.B. and LEE K.L., "Liquefaction of Sands During Cyclic Loading", Journal of the Soil Mechanics and Foundations Division, ASCE, Vol. 92, No.SM3, Proceedings Paper 4824, pp. 105-134, May, 1966.

SEED H.B. and MARTIN G.R., "The Seismic Coefficient of Earth Dam Design", Journal of the Soil Mechanics and Foundations Division, ASCE, Vol. 92, No.SM3, Proceedings Paper 4824, pp. 25-48, May, 1966.

SEED H.B. and PEACOCK W.H., "Test Procedures for Measuring Soil Liquefaction Characteristics", Journal of the Soil Mechanics and Foundation Engineering Division, ASCE, Vol. 97, No. SM8, Proceedings Paper 8330, pp. 1099-1122, August, 1971.

SEED H.B., LEE K.L, IDRISS I.M. and MAKDISI F.I., "Analyses of the Slides in the San Fernando Dams During the Earthquake of February 9, 1971", Report No. EERC/73-2, University of California, Berkeley, (NTIS No. PB 223 402), 1973.

SEED H.B., LEE K.L, IDRISS I.M. and MAKDISI F.I., "The Slides in the San Fernando Dams During the Earthquake of February 9, 1971", Journal of the Geotechnical Engineering Division, ASCE, Vol. 101, No. GT7, pp. 651-689, July, 1975a.

SEED H.B., LEE K.L, IDRISS I.M. and MAKDISI F.I., "Dynamic Analyses of the Slide in the Lower San Fernando Dam During the Earthquake of February 9, 1971", Journal of the Geotechnical Engineering Division, ASCE, Vol. 101, No. GT9, pp. 889-911, September, 1975b.

SEED H.B., TOKIMATSU K., HARDER L.F. and CHUNG R.M., "Influence of SPT Procedures in soil Liquefaction Resistance Evaluations", Journal of Geotechnical Engineering, ASCE, Vol. 112, No. 11, pp. 1016-1032, November, 1986.

SEED H.B., SEED R.B., HARDER L.F., and JONG H.L., "Re-evaluation of the Slide in the Lower San Fernando Dam in the Earthquake of February 9, 1971", Report No. UCB/EERC-88/04, University of California, Berkeley, 1988.

SEED R.B. and HARDER L.F., "SPT-Based Analysis of Cyclic Pore Pressure Generation and Undrained Residual Strength", Proceedings, of the H. Bolton Seed Memorial Symposium, University of California, Berkeley, Vol. 2, pp. 351-376, May, 1990.

SHERARD J.L., CLUFF L.S. and ALLEN C.R., "Potentially Active Faults in Dam Foundations", Geotechnique, Vol. 24, No. 3, pp. 367-428, September, 1974.

SIDDHARTHAN R. and FINN W.D. Liam , "TARA-2: Two Dimensional Nonlinear Static and Dynamic Response Analysis", Soil Dynamics Group, University of British Columbia, Vancouver, Canada, 1982.

SPENCER E., "Thrust Line Criterion In Embankment Stability Analysis", Geotechnique, Vol. 23, No.1, pp. 67-85, 1973.

TAYLOR D.W., Letter to South Pacific Division, Corps of Engineers, San Francisco, CA, 14 April, 1953.

TAYLOR D.W., Letter to Dana Leslie at the South Pacific Division, Corps of Engineers, San Francisco, CA, 20 May, 1953.

TERZAGHI K., "Erdbaumechanik auf Bodenphysikalischer Grundlage, Vienna, Deuticke, 399 pg., 1925.

THOMAS J., "Static, Cyclic and Post-Liquefaction Undrained Behavior of Fraser River sand," MASc Thesis, Dept. of Civil Engineering, University of British Columbia, October, 1992.

U.S. ARMY ENGINEER DISTRICT, VICKSBURG, "The Sardis Earthquake Study", Supplement No. 1 to Design Memorandum 5, Earthquake Resistant Remedial Measures Design for Sardis Dam, 1988.

VAID Y.P, CHUNG E.K.F. and KUERBIS R.H., "Stress Path and Steady State", Canadian Geotechnical Journal, Vol. 27, No. 1, 1989.

VAID Y.P. and THOMAS J., "Post-Liquefaction Behavior of Sand", Proceedings, 13th International Conference of Soil Mechanics and Foundation Engineering", New Delhi, 1994 (in press).

VRYMOED J., "Dynamic Analysis of Oroville Dam", Office Report, Department of Water Resources, State of California, 1975.

WAHL R.E., CRAWFORTH S.G., HYNES M.E., COMES G.D., and YULE D.E., "Seismic Stability Evaluation of Folsom Dam and Reservoir Project", Report No.8, Mormon Island Auxiliary Dam - Phase II, T.R. GL-87-14, U.S. Army Engineer Waterways Experiment Station, Vicksburg, Mississippi, October, 1988.

WANG W., "Some findings in soil liquefaction", Water Conservancy and Hydroelectric Power Scientific Research Institute, Beijing, China, August, 1979.

WILSON J.A., ENGEMON W.O., MCLEAN F.H. and HENSLEY P.J., "Utilization of Economical Slopes For Jordanelle Dam", Proceedings, ASCE Specialty Conference on Stability and Performance of Slopes and Embankments - II, Geotechnical Special Publication No. 31, Eds. R.B. Seed and R.W. Boulanger, ASCE, New York, NY, Vol. I, pp. 653-668, July, 1992.

YOUD T.L., "Ground Failure Displacement and Earthquake Damage to Buildings", Proceedings, Conference on Civil Engineering and Nuclear Power, 2nd, Knoxville, TN, 1980.

YOUD T.L., "Ground Failure Damage to Buildings During Earthquakes", Foundation Engineering: Current Principles and Practices, Vol. 1, Proceedings, Congress sponsored by the Geotechnical Engineering Division of ASCE, Evanston Illinois, pp. 758-770, 1989.

YOUD T.L., and PERKINS, D.M., "Mapping Liquefaction-Induced Ground Failure Potential", Journal of the Geotechnical Engineering Division, ASCE, Vol. 104, N. GT4, pp. 433-446, 1978.

ZIENKIEWICZ O.C., CLOUGH R.W. and SEED H.B., "Earthquake Analysis Procedures For Dams, State of the Art", ICOLD/CIGB Bulletin No. 52, International Commission on Large Dams, Paris, 148 pg., 1986.

ZIENKIEWICZ O.C., CHAN A.H.C., PASTOR M., PAUL D.K., AND SHIOMI T., "Static and Dynamic Behavior of Soils: A Rational Approach to Quantitative Solutions, Part I: Fully Saturated Problems", Proceedings, The Royal Society, London, pp. A429, 285-309, 1990a.

ZIENKIEWICZ O.C., XIE Y.M., SCHREFLER B.A., LEDESMA A., AND BICANIC N., "Static and Dynamic Behavior of Soils: A Rational Approach to Quantitative Solutions, Part II: Semi-Saturated Problems", Proceedings, The Royal Society, London, pp. A429, 311-321, 1990b.

Developments in Geotechnical Engineering, Balasubramaniam et al. (eds) © 1994 Balkema, Rotterdam, ISBN 90 5410 522 4

Civil engineering use of industrial waste in Japan

Masashi Kamon & Takeshi Katsumi
Disaster Prevention Research Institute, Kyoto University, Japan

Synopsis : Industrial waste management should be conducted rationally based on sound environmental geotechnology. Utilization of the industrial wastes as construction materials has been recommended, and many attempts on geotechnical waste utilization have been undertaken. This paper firstly introduces the recent governmental policy including the Japanese legal system and the present condition on industrial waste generation and management in Japan. An overview of researches and developments in last decade on utilization of various types of industrial wastes is provided systematically. Treatment techniques and utilization of surplus soil and waste sludge/slurry from construction works are discussed in detail. The paper concludes by mentioning the future outlook and our duties on waste utilization for a better environment.

1 INTRODUCTION

Japan has a dense population especially in the megalopolis, high degree economy and industry, and a large amount of consumption, while it is narrow in space and poor in natural resources. Subsequently, the problem on waste management has been very serious; Various types of wastes have been generated in quantity and a large amount of them are not being utilized, but disposed of in the limited disposal site which will be exhausted in the near future. Material balance in Japan shown in Fig. 1 exhibits that a large amount of virgin resources of both domestic and foreign are invested, and recycled resources are only 8% of total material investment. Especially, in the construction works, large quantities of materials are used, and we have many types of waste. During the excavation in urban area for the embedded lifelines, subways, etc. the surplus soil and the waste slurry/sludge are generated in quantities. Considering these situations, resource saving and waste disposal reduction must be undertaken by means of waste utilization from a standpoint of construction industry.

Utilization of waste as construction materials is advocated to be an effective strategy based on a viewpoint of the environmental geotechnology (Kamon 1989; Kamon et al. 1991) and a variety of researches and developments on waste application by proper treatment method has been carried out in Japan. This paper presents the current state on the waste utilization in construction works, governmental policy promoting waste utilization, utilization technic of various types of industrial wastes, and treatment method and applicability of surplus soil and waste sludge/slurry discharged from construction works in large quantities.

2 GOVERNMENTAL POLICY AND PRESENT STATE ON WASTE MANAGEMENT

2.1 *Legal System and Governmental Policy*

Public Nuisance Countermeasures Basic Law (PNCBL, 1967), which is the basis of Japanese legal system on environmental protection, prescribes the obligations of business operators, regulations on emission, preservation of nature, regulations of land use, subsidies and settlements, relating to environmental preservation. PNCBL dominates almost all of environmental laws, such as Water Pollution Control Law, Law for the Preservation of Natural Environment. Because the objectives of this law are passive and only for environmental protection from public nuisance, the enactment of a new substitute law named Environmental Basic Law has been planned which aims at establishment of the social responsibility that would reduce the burden to the environment and has an ability to promote sustainable development.

Waste Disposal and Public Cleaning Law (WDPCL, 1970) prescribes the frame on waste generation and management. The sphere of WDPCL is shown in Fig. 2. The law defines waste and classifies it into two groups based on its generating origin as shown in Fig. 3. Industrial waste and municipal waste are commanded to be managed by enterpriser and local government, respectively. Table 1 shows the classification of industrial wastes in detail. Two categories of Special Controlled Wastes were established based on the environmental impact when the law was revised fundamentally in 1992. The objective of the revised law is to secure a better public health by

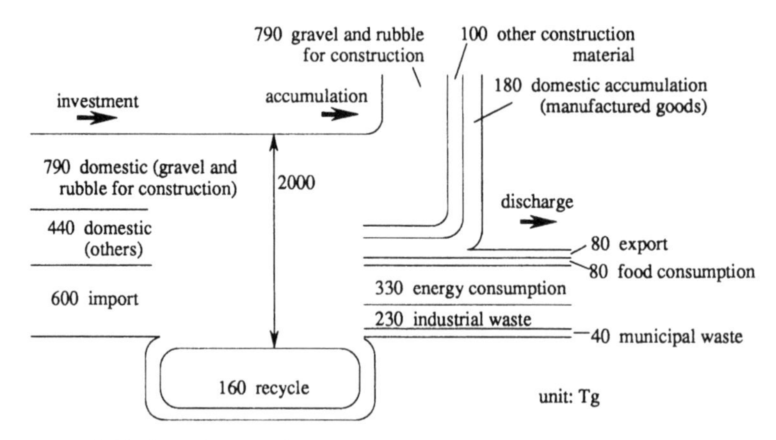

Fig. 1. Material balance in Japan in 1987 (Res. Comm. on Recycle, Environ. Agcy. 1991).

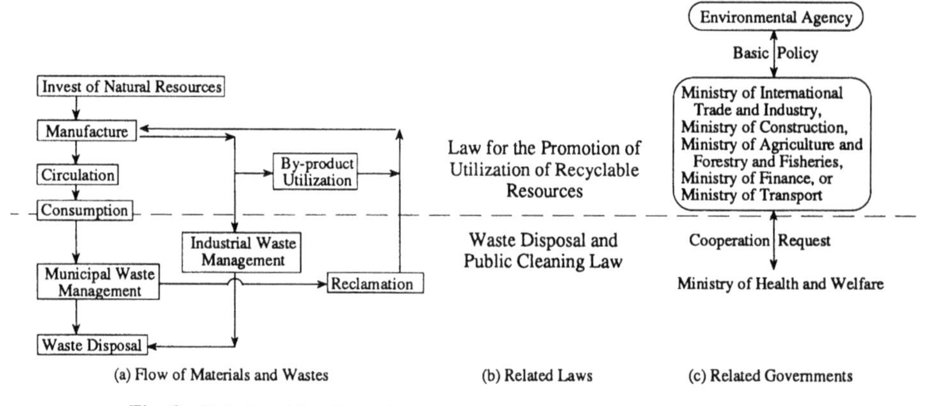

Fig. 2. Relationship of two laws on waste utilization and management.

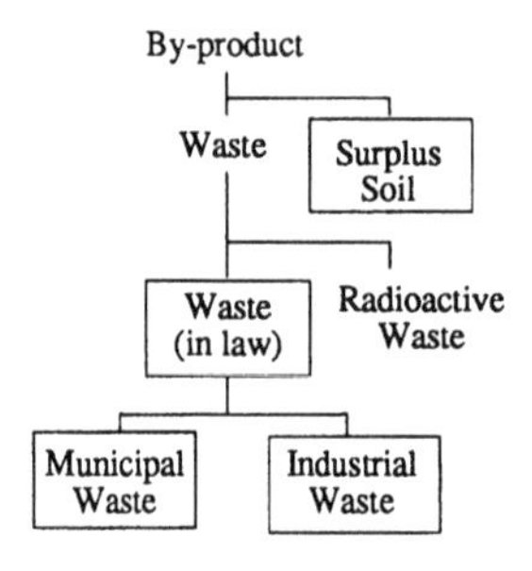

Fig. 3. Classification of waste in the law.

Table 1. Generation and recycle of industrial waste in 1990 (Environmental Agency 1993).

Type of Waste	Generation (Gg)	Recycling Ratio (%)
combustion residue	2,678	17.5
sludge	171,450	11.5
waste oil	3,471	24.5
waste acid	2,674	26.5
waste alkali	1,547	8.0
waste plastics	4,334	31.2
waste paper	94	63.2
waste wood	8,533	43.3
waste textile	5,295	51.1
animal and plant residues	3,543	31.3
rubber trash	1,193	12.5
metal trash	6,573	93.0
waste glass and ceramic	99	50.8
slag	42,507	84.7
construction waste	54,798	16.0
animal excreta	77,208	
dead animals	28	
dust	7,491	78.0
waste treated for disposal	1,218	7.4
Total	394,736	

means of reduction of waste generation and proper waste treatment. The problem of this law on waste utilization is that wastes improved for recycle must be named still wastes.

Law for the Promotion of Utilization of Recyclable Resources (LPURR) established in 1991 aims at sound development of the state economy by means of security of effective use of limited resources and promotive utilization of reclaimed resources. This law lays down Designated By-products (shown in Table 2), Specified Industry (paper industry, glassware manufacture, construction) and Designated Manufactured Goods (automobile, air-conditioner, TV-set, refrigerator, washing machine, can), because these by-products or goods from these industries should be reused in order to save limited resources. WDPCL and LPURR have a close relation and are inseparable for the creation of better environment by proper waste management, as shown in Fig. 2.

Table 2. Designated by-product in LPURR.

Type of By-product	Industry
Slag	Iron and Steel Industry
Coal Ash	Electric Power Generation Work
Surplus Soil Waste Concrete Waste Asphalt-Concrete Waste Wood	Construction Work

Other laws on waste management and utilization are introduced as follows: Act for Promotive Arrangement of Specified Facilities Relating to Industrial Waste Management (1992) intends to arrange the series of facilities in order to treat industrial wastes effectively and properly; Land disposal and ocean disposal must be undertaken to abide by Command on Technical Standard Relating to Waste Disposal Site (1977) and Law for the Prevention of Marine Pollution (1970), respectively. Considering the present state of waste generation from construction industry, Treatment Guideline of Construction Waste (TGCW) has been established by the Ministry of Health and Welfare in 1990, and about 160 local governments laid down the local laws of their own about the treatment and utilization of construction wastes.

Governmental ministries on waste management are shown in Fig. 2. Environmental Health Bureau, which is one main office of Ministry of Health and Welfare, carries out the tasks on waste management and prevention of pollution. Environmental Agency was established in 1971 to forward universal administration on environmental preservation. Ministry of Construction and Ministry of International Trade and Industry are expected to contribute resource-saving and by-product utilization, because they have to take responsibility in consuming a quantity of resources, manufacturing a mass of goods, and discharging a large amount of waste materials.

2.2 *Generation and Management of Waste*

Figs. 4, 5 and Table 1 exhibit the state of industrial waste generation. Total waste generation increases because of the remarkable increase of sludge and construction waste. While waste generations from iron and steel industry, mining, paper industry and chemical industry decreased from 1980 to 1985, construction works and lifeline works discharged much more wastes in 1985 than in 1980. Because the recent development and prosperity promoted construction works and many of the structures have been reconstructed for the redevelopment of urban areas these days. Table 3 shows the wastes from each industries. The main wastes are slag from iron and steel industry, sludge from chemical, paper and glass industries, and coal ash from electric supply works. From construction works, a large amount of waste sludge/slurry and surplus soil are also discharged from foundation works and dredging, as well as the construction rubbish.

As shown in Fig. 6, 30% of the discharged wastes disappears by intermediate treatment and 40% are reused owing to the technical development in waste management and utilization. The intermediate treatment methods are incineration, dehydration and melting. Incineration method which realizes volume reduction and sanitary resolution against harmful substances are spreading widely. Sludges which have high water content are

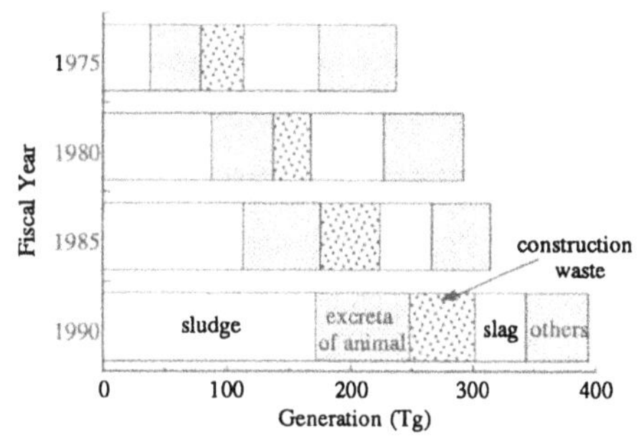

Fig. 4. Change of waste generation from industries (Environ. Agcy. 1993).

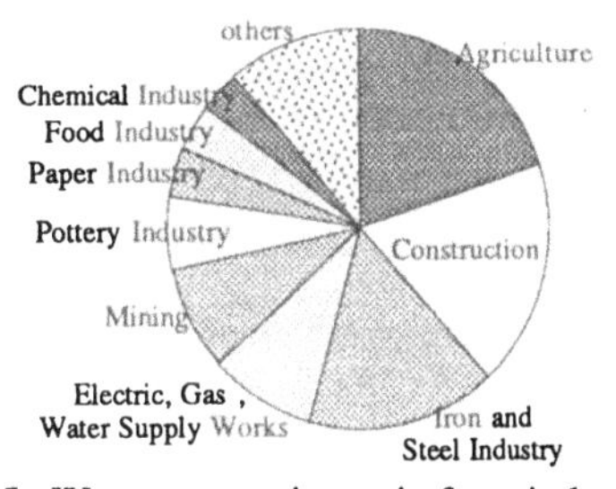

Fig. 5. Waste generation ratio from industries (Fiscal year : 1985, Total generation : 312.3 Tg, Environ. Agcy. 1993).

Table 3. Waste generated from manufacture (Ministry of International Trade and Industry 1991).

Industry	Waste Generation
Iron and Steel Industry	slag : 36160 Gg, dust : 4760 Gg, sludge : 520 Gg (recycling ratio, converter slag : 80%, electric furnace slag : 65%)
Paper Industry	sludge : 1150 Gg, boiler ash : 800 Gg, paper scale : 390 Gg (disposal : 2510 Gg)
Chemical Industry	sludge : 10810 Gg, waste acid : 2080 Gg, waste oil : 250 Gg (disposal : 2510 Gg)
Glass Industry	sludge : 2050 Gg (disposal : 120 Gg)
Textile Industry	waste textile, sludge
Nonferrous Metal Industry	slag : 2100 Gg (recycling ratio : 82%)
Electric Supply Work	coal ash : 3570 Gg, sludge : 1010 Gg (coal ash disposal : 1910 Gg)
Automotive Industry	slag : 540 Gg, waste oil : 80 Gg, waste plastics : 50 Gg (disposal : 320 Gg)
Semiconductor Industry	sludge : 50 Gg, waste alkali : 70 Gg, waste oil : 20 Gg
Oil Refinery Industry	sludge : 180 Gg, waste oil : 70 Gg, dust : 60 Gg

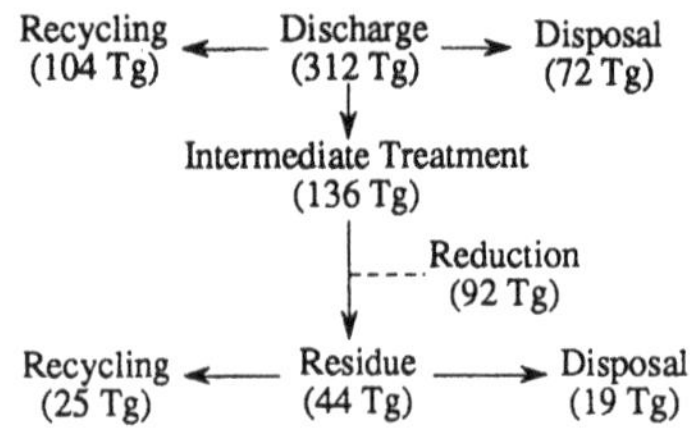

Fig. 6. Management flow of industrial wastes in 1985 (Ofc. Industrial Waste Management, Min. Health and Welfare 1992).

reduced in volume by dehydration as well as by incineration. Melting method has been mainly developed in treating sewage sludge for utilization purpose. However, construction rubbish from construction works are scarcely reduced by intermediate treatment. Table 1 shows recycling ratios of industrial wastes. Many of slag and dust are reused positively. But sludge and destruction rubbish generated in a large quantity have not yet been well utilized. Waste generation of 30%, except for the reused and the reduced, are disposed of in disposal sites (Fig. 6). Landfill sites in Japan are classified into three categories; Least Controlled Landfill Site for non-hazardous wastes, Controlled Landfill Site for organic refuses and non-hazardous wastes, and Strictly Controlled Landfill Site for hazardous wastes. The limited disposal sites are expected to be filled with waste materials in the near future and many local governments are struggling to secure a new disposal site, while the state government has planned the large-scale reclamation nearby Tokyo and Osaka Bays for disposals of the waste or surplus materials. Illegal dumping and improper treatment of waste, especially sludges and destruction rubbishes from construction are recent serious social problems.

3 UTILIZATION TECHNIQUE OF INDUSTRIAL WASTE

Industrial waste materials should be classified based on the process of generating industry. From the viewpoint of waste utilization, they needed to be grasped founded on their career and the relating characteristics. For example, from the treatment process, the waste materials are divided into three groups; wastes generated from incineration or melting (coal ash, iron slag, incinerated ash), wastes generated by crush (waste concrete powder, waste rock powder) and wastes generated as-is without any treatment (waste sludge, waste oil). Residues are always generated by incineration or melting which results in thermal power generation, the iron and steel refine work, or the incineration treatment of waste sludge and municipal waste. The characteristics of these wastes depend on their raw materials, the incineration temperature and time and the system of boiler. They are classified roughly into fly ash collected from fuel gas, bottom ash left at the bottom of boiler, and slag produced by melting. The second group wastes are generated from construction works in large quantities. The wastes categorized in the first and the second group are considered to be stabilized by compaction or chemical additives, like solidifiers, which are to be utilized as road materials or embankment according to the stabilization effect. The last group contains waste sludge, waste oil, waste plastics and so on. The treatment of these wastes are very difficult because of various technical and economical reasons.

Various cases of waste utilization have been researched and developed as shown in Table 4. Considering waste utilization, it is important to grasp the waste characteristics and the generating conditions. The former includes, whether the waste material is inorganic or organic, whether it contains heavy metals or not, and so on. The latter means when, where and how much the waste materials are generated. The main waste utilization cases are mentioned as follows.

3.1 *Coal Ash*

From the viewpoint of stable energy supply, electric supply workers have been reexamining the employment of thermal power plant. Therefore, the generation of coal ash which is by-product from the plant is about 5,000 Gg/year, and is expected to reach the level of 10,000 Gg/year in 2000s in Japan. Coal fly ash occupies 80% of coal ash generation, and recycling of coal fly ash is one of the main themes of all kinds of waste utilization. At present, 30% of coal fly ash is utilized in cement manufacturing, as the substitute raw materials of clay in making cement or the mixture of blended cement called fly ash-cement, but all the rest are disposed of without being reused. Recently, because the production of poor quality coal ash increases due to the use of various types of raw coal and the employment of

Table 4. Present grade of utilization of various wastes.

Type of wastes		Cement material				Road material			Soil material					Brick	Liner	Other waste treatment	Merits	Demerits
		Raw material	Blended cement	Soil stabilizer	Aggregate	Asphalt pavement	Base course	Subgrade	Filling-up	Embank-ment	Recla-mation	Caisson filler	Back-fill					
Coal ash	Pulverized coal fly ash	1	1	2	2	1	2	2	1	2	1	1	2	2		2	A2,A3,B1	
	Pulverized coal clinker ash			2	2		1	2	2	2		3	3	2			A1,A2	
	Fluidized bed combustion coal ash			2			2	2	3	2		3	3		3	2	A2,B1	U
Slag	Blast furnace slag	2	1	2	1		1	2	2	2				2			B1,B2	H
	Converter furnace slag	2		2		1	2	2	2	2							B1	E,H
	Electric furnace slag			2			2		3	3							B1	E,H
Sewage sludge	Sewage sludge incineration ash (by lime flocculants)	2		2		2	1	3	1	1	3			2			B1	H
	Sewage sludge incineration ash (by polymer flocculants)				2		1		1	3	3			1			B2	H
	Dehydrated sludge	2			2		4	4		3	3			3			B2	H,G
Waste sludge	Pulp sludge incineration ash					2	2	2		3	3	3	3	2		2	A2,B1	D
	Others	3		2										3			B1,B2	H
	Waste rock powder	1		2	2		3	3	1	3	3			4				
	Waste concrete powder			2			2	2	3	3	3			4				
	Waste soil	4	4	4	4	4	2	1	1	2	2							
	Waste slurry	4	4	4	4	4		2	2	2	2				3		A3	S
	Municipal waste incineration ash				4		3							3				H,D
	Cement kiln dust	1		2			2									2	B1	
	Waste oil	4	4	4	4	3	3			3	3	4	4	4	4		B3	G
	Waste plastic	4	4	4	4	2			4	4	4	4	4	4			B3	C
	Waste expanded polystyrol	4	4		4				2	2		3	2	4			B1	C

1) Grade of utilization, 1: utilized, 2: confirmed for utilization, 3: can be considered for utilization and 4: can not be considered for utilization.
2) Merit for utilization, A1: permeability, A2: light weight, A3: flow ability, B1: hydration characteristic, B2: baking characteristics and B3: containing oil.
3) Demerit for treatment utilization, H: containing heavy metals, D: containing dioxin, U: containing unburned carbon, C: chemical durability, G: production of gas or smell, S: soft condition and E: expansion characteristics.

the combust conditions which considers environmental impact, it is expected that coal ash should be reused as ground materials in large quantities.

Pulverized coal combustion system is most popular of coal combustion methods in electric generation plants. The pulverized coal fly ash is only used as a raw material for cement; Japanese Industrial Standard (JIS) has established the criteria for the selection of the coal ash. Because of the presence of silica which leads to the hardening reaction with lime, the utilization of coal ash as soil stabilizer with/without admixtures have been researched. One of the important characteristics of the coal ash is lightweight, and it is more expected to be used as lightweight soil materials, such as embankment or caisson filler. Table 5 shows actual results in construction of 67m-diameter and 15m-height man-made islands in Hakucho Ohashi (1380m-length of bridge) project (Kawasaki et al. 1992). Slurry disposal system by double mixing method, where coal fly ash is mixed with water doubly, is confirmed to be effectively used (Table 6).

Fluidized bed combustion method is spreading widely as independent electric power plant of chemical industries or iron and steel manufacturing plants. Because the system causes less air pollution than the ordinary method such as pulverized coal combustion system, and various kinds of raw coal can be used. The coal fly ash generated from fluidized bed combustion boiler, whose production is now only about 600 Gg/year, is expected to increase significantly. Table 7 shows the comparison of properties between pulverized coal fly ash and fluidized bed combustion coal fly ash. Because the fluidized bed combustion coal ash contains gypsum and lime due to the use of desulpherizer in the boiler, the utilization of the coal ash as soil stabilizer has been researched (Kamon and Katsumi 1994). Low combustion temperature at 800°C to 1000°C for prevention of air pollution leads to the generation of the coal ash containing a large amount of unburned substances, and the coal ash has the adsorptive ability of heavy metals and agricultural chemicals. Therefore, it is considered to be used as liner materials in the waste disposal site (Fujiwara et al. 1992).

Table 5. Properties of coal ash slurry.

Properties	Speci-fication	Slurry prepared Mean value	Standard deviation
Slump (cm)	8-13	10.30	0.69
Wet density (Mg/m^3)	1.60-	1.60	0.073
Strength (MPa) 28 days	0.62-	0.94	0.22
Bleeding ratio (%)	0-3	2.10	1.26

Table 6. Comparison of coal ash disposal system.

	Wet Disposal	Dry Disposal	Slurry Disposal
Slope of formed ground (degree)	0	0-5	0-20
Dry density (Mg/m^3)	0.75	0.85	1.07
ρ_d/ρ_{dmax} (%)	65	74	93
Compressive strength (kPa)	20	50	150
Leaching substances ratio (%)	65	50	6

Table 7. Comparison of chemical composition of coal fly ash.

	Pulverized Coal Fly Ash	Fluidized Bed Combustion Coal Fly Ash
SiO_2 (%)	50-55	25-40
Al_2O_3 (%)	25-30	15-25
Fe_2O_3 (%)	4-7	1-3
CaO (%)	4-7	10-30
MgO (%)	1-2	1-2
K_2O (%)	0-1	0-1
Na_2O (%)	1-2	0-1
SO_3 (%)	0-1	3-8
Ig-loss (%)	1-2	10-30

3.2 *Iron and Steel Slag*

The generation of Slags from the metallurgical industry has reached 40,000 Gg/year, and more than 85% of them are reused as road material, cement material, fertilizer, pottery material and soil stabilizer. Slags can be classified as blast furnace slag, converter furnace slag and electric furnace slag. The blast furnace slag and the converter furnace slag are produced through the process that iron is made of iron ore, and the electric furnace slag is generated from the steel making process using scrap iron as main raw material. While the production of the blast furnace slag (24,800 Gg) and the converter furnace slag (9,500 Gg) have decreased in recent years, the generation of the electric furnace slag (2,600 Gg) is at rise.

Blast furnace slag exhibits the hydration characteristics when mixed with lime or sulfate. 50% of blast furnace slags are positively used as materials of blended cement, namely blast-furnace-slag-cement, according to JIS. Blast-furnace-slag-cement has an excellent workability in constructing concrete and the concrete made of blast-furnace-slag-cement has more chemical durability against sea water, sulfate or acid. 25% of generated blast furnace slags are utilized as road subbase materials. The selection of the materials is prescribed in Asphalt Pavement Outline by the Road Association of Japan. The remaining 25% are reused as concrete aggregate or soil materials, so the by-product is not regarded as a waste but a resource.

Converter furnace slag exhibits the expansive characteristics in hydrating, which is the trouble in being utilized. It is well known that converter slags expand by the volume increase associated with the hydration of CaO. The slags with the expansion characteristic eliminated by aging have been reused as a subbase or base materials, as gravel in asphaltic concrete, and for soil stabilization.

Electric furnace slags also have the efflorescence and expansive characteristics like the converter

Table 8. Investigation of slag subbase by falling weight deflectmeter.

Case	Traffic Classi-fication	Age (month)	Thickness of Sub-base (cm)	Elastic Modulus (MPa)	
				Blast Furnace Slag	Electric Furnace Slag
1	B	10	15	1900	2500
2	C	14	15	3550	3400
3	C	12	25	2300	3900
4	C	11	10	1300	2000

Traffic Classification : B and C mean that truck traffic per day are 250-1000 and 1000-3000, respectively.

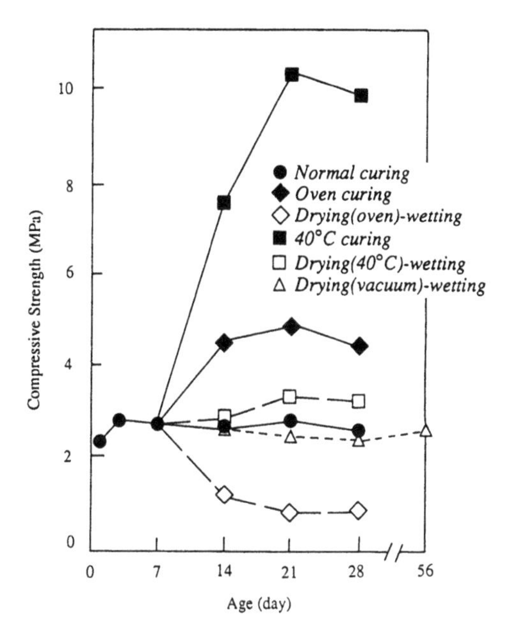

Fig. 7. Strength and durability of stainless-slag stabilized by 12 % stabilizer and 5 % kaolinite.

slags, so most of this production is disposed of in reclamation area. But recently, it has been proven that they can be reused as construction materials by treatment of aging (Kuwayama et al. 1992). Table 8 shows actual results when the electric furnace slags are reused tentatively as subbase materials in Osaka Prefecture (1993). Some types of electric furnace slags form the hydration products of calcium silicate hydrate and hydrated gehlenite in long term, so they do not only continue the volume increase but exhibit the hydraulic properties. It has been clarified that the slags are stabilized by cement admixture effectively enough to be utilized as road materials as shown in Fig. 7 (Kamon et al. 1993), and the slags blended with cement are proposed to be available for soil stabilization (Kamon and Nontananandh 1990).

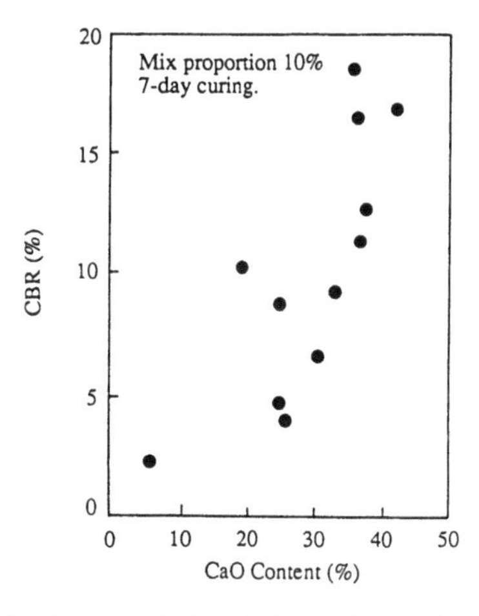

Fig. 8. Characteristics of clay soils stabilized by sewage incineration ash.

3.3 *Waste Sludge*

Various types of wastes sludge are generated from various kinds of industry. Among them is sewage sludge. Because sewage works are conducted in almost all parts of the city, a large amount, 240,000,000 m^3/year, of sewage sludge is produced. The sludge is reduced to 2,000,000 m^3/year of incineration ash and dehydrated residues by the intermediate treatment. Sewage sludge incineration ash are divided according to the flocculants used, such as lime and polymer flocculants. Some actual results exhibit that incineration ashes by lime flocculant is suitable for subbase, filling-up and embankment. This ash has similar characteristics to sandy soil, and has hygroscopic and hardening properties due to the remaining lime. Therefore, it has been clarified that the ash exhibits the improvement effect of soil stabilization in relation to lime content in Fig. 8 (Mazuda et al., 1991). On the other hand, incineration ashes by polymer flocculant are similar to clay soil so they are used only as subgrade and subbase, and has been tried as substitute of clay materials in brick-making.

Some processing methods have been researched for sludge utilization. One of them is the production of molten slag by means of melting method. The slag made from sewage sludge is suitable as construction material and compatible with the environment and is confirmed to be useful as interlocking blocks (Iwai et al., 1987). The other method is to make hardening materials from sludges by incinerating treatment. It has been proposed that by combining industrial waste sludge with lime, the material is good for soil stabilization, as shown in Fig. 9. These methods are applicable enough from the environmental and geotechnical points of view, but have economical problems to be settled. The methods of using stabilizers and/or additives, such as cement-based stabilizers or clays, are useful and produce materials which are suitable for subbase, subgrade or embankment.

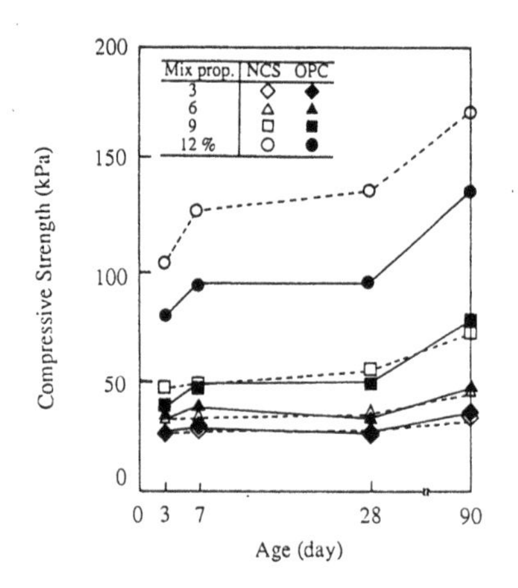

Fig. 9. Strength of loam soil stabilized with new cement like stabilizer (NCS) and ordinary Portland cement (OPC).

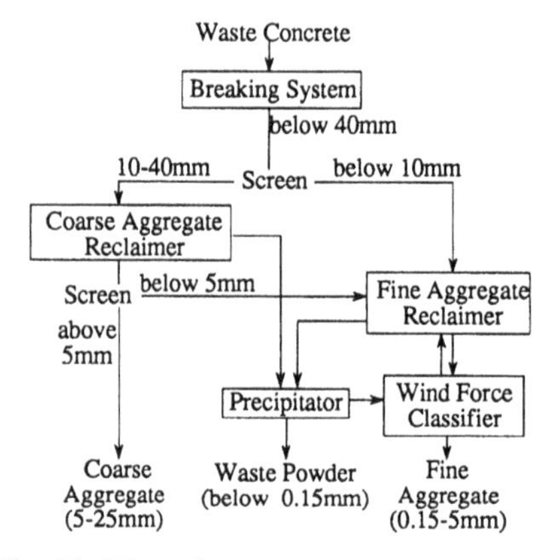

Fig. 10. Flow of waste concrete management.

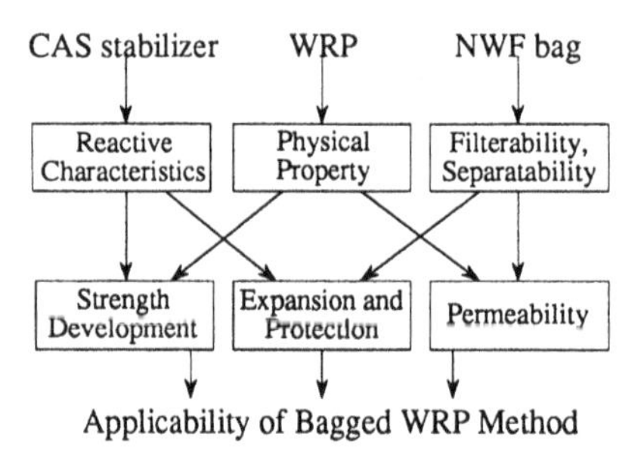

Fig. 11. Concept of Bagged WRP Method by waste rock powder.

3.4 *Waste Concrete and Waste Rock Powder*

Waste concrete is generated when concrete structures such as buildings and bridges are demolished, repaired or constructed; the production of concrete has reached the level of 150,000,000 m^3/year, and 25,000 Gg/year of waste concrete are generated. Waste concrete can be divided into waste concrete mass and powder. A little amount of waste concrete mass has been used as substitute materials of rubbles in road construction, but they must be used for concrete in large quantities in order to reduce the amount to be disposed of. Therefore, the system in which waste concrete mass can be reclaimed as concrete aggregate has been proposed (Fig. 10). In this system, it is difficult to produce the reclaimed aggregate suitable to JIS, and some kinds of equipment have been developed in order to remove the sticked cement component from the original aggregate (Honda and Yamada 1990). Another problem in this system is treatment of waste concrete powder generated also in producing the reclaimed aggregate. It has been proposed that waste concrete powder stabilized by hardening and additive materials can be utilized as road subbase (Kamon et al. 1992).

Due to the environmental constraint in the use of natural gravel, rubbles produced in crusher plants becomes alternative source, and 10,000 Gg of waste rock powder, by-product from the crusher plants, are generated annually. The characters of waste rock powder depend on the mother rock; the waste rock powder of limestone is used as a raw materials of cement due to its chemical composition, and silty sandstone is used as a filling material in exhausted diggings. The waste rock powder is non-hazardous in nature and valuable resource, however, suitable methods to utilize the waste rock powder in large quantities for construction are required and under discussion. The waste rock powder of sandstone, which has a large specific surface area and contains a large amount of amorphous materials increases the effect of lime stabilization of a soil in which a low amount of fine particle or amorphous materials is present (Nishida et al. 1992). The waste rock powder solidified by a newly developed stabilizer can have a potential utilization as permeable subgrade of road and back filling for retaining wall. A new method called "Bagged WRP Method" was developed. In this method, the non-woven fabric bags are filled with the dry mixture of the waste rock powder and the stabilizer which is solidified by soaking is proposed to be applied to sunk-levee materials or seafloor ground improvement, as shown in Fig. 11 (Kamon and Katsumi, 1994).

3.5 *Waste Plastics and Waste Oil*

Waste plastics and waste oil are difficult to be treated and disposed of, but recently variety of researches and developments have been carried out for utilization of the waste materials. Waste plastics are generated as both industrial waste and municipal waste. It is confirmed that some kinds of waste plastics are available effectively for asphalt aggregate of road pavement, as shown in Fig. 12 (Yamada and Inaba 1993). 15,000,000 m^3 of

Table 9. Amount of by-product from construction works (1990).

	Amount (/year)	Management Ratio (%)		
		Recycle	Reduction	Disposal
Surplus Soil	450,410,000 m^3	27.6	-	72.4
Waste Sludge	14,410,000 Mg	7.9	12.6	79.5
Waste Concrete	25,440,000 Mg	48.1	-	51.9
Waste Asphalt-Concrete	17,570,000 Mg	50.4	-	49.6
Mixed By-products	9,460,000 Mg	13.9	17.0	69.1

expanded polystyrene (EPS) are produced and they are abolished after use as much as two or three times. Small pieces of waste EPS are utilized as lightweight embankment materials by being mixed with soil and cement (Yamada et al. 1989). It has been shown that waste oil can be stabilized enough by using industrials wastes such as sludge incineration ash or iron slags with stabilizers, to be reused for ground materials (Sawa et al. 1994).

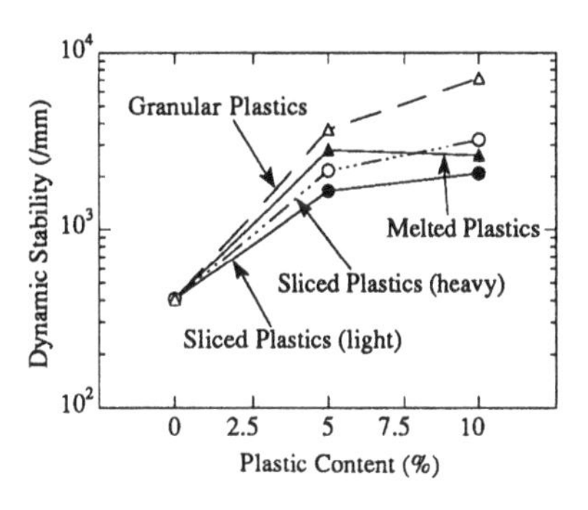

Fig. 12. Dynamic stability of asphalt mixture with waste plastics.

4 TREATMENT AND UTILIZATION OF SURPLUS SOIL AND WASTE SLUDGE/SLURRY FROM CONSTRUCTION WORKS

4.1 *Generation of By-products from Construction Works*

A large amount of by-product are generated from construction works, as stated above. Generation of the by-products, except for surplus soil and valuable coarse grain materials, have been considered as the wastes under WDPCL. Ministry of Construction started to carry out the examination of all by-products from construction works in 1990. Table 9 exhibits that the amout of surplus soil and waste sludge/slurry from foundation works leads to serious problems. Only 35% of generation are reused as reclaimed materials, but 60% of generation are disposed of not being utilized without any proper intermediate treatment (Kutara 1992).

Surplus soil of 450,000,000 m^3 and 14,000 Gg of waste sludge are generated from construction works and 70-80% of the generation are disposed of. Surplus soils have three kinds of destiny for utilization and disposal. Surplus soils which consists of soils in good quality such as sandy soil are utilized for filling-up or embankment without any treatment. Some kinds of surplus soils can be reused by means of soil improvement. The rest of surplus soil are disposed of because they can not be available as reclaimed materials with/without any treatment or effective treatment system are not established. Some amounts of waste sludge are reduced by intermediate treatment such as dehydration, but almost all of waste sludge and dehydrated sludge are reclaimed in disposal sites. One problem of waste sludge utilization is that the waste sludge treated for utilization purpose is regarded as waste in law and is obliged to be reclaimed in the area designated by laws. Another problem is that we cannot divide these by-products into valuable soil and waste sludge which leads to illegal dumping of large amounts of waste sludge. Nowadays, Treatment Guideline of Construction Wastes (TGCW, 1990) by Ministry of Health and Welfare has established the criteria (q_c = 2.0 kgf/cm² (= 196 kPa) or q_u = 0.5 kgf/cm² (= 49kPa)), according to which the classification of these by-products can be carried out.

4.2 *Utilization of Surplus Soil*

Criteria, methods and systems must be established in order to promote the by-products utilization as construction materials. For surplus soil utilization, utilization criteria were settled on LPURR, as shown in Table 10. Road embankment and reclamation embankment are the most general of utilization purposes. Recently, constructions of man-made island or big-scale levee are planned and a large amount of surplus soil is expected to be reused for their construction. Moreover, surplus soil of good quality can be utilized for road subbase, foundation of embankment, filling-up or back fill, and protection of slope, as well as some kinds of embankments.

Some soil stabilization methods are applied to surplus soil utilization. The most spreading method is by using chemical additives such as cement, lime and hygroscopic polymer and surface activator. Lime stabilization method is applied for soil utilization systems which was developed by Osaka City Government and Osaka Gas Co. Ltd. (Ninomiya et al. 1988). Another widely used is the

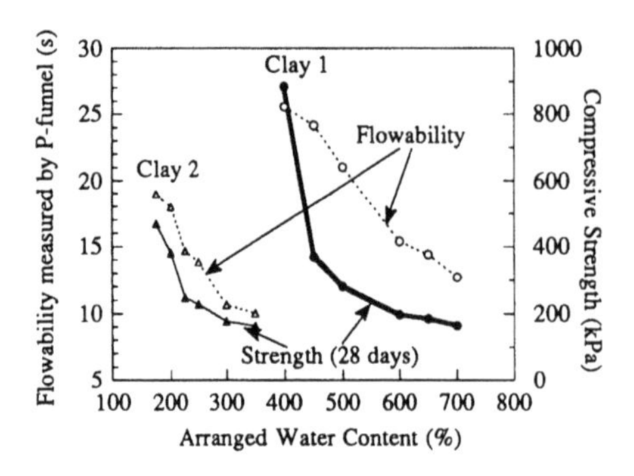

Fig. 13. Characteristics of soils stabilized by liquefied soil stabilization method.

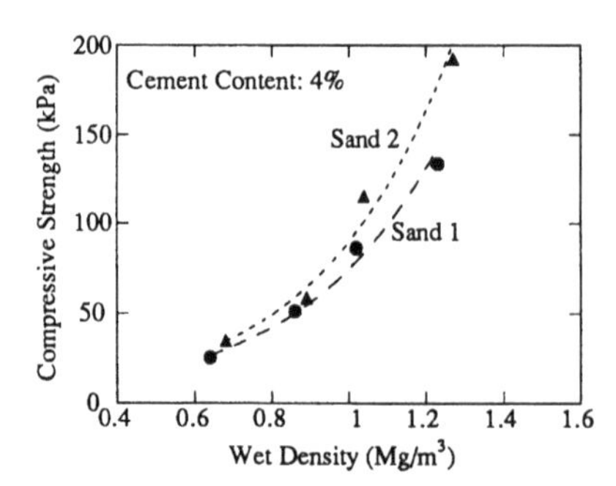

Fig. 14. Characteristics of soil stabilized with EPS and cement.

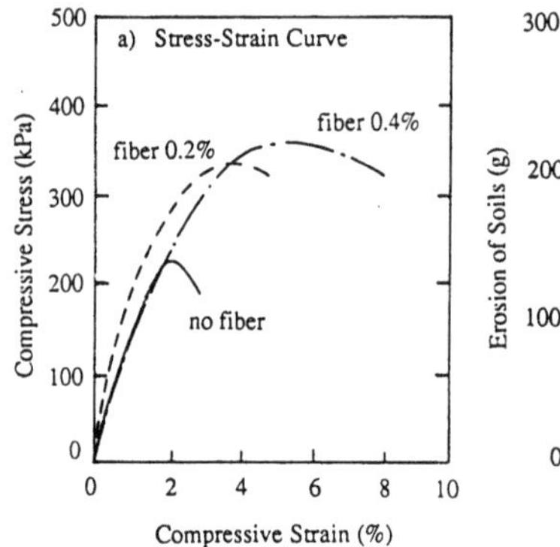

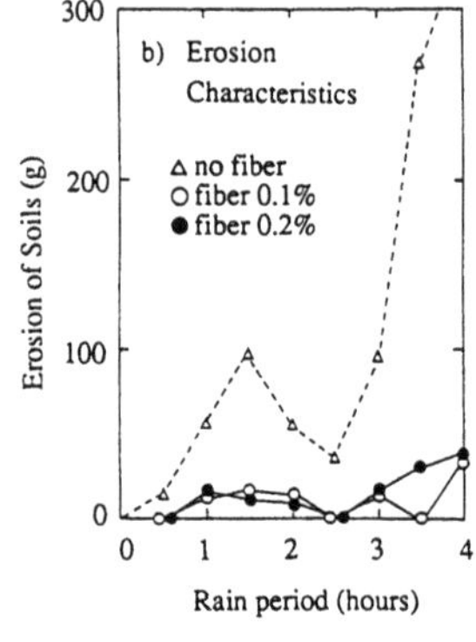

Fig. 15. Characteristics of sandy soil stabilized by fiber mixing method.

Table 10. Utilization criteria of construction surplus soil from LPURR.

Class	Content	Use
1st	sand, gravel, and the corresponding	back filling for construction work back-fill for structure road embankment fill for building lot
2nd	sandy soil, gravelly soil, and the corresponding	back-fill for structure road embankment river dike fill for building lot
3rd	clay soil which can be executed on, and the corresponding	back-fill for structure road subgrade embankment river dike fill for building lot water area reclamation
4th	clay soil, except for 3rd-class soil	water area reclamation

aging method where surplus soils are stored in a stockyard in long term and sometimes mixed up to be improved naturally. Recently some methods to increase the value of soil materials have been developed (Miki et al. 1992). Liquefied soil stabilization method, where the soil mixture blended with stabilizer and large amounts of water has flowability and hardening characteristics, are available for filling-up in underground pipe construction or backfill of retaining wall, in Fig. 13 (Kuno et al. 1992). Lightweight soil stabilization method, where surplus soils are mixed with lightweight materials such as EPS or foamed cement, is expected to be utilized for embankment and backfill. The soil mixture by this method has similar characteristics as general soil materials, differing from the use of EPS only. Utilization of waste EPS for the method are also discussed (Yamada et al. 1989). As shown in Fig. 14, the soil mixtures stabilized with the EPS and hardening materials have the wet density of 0.6-1.2 Mg/m^3 and the strength of 50-200 kPa. The method has been applied to lightweight embankment over soft ground and backfill of retaining wall. Fiber mixing method, where soils or stabilized soils are mixed with fibers (0.02-0.1mm of thickness and 3cm of length), can produce the persistent and durable soil materials against erosion or crack, as shown in Fig.15. Geotextile reinforced soil method is available for embankment construction by reclaiming surplus soil with high water content. Bagged soil method, where the non-woven fabric bags are filled with surplus soil, are being researched in order to construct underwater

Table 11. Effluent standard settled by law.

	Water Pollution Control Law	Sewage Law	Environmental Standard by PNCBL	
			River & Lake	Sea
SS (mg/l)	200	600	below 1 - below 25	-
pH	5 (5.8) - 9 (8.6)	5 - 9	6.5 (6.0) - 8.5	7.8 (7.0) - 8.3
BOD (mg/l)	160 (120)	600	below 1 - below 10	-
COD (mg/l)	160 (120)	-	below 1 - below 8	below 2 - below 8
Mineral oil (mg/l)	5	5	-	-
Animal oil (mg/l)	30	30	-	-

embankment or flexible bulkhead.

In order to manage the surplus soil utilization, the system must be established to have three components as follows; Setting up of plants for soil improvement in order to adjust the quality of soils, security of stockyard in order to regulate the amount and the time of supply and demand of soils, and arrangement of related data in order to conduct the optimum plan of utilization. The conceptual outline of utilization system is shown in Fig. 16. Some city governments and gas supply companies have developed the utilization system of surplus soil from 1980s. Since 1992, a utilization system, which has data-base system related to soil generation and demand, stockyard and plant for soil improvement, have been opened by Tokyo City Government (Maeda 1992).

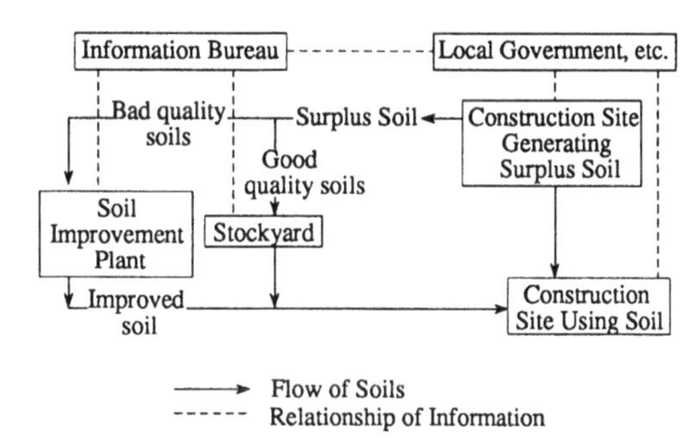

Fig. 16. Concept of management system of surplus soil.

4.3 *Utilization of Waste Sludge/Slurry*

Waste sludge/slurry or waste water, which can not be discharged into rivers and seas or can not be reclaimed for embankment as soil materials, are generated from foundation works in a large quantity. Waste slurry is a by-product of cast-in-place concrete pile method, continuous diaphragm walls method, shield tunneling method, and so on. Waste water is the water discharged from tunnel or the rain water collected in land development area. Waste slurry or water are generally treated by proper intermediate method, and the treated water can be released into river or sewage according to the environmental criteria in Table 11. Of the criteria, observance of SS (suspended solids), pH (potential of hydrogen), COD (chemical oxygen demands) and oil content is generally difficult. Soils and cakes produced by treatment are conveyed to be disposed of in landfill site, but the soils reaching the criteria of Treatment Guideline of Construction Wastes (q_c = 2.0 kgf/cm^2, or q_u = 0.5 kgf/cm^2) should be utilized as soil materials for embankment or backfill.

Fig. 17 shows the general flow of waste sludge treatment. Some processes can be omitted and some processes must be added according to the characteristics of waste slurry, environmental criteria and applicability of dehydrated soil or discharged water. The most important process in this flow is the dehydration, because the waste slurry or water is under the high water condition. It contains many fine particles and is difficult to dehydrate rapidly. In the case of slurry excavation methods in which bentonite or polymer such as Carboxy-methyl cellulose (CMC) are often used for regulating viscosity, these dispersants remain in the waste slurry and make it very difficult to dehydrate the slurry. In order to solve these problems, some kinds of flocculant chemicals and dehydration plants have been developed and utilized (Kita and

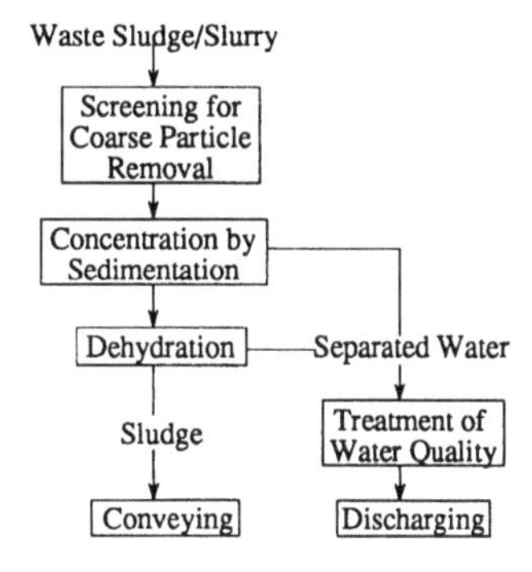

Fig. 17. Flow of waste sludge/slurry management.

Table 12. Various dehydration equipment.

Method	Dehydration mechanism	Working pressure	Management ability	Water content of dehydrated cake	Quality of discharged water
Belt-press	Mechanical	30-100 kPa	350-500kg/m·hr	100-200%	clear
Roller-press	Mechanical	30-100 kPa	350-500kg/m·hr	120-300%	clear
Filter-press	Sludge supply	300-500 kPa	7-25kg/m²·hr	50-100%	clear
High-pressure Filter-press	Sludge supply and Filtering	1.0-1.5 Mpa	5-20kg/m²·hr	40-80%	clear
Vacuum-dehydration	Suction	-30-60 kPa	15-20kg/m²·hr	150-300%	roughly clear
Screw decanter	Centrifuge	1500-3000 G	4-10t/hr	150-300%	not clear

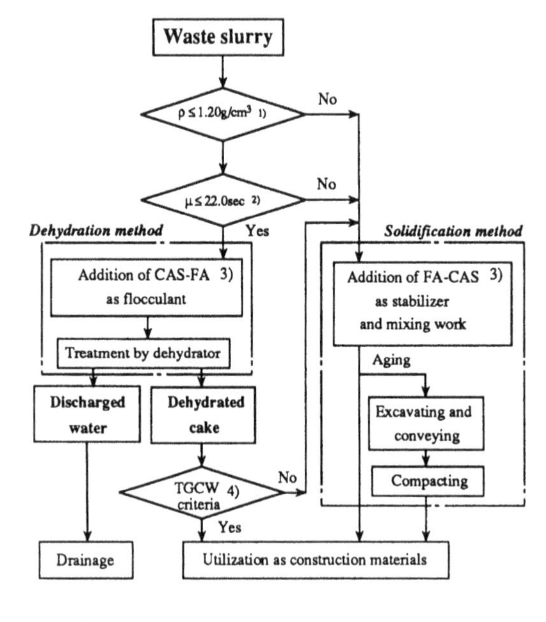

Fig. 18. Outline for utilization system of waste slurry.

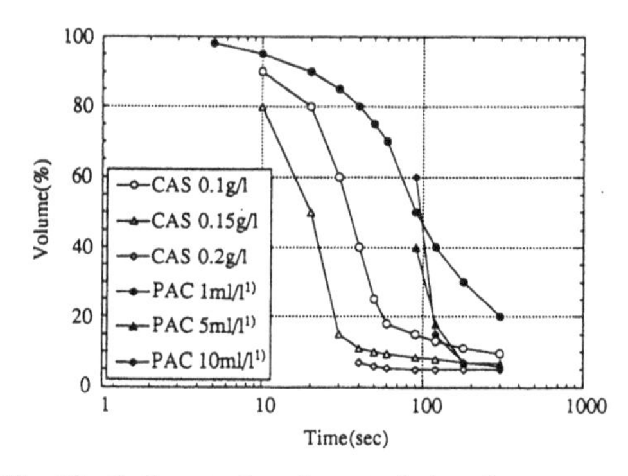

Fig. 19. Sedimentation characteristics of waste slurry with flocculants (Waste slurry; ρ = 1.003 g/cm³, '1)' ; with 10 ml/l addition of Polymer 0.01 % solution).

Tsuji 1981). Inorganic materials such as $Al_2(SO_4)_3$, PAC (Polyaluminium chloride) and polymer such as polyaclylamide are usually utilized as focculants. Dehydration plants developed are shown in Table 12. Filter-press method and Roller-press method are spreading widely.

A new utilization system of waste slurry is proposed by Kamon and Katsumi (1994) and the conceptual outline of which is shown in Fig. 18. It consists of dehydration and solidification methods and results in efficient treatment, decrease in volume, stabilization, and recycling as resources. The selection of treatment method is carried out based on the character of the waste slurry; the density (ρ) and funnel-viscosity (μ), universally measured to control the character of slurry at the excavation sites. The solid content which is indicated by the density, and the funnel-viscosity which is increased by the remains of bentonite and CMC, show the possibility or the effectiveness of dehydration treatment. Attempt on volume reduction by dehydrating a high solid content slurry is not always the best strategy from technical and economical aspects. The slurry with low density can be dehydrated easily, but the slurry with high viscosity is difficult to dehydrate though they have low density, because of the remains of dispersants . So the waste slurry which has higher ρ-value than 1.2 g/cm³ or higher μ-value than 22 seconds can be treated effectively in solidification method. In dehydration method, it is proposed that the waste slurry with Carbonated-Aluminate Salts (CAS) and fluidized bed combustion coal fly ash as flocculant should be dehydrated with the object of volume reduction (Fig. 19). CAS are newly developed and the mixtures of cement, $Al_2(SO_4)_3$, Na_2CO_3, $CaSO_4$, and so on. Especially by using high pressure dehydrator, the strength of dehydrated cakes can reach easily the criteria set by TGCW, therefore, they can be directly utilized as embankment and subgrade material (Fig. 20). Also the discharged

Table 13. Quality of waste water treated by CAS.

	Laboratry Test		Field Test	
	Untreated	Treated	Untreated	Treated
pH	7.0	6.9	7.6	7.4
COD (mg/l)	7.5	4.0	2.6	1.0
BOD (mg/l)	3.0	1.0	4.0	2.0
T-N (mg/l)	8.7	0.9	0.75	0.57
T-P (mg/l)	0.19	0.02	0.11	0.01
T-Fe (mg/l)	1.9	0.09	1.7	0.52
Turbidity (degree)	-	-	37	6

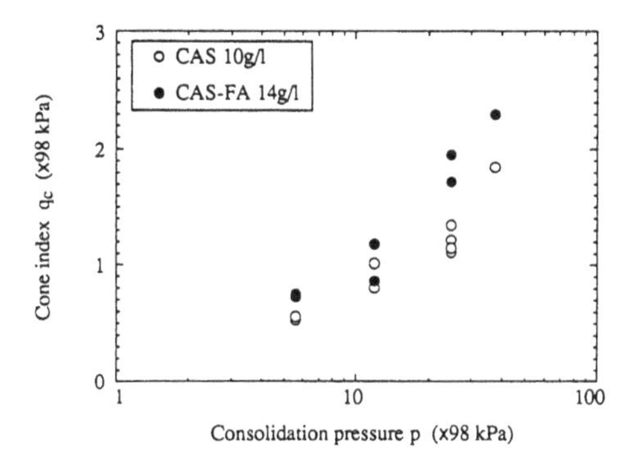

Fig. 20. Strength characteristics of dehydrated cakes (Waste slurry; $\rho = 1.060$ g/cm^3, $\mu = 19.2$ s).

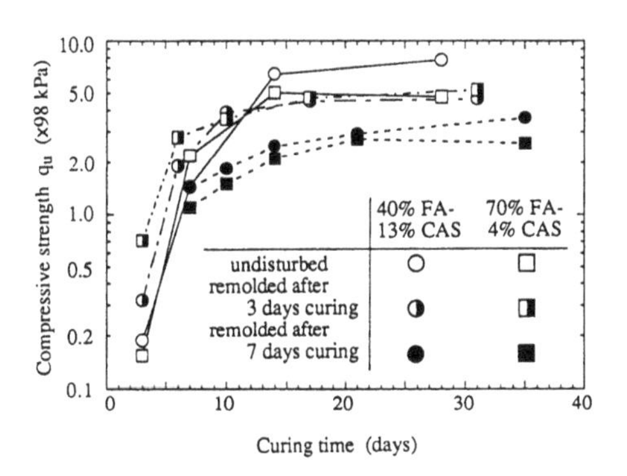

Fig. 21. Strengths of slurry-coal ash-CAS mixtures (Waste slurry; $\rho = 1.040$ g/cm^3, $\mu = 51.7$ s).

water satisfies the environmental standards in pH, SS and COD (Table 13). In solidification method, it is suggested that the slurry is stabilized by CAS and coal fly ash to increase the strength for embankment or subgrade purposes (Fig. 21). The use of coal fly ash can be very effective from both technical and economical point of view.

5 CONCLUSION - FUTURE ASPECT AND TASKS

Present situation on waste management and utilization in Japan were introduced and some attempts which have been developed and applied for by-product utilization were discussed in detail. Many technical methods are successfully achieved but some social problems still remain. The present social system, customs on construction and other industries, economical conception, etc, prevent many waste from being utilized in construction works. In order to constitute the social system for by-product utilization, we must arrange the related data systematically, establish the legal and tax system which gives advantage for by-product utilization, and courageously shift the present system to the ideal one.

REFERENCES

Civil Engrg. Ofc., Osaka Pref. 1993. *Application of Electric Furnace Slag to Subbase* (in Japanese).

Environ. Agcy. 1993. *A White Report on Environment* (in Japanese).

Fujiwara, Y., R. Miyazaki, M. Fukasawa & T. Sueka 1992. Contaminant Adsorption Properties of Coal Ash Containing High Content of Unburned Substance and its Utilization *Proc. 37th Symp. SMFE, JSSMFE*, pp.25-30 (in Japanese).

Honda, A. & M. Yamada 1990. *Treatment and Reutilization of Construction Wastes*: Energy-Saving Center (in Japanese).

Iwai, S., Y. Miura & T. Kawakatsu 1987. New Process of Sewage Sludge Treatment and Utilization of Slag for Construction Materials. *Environmental Geotechnics and Problematic Soils and Rocks, Balkema*, pp.447-459.

Kamon, M. 1992. Definition of Environmental Geotechnology. *Proc. 12th ICSMFE*, Vol.5, pp.3126-3130.

Kamon, M. & T. Katsumi 1994. Utilization of Waste Slurry from Construction Works. *Proc. 13th ICSMFE*, Vol.4, pp.1613-1616.

Kamon, M. & T. Katsumi 1994. Potential Utilization of Waste Rock Powder. *1st Inter. Cong. Environmental Geotechnics, ISSMFE and CGS* (in submitting).

Kamon, M. & S. Nontananandh 1990. Contribution of the Stainless-Steel Slag to the Development of Strength for Seabed Hedoro. *Soils and Foundations, JSSMFE*, Vol.30, No.4, pp.63-72.

Kamon, M. & S. Nontananandh 1991. Combining Industrial Wastes with Lime for Soil Stabilization. *Jour. Getech. Engrg. Div., ASCE*, Vol.117, No.1, pp.1-17.

Kamon, M., S. Nontananandh & T. Katsumi 1993. Utilization of Stainless-Steel Slag by Cement Hardening. *Soils and Foundations, JSSMFE*, Vol.33, No.3, pp.118-129.

Kamon, M., S. Nontananandh & S. Tomohisa 1991. Environmental Geotechnology for Potential Waste Utilization. *Proc. 9th ARC*, pp.397-400.

Kamon, M., S. Tomohisa, K. Tsubouchi & S. Nontananandh 1992. Reutilization of Waste Concrete Powder by Cement Hardening. *Soil Improvement, CJMR Vol.9, Elsevier Appl. Sci.*, pp.39-53.

Kawasaki, H., S. Horiuchi, M. Akatsuka & S. Sano 1992. Fly-Ash Slurry Island II. Construction in Hakucho Ohashi Project. *Jour. Mater. Engrg., ASCE*, Vol.4, No.2, pp.134-152.

Kita, D. & H. Tsuji 1981. Treatment and Utilization of Discharged Slurry from Slurry Excavation Methods. *Tsuchi-to-Kiso, JSSMFE*, Vol.29, No.11, pp.57-64 (in Japanese).

Kuno, G., M. Yoshihara, H. Ishizaki & Y. Omodaka 1992. Properties of Improved Surplus Soil by Liquefied Soil Stabilization Method. *Proc. 37th Symp. SMFE, JSSMFE*, pp.1-6 (in Japanese).

Kutara, K. 1992. The Present State and Future Trend on Disposal by By-products from Construction and Utilization of Recycled Materials. *Civil Engrg. Jour. Pub. Wks. Res. Inst., Min. Const.*, pp.32-39 (in Japanese).

Kuwayama, T., T. Mise, M. Yamada & A Honda 1992. Engineering Properties of Electric Furnace Slags. *Soil Improvement, CJMR Vol.9, Elsevier Appl. Sci.*, pp.85-100.

Maeda, M. 1992. Measures for Surplus Soil from Construction in the City of Tokyo. *Jour. JSCE*, Vol.77, No.6, pp.46-49 (in Japanese).

Masuda, T., T. Shiraishi, H. Miki, Y. Hayashi & Y. Ohshima 1991. Study on Soil Conditioning Effect of Incinerated Ash of Sewage Sludge. *Proc. Environ. & Sani. Eng. Res., JSCE*, Vol.27, pp.129-134 (in Japanese).

Miki, H., Y. Hayashi & N. Aoyama 1992. Development of New Techniques to Heighten the Value of Soil Materials. *Civil Engrg. Jour. Pub. Wks. Res. Inst., Min. Const.*, pp.58-65 (in Japanese).

Min. International Trade and Industry 1991. The Ideal of Waste Management and Reclaimed Resources in Future.

Ninomiya, T., R. Takamoto & T. Takano 1988. Reutilization of Waste/Surplus Soils in Road Construction. *Proc. JSCE*, No.397, pp.177-185 (in Japanese).

Nishida, K., S. Sasaki & Y. Kuboi 1992. Utilization of Waste Rock Powder for the Lime Stabilization of Residual Soil. *Soil Improvement, CJMR Vol.9, Elsevier Appl. Sci.*, pp.55-70.

Ofc. Industrial Waste Management, Min. Health and Welfare 1992. *Handbook of Industrial Waste Management* (in Japanese).

Res. Comm. on Recycle, Environ. Agcy. 1991. *New Age of Recycle* (in Japanese).

Sawa, K., M. Kamon, S. Tomohisa & N. Naitoh 1994. Waste Oil Hardening Treatment by Industrial Waste Materials. *1st Inter. Cong. Environmental Geotechnics, ISSMFE and CGS* (in submitting).

Yamada, M., & Y. Inaba 1993. Utilization of Waste Plastic to Asphalt Pavement. *Proc. 4th Ann. Conf. JSWME*, pp.385-388 (in Japanese).

Yamada, S., Y. Nagasaka, N. Nishida & T. Shirai 1989. Light Soil Mixture with Small Pieces of Expanded Polystyrol and Sand. *Tsuchi-to-Kiso, JSSMFE*, Vol.37, No.2, pp.25-30 (in Japanese).

NOTATION

1 Mg = 1 ton,
1 Gg = 1,000 ton, and
1 Tg = 1,000,000 ton.
PNCBL = Public Nuisance Countermeasures Basic Law,
WDPCL = Waste Disposal and Public Cleaning Law,
LPURR = Law for Promotive Utilization of Reclaimed Resources, and
TGCW = Treatment Guideline of Construction Waste.

Developments in Geotechnical Engineering, Balasubramaniam et al. (eds) © 1994 Balkema, Rotterdam, ISBN 90 5410 522 4

Seepage study for design of an offshore waste containment bund

S.L. Lee, S.A. Tan, K.Y. Ng, K.Y. Yong & G.P. Karunaratne
Department of Civil Engineering, National University of Singapore, Singapore

ABSTRACT: In land scarce coastal cities such as Singapore, the viability of an offshore waste disposal facility is seriously considered. For the design of such facilities, the potential of marine pollution from contaminant transport through the containment bund must be carefully studied. This paper describes some preliminary analysis of such nature to study the influence of tidal effects, liner thickness and permeability on the seepage and contaminant transport behaviour of a typical offshore bund.

1. INTRODUCTION

One of the major environmental problems facing land-scarce coastal cities with their high population density and robust economies is the lack of suitable sites for waste disposal by landfilling. In view of this, the creation of offshore waste disposal facilities are now being considered. In fact fore shores waste disposal facility has already been introduced at Yoshimien in Hiroshima, Japan (Aboshi et al, 1991) and is expected to be more common in years to come for land scarce countries. The proposed idea offers a relatively straight forward and affordable waste disposal option in view of the high demands for land. However, the disposal of waste in an offshore environment posed a potential pollution threat to the marine environment. This problem must be adequately addressed in the planning and design, from construction and operation to the closure phases of such facilities. Clearly, the design parameters of the perimeter containment bund cross section plays a very significant role in the control of possible contaminant migration out of the waste pond into the marine environment.

For economic and practical construction, it is necessary to use hydraulic sand filling as the means of constructing the perimeter bunds, with the possibility of incorporating geosynthetic or clay liners in the design. From careful site investigations, as far as it is possible, the location of bunds should be selected so as to rest on fairly impervious clayey soils. Therefore, it is likely that the bulk of the contaminant transport problem would arise from seepage flow across the perimeter bunds. A schematic figure of a hypothetical offshore waste disposal pond is shown in Fig. 1.

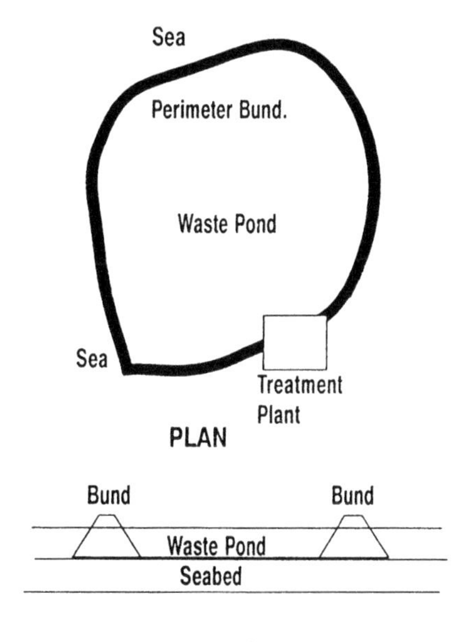

Fig. 1 Schematic of waste pond

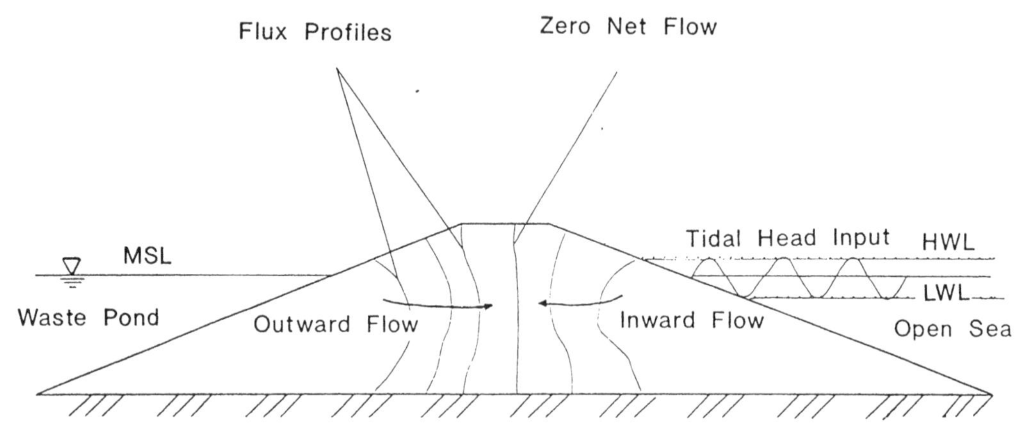

Fig. 2 Hydraulic trap

This paper describes the finite element simulation of the transient seepage flow through a proposed offshore waste containment bund, inclusive of the effects of outside tide variation. The first set of parameters examined in the modelling were the boundary conditions of waste pond level maintained at constant heads of mean and high water level, together with the downstream transient tidal movement which is a variable head. The other parameters studied were the effects of an upstream sloping clay liner with a possible range of thicknesses from 2m to 6m, and hydraulic conductivities from 5×10^{-7} to 1×10^{-8} m/s on the seepage flow in the bund. The modelling of transient seepage and contaminant transport is achieved through the use of SEEP/W and CTRAN/W developed by GEO-SLOPE with work done by Fredlund and his colleagues at the University of Saskatchewan, Canada, which incorporates the theory of flow through a saturated/unsaturated porous media as described by Freeze and Cherry (1979).

2. CONCEPT OF HYDRAULIC BARRIER

In the offshore environment, it is very difficult if not impossible to construct a leak-proof landfill which is typical of composite liner landfill designs used inland. Since the waste pond is situated in the sea, it would be more practical to create an inward gradient from the sea into the pond by control of pond water levels. A "hydraulic trap" refers to the use of a hydrodynamic gradient as a hydraulic barrier in containing leachate. A study of this phenomenon was discussed by Rowe (1988) and a case study on its feasibility for a sanitary landfill in a site of high water table at Saskatoon, Saskatchewan was reported by Haug et. al. (1989). They concluded that if a natural "hydraulic trap" can be designed into the system, a low cost effective seepage barrier against contaminant leakage into the groundwater can be achieved. In our study, it is believed that the presence of the outside tide with a constant waste pond level may provide a form of a "hydraulic trap" with the net flow constrained within the bund body or net seepage going into the waste pond.

The conceptual picture is shown in Fig. 2 for the case where the waste pond can be kept at mean sea level (MSL). When the tide is at high water level (HWL), there would be an inward flow into the pond due to the excess head from the sea to the waste pond. The opposite trend of flow would occur when the tide is at low water level (LWL). As the tide changes from HWL to LWL cyclically, the seepage gradient in the bund alternates with the tides. It can be envisaged that a phenomena could occur whereby the outward seepage cancels the inward seepage, thus creating a zone of "zero net flow" somewhere within the bund. When this occurs, a "hydraulic trap" is created which would prevent contaminants from escaping across the bund by advective transport.

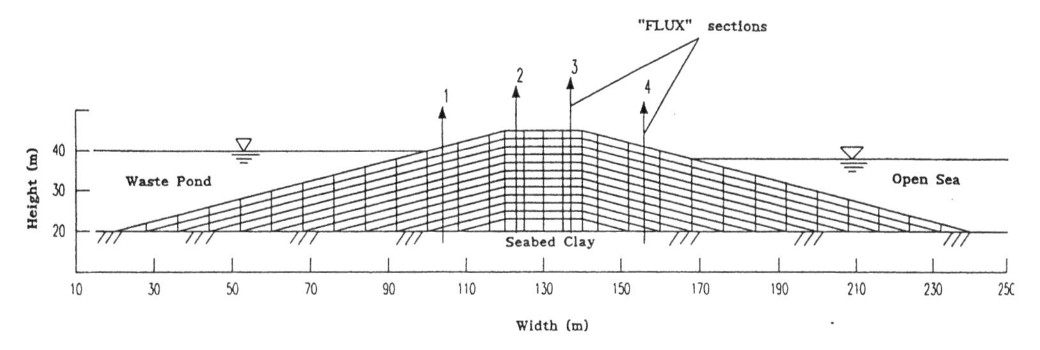

Fig. 3 Finite element mesh

3. FEM MODELLING OF TRANSIENT SEEPAGE

The above scenario is that of the filling up period during the operational life of the waste pond where there would be water inside the pond within the bund enclosure and the sea on the outside. Depending on the size of waste pond and the rate of waste disposal, the operational life can be as long as several decades. A finite element mesh of about 200 elements as shown in Fig. 3, is used to model a typical bund of 25m height (from Elevation 20m to 45m), 10m crest width, 240m base width, and a side slope of 1:4. The sea has a 12 hour tide period, with MSL at 38.75m elevation, HWL and LWL at 40.00m and 37.50m elevations, respectively. Boundary conditions were based on the geological conditions in the proposed offshore site, which showed the abundance of marine clay on the seabed (JICA, 1979). Thus the bund can be assumed to sit on a comparatively impermeable foundation ($k < 10^{-9}$m/s). The tide was modeled by a transient head boundary condition representing gradual changes from HWL to LWL in 6 hourly cycles.

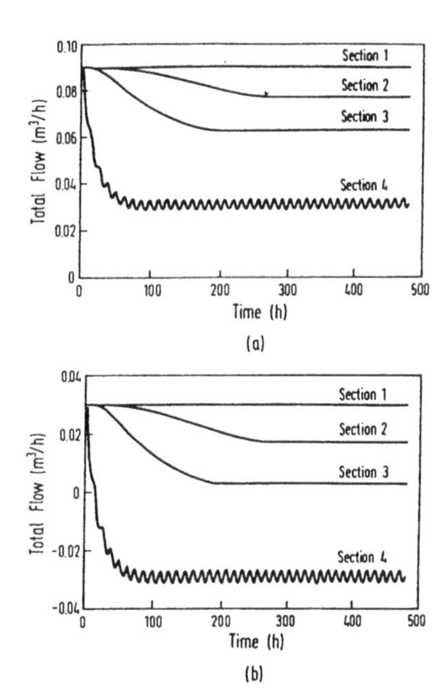

Fig. 4 Transient response of total fluxes in bund cross-sections

4. RESULTS OF PARAMETRIC STUDY

4.1 Operational condition

For the operational condition, it is feasible to maintain the waste pond level at mean sea level or lower by periodic pumping and treatment of waste pond water before disposing it back into the surrounding sea.

From the parametric study, it is demonstrated that the presence of the outside tide can create an impedance effect on the direction and magnitude of the seepage flow from the waste pond to the sea. Even with the use of a purely sand bund with a hydraulic conductivity as high as 10^{-4}m/s, a "hydraulic barrier" can be created in the bund for the condition of waste pond maintained at mean sea level or lower. This would be expected from steady state consideration, where with pond level at mean sea level, there will be zero flow across the bund. Whereas with pond level lower than MSL, there will be a net flow into the waste pond.

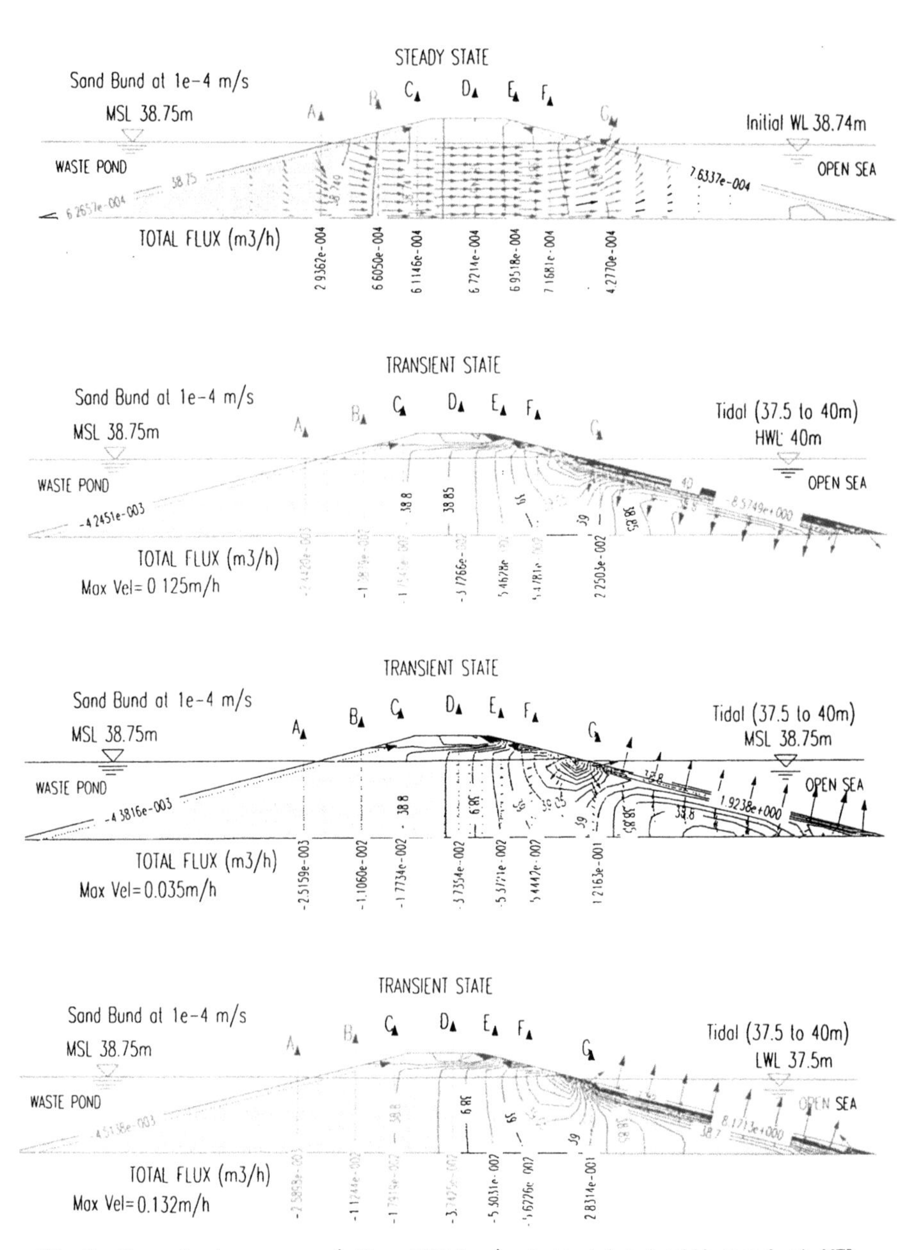

Fig.5 Transient seepage (after 300 hrs) of sand bund with pond at MSL and sea at MSL initially

The transient variation of total fluxes across four sections for an all sand bund and a clay-lined bund with waste pond at HWL, and the open sea at tidal boundary condition is shown in Fig. 4. This illustrates that it takes about 300 hours of transient cyclic seepage to achieve a steady condition across the bund. The flux section nearest the sea gave a cyclical response in consonance with the tidal effects, but sections 1 to 3 are far enough from the sea not to respond cyclically. The flux condition nearest the sea reaches a steady condition first followed by the sections further away from the sea towards the waste pond. Therefore, it is only necessary to show the results of the sea at HWL, MSL and LWL after 300 hours of simulation to obtain a clear picture of the state of seepage in the bund for any of the cases studied.

4.2 Case 1 - Sand bund (Pond at MSL)

For the case of an all sand bund of 1×10^{-4} m/s permeability and waste pond at MSL, the steady condition after 300 hours is shown in Fig. 5. The initial steady state condition is in waste pond at MSL and the sea at a level of 0.01m less than MSL. At this state there is a very small net flow (7×10^{-4} m^3/h) out of the pond into the sea. Subsequent transient flow simulations causes the flow to reverse into the pond, that after 300 hours, irrespective of the sea at HWL, MSL and LWL, there is a net flux of 4×10^{-3} m^3/h into the waste pond. Thus an active hydraulic barrier is produced which can be used to advantage to prevent pollutant from leaking out of the waste pond into the sea via advective transport. Therefore for the operational life of the waste disposal facility, from the seepage standpoint, an active hydraulic barrier can be created provided that the waste pond level can be kept at MSL or lower at all times.

4.3 Long term and closure condition

For the long term condition after the waste pond is filled up, it would be impossible to control the water level on the inside. The water table then would be a function of climate, inclusive of rainfall, infiltration, transpiration and evaporation together with the mounding of water table from the island effect. Under such condition, it can be expected that the water table will remain high, possibly at HWL over the long term. Therefore the crucial condition to study is that of the waste fill at HWL with the sea varying cyclically about MSL.

4.4 Case 2 - Sand bund (Waste at HWL)

Starting at the initial condition of waste at HWL, and sea at MSL, it is obvious that there will be a net outflow into the sea of about 0.095 m^3/h as shown in Fig. 6. Based on an earlier EPA recommendations for hazardous landfill liners, an annual leakage per acre of liner of 50,000 U.S. gals is permissible. This translates to a flux rate of 0.0004 m^3/h for the bund under study. Therefore the potential leakage across a sand bund is highly unacceptable in the long term. The transient conditions shown in Fig. 6 indicate that the tidal effect is to attenuate the flow going out into the sea as indicated by the reducing fluxes going from sections A to G, but there is no existence of a hydraulic barrier for this case, since there is a net flow from the waste pond into the sea.

4.5 Case 3 - 2m clay lined bund

When a 2m sloping clay liner with permeability of 5×10^{-7} m/s is added to the sand bund, the flux into the waste pond is reduced from 0.09 in the previous case to 0.009 m^3/h. Fig. 7 illustrates the transient seepage state of a 2m clay lined bund with waste fill at HWL and the sea initially at MSL. In this case, it is observed that the fluxes reduce from sections A to C in a direction towards the sea, with a reversal of fluxes from sections D to F. This indicate that there is a zero flux condition between sections C and D forming a "hydraulic trap" between these two sections. Thus no flow from the waste pond will go past section D to F and we have in place an active hydraulic barrier for the waste pond. The equipotentials between sections B and F are essentially at 39.25, indicating a zone of very little to no flow in the middle of the bund.

4.6 Effect of liner permeability

Further, it can be shown that reducing the clay liner permeability significantly reduces seepage magnitudes, and enhanced the effects of the

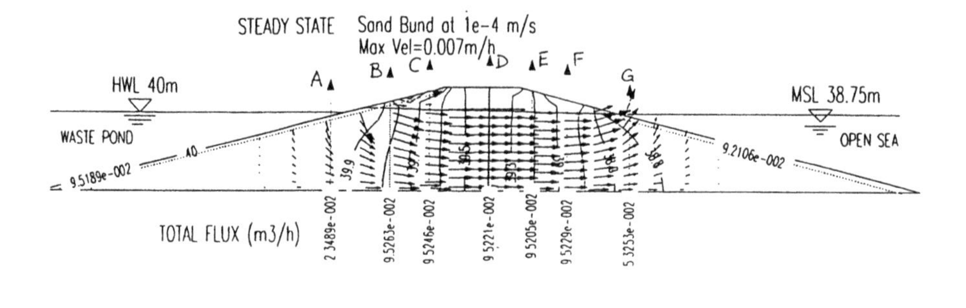

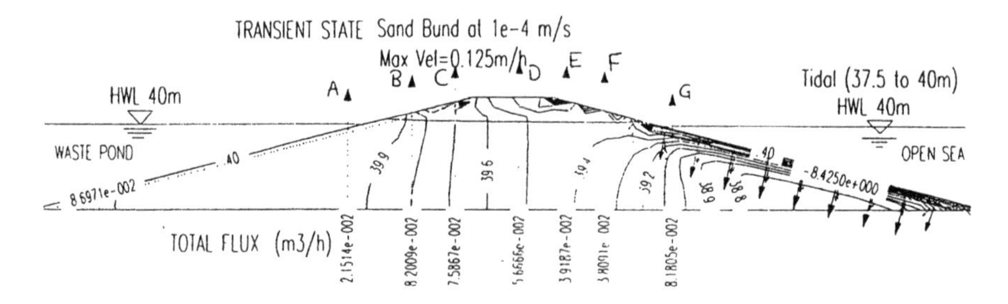

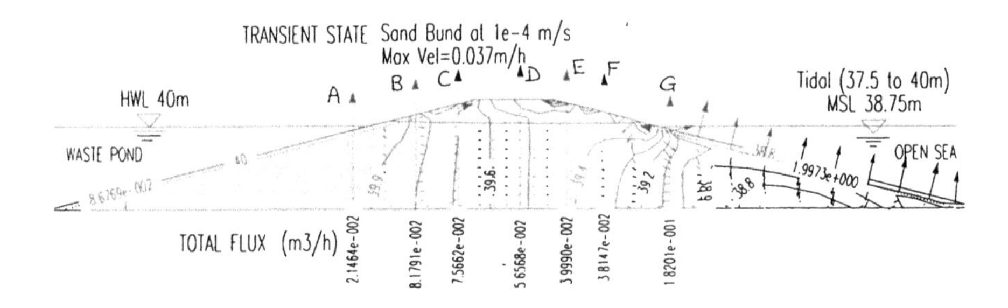

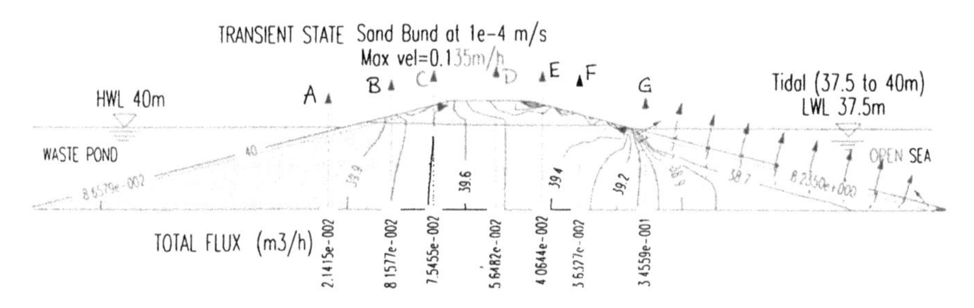

Fig.6 Transient seepage (after 300 hrs) of sand bund with pond at HWL and sea at MSL initially

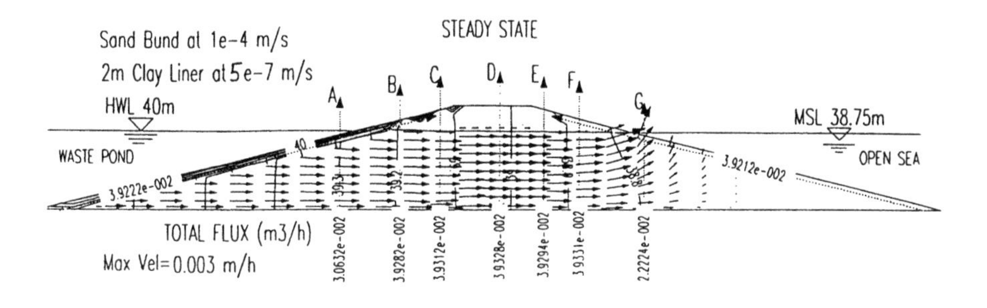

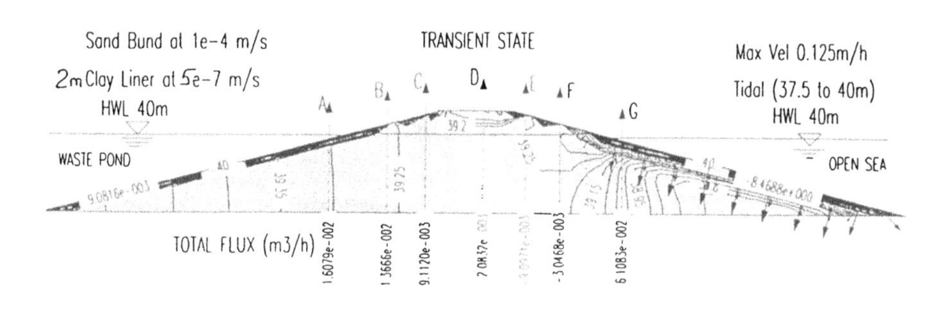

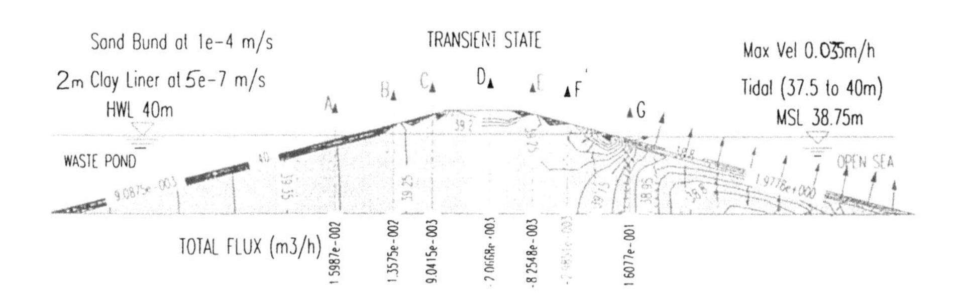

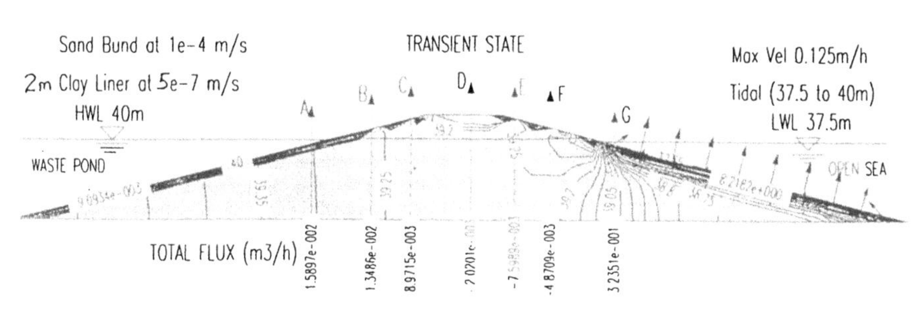

Fig.7 Transient seepage (after 300 hrs) of sand bund and 2m clay liner, with pond at HWL and sea at MSL initially

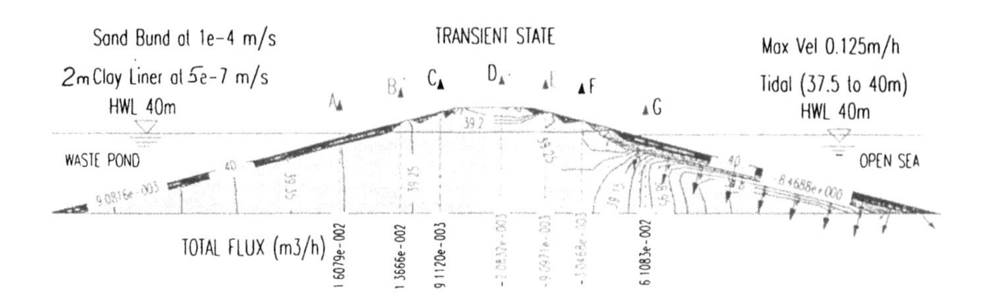

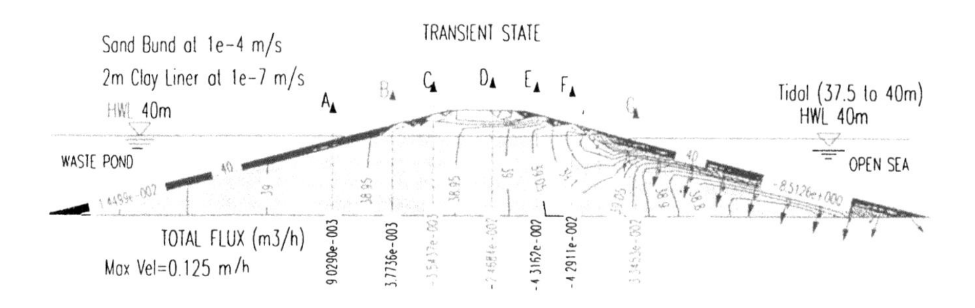

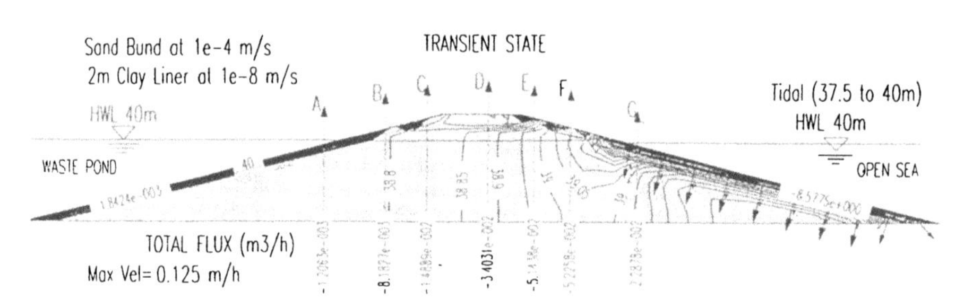

Fig.8 Comparison of Transient Seepage (after 300 hrs) for cases of 2m clay liner with different permeabilities

"hydraulic trap" achieved from the tidal movement of the sea. Using an upstream sloping clay liner of 2m thickness, with different hydraulic conductivities show the benefits that can be derived from clay liners. Figure 8 compares the three cases of a 2m upstream clay liner of 5×10^{-7}, 1×10^{-7} and 1×10^{-8}m/s, respectively, with pond at HWL and the instant when the sea is also at HWL after 300 hours of transient seepage. Reducing the clay liner permeability from 5×10^{-7} to 1×10^{-7}m/s shifts the "hydraulic trap" from between Sections C and D, to between Sections B and C. When the clay liner conductivity is reduced further to 1×10^{-8}m/s, this cause a shift of the trap zone inwards towards the waste pond between Sections A and the upstream face of the bund. Also, there is a further reduction in the seepage flux of one order of magnitude in these sections. Thus, the presence of a clay liner of low hydraulic conductivity is a very positive enhancement to the safety against pollutant leakage out of the waste pond into the sea.

4.7 Effect of liner thickness

The transient results with pond at HWL and the instant when the sea is also at HWL, for the cases of liners of 2m, 4m and 6m thickness at permeability of 5×10^{-7}m/s are shown in Fig. 9. The results implied that increasing clay liner thickness do have some beneficial effects of shifting the "hydraulic trap" upstream towards the waste pond. However, the revesal of fluxes still remains between sections C and D for all three thicknesses, with the zero flux line closer to section C as the thickness of liner is increased. Thus, the impact of clay liner thickness on the "hydraulic trap" is not as significant as the effect of liner permeability.

5. CONTAMINANT TRANSPORT MODELLING

The transport of non-reactive contaminants through a porous media is governed primarily by three mechanisms. They are the advection where the solutes are bodily transported along with the seepage flow, and the spreading out of the contaminant by dispersion. The dispersion is controlled by two processes, one is mechanical 'mixing' and the other is molecular diffusion. The first is dependent on the flow velocity, while diffusion is a function of concentration gradients. For the long-term condition of the waste facility after complete filling up and closure, it is assumed that the steady state condition could be represented by a water table that is effectively at HWL arising from the island effect of the island created by all the waste dumped up to the end of its operational life.

Under these conditions, if a constant contaminant source of 1 mg/l is used, the 10% plume development can be simulated for various conditions. The conditions studied are for clay liners of 2m, 4m and 6m thickness with conductivities of 1×10^{-7} and 1×10^{-8}m/s. It can be seen from Figs. 10 and 11 that the time for the 10% concentration plume to reach the open sea decreases with clay liner thickness and lower hydraulic conductivity. The results are summarized in Table 1 below:

Table 1 Travel time of 10% plume

Cases (k in m/s)	Time (yrs)
1. Sand Bund at 1e-4	0.7
2. 2m Clay at 1e-7	4
3. 4m Clay at 1e-7	7
4. 6m Clay at 1e-7	15
5. 2m Clay at 1e-8	30
6. 4m Clay at 1e-8	70

Thus a 4m clay liner of 1×10^{8} m/s will provide adequate protection against potential leakage from the waste island into the open sea for a waste pond with an operational life of 20 to 30 years.

The above is a very conservative scenario. The transient situation will result in lower seepage velocities than what is assumed in a steady state analysis. The actual situation would be better as waste with polluting potential will first be encased in cellular liners before disposing into the waste pond. Even if the cellular liners sprung a leak, these will only constitute contaminants of finite mass and not one of constant concentration throughout the waste pond. What the bunds must be able to provide is a secondary line of defense in case of failure of the internal control and treatment

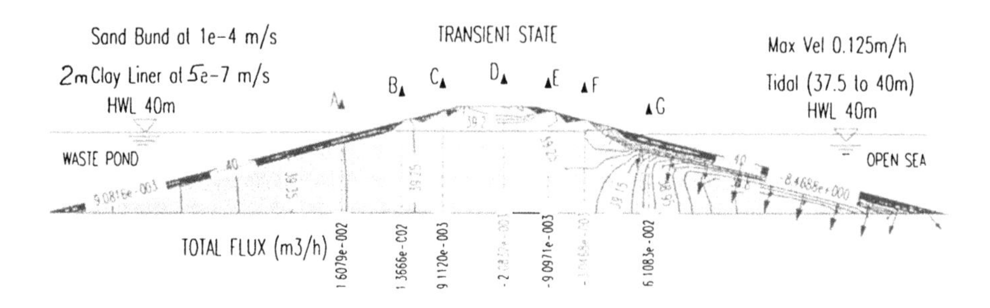

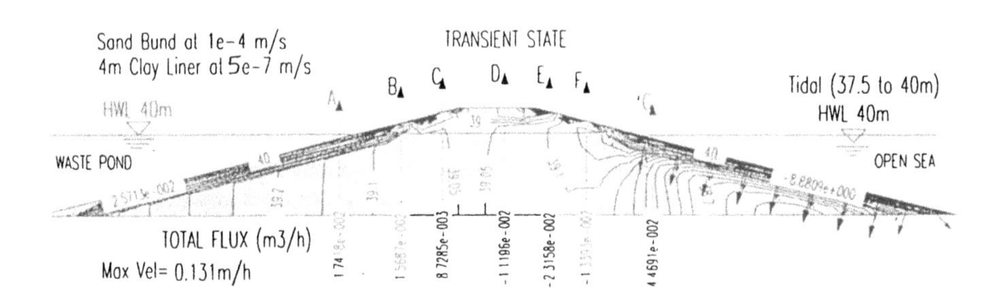

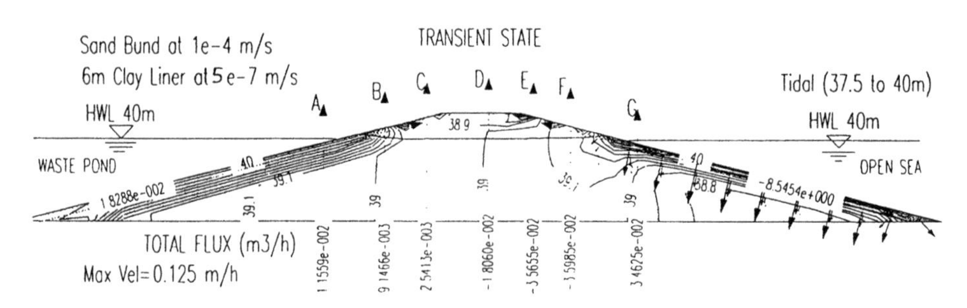

Fig.9 Comparison of Transient Seepage (after 300 hrs) for cases of 2m, 4m, and 6m clay liner thicknesses

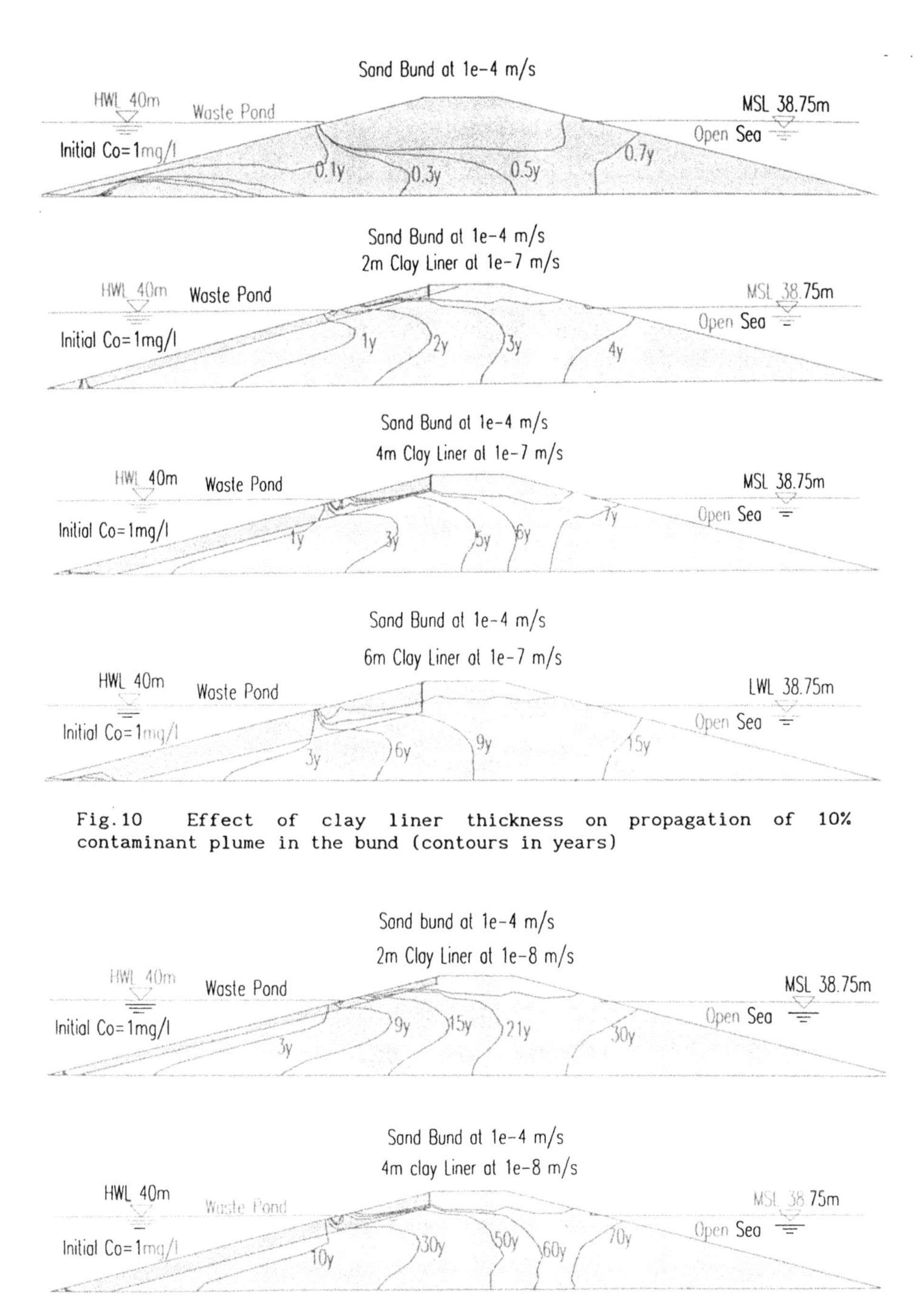

Fig.10 Effect of clay liner thickness on propagation of 10% contaminant plume in the bund (contours in years)

Fig.11 Effect of clay liner permeability and thickness on propagation of 10% contaminant plume in the bund (contours in years)

system that is designed to cope with the expected quantities of waste at the disposal facility.

6. CONCLUSIONS

It is possible to create a "hydraulic trap" in an offshore waste disposal facility by enclosing the waste site with a perimeter bund, maintaining waste pond at MSL or lower, and allowing the sea tides to keep the net seepage within the bund. The use of upstream sloping clay liner will enhance the safety barrier against leakage by shifting the "hydraulic trap" towards the waste pond, and reducing the seepage flux quantities.

For contaminant transport processes, the clay liner plays a significant role in the attenuation of many contaminants through the mechanisms of diffusion, adsorption and precipitation by cation exchanges. Thus, the combined use of an upstream clay liner and proper management of waste pond level to capitalize on the impedance effect of the outside tide variation can produce an economical safeguard against pollution of the marine environment in the operation of an offshore waste disposal facility.

7. ACKNOWLEDGEMENTS

The funding provided by the National Science and Technology Board, Singapore under RDAS Grant No. ST/86/05 is gratefully acknowledged.

REFERENCES

Aboshi, H., K. Fukuda, T. Ogura & T. Inoue 1991. The soil stabilization for a final disposal site. Geo-Coast '91, International Conference on Geotechnical Engineering for Coastal Development, Yokohama, Vol. 1, No. 2, 147-152.

Freeze, R.A. & J.A. Cherry 1979. Groundwater. Prentice Halls, Englewoods Cliffs, New Jersey, 604 pp.

Geoslope-International (Krahn and Fredlund) 1992. SEEP/W and CTRAN/W user's guide. finite element program for seepage and contaminant transport analysis.

Haug, M.D., D.J.L. Forgie & S.L. Barbour 1989. Design of a hydrodynamic leachate containment system. Canadian Journal of Civil Engineering, Vol. 16, 615-626.

JICA 1979. Study of fill material for reclamation projects in Singapore territorial waters. Report No. 32, Tokyo, Japan.

Rowe, R.K. 1988. Eleven Canadian Geotechnical Colloquium: Contaminant migration through groundwater: the role of modelling in the design of barriers. Canadian Geotechnical Journal, Vol. 25, 778-798.

Developments in Geotechnical Engineering, Balasubramaniam et al. (eds) © 1994 Balkema, Rotterdam, ISBN 90 5410 522 4

Liquefaction-induced earth movements and mitigation in an earthquake-prone area

Kuo-Ping Chang
Eastern Construction Company, Taipei, Taiwan

Tzyy-Shiou Chang
Memphis State University, Tenn., USA

ABSTRACT: Earthquake-induced settlement in sand deposits and lateral movement of slopes due to an underlying failed sand layer are the major damaging effects on structures resulting from either consolidation of unsaturated sand or liquefaction of saturated sand. Memphis is an important metropolitan area in Tennessee, USA, located within 40 miles south of the new Madrid seismic zone (NMSZ) in the central USA. Granular sand layers overlain by a layer of surficial clayey soils (silty to sandy clay) or loessial soils (clayey to sandy silt) are present throughout the Memphis area. In general, the surface of these sand layers are located at a depth as shallow as 10 feet to more than 40 feet with thickness ranging from about 5 feet to greater than 40 feet. In this study, earthquake-induced earth movements are estimated on the basis of engineering boring logs, previous seismic hazard and theoretical earth movement analyses, and on results of field observations of foundation failures and lateral slope movements following past earthquakes. The average and possible range of earth movements corresponding to assigned liquefaction risks are shown in a series of earth movement potential maps. Liquefaction mitigation techniques are also briefly discussed.

INTRODUCTION

Evidence from past earthquakes in the U.S. and abroad has shown that structural failures in the event of an earthquake are strongly site-related (Seed and Idriss 1969, Seed et al. 1972, Astaneh et al. 1989, Wyllie and Filson 1989). In general, earthquake damage to a structure in an earthquake-prone region is controlled by the following site-related factors:

(1) distance from the epicentral area; in general, the farther a structure from the epicenter, the weaker the shaking becomes,

(2) potential for amplified ground shaking due to localized soil conditions; usually, this is caused by a relatively thin layer of weak soils near ground surface. This ground amplification induces a significantly larger shear force on buildings than in other areas where ground shaking is not amplified by resonance,

(3) potential of earth movements (excessive vertical settlement and lateral slope movement) and loss of soil bearing capacity due to liquefaction.

Liquefaction of sandy soils has been observed as a very spectacular, astonishing phenomenon following earthquakes for a long time. However, the mechanism and damage consequences of liquefaction were not well understood until the Niigata, Japan and Alaska earthquake of 1964. Since then, many liquefaction-induced damage to structures and life lines were studied to develop empirical methods for evaluating liquefaction susceptibility of sites in earthquake-prone areas. More recently, many mitigation techniques have been developed and utilized for reducing damage risk in high liquefaction potential areas.

This paper focuses on potential earthquake-induced earth movements in Memphis and Shelby County (Located in the State of Tennessee, USA, as shown in Figure 1) based on an analysis of engineering boring data gathered throughout the study area from previous research (Chang et al. 1991a), the volume strain - shear stress ratio - (N1)60 relation developed from field observation of foundation failures, equations derived from theoretical analysis, and from slope movement monitoring in other earthquake-prone areas in the world after past earthquakes (Tokimatsu and Seed 1987, Hamada et al. 1987, Bryne 1991). Vertical settlements of sand layers caused by liquefaction or

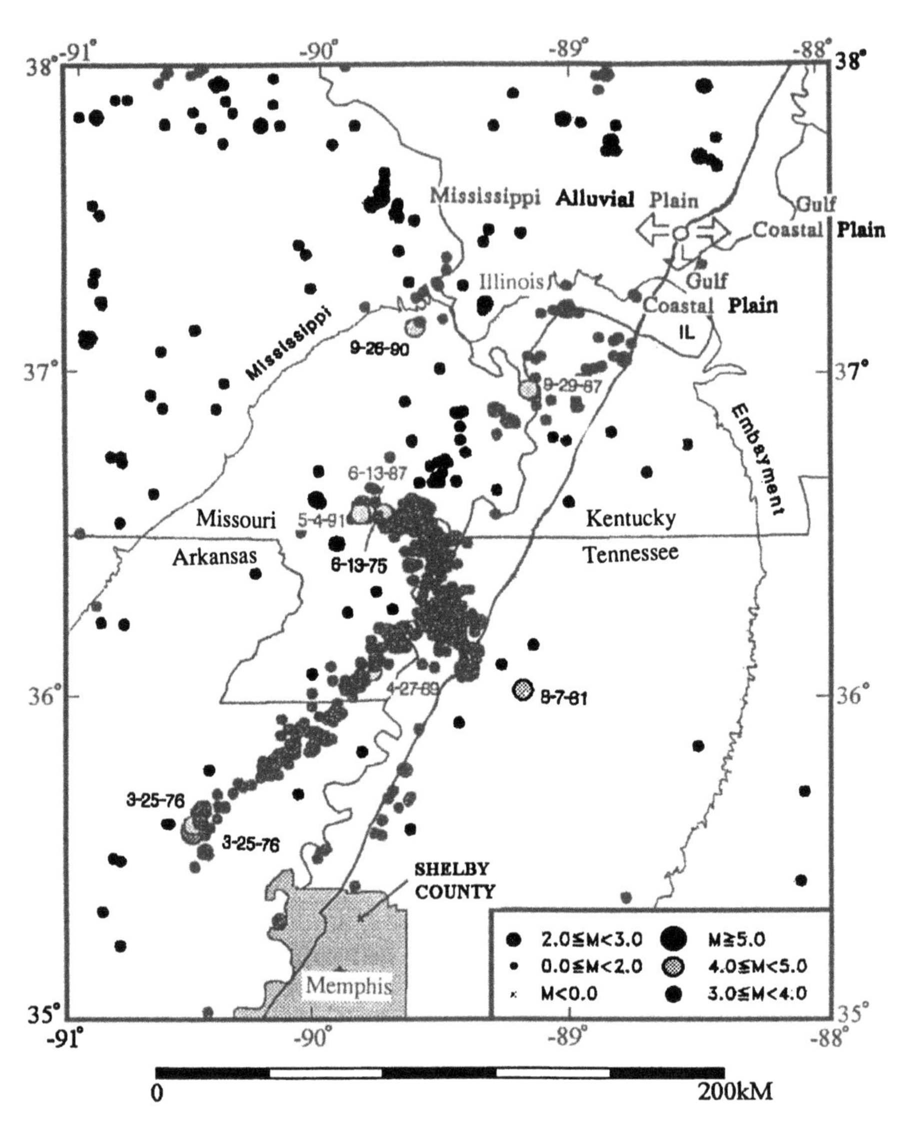

Figure 1 New Madrid Seismicity: 1974-1990

consolidation during earthquakes are also calculated in the study. On the basis of these study results, a series of earth movement potential maps corresponding to an assigned liquefaction risk are presented for the Memphis area.

DATA SOURCES AND RESEARCH METHODS

Data Sources

A subsurface database consisting of more than 500 representative soil profiles was compiled in a previous study and is based upon data from more than 8000 engineering boring logs (Chang et al. 1991a). These engineering boring logs were furnished to the CERI (Center for Earthquake Research and Information) project by local governmental agencies, private consulting companies, and other contributors. The computerized database summarizing these representative soil profiles contains the following engineering information:

(1) location and elevation of subsurface exploration,
(2) sequence of strata in profile, layer thickness and

soil description including classification,
(3) field and laboratory test results, and
(4) groundwater information at the time of drilling operation and other water level observations.

Data related to low-strain shear wave velocity and strain-dependent non-linear behavior of the soils were available from the results of a previous study (Chang et al. 1992). In that study, 78 dynamic tests were performed on 35 selected typical soil samples from the general NMSZ region which includes the current study area.

Research Methods

1. Liquefaction potential index PL: The sand layers revealed in the representative soil database were evaluated first for their liquefaction potential based on Seed's field relation and pore-pressure development criteria available from previous research conducted at the CERI (Seed and Idriss 1969, 1982, Chang and Chang 1993, Chang et al. 1991b). The liquefaction potential of the sites was established for constant common ground surface accelerations of 0.1g and 0.2g, and for ground surface accelerations estimated from a seismic hazard curve for an event of 0.001 annual probability of exceedance (Johnston 1988). This event results in an estimated ground surface acceleration which is highest in the northwestern part of the study area (>0.3g) and decreases toward the southeastern part of the study area where it is the lowest in the region (<0.2g). Liquefaction is judged a potential to occur if the safety factor, FL, is smaller than one in a sand layer. The safety factor, FL, is defined as the ratio of the stress ratio required to trigger liquefaction to the stress ratio induced by the driving earthquake accelerations at the ground surface. The FL value does not imply the spatial extent of the liquefiable layer such as depth or thickness. Once a liquefiable layer (or layers) is identified at the site, the depth and thickness of the layer(s) must be taken into account to evaluate the damaging effect that the whole profile has on the safety of a supported structure. The liquefaction damage effect at a site can be expressed in the form of the liquefaction damage potential index, PL, which is expressed as Equation 1 (Iwasaki et al. 1982):

$$PL = \sum_{i=1}^{n} G_i \times W_i \times H_i \qquad (1)$$

where

PL = liquefaction damage potential index
n = number of liquefied layers
G_i = severity of the i-th liquefied layer (G_i = 1 - FL)
W_i = a weighting function accounting for the influence of the depth of the i-th liquefied layer (W_i = 10 - 0.5 z)
H_i = thickness of the i-th layer in meters
z = depth of soil in meters

2. Estimation of earthquake-induced settlements: The primary result of vibration (cyclic loading) of saturated granular soils is the generation of pore pressure in the voids of the soil skeleton. The dissipation of this excess pore pressure with lapse of time can be accompanied by a net decrease in volume of the voids which can be translated into settlement at or near the ground surface. Thus, even though the saturated sand may not liquefy completely during earthquake motion, some excess pore pressures may still develop. The subsequent dissipation of this pore pressure can result in small amounts of settlement after cyclic loading ceases (Chang et al. 1991b). In addition, consolidation of loose dry sand deposits subjected to earthquake shaking can also result in some settlement depending upon the layer thickness, relative density of the sand layer and the shaking intensity.

Tokimatsu and Seed (1987) compiled laboratory test data from studies on sand settlement due to liquefaction and concluded that the primary factors affecting induced settlement in terms of volume strain are the cyclic stress ratio and the maximum shear strain induced by cyclic loadings in dry, partially or fully saturated sands. Based on these studies, Tokimatsu and Seed proposed a chart to estimate the settlement of saturated sand after complete liquefaction. Once the (N1)60 and the earthquake-induced cyclic stress ratio are obtained, the volume strain can be determined from the chart to estimate the settlement of the sand deposit. For settlements resulting from incomplete liquefaction, a normalized shear stress ratio (ratio of earthquake-induced shear stress ratio to shear stress ratio required to cause liquefaction) is used to characterize the extent of the settlement. On the basis of the volume strain, the settlement can be estimated for sites either (1) after complete liquefaction for liquefied sites, or (2) due to consolidation of the sand deposits during earthquake shaking for nonliquefied sites, as expressed in Equation 2:

$$\text{Total Settlement} = \sum_{i=1}^{n} e_i \times H_i \qquad (2)$$

where

Table 1 Summary of results of earthquake-induced estimated settlements in the Memphis area (A total of 518 sites).

Ground Surface Acceleration	Liquefaction Potential Index (PL)	Number of Sites (%)	Settlement Range (in)	Number of Sites (%)
0.1 g	PL = 0 (no liquefaction)	503 sites (97%)	< 0.5	432 sites (86%)
			0.5 - 2.5	72 sites (14%)
	PL > 0 (liquefaction)	15 sites (3%)	< 0.5	0 site (0%)
			0.5 - 12	12 sites (80%)
			> 12	3 sites (20%)
0.2g	PL = 0 (no liquefaction)	416 sites (80%)	< 0.5	359 sites (86%)
			0.5 - 2.5	57 sites (14%)
	PL > 0 (liquefaction)	102 sites (20%)	< 0.5	0 site (0%)
			0.5 - 12	92 sites (90%)
			> 12	10 sites (10%)
Estimated from Seismic Hazard Curve	PL = 0 (no liquefaction)	362 sites (70%)	< 0.5	305 sites (84%)
			0.5 - 2.5	57 sites (16%)
	PL > 0 (liquefaction)	156 sites (30%)	< 0.5	5 sites (3%)
			0.5 - 12	135 sites (87%)
			> 12	16 sites (10%)

Table 2 Earthquake-induced estimated settlement range for various liquefaction risks on the basis of liquefaction potential index PL for the Memphis area.

Liquefaction Potential	Liquefaction Risk	Settlement (in)	
		Average	Possible Range
PL = 0	very low (no liquefaction)	1.0 ± 1	< 0.01 to 3
0 < PL≥5	low	2.5 ± 1.5	1/2 to 9
5 < PL≥15	high	6.0 ± 2	3 to 14
15 > PL	very high	> 8	5 to 18

n = number of sand deposit
e_i = volume strain of the i-th sand layer
H_i = thickness of the i-th sand layer

3. Estimation of liquefaction-induced lateral movement of slopes: Several theoretical or empirical procedures have been developed since early 1960 that can be used to predict liquefaction-induced lateral displacements by taking into account the post liquefaction stress-strain-strength behavior and its dependency on (N1)60 value (Newmark 1965, Hamada et al. 1987, Bryne 1991). Bryne indicated that these procedures are in good agreement with field and laboratory observations. In this study, liquefaction-induced lateral displacement of slopes is calculated from an empirical equation based on field experience proposed by Hamada et al. (1987) as follows:

$$\text{Lateral earth movement} = 0.75\,(H)^{1/2}\,(\phi)^{1/3} \quad (3)$$

where
H = thickness of the liquefied sand layer
ϕ = slope of the ground surface in %.

RESULTS AND DISCUSSIONS

Liquefaction Potential

A total of 518 sites have been evaluated in this study to determine the liquefaction potential in terms of liquefaction damage potential index (PL) as defined by Iwasaki et al. (1982). This procedure considers the effect of liquefied sand layers on the total behavior of soil profile within the influence zone for the support of structures. Results of this phase of the study are summarized in Table 1.

Earthquake-induced Settlement

1. Common ground surface acceleration of 0.1g:

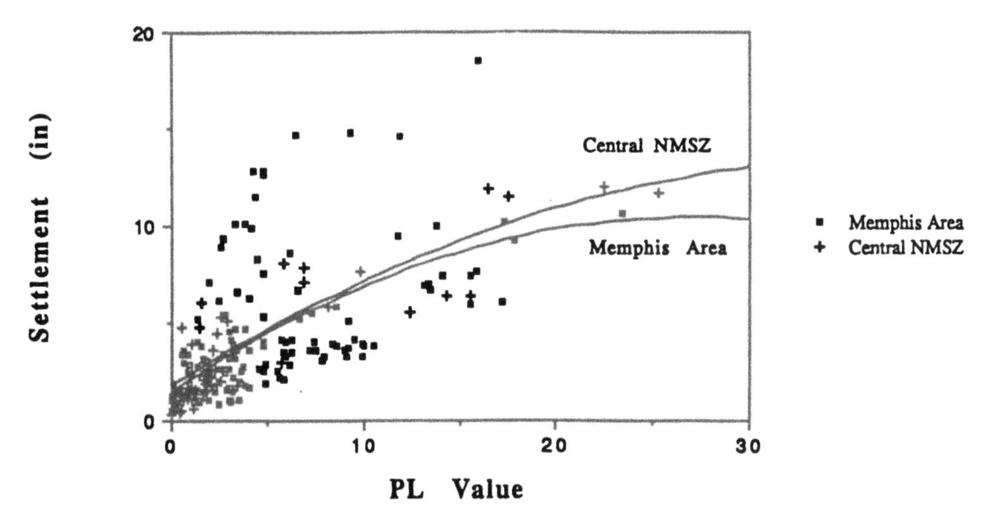

FIGURE 2 Earthquake-induced settlement potential as a function of liquefaction potential index PL for the Memphis Areas in comparison with that for the central NMSZ.

When liquefaction potential does not develop (PL = 0 for 503 sites, 97% of the evaluated sites), the possible settlement induced by consolidation of the sand layers ranges from a minimal value (less than 0.01 inch) to one in the order of 2 inches. A total of 72 non-PL sites (about 14% of the total non-PL sites) exhibit significant settlement potential (>0.5 in) due to consolidation because of great thickness of the sand layers even though the volume strain induced by earthquake shaking is small. When liquefaction potential does develop (15 sites, about 3% for 0.1g), all sites exhibit potential for settlement greater than 0.5 inch. Three sites (about 20% of the total liquefiable sites) may undergo very damaging settlements of greater than 12 inches because of a thick, loose underlying sand layer.

2. Common ground surface acceleration of 0.2g: When liquefaction potential does not develop (PL = 0 for 416 sites, 80% of the evaluated sites), the possible settlement induced by consolidation of the sand layers ranges from a minimal value (less than 0.01 inch) to one in the order of 3 inches. A total of 57 non-PL sites (about 14% of the total non-PL sites) exhibit significant settlement potential (>0.5 in) as a result of consolidation. When liquefaction potential does develop (102 sites, about 20% for 0.2g), all sites exhibit a settlement potential greater than 0.5 inch. Several sites (10 sites, about 10% of the total liquefiable sites) may experience settlements in excess of 12 inches. Results of analysis for ground surface acceleration estimated from the existing seismic hazard curve in the study area (0.001 annual probability of exceedance) are similar to those of 0.2g, therefore, no further discussion of this case is required. From these results, liquefaction-induced settlements plotted as a function of liquefaction potential index PL are presented in Figure 2. A least square curve fitting method was used to develop the curves on this graph. It clearly shows that earthquake-induced settlements generally increase with increasing PL.

The liquefaction risk categories with respect to various PL and their corresponding average and possible range of settlement in the Memphis area is summarized in Table 2. The information used to develop Table 2 was derived from Figure 2. By comparing the potential earthquake-induced settlements in Memphis from a previous study (Chang et al. 1992), the average settlements are almost the same for PL< 15 (very low, low, and high liquefaction potential), while the average settlements in the central NMSZ may be 1 to 2 inches greater than those in the Memphis area for very high liquefaction potential (PL >15). This is reasonable because soil conditions in the central NMSZ are somewhat worse than those in the Memphis area, i. e. with shallow, loose, more sandy soil layers that make up the Mississippi alluvial plain. The PL value and corresponding potential settlement were plotted for each site studied to develop maps showing potential earthquake-induced settlements for Memphis and Shelby County. As shown in Figure 3, these maps agree well with previous liquefaction potential maps (Chang et al. 1991b). Figure 3 provides useful information regarding the damaging effect of liquefaction for engineering applications in the Memphis area.

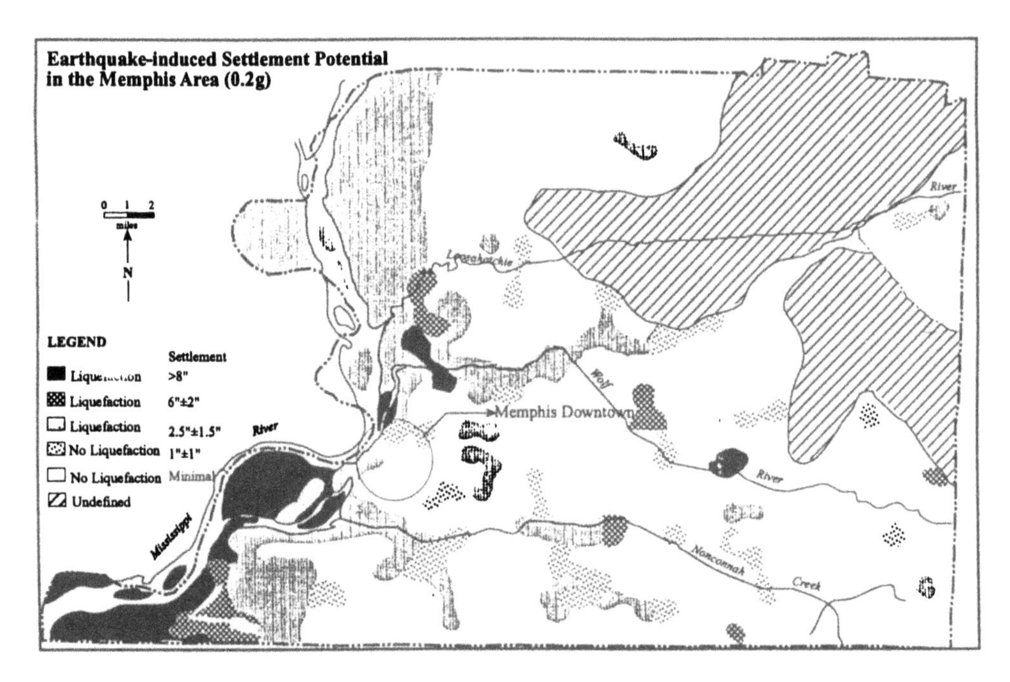

Figure 3 Earthquake-induced settlement potential for common ground accelaration of 0.2g in the Memphis area

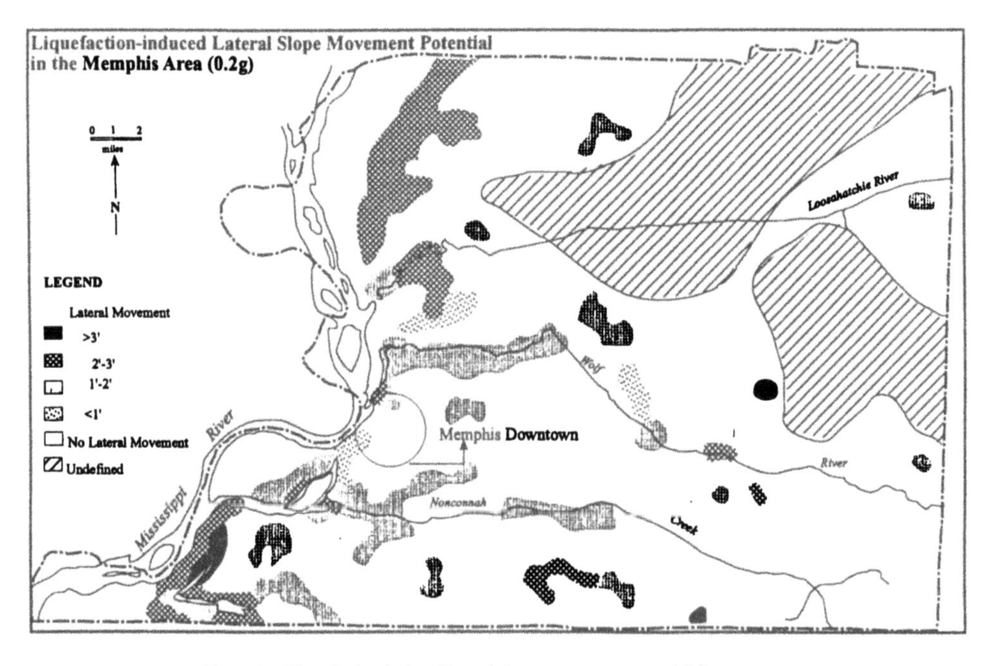

Figure 4 Liquefaction induced lateral slope movement potential for a common ground surface accelaration of 0.2g in the Memphis area

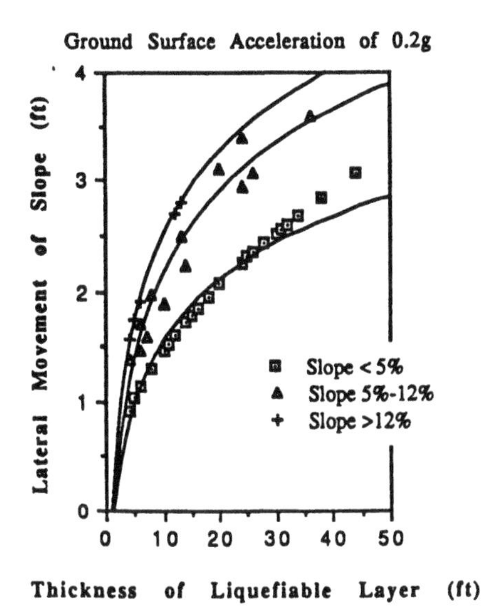

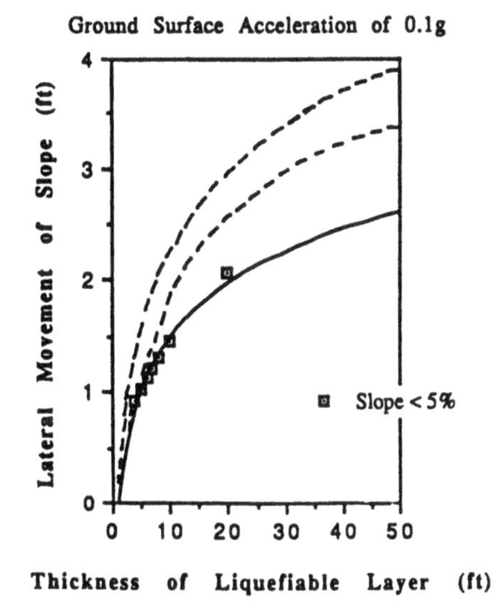

FIGURE 5 Liquefaction-induced lateral slope movement potential in the Memphis Area as a function of site slope angle, thickness of liquefiable layer and ground surface acceleration.

Earthquake-induced Lateral Movement of Slopes

As indicated by Bryne (1991), Hamada's empirical equation does not include a parameter that reflect the density of the sand layer. Bryne also indicated that the evaluation of lateral movement by Hamada's equation is very conservative. The results reported by Bryne have been studied through a comparison between the Hamada model and Bryne's model. Bryne's model considered residual strength and the limiting shear strain upon liquefaction and incorporated a safety factor determined from a static limit equilibrium slope stability analysis (Bryne 1991). The Hamada model is based on data collected from field observations of past earthquakes which are mainly applicable to a relatively thin layer of loose sand. Results from Bryne's model may be more accurate; however, Hamada's model is more convenient and applicable for an estimate in a region containing several hundred sites without detailed slope definition. The lateral displacement results reported by Bryne (1991) for a liquefied layer of 5 ft in thickness have been studied and concluded that the Hamada equation may be over-estimated with a factor of 3 to 5 depending upon the initial density of the liquefied sand layer. This paper presents Figure 4 as the map for liquefaction-induced lateral slope movement potential for 0.2g in the Memphis area on the basis of the Hamada equation modified by Bryne's model with a reduction factor of 3. Figure 5 shows the results of the analysis. The lateral slope movements are estimated ranging from less than 1 ft to in the order of 3 ft in the study area.

LIQUEFACTION HAZARD MITIGATION

If results of evaluation indicate that the liquefaction potential and the resulting damage and earth movement potential are not acceptable, remedial measures should be used to minimized the risk of foundation/structural failures. Significant advances have been made in the field of soil improvement over the past several decades (Welsh, 1987, Borden et al., 1992). At this time, the geotechnical techniques that are currently available for improving resistance of liquefiable soils include:

(1) appropriate foundation type and a greater safety factor,
(2) removal and replacement,
(3) dewatering,
(4) in-situ soil improvement: densification, drainage, chemical grout/modification, and
(5) relocation of the project, if necessary.

The improvement of liquefiable soils should be verified with the use of laboratory and/or in-situ tests to determine if the required liquefaction resistance of the soils are achieved. These tests include:

(1) laboratory testing: visual inspection, density, permeability, compressibility and static/dynamic strength.

(2) in-situ testing: geotechnical methods (SPT, CPT, DMP, etc.) geophysical methods (cross/down hole, probing radar, etc.)

CONCLUSIONS

1. Seismic hazard due to liquefaction can be evaluated with the use of existing soil data in the region, available theory in geotechnical earthquake engineering, and experiences from previous earthquakes. Appropriate mitigation techniques should be utilized to minimize the risk of casualty and property loss due to liquefaction in an earthquake-prone area.

2. For a common ground surface acceleration of 0.1g, only 15 sites (3%) of the evaluated sites exhibit any liquefaction potential (PL>0), while 102 sites (20%) exhibit significantly higher Liquefaction potential for 0.2g. When liquefaction does not develop (PL=0), potential earthquake-induced settlements range from a very minimal value (<0.01 in) to about 1.0 ± 1 inches. When liquefaction potential does develop (PL>0), the average liquefaction-induced settlement potential is 2.5 ± 1.5 in, 6 ± 2 in, and greater than 8 (to about 18 inches as the largest), for low, high, and very high liquefaction risk, respectively.

3. The charts derived in the study, showing the relationship between estimated lateral slope movement, thickness of liquefiable layer, slope angle, and ground surface acceleration, are very useful for engineering applications. The liquefaction-induced estimated lateral movement of slopes may range from less than 1 ft to about 3 ft in the Memphis area.

ACKNOWLEDGMENT

The authors express their appreciation to the U. S. Geological Survey for the financial support of the study (USGS Award #14-08-0001-G2000 and #1434-92-G-2197). The comments by Dr. Fry of Memphis State University and by Dr. Woods of the University of Michigan are also gratefully acknowledged. Special thanks are given to Ms. Valinda Stokes, Ms. Jill Stevens, Ms. Tanya George, and Ms. Hsueh-Jung Chris for preparing the paper.

REFERENCES

Astaneh, A., et al. (1989) Preliminary Report on the Seismology and Engineering Aspects of the October 17, 1989, Santa Cruz (Loma Prieta) Earthquake. Earthquake Engineering Research Center, University of California at Berkeley, Report No. UCB/EERC-89/14, 51 p.

Borden, R.H., Holtz, R.D. and Juran, I., editors (1992), "Grouting, Soil Improvement and Geosynthetics." ASCE Geotechnical Special Publication No.30, ASCE, NY, NY, 1453p.

Bryne, M. B. (1991) "A Model for Predicting Liquefaction Induced Displacement." Proceedings of Second International Conference on Recent Advances in Geotechnical Earthquake Engineering and Soil Dynamics, March 11-15, 1991 St. Louis, Missouri, 1027-1035.

Chang, T.S. and Chang, K. P. (1993) "Earthquake-induced Earth Movement Potential in the Memphis Area." Proceedings of 1993 National Earthquake Conference, May 2-5, Memphis, Tennessee, p. 15-24.

Chang, T.S., Hwang, H.,Ng, K.W. and Lee, C.S. (1991,a) "Subsurface Conditions in Memphis and Shelby County." Proceedings of the Second International Conference on Recent Advances in Geotechnical Earthquake Engineering and Soil Dynamics, St. Louis, Missouri, March 11-15, 1991, pp. 1305-1311.

Chang, T.S., Lee, P.S. and Hwang, H.M. (1991,b) "Liquefaction Potential in the Memphis Area." Proceedings of the Fourth International Conference on Seismic Zonation, August 26-29, Stanford University, California, 459-466.

Chang, T.S., Teh, L. K. and Zhang, Y. (1992) Seismic Characteristics of Sediments in the New Madrid Seismic Zone, CERI technical report submitted to the NSF, Memphis State University, Memphis, Tennessee. 141p.

Hamada, M., Towhata, I., Yasuda, S. and Isoyama, R. (1987) "Study on Permanent Ground Displacements Induced by Seismic Liquefaction." Computers and Geomechanics 4, 197-220.

Iwasaki, T., Tokida, K., Tatsuka, F., Watanabe, S., Yasuda, S. and Sato, H. (1982) "Microzonation for Soil Liquefaction Potential Using Simplified Methods." Proceedings of the 3rd International Earthquake Microzonation Conference, Seattle, 1319-1330.

Johnston, A. C. (1988) Seismic Ground Motions In Shelby County, Tennessee, Resulting From Large New Madrid Earthquakes. CERI Technical report

88-1, Memphis State University, Memphis, Tennessee. 36p.
Newmark, N. M. (1965) "Effect of Earthquakes on Dam and Embankments." Geotechnique, Vol. 15, No. 2, 139-160.
Seed, H. B. and Idriss, I. M. (1982) "Ground Motions and Soil Liquefaction During Earthquakes." EERI Monograph Series, Earthquake Engineering Research Institute, Oakland, CA, 134p.
Seed, H. B. and Idriss, I. M. (1969) "Influence of Soils Conditions on Ground Motions During Earthquakes." Journal of the Soil Mechanics and Foundation Engineering Division, ASCE, 95: (SM1) 99-137.
Seed, H. B., Whitman, R. V., Dezfulian, H., Dobry, R. and Idriss, I. M. (1972) "Soil Conditions and Building Damage in 1967 Caracas Earthquake." Journal of the Soil Mechanics and Foundation Engineering Division, ASCE, Volume 98, SM8.
Tokimatsu, K. and Seed, H. B. (1987) "Evaluation of Settlements in Sands Due to Earthquake Shaking." Journal of Geotechnical Engineering, ASCE, Vol. 113, No. 8, 861-878.
Welsh, J.P., editor (1987) "Soil Improvement - A ten Year Update." ASCE Geotechnical Special Publication No. 12, ASCE, NY, NY, 331p.
Wyllie, L.A. and Filson, J.R. (Eds.) (1989) Armenia Earthquake Reconnaissance Report. Special Supplement. Professional Journal of the Earthquake Engineering Research Institute, 175 p.

Developments in Geotechnical Engineering, Balasubramaniam et al. (eds) © 1994 Balkema, Rotterdam, ISBN 90 5410 522 4

Vertical barriers for municipal and hazardous waste containment

H. Brandl
Technical University of Vienna, Austria

ABSTRACT: For sealing off old landfills or new waste disposal facilities vertical cut-off walls have proved suitable. The efficiency of the containment can be increased significantly by installing a cellular cut-off system. The fundamental idea of cellular cut-offs is an isolating by using two parallel screens with linking cross walls at certain longitudinal distances. The groundwater level within these cells is kept lower than outside, but higher than inside the enclosure, i.e. the waste deposit. This hydraulic system does not allow an outward discharge of contaminated groundwater; it retards diffusion of pollutants. Furthermore it enables a permanent control and easy repair during its whole life-time. The paper deals with groundwater discharge evaluation, pollutant migration, sealing mixtures, cut-off wall construction, quality assurance control and long-term monitoring.

1. INTRODUCTION

In case of standard waste disposal units, monitoring is limited to leak detection in the secondary leachate collection and removal system, and to well/stand pipe controlling outside the site. If hazardous constituents are detected in the external monitoring system, the ground is already going to be contaminated. Consequently, for hazardous wastes with a great toxic potential, high safety disposal facilities have to be required. Because prevention of environmental pollution is better than remedial work, a multi-barrier system is preferred. This containment goal can be achieved by encapsulating wastes with vertical cut-off walls. Table 1 gives an overview of the currently used technologies and their capacity. For waste containment, diaphragm walls and vibrating beam slurry walls ("thin walls - Fig. 1) dominate. Jet grouting is commonly used for leak sealing; furthermore it enables the installation of inclined cut-off walls (Fig.2).

The maximum available depth of cut-off walls depends on several parameters: soil (rock) properties, technology and site equipment, required thickness and accuracy of the wall, allowable construction time and costs.

The efficiency of vertical screens increases significant-

Table 1. Overview of methods for cut-off wall construchtion
Approximate values for common width d (m) and currently maximum wall depth t_{max} (m)

TECHNOLOGY	CUT-OFF SYSTEM	GROUND PLAN (schematical)	DIMENSIONS d (m)	t_{max}(m)
permeability reduction of in-situ soil	compaction wall	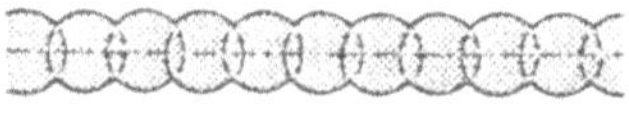	0,4-1,0 [1)]	10-20
	grouting wall		1,0-2,5	20-80
	soil freezing wall	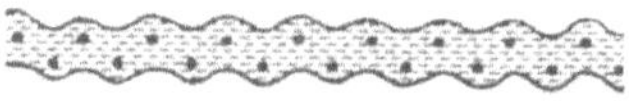	≥ 0,7	50-100
	jet grouting wall	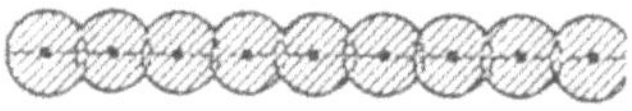	0,4-2,5	30-70
			≥0,15-0,3 [2)] (lamella)	20-30

soil displacement methods	sheet pile wall		≈ 0,02	20-30
	vibrating beam -slurry wall "thin (diaphragm) wall"	 	≥0,05-0,2 [3)]	10-35
	earth concrete driven- sheet pile wall	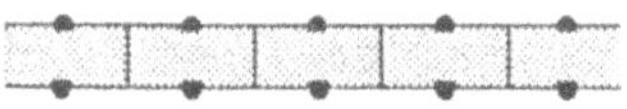	≥ 0,4	15-25
excavation methods	secant bored pile wall		0,4-1,5	20-40
	diaphragm wall (with hydrofraise)		0,4-1,6 [4)]	100-170
	diaphragm wall (with grab)		0,4-1,0	40-50 [6)]
	diaphragm wall with incorporated liner(s)		0,6-1,0 (0,4-1,6) [5)]	20-50

1) vibrocompaction, vibroflotation (vibrodisplacement, vibroreplacement)
2) total width of the lozenge-shaped jet grouting walls: ≥ 0,5 m
3) near the flanges of the vibrating beam significantly wider
4) up to 3,0 m in special cases
5) in special cases
6) 70-85m in exceptional cases

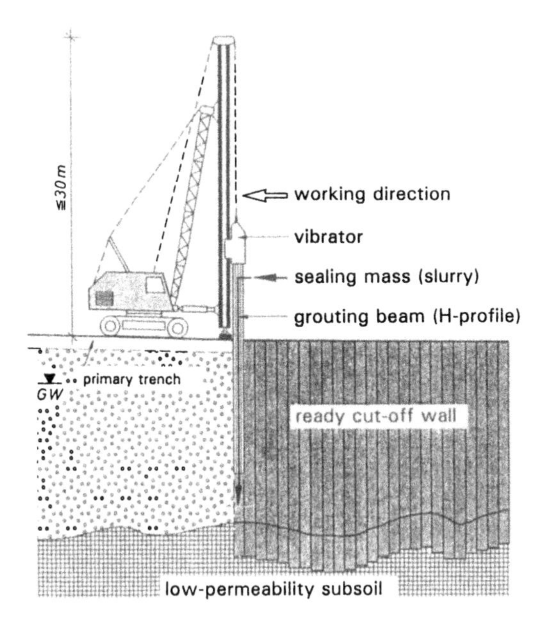

Fig.1:
Installation of a vibrating beam slurry wall ("thin diaphragm wall") - schematical.

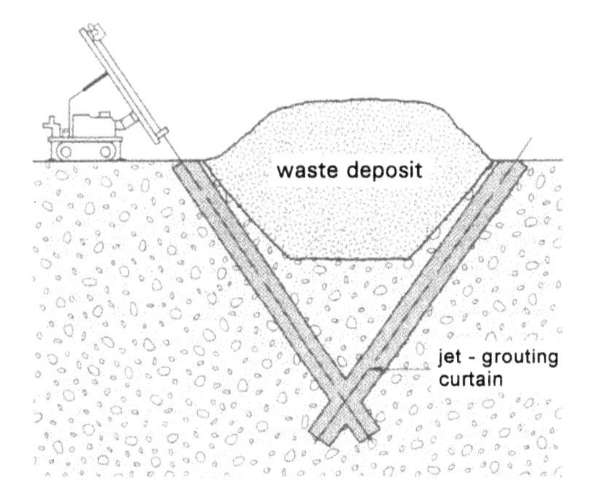

Fig.2:
Encapsulation of a waste deposit by inclined cut-off walls with the jet-grouting method.

ly if the groundwater level within the waste disposal area is lowered (Fig. 3). The cut-off walls should be embedded in a low permeability layer which may be a natural stratum or an artificial base (e.g. grouted zone). Otherwise the pumping discharge makes the method uneconomical for long-term containment.

2. GENERAL FEATURES OF CELLULAR CUT-OFF WALLS

From theoretical investigations and practical experience it is well known, that it is extremely difficult to localize leakages in vertical barriers which consist only of one wall. Furthermore a leak area of only 0,1 % of the screen area causes already a significant water discharge.

In case of a "membrane" the discharge is approximately 80 % of the full value (no sealing) for one leak and even more than 95 % for 10 leaks (of the same area as one large leak). But also for thicker cut-off walls the screening effect decreases strongly in cases of small leaks. Only if the surrounding soil fills these leaks, do the permeation conditions improve somewhat.

An accurate localization of leaks in horizontal base sealings is also rather problematical.

In order to avoid specific uncertainties of conventional sealing measures, a multi-barrier system was developed for hazardous waste deposits. It combines the advantage of vertical and horizontal cut-offs and fulfills all requirements of a high-safety disposal facility. Figures 4 and 5 show a solution that has proven successful even adjacent to large urban settlements. Generally, it may be used for all land disposal units

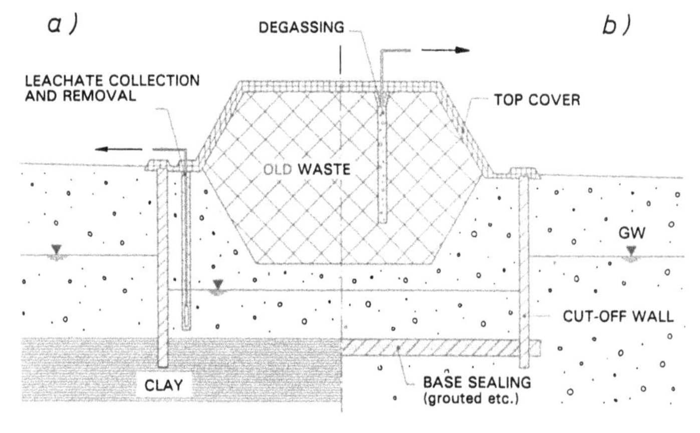

Fig.3:
Encapsulation of a waste deposit and groundwater lowering within the cut-off walls.
a) embedment of the cut-off wall in a natural low-permeability stratum
b) embedment of the cut-off wall in a grouted base sealing

(old and new ones; landfills, surface impoundments, waste piles), if the geological and topographical conditions are appropriate. The fundamental idea is to isolate the waste deposit by d o u b l e v e r t i c a l c u t - o f f s. These screens consists of two parallel walls being connected by cross walls at certain longitudinal intervals. About 4 to 8 m is recommended as the optimum distance between the longitudinal screens, according to the construction method. The transverse elements should not exceed a mutual distance of about 50 m. Thus a row of consecutive cells is formed surrounding the waste disposal site. The groundwater level within these cells is kept lower than outside the cells, but higher than inside the enclosure, i.e. the waste deposit. According to its first application in the year 1985, this method has been called the "Vienna cut-off double wall system".

This hydraulic system does not allow an outward discharge of contaminated groundwater. Rather a certain purification of the inner wall face is achieved. Furthermore the inward seepage pressure has a counter effect towards an eventual diffusion (e.g. chlorinated hydrocarbons) through the wall. Finally the double wall system acts as a gas barrier as could be proven on construction sites.
The cellular cut-off system has another advantage: It allows an accurate localization of leaks already during the construction period. As each cell contains a well and piezometer respectively, long-term monitoring and easy repairs are possible at all time. If measurements showed an increase of the groundwater level in a cell, this would indicate that the outer screen of this cell leaks. If the groundwater level in a cell fell to the level inside the encapsulation, the inner screen of this cell would obviously leak. This control system enables an exact local screen repair. Furthermore, groundwater being eventually contaminated by diffusion may be pumped off from the cells before polluting the surrounding subsoil. In all, the cellular arrangement of vertical cut-off walls, and the stepped groundwater lowering

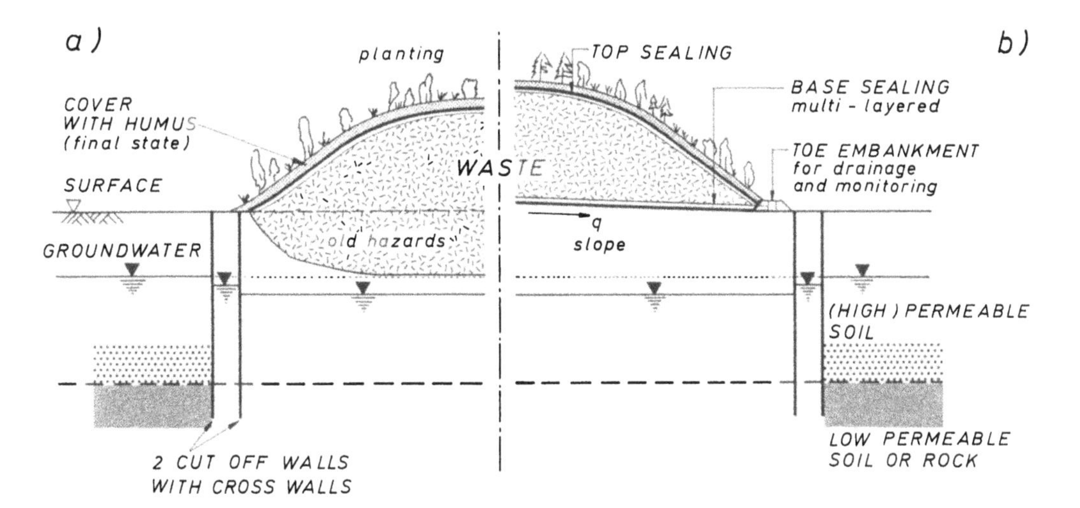

Fig.4:
High-safety encapsulation of waste deposits. Schematical sketch:

a) cut-off and top sealing of old waste deposits (or critical sites which might cause an environmental pollution)
b) scheme for designing new waste deposits with multi-barrier sealings.

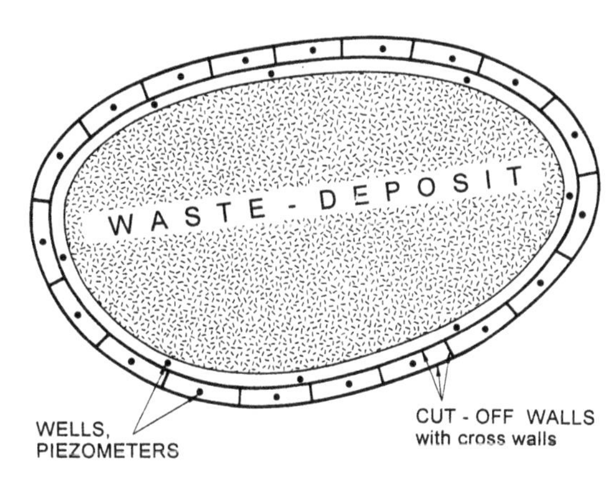

Fig.5:
Schematic ground plan to Fig.4 showing the "Vienna cut-off double wall system" with a cellular screen around a waste deposit site.

surmount the problems which may arise from conventional waste deposit designs.

The vertical screens need not reach in an "impermeable" stratum. A more or less horizontal layer with low permeability is sufficient, and the toe of the walls must be embedded well in it. Theoretical investigations and pracical experience have disclosed, that in most cases an embedding depth of 1,5 to 10 m is sufficient. Both parameters (k-value, embedding depth) are of great influence on the water amount which has to be pumped out from the encapsulated soil.

If the effectiveness of a base sealing (double-liner system) can be proven by a leak detection system and by groundwater monitoring beneath the bottom, the pumped water may be recharged outside the cut-off walls. However, contaminated water must be treated. A sewage sprinkling on top of the waste deposit is possible too - depending on the degree of the water contamination, on the local climate, and on the operation features of the waste deposit and the clarification plant.

The utilization or treatment of the pumped groundwater

depends on its contamination and on the waste disposal operation. In many cases the water can be used to dampen the wastes during filling and compaction (dust reduction), and to irrigate the greenery on the capping soil cover. On many sites only a minor part of the liquid has to be treated in purification plants.

Long-term experience has shown that a groundwater level difference of about 0,5 m (to 0,8 m) between inside and outside the vertical screens is sufficient even for large areas. Theoretically, even a value of $\Delta h \geqq 0,2$ m would be sufficient. In practice $\Delta h \leqq 0,5$ m is recommended for safety reasons as a reserve (failure of pumps, electric supply). Wells and piezometers within the enclosed waste disposal site must be installed only along the perimeter (indicated schematically in Fig. 5). Their optimal position depends on hydrological parameters and is therefore not as regular as that of the wells within the cut-off cells.

Inhomogeneities of the subsoil or geological pecularities within the enclosure or below the cut-offs are only of secondary importance: a local underseepage of the wall toes only increases the amount of water which has to be pumped from the enclosed area. But this risk can be minimized by appropriate additional soil investigations during the construction of the cut-off walls. Their depth is easily adaptable to the local soil conditions.

If such barriers are installed within a free groundwater field (above "impermeable" layers), the alteration of the ground water level outside the enclosure is negligibly small. In case of the Vienna central waste deposit (600.000 m^2 area) the maximum groundwater rise was 0,4 m (upstream), and the maximum drawdown was 0,25 m downstream the vertical barrier. Such magnitudes have been measured also on other construction sites; they are smaller than the natural seasonal fluctuations of the ground water table.
To summerize, the waste encapsulation acts as a pot-shaped structure being flown beyond without harming the surrounding ground water.

The cellular double wall system is clearly superior to simple cut-offs as could be realized not only theoretically but also in practice. A lowering of the groundwater table within the vertical screens of of a waste deposit in order to avoid an outward flow of contaminated water had been already designed in the year 1981 - in order to reclaim a contaminated area in Vienna. Difficulties in finding some leaks in the cut-offs already during the construction period (requiring an increasingly great pump capacity) finally led to the cellular cut-off scheme.

Fig. 4a and 4b indicate that this multi-barrier system is suitable for sealing off existing waste disposal dumps as well as for new waste disposal facilities. A top lining should be provided in any case when the deposit is closed. This reduces the vertical water inflow and finally enables a planting on the soil cover.

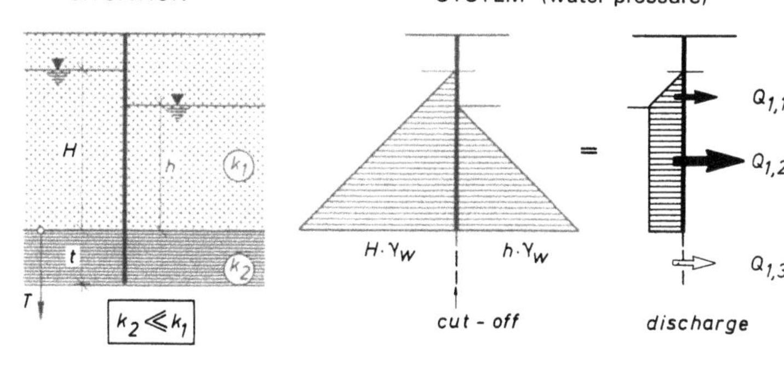

Fig.6:
Water discharge through a cut-off wall;

Fig. 4b shows a design with a self-sustaining leachate collection scheme. The outward gradient of the base causes drainage and waste waters to flow freely to the ring shaped toe embankment. This surrounding embankment contains the sewage collector and increases the static safety of the waste deposit. The only disadvantage of this cupola-shaped base is the reduced volume of deposits - if compared with the hollow-shaped construction according to Fig. 4a.

For vertical waste containment diaphragm walls and vibrating beam slurry walls ("thin walls") dominate. Secant bored pile walls have proved suitable only for small areas and depths, and in case of caverns or bulky wastes in the ground. Jet grouting is preferred for local sealings and leak repair.

3. GROUNDWATER DISCHARGE EVALUATION

Of particular importance for the long-term operation is the tightness of the vertical cut-off walls and the permeability of the stratum where their toe is embedded. Hence a water discharge evaluation is necessary for a prognosis of the pump capacity and for a detection of possible leaks.

According to the narrow mutual distance of the parallel walls of a cellular cut-off, the system approximately may be considered hydraulically a single cut-off.

In most cases the subsoil conditions of the described multi-barrier system with vertical screens are charakterized by a high-permeable layer which is underlain by a low-permeability stratum - as schematically indicated in Fig. 4. If the filter-loss in the high-permeability aquifer is neglected, the discharge i n t o the enclosed soil body may be evaluated as follows:

Discharge through a vertical cut-off, Q_1:
(along the high-permeable aquifer - see Fig. 6)

$$Q_1 = Q_{1,1} + Q_{1,2} + Q_{1,3} \quad (1)$$

$$Q_{1,1} = \frac{k}{d} . 0,5 (H - h) . \gamma_w . \quad (2)$$

$$. (H - h) . C = \frac{k}{d} . 0,5 . (H - h)^2 . C$$

$$\text{for } \gamma_w \doteq 1 \text{ t/m}^3$$

$$Q_{1,2} = \frac{k}{d} . (H - h) . h . C \quad (3)$$

$$Q_{1,3} \simeq 0 \quad (4)$$

This term may be neglected in practice. In case of local uncertainties about the low-permeability of the underlain layer, the value H should be increased properly.

Hence

$$Q_1 = 0{,}5.\left(\frac{k}{d}\right).(H^2 - h^2).C \quad (5)$$

C = circumference of the cut-off waste disposal site
k = permeability coefficient of the cut-off
d = thickness of the cut-off
k/d = "permittivity" (ψ) of the cut-off wall (should be $\leq 10^{-7}$ sec^{-1})
H,h = depth of groundwater in the high-permeable aquifer (Fig. 6)
k_2 = permeability coefficient of the low-permeable stratum
t = embedment depth of the cut-off in the low-permeable stratum
T = (ficticious) depth of the low-permeable stratum, relevant for the hydraulic calculation ("hydraulic influence depth")

Discharge from the low-permeability subsoil, Q_2:
("inverse underseepage" - see Fig. 7)

$$Q_2 = C.q_2 = C.\left(k_2.\frac{H-h}{2}\right) : \left[\sqrt[3]{\frac{t}{T-t}} + \frac{d}{2(T-t)}\right] \quad (6)$$

As the term $\frac{d}{2(T-t)}$ can be neglected in practice, the formula decreases to

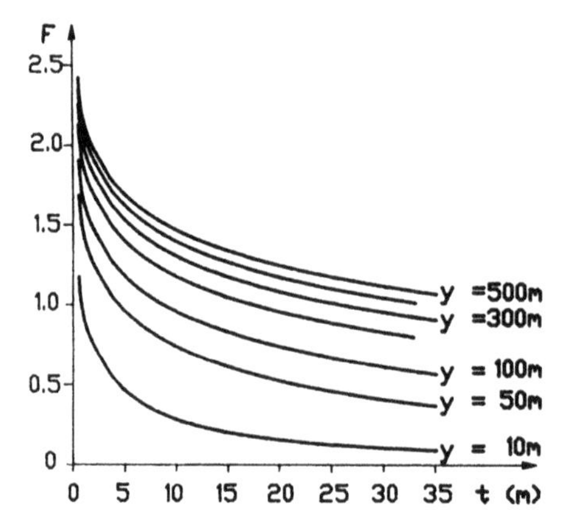

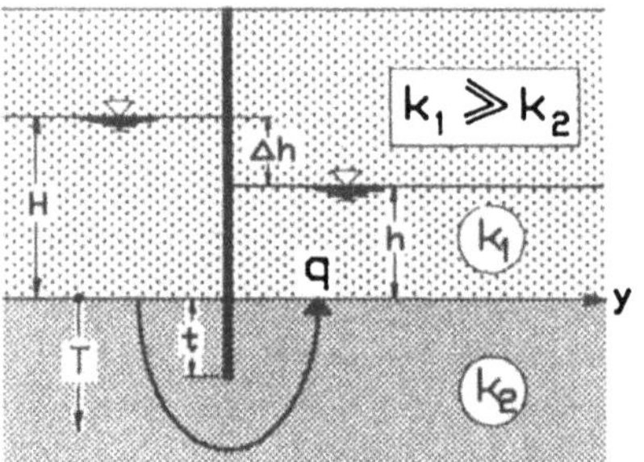

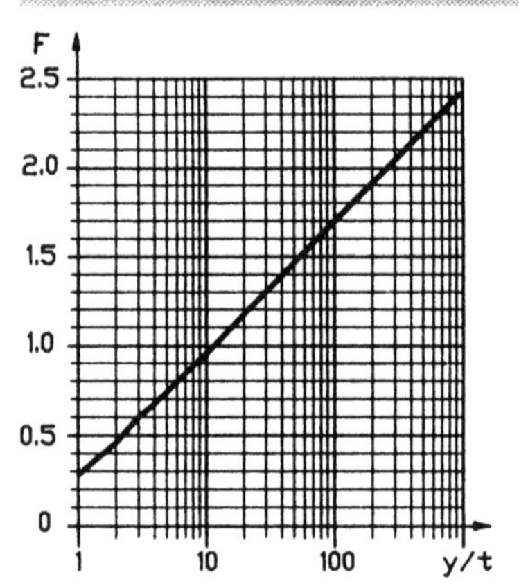

Fig.7:
Diagram for assessing the groundwater amount of underflown cut-off walls
F....shape factor of the underseepage (see equ.5)

$$Q_2 = C.k_2.0{,}5.(H-h).\sqrt[3]{T/t - 1} \quad (7)$$

Although the equations (6) and (7) are based on several approximations, they have proved sufficiently reliable in engineering practice. Commonly, the natural scattering of the soil parameters on a site is of much more influence on the results than a sophisticated theory.

The value T can only roughly be estimated by parametric studies, if the low-permeability stratum reaches very deeply. In most cases the increase of Q_2 may be neglected in practice for $T \geqq 3t$.

In case of a deep-reaching low-permeability subsoil ($T \rightarrow \infty$), the underflow of the vertical screens may also be evaluated according to Fig.7:

$$Q_2 = \sum_{i=1}^{n} \Delta h_i \cdot k_{2i} \cdot F_i \cdot \Delta l_i \quad (8)$$

$$F_i = \frac{1}{\pi} \cdot \int_0^{y} \frac{dy}{\sqrt{t_i^2 + y_i^2}} \quad (9)$$

$$\sum_{i=1}^{n} \Delta l_i = C \quad (10)$$

Δl_i = length of the element i
F_i = shape factor of the underseepage
y = horizontal distance from the underflown cut-off
t_i = embedding depth of the cut-off in the considered point i
Δh_i = effective total head difference in point i
k_{2i} = permeability coefficient of the low-permeability stratum in point i

The distances y have to be chosen in a way that the seepage lines don't intersect within the cut-off area. Hence a flow net over the whole cut-off site should be sketched. Fig. 7 gives the value F as a function of the depth t and the distance y. It clearly demonstrates that the maximum sealing effect is gained with $t \leqq 10$ m. For greater depths the relation between cut-off effect and economical aspects (construction costs) becomes rapidly worse.

In practice, equs. (7) and (8) provide the same results if proper parameters are chosen. Comparative calculations are recommended.

For evaluating the discharge Q_f through faults or leaks in the cut-offs, the following formula can be used:

$$Q_f = 2.(H - h).k_i.\sqrt{A/\pi} \quad (11)$$

k_i = permeability coefficient of the soil surrounding the leak
A = area of the leak

Equ. (11) is based on Dachler's theory (1936) referring to an artesian groundwater flow to a well or trench with a horizontal and plane bottom (Fig. 8):

$$Q = k.(H - h_o).4r_o \quad (12)$$

In case of a vertical cut-off, the hydraulic scheme of Fig. 8 must be turned by 90°, and a mathematical "reflection" is necessary to describe the hydraulic resistances through a leak. Contrary to a well or trench, there exists not only an inflow resistance but also an outflow resistance of about the same magnitude. Furthermore, the area of the leak or fault with a more or less irregular shape is roughly approximated by a circle. Hence, Formula (12) is modified by:

$$r_o \rightarrow \sqrt{A/\pi}$$

$$H - h_o \rightarrow H - h$$

$$\text{factor } 4 \rightarrow 4:2 = 2$$

$Q \rightarrow Q_f$, thus leading to equ. (11)

Experience has shown that equations (5), (7), (8), (11) provide fairly reasonable results. Sophisticated calculations by the FEM seam to increase the accuracy only within a range which is secondary for the practice, as parametric and comparative studies revealed. Of primary importance is a proper choice of the soil characteristics, combined with a detailed knowledge of local inhomogeneities (interlayers, lenses etc.). A scatter of the hydraulic conductivity by three to five decimal exponents even in geologically similar strata of one site is no rarity in practice.

Parametric studies and long-term site experience have revealed two essential results (Radl, 1991):

- A low hydraulic permittivity of cut-offs is necessary only to a certain extent (depending on the characteristics of the low permeability stratum). Additional measures do not reduce significantly the annually pumped water discharge.
- The embedment depth of the cut off walls reduces the annually pumped water discharge more than a reduction of the wall permeability, if the embedding layer is relatively permeable.

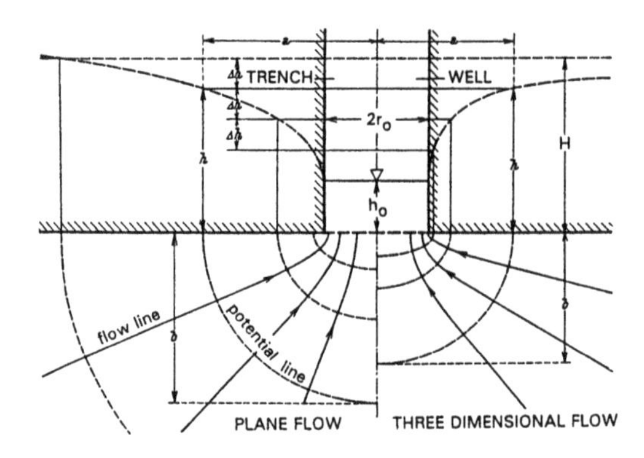

Fig.8:
Inflow of artesian groundwater to a trench or well with horizontal and plane bottom.

These results are demonstrated in Fig. 9 for a practical example with a cut-off ground area of 300 by 200 m. The intersecting point between the 15 % - line and the specific curve provides those k/d-values, where the pumped discharge (pumping amount) exceeds the theoretical minimum of k/d = 0 (impermeable wall) only by 15 %. According to Fig. 7 a hydraulic influence depth of $T = \infty$ was assumed for theoretical reasons. Actually the increase in Q for $T \geqq 3t$ is practically negligable.

Figure 9 is valid for a cut-off circumference of C = 1000 m. In case of $C \gtrless 1000$ m the pumping amount approximately can be derived by a simple conversion linearly to C. The result is sufficiently accurate for common site shapes. Extremely long and narrow areas have a relatively small "hydraulic radius" R = A/C resulting in a decreasing

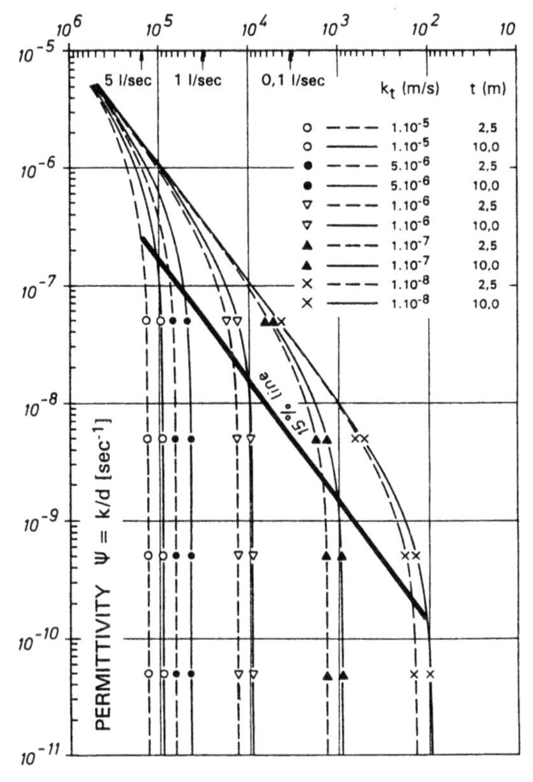

Fig.9:
Results of parametric calculations on the effectiveness of cut-off walls: pumping amount Q for lowering the groundwater within an isolated waste disposal area (300 x 200m) by $\Delta h = 0{,}5m$. Thickness of water bearing aquifer $H = 10m$ (see Fig.6,7)
$Q = f(k/d, k_t, t)$....discharge through the cut-off wall and underseepage for $T = \infty$
$k_t = k_2$ in Fig.6,7

discharge in relation to a circular or square shaped area. Consequently, equ. (7) and (8) lead to an overestimation of Q_2 in such cases.

Fig. 9 shows that even in case of a very low permeable stratum with $k_t = 10^{-8}$ m/s a permittivity of $\psi = k/d \leqq 10^{-10}$ sec^{-1} is not suitable. Commonly, the in situ large scale conductivity will be $k_t \geqq 10^{-7}$ m/s, thus justifying an allowable permittivity of $\psi \geqq 10^{-9}$ sec^{-1}. This value can be gained with cut off walls commonly used in ground engineering practice. The following range of application has proved suitable:

$$10^{-9} \; sec^{-1} \leqq \psi \leqq 10^{-8} \; sec^{-1}$$

for diaphragm walls ($d \geqq 60$ cm)

$$10^{-8} \; sec^{-1} \leqq \psi \leqq 10^{-6} \; sec^{-1}$$

for vibrating beam slurry walls ($d \geqq 5$ cm)

In practice the permittivity should be considered the dominating parameter describing the hydraulic conductivity of a cut-off wall. Consequently, a low k-value must be achieved for thin walls, and a higher hydraulic conductivity is allowed for thick diaphragm walls. Thus the wall thickness could be left to the choice of the contractor (according to his equipment etc.). Commonly $\psi \leqq 1.10^{-7}$ sec^{-1} should be required.

4. POLLUTANT MIGRATION

The migration of pollutants through a cut-off wall may occur by convection (hydraulic flow) and/or diffusion; sorption has a retardation effect (adsorption), but desorption is also possible.

The required hydraulic conductivity of diaphragm walls or "thin walls" commonly is about $k \leqq 5.10^{-9}$ m/s, but values down to $k \doteq 2.10^{-12}$ m/s have also been achieved frequently. In spite of such a low permeability, migration of hazardous constituents may occur by permeation, a process which involves both - diffusion and sorption. This migration can be described by the differential equation:

$$D.\frac{\partial^2 c}{\partial x^2} + k.(i - i_o).\frac{\partial c}{\partial x} - S.\frac{\partial c}{\partial t} = 0 \quad (13)$$

c = concentration in the place x at the time t
D = diffusion coefficient
k = hydraulic conductivity
i = hydraulic gradient (i_o = initial value)
S = sorption coefficient

Below the initial hydraulic gradient, i_o, convection is negligibly small or zero. It is defined as intersection point of the k-i-line with the i-axis.

The installation of an impermeable liner ("composite cut-offs") leads practically to an elimination of the second term of the differential equation as $k.(i - i_o) \rightarrow 0$. The diffusion coefficient is also reduced but can be still more improved by using slurry sealing mixes of a very small void ratio.
Both methods increase the resistance towards chemical aggression significantly and are a suitable measure to improve vapour containment. Generally the diffusion coefficient decreases with the void ratio. Consequently, mixtures have been developed which provide a very high density.

Another possibility of avoiding pollutant migration occurs when the groundwater table is lowered within cut-off walls (similar to Fig. 3, 4). By means of an inverse flow, a fluid-dynamic equilibrium is obtained which minimizes an outward diffusion of contaminants. The penetration of the CHC phase in a water-saturated pore matrix does not commence below a certain "initial pressure", created by the capillary depression at the interface between the non-wetting CHC phase and the water phase (Friedrich, 1989). Consequently, the existence of this limit value is somewhat similar to the "initial gradient" in case of hydraulic conductivity. It can be evaluated by the following physical parameters: surface tension and contact angle of the solid/CHC/water - three phase system, and effective pore channel radius of the solid matrix. The differential equation of pollutant migration changes to:

$$D.\frac{\partial^2 c}{\partial x^2} - k.i.\frac{\partial c}{\partial x} - S.\frac{\partial c}{\partial t} = 0 \quad (14)$$

For the stationary state a minimum ground water lowering Δh within the cut-off walls may be roughly estimated by the following formula which was originally developed by Friedrich (1989) for vertical CHC seepage within an encapsulated waste disposal:

$$(H - h)_{min} = \min \Delta h = \frac{D.n}{k} \cdot l_n\left(\frac{c_o}{c}\right) \quad (15)$$

(see Fig. 7)

D = diffusion coefficient
k = hydraulic conductivity of the cut-off
n = void ratio of the cut-off
c_o = pollutant concentration in

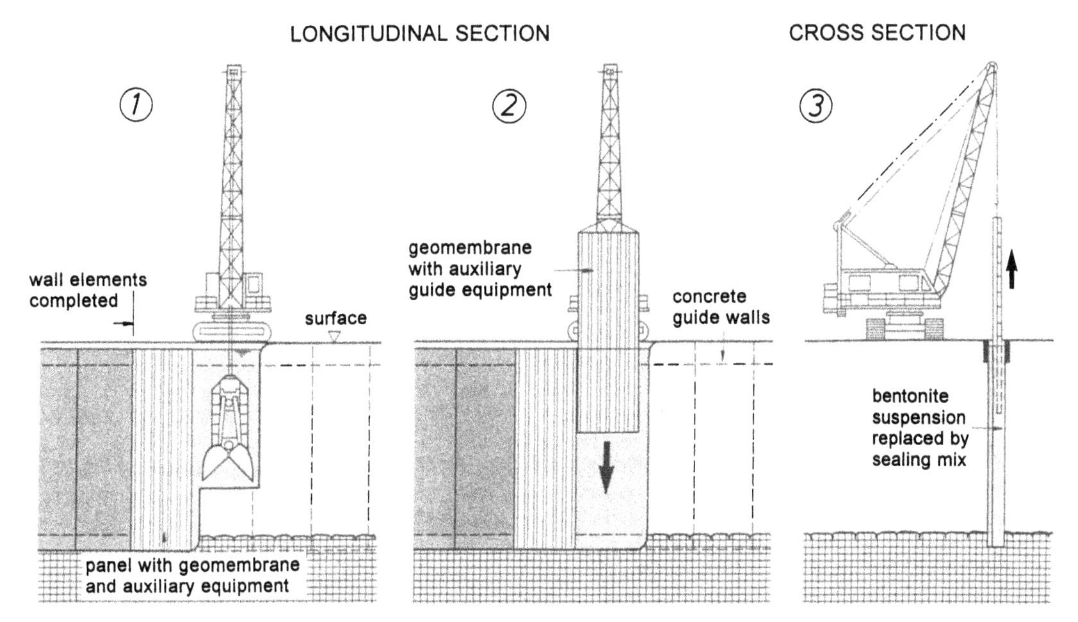

Fig.10:
Diaphragm wall with eccentricly incorporated geomembrane (Fischer 1988)
1......excavation of the slurry trench; two phase diaphragm wall technique (suspension replaced by a special sealing mix)
2......installation of the geomembrane
3......withdrawing of the auxiliary structures

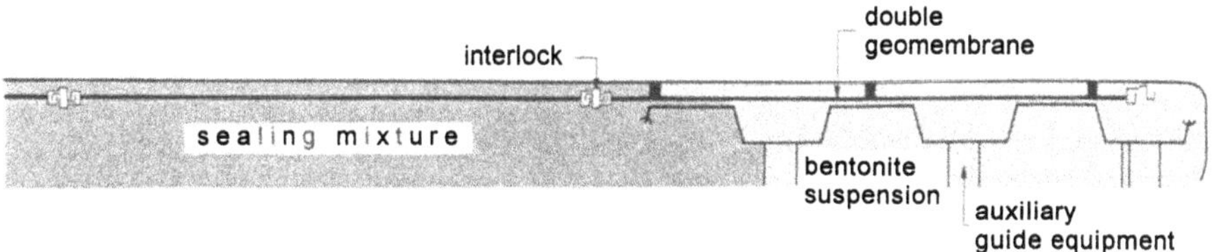

Fig.11:
Ground plan to Fig.10;
construction stage 3 and completed wall

the groundwater within the cut-off
c = pollutant concentration in the groundwater outside the cut-off

The value Δh is necessary to reduce the concentration from c to c_o. It should be emphasized that it does not depend on the thickness of the cut-off; hence thin vibrating beam slurry walls would have (theoretically) the same effect like thick diaphragm walls. The essential difference lies in the quantity of the permeating constituents; an increasing wall width reduces the concentration gradient and

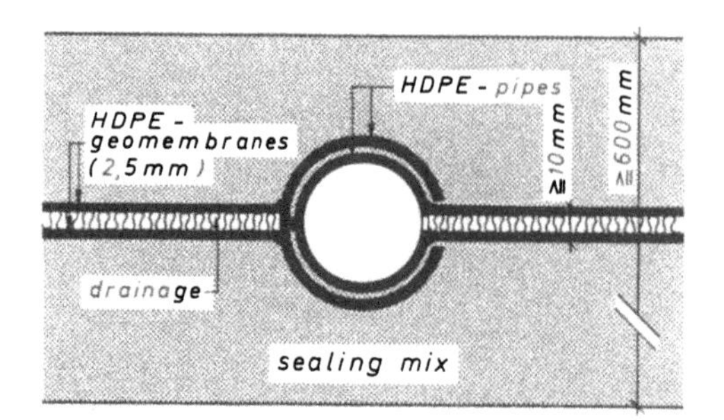

Fig. 12:
Composite cut-off wall with integrated geomembrane.
Central geocomposite lining system with two geomembranes and a drainage core as a leachate collection and removal system.

therefore reduces the diffusing discharge.
Long-term monitoring has shown that equ. (15) is only valid for a permeability coefficient of about

$$10^{-9}\ \text{m/s} \leqq k \leqq 10^{-7}\ \text{m/s}$$

From site experience, it is known that a groundwater lowering of about Δh = 0,5 to 1,0 m is normally sufficient to prevent pollutant permeation through cut-off walls.

5. COMPOSITE CUT-OFF WALLS

Diaphragm walls with an integrated second sealing screen are called "composite cut-off walls". The additional vertical elements may be geomembranes (with geocomposites), steel or special glass liners, or prefabricated reinforced concrete panels. Geosynthetics are preferred, the maximum depth hitherto achieved being about 30 m. Commonly, HDPE-geomembranes are mounted on a steel frame and lowered into the fresh suspension of the slurry trench; guide planks are also in use as auxiliary measures (Fig. 10, 11). For greater depths (up to 50 m) the geomembrane should be installed by unrolling it from a special endless conveyer and using toe weights.

Fig. 12 shows a geocomposite lining system with a drainage core. In case of a discharge through the wall or geomembranes, this drainage system serves as a leachate collection and removal system; hence it acts as a leak detection system. Furthermore, hazardous constituents, migrating by diffusion can be removed by the drainage system and the monitoring pipe. Further details on interlocking schemes to connect geomembrane liners are given in (Brandl, 1990).

Fig. 13 demonstrates a composite cut-off wall with a special glass liner; double or multiple glass screen systems are also available. The latest innovation is a special glass liner with incorporated pollutant indicators which enable a permanent monitoring.

Another alternative is the insertion of sandwich-like composite panels into the slurry trench wall (diaphragm wall). In this case a metal foil (usually aluminium) is placed between two HDPE-geomembranes. The method is theoretically useful if seepage or groundwater is heavily contaminated by chlorinated hydrocarbons, and also protects against certain vapour migration. Comprehensive practical experience does not yet exist however.

Usually the impermeable screen is placed centrically within the diaphragm wall. With regard to the standard construction equipment and the width of the interlocks the thickness of such composite cut-off walls is 600 mm at the minimum. A reduction to 400 mm may be possible but should be applied only in case of box-shaped cut-off systems or if the groundwater level is lowered by pumping within the enclosed area. The length of the wall sections depends on the soil properties, groundwater level and wall depth. It varies between 2,5 to 10,0 m, commonly being about 5 to 7 m.

The compatibility of the incorporated liner and the surrounding sealing mix must be tested initially and proven site-specifically.
A disadvantage of composite cut-off walls is the complicated construction, the high costs and a lack of long-term experience with their use. Consequently, they are used only in exceptional cases. Cellular cut-off systems with a lowered groundwater table are practically more effective than single cut-offs with an inner liner.

Efficiency control and long-term monitoring of single cut-off walls requires a close net of piezometers on either side of the curtain. The longitudinal spacing of the piezometers should be 50 m at the maximum; in case of leak detection even 10 m might become necessary. Moreover it is recommended to install two piezometer rows outside the cut-off in order to achieve a flow net. This facilitates the interpretation of measurement results significantly.

6. CONSTRUCTIONS

Figure 4 and 5 indicate that the cellular cut-off system may be used for enclosing existing waste as well as new disposal sites.
Cut-off walls should be installed, if possible, outside a highly contaminated

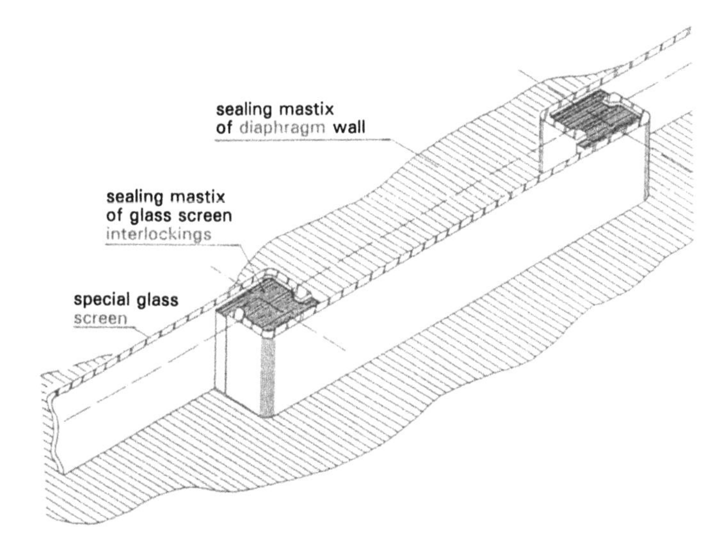

Fig.13:
Composite cut-off wall with integrated special glass liner. Interlockings are filled with a pollutant-resistant, volume-constant sealing mastix.

area. In case of an inverse groundwater seepage the life of the cut-off is significantely increased and the diffusion reduced. If the waste disposal facility is equipped with a base sealing and drainage system (e.g. Fig. 4b), the cut-off walls are completely protected from chemical attack by aggressive waters (leachates).
The embedment depth of single cut-off walls should be somewhat greater than for a cellular cut-off system under comparable conditions. The additional depth ranges between 1,0 to 2,5 m, depending on the subsoil characteristics and the cut-off wall geometry.

Depending on the particular circumstances (and the costs) the vertical cut-offs may be vibrating beam slurry walls ("thin walls"), diaphragm walls, injection walls (jet grouting) etc. - see Table 1. Usually the one-phase construction method is preferred for diaphragm walls, although their density is somewhat smaller than of two-phase diaphragm walls. For deep systems ($\geqq$ 40 to 50 m) rather the classical two-phase method with a bentonite suspension and hydrofraise excavators should be used.

The hydrofraise enables an excavation of slurry trenches even in fissured rock. Moreover a permanent monitoring with inclinometers provides a high accuracy during construction: presently about 0,1 to 0,5 %, depending on the subsoil, site equipment and depth. The latest innovation is hydrofraise excavation combined with a one-phase cut-off wall, using a special suspension (with 250 kg/m³ solid additives).

The installation of vibrating beam slurry walls is significantly cheaper and quicker than of diaphragm walls. The common H-beam and overlapping are indicated in Fig. 14a. In dense subsoil an overlapping length of about 50 % of the beam has proven suitable (Fig. 14b). The theoretical minimum width of d $\geqq$ 5 to 8 cm is commonly exceeded in practice, depending on the grouting pressure.

Driving H-beams of thin slurry walls in dense or even cemented soil may cause problems with the stability of the fresh slurry. Site experience and in-situ tests have shown that long driving periods in heavy ground lead to a significant heat development. This may desta-

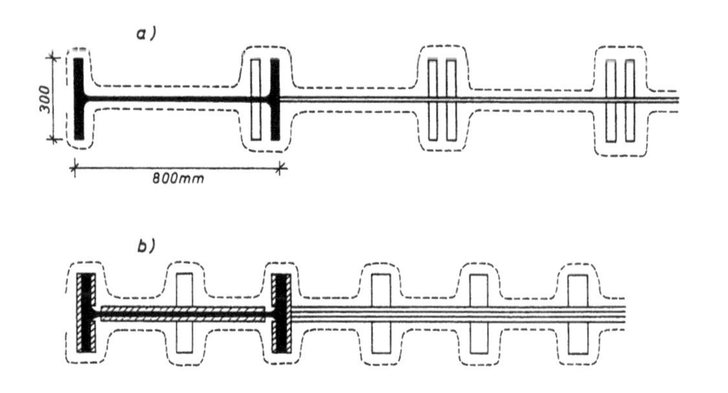

Fig.14:
Ground plan of a vibrating beam slurry wall ("thin wall") with overlapping wall units and infiltration zone (broken line); schematical
a) Common profile and 10% overlapping
b) Toe-stiffened beam and 50% overlapping (in dense soils)

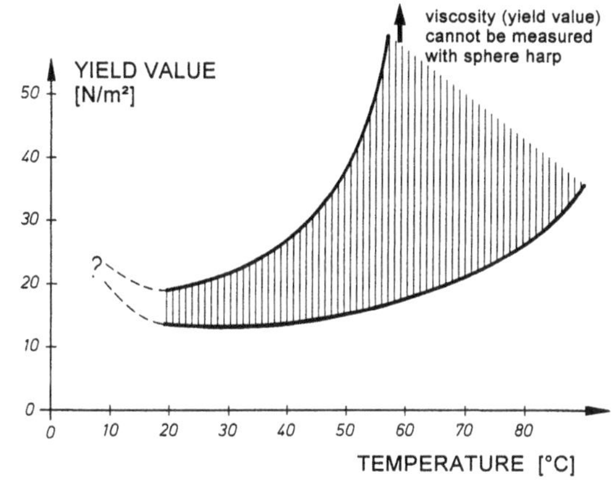

Fig.15:
Yield value versus working temperature for common sealing mixes of vibrating beam slurry walls ("thin diaphragm walls").

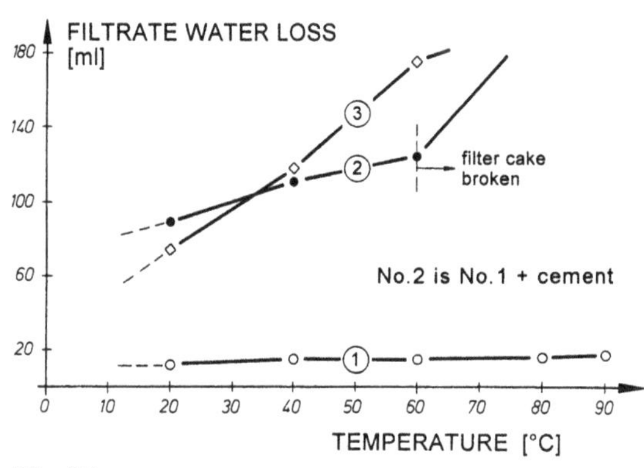

Fig.16:
Filtrate water loss versus working temperature for three typical sealing mixes of vibrating beam slurry walls.
Filter press (API) 700kN/m², 450 sec

bilize the slurry (depending on its mixture), and provide a highly permeable cut-off wall:
Fig. 15 shows how the yield value (determined by means of the sphere harp) increases with temperature. The hatched area indicates the scatter of commonly used slurries. On site, temperatures of about 80°C may thoroughly occur during vibrating in dense soil. Too high yield values lead to a significant loss of workability of the fresh slurry, and consequently to a lower quality of the cut-off wall.
Similar conclusions could be drawn by checking the filtrate water loss (Fig. 16). High temperature may cause an excessive loss of filtrate water, and consequently a rather permeable cut-off wall.
Even the stress-strain behaviour is influenced by the temperature (e.g. Fig. 17), though this effect is usually not so important.

The maximum depth of vibrating beam slurry walls frequently is limited due to boulders or cemented zones in the ground. Latest innovations overcome these problems by combining the vibrating beam system with jet grouting (e.g. "vibrosol-wall"). During vibrating downwards the penetration resistance is reduced significantly and when pulling the beam high pressure grouting is performed to achieve a close connection between adjacent curtain panels and a sufficient thickness of the cut-off. Thus, vibrating beam walls can be constructed also in cohesive silty soils. Other precutting methods also have been developed.

If the ground contains relatively large voids or hollow spaces, caverns etc. (e.g. artificial loose fills, leached out or karstic zones) a previous grouting is recommended before diaphragm walls or vibrating beam slurry walls are installed. Otherwise a high percentage of the slurry would get lost in the ground, and the cut-off wall would exhibit local leaks ("windows").

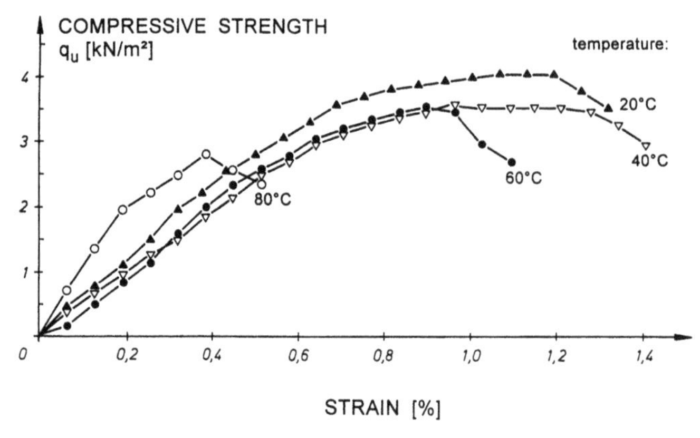

Fig.17:
Unconfined compressive strength versus strain for a typical sealing mix of vibrating beam slurry walls walls. Working temperature (during installation) as a parameter.
Ratio height: diameter of samples = 2:1, 28 days of curing (20°C)

The installation of cut off walls causes vibrations which locally may exceed allowable values in adjacent buildings (e.g. for sensitive print equipments, computer centres etc.). Vibrating beam slurry walls are especially critical. In this case it has proved suitable to place waste tyres beneath the crawlers of the vibrating equipment. This simple measure significantly reduces the maximum vibration velocity.

The advantages of diaphragm cut-off walls may be summarized as follows:

- well established technique;
- flexible structure with high earthquake resistance;
- dense soil, boulders and even rock are no definite obstacles;
- large depths;
- easy repairability;
- installation of additional sealing elements (e.g. Fig. 10 to 13).

The optimum mixture of diaphragm walls (as well as vibrating beam slurry walls) must be adapted to the specific requirements of the waste disposal and local subsoil, but also to the workability during construction. Therefore comprehensive compatibility tests in the laboratory and site controls are necessary. Numerous investigations have shown that the permeability and strength of the wall is influenced by various parameters as mixture composition (percentage of components), chemical additives, quality of bentonite (Ca^{++} increasingly preferred to Na^{+}), cement-grade and its fineness of grinding etc.

Generally, a unit weight of $\gamma_d \geq 12$ kN/m³ should be achieved after setting and hardening of the slurry wall. The high density provides an efficient barrier against diffusion of contaminants and a high chemo-physical resistance of the cut-off. In case of vibrating beam slurry walls the quality assurance should be based on in-situ samples from the hardened wall or on laboratory-samples (from the fresh in-situ slurry) which are cured under a lateral pressure exceeding the earth pressure at rest. Otherwise too small densities are measured.

Frequently used recipes of mixtures for one-phase diaphragm cut-off walls are (related to 1 m^3):

25 to 40 kg natrium-bentonite
150 to 250 kg blast furnace cement
900 to 940 kg water

Na-bentonite has a higher capacity of pollutant adsorption than Ca-bentonite. But on the other hand such cut-off walls are more sensitive towards long-term chemical attack, and they show a higher diffusion of several pollutants. Therefore Ca-bentonite is preferred in many cases.

At the central Vienna waste disposal facility the following (standard) mixtures were used for 1 m^3 sealing mass:

Vibrating beam slurry walls:
(area 145 000 m^2; depth 20 to 27 m)

115 kg calcium-bentonite
138 kg blast furnace cement
576 kg filler (pulverized limestone)
645 kg water

In case of Na-bentonite a higher quantity of filler (about 700 kg) would have been necessary to provide a sufficient density of the wall.

Diaphragm walls (one-phase method):
(area 8000 m^2; depth 27 to 45 m)

165 kg calcium-bentonite
144 kg blast furnace cement
826 kg water
3,8 kg chemical additives

Using Na-bentonite requires less bentonite but a higher quality of cement to provide a sufficiently low permeability of the wall (e.g. Fig. 18 and Meseck, 1987). Finally it should be emphasized that the cement properties are also of outstanding influence on the quality of the cut-off. Experience has shown that they often differ significantly.

For special cases (chemical attack), mixtures without cement have been developed for cut-off walls. Commonly this requires a two-phase construction, and the mixes consist of about 10 % of fluid and 90 % of solid mass: silanes, water

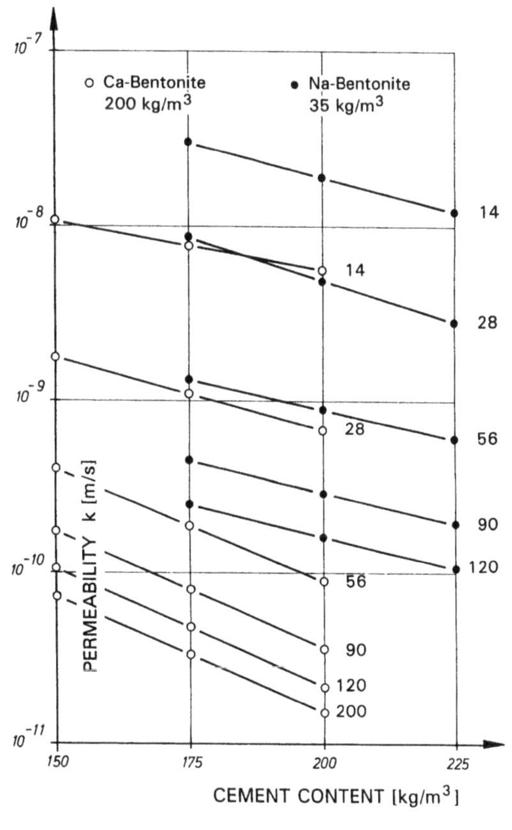

Fig. 18:
Permeability (hydraulic conductivity) versus cement content of sealing mixes with Na- or Ca-bentonite. Curing age (days) as parameter. Water/cement-factor decreases with increasing cement content (blast furnace cement).

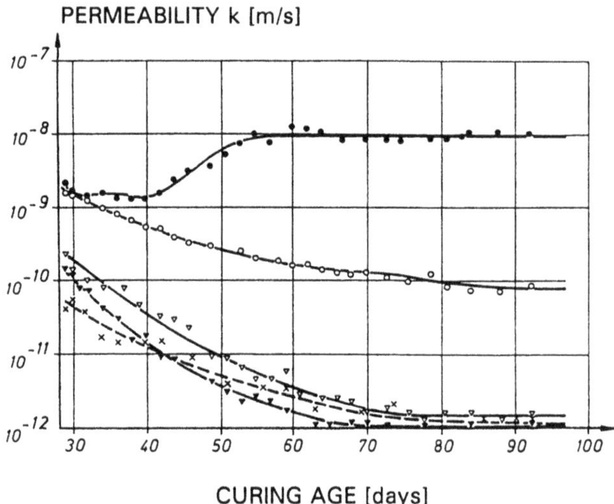

Additives to sealing mix	Test fluid	
	Aqua dest.	Leachate
Na-Bentonite	o	●
Ca-Bentonite	▽	▼
Diwa -mix	x	— — —

Fig.19:
Alteration of the hydraulic conductivity of cut-off wall sealing compounds with time.
Additives to the mixture and test fluid as parameters.

glass, water, clay, sand + gravel (0/32 mm). Bentonite is preferred as clay material, but illite and kaolinite are also in use, according to the specificly required resistance and retardation/sorption effect. Such mixtures reach a hydraulic conductivity of about $k \leqq 5.10^{-11}$ m/s and a very low void ratio (diffusion coefficient respectively). If no leaching or erosion occurs, the strength of the walls increases with time, and the permeability decreases.

The suitability test must include laboratory investigations of the long-term behaviour of the sealing compounds. Commonly, permeability tests and free storage tests with chemically treated test liquids are used for this purpose (e.g. Fig. 19). A modification is proposed by Müller-Kirchenbauer et al. (1991), taking into account a simultaneous convection and diffusion in one direction (worst case).

7. QUALITY CONTROL

The quality control of cut-off walls is begun by testing of the construction materials and mixtures (quality assurance), and followed by comprehensive construction testing, site supervising and (long-term) control. This procedure requires laboratory testing and in-situ testing, depending on the system of the cut-off walls. During and after construction, the following control measures are recommended:

- Taking specimens from the fresh site mix for determining density, filtrate water loss, yield value etc.
- Taking samples of slurry from the trench (before settling) using a discreet sampler; determining density, permeability, erosion stability, unconfined compressive strength, deformability etc.
- Taking core samples from the hardened wall for determining density, permeability,

unconfined compressive strength etc.

- In-situ permeability tests as recharge tests in boreholes (Fig. 20).
- Full-scale permeability tests with the whole cut-off system (Fig. 21).

Commonly a hydraulic conductivity of $k \leqq 10^{-8}$ to 10^{-10} m/s is required after 90 days.

For checking the erosion stability, the pinhole-test has proved suitable. This test was developed originally for identifying dispersive soils (ASTM, D 4647-87). For investigating cut-off wall mixtures a modified version is used. The hydraulic gradient should be i = 200 and the length of the pin-hole 15 cm.

A curing period of 56 days is recommended for the samples. The test itself should run at least five weaks: experience has shown that in many cases no indication of inner erosion may appear within the first three weeks, though the sealing mixture exhibits no long-term resistance.

The unconfined compressive strength is another test used to measure the erosion stability of cut-off walls. Theoretically a permanently plastic, i.e. flexible sealing mass would be optimal, but in practice, a very low strength favours inner erosion. Therefore the unconfined compressive strength of 7 days old samples should exceed 0,3 to 0,5 MN/m² (ratio of height to diameter h:d = 2:1). The short curing period of 7 days (instead of 28 days as common) is unavoidable in order to optimize the sealing mixture on the construction site in time.

The resistance against chemicals has to be investigated by long-term tests lasting over a period of 90 days at the minimum.

Further control refers to:

- accuracy of position of the cut-off panels,
- verticality of the cut-off panels,
- overlapping of excavation panels (diaphragm walls) or vibration panels (vibrating beam slurry walls),
- embedment depth of the cut-off panels,

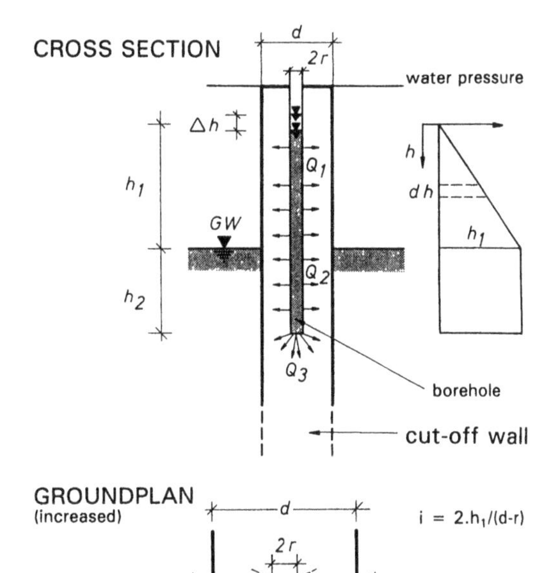

Fig.20:
In-situ permeability test in boreholes within cut-off diphragm walls (schematical).

- soil or rock characteristics along the cut-off wall, especially of the low-permeability stratum.

Depending on the trenching equipment the verticality of the panels can be supervised by:

- inclinometers
- ultrasonic control of the trench
- electronic measuring device mounted on the grap (constant monitoring of the angle of inclination and depth)

Of particular importance is - above all - a permament control of the wall toe embedment in the low permeable stratum. This must be checked continually during construction, i.e. for each unit. The following methods are recommended:

Diaphragm walls:

- Taking soil samples above and especially from the toe of the cut-off wall.
- Surveying the excavation resistance (especially writing recorder in case of hydrofraise equipment).

Vibrating beam slurry walls (thin walls):

- Drilling secondary exploration boreholes in front of the vibrating beam equipment (in addition to the primary boreholes during the designing stage). A borehole spacing of approximately 25 m (mean value) is recommended for heterogeneous subsoils.
- Taking soil samples above the toe of the cut-off; disturbed soil samples can be obtained with a special equipment on the H-beam.
- Surveying the penetration resistance (speed) of the beam, necessary capacity of the vibrator, and the suspension consumption; automatical registration by recorders is recommended.
- Physiological control: vibration caused sounds are noticed when the soil layers change significantly.

In many cases the results of permeability tests in the laboratory are widely scattered. Recharging tests within the diaphragm wall provide valuable additional information: After filling a borehole as indicated in Fig. 20, the saturation of the wall and a steady-state flow should be attained before taking measurements. If the test is repeated after certain periods (e.g. 28 days, 60 d, 90 d, etc.), data concerning the long-term behaviour are obtained. For the interpretation the following formulas are appropriate:

$$k = \frac{\Delta h}{\Delta t} \cdot \frac{1}{h_1 \cdot (h_1 + 2h_2)} \cdot \frac{r \cdot (d - r)}{2} \qquad (16)$$

$$k = \frac{\Delta h}{\Delta t} \cdot \frac{1}{h_1 \cdot (h_1 + 2h_2)} \cdot r^2 \cdot \ln \frac{2d}{\pi \cdot r} \qquad (17)$$

With Δh as the water level difference in the period Δt, the hydraulic gradient is

$$i = \frac{h}{l} = \frac{2 \cdot h_1}{d - r} \; . \qquad (18)$$

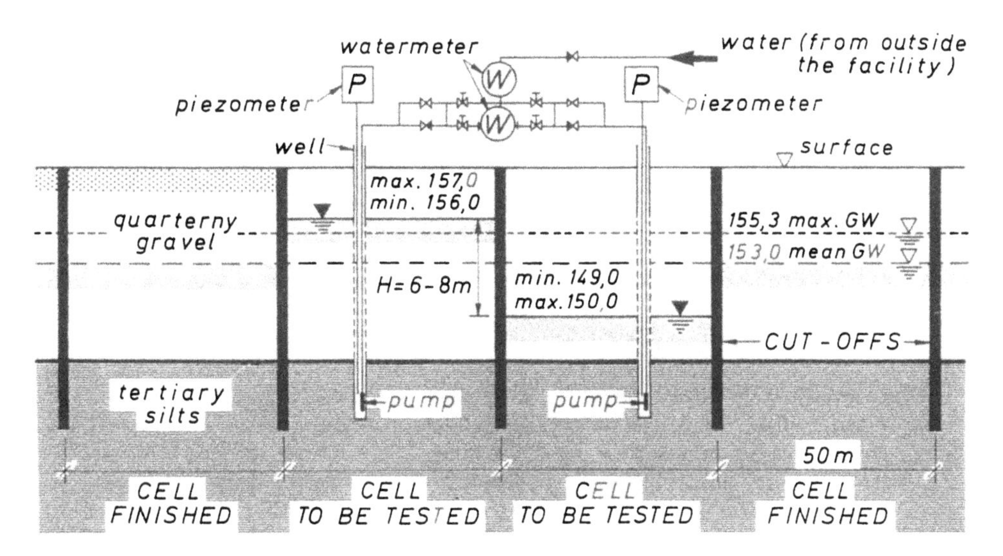

Fig.21:
Test pumping for determining the sealing efficiency of the cellular cut-off system. Suitable for quality assurance during the contruction period and for periodic long-term monitoring.
Data from the 600 000m² central waste deposit in Vienna: The hydraulic head between two adjacent cells is temporarily raised to 6 - 8m.

In equ. (16) the amount of Q_3 (Fig. 17) is commonly neglected. According to US Bureau of Reclamation Q_3 can be assessed by:

$$Q_3 = k.5,5.r.h_1$$

Hence equ. (16) becomes:

$$k = \frac{r.\Delta h}{\Delta t.h_1\left[\dfrac{2h_1 + 4h_2}{d - r} + 1,75\right]} \qquad (19)$$

Equations 16 and 19 respectively are preferred in practice because they provide somewhat higher permeability coefficients k. With regard to on-site uncertainties this conservative calculation can be taken as an additional safety factor.

The most reliable verification of a cut-off system is achieved by pumping tests. This test requires a lowering or raising of the groundwater table within the isolated area, if only single cut-off walls were installed. In case of cellular cut-off walls, each cell can be checked according to a system shown in Figure 21.

After completion of the cut-off walls the water level is lowered in one cell, the water being used to simultaneously fill the adjacent cell. For safety reasons, the lowering should be at least 3 m beneath the groundwater table outside the cut-off, and the filling $\geq$ 0,5 m above this value. After reaching a stationary stage, the water levels within the adjacent cells should be kept constant for 24 hours at a minimum. The water levels inside and outside the cells and the amount of pumped water have to be registered continuously as hydrographs which enables a comprehensive quality assurance. In total, each cell is tested by raising and lowering the water table alternately.

8. LONG-TERM OPERATION AND MONITORING

The equipment for permanent groundwater lowering and the monitoring instruments should be automatized without restricting an additional manual service and control. For very large waste deposit areas the control mechanism, remote recording etc. should be situated within "sub-stations" along the circumference of the disposal area. All data have to be submitted by a ring main to a central computer station which commonly is combined with the operation center of the waste disposal facility. The pumps within the vertical screens and the cut-off cells should operate automatically according to the required water level. The permanent recording should comprise:

- the water level inside and outside the cut-off walls;
- the water level within each chamber of the cellular cut-off system;
- the amount of pumped water (inside and outside the cells).

The control wells and stand-pipes provide a permanent integrity test under working conditions. The remote readings should be controlled by random manual supervision. Additionally, the long-term permeability of cellular cut-off systems has to be checked periodically by in-situ pumping tests similar those shown in Fig. 21. Thus, weak points can be localized and repaired in time. The conductivity test should be performed every year during the first 3-years period. Later, the testing intervals may be progressively extended unless the permanent water level readings indicate a possible fault of the structure.

Experience has shown that the hydraulic conductivity of properly designed and carefully constructed cut-off walls decreases with time. In case of the Vienna central waste deposit, this decrease was a factor of three within six years after the acceptance tests at the end of the construction period. On other sites, a long-term decrease of the permeability even by one to two orders of magnitude has been measured. The long-term improvement of the overall barrier effect of the cut-off system usually is less than derived from laboratory tests. Commonly, it occurs as an exponential-function, unless the cut-off walls are impaired locally by chemical attack:

$$\Delta k = a.(1 - e^{-b.t}) \qquad (20)$$

Δk = decrease of hydraulic conductivity
t = time
a, b = empirical factors (from laboratory tests and site experience), depending on the properties of the sealing mass, the groundwater and migrating substances (pollutants)

Random samples of the pumped water should be analyzed chemico-physically. If the waste deposit has an effective base sealing (e.g. Fig. 4b), pumping is necessary only temporarily or even not at all. In such cases, the cut-off containment provides only

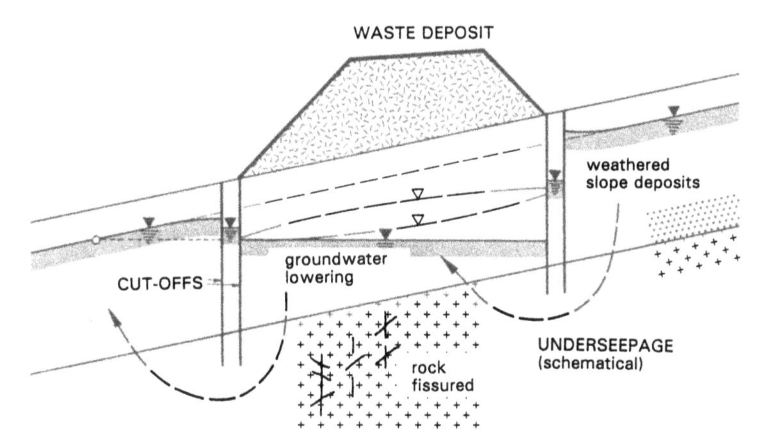

Fig.22:
Cellular cut-off wall system to isolate a waste deposit on a slope (schematical); risk of a large-scale outward underseepage of the downside double cut-off wall.

an additional safety. If the base sealing leaks, or if there is no base sealing (e.g. Fig. 4a), a permanent groundwater lowering within the cut-off walls should be carried out. According to the slow hydraulic response of the cut-off system and the low-permeability soil, intermittent pumping is sufficient in most cases. This could be experienced even on waste disposal sites where neither a capping cover nor a base sealing exists. The optimal operation scheme depends on the local conditions and can be found only by in-situ trial tests.

Long-term tests have revealed that a temporary failure of single pumps or energy supply does not reduce the safety factor of the cut-off system.

Even a stop of the whole groundwater lowering equipment over a period of several weeks has no relevant influence. The experience at the central waste deposit of Vienna shall demonstrate this as an example:

- cut-off area: 600 000 m²
- circumference: 3200 m
- natural groundwater table: 6 - 7 m beneath original ground
- subsoil (mean values):

quarternary sandy gravel	$k \doteq 3.10^{-3}$ m/s = k_1 in Fig. 7
tertiary silts (clayey, sandy)	$k \doteq 1,1.10^{-6}$ m/s = k_2 in Fig. 7 (scatter: 10^{-5} to 10^{-11} m/s)
tertiary surface	12,5 m beneath original ground (max. 30 m deep)

- permittivity of the cellular cut-off walls (mainly thin vibrating beam slurry walls; only locally diaphragm walls): $\psi \leqq 10^{-7}$ sec^{-1}
- embedment depth of the cut-offs in the tertiary layers (mean value): t = 8,4 m
- water level difference: Δh = 0,5 to 1,5 m
- no base sealing and no capping of the waste deposit;
- pumping discharge (total amount; annual mean value): q = 4 l/s

- pumping amount without rainfall (back calculated for final stage with sealing cap and a maximum water level difference of Δh = 1,3 m): q = 2,5 l/s

Long-term tests showed that it took 7 months until the groundwater level difference decreased from Δh = 1,2 m to Δh = 0,3 m, when pumping was completely stopped. After capping the waste deposit, this period will be significantly longer, and the pumping amount will be approximately $q \leqq$ 1,5 to 2,0 l/s (annual mean value for $\Delta h \geqq$ 0,5 m).

From spring to autumn the pumped water is used for irrigating the waste and site vegetation; in winter it is pumped to the adjacent municipal sewerage network and the sewage purification plant.

Accordingly, the operation and maintenance costs of such multi-barrier systems for waste deposits are very low. In case of the Vienna central waste deposit, the total annual expenditure is even less than the lighting costs for a medium-sized underground station. A further cost reduction is possible by using windpower to create the necessary energy which is rather small and is required only periodically. The energy is also provided by utilizing the landfill gas production on this site.

9. SPECIAL APPLICATIONS

Special applications of the cellular cut-off wall systems result from site-specific subsoil conditions and sometimes in connection with the encapsulation of existing waste deposits. The geomorphology and the required intensity of monitoring are also of influence on the scheme.

9.1. Inclined ground

If the original ground, and the groundwater level respec-

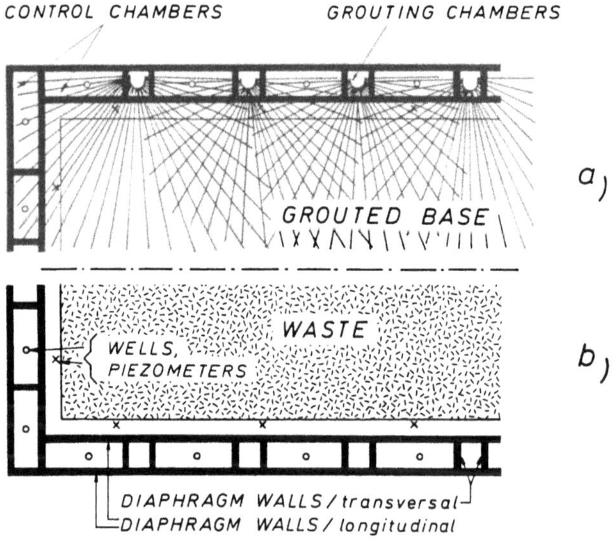

Fig.23: Ground plan of a cellular cut-off wall system to isolate a hazardous waste deposit in high-permeable subsoil. Not to scale (area: 110 x 750m).

a) Horizontal section through the grouted sealing base; grout pipes schematically indicated
b) Horizontal section through the waste

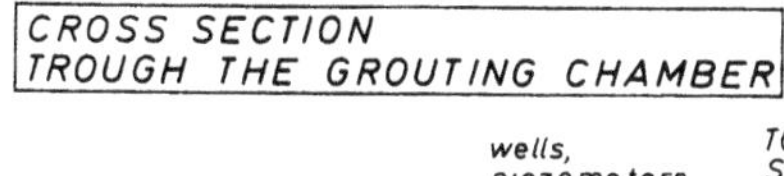

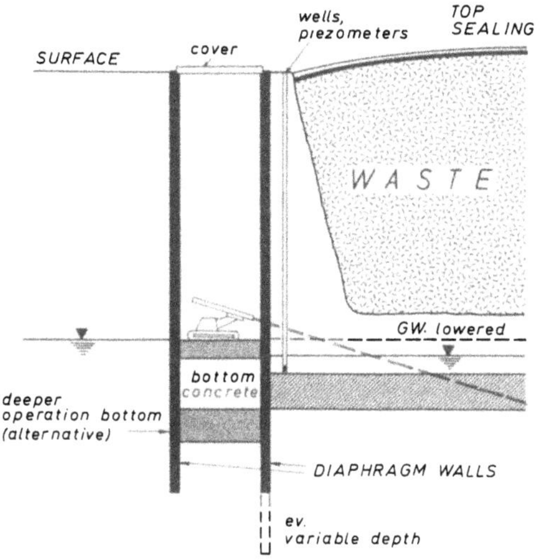

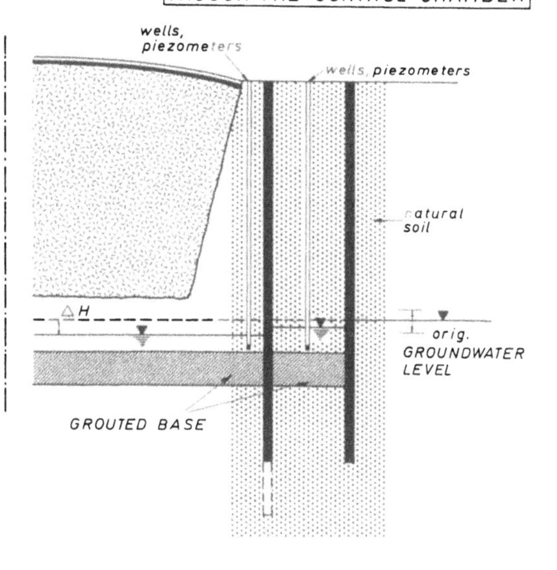

Fig.24:
Cross sections to Fig.23; also suitable for cutting off industrial plants which contaminate subsoil and groundwater. Possibility of permanent monitoring and repairability.

tively are highly sloped, the cellular double-wall cut-off system is applicable only to a limited extent. Fig. 22 shows a situation, where it should not be utilized:

The diaphragm wall toes are embedded in weathered rock. For horizontal ground (water)conditions its permeability would be sufficiently low, and only an increased water amount would have to be pumped from an inward underseepage of the cut-offs due to the rock fissures. But in case of a highly sloped ground-water table, an outward underflow is likely, in spite of a significant ground water lowering within the cut-off waste deposit (Fig. 22).

9.2 Low-permeability stratum in great depth

For an economical installing of vertical cut-off walls, the most suitable depth of the low-permeability stratum is about 8 to 25 m below the surface. Depending on the subsoil conditions, relatively cheap vibrated beam slurry walls can then be constructed.

For greater depths, the installation of diaphragm walls is unavoidable (and the risk of flaws increases).

If the low-permeability stratum lies in a great depth, and the waste disposal site is rather large, several alternatives are possible to construct a similar system as shown in Fig. 3 and 4. But all of them are rather expensive. Accordingly, the main application of such measures is limited to the following case:

- encapsulation of dangerous old wastes, or existing ground-water pollution;
- encapsulation of industrial areas where ground-water contamination might come from (e.g. atomic power plants, chemical industries, oil refineries).

Generally, new waste disposal units should not be designed in areas where the low-permeable stratum lies more than approximately 50 m deep, the overlying strata being aquifers.

The maximum depth of diaphragm walls was increased to 130 m and 170 m recently. In such cases, one should try to limit the vertical cut-off to one wall in order to save substantial costs. This method requires an extraordinary high quality work on the construction site (hydrofraise with inclinometers) and a continuous, strict survey. In case of leaks, it might become difficult to lower the groundwater table within the enclosure, because no double barrier system of screening cell exists.

Generally, the allowable deviation from the vertical of diaphragm walls should be limited to about 0,4 % per m depth. Actually, an accuracy of 0,1 % for 50 to 75 m deep walls could be achieved even in heterogeneous tertiary sediments (sand, silt, clay, with local faults but also rock-like cementations).

In spite of this high accuracy of wall installation, adjacent panels should overlap at least by 20 cm even for short cut offs with a depth less than 30 m. Experience and in situ measurements have shown that this overlapping increases the waterproofness and contaminant retention significantly.

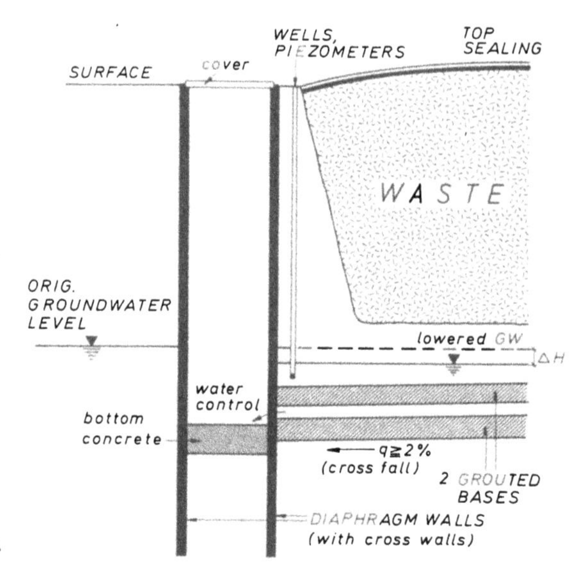

Fig.25:
Alternative to Fig.24 with a leachate collection and removal system within the grouted base sealing.

If the "impermeable" stratum cannot be reached, it is necessary to create a low-permeability zone within the subsoil beneath the hazardous waste deposit or the contaminating industrial facility. This can be gained by injections or (jet)grouting in all directions. Vertical borings are preferred generally, but should not be drilled through old wastes. In this case the subsoil might be polluted additionally by transferring hazardous substances. Therefore a previous installation of vertical cut-off walls around the area is recommended. Similar to Fig. 23 and 24 two diaphragm walls with cross elements at certain intervals should be installed. From certain cells, an approximately horizontal screen under the waste deposit can be constructed then. If this "horizontal" barrier is split into two low-permeability layers with a slope to the diaphragm walls, leachate from the waste, or seepage water from a (defect) industrial plant can be drained and monitored (Fig. 25).

The cut-off diaphragm walls must be embedded well in the soil improvement, their toe reaching beneath the grouted zone. Special care must be taken during construction to the connection zone between diaphragm walls and grouted soil, in order to avoid an uncontrolled local permeability: A soil penetration by the diaphragm wall suspension during excavation makes subsequent grouting more difficult. For long-term monitoring and eventual subsequent remedial works, some working cells should remain excavated and accessible. Those cells need a working platform consisting of a thick concrete bottom for strutting the parallel diaphragm walls and against uplift (Figs. 24, 25).

9.3 Underflown cut-off wall ("incomplete curtain wall")

Instead of the costly cut-off measures shown in Figs. 23 to 25, a significantly less expensive alternative may be possible under certain circumstances: Fig. 26 shows a hazardous waste deposit with a high toxic potential, threatening a large drinking water reservoir downstream.

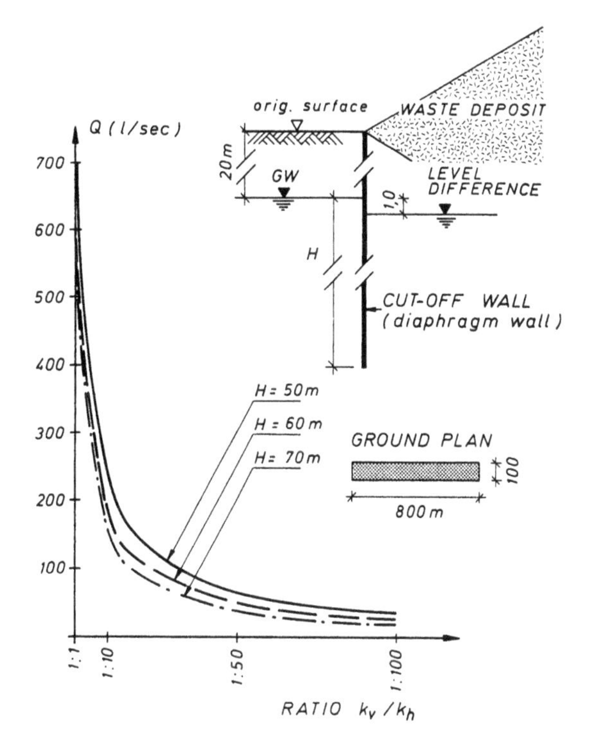

Fig.26:
Underflown diaphragm wall for a temporary cutting off a hazardous waste deposit during waste removal and replacement.
Influence of the ratio k_v:k_h on the pumping amount to lower the groundwater table within the isolation.
k_v..... vertical hydraulic conductivity
k_h..... horizontal hydraulic conductivity
Results from am FEM calculation with a mean value of $k_h = 1{,}0.10^{-3}$ m/s.

The disposal area is 100 x 800 m, and the depth of the waste bottom is about 20 m below original surface. Quaternary sand and gravel reach to a depth of 100 to 150 m. These high-permeable soils are interlain by bands of silt with medium to low hydraulic conductivity.

According to risk assessment, an excavation/replacement of the wastes (after bottom lining installation) could not be allowed without protective measure. Therefore, a temporary screening and a groundwater lowering within vertical cut-off walls were necessary.

Fig. 26 shows the result of parametric studies which were performed for a cut-off wall being underflown by the groundwater. The vertical permeability of a multi-layered subsoil is commonly much lower than the horizontal one. Consequently, a decreasing ratio $k_v : k_h < 1:1$ to $1:15$, due to the quaternary sedimentation and silty interlayers within the sandy gravel, is of greater influence on the amount of pumped water than an increase of the embedment depth of the cut-off wall.

The diaphragm walls must have a thickness of 80 cm as a minimum to provide a tight screen (with regard to possible inclination deviations of elements during installation). Their depth is determined by the amount of underseepage. The cut-off walls prevent a further groundwater pollution outside the enclosure, and the pumping beneath the waste disposal site enables a certain cleaning-up of the contaminated groundwater. After installation of the base sealing (mineral liner and geomembrane) at the bottom of the excavated landfill, the isolated area is filled again, but only with non-hazardous waste. Accordingly, the cut-off wall serves only as safety reserve in the final state,and a groundwater lowering is not necessary (unless monitoring reveals a leachate escape).

9.4 Unsymmetric systems

The parallel walls of a cellular screen need not have the same length in all cases. Sometimes the inner cut-off wall may be shorter - depending on the subsoil properties. Significant cost savings are possible, if this one-sided depth reduction allows the installation of a thin vibrating beam slurry wall there instead of a diaphragm wall (e.g. Fig. 27). The ratio of construction costs for vibrating beam slurry walls to diaphragm walls is about 1:3 to 8 for a depth up to approximately 20 to 30 m.

For deeper cut-offs only

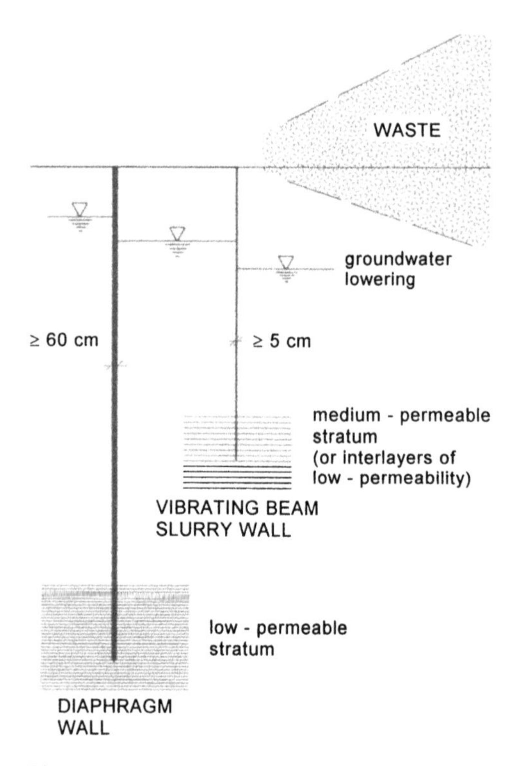

Fig.27:
Combined cellular cut-off system consisting of a diaphragm wall and a vibrating beam slurry wall. (schematical; not to scale)

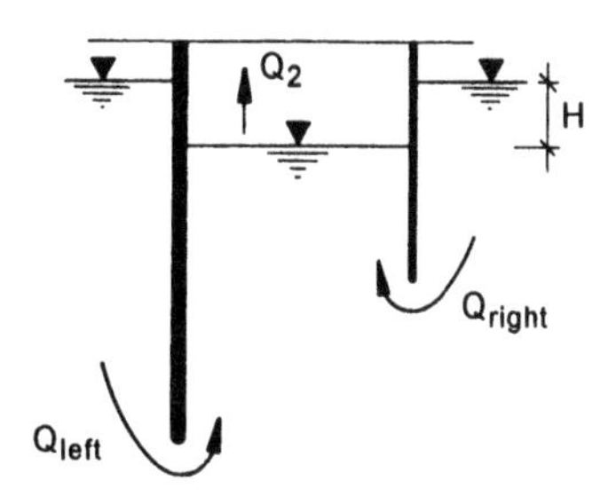

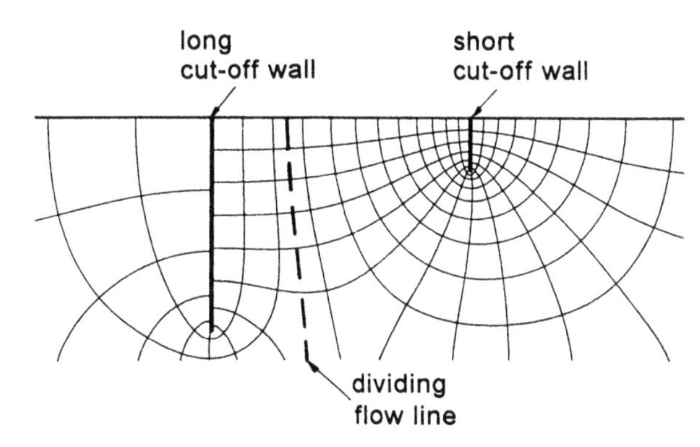

Fig. 28:
Cross section through an unsymmetric cut - off system with large wall spacing. Quality assurance test by lowering the ground water level within the cell.
Flow - net to demonstrate idealization as two seperate flow occurances (left, right side) in an isotropic medium.

diaphragm walls (or secant pile walls, or jet grouting screens) are appropriate for a permanent hazardous waste containment.

Jet grouting also has proved suitable for such portions of a site where the line of the cut-off wall crosses either underground or surface services which could not be disconnected or rerouted and removed. In those areas, it especially replaces the vibrating beam and the diaphragm wall technique.

The deep screen of unsymmetric barrier systems should be embedded in a low-permeable stratum. The shorter screen should reach in an at least semi-permeable stratum or in a subsoil with low-permeable interlayers.

From hydraulic point of view such unsymmetric cut-off systems with walls of unequal length are rather complicated. Especially, if their depth also varies in the longitudinal direction. This may be necessary to adapt the screen to local soil properties. The traditional method of analysis for the prediction of flow rates and exit gradients involves the use of flow nets. But this requires much practice and experience.

In Vienna cellular cut-off systems have proved suitable for waste containment and for groundwater-level regulating in connection with hydraulic power plants and flood protection. In total, some 50 000 m run of cut-off walls have been installed there. While analythic methods are available for symmetric cut-off systems (see chapter 3), the finite element method has been preferred to assess the steady seepage through cut-off walls of unequal length and through the subsoil beneath the wall toes. Numerous soil investigations and in situ measurements have been used to validate this method.

The FE-method easily can be validated or calibrated resp. when each cell of the cut-off system is checked separately by in situ measurements.

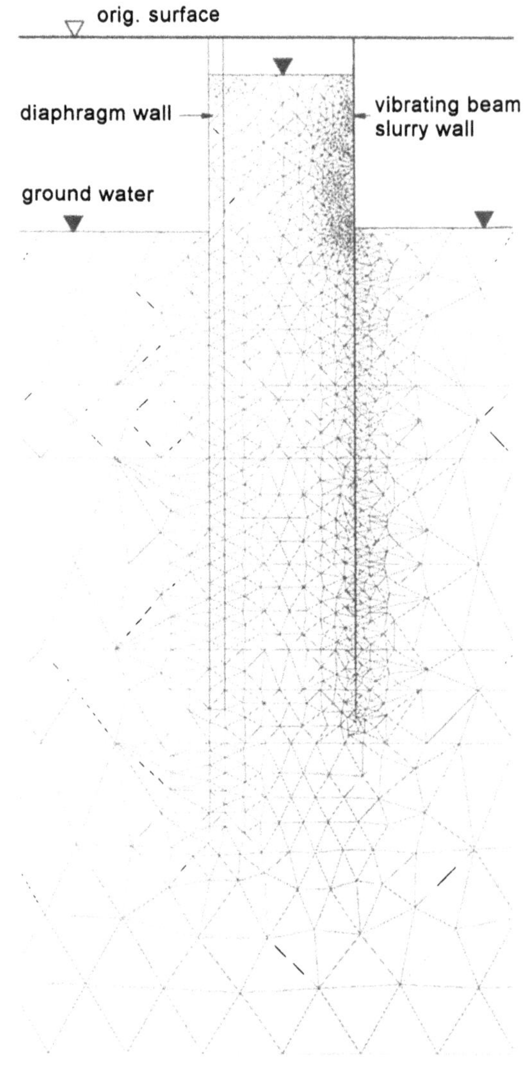

Fig. 29:
Cross section through an unsymmetric cut - off system with small wall spacing (clear distance = 5m). Quality assurance test by raising the ground water level within the cell.
Finite Element net for multi-layered, unisotropic subsoil and variable wall depths.

In order to optimize the cut-off system according to local soil properties and ground-water levels, parametric studies are necessary in which the wall depth and spacing are systematically varied. The length of the cells is less influenced by hydraulic than by control aspects. Short cells facilitate leak detection and the localization of wall sections with low quality.

In case of a small wall spacing, long cells and relatively homogeneous subsoil the calculation can be reduced to a two-dimensional problem. Otherwise a three-dimensional calculation is recommended.

Quality assurance tests at the end of construction and during long-term monitoring require a temporary rising or lowering of the groundwater level within the cells. The procedure is similar to symmetric cut-off systems (see chapter 7, 8), but the calculations differ widely.

In case of large wall spacings and lowered groundwater within the twin walls the system can be idealized to a reasonable degree of accuracy as two independent flow problems separated by a vertical impermeable barrier - assuming that the true dividing flow line is approximately vertical (Fig. 28). This approach is similar to seepage calculation beneath unsymmetric cofferdams (Griffiths, 1994).

Accordingly, the selection of the contour interval is based on the flow net equation

$$Q_2 = k \cdot H(n_f/n_d) =$$

$$= Q_{left} + Q_{right} \qquad (21)$$

where

Q_2 = flow rate by underseepage
k = isotropic permeability
H = total head loss
n_f = number of flow channels

n_d = number of equipotential drops

This flow rate Q_2 has to be superposed to the discharge through the cut-offs, Q_1 (see chapter 3).

Fig. 29 shows a typical example for a small wall spacing and multi-layered ground. The anisotropy was described by $k_h = 5\ k_v$, where k_h, k_v are the horizontal and vertical permeability coefficients. The hydraulic conductivity of the cut-off walls was taken into account by a permittivity of $\psi = k/d = 2.10^{-8}\ sec^{-1}$, derived from quality assurance tests (samples, thickness control). The fundamental geometry of the FE-net could be kept constant for variable wall depths.

Experiences has shown that a two-dimensional calculation is acceptable in most cases. Commonly, the cells of a twin-wall cut-off system have a length of about 50 m (ranging from 20 m to 100 m), and the typical width is about 4 to 6 m. A larger width and smaller length increases the costs but facilitates the localization of leaks or excessive local underseepage - especially if hydraulic disturbances exist locally in the ground (e.g. natural inhomogeneities or defect water pipes, wells etc.). Hence, the optimum geometry of unsymmetric cut-off systems depends on several factors and can be obtained only by parametric studies.

10. CONCLUSIONS

Vertical barriers have proved successful to cut off hazardous wastes in refused dumps or in case of new waste disposal facilities. The safety factor can be increased significantly by installing cellular cut-offs and lowering the groundwater table. When using thin vibrating beam slurry walls, this system is even cheaper than a single curtain of a diaphragm wall. It enables an easy control and localization of leaks. The parallel walls need not have the same length; in special cases the inner cut-off may be shorter - depending on the subsoil properties.

Incomplete curtain walls have proved economic in case of temporary cut-off measures. They do not reach in a low permeable stratum but take into account a stronger underseepage. If heavy groundwater pollutants prevail, such systems are not appropriate.

The results of calculation are commonly much more influenced by the natural scatter of the soil parameters (horizontal and vertical k-values) than by the calculation method itself. Accordingly, theoretical sophistications seem to be rather of academic interest than of practical importance. Nevertheless, a comparison of analyses based on different theoretical approximations is recommended. Moreover, a proper design should involve parametric studies with varying soil characteristics in order to check possible boundary values. Practice should focus on proper sealing mixtures and installation procedures. This represents a wide field of applied research.

REFERENCES

Brandl, H. 1990. Geomembranes for vertical waste containment sealing. 4th Internat. Conference on Geotextiles, Geomembranes and Related Products, The Hague, 511-516.

Brauns, J., Oliveira, L.M.Ch.de 1987. Abströmung aus Prüflöchern in Dichtwänden. Geotechnik 1987/1. Deutsche Gesellschaft für Erd-und Grundbau, Essen.

Esser, P., Staubermann, P. 1988. Dichtwandmassen mit geringer Durchlässigkeit.Straßen- und Tiefbau, Heft 9.

Fischer, J. 1988. Von der Basis bis zum Deckel. In U.Dettmann & A.Kirchne(eds.), VDI Tagungsbericht Moderner Deponiebau und Reinigung/ Verwertung durch Altlasten kontaminierter Böden. Hamburg.

Friedrich, W.1989.Ausbreitung chlorierter Kohlenwasserstoffe bei Einkapselung und Inversionsströmung. Mitteilungen des Institutes für Grundbau und Bodenmechanik, Universität Hannover; Heft 27.

Griffiths, D.V. 1994, Seepage beneath unsymmetric cofferdams. Géotechnique 44, No. 2, 297-305

Horn, A. 1987. Beiträge zur Deponietechnik. Mitteilungen des Institutes für Bodenmechanik und Grundbau, Universität der Bundeswehr, München; Heft 7.

Meseck, H. 1987. Mechanische Eigenschaften von mineralischen Dichtwandmassen. Mitteilungen des Institutes für Grundbau und Bodenmechanik;Techn. Universität Braunschweig, Heft 25.

Müller-Kirchenbauer,H.,Rogner, J., Friedrich, W. 1991. Estimation of a long-term resistance of sealing materials based upon laboratory tests.10.European Conference on Soil Mech. & Found. Eng.Florence, 897-902.

Radl, F., Kiefl, M. 1987.Umschließung einer Großdeponie in Theorie und Praxis. Mitteilungen für Gundbau, Bodenmechanik und Felsbau, Technische Universität Wien; Heft 4.

Radl, F., 1991. Dichtwandumschliessungen zur Sicherung von Deponien und Altlasten. Umweltbuch der Gemeinde Wien.

Developments in Geotechnical Engineering, Balasubramaniam et al. (eds) © *1994 Balkema, Rotterdam, ISBN 90 5410 522 4*

Decontamination and rehabilitation of contaminated sites

Raymond N. Yong
Geotechnical Research Centre, McGill University, Montreal, Ont., Canada

ABSTRACT

Rehabilitation of contaminated sites is examined in relation to contamination problems and land use or capability requirements. Site decontamination technology, processes, and techniques not necessarily common in the practice of classical geotechnical engineering. A brief review of the basic contaminant-soil interaction and bonding mechanisms is given, followed by an evaluation of the general requirements for the more "popular" site decontamination techniques. These are reviewed in relation to the applicability and capability of the technique to "remove" or "extract" the contaminants. The basic contaminant-soil interaction mechanisms are examined in relation to decontamination requirements from the point of view which reasons that if one understands how the contaminants are "bonded" to the soil substrate, the necessary techniques and technology can be developed to implement decontamination, i.e. "debonding" of the contaminants. Furthermore, it is also reasoned that this basic knowledge will also lead to determination of whether treatment in-situ or ex-situ would be more appropriate -- given the boundary conditions and site specifics.

INTRODUCTION

Rehabilitation of contaminated sites requires several critical decisions to be made involving various interacting parameters, not all of which are easily defined or characterized. The problem relates to the nature of the contaminants in the site, concentration (lethality or toxicity), extent of contamination, land use requirements, planning, zoning laws, cost-efficiency, best available technology, timing, etc. All these factors need to be considered, not necessarily in equal terms or proportions, before reaching the decision point which (a) dictates or limits the type of level of land use attainable as the rehabilitation process, or (b) specifies the land use and thus establishes the requirements and technologies for site decontamination.

As noted above, assessment of a contaminated site for decontamination requirements and procedures involves not only technical/scientific issues relating to the site and contaminants, but also the land use requirements and capabilities. By and large, the technical/scientific aspects of the problem include determination of the degree of contamination (extent and level of contamination), type of contaminants involved, and some information regarding site geology and substrate material (soil type). With respect to land use capabilities and/or requirements, consideration is given to required level of "clean-up", time frame allowed, costs or budget, processes, impact on immediate surroundings. The type of technologies to be used are by and large "dictated" as much as by "what is available" and by the preceding "issues", with the record showing very little consideration being given to the nature of the bondings and attachment mechanisms established between the contaminants and the various soil fractions constituting the substrate material in the determination of the technology to be used. It is not unusual to find that in many instances, the same decontamination techniques and operating procedures used for a new site -- as used previously to "similarly contaminated" sites and situations. This is a situation which one might label as "brute-force engineering". Thus for example, for sites contaminated with various kinds of organics, one observes that biological remediation techniques is overwhelming. As can be seen in literature, "bio-remediation" techniques and technology cover a wide range of possibilities, most of which require decisions concerning biological detoxification, which is generally treated as a "black box process", and assessment of treatabilities and efficiencies. Similar sets of issues can be raised with regard to sites contaminated by inorganic contaminants such as toxic metals, where "chemical washing" techniques are generally considered as likely candidates for contaminant removal.

So long as we continue to treat the various decontamination procedures as black box processes, the reasons for success or failure of the procedure to decontaminate partially or totally can never be assessed. The consequences of such an approach include: (a) limit improvement of decontamination capability, (b) development of inadvertent adverse reactions or treatment products that could be more toxic than the original contaminants, e.g. production of chlorinated catechols through biotransformation of some chlorinated aromatics.

In this presentation, a brief review of the more significant contaminant-soil interaction bondings and mechanisms is given to provide the basis for examination of the present capabilities and principal remediation "philosophies". The evaluation of some common procedures for decontamination of "polluted" soils is made, not with respect to the exact technology used, but in regard to the inter-relationships established between the contaminants (pollutants) and the soil medium. A knowledge of the nature of the bonding mechanisms established between the contaminants and the soil constituents is considered to be essential if one needs to determine the various parameters which are prominent in determination of the persistence and fate of the contaminants. This knowledge is also considered to be necessary as an a priori condition for development or structuring of the decontamination (remediation) process. Whereas the terms "pollutants", "toxicants", and "contaminants" have often been used in the literature with some ambiguity and often interchanged. One should be aware that these terms have specific meanings and connotations. Whilst one could perhaps consider all contaminants to be pollutants to a soil system, i.e. any change in the state of the soil material due to an "invasion" of contaminants will "degrade" or "pollute" the system, the term "pollutant" is generally used in relation to contaminants that pose threats to human health and the environment. The term "toxicant" is generally used to identify the health-threatening nature of the pollutant, and is sometimes used interchangeably with the term "pollutant" when it is known that such a pollutant produces toxic effects.

CONTAMINANT-SOIL INTERACTIONS AND ENERGIES

Contaminant Bondings and Energies of Interaction

The molecular interactions that govern adsorption processes are essentially electrostatic in nature, and involve the Coulombic interactions between nuclei and electrons. For removal of contaminant ions one needs to overcome the forces "binding them to the soil particles", i.e. overcome the energies of interaction developed between the contaminants and the soil particles. Forces of attraction between atoms and/or molecules have several sources. The strongest force, which is the Coulombic or ionic force between a positively charged and a negatively charged atom, decreases with the square of the distance separating the atoms. Forces of attraction between uncharged molecules which are neutral but in which the centres of positive and negative charge are separated are called *dipoles*. Certain orientations of adjacent dipoles are statistically preferred and a net attraction results. Even molecules which are not polar can be considered as instantaneous dipoles because the instantaneous positions of the electrons in the atomic shells change. This instantaneous dipole induces an in-phase dipole in an adjacent molecule, with a resultant net attraction. The maximum results when the dipoles are oriented in an end-to-end configuration. Combinations of these forces are also possible, e.g. ion-dipole attraction. The types of forces of attraction developed (Table 1) can be categorized into three components of *van der Waals* forces; (a) Keesom - forces developed as a result of dipole orientation, (b) Debye -- forces developed due to induction, and (c) London dispersion forces.

Determination of the energy by which contaminants are held within the soil matrix can be theoretically calculated using the modified DDL model as discussed by Yong and others (1992). Because the assumptions used require some idealization of particle arrangements and contaminant species, the calculated energies of

Table 1 - Intermolecular Forces

Type of bond	Nature	Mechanism	Force
Ionic	Balance of electrostatic attraction forces and Born repulsive forces.	Electron transfer occurs between the atoms which are subsequently held together by the opposite charge attraction of the ions formed.	$1/r^2$
Covalent	As above	Electrons are shared between two or more atomic nuclei.	
Coulombic	Electrostatic	ion-ion	$1/r^2$
van der Waals	Electromagnetic	Dipole-dipole (Keesom); dipole-induced dipole (Debye); instantaneous dipole-dipole (London dispersion)	$1/r^6$
Steric	Ion hydration surface adsorption	Strong attraction	$1/r^{12}$

interaction developed between contaminants and soil particles can at best be considered as "approximate". Nevertheless, the calculations are useful in that they provide guidance into the magnitude of effort required to counter the energies of interaction if one chooses to "detach" the contaminants from the soil-water system. Using the Grahame (1947) partitioning of the Stern layer into the inner and outer Helmholtz planes, the modified DDL model (Fig. 1) show two arrangements of cations in proximity with a soil particle. The (A) arrangement shows hydrated ions interacting with the electrified surface through a layer of water molecules, whereas the (B) arrangement shows partially hydrated ions in direct contact with the surface.

The potential at the inner Helmholtz plane (IHP) can be designated as ψ_{ih}. The potential at the outer Helmholtz plane (OHP) ψ_{oh} can be considered to be equal to the potential ψ_δ at the Stern layer boundary. One presumes that cations in the immediate vicinity of reactive soil particle surfaces will have a mixture of the two kinds of arrangements. i.e. some partly hydrated cations "sticking" to the electrified surface and some water molecules other than the cation hydration water {(A)} also "sticking" to the same surface. The interaction energy in the IHP, E_{ih}, is obtained directly as Coulombic interaction energy between the ions and the forces from particle

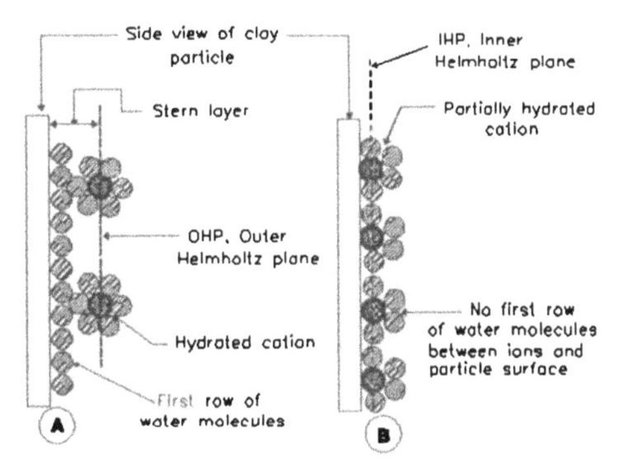

Fig. 1 - Two types of arrangement of ions at the particle surface-water interface, showing also the inner and outer Helmholtz planes.

as: $E_{ih} = E_c = \dfrac{z_i \varepsilon^2}{\epsilon R}$, where E_c = Coulombic interaction energy, R = distance between the centre of the ion i, and the negative charge site of the particle. The interaction energy in the OHP, E_{oh}, includes interaction with the Coulomb forces and also the short-range forces due to (a) ion-dipole interaction, (b) dipole-dipole interaction, and (c) dipole-site interaction. The ion-dipole interaction energy (E_{id}) is computed as (Barrow, 1973): $E_{id} = \dfrac{\mu \varepsilon}{\epsilon r^2}$ where the dipole moment of the water molecule μ can be taken to be 1.8×10^{-10} esu cm, and where r is the sum of the radii of the ion and water molecule. Interaction between water molecules in the region of the OHP and the soil particle surface reduces E_{id}, with the amount of reduction (of E_{id}) being dependent on the number of adsorbed interacting water molecules within the region of concern. The dipole-site interaction energy E_{ds} is similar in almost all respects to the ion-dipole interaction energy E_{id}, except that instead of considering r_1, the distance between the centre of the dipole and the negative charge site associated with the particle is used, i.e. $E_{ds} = \frac{\mu \varepsilon}{\epsilon r_1^2}$. The interaction energy in the OHP, E_{oh}, is obtained as the summation of the various interaction energy components, i.e. $E_{oh} = E_c + E_{id} + E_{ds}$. Thus, the interaction energies in the IHP and OHP can be given as follows:

$$E_{ih} = \frac{z_i \varepsilon^2}{\epsilon R}$$

$$E_{oh} = \frac{z_i \varepsilon^2}{\epsilon R} + \frac{\mu \varepsilon}{\epsilon r^2}\left(1 - \frac{D_n \mu^2}{\epsilon r^3}\right) + \frac{\mu \varepsilon}{\epsilon r_1^2}$$

Functional Groups and Interactions

The inorganic surface functional group OH^- and the surface charges associated with the soil solids develop as a consequence of association or dissociation of the hydroxyl groups with protons because of the pH of the soil solution. These are of considerable importance in the development of "bonding" relationships between the soil fractions and contaminants. The protons are important potential determining ions, as are the anions OH^- and $Al(OH)_4^-$. For non-mineral soil fractions such as soil organics, the main functional groups which include: (a) carboxyl (-COOH), (b) hydroxyl (-OH), {phenolic and alcoholic}, (c) carbonyl (C=O), and (d) amino ($-NH_2$), and (e) sulfhydryl (-SH), contribute most to the cation exchange capacity of these materials, and allow them to act as proton donors or acceptors. Organic matter contained in surface soils forms interparticle bonds. Sometimes electrical bonds are formed between the negative charge on the clay and a positive charge on the organic matter, e.g. protein, or between negative organic acids and positive charge on clay edges. Organic anions are normally repelled from the surface of negatively charged particles. They interact with clay surfaces through polyvalent bridges, and adsorption can be in the form of: (a) anion associated directly with cation, or (b) anion associated with cation via a water bridge -- referred to as a *cation bridge*. Organic anions can be associated with the oxides by simple Coulombic attraction, and adsorption of the organic anion is readily reversible by exchange with chloride or nitrate ions.

Cation exchange in soils refers to the positively charged ions from salts in the pore water which are attracted to the surfaces of the clay particles as part of the process which seeks to attain electroneutrality -- i.e. the process is stoichiometric, and electroneutrality requirements dictate that to balance the cations that leave the diffuse ion-layer, an equivalent amount of cations need to be "brought in" to satisfy the negative charges from the clay particle surfaces. Contaminant cations such as toxic metals are good examples of these exchangeable cations, and chemical washing for removal of contaminant ions is a good demonstration of this exchangeable cation mechanism, i.e. one cation can be readily replaced by another of equal valence, or by two of one-half the valence of the original one. The relative energy with which different cations are held at the clay surface can be found by determining the relative ease of replacement or exchange by a chosen cation at a chosen concentration. The valence of the cation has the dominant influence on ease of replacement. The higher the valence of the cation, the greater is the replacing power of the ion. Conversely, the higher the valence of the cation at the surface of the clay particles, the harder it is to replace. For ions of the same valence, increasing ion size gives greater replacing power. The number of exchangeable cations replaced obviously depends upon the concentration of ions in the replacing solution. Several equations have been derived with different assumptions about the nature of the exchange process. The simplest useful equation is that used first by Gapon:

$$\frac{M_e^{+m}}{N_e^{+n}} = \mathrm{K}\frac{\left(M_o^{+m}\right)^{1/m}}{\left(N_o^{+n}\right)^{1/n}}$$

where m and n refer to the valence of the cations and the subscripts e and o refer to the exchangeable and bulk solution ions. The constant **K** is dependent on specific cation adsorption effects and on the clay surface, and decreases in value as the surface density of charges increases.

In soil organic matter, two or three carboxyl are more selective towards multivalent cations when they are attached to adjacent carbon atoms in ring structures or in "straight chain" compounds, and two adjacent phenolic hydroxyl groups and mixed carboxyl-hydroxyl groups are more active in binding cations when they occur separately. The order for cation selectivity shows decreasing cation selectivity as follows:

carboxylic > carboxyl-phenolic hydroxyls
> phenolic hydroxyls

Exchangeable anions, i.e. those extractable with a replacing anion, include Cl^-, NO_3^-, ClO_4^-, SO_4^{2-}, are indifferent ions and electrostatically held in the diffuse ion-layers. They specifically adsorbed anions that are potential determining ions, and are adsorbed onto the clay particle surfaces through ligand exchange and not through electrostatic interaction. The chemisorption process involves the establishment of chemical valence bonds with the surface functional groups of the soil particles. By and large, the anions which fall into this category are oxyanions -- with the exception of *F*.

IN-SITU AND EX-SITU TREATMENT

The determination of whether the remediation process will be an ex-situ or in-situ process depends as much on site specifics and other externalities such as economics and regulatory requirements, etc. Obviously, the simplest procedure is to extract the contaminated soil and to replace the extracted soil with uncontaminated soil material, i.e. an ex-situ treatment procedure. The replacement uncontaminated soil may be fresh material, or may also be the decontaminated extracted soil. Treatment and implementation procedures are not likely to be exactly similar for ex-situ and in-situ conditions. In the former procedure (ex-situ), onc considers processes and technology applied, both on-site with soil material removed for treatment and either reused as rehabilitated (decontaminated) material, or off-site, where soil material is removed for treatment and final disposal. In the in-situ treatment case, remedial treatment is done within the substrate, with no physical removal of soil material. In some instances, when contaminated groundwater is pumped to the surface for treatment, the total remediation process is still identified as an in-situ procedure, although technically one could consider this to be a combined in-situ/ex-situ treatment.

The basic factors considered in determination of the appropriate technology and procedures for decontamination of contaminated sites include:

(Contaminants): Nature, extent, and degree of contamination, i.e. type of contaminants in the ground (site), extent of contamination (lateral and depth), and degree to which the site is contaminated, i.e.concentration and distribution of contaminants;

(Site): Site specificities, i.e. location, site constraints, substrate soil material, lithography, stratigraphy, geology, hydrogeology, fluid transmission properties, etc;
(Reclamation): Intended land use, land suitability/capability, regulations and requirements for clean-up remediation;
($$$$): Economics and compatible (feasibility?) technology, efficiency, time and penalties.

Whereas the first three factors are required in the evaluation of the technical feasibility for site decontamination and determination of the best available technology for site decontamination and rehabilitation, the final choice is generally made in accord with other governing considerations, e.g. risk, treatment effectiveness, benefits, and permanency of treatment. The question of "how clean should the soil or site be", i.e. "how clean is clean", needs to be answered in the context of the intended land use. Where good and effective clean-up can be achieved, the spectrum of land use can be widened. However, where effective clean-up cannot be achieved within the $$ and other constraints (technology and time), land use options are decreased, and initial land use choices may need to be re-examined. Thus regulations and requirements become very important considerations. However, it is not unusual to encounter the situation where the question of "how clean is clean" is replaced by the question "how dangerous is dangerous". The preceding notwithstanding, it is necessary to note that the choice of remediation/decontamination technique requires one not only to consider the many "scientific and technological aspects of the problem, but also "hazard identification", "toxicity and exposure" and "risk characterization or evaluation".

BASIC SOIL DECONTAMINATION TECHNIQUES

Decontamination of contaminated sites involve procedures which seek to achieve any or all of the following: (1) removal of the contaminants in the substrate, (2) detoxification of the contaminants, (3) immobilizing the contaminants in the substrate, thus rendering them "harmless" because of their inability to "move". Removal of the contaminants can be achieved either by treatment processes which will remove ("wash"/"strip") the contaminants from the substrate material in the site, or by physically removing the substrate material. However, it is not often clear that specification of the required contaminant-removal technology is made on the basis of a good working knowledge of how the contaminants are retained within the soil-water system, and whether it is feasible (i.e. technologically possible) to "destroy" these mechanisms to achieve contaminant removal.

In-situ detoxification of the contaminants can be achieved by biological and/or chemical means, and as stated previously, most of the processes involved are treated as "black boxes". Treatability experiments do not often require the determination of interaction processes and reactions resulting from the treatments, nor assessment or evaluation of the intermediate products formed. For example, reactor engineering studies with regard to reaction kinetics and multi-component species and biodegradation in an inhibitory substrate are not available to provide the guidance needed to determine effectiveness of biological detoxification of contaminated substrates.

Immobilization of contaminants is generally achieved by processes which "fix" the contaminants in the substrate, or by virtual thermal destruction. Using "end point objectives" as specified in regulatory requirements for remediation of the particular site in question, the required remediation technology can be developed in conjunction with geotechnical engineering input to produce the desired sets of actions. The general techniques which support the end-point objectives can be broadly listed in the following groups:

(Group 1) Physical-Chemical, e.g. techniques relying on physical and/or chemical procedures for removal of the contaminants, such as precipitation, desorption, soil washing,ion exchange, floatation, air stripping, vapour/vacuum extraction, demulsification, solidification stabilization, electrochemical oxidation, reverse osmosis, etc;
(Group 2) Biological, i.e. generally bacterial degradation of organic compounds, biological detoxification; bioventing, aeration, fermentation, in-situ biorestoration;
(Group 3) Thermal, e.g. vitrification, closed-loop detoxification, thermal fixation, pyrolysis, super critical water oxidation, circulating fluidized-bed combustion;
(Group 4) Electrical-Acoustic-Magnetic, e.g. techniques involving electrical, and/or acoustic, magnetic, procedures for decontamination, such as electrokinetics, electrocoagulation, ultrasonic, electroacoustics, etc;
(Group 5) Combination of any or all of the preceding four groups, e.g. laser induced photochemical, photolytic/biological, etc.

Contaminant retention mechanisms for the soil constituents differ, depending not only on the type and nature of the contaminants, but also on the nature of the soil fractions (mineral, amorphous material, soil organic, etc.). and their distribution. The distribution of the contaminants in the substrate soil is dictated by the soil fractions, the contaminant load boundary condition, and the local environmental conditions (particularly rainfall intensity and distribution). One needs to be reminded that contaminant transport in soils relies upon water as the carrying medium. The forces controlling water movement and retention will also control contaminant transport and distribution. Whereas contaminants can be conveniently grouped into classes such as acids, bases, solvents, etc., very few leachates or contaminated sites show contaminants consisting of only one single species. The presence of mixtures of various kinds of inorganic and organic contaminants makes it difficult to establish "single" treatment techniques that will effectively address all the contaminants in the contaminated material, -- assuming that in-situ decontamination is desired. To design combined techniques which incorporate multiple procedures, one needs to determine the principal processes which control retention of the various kinds of contaminants, so that the treatment system would selectively address the various contaminants.

PHYSICAL TECHNIQUES

Soil Removal and Treatment:

The simplest physical procedure for decontamination of a contaminated site is an ex-situ procedure which involves removal of the contaminated soil in the "affected" region, and replacement with "clean" soil. This procedure is most often considered, not only because of the obvious "simplicity" in site rehabilitation, but also it avoids all the complications in regard to procedures and requirements for in-situ or onsite contaminant removal. The "removed" contaminated soil is relocated in a prepared "waste containment" site, or is treated by any of the means covered under Groups (1) through (4) in the preceding discussion. To treat the removed soil, one either uses soil washing procedures to extract the contaminated water for treatment or incineration of the total soil volume. The typical soil washing procedure shown in Figure 2 works best for granular-type materials (sands and silts). For such types of contaminated soils, the amount of contaminants adsorbed onto the soil fractions can be considered to be very small, i.e. most of the contaminants is considered to be in the soil-water. Thus, through physical liquid-solid separation processes such as hydrocloning, and with density difference separation, one obtains sand and organic debris as the solid residue. The assumption of "clean sand" is predicated on the assumption that no contaminants are adsorbed onto the solids. Some contaminants will be attached to the organics however.

In clay soils, contaminant removal by soil washing is not very efficient because of contaminants adsorbed onto the soil solids, and because of the need to introduce dispersants at the *Wet Screening* stage to disperse the soil solids for chemical washing to achieve effective contaminant removal. For soils contaminated with organics, incineration of the soil is most often recommended -- for "destruction" of the contaminants. Relocation of the contaminated soil in a waste containment site without treatment is not an appropriate solution since this effectively "postpones" the problem/solution. For small soil volumes, contaminated soil removal and replacement with "clean" soil can be an effective solution. However, in situations where the soil volumes are large, this type of decontamination procedure is generally not cost-effective.

Vacuum Extraction:

Whereas vacuum extraction is generally applied as a physical in-situ soil decontamination process, this technique only addresses the contaminated groundwater. One presumes that contamination of the soil solids is not particularly important, or that

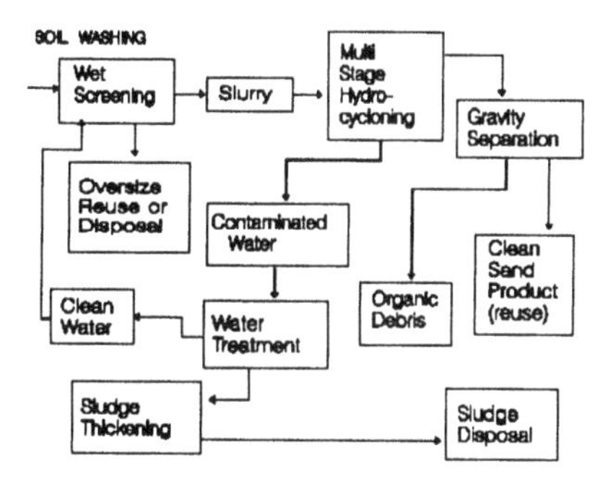

Figure 2 - Typical soil washing sequence.

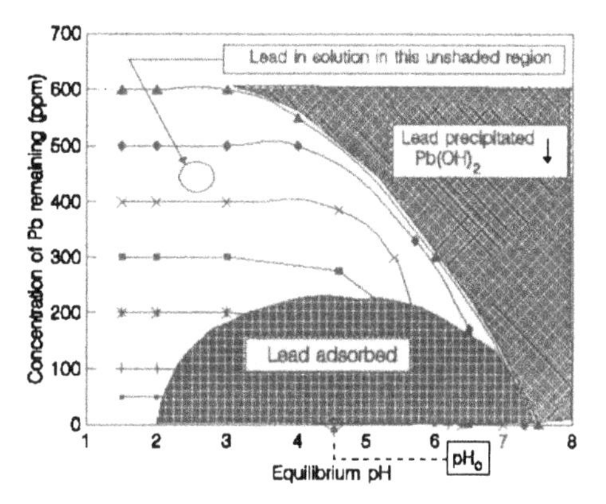

Figure 3 - Results from Yong and Galvez-Cloutier (1993) indicating partitioning of lead (Pb) contaminant.

this part of the problem will be handled by other means. Contaminant-soil interaction theory and experience show that when contamination of groundwater occurs, the soil solids will also be contaminated, i.e. there will be contaminants "bonded" to the soil solids via sorption mechanisms such as specific adsorption, chemisorption, and chelation. As an example of the problem at hand, we can consider the results from Yong and Galvez-Cloutier (1993), shown in Figure 3 where the results of interaction of a lead (Pb) contaminant leachate in a kaolinite soil are portrayed. The partitioning of Pb is seen to depend upon the pH of the total soil-water system and the concentration of Pb initially applied to the system. If vacuum extraction of the water is applied when the soil-water system is at pH levels above 7, there will be little success in obtaining Pb from the water extracted -- because of the separate phases formed from Pb precipitation. At levels of pH below 7, various amounts of Pb will be extracted depending on the pH level, with the rest remaining as Pb adsorbed to the soil solids

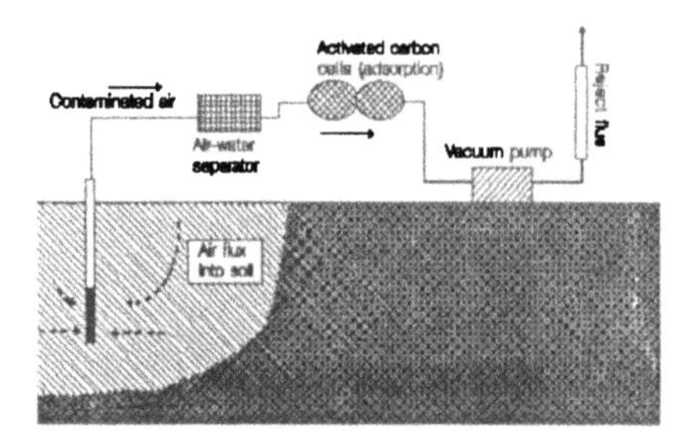

Figure 4 - Typical in-situ respiration and bio-venting systems for hydrocarbon contaminated sites.

and as precipitates associated with the soil solids or as separate phases. One can project from this type of results for application to other heavy metals. The relevant information obtained from this type of examination relates to the importance of the chemistry of the soil-water system in the structuring of techniques for extraction of pore water from a contaminated soil.

If vacuum extraction is used to obtain contaminated groundwater, one can consider this as a physical technique, in the same manner as physical removal of contaminated soil. The treatment of the extracted groundwater which is required before discharge, can be achieved by several means, not the least of which are the standard wastewater chemical and biological treatment techniques and "air stripping". Standard wastewater treatment will not be discussed herein. In the case where the vacuum technique is designed for soil vapour extraction, this is best suited for treatment of VOCs where there is the tendency of the VOCs to volatilize from water into air. Whereas the first part of the technique is physical (i.e. soil vapour extraction), the second part of the technique where cleaning of the soil vapour occurs is not necessarily "physical" since one generally uses a packed tower containing activated carbon or synthetic resins to facilitate interphase mass transfer. A typical procedure of soil vapour extraction is shown in Figure 4 for a VOC contaminated site where activated carbon cells are used as the after-stage following the air-water separation process.

Electro-Kinetic Application

The use of electro-kinetic phenomena in containment or treatment of sites with inorganic contaminants has attracted considerable attention, partly because of previous experiences with electro-osmotic procedures in soil dewatering, and partly because of the relative "simplicity" of the field application method. Strictly speaking, this is a physico-chemical technique, but is most often considered as a physical technique because of the field application methods, i.e. the use of electrodes and "current energy". Whether the procedure is effective depends on both the types of inorganic contaminants contained in the site and also the type of soil affected. The procedure relies heavily on the fundamental aspects of electro-chemistry which governs the movement of contaminant cations and anions to the respective electrodes. For the more granular types of soils (silts), the procedure can be effective. However, in the case of clay soils, diffuse double-layer mechanisms developed in the soils can pose several problems, not the least of which are the energy requirements needed to maintain ionic movement. Figure 5 shows the scheme which illustrates the interpenetration of diffuse double layers between adjacent soil particles and the development of the resultant zeta potential ζ. In application of electro-kinetic technology, one introduces similar procedures used in electro-osmotic dewatering, i.e. anodes and cathodes are inserted into the soil to produce movement of cations to the cathodes and anions to the anodes. In soils that have significant surface activity where interpenetration of diffuse double layers are prominent (as in Fig. 5), one needs to move contaminant cations from the region dominated by interpenetrated diffuse double layers -- if "dislocation" of the contaminant ions is to be achieved. The amount of energy required will need to be greater than the interaction energies established between the contaminant ions and the soil particles.

A typical situation is shown in Figure 6 where a waste containment slurry wall system surrounded by the appropriate electrodes which provide the basis for electro-kinetic activity. The governing conditions or parameters controlling movement of the contaminant ions are often reduced to considerations of permeability of the soil, diffusivity of ions, and electrode clogging or "fouling". Treatment of the plume contained between the slurry walls constitutes another separate set of procedures. A variant of this type of procedure has been used in containing the potential leachate plume under a waste landfill in Austria. In this instance, the slurry wall which was placed around the site was made with a bentonite-cement mixture, and the purpose of the wall was to provide for an impermeable barrier.

In other kinds of application, when the contaminant source "leaks" contaminants and where the appropriate soil conditions are met, it is possible to use judiciously placed electrodes to intercept the contaminant plume. However, whereas the systems have been found to be feasible in laboratory studies, it remains to be determined if the laboratory experiments can be translated into field applications under cost-effective terms, particularly because of the types of contaminants found in the various types of contaminated sites, and also because of the problem of electrode fouling.

CHEMICAL TECHNIQUES

Whilst a shopping list for chemical techniques can be made, it is not pertinent to produce such a list since innovative chemical technologies are continuously being developed, thus making it difficult to develop a comprehensive or all-inclusive list. Instead, it would be useful to examine the chemical techniques in a generic sense. To do so, one needs to first consider the type of "bonding" established between contaminants and soil constituents, whereupon the "type of chemistry" needed to "de-bond" might be better prescribed. If the contaminant-soil interactions which control bonding indicate that chemical de-bonding is not feasible, one could then entertain either partial removal via chemical means, or other methods for contaminant removal. Application of a remediation technique vis-a-vis in-situ or ex-situ can be better structured with this a priori knowledge of contaminant-soil interaction and retention knowledge.

Experience shows that whilst contaminants can be conveniently grouped into classes such as acids, bases, solvents, etc. very few waste leachates or contaminants found in polluted sites are single species. It is not uncommon to find various kinds of organics or inorganics, or mixtures of these, in contaminated soil, thus making it difficult to structure a one-step remedial treatment technique which can effectively address the spectrum of contaminants. By determining the principal sorption processes which control the retention, persistence and fate of the various individual species of contaminants in soils, one could develop techniques which could selectively address the various types of contaminants.

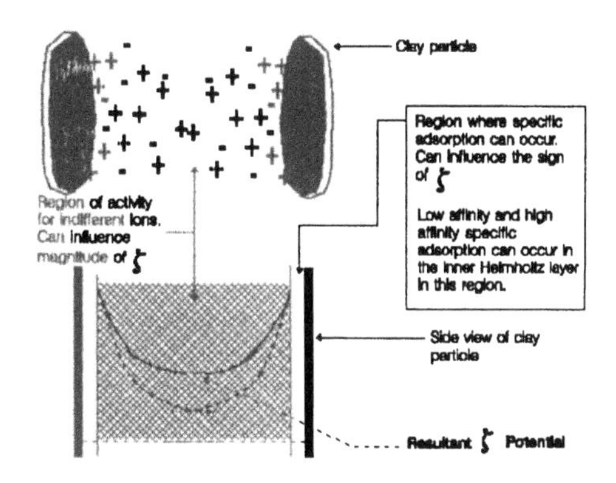

Figure 5 - Interacting charged surfaces and interpenetration of diffuse double layers.

Inorganic contaminants -- the heavy metals example

A very good example of the need to determine the inter-relationships established between soil constituents and contaminants can be seen in the selective sequential extraction (SSE) of heavy metals (h.m.) from h.m.-contaminated soils (Yong et al, 1993). In this technique, one chooses appropriate chemical reagents to destroy the "binding phase" between the h.m. and the soil constituents, thereby releasing the different h.m. fractions from the soil, and allowing the h.m. to be removed from the soil by washing techniques. It is important to determine the compositional nature of the soil since the various kinds of soil constituents, e.g. clay minerals, soil organics, carbonates, oxides and hydroxides, etc. establish different bonding relationships with the h.m. In addition, one notes from Fig. 3 the importance of the "local" chemistry of the soil-water system, i.e. pH control of the partitioning of the h.m.

Recognition of the differences in interactions established between the heavy metals and the soil constituents will lead one to the realization that different chemical reagents will be required to specifically address the correspondingly different binding mechanisms established between the h.m. and the particular soil constituent. The typical results reported by Yong et al (1993) in Fig. 7 demonstrate the proportioning of the adsorbed Pb by the soil solids. The four different soils, ranging from a natural clay soil to the pure clay mineral soils (montmorillonite, kaolinite), contain varying proportions of soil organics, carbonates and clay minerals, have been assessed for Pb retention (adsorbed) in relation to the local pH.

A knowledge of the types of soil constituents

present in the contaminated soil, and the various mechanisms of h.m. adsorption by the constituents will serve to indicate the most appropriate type of extractant. In general, the range of extractant reagents that need to be used will include concentrated inert electrolytes, weak acids, reducing agents, complexing agents, oxidising agents, and strong acids. These are generally applied in sequential format, depending on the type of soil constituents present, and the proportions of the various constituents. As an example, metals held by exchange mechanisms will generally be non-specifically adsorbed and ion-exchangeable, thus they can be replaced by competing cations. By and large, the soil fractions (constituents) that participate well in development of exchange mechanisms are the clay minerals, natural soil organics, and amorphous materials (hydrous oxides). Neutral salts such as $MgCl_2$, $CaCl_2$, and $NaNO_3$ can be used as flushing fluids with concentrations higher than 1M. It is important to appreciate that many other types of salts can be used, but one needs to consider that some of these may dissolve considerable amounts of carbonate and sulphate compounds. It is arguable that this may not be a detriment since metals precipitated or co-precipitated as natural carbonates will most often need to be extracted from the soil in a "clean-up" process.

In the various diagrams shown in Fig. 7, one needs to note that the portion of heavy metals held as "residual" indicate the portion that cannot be readily removed in a normal "clean-up" procedure -- unless complete dissolution of the soils fractions is contemplated. Changing the pH environment can be very useful in obtaining release of the heavy metals. Precipitated metals can be released by application of an acid, whereas metals retained by the soil organic matter via complexation, adsorption and chelation mechanisms require release through oxidation of the organic matter.

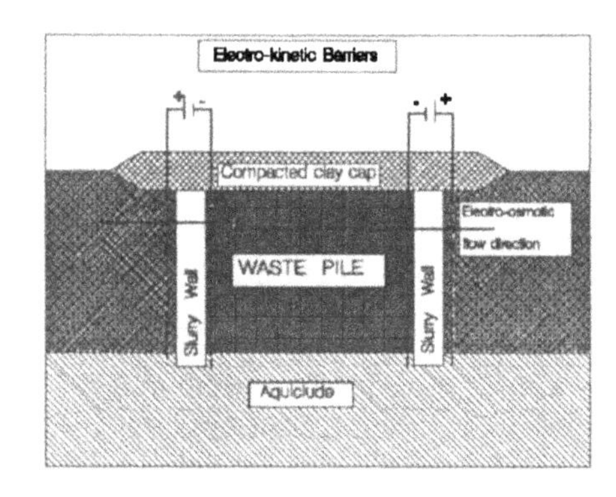

Figure 6 - Electro-kinetics applied to a slurry wall system for waste containment.

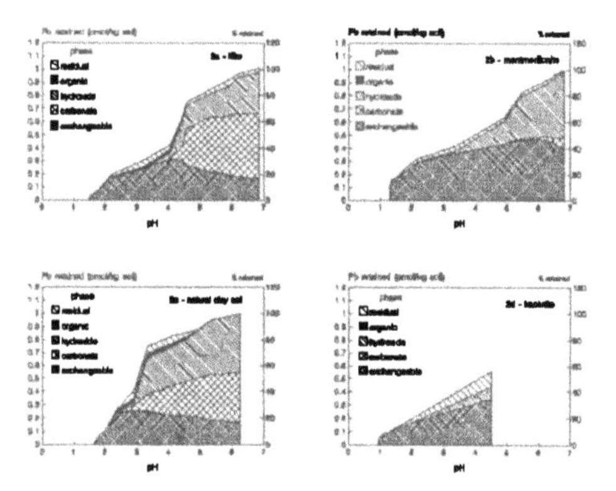

Figure 7 - Effect of type of soil constituents on adsorption of lead (Pb) contaminants in relation to pH of the soil-water system. (Yong et al., 1993)

Organic Contaminants

Application of chemical procedures for removal of organic contaminants can benefit from the knowledge of the interactions between the functional groups of the organic contaminants and the soil fractions. Chemical properties of the functional groups of the soil fractions will influence the acidity of the soil particles -- a significant property of the soil since surface acidity is very important in the adsorption of ionizable organic molecules of clays. Surface acidity is a major factor in clay adsorption of amines, *s*-triazines, amides and substituted ureas where protonation takes place on the carbonyl group. Many organic molecules (amine, alcohol and carbonyl groups) are positively charged by protonation and are adsorbed on clays, the extent of which depends on the CEC of the clay minerals, the amount of adsorbed surfaces, and the molecular weight of the organic cations. In addition, large organic cations are adsorbed more strongly than inorganic cations because they are longer and have higher molecular weights.

Polymeric hydroxyl cations are adsorbed in preference to monomeric species not only because of the lower hydration energies, but also because of the higher positive charges and stronger interactive electrostatic forces. The hydroxyl groups in organic contaminants consist of two broad classes of compounds, alcohols (ethyl, methyl isopropyl, etc) and phenols (monohydric and polyhydric), and the two types of compound

functional groups are those having a *C-O* bond (carboxyl, carbonyl, methoxyl, etc) and the nitrogen-bonding group(amine and nitrile).

Carbonyl compounds possess dipole moments as a result of the unsymmetrically shared electrons in the double bond, and are adsorbed on clay minerals by hydrogen bonding between the *OH* group of the adsorbent and the carbonyl group of the ketone or through a water bridge. On the other hand, the carbonyl group of organic acids such as benzoic and acetic acids interacts directly with the interlayer of cation or by forming a hydrogen bond with the water molecules (water bridging) coordinated to the exchangeable cation of the clay complex.

The NH_2 functional group of amines can protonate in soil, thereby replacing inorganic cations from the clay complex by ion exchange. The phenolic function group, which consists of a hydroxyl attached directly to a carbon atom of an aromatic can combine with other components such as pesticides, alcohol, and the hydrocarbons to form new compounds. The sulfoxide group, which is a polar organic functional group forms complexes through either the sulphur or oxygen atom. Complexes are formed with transition metals and with exchangeable cations.

From the preceding brief and selected summary of possible bonding mechanisms, the chemical treatment processes for removal of organic contaminants that appear to be feasible include:

(a) Base-catalyzed hydrolysis, using water with lime or *NaOH*: for application to contaminants such as esters, amides, and carbamates;

(b) Polymerization, using catalyst activation, i.e. conversion of the compound to a larger chemical multiple of itself: for application to aliphatic, aromatic and oxygenated monomers;

(c) Oxidation: for application to benzene, phenols PAHs, etc.;

(d) Reduction: for application to nitro and chlorinated aromatics.

BIOLOGICAL TECHNIQUES

The use of soil microorganisms which exist in the soil substrate, in a contaminated site, to metabolize (biodegrade) synthetic organic compounds (SOCs) benefits considerably from the similarity between many of the SOCs and naturally occurring organic compounds (NOCs). Some examples of these similarities (Hopper, 1989) are:

Aromatic NOCs	*Aromatic SOCs*
Phenylaianine	Benzenes, toulenes
Vanillin	Xylenes
Lignin	Chlorophenols
Tannins	PAHs, phenols, napthalenes, phthalates
NOCs (Sugar)	*SOCs (Sugar)*
Glucose	Cyclohexane
Cellulose	Cyclohexanol
Sucrose	Chlorocyclohexanes
Pectin	Heptachlor
Starch	Toxaphene
Aliphatic NOCs	*Aliphatic SOCs*
Fatty acids	Alkanes
Ethanol	Alkenes
Acetate	Chloroalkenes
Glycine	Chloroalkanes
Cyanides	Cyanides, nitriles, paraffins

The use of bio-remediation techniques for the clean-up of organics-contaminated sites has received considerable attention to a very large extent because of its capability for treatment in-situ using the in-place contaminants in the respiration process of microorganisms. Proper accounting for the many field and local environmental interactive factors must be given if the treatment procedure is to succeed. The use of bioventing techniques for enhancement of the respiration rates of the indigenous microorganisms works well for organic contaminants that show biodegradability, particularly with the more granular-type soils. Controls on the rate of air entry in the system and the composition of the "air" are issues that need to be "sorted out". The use of soil venting as a separate type of venting technique is useful for soil vapour extraction. The technique is both a physical-chemical process that could also contribute to biological activities in the system. The typical systems for in-situ respiration and bio-venting are shown in Figure 8.

Controls on air entry rate into the substrate system and the composition of the "air" are significant issues. The more air permeable soils are good candidates for treatment. However the experience reported by Yong et al.(1991) with regard to bio-remediation studies of soil samples obtained from a site contaminated by PAHs, PCBs and heavy metals, using bioventing is a good example of the requirement for laboratory treatibility evaluation of the interacting physical-chemical and biological parameters. In the report, the extent of disappearance of PAHs under forced aeration was measured, with

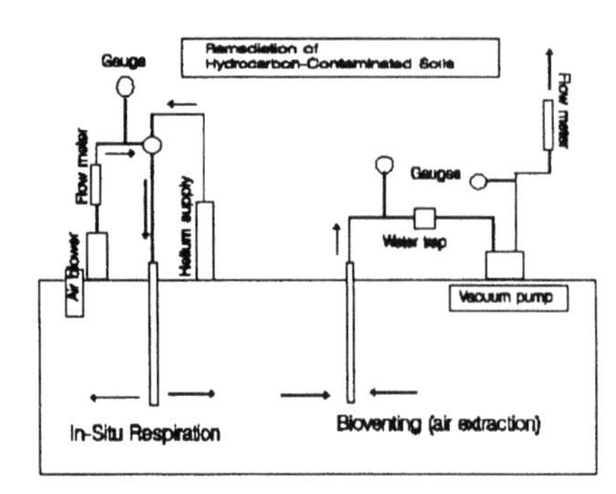

Figure 8 - Typical in-situ respiration and bio-venting systems for hydrocarbon contaminated sites.

different parameters used to distinguish biological degradation from chemical/physical degradation. The use of sterile and non-sterile samples demonstrated that biological degradation was responsible for 20% of PAH disappearance over 20 days of forced aeration. Various soil microorganisms isolated in soil extract media were studied, the results of which showed that bioremediation as the sole decontamination technique may not be as effective as anticipated, and that chemical and physical mechanisms such as volatilization and masking (i.e. transformation into an altered compound) are also mechanisms of PAH disappearance need to be considered.

CONCLUDING REMARKS

Experience shows that a combination treatment technique is generally more beneficial in site rehabilitation - i.e. remediation of a contaminated site. This is primarily because most contaminated sites consist of a whole variety of organic and inorganic contaminants. Laboratory treatability studies are mandatory, and pilot testing should always be implemented if circumstances permit. Scaling from laboratory and pilot tests will always remain as the most challenging task, particularly if new technology is to be developed. The scale-up procedures suffer not only from scale effects, but also from lack of control of soil and contaminant compositions and uniformity, and local physical/chemical control. Biological, chemical and physical reactions do not appear to scale linearly, and interactive relationships are likewise affected.

Treatment in-situ can become complicated when complex mixtures of contaminants are encountered. Aeration or air stripping, steam stripping, soil vapour extraction and thermal adsorption are techniques that are suited for removal of volatile organics. Chemical precipitation and soil washing can be used for "removal" of many of the heavy metals in the soil-water in in-situ treatment procedures. However, "complete" removal will be difficult because of high affinity and specific adsorption of the contaminant ions. A good working knowledge of contaminant-soil bonding would provide for better structuring of appropriate options and compatible technology for soil decontamination and site remediation.

REFERENCES

Hopper, D.R., (1989), *Remediation's goal: Protect human health and the environment, Chemical Envineering,* August, pp. 94-110

Yong, R.N. and Galvez-Cloutier, R. (1993); *pH control on lead accumulation mechanisms in kaolinite-lead contaminant interaction,* Proc. Environnement et Geotechnique: de la decontamination a la protection du sous-sol, Presses de l'ecole nationale des Ponts et Chausses, April 1993, Paris, pp. 309-316.

Elektorowicz, M. and Yong, R.N. (1993); *Processus retardant la biorestauration des sites contamines par des hydrocarbures,* Proc. Environnement et Geotechnique: de la decontamination a la protection du sous-sol, Presses de l'ecole nationale des Ponts et Chaussees, April 1993, Paris. pp. 373-380.

Yong, R.N., Mohamed, A.M.O., and Warkentin, B.P., 1992; *Principles of contaminant transport in soils,* Elsevier Publ. Co., Holland, 327p.

Yong, R.N., Galvez-Cloutier, R. and Phadungchewit, Y. (1993); *Selective sequential extraction analysis of heavy metal retention in soil,* Can. Geotech. Jour. October.

Yong, R.N., and Phadungchewit, Y. (1993); *pH influence on selectivity and retention of heavy metals in some clay soils,* Can. Geotech. Jour. October.

Yong, R.N., Tousignant, L., Leduc, R., Chan, E.C.S., (1991); *Disappearance of PAHs in a contaminated soil from Mascouche, Proc. Int. Symp. on In-Situ and On Site Bioreclamation,* Hinchee (ed). pp. 377-395.

Proceedings, Environnement et Geotechnique: de la decontamination a la protection du sous-sol, Presses de l'ecole nationale des Ponts et Chausses, April 1993, Paris,

Proceedings, Joint CSCE-ASCE National Conf. on Environmental Engineering., July 1993, Montreal, Canada.

Developments in Geotechnical Engineering, Balasubramaniam et al. (eds) © 1994 Balkema, Rotterdam, ISBN 90 5410 522 4

Geotechnical risk assessment – Different perspectives

R.N.Chowdhury
Department of Civil and Mining Engineering, University of Wollongong, N.S.W., Australia

ABSTRACT: Rapid progress in soil mechanics over the last half century has facilitated the solution of challenging problems in geotechnical engineering. The profession is currently assessing the needs and priorities for reliable assessments of risk and there are major challenges posed by complex structures as well as by geo-environmental problems. This paper highlights the different approaches to risk assessment. Traditional approaches include deterministic analyses and observational perspectives. These can now be supplemented by probabilistic approaches. The role, scope and limitation of different approaches are briefly explored.

1. INTRODUCTION

There has been a remarkable progress in the development of soil mechanics and foundation engineering over the last fifty years. Research at a fundamental level has been carried out in many countries to understand the behaviour of different soils under laboratory and field conditions. There have been many developments in equipment and methods for laboratory and field testing and in instrumentation for monitoring pore water pressures, deformations and stresses within natural and man-made soil masses. Classical theories of soil mechanics have been reassessed and their limitations recognised. New concepts and theories have been developed and their implications explored. With the rapid development of modern computers, the application of advanced numerical methods of analysis to geotechnical engineering problems has become increasingly feasible and attractive. Simple and sophisticated models of soil behaviour have been developed to facilitate geotechnical analysis under different loading and environmental conditions. Centrifuge testing of soils proved to be useful for a variety of field problems such as those concerning the behaviour of slopes, embankments and offshore structures. There have also been many developments in geotechnology enabling the construction of increasingly tall buildings, earth dams of increasing height, offshore structures, foundations of nuclear power plants and other complex projects.

Increasingly it has become necessary to construct structures at sites with unfavourable or poor ground conditions, to implement significant land reclamation projects and to deal with contaminated or polluted land. There is now increasing pressure to consider the environmental impact of all development projects. Geo-environmental studies are assuming a special significance all over the world. For example, the development of reliable waste containment systems and the control of containment transport are posing new challenges to geotechnical engineers.

Along with these developments and challenges there has been a growing recognition of the need to consider the effect of natural events, such as rainstorms and earthquakes, on the behaviour of natural earth masses, earth dams, foundations and other geotechnical structures. Hardly a month goes by without the report of a landslide. Landslides often occur without any warning or immediate trigger. Of course, many landslides are very clearly associated with rainstorms or earthquakes. Before the event of failure, it is often difficult to distinguish between the potential for failure as a long-term process on the one hand and the potential for failure associated with natural events on the other.

Several phenomena are related to the effect of ground motion associated with earthquakes. The liquefaction of soil masses has caused major damage to structures during many earthquakes including the collapse or sinking into the ground and tilting of buildings. Excessive deformations of earth dams have occurred, some dams have failed partially and a few completely as a result of earthquakes.

Geotechnical engineers may well ask the question:

Given the current state-of-the-art of soil mechanics and foundation engineering, how well is the profession able to recognise, understand, assess and manage risks?

2. RISK IN GEOTECHNICAL ENGINEERING

The use of terms such as *risk analysis, risk evaluation* or *risk assessment* in the context of geotechnical engineering is relatively new and, in fact, many engineers still do not use these terms in relation to geotechnical performance. Yet geotechnical engineers have always been dealing with risk as part of their work related to analysis, design or construction of geotechnical facilities or projects.

It is instructive to recall that Casagrande (1965) chose for his Terzaghi lecture the title *'The role of the "calculated risk" in earthwork and foundation engineering'*. He gave the following explanation of the term "calculated risk" which, in his opinion, was to be used in relation to qualitative rather than quantitative assessment of the performance of a geotechnical project.

"(a) The use of imperfect knowledge, guided by judgement and experience, to estimate the possible ranges for all pertinent quantities that enter into the solution of the problem.

(b) The decision on an appropriate margin of safety or degree of risk, taking into consideration economic factors and the magnitude of losses that would result from failure".

Eighteen years later Whitman (1984) summarised the rapid developments which had occurred with regard to the quantitative evaluation of risk and gave examples of first attempts to apply probability theory in geotechnical practice. He also made the point, which has been reiterated in different ways by others, that systematic analysis within a probabilistic framework facilitates the understanding of major sources of risk even when a precise quantitative assessment cannot be made.

There has been a dramatic increase in the range and complexity of problems and issues with which geotechnical engineers have to grapple. Moreover, the subject of risk has often to be considered in a wider context than that of the design of individual elements in a system. In order to deal with risk, it is necessary to understand its nature and its sources. The main cause of risk is uncertainty and there is a direct relationship between dealing with uncertainty on the one hand and managing risk on the other. Risk may relate to the economic aspects of a project, to its performance and function or to its environmental impact. Although overall assessment should consider all these aspects, geotechnical risk assessment is generally or predominantly concerned with performance evaluation.

The risk of failure of a geotechnical structure may be defined as the probability of inadequate performance. More appropriately, however, the term risk includes the consequences of failure as well. Conversely, reliability may be defined as the probability of adequate performance or success. In conventional deterministic soil mechanics, the factor of safety, F, is an indicator of performance in problems of stability. In reliability analysis, one may use the reliability index, ß, which includes the mean value of the factor of safety as well as its standard deviation according to the following simple definition:-

$$\beta = \frac{\bar{F} - 1}{F} \qquad (1)$$

In general, ß reflects the probability distribution of the factor of safety which is regarded as a random variable or a stochastic parameter.

Often a typical value of F = 1.5 is considered acceptable for slope stability problems. According to Whitman (1984), a value of reliability index ß = 2 represents "much commonly accepted geotechnical practice". He noted that there are examples where ß is larger or smaller than 2. If average shear strength along a slip surface has a coefficient of variation of 30% then a mean safety factor $\bar{F}$ = 2.5 would be required to obtain a reliability index of 2; a mean safety factor $\bar{F}$ = 1.5 would imply a reliability index of only 1.1.

It can be misleading to adopt a typical value of the factor of safety without considering the nature of uncertainties and the type of the problem. It is more appropriate to have a target value of the reliability index ß. However, a reliability index ß = 2 may be suitable for embankments but such a value may be too low for those foundations and earth structures the failure of which would be associated with severe or potentially disastrous economic consequences or with the loss of life. It is also important to note that the average coefficients of variation of shear strength are usually smaller than those calculated from point values based on test results. Moreover, engineers tend to underestimate the shear strength and this conservatism introduces bias in the estimation of the factor of safety as well as that of the reliability index.

The probability of failure, p_f, is defined as the probability that the factor of safety is less than one, ie.

$$p_f = P\,[F \leq 1] \qquad (2)$$

The calculation of the probability of failure requires methods for the calculation of statistical moments of the performance function such as F or a simulation method for determining directly its probability distribution. Commonly used methods include (a) the first order - second moment method based on Taylor series approximation; (b) the point estimate method (Rosenblueth, 1975); and (c) the Monte Carlo simulation method.

It is pertinent to ask whether one should adopt a deterministic or probabilistic approach for evaluating risk. The controversy about alternative perspectives has been going on for decades. It is of interest to note that even 28 years back Sherman (1965) referred to the use of formal probabilistic analysis and to the relationships between the factor of safety and the probability of failure.

Traditionally, engineers have incorporated subjective probability in decisions through the exercise of judgement based on experience and precedent. The observational approach developed by Terzaghi and presented in a systematic way by Peck (1969) has generally been regarded as a design philosophy. Alternatively, however, it could be regarded as an approach for risk evaluation and management within a deterministic framework.

There is an increasing acceptance of the role of probabilistic analysis especially if it is used to complement deterministic analysis and/or to aid and develop engineering judgement.

Whatever the type of analysis or the framework of analysis, performance often relates to deformations rather than stability. Similarly, performance may relate to any other aspect of a project which determines its success or failure, e.g., the control of leakage in a waste-containment system. Again it may be necessary to define performance in terms of environmental impact especially in geo-environmental projects.

Whatever the vision of an owner or a geotechnical engineer or the public concerning performance and what it signifies, it is often necessary for analysis to define a performance function such as F, the factor of safety. From the formulation of a performance function one can develop appropriate models and devise methods of analysis and assessment for evaluating risk or reliability. The formulation of the performance function must include all the parameters that can influence performance. Therefore, a clear and comprehensive understanding of soil behaviour is a pre-requisite for successful formulation of the performance function.

A performance function may be an explicit function of the basic parameters (random variables and constants). More often, however, it will be an inexplicit and non-linear function requiring an efficient numerical method for its solution.

Risk may be assessed either on a qualitative or quantitative basis. Even if the final assessment is expressed in qualitative terms, quantitative studies may be carried out as part of the assessment process. Whether a qualitative or quantitative assessment is required, the understanding of the essential geological, geomorphological and geotechnical aspects of the project is a pre-requisite for recognising risk and understanding its nature and scope. A geotechnical engineer has to visualise all potential mechanisms of behaviour and modes of failure. Carrying out even the most sophisticated analysis is futile if an important failure mode has not been recognised.

Geotechnical engineers have had variable success in prediction of field performance even under controlled conditions where considerable data were made available. For the Muar test embankment fill height prediction (Brand and Premchitt, 1989), there were four expert assessments and a total of 30 participants. Histograms of predicted fill height at failure showed a coefficient of variation of 32% considering all the participants but only 14% considering only the experts. It has been argued by Kay (1993) that these values provide initial guidelines for the relevant uncertainty components in probabilistic analysis. A more important conclusion from such exercises, which are certainly very desirable, is that geotechnical engineers should focus more on risk assessment rather than on precise predictions.

3. SOURCES AND TYPES OF UNCERTAINTY

The major sources of uncertainty in geotechnical engineering relate to the nature of soil masses which are often heterogeneous and anisotropic but may be idealised as homogeneous and isotropic. The values of parameters to be used in any geotechnical analysis are subject to several kinds of uncertainty. Firstly, a finite number of samples from a small number of borings cannot accurately represent the whole of a soil mass. This introduces *statistical uncertainty.* Secondly, there is always the likelihood that the value of a parameter estimated by a laboratory or field test or by back-analysis is different from its real value. This introduces *systematic uncertainty or bias.* Thirdly, variability of a parameter inferred from point values (tests performed at individual points or from individual samples) does not accurately reflect the *natural variability* or *spatial variability* which depends on correlation characteristics and which may be different in vertical and lateral directions. The variability inferred from point values is always an overestimate and, therefore, a spatial averaging process is necessary. This process can be carried out if the correlation characteristics of the parameter in different directions are known. Slope stability analysis with consideration for correlations between parameters was demonstrated by Alonso (1976) and, using a different approach, by Vanmarcke (1977).

There may be significant uncertainties associated with geotechnical models and such uncertainties must be recognised. The risk of using the wrong geotechnical criteria (McMahon, 1985) may be regarded as part of this uncertainty which may be denoted as *model uncertainty.* In addition, one must recognise that there is always *uncertainty due to human error* in geotechnical planning, analysis, design or construction. The importance of these factors has been borne out by the higher rate of failures associated with offshore structures in the North Sea than the rate predicted by reliability analysis which only accounts for uncertainties in loads and material properties (Whitman, 1984).

One may also distinguish between known and unknown types of uncertainty. *Geological anamolies* represent one kind of unknown uncertainty described by McMahon (1985) as the risk of encountering an unknown geological condition. Recent attempts to model anamolies probabilistically have been summarised by Tang (1993).

It is important to note that many failures of slopes are due to undetected weak subsurface strata or due to pore water pressure distributions quite different from those assumed. According to De Mello (1977) safety is determined more by the adequacy of our hypothesis than by the accuracy of our calculations.

The frequency of geotechnical problems during construction and mining was analysed by McMahon (1985) on the basis of data on 130 major projects supplied by several engineers. He noted that 64% had no significant problems, 12% encountered problems due to unknown geological conditions, 8% due to bias and/or variation in the design parameters being greater than estimated, 5% due to human error, 4% due to design changes (some projects encountered more than one type of problem and, therefore, the above frequencies add to more than 100%).

He also analysed the frequency of post-construction failures or disasters and found that only 5% encountered problems which were not corrected during construction or mining and subsequently led to failure of the finished structure (7 out of 99 civil projects) or a mining disaster (1 out of 31 mines).

More frequency analyses of different types of geotechnical problems and failures would be of immense value to the geotechnical engineering profession, especially if such studies were carried out at regular intervals and subject to peer review. Such studies would contribute to development of research priorities in general and more realistic risk assessment procedures in particular.

In geo-environmental problems, such as toxic-waste clean-up, risk assessment includes data collection and evaluation as well as assessment of exposure and toxicity. Data validation is a part of the evaluation phase and the process consists of checking for accuracy, representativeness, comparability and completeness through the use of data quality indicators. According to Balasundaram and Shashidara (1993) the current practice in the United States consists of blind evaluation in which data validators are kept in the dark about the circumstances of data collection and how the data will be used. They suggest that this practice should be improved so that validators can judge the worth of the data and the context in which it is to be used. For example, data validators should be given details of the sampling plan and the methods used for sampling because the results are significantly influenced by decisions and processes involved in developing these aspects of the project. They also point to the lack of communication between the collectors, analysers, users and validators of data so that those judging the validity of the data often have no real appreciation of the importance of their work in relation to the project as a whole.

Lack of communication between those responsible for investigation, analysis, design and construction contributes to uncertainty in any geotechnical project. This uncertainty could be included as part of the category of human error. However, it is suggested that considering it as a separate category would be far more effective for risk assessment purposes.

4. ASSESSMENT OF RISK - DIFFERENT PERSPECTIVES

Generally speaking one may classify risk assessment perspectives as those dominated by the following:

- Deterministic framework of analysis;
- Observational approach;
- Probabilistic framework of analysis; and
- A combination of analysis and observation.

4.1 *The Conventional Deterministic Perspective*

This approach relies on experience and the use of engineering judgement for the selection of the design approach. From available data the values of parameters for analysis and design are also selected on the basis of judgement. Peformance evaluation is carried out in terms of the best estimates of the important parameters without any formal analysis of uncertainty. Values of the factor of safety, total settlement and differential settlement acceptable in different situations are also selected from precedent and experience. While the likelihood of failure or of inadequate performance is recognised, no attempt is made to quantify the relevant probability in any given situation. In order to facilitate realistic and efficient design, sensitivity studies of the performance function are made with respect to individual parameters on which the function depends.

There are obvious limitations of conventional methods because of uncertainties related to geological, geotechnical and other parameters. More importantly, however, many phenomena are not fully understood and conventional geotechnical models are, therefore, not effective in performance prediction. For example, the dynamics of landslides has defied modelling and analysis. Uncertainties related to landslide mobility and forecasting of behaviour have been discussed by Salt (1988) and Oboni (1988) among others.

4.2 *The Observational Approach*

This approach is used as the dominant perspective for risk assessment in situations where geotechnical models and analytical procedures have not been successful. The most notable examples concern landslides in different geological formations. It is difficult to model the process of landsliding completely and accurately. As stated above, the deformations, velocities and accelerations associated with natural slope masses cannot be predicted on the basis of the current state-of-the-art of geotechnical engineering. Consequently, these quantities are measured in the field continually over a long period of time. The variation of these quantities with time provides important data for decisions concerning risk assessment. Data from other kinds of observation may, of course, be used to supplement the data on movements before interpretation, and decisions concerning risk are carried out. For example, increase in pore water pressures and stresses within the elements of a geotechnical system may provide a very useful basis for evaluating risk.

4.3 *The Probabilistic Perspective*

The probabilistic perspective recognises explicitly that failure can occur regardless of the calculated value of the factor of safety. The risk of failure or inadequate performance can be reduced but never eliminated and, therefore, it is important to quantify and express the risk in terms of probabilities. Based on this perspective the performance function is considered to be a stochastic parameter depending on basic random variables or a combination of random processes, random variables and constants. The probability distribution of a performance function is assumed or derived from the probability distributions of the basic random processes and variables.

The calculated probability of failure or of inadequate performance depends on the assumed or derived distribution of the performance function and is very sensitive to the distribution as the threshold value (e.g. $F = 1$) approaches the tails of the distribution. Consequently, absolute values of calculated probability of failure are subject to greater uncertainty for those projects in which it is necessary to aim for very low risk, e.g., dams, nuclear power plants, major offshore structures. For dams the order of observed failure probabilities is 10^{-4} per dam year. Peck (1980) referred to our increasing knowledge and experience concerning the reasons for dam failure and methods for preventing the causes of such failures. According to him, "... we should be able to justify a base level probability no more than 10^{-5} per dam year. Such an improvement is now within the state-of-the-art. Its achievement does not depend on the acquisition of new knowledge ...". Considering the consequences of the failure of a nuclear power plant, much higher reliability should be required than that for dams. Unfortunately, observed or historical data on successes and failures is limited and, therefore, experience should be supplemented by judgement based on consideration of uncertainties, known and unknown.

It is necessary to acknowledge that there may be not direct relationship between the observed (historical) failure probability of a class of geotechnical structures, such as dams, and the calculated failure probability of individual structures in that class. The former, i.e. observed (historical) failure probability, provides the engineer with a guide to the order of magnitude of the calculated failure probability that one may consider acceptable.

Traditionalists would, of course, argue that it should be the aim of good design to reduce the calculated failure probability to zero and achieve one hundred percent reliability. Indeed, such a goal may be valid in principle within a deterministic perspective. On the other hand, a zero probability of failure or one hundred percent reliability is not achievable in principle considering a probabilistic framework.

Considering now the same issue at a practical level, it may be valid to aim for 100% reliability in the case of earth dams because of considerable experience of analysis, design, construction and observed performance. However, experience tells us that the behaviour of natural slopes and landslide areas is subject to far greater uncertainty and is often totally unpredictable. It would hardly be a realistic goal to set the acceptable failure probability at or close to zero.

Another practical consideration is, of course, cost. Assuming that the probability of failure of an earth structure or geotechnical project can be reduced to zero, the cost required to achieve that goal may be huge or, at best, unrealistic. Consequently, it is far more practical to optimise on the basis of total cost, C_T, which may be defined as the sum of (a) initial project cost, C_o; and (b) the product of the failure probability and the cost of damage and reconstruction/repair, C_D, in the event of failure. Thus one may write simply:

$$C_T = C_o + p_f C_D \qquad (3)$$

As one attempts to optimise C_T it becomes obvious that a certain level of failure probability is implied for a cost-effective design. Note that C_o increases and the second term decreases as p_f decreases.

A more detailed or refined analysis for optimum cost may, however, be made. For example, the cost C_D would, in general, depend on the extent and severity of the failure. Again, there may be environmental damage associated with the failure which must be considered. The question of loss of life must be considered since it is not easy to provide for loss of life in monetary terms. The detailed discussion of all these aspects and of cost-benefit studies generally is outside the scope of this paper. A system reliability approach is also feasible within a probabilistic framework and the basic principles for calculating system reliability bounds have been presented by Ang and Tang (1984). Application of system reliability approach to slope stability problems has been discussed by Oka and Wu (1990) and Chowdhury and Xu (1993a).

4.4 *Combination of Analysis and Observation*

Engineers often rely on more than one approach during the decision-making process. Even within a deterministic framework of analysis, the use of more than one method of analysis and consideration of alternative geotechnical models is not unusual. More importantly, geotechnical engineers have increasingly adopted the use of observational approaches not only qualitatively, but in terms of detailed monitoring of stresses, strains, deformations and pore pressures. A systematic approach developed by Terzaghi and presented in his lucid and inimitably organised style by Peck (1969) is a well known example of combining conventional deterministic analysis with observation. The use of back-analyses of failures to derive more reliable geotechnical parameters for analysis is another example of combining deterministic analysis with observed performance. These approaches can be very effective in the evaluation and management of geotechnical risk. The use of back-analyses is only applicable if failures occur frequently. However, one cannot accept the view that improved design and risk assessment should await the failure of major structures. To keep pace with modern developments it is necessary that other approaches be used where historical data and experience of successes and failures is limited.

The use of instrumentation and monitoring during and after construction of projects, however novel and complex, is certainly desirable and often very valuable. Nevertheless, such an observational approach implies that there is little uncertainty about what parameters are to be measured and monitored. Such may not be the case and, therefore, the quality and accuracy of risk assessment will suffer if the unknowns and uncertainties are ignored. More importantly, decisions have to be made before the period of construction and, therefore, of observation begins. One may argue that such decisions should rely solely on a deterministic perspective aided by engineering judgement based on experience. On the other hand, there are unique situations to be managed by geotechnical engineers where experience is limited and where there may be few precedents. Under these circumstances it would be useful to explore the combination of a probabilistic framework and observation during and after construction.

The Bayesian approach is well known for updating probabilities on the basis of additional information. The initial or prior probabilities and additional information or likelihood function are combined on the basis of Bayes' theorem to calculate the updated or posterior probabilities. In principle, both the success and the failure of a project yield new information and, therefore, can be used to modify prior probabilities. The successful implementation of a stage of excavation would obviously yield a posterior probability lower than the prior. On the other hand, failure at any stage of a project or failure of part of an earth structure would yield a posterior probability higher than the prior for the remaining stages or the next phase of the project or for similar structures yet to be constructed.

The combination of probabilistic framework in combination with observation need not be restricted to a Bayesian approach. The probabilistic approach can provide engineers with new perspectives of risk perception and assessment not generally available within a deterministic framework. For example, the progression of failure in space and time is rarely modelled within a deterministic framework. Although one may carry out calculations of the factor of safety for different stages or periods of time, these calculations are mainly considered to be valid for $F > 1$ only, i.e, state of success. Failure is generally regarded as simultaneous rather than progressive. Therefore, it is difficult to ask a question in the following form:

"If failure has progressed to a stage A in spatial terms (where A is incomplete or partial failure), what is the probability of failure progressing to a more advanced stage B or what is the probability of complete failure?"

For example, in slope stability the local factor of safety is considered to be equal to the overall factor of safety in almost all the simplified and rigorous methods of slope stability analysis. Yet it is well known from stress-deformation studies that failure is a progressive process and, in fact, the development of the slip surface itself can be simulated in certain situations by sensible use of these methods (eg. Indraratna *et al*, 1992).

The calculation of the probability of progressive failure on the basis of bivariate probability distribution of two elements or segments of a system was demonstrated by Chowdhury and A-Grivas (1982). A model for the simulation of the probability of complete failure of a slope, the sliding probability, was presented by Chowdhury *et al* (1987). The models for calculating conditional probability of complete failure given that partial failure has been observed was presented by Chowdhury (1992) and Chowdhury and Zhang (1993). The increase of the sliding probability with the increase of proportion of slip surface observed to have failed can be simulated with these models. Consequently, these approaches can be used for decision-making concerning landslides based on probabilistic analysis and observational approaches.

5. APPLICATION OF RISK EVALUATION APPROACHES

The selection of a particular approach for risk assessment depends on a number of factors including (a) the type of project; (b) the extent and quality of information and data which is available or which can be obtained; (c) the availability of experience and precedent (d) the feasibility of significant observational procedures and the availability of time for such procedures; (e) the expertise of the engineers; (f) the wishes of the owners; (g) the requirements of regulatory agencies and codes of practice; (h) the economic constraints; (i) the consequences of failure; (j) the environmental impact of the project; (k) the technical judgement of the engineer; (l) the extent to which there has been effective transfer of knowledge and technology concerning new approaches.

Considering the large number of factors which influence the choice, it is not surprising that there has been reluctance among engineers to adopt a radically new frame of reference for assessment of risk. There are other difficulties such as the one pointed out by Oboni (1993) in the following words:

"From a classical point of view one could say that statistics is the tool necessary to characterise the stochastic variability of the input parameters of a probabilistic analysis. Unhappily this path very often presents a dead end in geotechnical applications due to excessive costs of sampling and testing required". A "pragmatic probabilistic approach" may, however, be adopted where coefficients of variation of different parameters are selected on the basis of experience and judgement and then a first order - second moment analysis or Monte Carlo simulation is carried out. Even then geotechnical engineers may not be willing to "... submit themselves to the conceptual discipline required for a probabilistic analysis, ie., a probabilistic analysis that is (a) feasible; (b) affordable; and (c) leads to results that can actually deserve our client's interests".

It has been observed by Whitman (1984) that reliability theory is of little use in selecting an appropriate safety factor for problems involving soils for which there is very little direct experience such as some problems of soft clays discussed by Leonards (1982) and many slopes in residual soils. This observation should not be interpreted to mean that probabilistic analysis is irrelevant in such problems. In fact the assessment of risk based on uncertainty analysis would be the best form of assessment in these cases especially if one is willing to adopt a pragmatic approach. The selection of appropriate safety factors is just one type of use, among many others, to which reliability analysis can be put.

Consider a geotechnical problem in a highly urbanised area where deformations play an important part during foundation excavation, support and construction. Often there are economic constraints to the extent of site investigation that can be undertaken. Moreover, in these situations there are many physical constraints and the geotechnical engineer has to rely largely on data from adjacent projects which may become available. Based on these sources and whatever new investigations and tests are undertaken, site characterisation can be carried out including decisions on the average properties of significant geotechnical parameters and their coefficients of variation. Such decisions would, of course, be influenced by experience and engineering judgement. After making these decisions, the performance functions for lateral and vertical deformations at different points of interest may be defined based on the appropriate geotechnical models. Using reliability analysis, the average values of deformation components and their coefficients of variation can be calculated. On this basis the engineer can make judgements concerning the maximum and minimum values of deformation to be expected and these can be checked against observed values at points which are instrumented. Essentially, the use of reliability analysis allows the engineers to make effective risk evaluation and to manage the jobs within constraints of time, economy and site location. Such approaches have been used with considerable success for some projects in Switzerland according to Oboni (1993). Similarly pragmatic risk approaches have been used successfully in geo-environmental problems.

The need for continual review and improvement of geotechnical models cannot be over-emphasised in relation to improved risk assessment and management. For example, the simulation of the behaviour of embankments and earth dams during earthquakes is of significant interest. The unpredictability of earthquakes leads to significant uncertainties concerning the duration and intensity of ground motions for which earth structures should be designed. However, even if a decision is made concerning the particular time-acceleration history for which an earth structure is to be analysed, the choice of a comprehensive model for simulation of its response is very important. An adequate model must consider the change in the static factor of safety, F, during dynamic loading and this can only be achieved by taking into account mechanisms of soil strength degradation due either to strain-softening or development of excess pore water pressures. Such a model has been developed by Chowdhury and Xu (1993b, 1994) on the basis of the well known Newmark approach. Using this improved model, risk assessment can be carried out either on a deterministic or a probabilistic basis. Moreover risk assessment in such cases can include not only overall safety or reliability as a function of time but also relative deformations along a potential slip surface.

6. CONCLUSIONS

The main conclusions of this paper are the following:

- There are several perspectives about the methods and procedures to be used for risk assessment in geotechnical engineering. However, it is necessary for geotechnical engineers to appreciate the role, scope and limitations of each of these approaches.

- The central issue concerning risk assessment concerns the impact of various types of uncertainties on decisions concerning a geotechnical engineering project. The multi-disciplinary nature of most projects and the fact that risk may relate to economic, performance and environmental aspects in varying degrees should be appreciated.

- Conventional deterministic approaches rely heavily on engineering judgement in the selection of geotechnical models and the values of individual geotechnical parameters. The importance of reviewing and improving geotechnical models on a continual basis should be emphasised.

- Probabilistic approaches require systematic analysis of uncertainty and, therefore, require the engineer to explore all significant sources of uncertainty. Consequently, realistic assessments concerning risk can be made on the basis of available data. Even if detailed and comprehensive data are not available, pragmatic probabilistic approaches can be used to facilitate risk evaluation.

- The judicious use of observational approaches has facilitated risk assessment in many geotechnical engineering projects. However, while observation may play a dominant role in such approaches, some form of analysis is always useful, if not absolutely necessary.

- The use of probabilistic approaches to complement traditional practice is the best way to progress greater awareness, among geotechnical engineers and others, of the value of these approaches in decision-making.

- Major issues concerning risk assessment concern acceptable values of factor of safety, reliability index and probability of failure for different projects and the handling of unknown uncertainties and anamolies.

7. ACKNOWLEDGEMENTS

The support of the Australian Research Council and the University of Wollongong for research and scholarly activities is acknowledged. The author is Coordinator of the Water Engineering and Geomechanics Research Program and research work is carried out by the author and twelve other members as part of the activities of this group.

8. REFERENCES

Ang, A.H-S. and Tang, W.H. (1984), *Probability Concepts in Engineering Planning and Design*, Vol II, John Wiley, 562 pp.

Balasundaram, V. and Shashidhara, N. (1993), Data Validation Practice and Risk Assessment, *Civil Engineering,* ASCE, pp 60-61.

Brand, E.W. and Premchit, J. (1989), 'Comparisons of Predicted and Observed Performance of the Muar Test *Embankment', Proceedings, Intl Symp on Trial Embankments on Malaysian Marine Clays*, Kuala Lumpur.

Casagrande, A. (1965), The Role of the "Calculated Risk" in Earthwork and Foundation Engineering, *Journal of Soil Mechanics and Foundations Division,* ASCE, 91, SM4 reprinted in Terzaghi Lectures (1963-1972),ASCE, pp 72-111.

Chowdhury, R.N. (1992), Modelling the Risk of Progressive Slope Failure, *Canadian Geotechnical Journal,* 29, 1, 94-102.

Chowdhury, R.N. and A-Grivas, D. (1982), Probabilistic Model of Progressive Failure of Slopes, *Journal of Geotechnical Engineering Division,* ASCE, 108 (GT6), 803-19.

Chowdhury, R.N., Tang, W.H. and Sidi, I. (1987), Reliability Model of Progressive Failure, *Geotechnique,* 37(4), 467-81.

Chowdhury, R.N. and Xu, D.W. (1992), Reliability Index for Slope Stability Assessment - Two Methods Compared, *Reliability Engineering and System Safety,* 37, 99-108.

Chowdhury, R.N. and Xu, D.W. (1993a), System Reliability Bounds for Slopes, (unpublished).

Chowdhury, R.N. and Xu, D.W. (1993b), Earthquake Analysis for Embankments, Environmental Management, Geo-Water and Engineering Aspects, R.N. Chowdhury and S. Sivakumar (eds), Balkema, 753-759.

Chowdhury, R.N. and Xu, D.W. (1994), Embankment Dam Deformation during Earthquakes, Geotechnical Engineering - Emerging Trends in Design and Practice, K.R. Saxena (ed), Oxford and TBH Publishing, New Delhi, 15 pp.

Chowdhury, R.N. and Zhang, S. (1993), Modelling the Risk of Progressive Slope Failure: A New Approach, *Reliability Engineering and System Safety,* 40, 17-30.

De Mello, V.F.B. (1977), Reflections on Design Decisions of Practical Significance to Embankment Dams, 17th Rankine Lecture, *Geotechnique,* 27, 3, pp 279-355.

Indraratna, B., Balasubramaniam, A. and Balachandran, S. (1992), Performance of Test Embankment Constructed on Soft Clay, *Journal of Geotechnical Engineering,* ASCE, 118, 1, 12-33.

Kay, J.N. (1993), 'Probabilistic Design of Foundations and Earth Structures', Probabilistic Methods in Geotechnical Engineering, Li and Lo (eds), Balkema, 49-62.

Leonards, G.A. (1982), Investigation of Failures, *Journal of the Geotechnical Engineering Division,* ASCE, 108, GT2, 185-216.

McMahon, (1985), 'Geotechnical Design in the Face of Uncertainty', E.H. Davis Memorial Lecture, Australian Geomechanics Society, 19 pp.

Oboni, F. (1988), Analysis Methods and Forecasting of Behaviour - General Report, Proc Fifth Intl Symp on Landslides, C. Bonnard (ed), Balkema, 1, 491-499.

Oboni, F. and Holbil, Z. (1993), 'Course Notes for Geotechnical Risk Assessment', Short Course, University of Wollongong, Australia, February, 15 pp.

Oka, Y. and Wu, T.H. (1990), System Reliability of Slope Stability, *Journal of Geotechnical Engineering,* ASCE, 116, 8, 1185-89.

Peck, R.B. (1969), Advantages and Limitations of the Observational Method in Applied Soil Mechanics, Ninth Rankine Lecture, *Geotechnique,* 19, 171-187.

Peck, R.B. (1977), Pitfalls of overconservatism in Geotechnical Engineering, *Civil Engineering,* ASCE, 47, 2, 62-66.

Peck, R.B. (1980), Where has all the Judgement Gone?, *Laurits Bjerrum Mineforedrag* No 5, Norges Geotekniske Institutt, Oslo, 5 pp, also *Canadian Geotechnical Journal,* 1980, 17, 4, 585-590 and *Norwegian Geotechnical Institute Publn, 134 (1980), pp 1-5.*

Rosenblueth, E. (1975), 'Point Estimates for Probability Moments', Proc Natl Acad Sci, USA, 72, 10, 3812-3814.

Salt, G. (1988), 'Landslide Mobility and Remedial Measures', Proc Fifth Intl Symp on Landslides, C. Bonnard (ed), Balkema, 1, 757-762.

Sherman, I. (1965), Discussion on 'The Role of the "Calculated Risk" in Earthwork and Foundation Engineering', reprinted in Terzaghi Lectures (1963-1972), ASCE, 119-122.

Tang, W.H. (1993), Recent Developments in Geotechnical Reliability, Probabilistic Methods in Geotechnical Engineering, Li and Lo (eds), Balkema, 3-27.

Whitman, R.V. (1984), Evaluating Calculated Risk in Geotechnical Engineering, *Journal of Geotechnical Engineering,* ASCE, 110, 2, 145-187.

Developments in Geotechnical Engineering, Balasubramaniam et al. (eds) © 1994 Balkema, Rotterdam, ISBN 90 5410 522 4

Environmental effects on critical strain of rocks

S. Sakurai, I. Kawashima & T. Otani
Kobe University, Japan

ABSTRUCT Critical strain, defined as the ratio of the uniaxial strength and Young's modulus, has practically been adopted as a criterion to assess the stability of underground structures such as tunnels. It has already been demonstrated that the critical strain is more or less independent from the specimen size. This implies that the critical strain of large rock masses can be determined from a small specimen tested in the laboratory. In this study some laboratory tests were carried out to investigate environmental effects on critical strain such as confining pressure, moisture content and temperature. In this paper the results of the tests are shown, and the environmental effects are discussed.

1. INTRODUCTION

For assessing the stability of underground structures like tunnels, Sakurai (1982) proposed the "Direct Strain Evaluation Technique (DSET)." According to this technique, the stability of tunnels is assessed by comparing strain occurring in the ground around the tunnels with its allowable value. The strain should be maintained within the allowable value for stabilizing the ground. Sakurai (1982) also proposed critical strain* defined as Eq. (1), which is always smaller than failure strain, and it can be adopted as the allowable value of strain.

$$\varepsilon_0 = \sigma_c / E \qquad (1)$$

where σ_c: uniaxial compressive strength
E : Young's Modulus

It is well known that uniaxial strength and Young's modulus of in-situ rock masses are usually smaller than those of small specimen. They can be expressed as:

$$\sigma_{cR} = m\,\sigma_c \qquad (2)$$

$$E_R = n\,E \qquad (3)$$

where σ_{cR} and E_R are uniaxial strength and Young's modulus of in-situ rock masses, respectively. σ_c and E are the values for small specimens. m and n are coefficients which are always less than one. Thus, critical strain of in-situ rock masses is expressed as:

$$\varepsilon_{0R} = \frac{\sigma_{cR}}{E_R} = \frac{m\,\sigma_c}{n\,E} = \left(\frac{m}{n}\right)\varepsilon_0 \qquad (4)$$

In Eq. (4) m/n is a modification factor for relating the critical strain of small rock specimens with that of large in-situ rock masses. It has already been demonstrated that it ranges approximately 1.0 - 3.0 depending on rock types [Sakurai (1983)]. This implies that, if we adopt critical strain as an allowable value for assessing the stability of rock masses, the safety factor may increase slightly up to 3.0.

It is worth mentioning that critical strain is more or less identical for both small specimens and large in-situ rock masses. This is a rather surprising fact, and it may be interpreted as both the uniaxial strength and Young's modulus of rocks being strongly influenced by the size of specimen. If, however, the uniaxial strength is divided by the Young's modulus, that is the critical strain, then the effects of specimen size cancel each other out. Thus, the critical strain shows almost identical values for both small specimens and large scaled in-situ rock masses.

Based on the above mentioned discussions on the effects of specimen size on the critical strain, it may be predicted that the effects of environmental factors such as confining pressure, moisture content, and temperature also do not seriously influence the critical strain, although the strength and Young's modulus are very much affected by these environmental

*Deere (1968) defined "Modulus ratio," which is a reciprocal number of the critical strain defined by Eq. (1). However, the physical meaning of the modulus ratio is unclear.

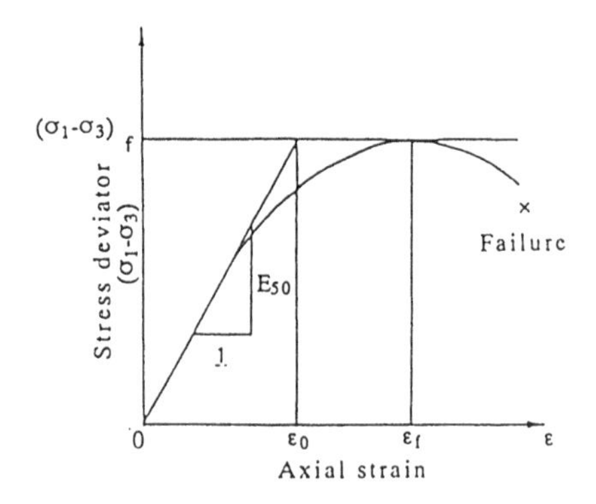

Fig.1 Definition of critical strain in triaxial condition

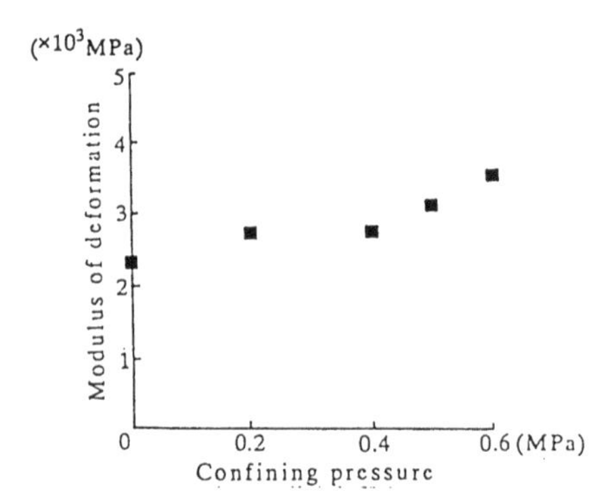

Fig.3 Effect of confinig pressure on modulus of deformation

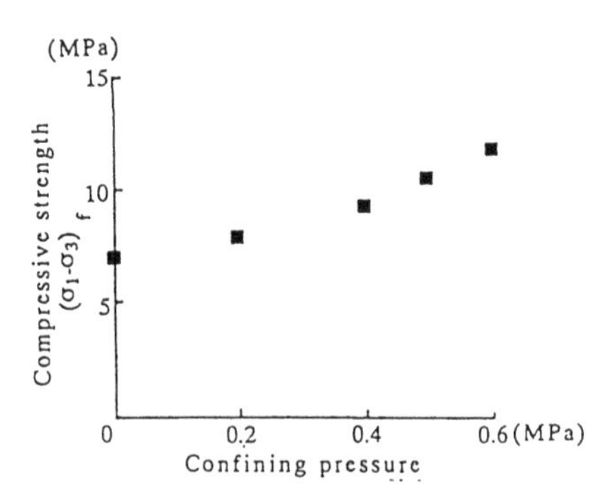

Fig.2 Effect of confinig pressure on compressive strength

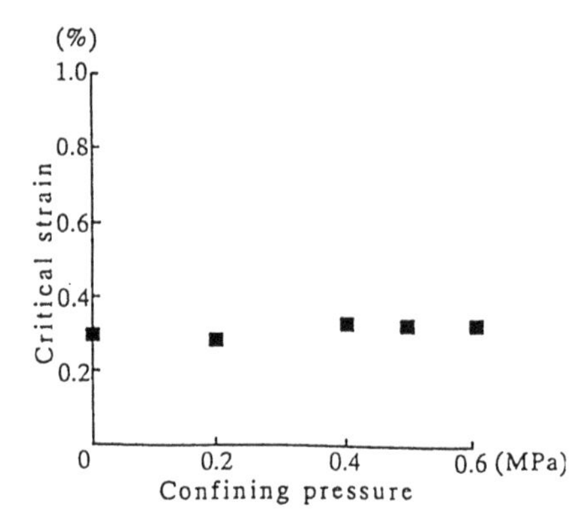

Fig.4 Effet of confinig pressure on critical strain

factors. To prove this prediction, some laboratory tests were carried out. In this paper, the test results are shown, and the effects of the environmental factors on critical strain are discussed.

2. CRITICAL STRAIN IN TRIAXIAL CONDITION

The critical strain is originally defined under a uniaxial condition. In engineering practice, however, geomaterials are usually loaded under triaxial condition. Thus, the critical strain in triaxial condition is defined as follows (see Fig.1).

$$\varepsilon_0 = \frac{(\sigma_1 - \sigma_3)_f}{E} \qquad (5)$$

where $(\sigma_1 - \sigma_3)_f$ denotes maximum stress deviator at failure, and E is the modulus of deformation.

3. EFFECTS OF CONFINING PRESSURE

Porous tuff was tested under triaxial conditions in the laboratory. The size of a cylindrical specimen was ϕ50mm x H100mm. The results of tests under both fully saturated and drained conditions are shown in Figs. 2,3 and 4, where maximum stress deviator, modulus of deformation and critical strain are plotted with respect to confining pressure, respectively. It is clearly understood from these figures that the maximum stress deviator and the modulus of deformation are strongly influenced by confining pressure, and they increase with any increase in confining pressure. However, the critical strain is fairly independent from confining pressure, and indicates an almost constant value.

It is worth mentioning that the critical strain under the triaxial condition is almost identical to that of the uniaxial condition (confining pressure is zero).

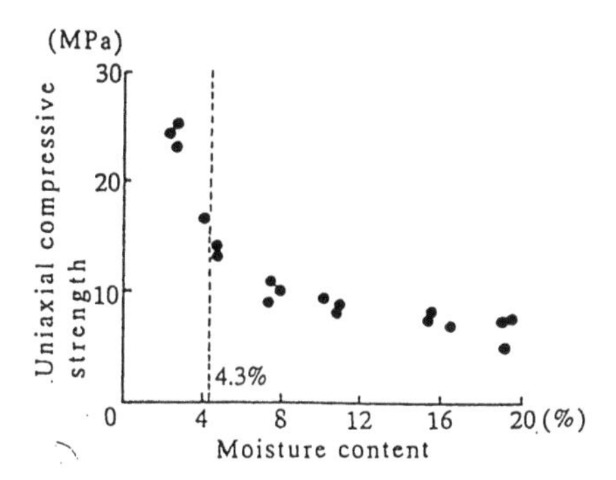

Fig.5 Effect of moisture content on uniaxial compressive strength

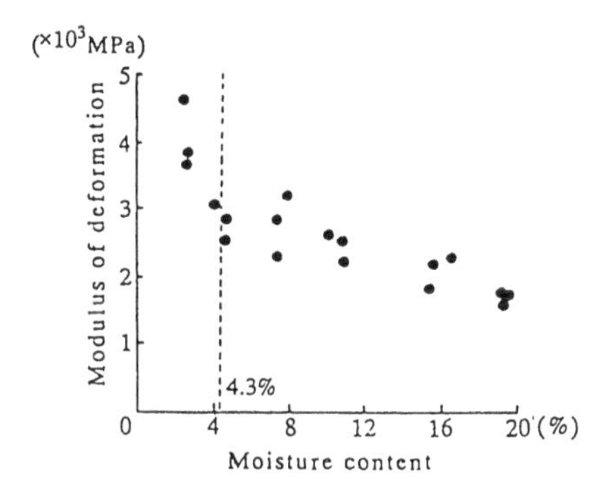

Fig.6 Effect of moisture content on modulus of deformation

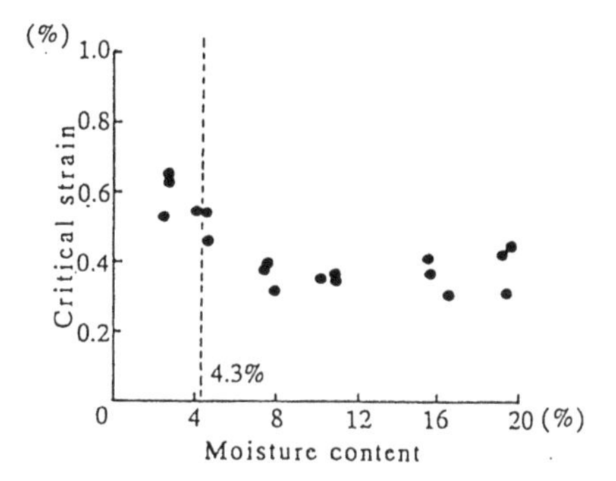

Fig.7 Effect of moisture content on critical strain

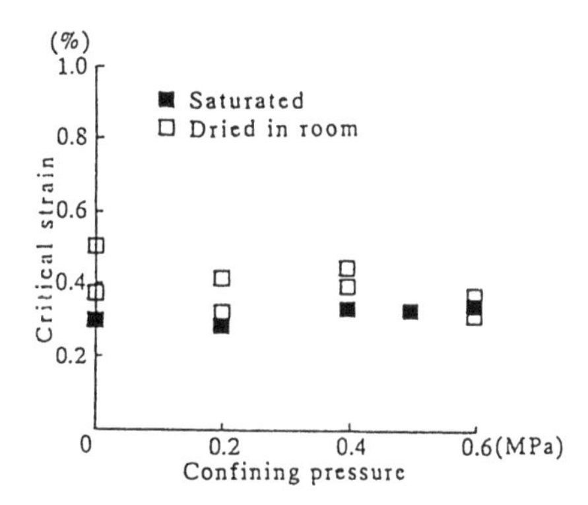

Fig.8 Effect of confinig pressure on critical strain

4. EFFECTS OF MOISTURE CONTENT

Cylindrical specimens of porous tuff were also used. The tests were carried out under a uniaxial compressive state by changing moisture content ranging from 0% to 100% saturation with 20% interval. Three tests were repeatedly conducted for each moisture content.

The relationship between uniaxial compressive strength and moisture content is shown in Fig. 5. The modulus of deformation is shown in Fig. 6 for different moisture content. It can be observed from these two figures that both the uniaxial compressive strength and modulus of deformation decrease with an increase in moisture content. On the other hand, however, it is understood from Fig. 7 that the critical strain seems to be approximately constant, with the exception of a low moisture content condition. It may be interesting to note that the critical strain tends to increase rapidly as moisture content becomes less than the natural moisture of the laboratory. (The natural moisture was 4.3% during tests).

In order to verify the reason why critical strain increases when moisture content is less than natural moisture, micro-structures of the rocks were investigated by using a scanning electron microscope. The microscopic structure of rocks are shown in photos 1(a), (b) and (c). The photos 1(a) and (b) are taken for the rocks under fully saturated and natural moisture conditions, respectively. It is seen from these photos that there seems to be no difference between the two. On the other hand, photo 1(c) is for a rock in an absolutely dried condition. This photo indicates clearly the change of micro-structure of the rocks due to an absolute dryness.

The comparison of these three photos may conclude that the physical constitution of the rocks is no more the same as the original one, when moisture content of the rocks becomes less than the natural moisture in a room. In other words, the rocks become a different kind, so that critical strain is no more constant.

Fig. 8 shows a comparison between critical strains of both fully saturated and dried specimens determined under different confining pressures. It can be seen

(a) Saturated

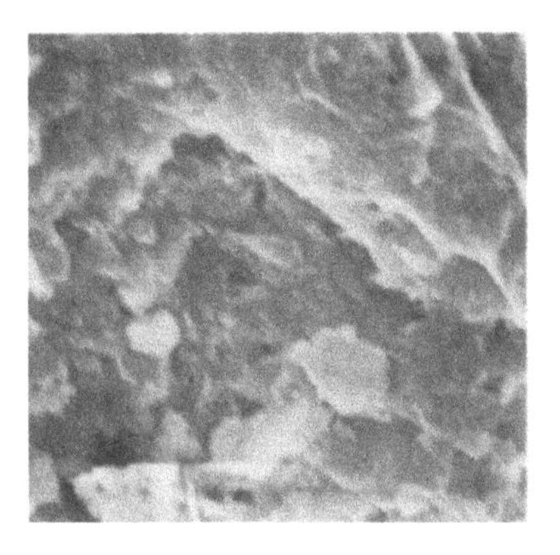

(b) Dried in room

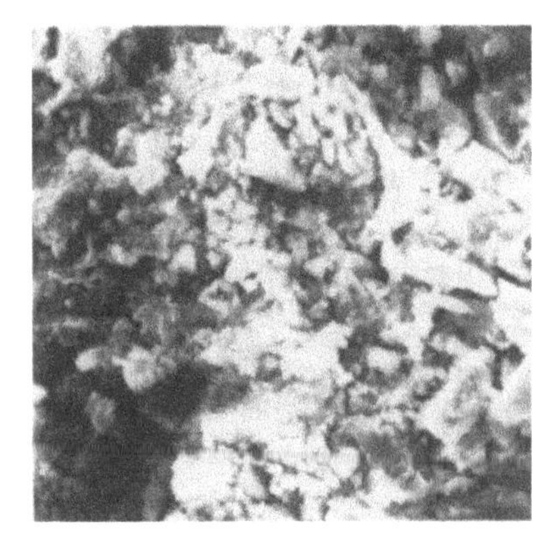

(c) Absolutely dried

Photo 1 Microstructure of specimens (x2000)

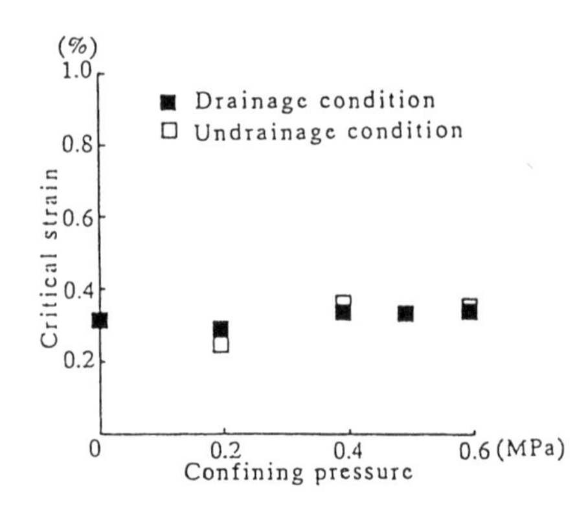

Fig.9 Effect of confining pressure on critical strain

here that both results are almost identical to each other. This indicates that the effect of confining pressure is negligibly small even for saturated rocks.

Triaxial tests were also carried out under both drainage and non-drainage conditions. The results are shown in Fig. 9. It can be deduced from this figure that pore water pressure does not cause any influence on critical strain. Therefore, it may conclude that critical strain is almost independent from the change of both pore water pressure and confining pressure.

5. EFFECTS OF TEMPERATURE

The effects of temperature on critical strain were investigated by using data already published in literature. Inada et al.(1992) carried out uniaxial compression tests on granite, andesite, sandstone and tuff under different temperatures. Their test results have been used for this study. Critical strains are determined and plotted with respect to temperatures, as shown in Figs.10 and 11. Fig.10 is for fully saturated specimens, and Fig.11 for dried specimens. It may be interesting to note that critical strain is also, for the most part, independent from temperature as well as moisture content.

6. CONCLUSION

Critical strain has been defined for triaxial compression conditions. The effects of confining pressure, moisture content, pore water pressure and temperature on the critical strain have been studied, and the following conclusions are derived:

(1) The effect of confining pressure on critical strain is negligibly small. Thus, critical strain obtained from uniaxial compressive conditions can represent a value even for triaxial conditions.

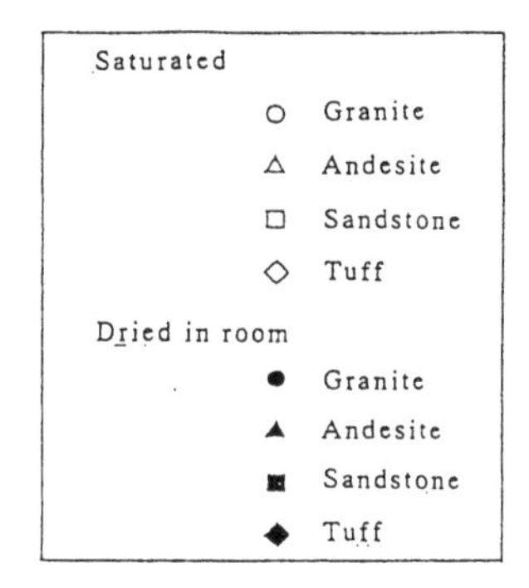

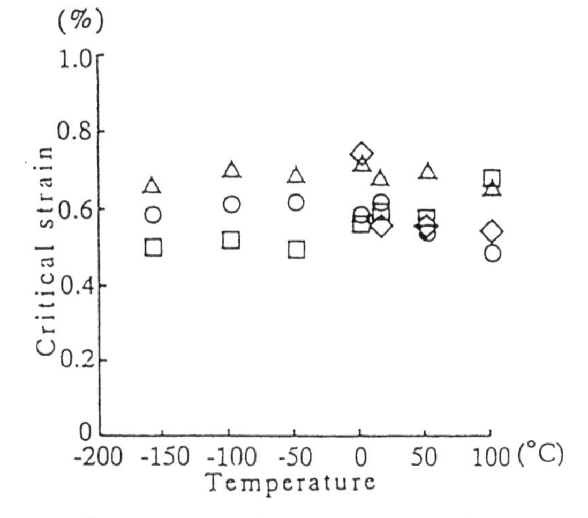

Fig.10 Effect of temperature on critical strain (Saturated specimens)

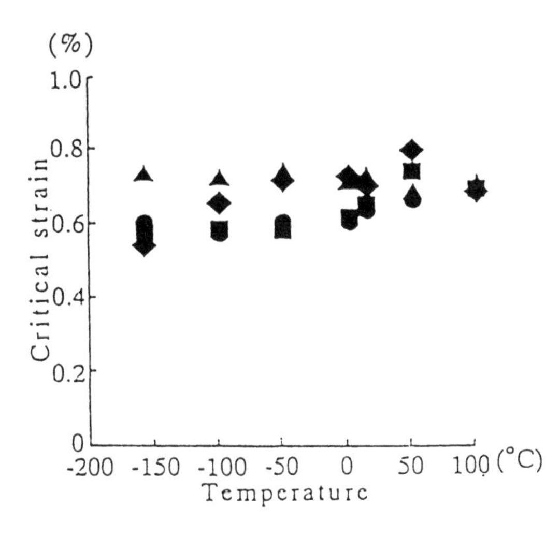

Fig.11 Effect of temperature on critical strain (Dried specimens)

(2) Moisture content does not effect critical strain, provided that moisture content is greater than natural moisture of the laboratory. When moisture content decreases and becomes less than natural moisture, the physical constitution of the rocks is no more the same as the original one.

(3) Critical strain is also, for the most part, independent from temperature, and is assumed to be a constant value depending on the kind of rock.

It has already been proved that critical strain of large scale rock masses is approximately the same as that of small sized specimens. Therefore, all the laboratory test results with respect to critical strain obtained in this study can easily be extended to large scale in-situ rock masses.

Aknowledgement

The authors wish to thank Prof. K. Nakamae and Mr. A.Naohara, Graduate Student, Dept. of Chemical Science and Engineering, for their help to take photos by using a scanning electron microscope.

Reference

Deere,D.U. (1968) Geological considerations, Rock Mechanics in Engineering Practice, Edited by K.G.Stagg and O.C.Zienkiewicz, pp.1-20

Inada,Y. and Kokudo,Y. (1992) Effect of high and low temperature on failure characteristics of rock under compression, J. Soc. Mat. Sci., Japan, Vol. 41, No. 463, pp.410-416.(In Japanese)

Sakurai,S. (1982) An evaluation technique of displacement measurements in tunnels, Proc. of JSCE, No. 317, pp.93-100.(In Japanese)

Sakurai,S. (1983) Displacement measurements associated with the design of underground openings, Proc. Int. Sympo. Field Measurements in Geomechanics, Zurich, Vol. 2, pp.1163-1178.

Developments in Geotechnical Engineering, Balasubramaniam et al. (eds) © 1994 Balkema, Rotterdam, ISBN 90 5410 522 4

Application of satellite technology to environmental geotechnology

Za-Chieh Moh & Chung-Tien Chin
Moh and Associates, Inc., Taipei, Taiwan

ABSTRACT: Ground subsidence is one of the most serious environmental geotechnology problems in Taiwan. It has been reported that about 1170km^2 of land has settled during the past few decades due to extraction of groundwater. Recently, the Government sponsored a research which is aimed at providing guidelines for better control of ground subsidence. One of the major tasks of this research is to introduce satellite technology in the study of ground subsidence. Global Positioning System (GPS) has been successfully applied in this study. The major concern of the GPS application is its accuracy in vertical direction. This study indicates that the root-mean-square difference between level survey and GPS can be controlled to be within 2cm which is acceptable for study in large areas with significant subsidence. Since satellite image can now be relatively frequently and economically obtained, it has become an extremely useful data in the control of land use. This study introduced the use of neural network concept to process the images from Satellite SPOT. It can identify the characteristics of land with reasonable accuracy. The processed image can then be incorporated in a Geographic Information System (GIS) for land use control. The successful use of satellite technology in this study indicates that further applications of these techniques in environmental geotechnical problems are very encouraging.

INTRODUCTION

Ground subsidence is one of the most serious environmental geotechnology problems in Taiwan. The total area of Taiwan Island is 36000km^2. It has been reported that about 1170km^2 of land has settled during the past few decades due to extraction of groundwater. Many coastal areas have even been seriously flooded during typhoon seasons. Considering two thirds of the land in Taiwan is mountainous, the ground subsidence area does occupy a significant portion of very valuable land which can be used for living, agriculture and industry. This ground subsidence issue has caught more attention in the last few years since the Government announced the "Six-Year Development Plan". The total budget of this plan is approximately 300 billion US dollars of which more than 100 billion dollars will be spent on infrastructures. These infrastructures include highways, rapid transit systems, high speed railway, and other engineering facilities. Some of the major constructions will be located in ground subsidence areas.

Many studies on ground subsidence have been conducted in different areas in Taiwan. Most of these studies use conventional survey methods and monitoring instrument to measure ground level and groundwater pressure. Since the impact of ground subsidence becomes more significant to the future development in Taiwan, the Public Construction Supervisory Board of the Executive Yuan (Cabinet) in 1991 commissioned Moh and Associates, Inc. to carry out a study which is aimed at providing guidelines for better control of ground subsidence (MAA, 1992; MAA, 1993). This research is methodology-development in nature. One of the major tasks of this research is to introduce the state-of-the-art technology to future study on ground subsidence problems. Of particular interest to this paper is the application of satellite technology: Global Positioning System (GPS) is used to measure the ground elevation and the French Satellite SPOT images are used to characterize land use.

SITE CONDITION

The study site is Yong-An area in Kaohsiung County. It is located in the southwest coast of Taiwan (Fig. 1). Yong-An is one of the most extensively monitored ground subsidence areas in Taiwan. Because of the deep well pumping for fishery, Yong-An area has experienced significant ground subsidence since early 1980s. In addition to this regional ground subsidence, settlement problem is more severe in two major industrial facilities in this area. They are the Hsinta Power Plant and the Liquified Natural Gas (LNG) Receiving Terminal located in the north and south sidc of Yong-An, respectively. Both of these two facilities were built on reclaimed land. For successful operation of these facilities, settlement of the reclaimed land itself and settlement caused by surcharge load of the heavy industrial equipment should be taken into account. The settlement problem is even more complicated in the LNG Receiving Terminal. In order to build 7 underground storage tanks of 100,000kl capacity each, diaphragm walls of 1.2m thick and 54.5m long have been constructed. The groundwater has to be lowered during excavation. In 1986, the pumping used to lower the groundwater in the excavation has caused significant groundwater drawdown and consequently serious settlement of the site.

Figure 2 presents the typical soil profile of the study

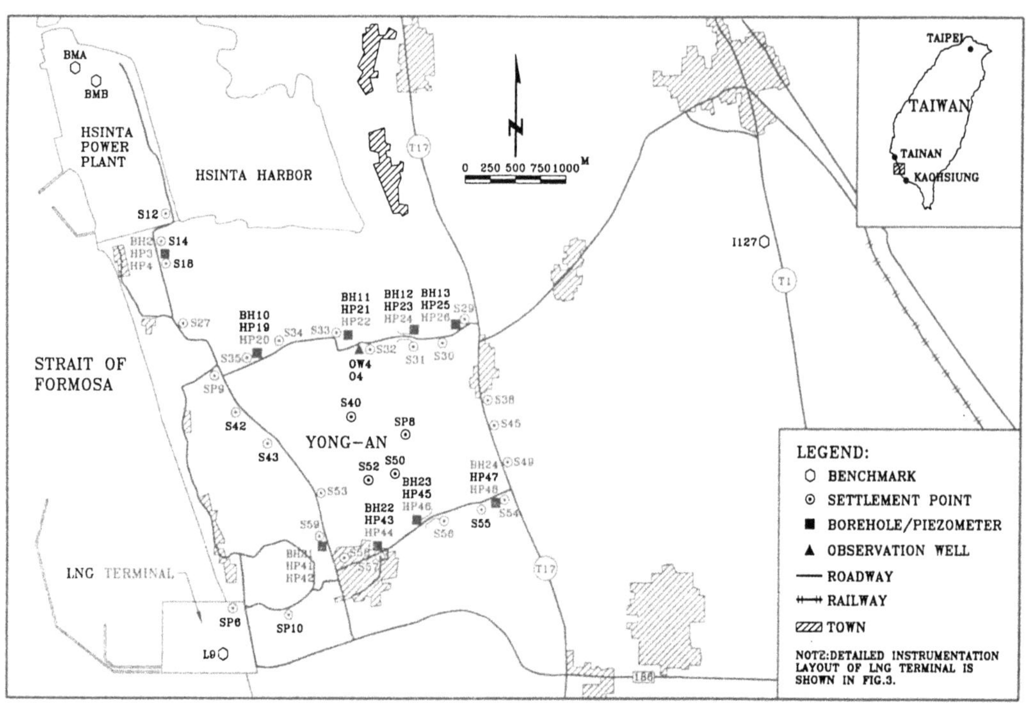

Figure 1. Typical Soil Condition of the Study Area

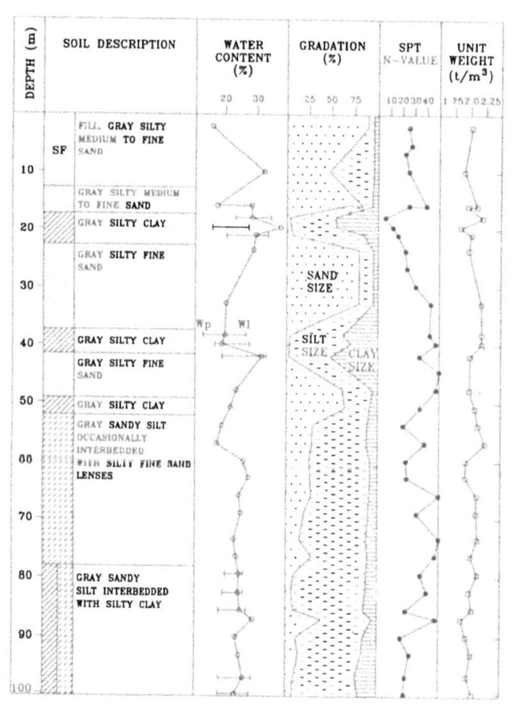

Figure 2. Soil Profile of the Study Area

area from ground surface to the depth of 100 m. Underlying surface soils are alternating silty sand and silty clay layers with depth down to approximately 52 m. Between the depths of 52 m and 78 m is a sandy silt layer with silty fine sand lenses. Below this layer is a thick interbedded sandy silt and silty clay layer. Other deep borehole information indicates that there is a gray silty clay stratum located between the depths of 100 and 200 m approximately. The groundwater table is very close to the ground surface, but the groundwater pressure in deep layer is significantly lower than the hydrostatic pressure.

MONITORING INSTRUMENTS

The general layout of the monitoring instruments in Yong-An is shown in Fig. 1. Figure 3 presents the detailed layout of the instruments in the LNG Receiving Terminal. Monitoring of ground subsidence and groundwater variation has been carried out in many stages by various engineering and research institutes in the last few years. Total settlement at ground level, compression of each soil stratum, and variation of groundwater pressure are monitored.

Level survey has been conducted to measure the elevation change of the shallow settlement points as well as the deep benchmarks. In this study, Wild NA3000 Electronic Level was used. In addition to the conventional level survey, the ground elevation has also been measured by the use of GPS. It will be further discussed in the following section.

Compression of each stratum has been measured by extensometers. Both rod extensometers and SONDEX

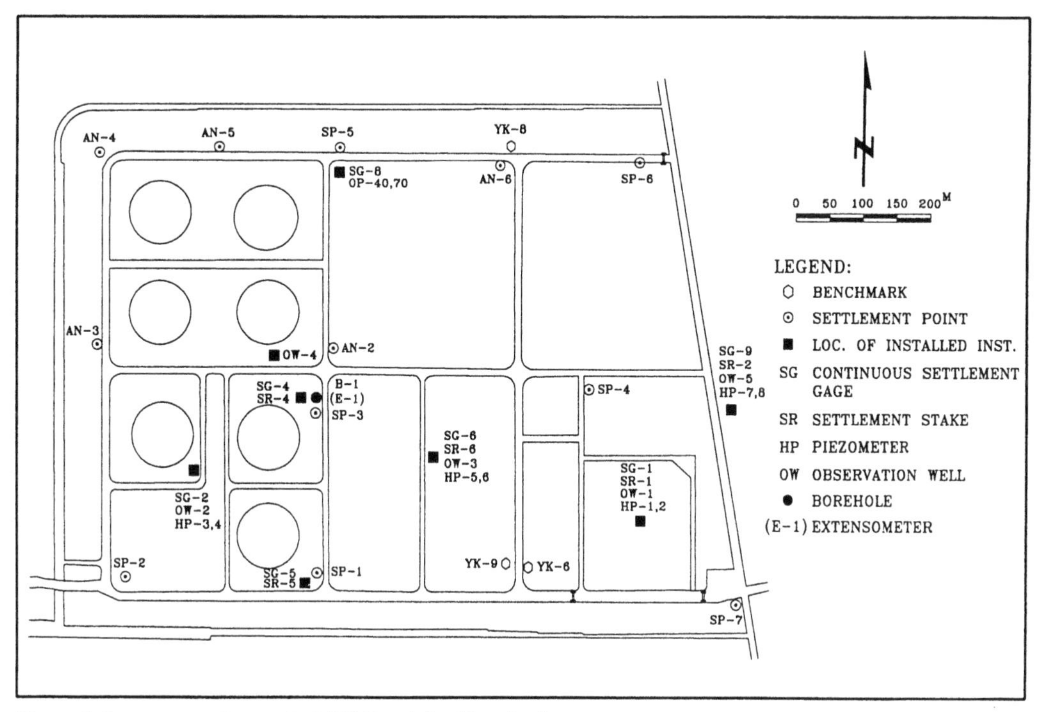

Figure 3. Instruments Layout in LNG Receiving Terminal

system have been used. Measurement of the compression of each stratum is critical to the prediction of the amount and duration of the settlement. With the long-term monitored compression and groundwater pressure data of each layer, back analysis can be conducted. Soil properties of each layer can be estimated, as well as the analysis model can be calibrated. The back-analysis results can then be used for prediction of future settlement of this site.

Groundwater pressure was measured by means of observation wells and piezometers. In this area, hydraulic-type, pneumatic-type and electronic-type piezometers have all been used. Groundwater monitoring results provide extremely useful information in understanding the ground subsidence mechanism of the region.

GLOBAL POSITIONING SYSTEM

The Global Positioning System (GPS) is developed by the U.S. Department of Defense. It is a satellite-based positioning system to support military navigation and timing needs. Once fully operational, it will consist of 24 satellites arranged in six orbital planes. The orbit is at an altitude of 20,200km above Earth with a 55° inclination. Currently there are 22 operational satellites which can allow for use of the system in Taiwan for almost 24 hours a day with good geometric distribution. During the surveying period, there existed 18 operational satellites. The absolute position of a ground point can be determined from a GPS receiver at this selected point receiving signals from at least 4 simultaneously observed satellites. With only one GPS receiver, the accuracy of the absolute positioning is not sufficient for engineering surveys. If two or more ground receivers are used simultaneously, then relative positioning between the two stations can be performed. The error can be reduced to the range of a few centimeters which can be acceptable for some specific uses in engineering practice. Compared to the conventional level survey, GPS is much faster and less expensive. Besides that, GPS survey has several other merits compared with terrestrial surveys (Chrzanowski et al., 1991). Intervisibility between stations is not needed, which simplifies the design of monitoring schemes and allows for the optimal location of points. It also makes it easier to select the reference points outside the subsidence area. In addition, GPS surveys provide simultaneously three-dimensional information on the deformation status. Over the past few years, GPS has been tested and applied in ground movement measurements in various projects such as ground subsidence study in the oil fields in Venezuela (Chrzanowski et al., 1991). The application of GPS to monitor ground subsidence due to water withdrawal (Strange, 1989) and the determination of tectonic movements (Bock & Murray, 1988) have also been reported.

In this study, GPS surveys have been conducted in September 1992, January 1993, and April 1993. Two ground receivers have been mobilized. "Rapid-static" survey method has been adopted in this study. The receivers occupied the station only for a few minutes to collect the data. A commercial software package LEICA SKI was used to process received signals. The working hours selected were dependent on the relative location among the ground receivers and satellites. The index to reflect the effect of the geometrical distribution of the satellites on the overall positioning accuracy is Geometrical Dilution of Precision (GDOP) which is given as a ratio of the actual positioning/time precision to the measurement precision. The smaller the numerical value of GDOP the better the geometry. In this study, all measurements were taken at GDOP less than 6. Level survey was carried out at the same time as the GPS survey was conducted so that the accuracy of GPS survey can be evaluated. The survey covers an area of approximately 30km². The elevations of 26 settlement points have been surveyed. Tables 1 to 3 compare the measured elevation from the leveling survey and GPS survey. They indicate that the average differences in these 3 sets of comparisons are 0.6cm, 1.0cm and 1.1cm. The average of all the measurements is 0.9cm. The root-mean-square errors of these 3 sets of measurements are 0.9cm, 1.4cm and 1.7cm. The accuracy of GPS survey of this study is comparable to the engineering level survey. It should be noted that the level survey carried out in this study took 7 days each time, but the GPS survey only required two working days at most.

GROUND SUBSIDENCE INFORMATION SYSTEM

In order to make the best use of all available information in ground subsidence area, Ground Subsidence Information System was established in this study. With its emphasis on the treatment of ground subsidence related problems, this system functionally is a Geographic Information System (GIS) which can be generally characterized as an organized collection of computer hardware, software, and

Table 1. Comparisons between GPS and Level Survey Measurements (Sept., 1992)

September, 1992 Measurement				
SETTLEMENT POINT	ELEVATION (Level Survey H (m)	ELEVATION (GPS Survey) Hi (m)	H-Hi (m)	GDOP
YK8	3.937	3.940	0.003	2
YK9				
S55	2.061	2.061	0.000	3
HT1	2.095	2.095	0.000	2
HT2				
AN3	4.087	4.074	-0.013	2
AN2	6.420	6.417	-0.003	2
AN4	4.777	4.776	-0.001	2
AN5	6.040	6.037	-0.003	2
AN6	3.659	3.658	-0.001	2
SP1				
SP2				
SP3	3.857	3.865	0.008	3
SP4	3.746	3.748	0.002	3
SP6	3.981	3.978	-0.003	3
SP7	4.526	4.529	0.003	4
SP8	2.832	2.819	-0.013	2
SP9	1.279	1.280	0.001	3
SP10	1.637	1.659	0.022	3
SP27				
SP31				
SP38	4.110	4.126	0.016	3
SP43	1.452	1.455	0.003	3
SP53	1.574	1.583	0.009	3
SP54				
SP57				
Average Error (m)		0.006		
Root-Mean-Square Error (m)		0.009		

Table 2. Comparisons between GPS and Level Survey Measurements (Jan., 1993)

January, 1993 Measurement				
SETTLEMENT POINT	ELEVATION (Level Survey H (m)	ELEVATION (GPS Survey) Hi (m)	H-Hi (m)	GDOP
YK8	3.928	3.928	0.000	
YK9	4.039	4.044	0.005	2
S55				
HT1	1.067	2.068	0.001	1
HT2	9.481	9.481	0.000	1
AN3	4.047	4.049	0.002	1
AN2	6.378	6.383	0.005	3
AN4	4.752	4.753	0.001	4
AN5	6.014	6.021	0.007	2
AN6	3.633	3.657	0.024	4
SP1	3.642	3.618	-0.024	3
SP2	4.258	4.235	-0.023	4
SP3	3.829	3.821	-0.008	3
SP4	3.723	3.741	0.018	5
SP6	3.961	3.977	0.016	3
SP7	4.506	4.505	-0.001	2
SP8				
SP9	1.249	1.263	0.014	2
SP10	1.616	1.630	0.014	2
SP27	0.327	0.325	-0.002	1
SP31	1.225	1.239	0.014	2
SP38	4.080	4.062	-0.018	1
SP43	1.425	1.443	0.018	2
SP53				
SP54	5.143	5.144	0.001	3
SP57	1.826	1.846	0.020	3
Average Error (m)		0.010		
Root-Mean-Square Error (m)		0.014		

Table 3. Comparisons between GPS and Level Survey Measurements (April, 1993)

April, 1993 Measurement				
SETTLEMENT POINT	ELEVATION (Level Survey H (m)	ELEVATION (GPS Survey) Hi (m)	H-Hi (m)	GDOP
YK8	3.929	3.929	0.000	
YK9	4.039	4.039	0.000	2
S55				
HT1	2.067	2.067	0.000	1
HT2	9.476	9.476	0.000	
AN3	4.039	4.045	0.006	2
AN2	6.360	6.378	0.018	2
AN4	4.740	4.745	0.005	2
AN5	6.004	6.010	0.006	1
AN6	3.622	3.634	0.012	2
SP1	3.635	3.637	0.002	1
SP2				
SP3	3.817	3.824	0.007	1
SP4	3.712	3.733	0.021	2
SP6	3.953	3.979	0.026	2
SP7				
SP8				
SP9				
SP10	1.609	1.609	0.000	2
SP27				
SP31	1.221	1.246	0.025	1
SP38	4.072	4.081	0.009	2
SP43	1.420	1.464	0.044	2
SP53				
SP54	5.138	5.139	0.001	2
SP57	1.820	1.852	0.032	2
Average Error (m)		0.011		
Root-Mean-Square Error (m)		0.017		

geographic data designed to efficiently capture, store, update, manipulate, analyze and display all forms of geographically referenced information (Dangermond, 1992).

In addition to the general features of GIS, this Ground Subsidence Information System was designed to satisfy two other major requirements: (1) all borehole information and monitoring results have to be incorporated in the system and which can effectively processed so that the fundamental subsidence mechanism can be understood and the analysis on future settlement can be conducted; and (2) images of land-cover and processed land-use map should be included in the system and which can easily be updated so that the government agencies can use this system to control the land use.

Details of this system is presented in Moh and Associates (1993). Data in Ground Subsidence Information System are organized as shown in Fig. 4. The major hardware used in this system is the INTERGRAPH Workstation InterPro 2730. The basic software used is the MGE System which is the platform and front-end setup for the Modular GIS Environment. Most of the other GIS applications are based on MicroStation 32, such as MGE Analyst, which can be added to MGE for classic GIS analysis capabilities.

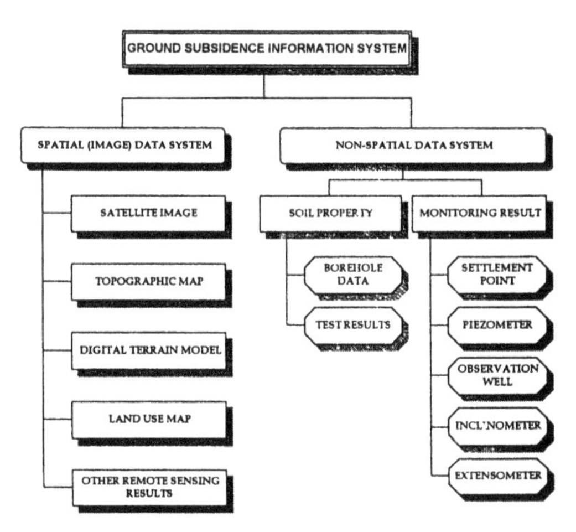

Figure 4. Data Structure of Ground Subsidence Information System

APPLICATION OF SATELLITE IMAGES

Information on land-cover and land-use are very important to environmental planning and control in using Geographic Information Systems. Since earth resource satellite can provide relatively frequent data over large areas, it has increasingly become a major source of land-cover and land-use information. In Taiwan, Satellite Ground Receiving Station at the National Central University is now routinely receiving data from satellites Landsat and SPOT.

A significant amount of ground subsidence in very wide areas in Taiwan is caused by deep pumping from illegal fish farming. Government needs to use this Ground Subsidence Information System to monitor the land use. Satellite images are very helpful in carrying out this task. But unless a reliable automated classification system can be implemented in this system, the advantages of using satellite images cannot be fully developed. Many classification algorithms have been developed since the launch of Landsat MSS and relatively reasonable results can be obtained (Lillesand and Kiefer, 1987). However, the statistical approach, commonly used by most of the traditional classification algorithms (Swain and Davis, 1978), is not designed to handle high spatial resolution satellite data, such as 20m resolution of SPOT which is used in this study.

This study introduces the use of neural network to perform land classification from SPOT images (Heermann and Khazenie, 1992). The neural network is the model that attempts to simulate the learning and recognition processes of the human brain. One of the most popular and most widely used neural network models is known as back-propagation (Chester, 1993). The topology of this type of network typically consists of three or more layers: an input layer, an output layer, and one or more hidden layers in between. Figure 5 illustrates a three-layer neural network. Each layer contains a number of nodes, which correspond to the human biological neurons. The input layer accepts input signals, and the output layer stores the results of the network. The hidden layer performs activation of the input signals and transfers the activation to the output layer through a nonlinear procedure. A complete processing of the neural network normally begins with pattern learning and ends with pattern recognition. The purpose of learning is to use training data to train the neural network to recognize a set of desired output patterns. As a result of learning, the connection between layers will contain a set of weights which can be used to perform pattern recognition. After the neural network has been trained, the unknown input can be processed and recognized as proper pattern by the derived weights in the neural network.

In this study, the training data is represented by the brightness values of SPOT multispectral image and the desired output patterns are corresponding land-cover/land-use classes. SPOT images of 3rd October 1986 and 30th January 1988 were processed. Comparisons between the processed data can clearly identify the change of fishing farm areas. The results are in very good agreement with the change of land use.

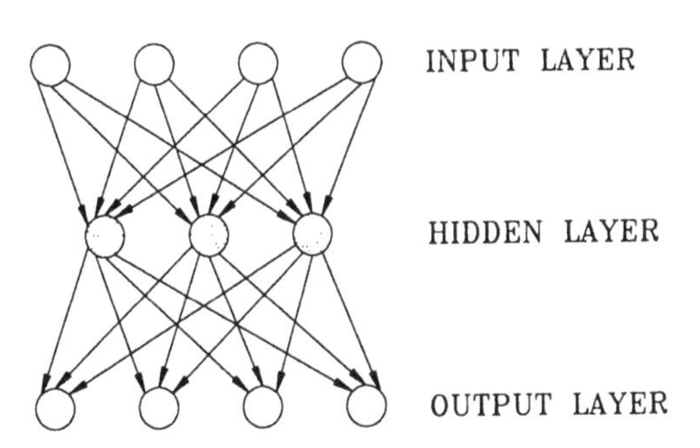

Figure 5. A Three Layer Neural Network

CONCLUSIONS

Environment geotechnology is a relatively new field. Many of environmental geotechnical issue did not exist in the days of Terzaghi. It had not been included as a specialty session in the International Conference on Soil Mechanics and Foundation Engineering until the IXth Conference in Tokyo in 1975 (Moh, 1975). But now environmental geotechnical problems have become one of the most seriously concerned topics to geotechnical engineers. Today while many high-techs are making dramatic progresses, geotechnical engineers must be equipped with those advanced technologies to tackle those new environmental challenges. This paper briefly describes the application of satellite technology to ground subsidence problems. The successful result indicates that further applications of these techniques to other environmental geotechnical problems are very promising. More importantly, it exemplifies a situation that how a geotechnical engineer solves a problem with non-conventional tools - it may be quite frequently the case in the next century.

ACKNOWLEDGEMENTS

The authors wish to thank the Public Construction Superviorsy Board, Executive Yuan, R.O.C. for sponsoring this study. The authors are grateful to Prof. L. C. Chen, Prof. J. Wu and Prof. C. F. Chen of the Center for Space and Remote Sensing Research, National Central University for their participation in the project. Their assistance in preparation of this paper is also greatly appreciated.

REFERENCES

Bock, Y. & M. H. Murray 1988, "*Assessing the Long-term Repeatability and Accuracy of GPS: Analysis of Three Campaigns in California*" Proc., 5th Int Symp. on Deformation Measurements, Univ. of New Brunswick, Fredericton, Canada, pp. 91-102.

Chester, M. 1993, "*Neural Networks: A Tutorial*" PTR Prentice Hall, New York, N.Y.

Chrzanowski, A., Chen, Y. Q., Leal, J., Murria, J. & T. Poplawski 1991, "*Use of Global Positioning System (GPS) for Ground Subsidence Measurements in Western Venezuela Oil Fields*" Proc., 4th Int'l Symp. on Land Subsidence, Houston, Texas, U.S.A., pp. 410-431.

Dangermond, J. 1992, "*What is a Geographic Information System (GIS)?*", Geographic Information Systems (GIS) and Mapping - Practices and Standards, Edited by A.I. Johnson, C.B. Peterson, and L.J. Fulton, ASTM Publication Code Number 04-011260-38, pp. 1-17.

Heermann, P. D., & N. Khazenie 1992, "*Classification of Multispectral Remote Sensing Data Using a Back-Propagation Neural Network*", IEEE Transactions on Geoscience and Remote Sensing, Vol. 30, No. 1, pp. 81-88.

Lillesand, T. M., & R. W. Kiefer 1987, "*Remote Sensing and Image Interpretation*", 2nd edition, John Wiley & Sons. Inc., New York, N.Y.

Moh, Z. C. 1975, "*Session Report on Geotechnical Engineering and Environment Control*", Proc., IX Int'l Conf. on Soil Mech. and Found. Engrg., Vol. 3, pp. 559-561.

Moh and Associates, Inc. 1992, "*Ground Subsidence in Taiwan (Phasse I)*", Research Report sponsored by the Public Construction Supervisory Board, Executive Yuan, R.O.C.

Moh and Associates, Inc. 1993, "*Ground Subsidence in Taiwan (Phasse II)*", Research Report sponsored by the Public Construction Supervisory Board, Executive Yuan, R.O.C.

Strange, W. 1989, "*GPS Determination of Groundwater Withdrawal Subsidence*", J. of Surveying Engineering, Vol. 115, No. 2, pp. 198-206.

Swain, P. H., & S. M. Davis 1978, "*Remote Sensing: The Quantitative Approach*", McGraw-Hill, New York, N. Y.

Developments in Geotechnical Engineering, Balasubramaniam et al. (eds) © 1994 Balkema, Rotterdam, ISBN 90 5410 522 4

In-situ testing methods for groundwater contamination studies

R.G.Campanella, M.P.Davies, T.J.Boyd & J.L.Everard
In-situ Testing Group, Civil Engineering Department, The University of British Columbia, Vancouver, B.C., Canada

ABSTRACT: *In-situ* testing where groundwater contamination is either documented or suspected is carried out to provide GeoEnvironmental site characterization. GeoEnvironmental characterization refers to the surficial and subsurficial representation that approximates the actual *in-situ* conditions. Surficial characterization, although often not straightforward, does allow visual appraisal of essentially all of the materials involved. Subsurface characterization,on the other hand, rarely allows even a micro-percentage appraisal. From a statistical significance standpoint, subsurface characterizations for groundwater studies are almost always inadequate and therefore large assumptions based on the best possible information are required.

In-situ testing methods, such as the piezocone penetration test, often provide the best possible information at selected locations for groundwater contamination problems where stratigraphic, geotechnical, hydrogeological and some specific environmental parameters are required on a specific and/or screening basis. The addition of a resistivity module to the basic piezocone has applications for both organic and inorganic contaminants. This paper presents a review of the currently available *in-situ* methodologies for groundwater contamination problems. The methods reviewed include conventional drilling techniques, piezometer cone penetration test technology with resistivity measurements and discrete water sampling systems. The advantages and disadvantages of using each methodology are presented and include comparisons of testing cost, data reliability and regulatory acceptance. Brief case histories are used to demonstrate the preferred technologies. The paper concludes with an overview of current research including recent advances in self-grouting technology for cross-contamination concerns.

1 INTRODUCTION

GeoEnvironmental is a relatively new term in the geoscience field. There are currently several interpretations of meaning for GeoEnvironmental, and the definition used herein is: "the field of study that links geological, geotechnical and environmental engineering and engineering sciences to form an area of interest that includes all physical and environmental concerns within geological media."

Groundwater contamination is an excellent example of an area of interest within the scope of the GeoEnviornmental field. With the basic definition established, GeoEnvironmental subsurface site characterization for groundwater studies can be defined as: "the formulation of the physical, chemical and/or biological information for the site that allows adequate three-dimensional representation for the prevailing subsurface conditions and the engineering requirements of the data."

Modern *in-situ* testing instruments provide the best current technology for developing this three-dimensional representation for many site conditions. This paper presents an overview of the currently available *in-situ* methods for site characterization. In-situ testing embodies all of the prevailing field conditions on material behavior whereas laboratory testing, usually on disturbed samples, attempts to estimate these many factors. The value of a given *in-situ* test is directly related to how much non-quantifiable disturbance is caused during tool insertion; e.g.,the degree of data repeatability.

2 GROUNDWATER CONTAMINATION: SITE CHARACTERIZATION STUDIES

As will be shown in the following sections, what makes modern *in-situ* testing methods attractive for groundwater site characterization studies is that, it is accurate, repeatable, and direct results can be obtained in a fast, cost-effective manner. Traditional methods of *in-situ* site characterization, like conventional drilling and sampling, cannot match the speed, accuracy or economy achieved by the tools described in this paper. However, in cemented and fractured sedimentary formations and in compact, thick gravel layers (where penetration tools are restricted) drilling and coring likely have an advantage. Surface geophysical methods are least disruptive, but they lack ground-truthing capability and therefore the results are difficult to interpret.

Modern *in-situ* testing also provides a huge advantage over conventional drilling in that soil cuttings are not produced. On many contaminated sites, the cost of preparing for, handling and disposing of, these cuttings can be extremely high and tends to limit the amount of necessary subsurface investigation ultimately carried out.

The key items that must be evaluated in order to formulate the subsurface characterization for appropriate groundwater studies are:

- the stratigraphic profile;
- the relative geotechnical properties of each stratigraphic unit;
- the hydraulic properties of aquifer and aquitard materials; and
- the nature of the pore fluid/gases that are present in the materials.

Therefore, the assessment of any given *in-situ* technique must include an evaluation of whether it will provide the requisite items and how comprehensive and accurate the results provided will be.

3 *IN-SITU* TESTING

As defined above, *in-situ* testing involves evaluating material properties "in place;" where physical and biochemical stresses are evaluated as is. Conventional drilling and discrete sampling represents a quasi *in-situ* test method although laboratory evaluations are typically required for samples retrieved. However, drilling is included in the definition for this paper to allow a comparison of the current industry standard with the more sophisticated tools and evaluation procedures available. On the other hand, surface geophysical methods have not been included as most of these approaches are in the early stages with respect to assessing environmental parameters.

3.1 *Conventional Drilling and Discrete Sampling*

The most common manner in which site characterization for groundwater studies is carried out remains to be conventional drilling with discrete sampling. The large experience base and the retrieval of a physical sample are important advantages of the method whereas lack of standards, relatively high cost and questionable sample quality represent the key disadvantages.

Drilling with reasonable, albeit far from continuous, sampling frequency will result in a production rate for a 30 meter (100 ft) hole from 10 to 20 hours and more. The actual production rate will depend upon soil conditions, equipment and the operator quality, the method of sampling used and decontamination and grouting requirements. Although specifying conventional drilling on a meterage basis is never recommended for obvious reasons, we can apply a generic $150 hourly rate with modest consumable to provide an approximate $50 to $100 per meter for this discrete sampling.

Once a sample is obtained, its quality must be assessed. Table 1 presents the steps and sources of error in taking a sample from its *in-situ* condition to the laboratory. Table 1 does not include the errors introduced by the laboratory itself, such as sample disturbance, loss of *in-situ* stresses, and chemical alterations, etc.

TABLE 1 -- Steps in groundwater sampling and their related sources of error

	Step	Sources of error
	In-Situ Condition	--------
1.	Establishing a sample location	Improper borehole completion/placement; poor location choice; poor material choice
2.	Sample collection	Sampling mechanism/procedural bias; operator error
3.	Sample retrieval	Sampling mechanism/procedural bias; sample exposure; degassing; oxygenation; cross-contamination
4.	Preservation/storage	Handling/labeling errors; temperature, etc. control
5.	Transportation	Delay; sample loss

Drilling and sampling are therefore shown to be relatively expensive and provide potentially suspect reliability on a non-continuous basis. However, due to its popular use, regulatory acceptance has traditionally been high although recent trends in data quality management and geostatistical evaluations are tending to make it more difficult to convince all regulators that data from drilling and sampling programs are either comprehensive or accurate enough.

3.2 *Piezometer Cone Penetration Test*

The piezometer cone penetration test (CPTU) is increasingly being used by practicing engineers as an effective means of geotechnical classification. The method provides a fast, economical, and accurate means of delineating soil stratigraphy and determining geotechnical parameters. The cone has a standard 10 cm^2, 60° conical tip, friction sleeve with an area of 150 cm^2 and one or more pore pressure sensors. The CPTU measures tip resistance (q_c), friction sleeve stress (f_s) and pore pressure at up to three locations on the cone. All channels are continuously monitored and reported at either typically 25 or 50 mm intervals, thus providing essentially continuous *in-situ* data sampling. There are extensive relationships available between the various CPTU channels, and combinations thereof, that provide soil behavior type (often equivalent to stratigraphy), geotechnical strength parameters and hydraulic parameters. Temperature (t) and instrument inclination (i) are usually measured simultaneously. In addition, a seismic model of the cone has been developed that allows the measurement of low strain dynamic properties of the soil.

CPTU is generally regarded as the most effective and cost efficient tool for stratigraphic logging of the soils (Robertson et al., 1986). The cone is pushed into the ground at a constant 2 cm per second (or about 1 m per minute) by a hydraulic pushing source; often a drill rig or a specially outfitted cone truck or penetration vehicle. As the cone is advanced, pore pressures are generated around the cone tip and sleeve. The measurement of the excess pore pressures which are generated during penetration, and their subsequent dissipation, provide insight into the soil type and its hydraulic parameters.

The CPTU can and should be specified on a meterage basis. The rigorous ASTM and International standards ensure high repeatability and reliability, thus eliminating traditional concerns such as the speed of testing being inversely related to data quality. Although there are regional variances, CPTU soundings are typically carried out for between $20 and $50 per meter for the continuous data and plots provided. Due to its high standard and operator requirements, engineering supervision to the degree required for the conventional drilling is not necessary. Its continuous data, speed, relatively low cost and excellent repeatability are its main advantages. The main disadvantage is the lack of physical samples but for screening assessments, confirmatory investigations or routine monitoring, the requirement for a physical sample is becoming less important as confidence in the method grows. Regulatory acceptance is variable with the knowledge of the regulator being the key discriminator as to the success of approval. The trend, however, is one of acceptance with the provision of at least an occasional confirmatory/collaborative bore-hole; an approach with which we fully concur.

3.3 *Resistivity Piezometer Cone Penetration Test*

The resistivity piezometer cone penetration test (RCPTU) is a recent modification of the standard piezocone test (CPTU). The ability to measure the resistance to current flow in saturated soils on a continuous basis is extremely valuable due to the large effects that dissolved and free product constituents have on soil resistivity (conductivity).

Table two presents typical values of bulk soil resistivity measurements with the RCPTU and corresponding measurements of pore fluid resistivity. Since conductivity is the reciprocal of resistivity, it is easy to convert according to:

$$\text{Conductivity}[\mu S/cm] = 10{,}000 \div \text{Resistivity}[\Omega\text{-m}] \quad (1)$$

Table 2 also gives corresponding values in units of conductivity due to their popular use in the geochemical field. The results in Table 2 show that the range of resistivity (conductivity) values is very large from about one (10000) to about 7000 Ohm-m (14 μS/cm) and is very sensitive to both soluble salts and low solubility organic contaminants.

Thus, the RCPTU has the advantage of providing continuous resistivity measurements (which directly correlate on a site specific basis with contaminant concentration) in combination with all of the traditional CPTU data. The current electrode separation used by researchers and practitioners varies from about 10mm to over 150 mm providing different depths of lateral penetration. The more advanced tools have several electrode spacings so that both profiling and sounding resistivity tests are carried out simultaneously. Figure 1 shows a schematic representation of one of UBC's RCPTU tools. Campanella and Weemes (1990) provide a complete summary of the RCPTU and its developments.

Therefore, the RCPTU meets all of the key items identified for formulating a subsurface site characterization. The actual quantification of non-background geochemical processes requires site specific calibration if more than "screening" evaluations are intended. This site specific calibration involves accurate pore fluid-gas sampling which is described in the following section. The cost per meter of RCPTU testing programs is essentially the same as for the CPTU with an allowance of an additional 10% for data reduction. Regulatory approval is variable due to the relative newness of the tool but in our experience, once the technology is demonstrated, the acceptance is typically rapid.

3.4 *In-Situ Water Sampling*

The main goal of any water sampling program is to rapidly and economically collect representative samples of the *in-situ* regime. Conventionally, bailers and pumps are used to obtain samples from drilled or installed wells or piezometer. To obtain high levels of sample integrity with conventional methods, costly and time consuming installations and procedures are typically required.

In recent years there have been significant developments in *in-situ* discrete depth water samplers. These samplers, such as the BAT Enviroprobe and QED Hydropunch II, minimize purging and exposure to the sample, allow for rapid, high quality, representative samples and are easy and economical to use. (Blegen, et. al, 1988, Zemo, et al, 1992)

A key advantage of these discrete depth samplers is that they can be used effectively in screening studies to better delineate contaminant and thus optimize the location(s) of permanent monitoring wells. Regulatory approval of these new *in-situ*

TABLE 2 -- Summary of typical resistivity (conductivity) measurements of bulk soil mixtures and pore fluid(UBC experience)

Material type	Bulk Resistivity ρ_b, Ω-m	Fluid Resistivity ρ_f, Ω-m	Bulk Conductivity μS/cm	Fluid Conductivity μS/cm
Sea water	---	0.2	---	50000
Drinking water	---	>15	---	<665
McDonald Farm clay	1.5	0.3	6700	33300
Laing Bridge site clay	20	7	500	1430
Colebrook site clay	25	18.2	400	550
TC @ 232 Ave clay	8	---	1250	---
Strong Pit clay	35	---	285	---
Kidd 2 site clay	14	12.5	715	800
McDonald Farm site sand	5-20	1.5-6	2000-500	6700-1670
Laing Bridge site sand	5-40	1.5-10	2000-250	6700-1000
Colebrook site sand	70		143	
Strong Pit site sand	115		89	
Kidd 2 site sand	1.5-40	0.5-21	6700-225	20000-475
Typical landfill leachate	1-30	.5-10	10000-330	20000-1000
Mine tailings site sand with acid drainage leachate	1-40	2-27	10000-250	5000-370
Mine tailings site sand without acid drainage	70-100	15-50	145-100	665-200
Industry site-inorganic contaminants in sand	0.5-1.5	0.3-0.5	20000-6500	33000-20000
100% ethylene dichloride (ED)	---	20,400	---	0.5
50% ED/50% water in sand	700	---	14	---
17% ED/83% water in sand	275	---	36	---
Industry site - organic contaminants in sand	125	---	80	---
BC Place Parcel 2, PAHs (coal gas plant)	200-300	---	50-33	---
BC Place Parcel 2 (wood waste)	300-600	---	33-66	---

sampling methods has been widespread and rapid. As use and experience increases, the methods will likely become industry standard for appropriate sites.

Figure 2 shows a schematic sketch of the rapid sampling probe used at UBC which uses the BAT groundwater system (Torstensson, B-A, 1984). This 50 mm diameter probe with a central HDP filter is pushed down the hole previously made by the RCPTU which left a 44 mm diameter hole.The probe is pushed to a specific depth to obtain a pore water sample which is then chemically analyzed to provide site specific correlations of contaminant concentrations with bulk resistivity measurements. The preceding RCPTU sounding provides soil classification information with the use of CPTINT (a cone interpretation program developed at UBC) to indicate the depth where high hydraulic conductivity zones exist so rapid water sampling can be achieved. Soil classification is reliably predicted using either friction ratio or pore pressure ratio.

Current studies with this continually advancing water sampling probe indicate that it takes representative samples of volumes of about 250 ml. A 250 ml sample can often be obtained in less than 10 minutes as long as its position is carefully placed in the coarse zones as indicated by CPTINT soil classification. A complete RCPTU sounding to 30m depth followed by 10 or more water samples (including a purge sample at each depth) can be completed in one working day.

4 CASE EXAMPLES

4.1 *Organic Contaminants: Heavy Oils*

RCPTU and discrete depth water sampling were used in a research capacity to confirm and augment data obtained through a conventional drilling and sampling program at a contaminated site in the Lower Mainland, B.C. The aim behind the research program was to demonstrate the capabilities and limitations of using this technology as a screening tool on organically contaminated sites.

The field program consisted of 7 RCPTU soundings (3 were used as water sampling holes) located to make optimum use of existing information. Figure 3 shows a typical RCPTU

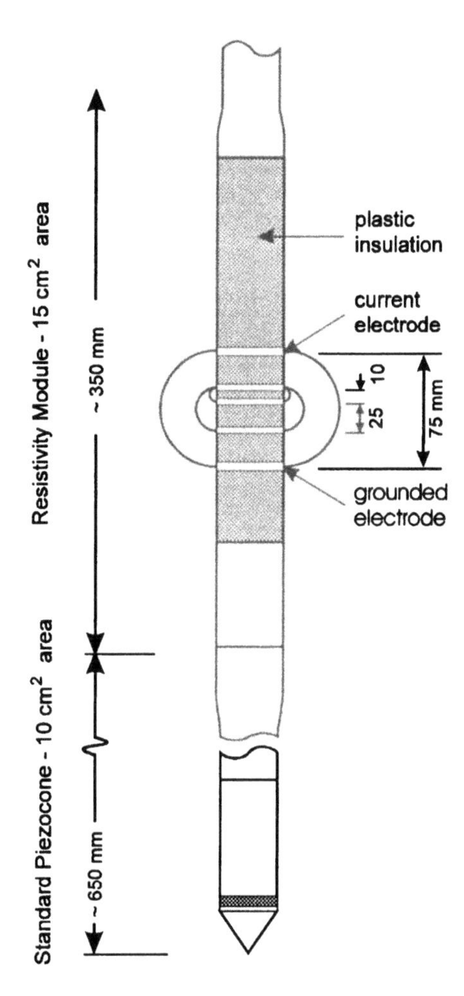

Fig. 1 UBC Resistivity Piezocone (RCPTU)

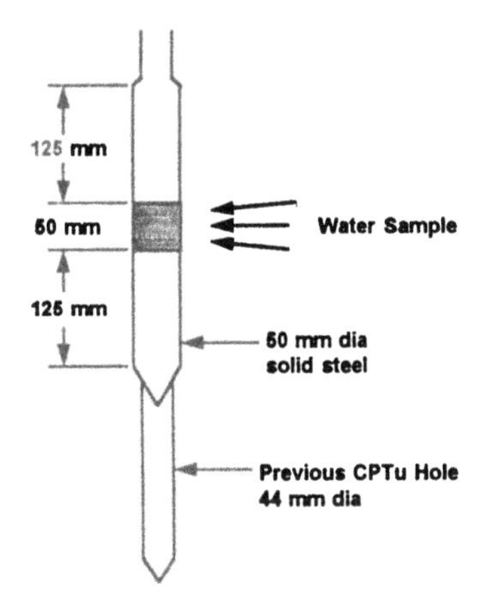

Fig. 2 UBC Modified BAT Pore Water Sampler

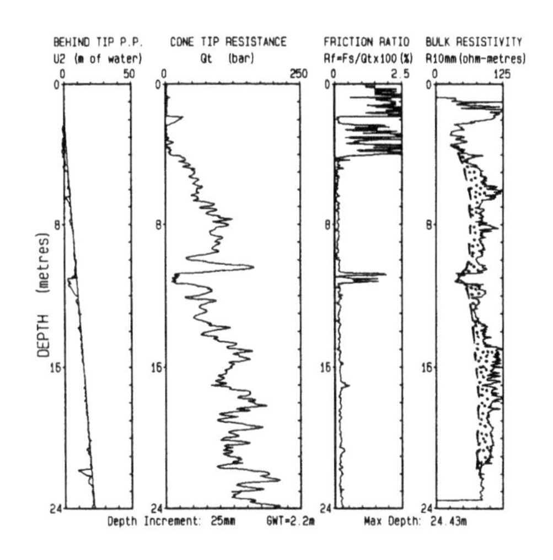

Fig. 3 RCPTU Sounding:
Organic Contamination by Heavy Oils

profile of cone tip resistance, pore pressure, friction ratio and resistivity with depth. Contrasts in bulk resistivity showing elevated values (due to the insulating quality of the organic contaminants) compared to "background" values are indicated by the shaded zones in Fig.3. The changes in bulk resistivity are dominated by pore fluid chemistry which was verified by discrete depth water sampling and lab testing.

4.2 *Industrial Inorganic Pollutants - Western Canada*

RCPTU and discrete depth water sampling were used to characterize an individual site in Western Canada with respect to the geological, hydrogeological and geochemical setting. The purpose of the investigation was to delineate the plume of inorganic contaminants (sulphates, chlorides, and phosphates) and to speculate on the potential for offsite migration. RCPTU and discrete depth water sampling were chosen for their technical and cost advantages. Geotechnical and hydrogeological parameters (e.g. soil type, relative density, hydraulic conductivity, etc.) were readily obtained from the investigation. Chemical analyses conducted on groundwater samples correlated reasonably well with bulk resistivity contours. A trend between total dissolved solids (TDS) and bulk resistivity was noted to be similar to Ebraheem, et al (1990). In contaminated zones, bulk resistivity varied between 0.5 and 1.5 Ohm-m (20,000 - 6500 μS/cm). Pore fluid resistivities within contaminated zones varied between 0.3 and 0.5 Ohm-m (30,000-20,000 μS/cm). Contaminated zones and their direction and rate of movement were subsequently identified through on-going RCPTU studies.

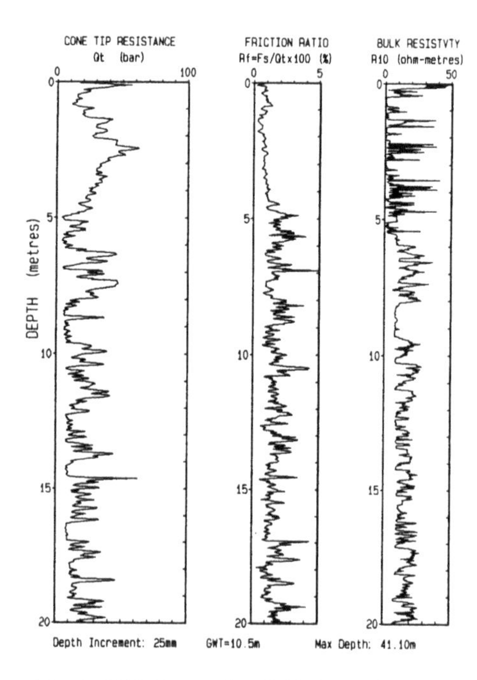

Fig. 4 RCPTU Sounding: Acid Generating Mine Tailings

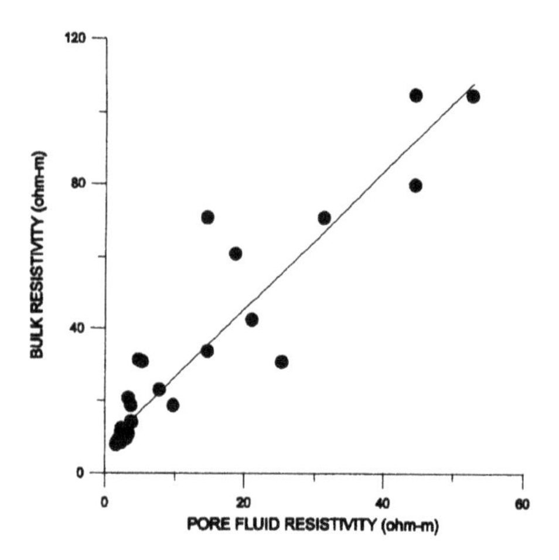

Fig. 5 Bulk versus Fluid Resistivity for Mine Tailings

4.3 *Acid Rock Drainage (ARD) in Mine Tailings - Eastern Canada*

A field program was conducted at a sulphide tailings impoundment in eastern Canada, at which a well documented case of ARD is occurring. The field program consisted of more than thirty RCPTU soundings with calibrative discrete groundwater sampling using the UBC modified BAT System, previously described, at selected holes.

A typical RCPTU sounding in acid generating tailings is shown below in Fig.4. Note that these tailings below 5 m depth are very mixed and stratified as indicated by the cyclic nature of the cone resistance and friction ratio with depth and are fairly fine and quite loose. However, the soil to 5 m depth is fairly uniform fine sand with little fines as shown by the consistent value of 1% and less for friction ratio.

The extremely low bulk resistivity measurements of less than 1 Ohm-m (or bulk conductivity more than 10,000 μS/cm) are very interesting and suggest that from ground surface to about 5 m in depth there is very high accumulation of ions in solution. This is a zone of high capillary pore water tension in fairly fine sands where saturation levels are high (causing lower resistivity), but, more important, this is a zone of high oxidation which gives rise to sulphate and acid generation which is the main cause for the very low resistivities. Note that from 5 m to well below the water table (10.5 m) the lowest bulk resistivity values are about 10 Ohm-m indicating a dilution of dissolved solids compared to the top 5 m

The results of resistivity logging and pore water sampling at another tailings impoundment are summarized in Fig.5 where bulk resistivity is plotted against fluid resistivity at all depths where discrete water samples were taken. The linear relationship in Fig. 5 gives a ratio of bulk to fluid resistivity (called Formation Factor) of about 2 in these tailings. Resistivity of soil particles is much higher than that of the pore fluid, and therefore it is expected that the bulk resistivity values would be significantly higher than that of the pore water. Typically the Formation Factor for clean quartz sands is between 2 and 5, but can be much lower with the clay minerals and particles with low resistivity surfaces. The agreement between the data in Fig. 5 demonstrate the effectiveness of the RCPTU in identifying zones of lower resistivity and potential groundwater contamination in mine tailings which contain mixed and often conductive fine grains.

Chemical testing has been conducted to determine sulphate and metal levels in the water samples. The results of the sulphate tests are presented in Fig.6 which shows a good correlation between sulphate concentration in mg/l and bulk conductivity in μS/cm. With this average correlation the, concentration of sulphates in the pore water can be mapped, three dimensionally, using the bulk resistivity measurements from the rapidly performed RCPTU soundings. This work is ongoing and promising results are indicated.

5 CURRENT RESEARCH

One of our highest research priorities is the development of a grouting and decontamination system for the cone penetrometer. This system must easily and economically grout and seal the cone hole to prevent potential cross contamination from occurring. Conventionally, the hole is grouted after the cone is removed and requires a special grouting pipe to be pushed back down the cone hole and withdrawn slowly while introducing grout. This procedure takes at least as long as the RCPTU soundings itself and is therefore time consuming and requires extra equipment, thereby increasing the cost substantially. It is therefore desirable to develop a "self"-grouting system.

Figure 7 shows a diagrammatic sketch of a self-grouting system where the CPTU is grouted while it is pushed in and when it is withdrawn. During the penetration, the grout is "sucked" down into the annulus (about 5 mm thick zone) behind the friction reducer. In the case of the resistivity-piezocone, the friction reducer is larger than the resistivity module and behind it, so the grouting does not interfere with the resistivity measurements. When the cone is withdrawn the friction reducer sleeve (which is about 50 to 75 mm long) is sacrificed and left in the ground so that the grout can fill the cone hole while it is withdrawn. Other than the time to premix the bentonite grout, little extra time is required. We use a recirculating Moyno pump and hopper to mix about 75 l of grout which is enough for a 30 m deep penetration.

To facilitate the cleaning and decontamination of the rods, multiple jet high pressure "steam," washed with detergent, all housed in a container fitted with a double rod wiper and under the cone truck. The high temperature wash uses a propane RV heater, a delivery pump and a high pressure washer. A recovery tank is also required so wash water is taken off site. The decontamination system is automatic and remote to the operations in the vehicle, hence requires no addition to the time of testing.

Initial results with this self grouting and decontamination system indicate that the system works efficiently and as expected. It is not clear who should be given credit for this idea for self grouting, but we first heard of it from Dr. Alan Lutenegger (1992), during his review of various grouting methodologies. We continue to evaluate and improve the performance characteristics of this system.

Other current priorities include the following:

1. To develop improved methods of rapid water sampling for screening studies;

2. To develop new techniques to measure the hydraulic conductivity, K, of coarse soils *in-situ* and to correlate measurements with CPTU parameters to allow RAPID estimate of K vs. depth;

3. To evaluate and develop other geophysical techniques applied to cone penetration technology, such as the induced polarization technique (IP) and ground probing broad band radar (GPR);

4. To evaluate and apply other sensor technology, especially in the vadose zone and in the detection and quantification of 'floaters' and specific DNAPLs and LNAPLs. The recent "explosion" of new developments in the sensor technology is evident when one reviews the proceedings of "Field Screening Methods for Hazardous Wastes and Toxic Chemicals," (1993), which is in two volumes and covers 1,323 pages.

5. The importation, application and use of high density, *in-situ* measurements of contamination levels directly into SQL databases, data worth models and contaminant transport models.

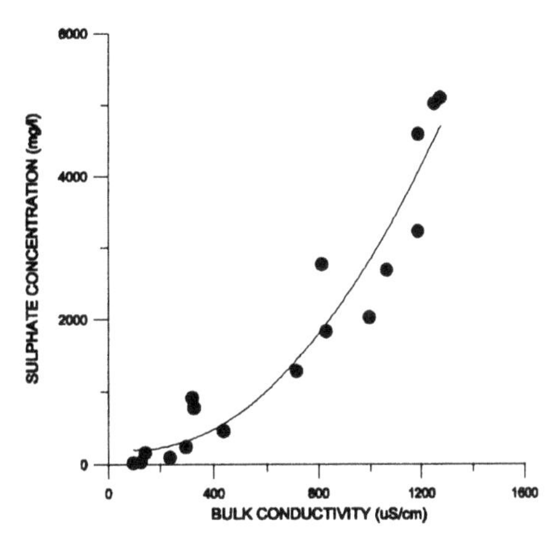

Fig. 6 Sulphate Concentration versus Bulk Conductivity: Mine Tailings

Specific groundwater contamination problems currently under investigation include:

1. Hydrocarbon detection and mapping of DNAPLs and LNAPLs;

2. Accumulation of nitrates and pesticides due to excessive agricultural activities;

3. Develop rapid methods of detection, mapping and characterization of metals and sulphates in mine tailings and waste dumps coupled with characterizing acid generation and drainage mechanisms;

All of our current research thrusts are related and applied to industrial and public needs concerning rapid and economical screening methods to measure groundwater quality.

6 CONCLUSIONS

The use of rapidly deployed logging tool like the Resistivity Piezocone (RCPTU) to identify subsurface stratigraphy and estimate groundwater flow parameters to guide the location for specific depth pore water sampling has been presented and described. The specific depth pore water sampling is also rapidly done with the use of a modified BAT groundwater system. The case histories relating to both organic and inorganic contamination demonstrate the application of the resistivity

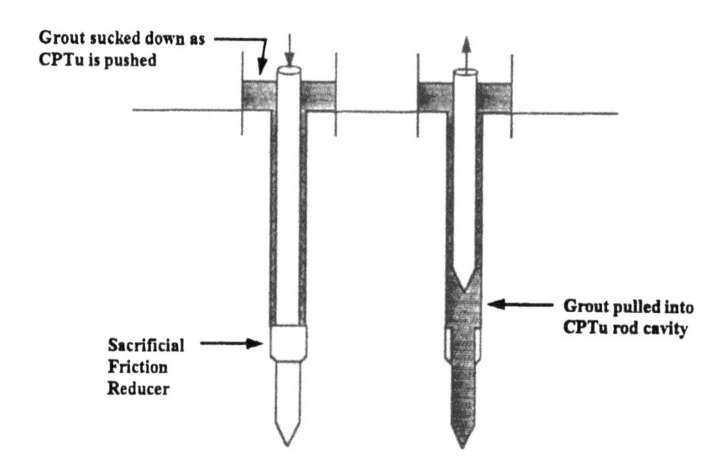

Fig. 7 Simplified Self-Grouting System for a Cone Penetration Sounding

penetration technology to identify levels of contamination. The results clearly show that these methods give a better definition of the lateral and vertical extent of the stratigraphy and contaminant zones than the conventional methods and at substantially lower cost and are especially useful in screening applications.

These *in-situ* testing tools are well past the research and development stage and should be made use of by practicing geoenvironmental engineers for site characterization of soil deposits. As of 1994, there are many cone contractors operating throughout the world who can demonstrate the effectiveness of modern penetration technology even on relatively small projects. Furthermore, the interpretation of piezocone test results can quickly be done by engineers, especially those new to the field, with the aid of user oriented software like CPTINT developed at UBC.

ACKNOWLEDGEMENTS

We take this opportunity to acknowledge the continuing support of the Natural Sciences and Engineering Research Council, Canada and the B.C. Science Council.

Also, the support from cone and penetration system manufacturers like Hogentogler and Co.; drilling and cone contractors, like Foundex Exploration Ltd. and ConeTec Investigations Ltd.; from governmental agencies like the Canadian Geological Survey, Environment Canada, BC Ministry of Highways and the BC Ministry of the Environment; from initiatives like MEND and the BC Task Force on Acid Drainage; from industries like BC Hydro, INCO, Cominco, Placer Dome, Domtar; and consulting companies like Klohn-Crippen, McLeod Geotechnical and HBT Agra are also much appreciated.

Our research efforts would not be possible without the expertise of technicians Scott Jackson and Harald Schrempp and programmers Jim Greig and Thomas Wong.

Finally, the collaborations with Dr. J.M.O. Hughes and the efforts of former research students are much appreciated.

REFERENCES

Bear, J. and , A Verruijt, A. (1987). "Modelling Groundwater Flow and Pollution", D. Reidel Publishing Co., Dordrecht, pp. 196-198.

Blegen, R.P., Hess, J.W. and Denne, J.E. (1988). "Field Comparison of Ground-Water Sampling Devices," North West Waterworks Assoc., 2nd Annual Outdoor Action Conference, Las Vegas, Nevada, 23 pages.

Campanella, R.G. and Weemes, I. (1990). "Development and Use of an Electrical Resistivity Cone for Groundwater Contamination Studies", Canadian Geotechnical Journal, 27:5, pp. 557-567 Oct.

Campanella, R.G., Davies, M.P. and Boyd, T.J. (1993). "The Use of *In-Situ* Testing to Characterize Contaminated Soil and Groundwater Systems", Second International Joint ASCE-CSCE Conference on Environmental Engineering, Publ. by Geotechnical Research Centre, McGill Univ., Montreal, Vol. 2, pp. 1497-1505

CPTINT 5.0, (1993). Piezocone Interpretation Software for the IBM-PC, Department of Civil Engineering, University of British Columbia, Vancouver, B.C., Canada, V6T 1Z4.

Ebraheem, A.M., Hambeurger, M.W., Bayless, and Krothe, N.C. (1990). "A Study of Acid Mine Drainage Using Earth Resistivity Measurements", Groundwater, Vol. 28, No. 3, pp. 361-368, May.

Field Screening Methods for Hazardous Wastes and Toxic Chemicals (1993). Proc. of International Symposium, U.S. Environmental Protection Agency and Air and Waste Management Association, P.O. Box 2861, Pittsburgh, PA 15230, Las Vegas, Nev., February.

Kokan, M.J., (1990). "Evaluation of Resistivity Cone Penetrometer in Studying Groundwater Quality", unpublished B.A.Sc. Thesis in Geological Engineering, University of British Columbia.

Lutenegger, A.J. (1992). Univ. of Massachusetts at Amhurst, Dept. of Civil Engineering, personal communication.

Robertson, P.K., Campanella, R.G., Gillespie, D.J. and Grieg, J. (1986). "Use of Piezometer Cone Data," Proc. of *In-Situ*-'86, ASCE, Geotechnical Special Publication No.6., pp. 1263-1280.

Telford, W.M., Geldart, L.P., Sheriff, R.E., and Keys, D.A. (1976). "Applied Geophysics". Cambridge University Press. Cambridge, pp. 442-457.

Torstensson, B-A. (1984). "A New System for Ground Water Monitoring", Ground Water Monitoring Review, Fall, pp. 131-138.

Zemo, D.A., Pierce, Y.G. and Gallinatte, J.D. (1992). "Cone Penetrometer Testing and Discrete-Depth Groundwater Sampling Techniques: A Cost-Effective Method of Site Characterization in a Multiple-aquifer Setting", Proc. 6th Outdoor Action Conference, National Ground Water Assoc., May, Las Vegas, Nevada.

4 Embankments, excavations and buried structures

Developments in Geotechnical Engineering, Balasubramaniam et al. (eds) © 1994 Balkema, Rotterdam, ISBN 90 5410 522 4

Embankments on soft clays, a continuing challenge of misspent efforts

Victor F. B. de Mello
São Paulo, Brazil

ABSTRACT: A few very important cases of embankment test fills are reviewed, and lessons are extracted confirming to a surprising degree the fact that any past effort, no matter how good at the time, inevitably calls for ulterior revisions, both of mistakes and bias, and of dispersions. However, using as mere examples the world-publicised cases of Vasby, Sweden 1946, the M.I.T. Performance vs. Prediction 1974, and the corresponding Kuala Lumpur 1989, one concludes that there is little cause for elation regarding ENGINEERNG, despite the notable advances in geotechnique's scientific bases.

1. INTRODUCTION

In surveying the developments in Geotechnique from Harvard 1936 to New Delhi 1994, we cannot but emphasize two very important points: Firstly, the difference between geotechnical "science" and the dominant routine applications in geotechnical "engineering": thereby, we can perceive somewhat of effective reality within different degrees of shadows, despite the dazzling brilliance of the advances of scientific methods and results. Secondly, the fact that for the needs of ENGINEERING, which is decision and action despite doubts and dispersions, one must respect an intermediate rate of change of available tools, to permit establishing "factors of adjustment" of nominal-truth to reality, and thence the experience and judgement. Too fast a rate of introduction of novelties leads to confusion and sterility; on the other hand, too slow an incorporation of advances must also be rejected as frustrating the very principles of progress and engineering optimization.

2. DESIGN REQUIREMENTS, AND QUEST FOR KNOWLEDGE. PREDICTIONS AND PERFORMANCE IN FOUNDATION ENGINEERING

Very rough estimates taking into account growths of population and, principally, of irreversible social requirements, have suggested that the annual expenditures in foundations in the world might well reach figures of the order of US$(100-400)$10^9$. It is, indeed, an industry in which investment well directed should result in great savings, all the more desired because costs are buried unperceived. There is a significant well-recognized bias in decisions, of much higher preference for avoiding loss than for seeking gain. Without any effort of imagination one can estimate that this bias becomes much greater when the loss is buried and distributed among millions unaware, than when it is blatantly concentrated on singular cases, and exposed to criticisms and highly punitive law suits. It would stand to reason, therefore, that Foundation Engineering Practice should have grown progressively more conservative and expensive (buried, generalized, unnecessary incremental costs) in proportion to the singular cases of courageous design, and the consequent doubly singular cases of blatant failures. The thousands of uneconomically and conservatively conducted routine cases establish a chronic epidemic, while deriving a vicarious pride and prestige from the occasional publicized over meticulously conducted big projects... much as in many a society and religion the dearths of the multitude are sublimated in the ostentatious wealth of the leader.

Such trends could only be aggravated by the complex of disparaging comparison between concepts of successful research and development in Academia and the synthetic Industries, versus the risks of Engineering in Civil and Geotechnical interaction with Nature. The exaltation of KNOWLEDGE as static-deterministic-scientific-mathematical pushes towards indefinite search for "knowledge of Nature" in her micro-tendencies (despite the admission that we never know the "status quo" but only alterations or differences thereof). If knowledge is rightly exalted, why should knowledge of ENGINEERING (i.e the artisan pursuit of economic moulding of Nature) be any less prestigious than the infinite search for the infinitesimal behavior trends in IN SITU INTACT SOIL ELEMENTS? The atrophy of the very concept of Engineering beckons us to re-evaluate some major milestones past, and their undercurrents, occasionally misdirected.

Foremost among these milestones are the rare and expensive prototype tests and Prediction vs. Performance Challenges, which merit summary cross examination.

T.W. Lambe's Rankine Lecture, 1973, rightly emphasized the preferences for Type A Predictions, and raised some possible (and all--too-frequent) suspicions against types B (really the basis of the Observational Method of design adjustments) and C ("one must be suspicious when an author uses type C predictions to 'prove' that any prediction technique is correct"). Systematic regretable simplifications and misunderstanding of those proposals, together with the psychology of seeking laurels at a professional Olympiad, have done a great and growing harm to our profession, which relies entirely on a patient progressive adjustment of estimates TOWARDS REALITY, at MINIMIZED INCREMENTAL COST AND WASTE, by Bayesian prior to posterior probability adjustments.

Any type C condition can be re-established as a renewed type A case, merely by making the existing case anonymous, with all identifying characteristics well altered (without altering the essentials of the geotechnical data), and with the known end-result kept secret.

Moreover, if we are honestly seeking systematic advance of our technology, there are irrefutable arguments for REVISITING OVER AND OVER AGAIN the type-C field cases, transformed by disguise and anonymity into periodically repeated type A prediction and Design--test cases on the self-same documented NATURAL BEHAVIOUR.

In any process of adjusting ourselves to a goal (by skew--Bayesian successive adjustments of prior and posterior probabilities of improving the aim at the target--center as well as narrowing the dispersion around the dead-center) the starting obligation is to maintain the WELL-DEFINED GOAL FIXED, IDENTICAL. In principle in the face of such cases there are 4 principal tests involved: (1) NATURE'S BEHAVIOUR,indelible, an asset invaluable as a single crown jewel, the HOPE DIAMOND, not only because of high costs already spent, but much more, because of time irretrievable; (2) Our capacity to investigate and observe; (3) Our capacity to analyse, forecast, and decide, with justifiable confidence in our consequent results and decisions; (4) Our capacity to educate ourselves, measurable by systematic evolution of improved procedures, ever more widely applicable and convincedly accepted. It is indeed a slur on us that in a profession most deprived of the conveniences of adequate-size model and prototype testing, and in a world dominated for over 50 years by the "cybernetics" of rapid yes-no refining of choices, we have not absorbed "into our groins" the lesson of such Bayesian evolution of experience.

In fact we are obliged to conclude that by having failed to draw the psychological and sociological lessons from such Type A field trials, which obviously had to give frustrating and disperse results, the net effect has been unfavourable, and detrimental to Engineering's service to Society. The incentive to search for the scientific "philosopher's stone" solution, the EUREKA COMPLEX, has only been spurred by the inabilities disclosed. Easier and more attention-attracting than to work at gradually improving our existing instruments, parameters and methods, has been to hasten to open more novel proposals, each and all inevitably born naked.

By references to Figs. 1, 2, 3, 4 we should emphasize that every major field test trial should be used not merely as a Prediction Challenge case (ability to hit the Average Predicted into equivalence with Performance Reality, within a minimal dispersion) but even more as a check on our benefit/cost Design Decision ability. For the latter the decrease of Dispersion is much more profitable than the improvement of the Average: inasfar as possible the data from the Prediction Challenges discussed below exemplify pointedly how our geotechnical engineering has foregone the dominant obligation of concentrating on both technical and economic improvement to Society.

3. EMBANKMENTS ON SOFT CLAYS

This is a very significant technical and economic problem, with mankind mostly settled around water: it has been faced since early geotechnical history as requiring our concentrated developmental attention. Sundry directions have been, and can be, taken in this discussion: among them, the many reinforcing treatments, and the different serviceability criteria on tolerable settlements and displacements. I choose to concentrate on the oldest geotechnical field test, Väsby, Sweden (Terzaghi Jan.16, 1946), the 2 internationally publicised Prediction vs Performance "challenges" (1974, 1989) on the limit height causing failure, and the vanguard investigations conducted at the Bothkennar Test Site, U.K. (Geotechnique, June 1992). Failure has always been abrupt, and undisputably observed: therefore the ENGINEERING AIM of minimally averting failure can be well defined, whereas most other serviceability and computational parameters are rather undesirably intangible for this presentation's purpose. Meanwhile the soft clay generally has such low ACTIVELY EFFECTIVE STRENGTH, so close to zero, that one can pardoningly justify the historic concentration of interest only on the clay: yet, on revisiting the issues one should correct this understandable error, because under scientific-technological principles, well-proportioned attention should be given both to the CAUSATIVE FACTOR (the fill) and to the AFFECTED/ REACTIVE CONTRIBUTOR (the clay), all the more so since both are obliged to behave within their ranges of wide statistical dispersions at close to zero.

3.1. Väsby, Sweden, 1946

Terzaghi's report recommended the field tests at Väsby "in such a manner and on such a scale that they will inform us on all the factors which determine the behaviour of soft clay under the influence of temporary and permanent surcharges. Foremost among them is the secondary time effect. Once this knowledge is available the preliminary investigations for the construction of a flying field on soft clay in any part of the country can be reduced to routine soil tests which can be performed for a short time". Rather deterministic and confident regarding "all the factors", "any part of the country", the credence to "routine soil tests", and the important professional problem of extrapolating from short-term to long-term behaviours to be predicted. Every such point quite understandable in historic retrospect.

However, on revisiting this milestone effort after 47 years of the possible infinite benefit/cost

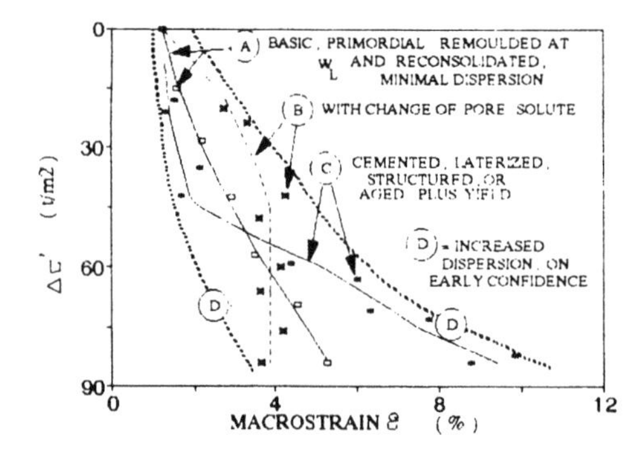

FIG. 1 SYSTEMATIC IMPACT OF LAB. RESEARCH FINDINGS ON THE SPECTRUM OF DISPERSIONS TROUBLING THE PROFESSIONAL (SCHEMATIC)

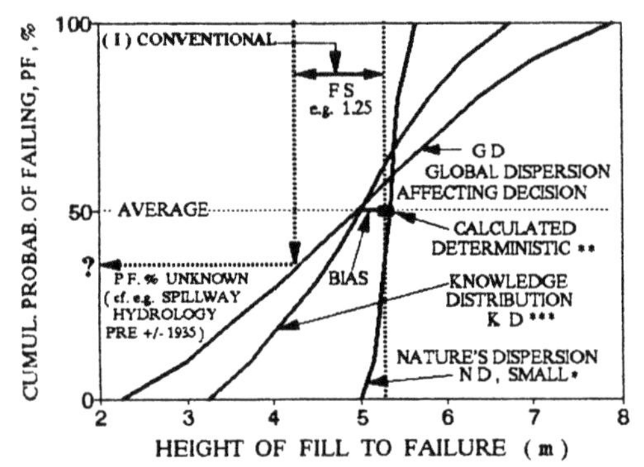

* WITNESS " PERFECT FAILURES "

** RESEARCH DEVELOPMENT AIM AVERAGE = REAL

*** GREATER , BIASSED, SKEW

(I) IGNORANCE OF BIAS, DISTRIBUTIONS, P F s, etc. HINDERS PROGRESS BY EFFICIENT SUCCESSIVE DECISIONS

FIG. 2 DIFFERENCES BETWEEN CHALLENGE OF SHARPSHOOTING AT MEDIAN MARK, AND ENGINEERING OF AVOIDING GREATER RISK THAN TOLERATED

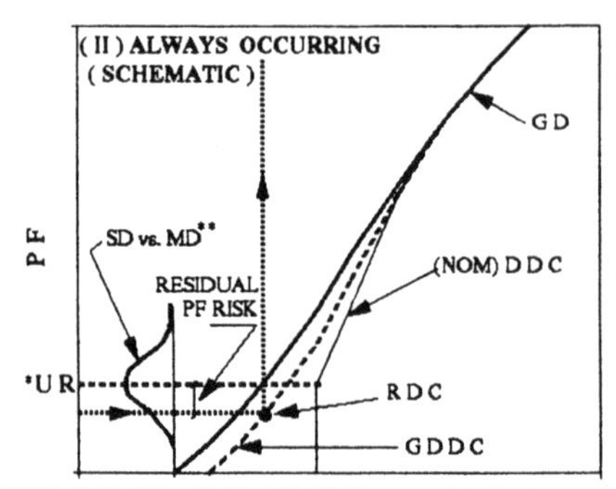

** FIRST - TRY AND SINGLE DECISIONS, SD, INEVITABLY VERY SUBJECTIVE, UNACCEPTABLY OPTIMISTIC OR PESSIMISTIC AND SKEW . FOR ASSESSING RESIDUAL RISK OF G D D C, SHOULD REALLY USE GAUSSIAN DISTRIBUTION OF MULTIPLE DECISIONS MD, OF SAME INDIVIDUAL + METHOD, IN " SAME CASE ". WE ASSUME ACCEPTABLE USING AS SIMILAR THE P D F OF " MANY EXPERIENCED "

G D - GLOBAL DISTRIBUTION, ASSUMED GAUSSIAN
G D D C - G D DECISION CORRECTED
R D C - P F OF REAL DECISION CUTOF
U R - ASSUMED UNACCEPTABLE RISK, FOR SINGLE PREDICTOR' S DECISION

FIG. 3 RECOGNITION THAT BECAUSE OF P D F, THE SINGLE DECISION CUTOFF AT GIVEN P F ALWAYS IMPLIES SOMEWHAT HIGHER RESIDUAL RISK P F THAN USED

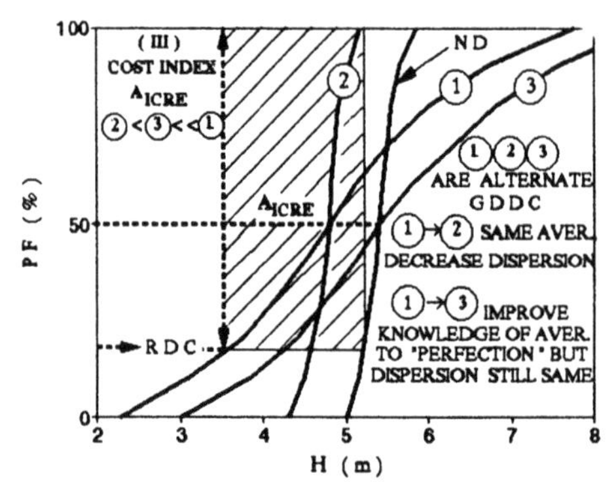

ASSUME GIVEN PHYSICALLY POSSIBLE FAILURE OF 50 m LENGTH, EMBANKMENT OF INCREASING H, 1000 m LONG ∴ ≈ 20 POSSIBLE PHYSICAL FAILURES. FAILURE CASES ARE NOT STATISTICALLY INDEPENDENT, BUT PROGRESSIVELLY INSTRUCTIVE, TO QUICKLY APPLIED EXPERIENCE INDEX.

A_{ICRE} = AREA INDEX OF INITIAL COST OF RISK EXCLUSION. (BY GIVEN DECISION)

ON ① LOWER AVER. AND GREATER DISPERSION , FOR FILL OF H > 3.4 m REINFORCED FOUNDATION USED BY R D C. HOWEVER , FILL RISING UP TO 5.2 m WOULD NOT REALLY REACH R D C ON N D.

∴ A_{ICRE} GOES FROM R D C TO 100 % AND 3.4 m TO 5.2 m CASE OF INCREASING DISPERSION OBVIOUSLY MUCH WORSE

FIG. 4 SCHEMATIC EXAMPLE OF CALCULATION OF COSTS OF RISKS. DEMONSTRATION THAT FOR ENGINEERING, DISPERSION IS MUCH MORE COSTLY, EVEN IF CHALLENGE MEDIAN MARKSMANSHIP ACHIEVES " PERFECTION "

ratio, because of being the singular case and because of the elapsed time irretrievable, what do we find? Shall we repeat the prototype test to be authentically Type A, and await till the year 2040 to be in a better position? How depressingly unscientific to repeat the starry-eyed belief that NOW, yes, WE do have the right to claim a grip on "all the factors". Moreover, in the oblivion gradually sentenced to the secondary time effect, declared as "FOREMOST among... factors..." (by the father of primary consolidation theory) how bitter to reflect that academia cannot devote interest to really long-term problems, while design professionals on their side can defend themselves all the better from liability suits and guilty consciences behind the mysticism curtain of collective ignorance.

The Väsby test fill is eloquent in proclaiming the obligation to repeated revisiting. A careful examination of the records serves as a most eloquent lesson on three facets: the importance of viewing our endeavours historically; time irretrievable in prototype observat ion; the great cost and value of Nature's behaviours well evidenced and remaining available for successive reanalyses while our methods undergo changes. During the recorded trajectory every single revisitation has taught something technical, but, above all, it should have taught the message of our need to return over and over with our erroneous and dispersive visions, to try to improve rational adjustment to the crystal clear course of Nature's behaviour. Some of the most illustrious institutions and geotechnical leaders have been involved, both in the initial effort, and in three important revisitations, around 1966-69 (after 20 years), and 1979-81 (35 years), and Apr.85, Sept.87 ASCE, and it behoves us to emphasize the obvious and MAYBE reasons why we have to continue correcting ourselves. Facts and questionings require being proclaimed aloud: the persons behind them dispense identification as mere laudable instruments of our cumulative service to an unfathomable destiny.

The initial program envisaged simultaneously two very distinct purposes: the practical engineering purpose of observing long-term settlements (as subject to viable accelerated anticipation and control, or not); the theoretical purpose of interpreting the settlement behaviour via the original idealized consolidation theory, or a generalizable revision thereof.

It is impossible to recount herein the series of insufficiencies and deficiencies reported, as resulted in the 20-year and 35-year Revisitations. Many are the lapses of investigational logic, associated mostly with the vicious circle of begging the question (lifting oneself by one's shoe laces) under wishful thinking. For instance, the interpretations on the presumed separation between primary and secondary consolidation are tied to the historic first-order pragmatic procedures of Taylor and Casagrande graphical interpretations in oedometer tests without pore pressure monitoring.

In short, and principally, as is inevitable, there is always a lack of superabundant redundancy of tests and/or instrumentation-monitoring, to establish statistical dispersions. And there is always a lack of superposition, at the same moment (same operators etc.) of the HISTORIC vs PRESENT optimized sampling-testing-interpreting.

If the theory of "self-induced primary consolidation process" has been firmly hypothesized, and "this process is likely to continue until the clay structure reestablished itself", the controlled experimental avenue should have been promptly followed. The theory is much more profitably confirmed under precise laboratory control.

Meanwhile, in the field the theory postulated a process likely to continue at so significant a constant rate as 6mm settlement per 100 days (!) until geologic reestablishment of stability, one is forced into incredulity; discrepancies are to be noted, inasfar as:

1. the natural ground is concluded to be stable (N.B. the 120-day observation would be unconvincing in terms of geologic or "secular" time);

2. the 30cm undrained fill is monitored to have "experienced no settlement during the life of the

load test" (22 years, June 1946 - Sept. 1968);

3. but regarding geologic dating, only the "lower clay's bottom" is dated, as 7900 B.C., while the more relevant dating attributable to the "upper clay", which is merely described as "post-glacial... relatively young", should be indispensable, and easily obtainable.

The fact is that after installing new highest precision modern piezometers, duly calibrated as to controlled responses IN SITU AS INSTALLED (by external tubes permitting injection or extraction of water at cell-tip), it should be highly profitable to add a thickness of fill on top of the existing one, to check on incremental behaviour, now better documented, regarding fill pressure and the developed excess pore pressure.

In retrospect we find ourselves most surprised at how little research data was collected on the fill, and on the theoretical vehicle for interpretation of secondary compression, which is the excess pore pressure: the explanation would seem to lie in the expectation that the field test could be based merely on comparison of analogous fills for drained vs. undrained behaviours of the clay, and on the observation of the end-result of settlement rates after fulfilled theoretically anticipated "total consolidation dissipation" of excess pore pressure. Research dominated by confident deterministic expectations could be pardoned in the infancy of the profession, but should have been recognized and corrected by now, in appropriate revisitations.

Having defaulted on some fundamental principles of progressive investigative adjustment of our transitory methods to the observed SIGNIFICANT REALITY, and the research having been temporarily finalized with an unusual theoretical conclusion classifiable as a THEORY OF A SINGULAR CASE, this field test presently stands at the extreme of a ZERO benefit/cost ratio, instead of being deservedly taken to the very high benefit/cost ratio corresponding to its AGE IRRETRIEVABLE. I dare pronounce that the many internal inconsistencies call for redress and for one or more additional revisitations.

As a startling beginning I venture a suspicion that the very poor definition of the quality of the gravel fill, coupled with a wrong intuition on changes of stresses accompanying the increasing settlement, can implode the very basis of the published hypothesis. The entire interpretation arises from an assumed calculation that as the fill settles below groundwater level (taken as fixed) the "submergence" would progressively REDUCE the applied (would-be) effective stresses causing excess pore pressure and ulterior settlements. The intuitions regarding the principles of Archimedes are ingrained: who would stop to reconsider the specific case, in face of so undisputable and elementary a calculation?

Are we not compelled to such reconsideration in the face of so very important a theoretical and professional problem as the secondary or "secular" compression in field conditions?

In the published Report (1981) the "causative factor", the gravel fill loading, is too minimally described: "The western half of the fill was placed by free dumping without compaction, while the eastern half was compacted after dumping. As a result of the method of placement, the western half of the fill was slightly higher than the eastern half. However, the magnitude of the load on the whole area was believed to be the same. The unit weight of the gravel fill in its uncompressed state was determined to be 1.7 t/m3". From the thesis that generated the Report one does not extract additional significant information.

I would conclude that presumably not only the gravel fill was rather loose, but understandably almost dry. Let us assume percent saturation and water content of the order of Sr=15% and 3.75. Obviously as such a fill submerges, its Sr would increase to about 95%: thus the unit weight of the submerged thickness would increase to 2.01 t/m3. Note that the intuition on decreasing pressure with submergence arises from compacted clayey fills that start at Sr≃90% and would hardly increase in Sr at all, or not more than 2-3% (cf. the need for back-pressure saturation).

Thereupon we refer to Fig.5 and go back to first principles of "prospective" effective stresses as total stresses minus pore pressures. With a constant ground-water level, at any depth z of a soil element the pore pressure remains constant. Assuming "no" lateral displacement of the clay above a given point, as compression occurs the total stress due to the clay remains constant because of the $\Delta\gamma$ increase compensating the ΔH compression. As far as concerns the gravel, repeating for times t=o and T, corresponding to x settlement, the applied total stress only increases linearly until the entire gravel fill is submerged (2.5m settlement). The comparative profiles (A) and (B) of Fig.5 should clarify the reasoning, and the graph (C) indicates the changing total stress with settlement: thereafter the changing tendency to generate compressing effective stress while the hydrostatic pore pressure due to constant ground-water level remains constant. In the 1985-87 revisitations this item was recognized as important, but treated lefthandedly, and left inconclusive and erroneous (?!).

Incidentally, the same MAYBE REVISITATIONS would apply also to the invaluable Skä-Edeby test fill, 1957. One would conclude that, to begin with, the data on the fill should be greatly incremented and improved; a very easy task. Moreover, considering the but modestly successful piezometric data of the past, and the enormously improved modern instrumentation, I vouch that Terzaghi must be asking that we should add another meter or so of fill, to confirm or dispell the maybe theories.

3.2. The M.I.T. 1974 "Foundation Deformation Prediction Symposium"

It may seem unfair and sterile to return after 20 years to that milestone case, but it is from such markers of the past, freely re-analysed, that we must develop our collective Experience, especially when as in any first-try there is the greatest tendency to misjudged orientations. Those were the days of concentrated faith and effort on effective stress analyses, computational modelling, finite elements, normalized behaviour generalizations and constitutive equations, greatly improved testing and instrumentation precisions, and the PROJECT SERVICEABILITY AIMS focussing on deformation. In fact, the Symposium's name was Prediction of Foundation Deformation, although inevitably the most salient feature shifted to being the neat FAILURE, the only significant and clear-cut behaviour.

Once again the oft-mentioned clear description of perfectly defined "brittle" FAILURE stood out as the fly for a sharp-shooter's marksmanship (1). It is clear that in this case of homogeneous clay deposits Nature's behaviour is "theoretically" crystalline as regards failure, whereupon any discrepancy or dispersion in our prediction lies squarely and only on our shoulders, and not on the oftslandered geologic erraticities. In fact, as was shown on Fig.2, Nature's behavioral dispersion is very much smaller than our capacity to quantify it; our task is both to approach the Average reality of PREDICTION ≡ PERFORMANCE, and, for economic design decisions, to decrease our much wider dispersions.

Meanwhile, both the aims and the conduct of the field test were too broadly-embracing and undefined as regards "performance of the foundation during and after

---

(1) "Early in the morning... a failure of extraordinary proportions occurred. Within minutes... crest to drop about 30 feet and the sides to heave as much as 14 feet. ...No surface cracking was noticed the previous day, nor was a clear indication of impending failure obtained from the field instrumentation. ...Failure occurred to both sides..."

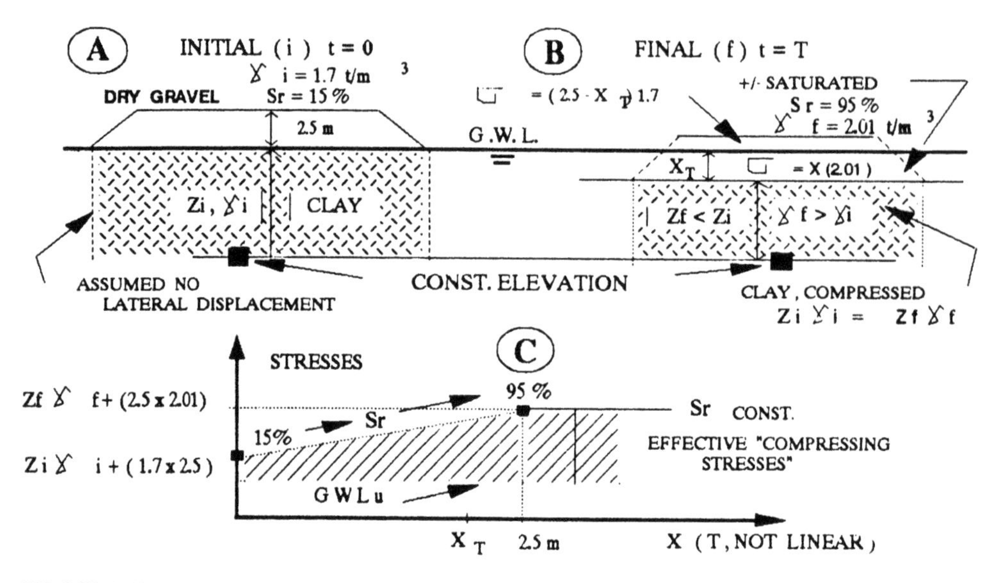

FIG. 5 CHANGE OF EFFECTIVE " COMPRESSING STRESSES WITH SETTLEMENT, GRAVEL FILL.

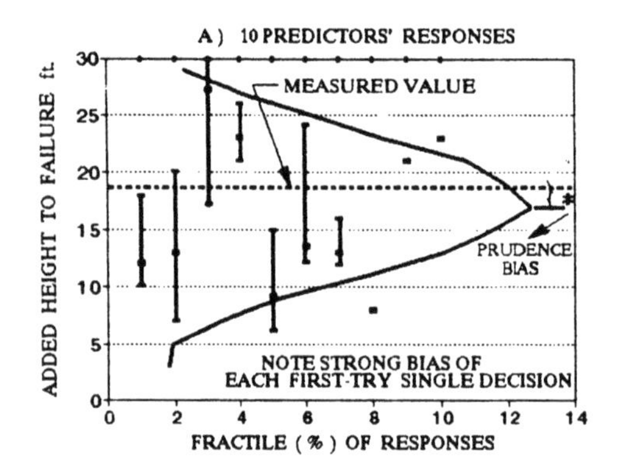

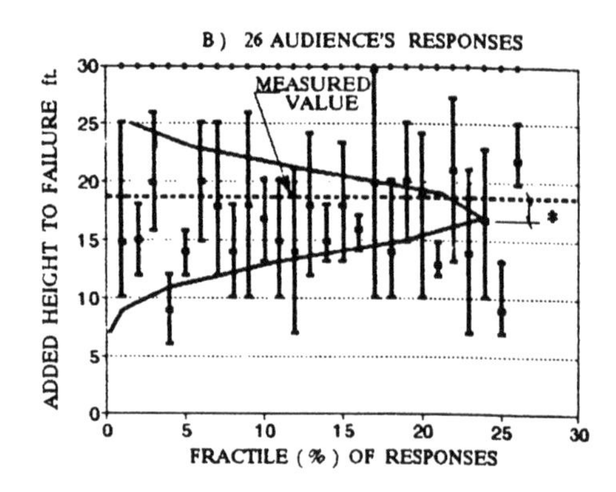

FIG. 6 M. I. T. 1974 CHALLENGE " PREDICTION vs. PERFORMANCE "

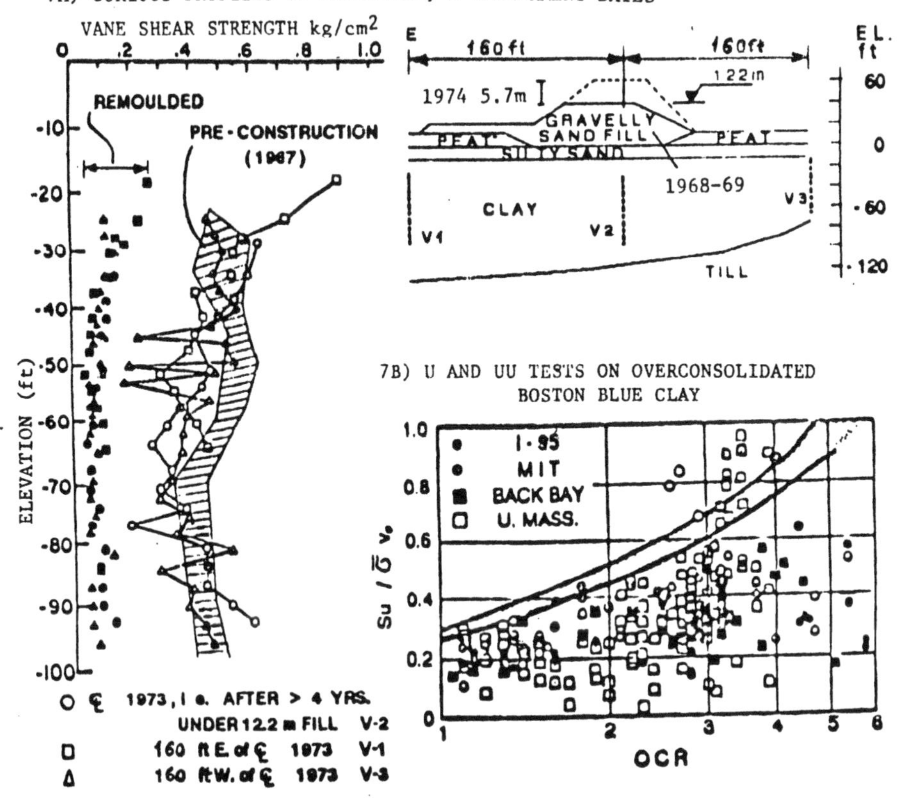

FIG. 7 M. I. T. 1974 TESTS, EXAMPLES OF EXTREME ERRACITY OF DATA, SOME CONTRARY TO LOGIC. INTERFERENCIES OF EQUIPMENT, PROCEDURES, PERSONNEL TO UNUSUAL DEGREE.

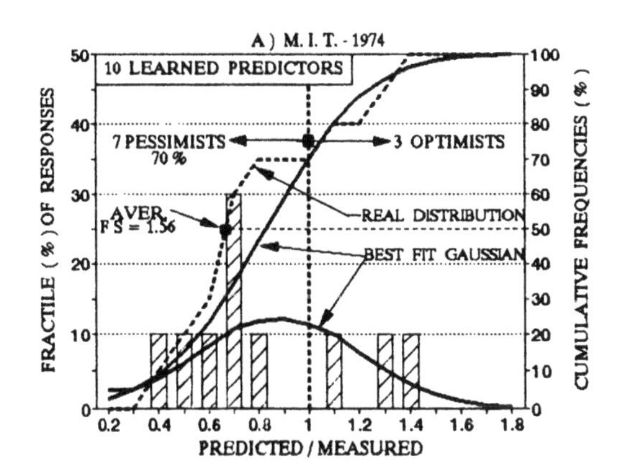

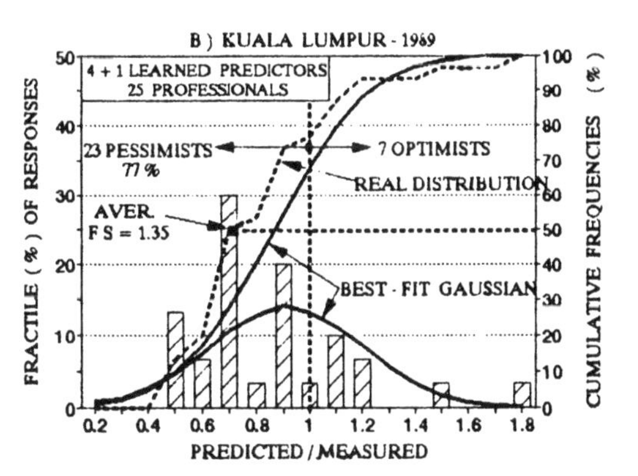

FIG. 8 COMPARATIVE STATISTICAL, ANALYSES OF 1974 AND 1989 CHALLENGES

construction": scientifically one should ever remember the partial--differential-equation principle, of aiming at one target at a time, and significant; professionally one tunes in on experience at what matters, which would be, in a nutshell, end-of-construction transitory destabilization potential, and/or long-term after-construction deformations, and not a confusing mixture of the two, that can only hint at a field test aimed at matching an idealized theoretical thesis, with but left-handed attention to the typical professional engineering problem and the need to tie back to digested experience.

For the present purpose of submitting how very much was lost in that case and could still be progressively regained by re-visitations, the results summarized in Fig.6 and 7 should suffice. Some striking facts of importance to ENGINEERING DECISIONS (the accept--reject prior cutoff in the knowledge distribution, cf. Figs. 2, 3, 4) may be summarized:

a) The 10 learned predictors (more documented than, say, 98% of typical similar professional cases (2)) used widely different personalized theoretical approaches, none of them adjusted to practice via case histories, and essentially all with such deterministic unfounded bias (optimism or pessimism) that mostly they did not individually straddle across the average or the observed result (Fig.6A).

If the client had decided to pay 10 times the (rather exceptional) design cost, and to average the 10 recommendations, by fluke he should have ended up with a good project.

As shown in Fig.6B, a cheaper design, of equivalent average and lesser dispersion, would have resulted from a few hours of "feel" by all the 26 members of the audience: strictly speaking, however, this should also be recognized as another fluke, because of other factors, some important and singular.

b) As regards prior professional experience, it should be noted that the proposed case was quite NOVEL, since it would not appear that any previous (or ulterior) embankment on soft clays had been designed on any basis other than FS with respect to FAILURE. No "end-of--construction deformations" had ever been of interest (in comparison with long-term settlements, secondary compression, maintenance etc., cf. Väsby), and no designer had ever considered monitoring construction-period deformations and piezometers, to accompany pre-failure indications. The Type A prediction was thus a challenge on untested and unadjusted theoretical presumptions, suggesting acceptance of "data" as factual, at stationary face value, stripped of historical transience.

c) Regarding such acceptance of test data (e.g. undrained strengths) at face value, Fig.7 summarizes two extreme graphs of heterogeneities quite beyond reason or acceptability. One notes the lack of any consistent attempts to "correct" for Sensitivity-remolding, boring-sampling-handling disturbances, sample and specimen quality as reflected in stress--strain curves etc.. In qualifying a sample merely as a (e.g.) "5-inch diameter undisturbed sample" the concern for such historic dictates as in Hvorslev "Subsurface exploration and sampling of soils for civil engineering purposes" (1940, ASCE) were neglected. Incidentally, the predictors did not express advance complaints, or desire for the conventional samples-tests (however poorer) to which their experience would have been adjusted.

d) The 2-step embankment filling, firstly of 12.2m height (Apr.1968 to May 1969, with winter interruption Nov. 15-Apr.15), and finally, five years later, of the 5.7m increment in "late summer 1974" (to failure,20/Sep/1974) constituted

---

(2) One opening statement was "The major cause of inaccurate predictions is faulty and insufficient data". Faulty they always are, to greater or lesser degrees, and intimacy and experience are called to compensate. In the face of professional practice "determinedly misdirected" might be a more realistic qualification than "insufficient".

another unusual complicating factor, obviating any "model-to--prototype" Bayesian adjustments. Moreover such adjustments could only be viable if the monitored parameters were significant, and pursued the same "laws" of phenomena in model-to-prototype evidenced behavior.

e) From an engineering standpoint the most striking fact was the absolute lack of attention to the fill itself, both as the basic causative factor, as having reached a thickness of up to 17.9m, and as having nevralgic strength and "brittle stress-strain" behaviors at overburden stresses close to zero, poorly quantifiable except in UU "quick" tests.

The 8 (only!?) field density tests varied between 1.74 and 2.20 t/m3, a ± 11% variation around the mean, leading to the same variation in applied pressure: however, the denser conditions are coincident with much higher strengths (at low stresses). And the fill's strength testing was limited to six CD(!?) triaxial tests, with possibly nominal effective stresses depending on suctions. Many more points may be made, calling for profitable reassessments (not all of them criticizable as of hindsight) of this case in which Nature's behavior was so definitive, and ours so very poor, and passable by fluke. It would be unfortunate if different "schools" should pursue their separate paths, heedless of each other's comparative advantages,and, especially most regrettably, heedless of the need to adjust to the only valid test, which is to improve technical-economically on the design solution for Society.

3.3. Kuala Lumpur K.L. 1989 trial embankments.

The type A prediction challenge in this case was better oriented with regard to typical design decisions. Firstly, the limiting height to failure, necessary for a cutoff decision on PF%. Secondly, for the situations considered beyond the acceptable height with its risk, the challenge to specialized ground treatment organizations (consultants, specialist contractors, and suppliers of proprietory products) to design and conduct alternative treatments to meet well-defined performance criteria of magnitudes and rates of settlement avoiding expressway surface regulation more than twice a year (by pavement experience the limit set of 100mm settlement over 2 years after commissioning).

Specially praiseworthy is the fact that COSTS are submitted, the indispensable second leg of ENGINEERING besides TECHNICAL EXPERTISE. In passing I submit my doubt that in my intense worldwide coverage of geotechnical papers over the past 40 years, more than 2 or 3 papers per thousand ever mention costs: a disparaging observation.

The treatment included: electrochemical injection; sand sandwich; preloading, geogrid reinforcement and prefabricated vertical drains (two different enterprises); well--point preloading; electrosmosis; prestressed spun piles; sand compaction piles; vacuum preloading and prefabricated vertical drains; preloading and prefabricated vertical drains. No further mention will be made herein on these treatments except that (1) dispersions and rushed novelties abounding are suffering, and taking from Society, the inevitable much higher toll of more frequent failures and disparaging comparisons; (2) more than 50% of the cases incurred in failure during the construction sequence or were abandoned (3); (3) the cost data permit shockingly revealing comparisons. At any rate, despite the insufficiencies and

---

(3) This should be recognized as unusual in the face of the dictum that generally a good creative solution should be superabundant in its achievement in order to be noticed and increasingly used. The explanation for the exception is simple: on the one hand the solutions hovered around the indeterminancies "close to zero", in the aim for economic competitiveness; on the other hand, they were solutions subconsciously pushed by vested interests.

failures that occurred, in order to avoid increased complexities and confusions, in my present purpose I adopt the reinforcement treatments as "perfect, no risk", and each at its minimum cost as published in the Proceedings.

In Fig.8 we present the comparative probability distribution curves and bar diagrams of predicted/observed failure heights as ratios, for comparison. From the best-fit Gaussian distributions there appears to have been in the 15-year interval a slight improvement both in the academic aim of the median coinciding with 1.0, and also in any typical design decision cutoff (e.g. 20% cumulative probability risk of failing). This impression needs correction, however.

The results of this additional geotechnical milestone have already been ably summarized and discussed. For my purpose of viewing the advances for the profession deriving from the historic ties and reappraisals, the geotechnical comments are minimized, while the cost implications to Society call for emphasis:

a) The fill's field density (given as associated with percent compactions of 91-100%) merited more attention: 365 tests averaged 2.04 t/m3, still with a dispersion of roughly ±9%. The fill's conditioning strength parameters were yet offered in terms of effective stresses, notwithstanding the very low stress range and the sandy-clay CH soil of $16 \leq$ hopt $\leq$ 18% and max. $1.75 \leq \gamma_d \leq 1.83$ t/m3. Predictors were cautioned as to discrepancies and low credibility of the strength parameters although determined from block samples.

b) Once again, essentially no comment on greatly disperse sample qualities, sensitivities, stress--strain curves, etc., the test results being taken automatically at face value. Incidentally, with the baptismally-blessed "undisturbed stationary-piston thin-wall" samples, simply described as such, and indiscriminately used to great and constant lengths (e.g. 100, 130cm) without judicious adjustments, we should reexamine if the intent of sampling with MINIMAL STRESS AND STRAIN DISTURBANCES AND READJUSTMENTS is not being disguised under the indexrobe of automatic imposed control of length changes, via compensating internal stresses and strains. With disturbance principally controlled by minimized Area Ratio and static penetration, the further principle was classically emphasized that the tendency to expand of each sample due to stress release (obviously increasing with depth of hole) should be essentially controlled at the cutting-edge bevel, and via the inside clearance; the stationary--piston should be but a complement of prudence, the sample's 100%(±) recovery ratio having already been reasonably adjusted by a sampler penetration that balances friction compression vs. tendency to expansion. One notes that the compartmentalized distance between field and office has increased so much that not only are such details not furnished, but neither are they demanded by the predictors.

The attitude of accepting "data" at face value extends to the piezometric records on unexplained hydrogeology, and essentially all parameters. A far cry from the indispensable approach that all data are always wrong, possibly to different degrees, and to estimable values of bias and dispersion. For instance, it is difficult to reason on "average" in situ strength profiles, when in most conventional testing disturbances only tend to decrease sensitive strengths: decreases of preconsolidation (or "yield") ' are the principal index, the in-situ values having to be along the upper limit profiled; but the concomitant logical trends have to be used also, confirming the decreased nominal C_c and peak δu, and correspondingly increased strains at failure. Logical trends should condition judicious choices of parameters.

c) One notes that a fair proportion of the analyses emphasizes the importance of "cohesion" strength of the fill, up to one extreme postulation that beyond a certain fill height the FS remains constant because each incremental thickness incorporates exactly the additional resisting force to compensate the unstabilizing increment. The cracking of the fill is also mentioned. Note that the added layer's cohesion is not acquired by fairy wand, the destabilizing weight becoming effective before compaction is completed.

The principal conclusion derived from the analyses submitted is the confirmation of the trend (schematically postulated in Fig.1) of increasing dispersions of methods and parameters that have spread across the world, even in so continually repeated a professional problem. Just as opposite examples one notes that in one case preference is given to unconfined compression strengths (the ±1945 practice, but with what sampling--handling?) whereas in another, success is hinged on the ever--elusive in situ K'o parameter.

It is not surprising that once again the knowledge Probability Distribution was somewhat pessimistic-prudent, and very dispersed, whereas Nature's behavior repeated (at the position of the test) the essentially clear-cut failure condition, almost deterministic, with but some longitudinal cracking the previous day. Incorporating some inevitable small dispersion (unknown, in any part of the world, because of prevailing single deterministic fail-don't fail approach, which is most unfortunate for engineering progress) along the longitudinal, and adopting the construction reality of a fill rising layer by layer, we now proceed to the key lesson to be extracted by revisiting this case. Figs.9 and 10 have been prepared based on the published costs, not to be discussed, but accepted as nominal and quantitatively comparative. The method of analysis refers back to Fig.2.

For the sake of simplicity(4) in the comparative nominal cost computations we adopt the hypothesis that any specific reinforcement is "perfect, no-risk": the same is applied, much more justifiably, to the hypothesis of reconstituting any failed pure embankment section additional fill, as much as necessary as a berm, and the rest to get back to fill height.

The increase of prudent pessimists from 70% in 1974 to 77% in 1989 represents an increased cost to Society (each project employs one designer only, that is, one decision, not the average of 30 opinions). If one designer has concluded that the failure height is 3.5m (say), he would really use a FS (say 1.25) limiting his design to acceptance of 2.8m without reinforcement; all the remaining length, of higher embankment, is forced to use some reinforcement, more expensive (Fig.9). However, for simplicity and on the conservative side we can assume that similar Design Decisions would arise from a subparallel Decision Distribution Curve at FS=1.0, which is analogous to the distribution curve reached by the 30 predictions(5) aiming at the bull's-eye of coincident average failure PREDICTION ≡ REALITY.

Along a long embankment of gradually increasing grade elevation, the lengths of stretches reinforced or not will vary from designer to designer. However, for the present we are well documented to imagine a case of a long (say 1000m stretch) of constant 6m height of embankment, for which the costs, for presumed perfect no-risk reinforcement solutions, derived from the conjunction of the varied pessimism (greater intensity of reinforcement) plus costs of the specialized services.

While we have concentrated on site-and component-issue of methods a, b, c vs. k, l, m, n what we have failed to realize is that the most important information of all, which is Nature's Distribution

---

(4) The more complicated situations are quite as straightforward, but lengthy, detracting from this presentation's purpose of emphasizing principles.

(5) In fact we are discussing an utopian condition of collective decision probabilities of our worldwide community. In unfortunate reality, since each client tends to rely on only one designer at a time, and each designer has his bias plus dispersion (the former much more dominating because of lack of repetitive cases for tuning in) the most uneconomical project would result from the most prudent pessimist.

curve (in this problem) is what we do not have (but the "experienced designer" with many repetitive cases begins to feel, if develop-mental academia will permit using the same method over and over). The most important embankment test would be just to face a long project with optimism (or repeated Type C-DISGUISED trials). Let us imagine such a trial, assuming a reasonable ND curve as shown in Fig.10 A.

If we are dealing with an optimist over the 1000m length of 6m embankment we would have 5,20%, 30% etc. cumulative probabilities of failure on reaching heights of 4.5, 4.7, 4.9m respectively. IT MUST BE EMPHASIZED THAT THIS RISK IS INSTANTANEOUS, WELL WORTH TAKING, BECAUSE STABILITY ONLY IMPROVES THENCEFORTH WITH TIME(6). The real failure data of the K.L. 1989 test were of a failure over essentially the entire short length of embankment, therefore pin-pointing a roughly 99% probability of failure on reaching the 5.7m height (the test was of too short a length).

For the sake of simple cost comparisons we assume that: (a) the fill rises by 0.2m lifts simultaneously over the entire 1000m length; (b) the physically viable failure lengths are ≥ 50m; (c) the drop of the crest, will be (1/3)H; (d) the volumes for re-constituting any failed section include completing the heave with an added 2m thick berm, plus going back to grade; (e) a reconstituted failed section is risk-free for the required additional height; (f) the ND data continue to apply to the remaining still unfailed lengths; (g) the cost per cubic meter of fill for reconstituting failed sections is 5 times the initial cost of fill.

The cost of such a "shameless" non-Bayesian embankment-construct-ion test is represented in Fig.10B. The conclusion should be absolutely startling, but irrefutable: the acceptance of up to 60% probability of failure roughly matches in cost with the cheapest of the perfect no-risk reinforcing treatments. In other words, are we not really failing to optimize engineering for society, while really minimizing cost of our prestige, at consider-able expense to society? (7).

The value of such a physical test (as above mentalized) to determine ND is absolutely inestimable, and at very low cost. Above all, along the kms of foundation clay reason-ably adopted as uniform (fixed statistical universe), no matter how much sophistication is incrementally introduced for the progress of geotechnical science, the starting principle is that the gross of the investigation must be logical, simple and very re-petitively usable, and the "monitoring" basically of facts flagrant in the engineering scale.

3.4. Bothkennar soft clay test site, U.K. 1992

To mention this remarkable additional MILESTONE still in the making can only be envisaged as a MAYBE CONTRIBUTION in the line of the present thesis. Much more than another trial embankment on soft clays, the farsighted and noble intent of the U.K. Science and Engineering Research Council (SERC) has been set towards developing one soft clay engineering research site for uninterrupted long-term research. And, besides counting on the greatest specialists in geo-technique, an earnest call has

---

(6) Consider, in comparison, the short-term risk that any dam engineer HAS TO ACCEPT in a cofferdam and diversion, and ponder on how we have been betraying the principles of Civil Engineering.

(7) Of course it must be recognized that prestige does have its fundamental "value" to be preserved, for the very sake of society also. There should be a concerted effort of educational communication to lead society to recognize ingrainedly that engineering is not deterministic right-wrong, and that in such problems of cost of risk close to nil, radical changes of attitude must be implanted into clients, media, and society.

been put out for enhancing it as an international test bed site, by promoting joint research in which outside bodies would collaborate with the U.K. group(s). The call for a worldwide cooperative effort prods me to use this international podium to extend the suggestions and appeal, because, as will be expatiated, much of the input for broader professional applicabilities will have to be contributed by the distinct past participants in less ambitious field tests across geography and time.

Engineering must straddle judiciously between singular sophisticated cases, and multitudinous roughly assessed similarities/variabilities. With regard to indispensable historic ties surely we need not remind ourselves of the feeling that each SINGLE MILESTONE, too widely separated for direct vision of others, could be seduced into fancying itself as the ultimate NIRVANA. Would one need to be guarded against "scorning the base degrees, By which he did ascend" i.e. forgetting that for all mortals there must be a ROAD (evolving practice) already spot-marked by other milestones, of which there will be more forthcoming, of course progressively altering course ?. A ROAD and GOAL are real, while the arrival is illusive: such is the concept of "uninterrupted long-term research" into past and future.

For instance, could it be that secondary compression testing and the milestone of Väsby have been consummately explored or have deservedly lost interest? So also such widely used indices as unconfined compression strengths and Sensitivity (as conventionally defined) etc., in statistical comparisons with ulterior "improved" substitutes?

The scientific conscience is markedly evidenced. Merely as an example I pick on the question of sampling and sample quality, to employ the researchers' own logic in benefit of worldwide "routine practices" and of tying-in with historic experience. The aim "to use the sampling and testing techniques that were regarded as the best available in current practice" is stated, and is meritorious for a spearhead. But, were not past efforts admissibly intentioned in like fashion? And what percentage of professional cases is (or will be, in foreseeable future) able to use similar spearhead practices?

Meanwhile, in the effort to preserve the legacy of past evidences of Nature's behaviour, two avenues are available, and in at least some significant cases should be used in complement. One is to repeat, for past cases, the current "best available practices" for due comparison. Despite the transience of such past involvements in analogous efforts by important institutions across the world, an earnest call must be made along this avenue, because it is the only way to add some statistical credence to the Bothkennar single-clay findings. The other, more feasible immediately, is to repeat in the case under current study some of the dominant practices associated with the past cases, so that, assuming moderate similarity, some adjustment factors may be quantified for present parameter estimates vs. the erstwhile adopted ones.

The very significant differences signalled (cf. examples summarized in Fig.11) between key results as obtained from the three "presently ideal samples" should reinforce our recognition of the need to compare also the results of the WIDELY VARIED SAMPLING AND TESTING TECHNIQUES that were spread across the world, and are still in duly respectful use by good disciples and acolytes. When results were poor, tending towards significant disturbance-remoulding, obviously the differences had been greatly attenuated, FAVOURING A COMMON LANGUAGE AND PRESCRIPTION; but they were made sufficient for each start, inescapably humble.

I would venture the guess that due to lags in time, geography, economics, and composite factors, surely more than 98% of geotechnical past-and-present experience and judgment is tied to much cruder sampling-handling--testing-interpreting practices than used in the Bothkennar research publications. It cannot escape notice that peak strength results differ by as much as 45%, and preconsolidation pressures

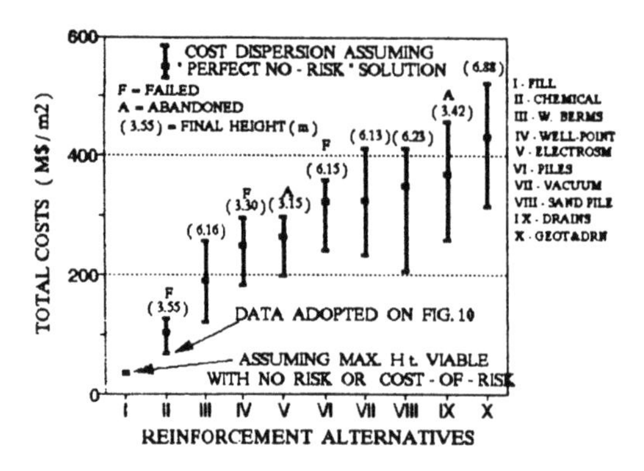

FIG. 9 ESTIMATED COSTS FOR 6 m HIGH EMBANKMENT

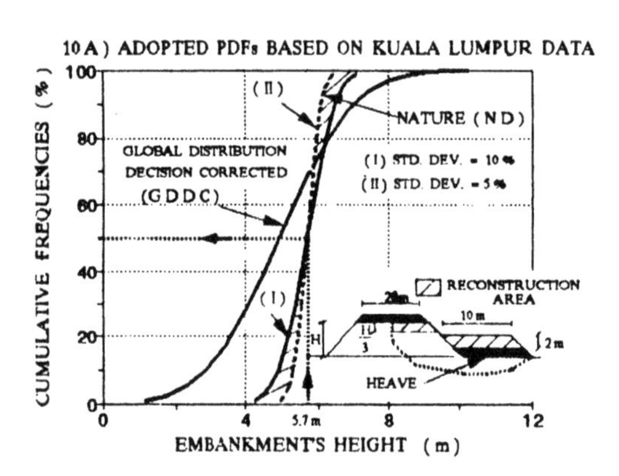

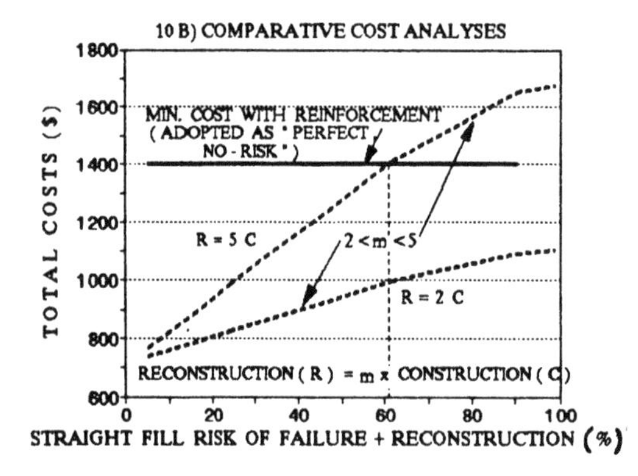

FIG. 10 COMPARATIVE COST ANALYSES OF OPTMIST RISKING FAILURES AND RECONSTRUCTIONS vs. MIN. COSTS OF REINFORCEMENTS FOR PESSIMIST.

determined by as much as 200% ! If we change (under the best and most laudable scientific intentions) our MEANS so very significantly, should it not automatically require proportionally significant adjustments of our EXPERIENCE-

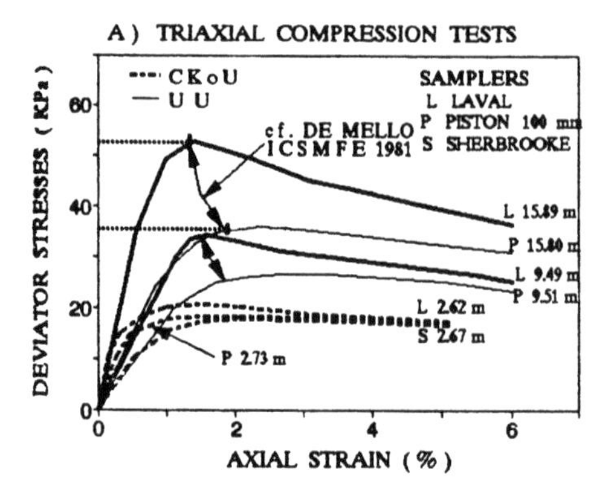

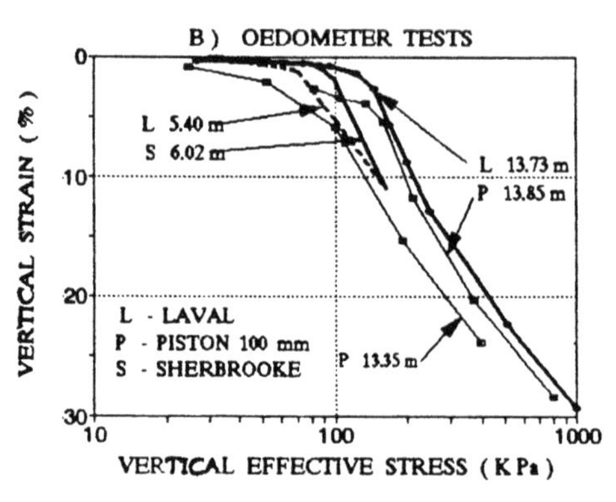

FIG. 11 MARKED DIFFERENCES IN KEY PARAMETERS DUE TO SAMPLERS (cf. BOTHKENNAR 1992)

-ADJUSTMENT COEFFICIENT towards the only point that is, in the final judgment, the PURPOSE of geotechnical engineering RESULTS?

Can we countenance disregard for the price paid for countless past field tests, and the immensity of project evidence of over-spending in totally non-misbehaved cases, spot-marked by failures questioningly analysed? And can we also disregard the vast majority of endeavours across the world that are still (and will always inevitably be) out of phase with any single spearhead of development? Surely not : and that is where a concerted worldwide effort, technical and financial, must be mustered around these presently final sprinters, to hand them the batons from across geography and time.

4. CONCLUSION

The classic problem used for these

analyses involves homogeneous, saturated, idealized sediments akin to primeval "text-book cases". It is a problem faced with great frequency in professional life, and has been subjected to over 300 test-fill studies. We have progressed very much indeed in quantifying a number of "additional" parameters, presumed relevant to engineering works. Clients and Society are called to pay much higher costs of investigations and analyses, if induced into most advanced current practices. It would seem, however, that the greatest proportion of professional practices across the world, even in well-developed areas, lags considerably behind the level of advances available for incorporation. Incidentally, in Engineering is there an advance if it is in methods, but with no perceptible advance in net results?

Two fundamental challenges to geotechnical CIVIL ENGINEERING have been neglected under the prestigious avalanche of the published WORD in scientific quantifications. One is the nurturing of past experience of individual cases, never entirely repeatable, but ever judiciously retrievable with estimated quantifications of the new parameters erstwhile disconsidered. The other is the global resulting comparative cost to Society of the constructed facility, with due inclusion of the costs of risk and of discredited professional prestige.

It takes relish of KNOWLEDGE to keep abreast of the most updated geotechnical-science refinements, but even more thirst of the WISDOM not to use recent knowledge any faster than it becomes tempered to PROJECT EXPERIENCE. Since the tasks of Civil Engineering remain in essence the same, with but changes of measures and methods, let it not be justly said of us that "...when he once attains the upmost round, He then unto the ladder turns his back, Looks into the clouds, scorning the base degrees By which he did ascend."

(Shakespeare, Julius Cacsar).

Developments in Geotechnical Engineering, Balasubramaniam et al. (eds) © 1994 Balkema, Rotterdam, ISBN 90 5410 522 4

Stage construction of embankments on soft soils

E.Togrol
Faculty of Engineering, Istanbul Technical University, Ayazaga, Turkey

S.F.(Ozer) Cinicioglu
Istanbul University, Turkey

ABSTRACT: The method of construction of embankments on deep deposits of soils is based on the utilization of the increase in shear strength due to consolidation. In this paper a reliable and useful method is proposed. The determination of stress paths is based on the basic principles of critical state theory. The design procedure allows for adjustments and modifications in accordance with field measurements during the construction.

INTRODUCTION

The construction of embankments on soft soils is associated with a number of problems. The loads imposed by an embankment bring the underlying soil very close to undrained failure and give rise to shear and volumetric deformations. The instability of an embankment is usually caused by the low shear strength of the soft and often normally consolidated subsoil.

The method of construction of embankments on deep deposits of soft soils is based on the utilization of the increase in shear strength due to consolidation. In this context it is important to predict, (1) the safe load at each stage of construction, (2) the necessary waiting time before subsequent loading, and (3) the amount and effect of soil volume change which is usually defined as settlements.

Until the recent years, designs of stage-constructed embankments were carried out by employing stability analysis based on total stresses. Even though the behaviour of embankments have been extensively studied during the last two or three decades the choice of total or effective stress analysis for the design of embankments is still debated. Brand and Premchitt (1989) stated that total stress analysis is the only means of making a safe design for embankments on soft clays. They argue that this is largely because of the difficulties encountered in predicting excess pore pressure distribution caused by the construction. On the other hand, Indraratna and Balasubramaniam (1993) in their reply to a discussion by Brand (1993) have rightly pointed out that the complete behaviour of a soil structure cannot be interpreted satisfactorily unless the development of excess pore water pressure and deformation mechanisms are properly investigated. Therefore, a stress-strain analysis that is truly representative of in-situ behaviour is of major concern in the evaluation of the stability of embankments on soft soils.

The three components of the prediction of excess pore water pressures developed in the soft soils under embankment loading are listed by Brand and Premchitt (1989), (1) the total stress changes in the ground caused by the embankment loading, (2) the relationship between the changes in total stresses and pore pressures, (3) the rate of dissipation of the induced excess pore pressures. Being effective on both the first and second components of prediction, total stress has a major influence on the reliability of effective stress analysis. Whereas the method of total stress calculation constitutes another source of error against the reliability of effective stress analysis. The only method of calculation of total stresses induced in a clayey soil deposit by embankment loads is by using the elastic theory whereas the validity of the method is highly questionable in relation to plastically behaving strained normally consolidated clay media.

In addition, the shear strength values mobilized in-situ may substantially differ from those found by either in-situ or laboratory tests. The mobilized value of the undrained shear strength depends on the stress path followed and also on the rate of loading. Reported cases showed that the application of Ø=0 analysis results in overestimation of values of the safety factor (Bjerrum, 1972; Pilot, 1972; Pilot et al, 1982).

A critical appraisal of the present state-of-the-art implies that the use of the effective stress methods has been justified so long as they are based on the true stress paths and effective stresses. Therefore, if a reliable, theoretically justifiable and easy to apply method based on effective stresses becomes available, the problem of embankment design will reduce to a straightforward application of that method. Until such a breakthrough is achieved we have to contend with corrections and modifications applied to the misleading factors in the present practice in a way that fundamental behaviour of soils will be accounted for more comprehensively.

The attempts of Bjerrum (1972) and Pilot (1972) to improve the usual stability analysis by introducing a correction factor for the undrained strength values obtained by field vane and cone penetrometer tests have not been justified by further reported cases. Later Trak et al.(1980) basing their argument on the results achieved by Mesri (1975) stated that "c_u" is independent of plasticity index, but is a function of preconsolidation pressure of the soil. Following a similar reasoning Chapuis (1982) proposed a method related to effective stress concepts. Chapuis's method makes use of conventional total stress analysis corrected to take account of the overconsolidation pressure and undrained strength profiles. Therefore, Chapuis's method can be considered as a transition towards effective stress methods.

Nevertheless, Schmertmann's (1975) recommendation in relation to the development of methods using the more fundamental effective stress behaviour concepts, still remains to be one of the basic requirements for achieving a realistic solution to the problem. Concurrently, the new methodology presented in this paper has been developed to meet the requirement for a design procedure which is based on the fundamental behaviour of soils and which can be controlled by field measurements during construction and even be modified if need arises.

Theoretical justification of the proposed design approach and the basic concepts related to the development and application of the method are discussed in detail by Cinicioglu and Togrol (1991). Togrol and Dadasbilge (1977) basing their argument on the design and construction experience of İskenderun breakwater in Turkey suggested that, the time required for consolidation periods should better be determined in the field for a successful application of stage-construction technique. Accordingly, the dimension of time is added to the application of the proposed design method in this paper and also the scope of the method is improved in such a way that means of controlling and modifying the design consistent with the field measurements are provided.

FURTHER CONSIDERATIONS RELATED TO VOLUME CHANGE-STABILITY RELATION

Chapuis (1982) denoted that conventional stability analyses make use only of stresses or elementary forces, whereas a failure is identified by abnormal displacements which are not considered in these analyses. Stability is being one of the main geotechnical consideration in the design of embankments on soft clays; settlement is the other criteria which has primary importance as pointed out by Loganathan et al.(1993). Field deformation analysis (FDA) is introduced by them as a new methodology to seperate and quantify settlement components for both loading and consolidation stages. Being applicable especially for the stage-constructed embankments, FDA with its formulation based on volumetric deformation concept is an important contribution to the present design practice. It can be argued that stability and strength changes in the foundation soils are determined by volume changes.

If the volume change affects the soil in the stiffening direction as it occurs during consolidation periods, this can be utilized, during stage-construction of embankments. As a number of researchers (Yamaguchi, 1984; Marche et al.,1974; Tavenas et al., 1980) pointed out, lateral deformations can be considered as indirect measures of stability criterion. Most of total deformation and thus major portion of lateral deformation occur during the construction (Mochizuki et al., 1980; Ortiz, 1967). This behaviour is also an indication of the close relation between embankment instability and lateral flow. Therefore, the main consideration for the safe design of embankments on soft soils should be to minimize lateral flow during loading stages. Suzuki (1988) pointed out that in construction of embankments on soft clay, the shear stress caused by the embankment load must be large enough compared with the in-situ shear strength to allow the underground lateral displacement to take place. Hence, if stresses in the ground can be kept at a safe value so that failure will not be initiated at any location beneath the embankment, lateral flow will be minimized which in turn contributes to stability.

In the proposed design procedure, stresses are not allowed to reach critical values and it secures the desired value of safety factor to be preserved at each stage of loading. Therefore the consideration of minimization of lateral deformations is satisfied.

Field deformation analysis can be used as an aid to the control mechanism of the design approach proposed in this paper. In this way, by calculating the settlement components, the effects of volumetric deformation changes on the soil properties can be predicted during construction and necessary modifications can be made.

A NEW APPROACH TO EMBANKMENT DESIGN

Methodology

Stress paths together with strength and time relations which are used in the proposed approach are given in Fig.1 in the form of a flow chart application of the proposed design procedure. The application is outlined below on a step by step basis. The symbols, N, M, Γ, and λ used in the equations are soil constants and can be determined by conventional laboratory and field soil mechanics experiments. The proposed procedure is developed for centre line region below embankments.

1. Initial stress state of the soil is represented as point A in Figs. 1a and 1b and can be determined from the data of saturated unit weights and from the knowledge of K_0. The initial isotropic normal pressure is expressed as,

$$p_{0_1} = \exp\left(\frac{N-v_1}{\lambda}\right) \quad (1)$$

where v_1 is the initial specific volume of the soil. If soil is normally consolidated, $p_c = p_{0_1}$, where p_c is preconsolidation pressure. If v_1 is not known, p_{0_1} can be determined from equation (2) and then is used in equation(1) in order to determine v_1. The equation of the yield surface according to the Modified Cam-clay theory is used to calculate p_{0_1}:

$$p_{0_1} = p_1 + \frac{q_1^2}{M^2 p_1} \quad (2)$$

where p_1 and q_1 are normal and shear stresses corresponding to the initial stress state and are found for normally consolidated soil with a stress anisotropy of K_0.

2. Normal and shear stress values corresponding to failure state can be found by:

$$p_{f_1} = \exp\left(\frac{\Gamma-v_1}{\lambda}\right) \quad (3)$$

$$q_{f_1} = M.\,p_{f_1} \quad (4)$$

3. L and m values which are required to construct intermediate (design) line in q-p-v stress space can be found by inserting a desired value of safety factor into the equations developed by Çinicioğlu and Togrol (1991).

$$L = \lambda \ln \frac{e^{N/\lambda} \pm [e^{2N/\lambda} - 4e^{(2T/\lambda - 2\ln F_s)}]^{0.5}}{2} \quad (5)$$

$$m = M\left[\exp\left(\frac{N-L}{\lambda}\right) - 1\right]^{0.5} \quad (6)$$

The projections of the design line on q-p' and v–lnp' planes are defined by the equation $q=mp'$ and $v=L-\lambda \ln p'$.

4. Up to this point, the boundaries of undrained loading stages are determined. From this point on, it is aimed to apply the embankment loads in such a way that shear stress level in the ground should always be kept at or below the design line. Thus,a desired value of safety factor will automatically be provided at each stage of loading by keeping the effective stress states on the design line. The value of normal design stress for the first stage of loading is predicted as:

$$p_{int_1} = \exp\left(\frac{L-v_1}{\lambda}\right) \quad (7)$$

and the value of design shear stress is found by inserting the values of p_{int_1} and p_{0_1} into the modified Cam-clay yield surface equation

$$q_{int_1} = \pm M\,[p_{int_1}(p_{0_1} - p_{int_1})]^{0.5} \quad (8)$$

or in an easier way from:

$$q_{int_1} = m.p_{int_1} \quad (9)$$

5. First loading stage is stopped on the design line at the point represented as B in Figs. 1a and 1b. Line BC represents the stress path during consolidation period. Point C representing the end of first consolidation stage lies on anisotropic loading line termed K_0-Line which passes through the origin and is characterized by the slope of $q/p' = 3(1-K_0)/(1+2K_0)$ on q-p' plane. The value of K_0 is a constant and will remain constant as long as the soil remains in normally consolidated condition. Since the stresses applied will not be relieved at any stage for the design problem analysed in this paper, the effective stress state of the soil will be carried back onto the normally consolidated K_0-line after complete dissipation of excess pore water pressures. Stress states at the beginning and end of the first consolidation period are represented as points B (P_{int_1}; q_{int_1}; v_1) and C (p_2; q_2; v_2) respectively. Values of stresses corresponding to point C are; $q_2=q_{int_1}$ and $p_2 = [(1+2K_0)/3(1-K_0)]\,q_{int_1}$. Thus, the amount of excess pore water pressures which should be dissipated during consolidation period is $u_2=p_2-p_{int_1}$.

6. The specific volume of the soil decreases from v_1 to v_2 during the first consolidation period as excess pore pressures dissipate. The value of v_2 which will be attained at the end of the consolidation period can be found as $v_2=N-\lambda \ln p_{0_2}$ at the point of intersection of constant volume contour corresponding to this state and the normal consolidation line. The value of preconsolidation pressure corresponding to the effective stress state represented as point C in Figs. 1.a and 1.b is:

$$p_{0_2} = p_2 + \frac{q_2^2}{M^2 p_2} \quad (10)$$

7. The time required for consolidation may be considered too long in regards of economical and managerial considerations. Accordingly, it may be desired to start a new loading period at any time during a consolidation pause period. In such a case, the stress state at the beginning of the new loading stage and the amount of consolidation achieved at this point should be determined. The assumptions of one-dimensional consolidation is highly applicable for the problem of embankments on soft clays, because of its geometry with a large loaded area in relation to the depth of soft layer. In addition to stress control mechanism provided in the proposed design procedure, a time control property has also been introduced by making use of "deformation-time" curves which are usually produced from oedometer test results. "Unit vertical strain - time" curves are converted to "v-time" curves by using the relation $v=v_0(1-\varepsilon)$. "v-time" curves are produced in such a way that load increments applied in consolidation tests simulates the loading programme of the design.

For practical purposes, consolidation curves corresponding to each consolidation period succeeding a loading period will be chained successively. This method of construction is developed with the assumption that duration of loading periods are short enough not to have a pronounced effect on soil consolidation. Then, by finding the value of "t_0" from the first consolidation curve and "t_∞" from the last one, a tangential curve which comprises the chain of consolidation curves is drawn and in this way an equivalent consolidation curve is obtained. This approach inherits the adoption of a single value for

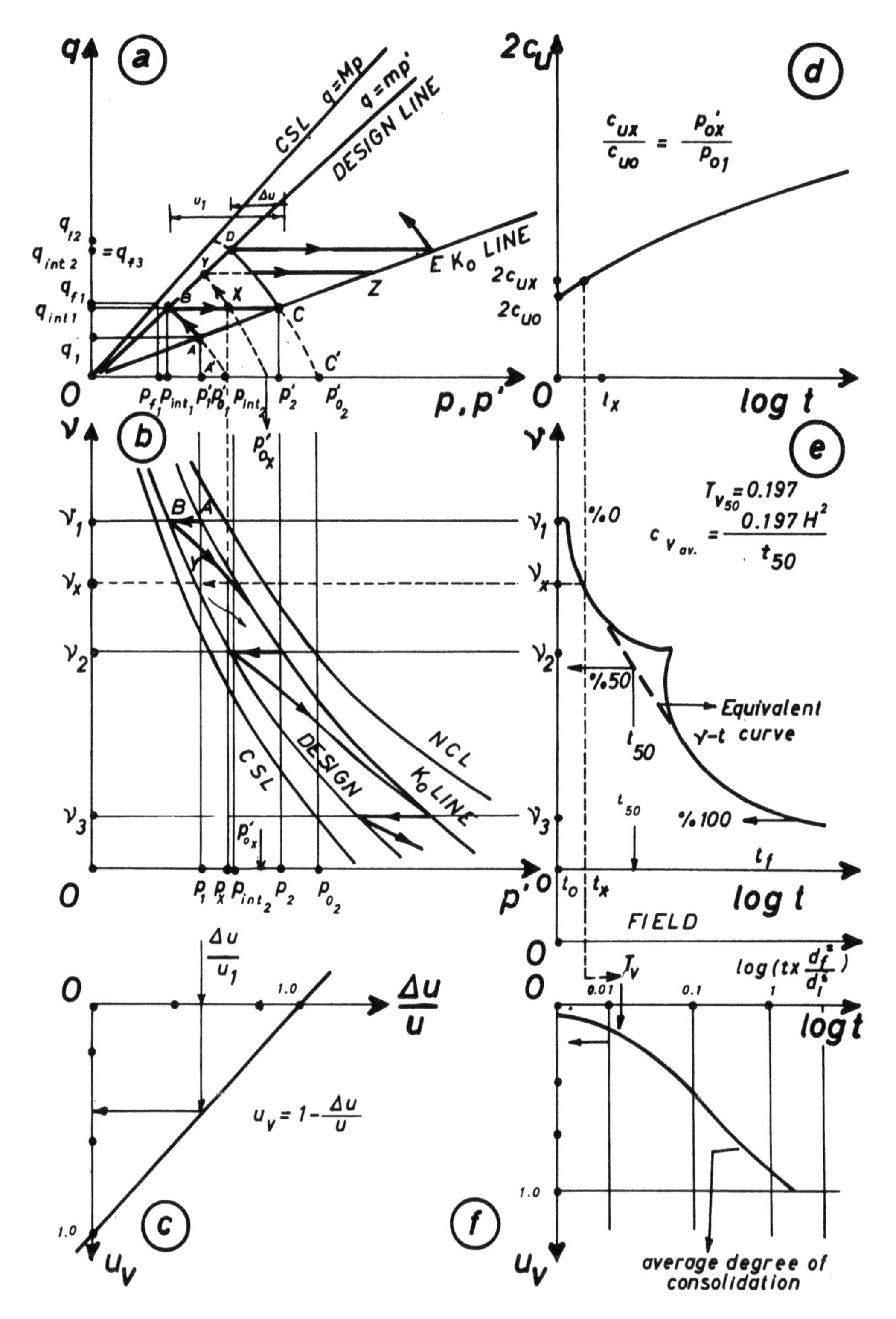

FIG.1. Flow chart of figures used in the proposed procedure

"c_v" and this is an acceptable approach for normally consolidated clays. Because remarkable variations in "c_v" values can only be expected as the stress state of the soil reaches the yield point, but after yielding, the variation in "c_v" is negligible. Barry and Nichols (1982) determined that the variation of "c_v" values is in the order of 1×10^{-8} m^2/sec for an increase of vertical stress from 100 kPa to 1000 kPa. Therefore, the error introduced by the adoption of a single "c_v" value for all stress levels is negligible.

8. The amount of load which will be applied during undrained loading stages can be found by the relation $\Delta q = \Delta\sigma_v(1-K_o)$. On the other hand, because there is no sufficient time for horizontal forces to mobilize during undrained loading periods, the relation of $\Delta q = \Delta\sigma_v$ can be used for computing the additional load. "ν-time" curves are developed with the assumption of achieving complete consolidation which means that stress state of the soil returns to the "at rest" stress state when the consolidation is completed. After achieving a single representative "ν-time" curve, a second axis of time is constructed for this curve by multiplying laboratory times by a conversion factor of $d^2_{field} / d^2_{sample}$. Thus consolidation curves are also fitted in an axes system relevant to field times. "ν-time" curves are given in Fig.1.e and "T_v-U" relation is given in Fig.1.f.

9. So long the curves and figures produced in this paper are achieved by an interpretation of laboratory strength and consolidation properties of foundation soils. From this stage on, a mechanism to control the loading and to take care of achieved conditions in the field has been introduced into the design and this new property enables the designer to modify the design according to the conditions in the field. This controlling mechanism makes use of field measurements and is applied on the developed figures presented in Fig.1. If it is desired to start a new loading stage at a time "t" the value of "ν" corresponding to this "t" is found from Fig.1.e. This "ν" value is used to pass from Fig.1.e. to Fig.1.b. The point of intersection of this ν value with the corresponding drained path is found as (ν; p'). ν is already known and p can be found by writing the equation of drained path between the points (ν_1; ln p_{int_1}) and (ν_2; ln p_2) on ν-lnp plane. Thus, p is obtained from:

$$p = \left(\frac{p_{int_1}}{p_2}\right)^A \cdot p_{int_1}$$

where $A = (\nu-\nu_1)/(\nu_1-\nu_2)$.

Preconsolidation pressure p_{o_2} corresponding to a normal pressure p_2 can be obtained from equation (10) and design procedure is continued in the same way using the procedure and equations given above (Cinicioglu and Togrol, 1991; Cinicioglu, 1993).

10. The value of p and q can be controlled by field measurements before starting a new loading stage. A c_u-time relation such as that produced in Fig.1.d can be obtained by application of field vane tests with appropriate time intervals. The ratio between the measured value of c_u at the beginning of a consolidation period and a c_u value measured at any time "t" before starting a new loading period equals the ratio of preconsolidation pressures corresponding to these stress states. It is a generally accepted opinion that undrained strength of natural soils are primarily dependent on their preconsolidation pressures (Chapuis, 1980; Trak et al. 1980; Bergdahl et al. 1987). Thus for a measured value of c_{u_x} at time t_x during consolidation, it can be written $c_{u_x} / c_{u_o} = p_{o_x} / p_{o_1}$ where c_{u_o} corresponds to the strength measurement at the beginning of consolidation period and p_{o_x} and p_{o_1} are the preconsolidation pressures corresponding to the stress states at time t_x and t_o . t_o denotes the time corresponding to the beginning of a consolidation period. This relation can be made use of to determine the yield envelope relevant to the current effective stress state of the soil by calculating $p_{o_x} = (c_{u_x} / c_{u_o}) . p_o$. Any necessary modifications or corrections can then be applied on calculated effective stress state of the soil and the procedure is continued. In this way, the design procedure based on effective stresses is controlled by field strength measurements.

The difference in the values of specific volume, $\delta\nu$, is used to calculate consolidation settlements by the relation $s = \Sigma(\delta\nu/\nu_1)h$.

11. The degree of consolidation achieved at any time during a consolidation period can be found by using Fig.1.c which is produced by using the equation $U_v = 1-\Delta u/u$. Mid-depth of soft clay layer for double-drainage conditions is the critical depth at which consolidation proceeds most slowly. If this property is not taken into account, the existence of a weak layer in the middle causes the formation of slip surfaces along this layer. If "T_v-U" relations are investigated, it is seen that the values of T_v which corresponds to 50% consolidation at mid-depth should at least be around T_v=0.30~0.40. This condition should also be checked during the application of the proposed design procedure. This can be done, with Fig.1.c by either using $\Delta u/u$ values calculated from Fig.1.a or employing $\Delta u/u$ values measured by piezometers.

CONCLUSIONS

The design of embankments on soft soils is a major concern of soil engineers in regards of the growing demand in harbour and road construction. In this paper, a reliable and useful method is proposed to achieve successful results in the field.

The construction of an embankment on a thick stratum of soft soil is usually made in stages. At each loading stage, increases in the shear strength and the bearing capacity are expected. The loads imposed by an embankment bring the underlying soil close to undrained failure, and careful monitoring of settlement and increase of the shear strength are required when stage loading is used.

Critical state theory is employed in the development of the proposed method and thus, the whole process is founded on an appropriate base.

The method is based on fundamental behaviour

of soils defined in terms of stress paths below the centre-line of embankment. The determination of stress paths is based on the basic principles of critical state theory. Measurements of pore pressures which are often unreliable are avoided in the basic construction of the method.

The settlement rate is generally calculated from Terzaghi's one-dimensional consolidation theory. However, the actual time required to drain out the excess water is less than that predicted on the basis of laboratory tests.

In order to adopt the method according to practical situations and field measurements, control mechanisms which can make use of strength and/or pore pressure measurements are also provided.

A comprehensive soil investigation is generally necessary if a large area is to be loaded.

In cases where the soil profile is erratic it would be difficult to obtain accurate data regarding the physical properties of individual soil stratum. The analytical approach, however rigorous, may be justified for fairly homogeneous normally consolidated soils.

REFERENCES

Balusubramaniam, A.S., Phin-wej, N., Indraratna, B., and Bergado, D.T. 1989. "Predicted behaviour of a test embankment on a Malaysian marine clay." *Proc. Int. Symp. on Trial Embankments on Malaysian Marine Clays, Kuala Lumpur, Malaysia*, 2, 1(1)-1(8).

Barry, A.J., and Nicholls, R.A. 1982. "Discussion" *in vertical drains* (London: Thomas Telford), 143-146.

Bergdahl, U., Hartlen, J., Larsson, R., Lechowicz, Z., Szymanski, A. and Wolski, W. 1987. Shear strength increase in organic soils due to embankment loading. *8 Krajowa Konferencja Mechaniki Gruntow i Fundomentowaina, Wroclaw.*

Bjerrum, L. 1972. "Embankments on soft ground" *Proc., ASCE speciality Conf.* on Performance of Earth and Earth Supported Struct., ASCE, New York, N.Y., Vol.2, 1-54.

Brand, E.W., and Premchitt, J. 1989. "Moderator's report for the predicted performance of the Muar test embankment". *Proc. Int. Symp. on Trial Embankments on Malaysian Marine Clays. Kuala Lumpur, Malaysia*, 2, 1(32)-1(49).

Brand, E.W. 1993. Discussion on Performance of Test Embankments Constructed to Failure on Soft Marine Clay. *J. Geotech. Engrg. ASCE*, 119(8), 1321-1323.

Chapuis, R.P. 1982. "New stability method for embankments on clay foundations". *Canadian Geotech, J.*, 19, 44-48.

Cinicioğlu, S.F. and Togrol, E. 1991. "Embankment design on soft clays". *Journal of Geotechnical Engrg., ASCE*, 117(11), 1691-1705.

Cinicioglu, S.F. 1993. "Stage Construction of Embankments with stress and time controlled loading procedure", *Proc. of 1st. Tech. Congress in Civ.Engrg. organized by ODTU and DAU in Northern Cyprus. (in Turkish).*

Indraratna, B. and Balasubramaniam, A.S. 1993. Response to Discussion by E.W.Brand. *J. Geotech. Engrg. ASCE*, 119(8), 1327-1329.

Loganathan, N.: Balasubramaniam, A.S., and Bergado, D.T. 1993. "Deformation Analysis of Embankments", *J. Geotech. Engrg., ASCE, 119(8), 1185-1206.*

Marche, R., and Chapuis, R., 1974. "Control of stability of embankments by the measurement of horizontal displacement". *Can. Geotech. J.* 11(1), 182-201.

Mesri, G. 1975. Discussion on new design procedure for stability of soft clays. *ASCE J. of Geotech. Engrg. Div.*, 101 (GT4) 409-412.

Mochizuki, K., Hiroyama, T., Moritz, Y. and Sakamaki, A. 1980. "Lateral deformation of soft ground -Case of Kurasiki". *Proc. 15th Japan National Conference on Soil Mechanics and Foundation Engineering.* E-2 216, pp.861-864 (in Japanase).

Ortiz, J.S., 1967. "Zumpango test embankment". *Proc. of ASCE*, Vol.93, No.SM-4, pp.199-209.

Pilot, G. 1972. "Study of five embankment failures on soft soils. *"Proc. ASCE Conf. on Performance of Earth and Earth Supported Struct., ASCE,* New York, N.Y., Vol.I, Part I, 81-100.

Pilot, G., Trak, B. and La Rochelle, P. (1982). "Effective stress analysis of the stability of embankments on soft soils". *Canadian Geotech. J.*, 19(4), 433-450.

Schmertmann, J.H. 1975. "Measurements of in sute shear strength. *"Proc. ASCE Speciality conf. on in situ Measurements of Soil Properties, Raleigh,* Vol.II, pp.57-138.

Suzuki, O. 1988. "The lateral flow of soil caused by banking on soft clay ground". *Soils Found.*, 28(4), 1-18.

Tavenas, F. and Leroueil, S. 1980. "The behaviour of embankments on clay foundation". *Can. Geotech. J.*, Vol. 17, 236-260.

Togrol, E. and Dadasbilge, K. 1977. "Stability of Embankments on Soft Submarine Sediments". *Proc. of 9th Int. Conf. on Soil Mech. and Found. Eng. Tokyo*, Vol.1, pp.773-776.

Trak, B., La Rochelle, P., Tavenas, F., Leroueil, S., and Roy, M. 1980. "A new approach to the stability analysis of embankments on sensitive clays", *Canadian Geotech. J.*, 17(4), 526-544.

Yamaguchi, H. 1984. "Effect of depth of embankment on foundation settlement". *Soils Found*, 24(1).

NOTATION

The following symbols are used in this paper

c_u = *undrained shear strength;*
D = *large increment of;*
F_s = *factor of safety;*
h = *(subcript) horizontal;*
K_0 = *coefficient of earth pressure at rest;*
L = *constant determining position of intermediate (design) line;*

M = *slope of critical state line when it is projected onto constant volume plane;*
m = *slope of intermediate line on the q-p plane;*
N = *specific volume of isotropically normally consolidated soil at* $p = 1.0\ kN/m^2$;
p = $(\sigma_1 + \sigma_2 + \sigma_3)/3$;
p = $(\sigma_1 + \sigma_3)/3$ *stress parameters for* $\sigma_2 = \sigma_3$;
p_c = *preconsolidation pressure;*
p_f = *value of p at failure;*
p_{int} = *intermediate value of mean normal pressure;*
p = $(\sigma_1 + \sigma_2 + \sigma_3)/3 = p - u$;
p_o = *value of p at start of test;*
q = $(\sigma_1 - \sigma_3)$;
q_{int} = *intermediate value of shear stress;*
q = $(\sigma_1 - \sigma_3)$;
q_f = $(\sigma_1 - \sigma_3)$, *value of q at failure;*
v = *(subscript) vertical;*
Γ = *specific volume of soil at critical state with* $p = 1.0\ kN/m^2$;
δ = *small increment;*
κ = *slope of overconsolidation line;*
λ = *slope of normal consolidation line;*
σ = *normal stress;*
τ = *shear stress;*
v = *specific volume;*
v_i = *initial specific volume; and*
Tv = *time factor*
t = *time*
u, Δ_u = *excess pore pressures*
U_v = *Degree of consolidation*
v = $1+e$ = *specific volume*

Note: Effective stresses are used in this paper.

Developments in Geotechnical Engineering, Balasubramaniam et al. (eds) © 1994 Balkema, Rotterdam, ISBN 90 5410 522 4

Unmanned excavation systems in pneumatic caissons

S. Shiraishi
Shiraishi Corporation, Tokyo, Japan

ABSTRACT: Before the excavation work of the ground in caissons was mechanized, pneumatic caissons were inefficient, labor-intensive means of subsurface construction. Excavating machines called *caisson shovel* were developed and they have been used efficiently for excavating the ground in large caissons since 1971. Through the improvements made to the various appliances associated with the use of caisson shovels, pneumatic-caisson technology has been remarkably renovated. Innovative unmanned excavation systems have been developed recently and they were utilized successfully in deep caissons. As a result of those renovations realized in pneumatic-caisson practices, the dreadful image associated with the cruel labor in the compressed-air work of the conventional pneumatic caissons was wiped out.

1 PNEUMATIC CAISSON

A pneumatic caisson is a huge box of reinforced concrete, a typical example of which is as shown in Fig. 1. The ground beneath the caisson is excavated and the caisson was sunk into the ground by gravity.

The bottom space of a pneumatic caisson which is commonly called *working chamber* is filled with compressed air. The pressure of the air in the working chamber is adjusted to keep balance with the pressure of the ground water at the depth level of the bottom end of the caisson. The ground water does not intrude into the working chamber as the ground water is held back by the compressed air. The hollow space in the working chamber is normally kept in dry condition.

During the old days in the early 20th Century, there was scarcely no other means for constructing deep underground structures than pneumatic caissons. Therefore, pneumatic caissons were used almost exclusively for constructing deep underground structures of large scale.

In those old days, men dug the ground in the working chamber of caissons with hand tools under high air pressure. Pneumatic caissons then were heavily labor-intensive means for constructing underground structures. The working condition under the high air pressure in pneumatic caissons then was very hazardous and the men working inside are endangered of being affected by the dreadful decompression disease which

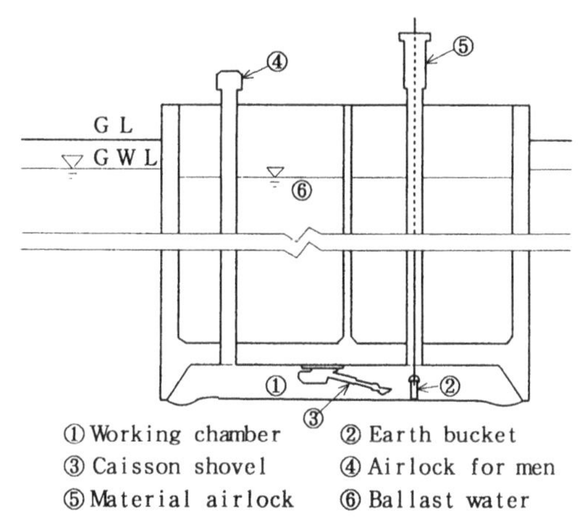

① Working chamber ② Earth bucket
③ Caisson shovel ④ Airlock for men
⑤ Material airlock ⑥ Ballast water

Fig. 1 An example of pneumatic caisson

is commonly called *caisson disease*.

The pneumatic caissons then were very expensive not only because the labor cost for the compressed-air work was high but also because it took much more time to build a caisson than to build a pile foundation of equivalent magnitude.

Becasue of the above mentioned drawbacks, pneumatic caissons had gradually been replaced with long piles and partly with large dredge caissons since long-size steel were biginning to be placed on production lines in the early 1930's.

The dredge caisson is a type of steel open caisson in which the ground is excavated mainly with dredging equipments.

Thus the use of pneumatic caissons diminished and ceased in the early 1940's in most of the developed nations except in Germany and Japan. Both of these two countries then had little reserve of steel to be allocated for domestic use and pneumatic caissons were continually built in these two countries because not very much steel was required to build them.

Therefore, the pneumatic-caisson technology was steadily preserved in these two nations throughout the particular period in the 1940's and more than a decade thereafter.

The technology of pneumatic caisson in Japan has remarkably been improved in a few recent decades and the pneumatic-caisson technology in there now has been advanced to the high level comparative to such a newly developed method as the cast-in-situ concrete piles or subsurface diaphragm walls.

2. CAISSON SHOVEL

A caisson shovel is a compact power shovel designed to be operated in the limited space in the low-head working chamber of a caisson.

As shown in Fig. 2, a caisson shovel is suspended from a small trolley tracked on a pair of I-shaped rails fastened up to the heavy reinforced-concrete ceiling slab of the working chamber. It runs back and forth along under the I-shaped rails like a suspended monorail wagon. Because the caisson shovel is suspended and unrestrained from below, it travels smoothly even at such an extremely soft ground where any crawler-mount equipment might have been stucked.

The maximum working radius of a caisson shovel is 4 m when its dipper boom is fully extended and it covers an 8-m wide strip of the ground below the I-shaped rails.

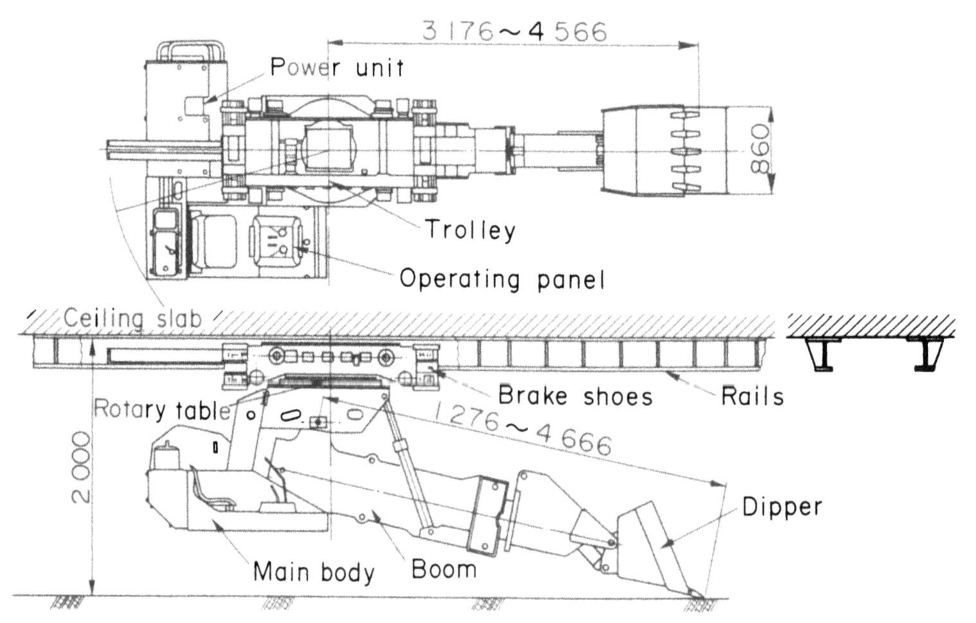

Fig. 2 Caisson shovel

The suspended body of a caisson shovel rotates 360 degree around its vertical axis. An 150-liter dipper bucket is attached to the front end of its telescopic boom. As its shoving thrust is as heavy as 10 t(98.1 kN) in maximum, very hard ground like stiff hardpan may be shoved out efficiently with the dipper.

A caisson shovel may be disassembled into five pieces each piece being designed to clear through the narrow passages in an air lock when it is brought down into the working chamber or brought up out of the working chamber through the airlock.

Since the first large-scale use of the caisson shovels was made in 1971, the accumulated total volume of the ground excavated with the caisson shovels in more than 500 caissons exceeds 2.5 million cubic meters.

The principal part of the caisson shovel had been patented but the effective period of the patent expired a long time ago. However, the great store of knowhows gained through the long-term experience in building those more than 500 caissons is indispensable to the efficient use of the caisson shovels.

Otherwise, without the support of the particular knowhows, any kind of the attempt to copy the design of the caisson shovels and to use them would not be successful and it might end up in vain.

The use of caisson shovels resulted in much increased efficiency of the excavation in caissons. And it realized remarkable man-power savings and cost-down of the works in pneumatic caissons.

3 UNMANNED EXCAVATION SYSTEM

A revolutional improvement of the excavation in caissons was realized by the newly developed unmanned excavation system(UES). In an UES, the operator of a caisson shovel sits in an atmospheric capsule installed in the working chamber of a caisson. He maneuvers his caisson shovel with a wireless remote-control kit in a manner similar to the operation of a radio-controlled model car.

As shown in Fig. 3, an atmospheric capsule is a cylindrical steel cabin suspended from the piston head in a vertical air-cylinder.

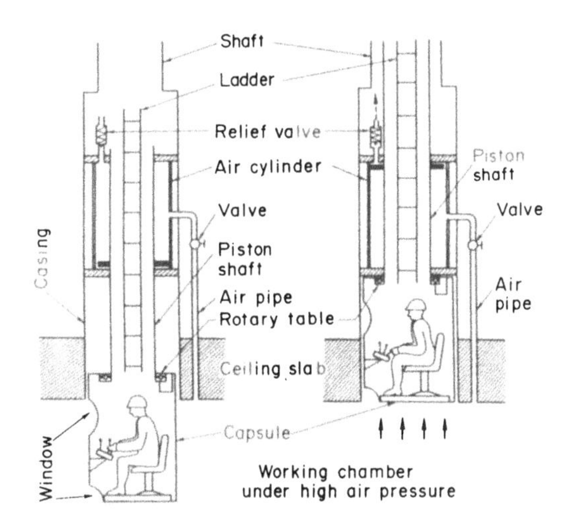

a. Lowest position b. Highest position

Fig. 3 Atmospheric capsule

The air-cylinder is firmly anchored down to the reinforced-concrete ceiling slab of the working chamber.

The capsule may be turned around at the operator's will and he may have its windows face his caisson shovel so that he always has a good view of the caisson shovel being operated by him.

The capsule normally takes its lowest operating position as shown in Fig. 3, a. If the capsule had been threatened to be flattened by accidental settlement of the caisson, the upward pressing of the piston-head rising together with the capsule would have induced excessively high air pressure in the air-cylinder. Then, the relief valve of the air-cylinder would be opened to release air causing the capsule to rise up to its highest position in a cased hole and out of harm's way, as is shown in Fig. 3, b.

The average hourly production rate attained by the operators of the UESs is lower than the rate attained by the operators driving their caisson shovels directly in the compressed air. The decline of the rate is due to the former's awkward and slow indirect operation from inside the capsule.

The recent records obtained in large, deep caissons indicate that the hourly production rate attained by the operator of an UES varies in

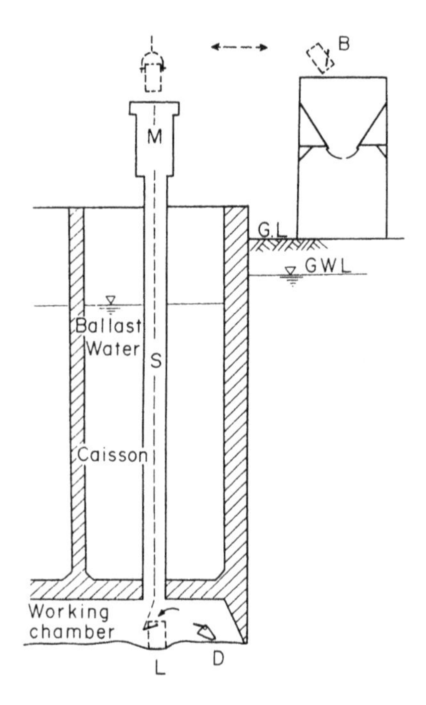

Fig. 4 Bucket handling process

a wide range from 3 cubic meter to 10 cubic meter. It depends on the skill of the operator and on such other factors as the size of caisson, soil type as well as the cycle time of earth-bucket handling process for hauling the excavated soil out of the caisson.

The average hourly production rate attained by the operators of UESs is generally about 50% lower than the average rate attained by the operators directly driving their caisson shovels in compressed air. However, the operators in the atmospheric capsules are able to work most of their work-shift hours in a day while the operators exposed to the high air pressure in the working chamber are not able to work for more than two hours in a day when the air pressure is higher than 3.5 kg. per sq.cm(3,434 hPa).

Therefore, even if the average hourly production rate declines by 50% in UES as compared to the average hourly rate attained by the direct driving in compressed air, the average daily rate attained by the operators of UESs may be in excess of doubly higher than the average daily rate attained by the opertors in compressed air at the above mentioned high pressure.

The meritorious results brought forth by the use of UESs are not only limited to the cost-saving effect realized by the increased productivity. The improved working environment in the atmospheric capsules resulted in a much better effect to the health and welfare of the workers. In the case of the world's largest pneumatic caisson, to be quoted later in this paper, where more than 20,000 entrances into the working chamber were recorded, the occurrence rate of the *caisson disease* was only less than 0.2%. The patients of the caisson disease in this case were medically treated in hospital air locks and all of them were cured completely.

In contrast to the above mentioned very low occurrence rate of caisson disease where most of excavation work was performed with the UESs, in the conventional caissons, where the digging of the ground was made by human labor in compressed air, the occurrence rate was generaly higher than 3%.

The work in the conventional pneumatic caissons is a dirty hazardous task. In the modern caissons equipped with UESs, however, the caisson job is a safe and clean task for anybody who joins it.

4 QUICK SOIL LOADER

The bottleneck of in-caisson excavation does not lie in the digging work itself. But it lies in the process of hauling the excavated soil out of the caisson for disposing it.

As shown in Fig. 4, the excavated soil taken into the dipper bucket of a caisson shovel is hauled from the digging point D to the loading point L where the soil is loaded to the bucket, which is customarily called *earth bucket*, put down below the shaft S. The bucket then is pulled up through the shaft S and the air lock for materials M and out above the air lock.

The bucket is further moved sideway to the point B above the soil storage bin where the bucket is turned upside-down for dumping down the soil into the storage bin.

After dumping the soil, the empty bucket is returned to the loading point L reversely through the airlock M and the shaft S. The cycle time of the above bucket handling process varies in a wide range, depending on the total length of the round-trip passage of the bucket, the speed of wire moving the bucket and some other pertinent factors. In an example of a large caisson, where the vertical head from the point L to the point M was 50 m and the distance between the point M and the point B was 20 m, the average cycle time was 6 minutes. In this example, every step of the bucket-handling process was automated.

The bucket-handling process was automated by utilizing a series of computer-controlled sequense programs composed for every action in the bucket-handling process.

For filling up a bucket of one-cubic-meter capacity by repeated feeding of soil from a caisson shovel, several minutes of time is required while the caisson shovel shuttles back and forth between the digging point D and the loading point L.

The time required for fully loading an one-cubic-meter bucket with soil varies depending on the soil type, the distance between the points D and L, as well as the skill of the operator. In an example in the ground of very soft clay where the digging was easy, the average distance between the points D and L was 12m, and the operators were unskilled, the average cycle time for loading fully the one-cubic-meter bucket was 6 minutes.

The time spent for loading soil into the bucket constitutes a loss time in the bucket-handling cycle as the bucket may not be moved up until the loading of soil is over.

Therefore, the cycle time of the bucket handling increased by the loss time of 6 minutes for loading soil into the bucket. Altogether, with the additional loss time of 6 minutes, the total cycle time of the bucket-handling process increased doubly to 12 minutes.

A quick soil loader is designed to eliminate the loss of time in the bucket-handling cycle consumed for loading soil into the bucket.

Instead of feeding soil repeatedly with a caisson shovel into the bucket, the soil is loaded into the storage vessel of a quick soil loader while the bucket is moved for discharging the soil out of the caisson and returned to the point L.

The oblong storage vessel of a

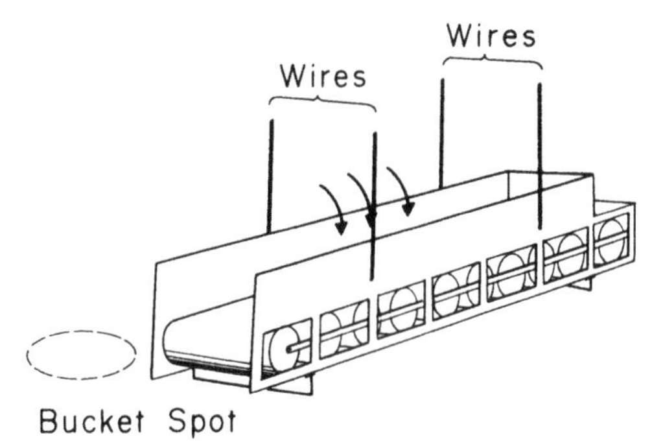

a. Storing position

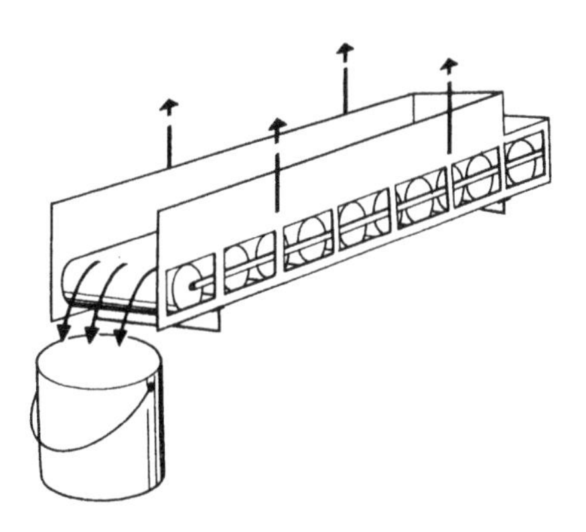

b. Dumping position

Fig. 5 Quick soil loader

quick soil loader with its built-in belt-conveyer bottom has a capacity of storing one-cubic-meter bucket full of soil.

During while the soil is fed by a caisson shovel, the quick soil loader is hung down to its lowest storing position as shown in Fig. 5, a.

When the storage vessel is fully loaded with soil, it is pulled up and moved horizontally over to the empty earth bucket and all the soil stored is dumped with a rush into the bucket in a matter of seconds by moving the belt conveyer for a short stroke, as shown in Fig. 5, b.

5 EXAMPLE OF UES USED IN WORLD'S LARGEST PNEUMATIC CAISSON

The world's largest pneumatic caisson of 160,000 cubic meter in total volume was built and sunk down to its final depth of 46.5 m below the mean sea level into the submarine ground of Tokyo Harbor in 1989.

This caisson is the foundation of the off-shore anchorage for the 798-m long suspension bridge crossing over the main navigation channel of the harbor.

This bridge was completed, nicknamed *Rainbow Bridge* and opened for traffic on August 27, 1993.

The anchorage caisson is 70 m by 45 m in plan, 51 m high and the volume of the submarine ground excavated in it 109,000 cubic meter.

The soft ground beneath the 12-m deep sea water was dug mainly with ten caisson shovels down to the depth of 15 m below the sea bed, while the operators drove their caisson shovels in the compressed-air chamber. Eight quick soil loaders were provided in order to expedite the soil discharging process with automated bucket-handling systems.

Those ten caisson shovels were converted to the unmanned excavation systems(UESs) when the ground was excavated down to the depth of 15 m below the sea bed. Those ten UESs were used thereafter for digging about 60,000 cubic meter of the ground down to the final depth of 34.5 m below the sea bed.

The maximum daily production rate attained by the 3-shift-a-day work by the operators of the ten UESs digging the soft clay in the depth range from 15 m to 27 m below the sea bed was 790 cubic meter.

The average hourly production rate attained by an operator of UES there was 3.8 cubic meter.

Because it took much more time for building up a 3-m high lift of caisson body involving more than 3,000 cubic meter of reinforced concrete, than for excavating 9,478 cubic meter of ground in the depth range of 3 m, the work of excavation was not on the critical path in the progress of over-all caisson works until the caisson sunk to the depth of 27 m below the sea bed through the soft clay.

In the deepest layer of 8-m thick hardpan, however, the progress of excavation slowed down due to hard digging and it became a critical factor to determine the overall progress of the entire caisson work.

Eight additional manned caisson shovels as well as four jib cutters were used to expedite the cutting of the hardpan. The jib cutter is a giant chainsaw for cutting out the hardpan from beneath the thickness of the external walls of the caisson. The UESs, the quick soil loaders and other refined appliances like automated bucket-handling systems made a great deal of contribution to the successful completion of the unprecedented huge caisson well in advance of schedule.

This work of the huge caisson was accomplished with the record-making rapid progress in excavation and the highly accurate control of the sinking operation of caisson body.

6 UNMANNED EXCAVATION SYSTEMS FOR CIRCULAR CAISSONS

The UESs as quoted before may be applicable to the large caissons wide enough to accomodate several sets of UESs, quick soil loaders and other pertinent appliances in the caisson's working chamber. However, because the UESs are much too bulky, they may not be suitably applied to the caissons smaller than 16 m in diameter or in width.

A modified type of unmanned excavation system for circular caissons (UES-C) has been newly developed.

The UES-C is applicable to the circular caissons not larger than 16 m in diameter. In an UES-C, a caisson shovel is set suspended from a small trolley tracked on a pair of circular running rails fastened up to the ceiling slab of the caisson's working chamber as shown in Fig. 6, a.

The caisson shovel is remotely operated from the control room installed on the ground surface near the caisson.

The motion of the caisson shovel in the working chamber is monitored by watching the graphic displays as shown in Fig. 6, a and b. Furthermore, the TV images and sounds transmitted from the TV cameras set in the working chamber and attached to the caisson shovel are also monitored for controlling the action of the caisson shovel.

The graphic display as shown in Fig. 6, a is composed by processing the electric signals issued from the sensors attached to the caisson shovel and it indicates the position and the posture of the caisson shovel.

The graphic display as shown in Fig. 6, b is composed by processing the electric signals transmitted from the supersonic sensors attached to the main body of the cais-

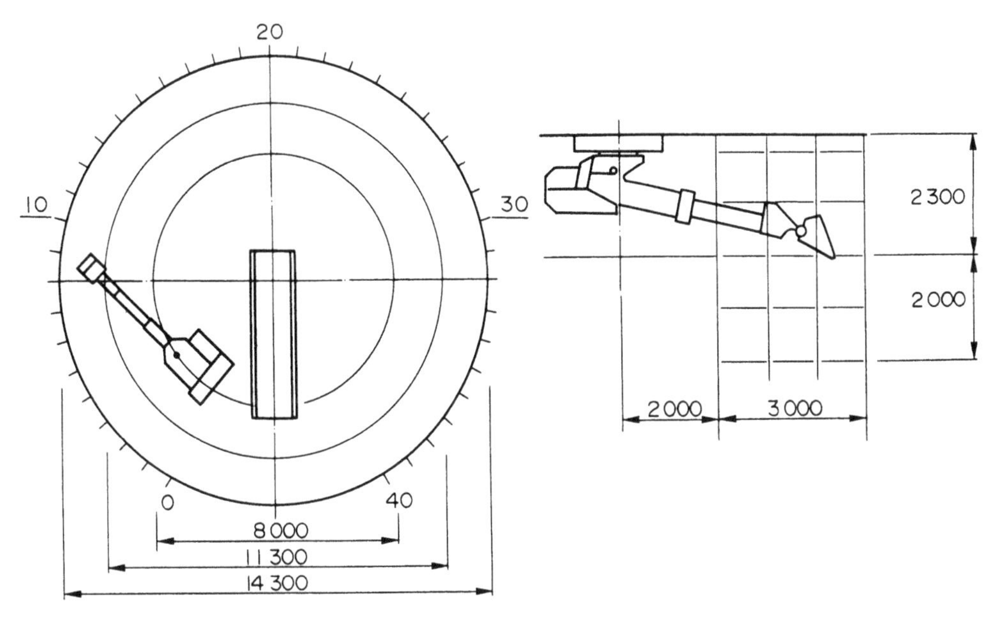

a. Graphic display of position and posture of caisson shovel

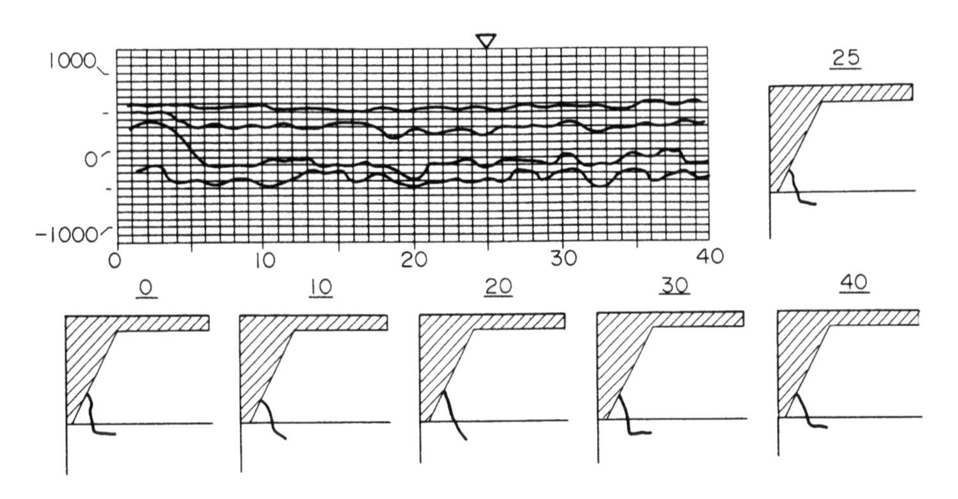

b. Graphic display of the ground suraface being excavated

Fig. 6 Examples of graphic displays

son shovel. The supersonic sensors catch the pulse of the supersonic wave reflected back from the ground surface and send electric signals to the micro-computer where the pulse-signals are processed to compose the graphic display indicating the undulation of the surface of the ground being excaveted.

In the UES-C, certain simple routine actions of the caisson shovel are automated. For those routine actions, the series of actions performed by a skilled operator are teached into the computer program.

The automatic control by the teached-in program is generaly applied to such simple routine actions like the movement of the caisson shovel from the digging point to the loading position, the feeding of soil into the quick soil loader and the returning run to the digging point. Digging action is manually controlled as it requires the operator's judgement by monitoring the graphic displays and TV screens showing the changing states of the ground being excavated.

7 CIRCULAR QUICK SOIL LOADER

The circular quick soil loader is designed for the use in the UES-C. As shown in Fig. 7, a circular quick soil loader consists of an internal tube and an external ring. The internal tube is fixed up to the ceiling slab of the working chamber at its top end. The hollow space inside the internal tube is to accomodate an earth bucket hung down into it. Three wide windows open on the upper part of the internal tube. The external ring consists of an annular base plate and a rotary spill-guard ring wall.

A pair of ejector blades are fixed to the ring wall from inside. The ring wall is revolved by an oil motor through chain transmission systems. The inside edge of the annular base plate is closely surrounding the outside surface of the internal tube. The gap between them is very small so that the soil fragments may not spill down too much through the gap.

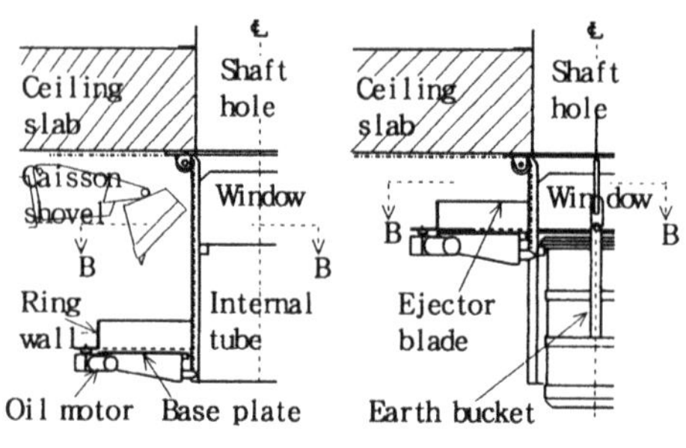

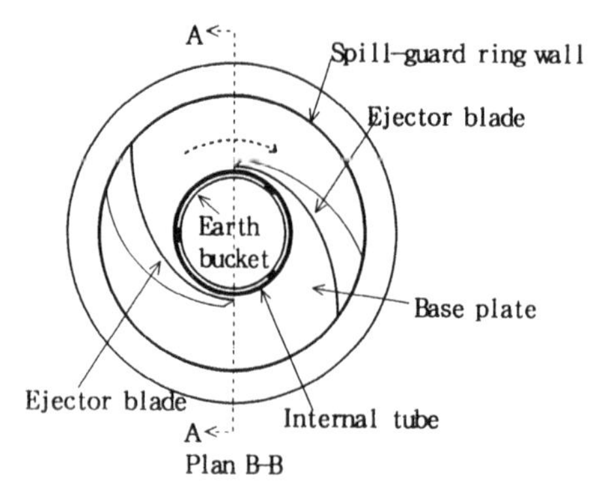

Fig. 7 Circular quick soil loader

8 COMBINED USE OF UES-C AND CIRCULAR QUICK SOIL LOADER

The combined use of a UES-C and a circular quick soil loader, as shown in Fig. 8, was made recently in a caisson of 15.2 m in outside diameter and sunk to the depth of 45.0 m below the ground surface. The average daily rate of production by 3 shift-a-day work of excavation into the ground of sandy silt in the range of depth from 25 m to 45 m below the ground surface was 112.3 cubic meter. The average hourly rate in this example was 5.3 cubic meter. The average length of cycle time for bucket-handling process was six minutes.

9 INSTRUMENTED CONTROL SYSTEM

Not only the excavation phase but also several other phases of caisson works are remarkably improved.

For controlling the sinking process of caissons, such a disorder as slanting or dislocation of the caisson body had formerly been observed by eye measurement or by reading measuring instruments. Such manual observations were not only inaccurate but also too slow to be ready promptly for timely remedy.

In order to keep pace with the faster progress of in-caisson excavation than it was before, large caissons are often instrumented and they are controlled by computer-aided systems. The inclination or the dislocation of the caisson body may be monitored with the graphic displays indicated on the TV screens placed in the control room with the accuracy of millimeters.

Disorder, such as inclination or dislocation of the caisson body, may be remedied promptly before the disorder grows too much.

The disorder may be remedied by such a measure as relocating the positions of ballast-water loading or by changing the unbalanced pattern of excavation.

10 UES IN ULTRA DEEP CAISSON

The UESs are applicable to very deep caissons where the maximum air pressure exceeds the ruled limit of 4 kg per square cm.(3,924 hPa).

However, men have to enter into the compressed-air chamber at very high pressure for occasional inspections or maintenance purposes.

In this connection, efforts were made to establish the authentic procedure for letting men enter the ultra high air pressure without causing any physical hazard.

There had been an established technology of deep-sea diving.

Helium-mixed air is to be used for breathing instead of natural air in the deep-sea diving practice.

If the natural air at very high pressure were breathed, symptoms of *nitrogen narcosis*, which are the conditions similar to the serious intoxication by liquors might have appeared to disable both mental and physical abilities of men.

Experimental research program based on the knowledge and experience reported in the references on deep-sea diving technology had been prepared.

Comprehensive experiments, including a series of physical and mental tests on animals and human bodies, were performed under very high air pressure in simulation vessels.

Those experiments were performed for these several years under close guidance by medical experts.

As the outcome of those experiments, the specifications for the procedure of entering into the high air pressure up to the gage pressure of 7 kg per square cm were compiled. In compressed-air works, it is customary to express the value of pressure as indicated on pressure gages.

The gage pressure of 7 kg per square cm corresponds to the absolute pressure of 8 kg per square cm(7,848 hPa) which is 8 times the atmospheric pressure.

For the next stage, aiming at the practical applications, the design of a prototype setup for proving the compiled specifications was prepared. Every necessary appliance, for letting men enter the compressed air cabin at very high pressure, will be provided for the performance tests in the prototype setup.

When the compiled specifications would be proved right by the performance tests in the prototype setup and the training program thereafter, applications will be filed to the government agencies concerned. And whenever the applications are authorized, the UESs may be applied to ultra deep caissons.

The first possibe application of UESs to the ultra deep caissons is scheduled to be made at the main-tower foundations of the Second West Port Bridge in Nagoya in the late 1995. This bridge will be a suspension span whose main towers will be founded on very deep caissons. The maximum depth of the caissons will be 45 m below the mean sea level.

Before the above application of the UESs to the ultra-deep caissons is realized, several barriers of government examination should be cleared through.

11 FUTURE DEMAND FOR PNEUMATIC CAISSONS

In most countries except in Japan and Germany, pneumatic caissons had been abandoned many years ago because they became outdated and unsuitable for modern needs.

For building pneumatic caissons, quite a deal of investment for such equipment as air compressors, air

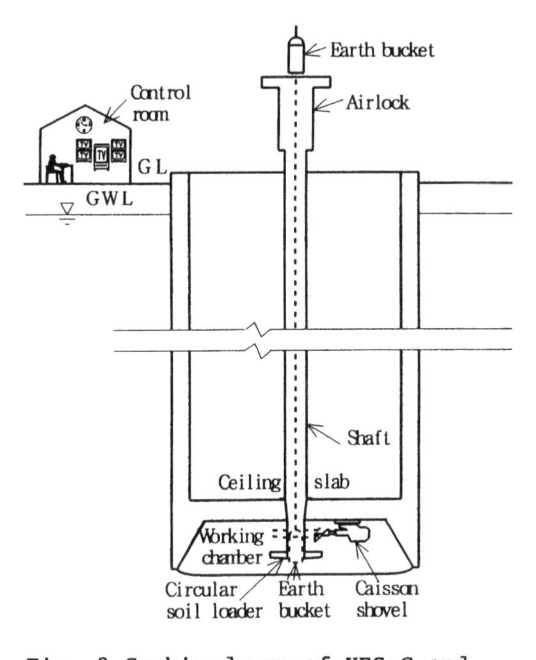

Fig. 8 Combined use of UES-C and circular quick soil loader

locks, caisson shovels and the like is required.

Because Japanese caisson contractors hold those pertinent equipments sufficiently for building a large amount of caissons, their equipment costs for caisson works are appreciably low because of bulk merit.

Therefore, they may be able to furnish necessary equipments and knowhows at reasonably low price to any project of pneumatic caissons not only in Japan but also in the overseas areas.

Some recent survey indicates that there are quite a few sites of the future projects planned in the Southeast Asia where pneumatic caissons may be suitably applied.

For those projects, the knowhows and equipment furnished by the experienced Japanese contractors at reasonable price may be profitably utilized and it may contribute a good deal of merits to the project.

12 CONCLUSION

The recent developments in the unmanned excavation systems(UESs) in pneumatic caisson may be summarized as follows:

1. By the use of UESs in pneumatic caissons, the productivity of excavation has been remarkably improved.
2. By the combined use of the UES-C and the circular quick soil loader, the productivity of excavation was raised appreciably.
3. The working environment in the caissons has been much improved so that the occurrence rate of caisson disease was minimized.
4. It is made possible to apply the UESs in ultra deep caissons.

ACKNOWLEDGEMENTS

Kashima Corporation and Shiraishi Corporation, the co-sponsors of the engineering program for developing the UES-C, were honored to receive the award of the Presidential Prize of the Japanese Construction Mechanization Association of 1990.

The writer is highly indebted to the key engineering staff of the above two firms for their efforts and contributions in developing unmanned excavation system for circular caissons(UES-C).

The writer is grateful also to the key personnel of the Metropolitan Expressway Public Corporation for their guidance and assistance in building the large caissons in Tokyo Harbor.

REFERENCES

Shiraishi, S.1989. Excavation of submarine ground in the world's largest pneumatic caisson in Tokyo Harbor, Symposium on Underground Excavation in Soils and Rocks, Sponsored by Asian Institute of Technology, Bangkok, Thailand, November 27-29, 1989.

Maeda, K. & Shiraishi, S. 1990. World's largest caisson, Civil Engineering, American Society of Civil Engineers, Vol. 60, No.8, August 1990, p. 47-49

Shiraishi, S. 1990. History and recent development of pneumatic caissons, Conference on Deep Foundation Practice, CI-Premier Conference Organization, Singapore, October 30-31, 1990,

Oda, S. et al 1992. Application of deep-sea diving technology to pnematic works in very deep caissons, Conference on Engineering and Health in Compressed Air Work, Construction Industry Research and Information Association, London, September 1992

Shiraishi, S. et al 1992. Unmanned excavation systems in pneumatic caissons, Conference on Engineering and Health in Compressed Air Work, Construction Industry Research Information Association, London, September 1992.

Shiraishi, S. 1993. Unmanned excavation systems in compressed air caissons, Tunnels & Tunnelling, British Tunnelling Society, Morgan-Grampian(Construction Press) Ltd., Vol. 25, No.7, July 1993, p. 27-29

Developments in Geotechnical Engineering, Balasubramaniam et al. (eds) © 1994 Balkema, Rotterdam, ISBN 90 5410 522 4

Design methodology and experiences with pile supported embankments

W.H.Ting, S.F.Chan & T.A.Ooi
Kuala Lumpur, Malaysia

ABSTRACT: Pile supported embankments have been constructed in Malaysia in soft ground areas. The success of the construction is mixed; it is successful with piles smaller than 300mm diameter but failure occurred when piles from 400mm to 500mm sizes were employed. The Swedish Method of design is adequate for the smaller piles provided that horizontal forces are catered for, the Method is inadequate for the larger piles as it results in spacings on the high side. Various models representing pile supported embankment behaviour are reviewed. A preferred design method based on the stone column analogy model is then proposed catering for the larger size pile supports.

1. INTRODUCTION

The pile supported embankment is a construction where the embankment is supported by a grid of piles with individual reinforced concrete caps. Such embankments have been constructed in several soft ground sites in Malaysia as approaches to abutments for bridges. The heights of the embankments range from about 4m to 6m. The successes of the pile supported embankments have been mixed. They have been successful where smaller piles less than about 300mm size have been used, whilst in more than one project where pre-stressed spun concrete cylindrical piles of diameters ranging from 400mm to 500mm have been used problems ensued.

This paper describes briefly the current design method and how it was applied successful when smaller piles were used provided some important design details were adhered to.

Design methods based on assumed mechnisms and various observed and theoretical mechanisms are discussed. The mechanisms deduced from small-scale models are first presented. Case histories on the performance of the embankment and piles in full scale problems are also presented. A stone column analogy model is then proposed for application in the design of large pile supported embankments where the current design method has proven to be inadequate.

2. CURRENT DESIGN METHODOLOGY

2.1 General

In a pile supported embankment on soft ground, the primary role of the piles is to enable the embankment to be built to a height that would otherwise be impossible to achieve in view of the weak soils. A typical arrangement of piles is shown in Fig. 1. As such the piles have important functions in controlling the stability and settlement of the embankment. In the usual case of an approach embankment to a bridge, the embankment is connected to the bridge structure at the high end; whereas at the low end of the pile supported embankment, it is connected to a conventional road embankment which is supported directly on the ground. Thus the settlement of the pile supported embankment must be compatible with those of the two above mentioned connecting structures, each of which has a completely different settlement behaviour, e.g., small settlement at the high end and comparatively large, time-dependent settlement at the low end.

In addition, the caps and pile spacing need careful consideration to ensure that the embankment loads are transmitted satisfactorily to the piles without causing uneven settlement ('mushroom' effect) at the road surface. The side slopes of the embankment also require attention both from the stability and settlement standpoints. The process of construction

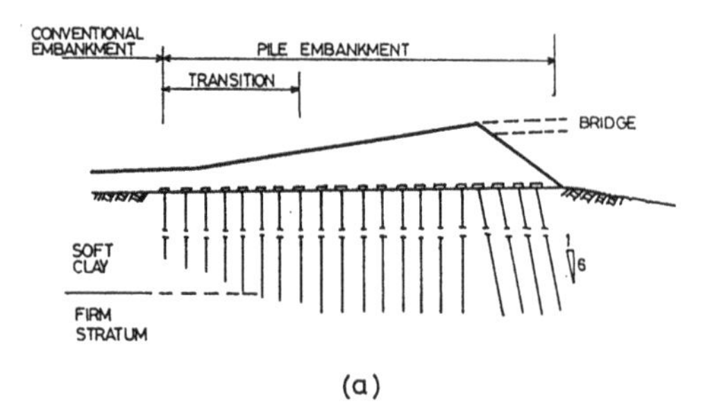

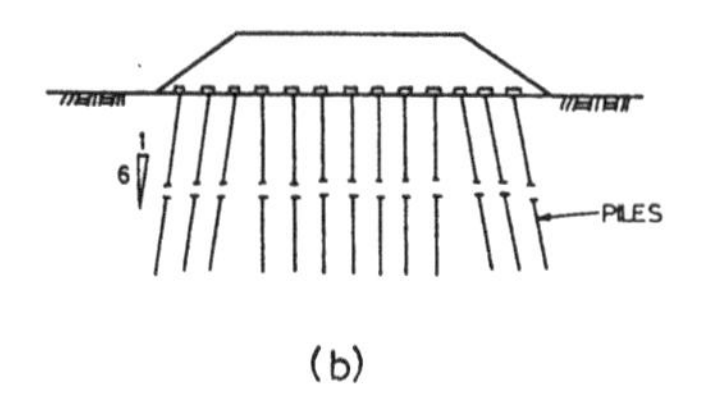

FIG. 1 PILE SUPPORTED EMBANKMENT: (a) LONGITUDINAL PROFILE, (b) CROSS-SECTION.

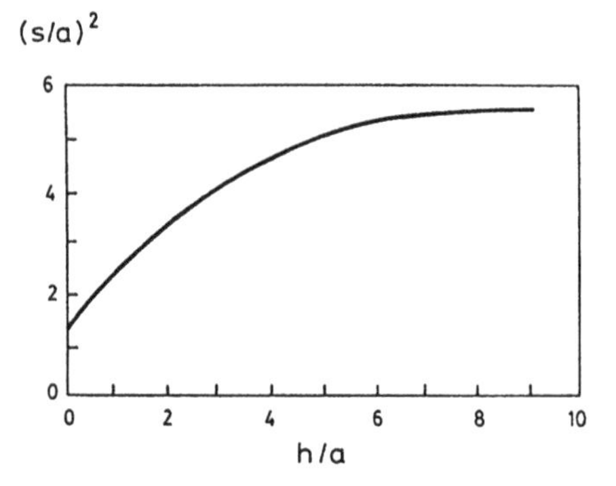

FIG. 2 RELATIONSHIP BETWEEN SIZE OF PILE CAP (a), HEIGHT OF EMBANKMENT (h) AND SPACING OF THE PILES (s) (after Swedish State Road Department, 1974)

of a pile supported embankment usually generates significant unbalanced temporary forces in the piles and caps and these should be accounted for in the design.

In Malaysia, the current design method is based generally on the recommendations of the Swedish State Road Department (1974), with some modifications based on local experience.

2.2 Piles

Precast reinforced concrete piles are commonly used in highway embankments. The piles are designed to carry the full weight of the embankment together with the traffic loads and the ground between caps is assumed to carry no load. Thus the benefit of soil arching between the pile caps is ignored in the calculation of pile load. These loads are transmitted to the piles through individual pile caps. The geotechnical design of the piles follows the usual method adopted, for example, as for buildings. In the structural design, the Swedish State Road Department (1974) method allows the safe structural load of pile to be 1.5 times that commonly used for buildings. Before adopting this provision, it is worth noting that the piles are usually not designed to cater for bending and other incidental forces that may be induced during construction or under service state. Except in the transition zone, which will be discussed in Section 2.5, the settlement of the piles should be small in order to be compatible with the settlement of the connecting bridge structure.

The spacing between piles, which is partly governed by the pile size, is a crucial factor in determining the success or failure of a pile supported embankment. The spacing should be chosen such that soil arches can form within the height of pile supported embankment. Large pile spacing may result in collapses of the soil arches under service state. These will be reflected as depressions in the unsupported areas of the piled embankment between the pile caps, thus adversely affecting the riding quality of the road surface.

2.3 Caps

The individual concrete caps, which are either precast or cast-in-situ, are designed to carry the full embankment weight together with the traffic loads. The loading on each cap is assumed to be uniform. The cap size can be calculated using Fig. 2 which is based on the results of model tests (Swedish State Road Department, 1974). In general, the total area of the caps amounts to 20% to 30% of the total area of the embankment.

The size of cap in relation to the pile spacing is crucial to the satisfactory performance of the embankment. In particular, the clear distance between the edges of two adjacent caps should not be too large as this will cause excessive settlement in the original ground between caps.

Another problem in this connection is related to construction. It is essential that a suitable working platform is constructed using selected material usually of 600mm thick or more. This layer will provide some form of lateral restraint for the piles. The space between the pile caps must be filled with selected frictional soil and carefully compacted. This will ensure that adequate restraint is provided to the pile caps so that the problems of tilting of pile caps and bending or shifting of piles are minimised. Filling of embankment can then proceed with the placement of the first layer of fill above the pile caps using light tractors.

In high embankments, horizontal concrete ties are required so that the horizontal forces in the pile caps are distributed and balanced, see Fig. 3. These ties will also help to minimise the problem of shifting, tilting and bending of piles.

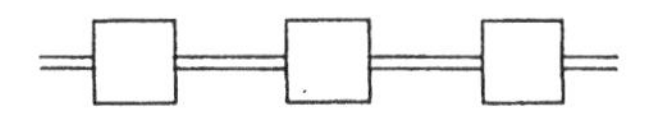

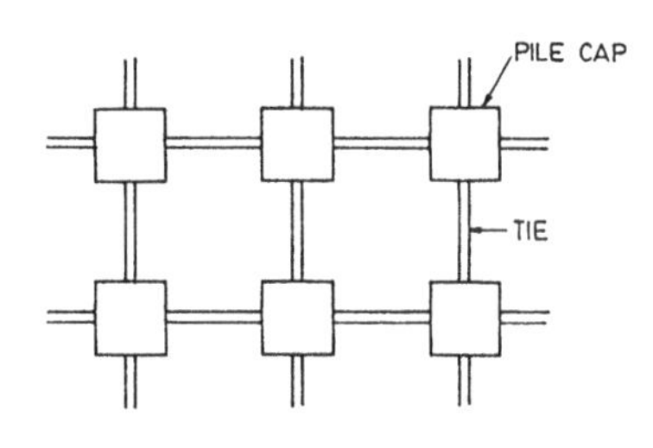

FIG. 3 CONCRETE TIES FOR PILE CAPS

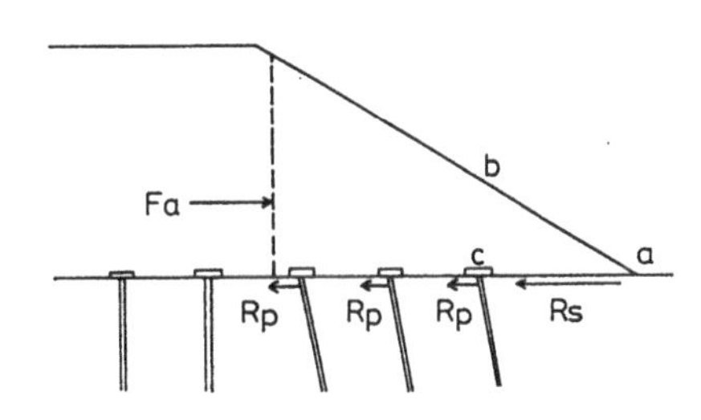

Rs = Shear resistance of soil
Rp = Horizontal resistance of pile
Fa = Force due to active pressure

FIG.4 SIDE SLOPE STABILITY.

2.4 Side Slopes

At the side slopes, in addition to vertical stability, there is also a need to provide for horizontal stability. The overall stability can be attained by providing raking piles. In the service state the destabilizing horizontal force arising from active earth pressure is resisted by horizontal shear resistance of the ground and the horizontal components of axial pile forces, as shown in Fig. 4.

At the outer edge of the embankment, there is a triangular block of fill material, marked abc in Fig. 4, which is not supported on piles. If the outside row of piles is stopped too far from the edge, this block will settle significantly and a crack will develop along bc.

2.5 Transition Zone

At the low end of the pile supported embankment, there is a need to provide a transition zone (see Fig. 1). This is where the pile supported embankment meets the conventional embankment which is supported directly on the soft ground. Since the settlement of the conventional embankment is relatively large, the transition zone should be designed such that its settlement is increasing gradually towards the low end. This requirement can be met by decreasing the pile length as the low end is approached. However, the design of piles to meet a specific embankment settlement is not an easy matter. Furthermore, in order to avoid large pile spacing it may be necessary to use smaller piles in the transition zone.

3. MODELS AND PERFORMANCE

Results from model tests and full scale field performance measurements are briefly described. Some theoretical models are presented from which design methods can be evolved.

3.1. Small-scale Models

3.1.1. Swedish Experiments

Berghdahl et al.(1979) carried out studies on 1:10

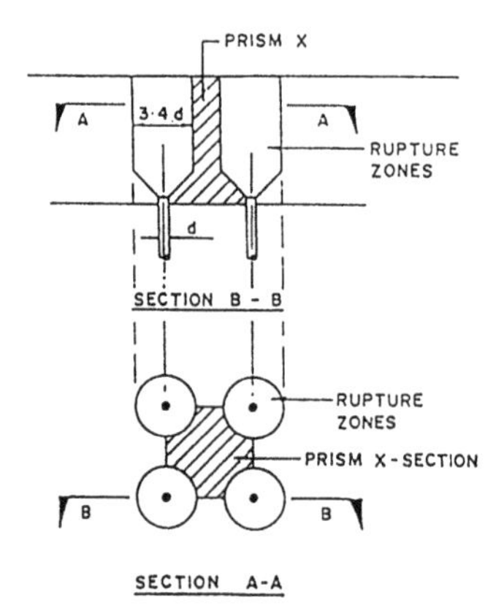

FIG. 5a LOAD SHARING MODEL

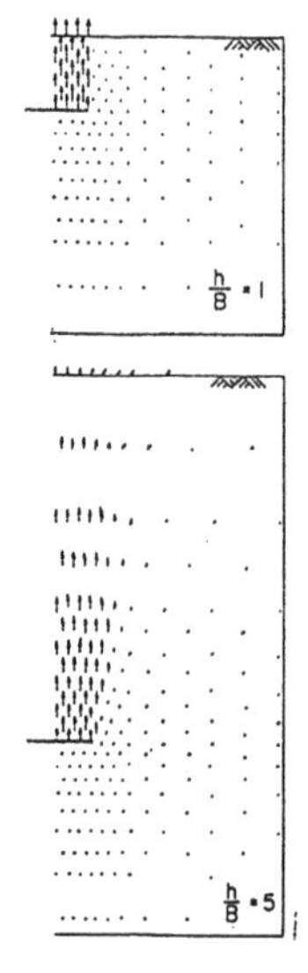

FIG. 5b VELOCITY FIELDS AT COLLAPSE;
$\phi = 30^{\circ}$, $\psi = 0$
(After Rowe & Davis, 1982, Fig.3)

scale models of pile supported fills with plates as pile caps to gain a better understanding of the behaviour of such systems. They concluded that for low embankments below a certain critical height, when settlements between plates are high, the average pressure on the ground (between plates) can reach 90% of the fill pressure. Above the critical height it is possible to support nearly all the fill load by the plates.

3.1.2. Model Experimental Results of Ting et al.

Ting et al. (1983) presented the findings of their model tests using loose and dense sand as fill. Piles of 38mm diameter (d) at spacing of 3d, 4d and 6d were studied. Fill heights were 9.4d and 19d.

A physical model (Fig.7 in above-mentioned paper) deduced from the tests is reproduced herein as Fig. 5a.

3.2. Full-scale Performance

3.2.1. Penang Bridge Approach

The eastern approach embankment was supported by 500mm diameter prestressed concrete cylindrical spun piles with 1.8m square pile individual caps at grid spacing of 3.4m to 4.2m. The embankments reached a height of 6m and the larger pile spacing was for low embankment areas.

Within months of the opening of the highway, differential settlements were observed on the pavements, between the areas of the embankment over the pile caps and those spanning the caps. The results of the differential settlements have the appearance of `mushrooms' over each pile position.

An investigation was initiated. It was estimated that approximately one and a half years after completion of the embankments, the total differential settlement was at least 225mm. Typical differential settlements measured at the commencement of the investigation are as shown in Fig 6. Significant lateral movements were also observed over the investigation period.

The formation of the `mushrooms' may be attributed to the selection of pile cap spacing specially in low embankment areas. Other factors that may be involved are the absence of lateral ties and inadequacies in pile support at embankment toes. A further factor to consider which is specific to the site is the consolidation arising from an old fill which underlies the new platform fill.

3.2.2. Tangkak Trial

The Malaysian Highway Authority conducted full-scale trial embankment performance studies at Tangkak. A section of the trial (Malaysian Highway Authority, 1989) was allocated to the embankment supported by 400mm diameter prestressed concrete spun piles fitted with 1.8m × 1.8m precast pile caps at grid spacing of 3.5m. The configuration of the intended embankment is shown in Fig. 7.

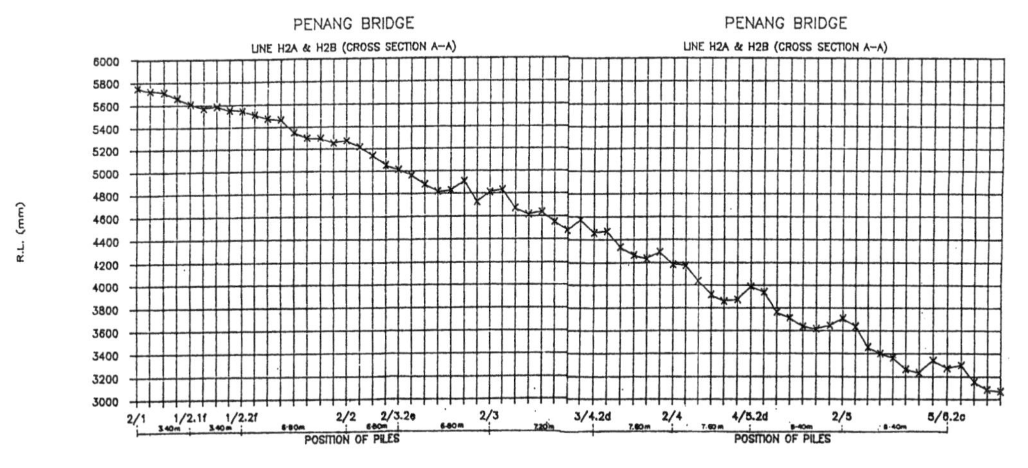

FIG. 6 SETTLEMENTS

In the course of the construction, cracks first appeared on the embankment when the height was about 3m. When the embankment neared the intended height of about 6m longitudinal cracks initially appeared on the top of one side of the embankment and after a lapse of about 21 days failure took place. A view of the failed embankment is as in Fig.8. On excavation of the embankment the configuration of the failed piles were surveyed and are as shown in Fig. 9.

The embankment failure which appears to involve a slip surface mechanism may be attributed to inadequacies in lateral support as well as the selected spacing of the pile caps.

Further details of the embankment and observations on its performance can be found in the Highway Authority Report.

3.2.3. *Small Bridge Approach Embankment*

Baskaran and Sundaraj (1993) reported the movements of the pile supported embankment serving as an approach to a small bridge in East Malaysia. The embankment and pile support configuration and contours of movements are shown in Fig. 10. It can be seen from the contours that lateral spread of the embankment has taken place giving rise to a settlement of up to 6mm over the measurement period of about three months. Such movements could have been reduced if lateral ties and raking piles had been incorporated in the designed configuration.

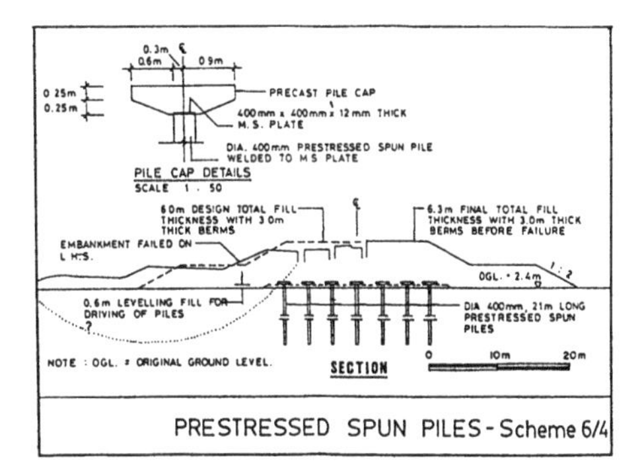

FIG. 7 PILE SUPPORTED EMBANKMENT

3.3. *Theoretical Models*

3.3.1. *Vault Concept*

Hewlett & Randolph (1988) proposed the vault model suggested by laboratory model tests. A representation of their model is reproduced here as Fig. 10. They concluded from their analysis that the vault action leads to lower bound values of fill loads transferred to pile supports when compared with field measurements.

3.3.2. *Inclusions Concept*

Combarieu (1988) considered the pile supported embankment from the point of view of the effects of the embankment pressure on the ground between shafts. This approach differs from those proposed so far which are concerned with the interaction of

FIG. 8 : AFFECTED PILE SUPPORTED EMBANKMENT

the embankment with the pile caps. The underlying principle involves the 'silos' type concept. The proposed model showing the piles as inclusions is as shown in Fig. 12. The pressure from the embankment is first calculated. The embankment pressure then acts on the original ground between pile shafts and the ensuing settlement is controlled. The results of the analyses are compared with the recommendations of the Swedish Method in Fig. 13.

The inclusion concept has the attraction that it can deal with the perplexing problem of the transfer of soil pressures from the embankment fill between pile caps to the original ground between pile shafts.

3.3.3. Proposed Stone Column Analogy Model

The physical model presented in Section 3.1.2 (Fig. 5) conceptualised from the small-scale model and finite element analysis can be developed further as a model represented analytically by a stone column analogy.

3.3.3.1. The Stone Column Model

Fig. 14 shows the proposed analogous stone column model as applied to the pile supported embankment situation. A solid cylinder of soil mass is tributary to each pile. An idealised column of soil is formed over the pile caps, as the soil mass begins to move down due to absence of support between pile caps. The column of soil is assumed to have the same equivalent diameter as the pile-cap. As the soil column is prevented from moving downwards by the pile-cap the rest of the moving soil mass continues its movement as a hollow soil cylinder. A similar mechanism has been applied to problems of soil mass reinforced by stone columns.

In the column model the soil between the pile caps tributary to a pile is assumed to be a part of a hollow soil cylinder supported by the shaft resistance of the column which in turn is supported by the pile-cap. In contrast, in the 'vault' model the soil between the pile caps is supported by the vault across two pile caps.

The pull-out of anchor plates involves the same mechanism in reverse to that of the stone column model proposed above. The studies of Rowe & Davis (1982) on the anchor plates can also be applied to show that the column model mechanism is reasonable (Fig. 5b).

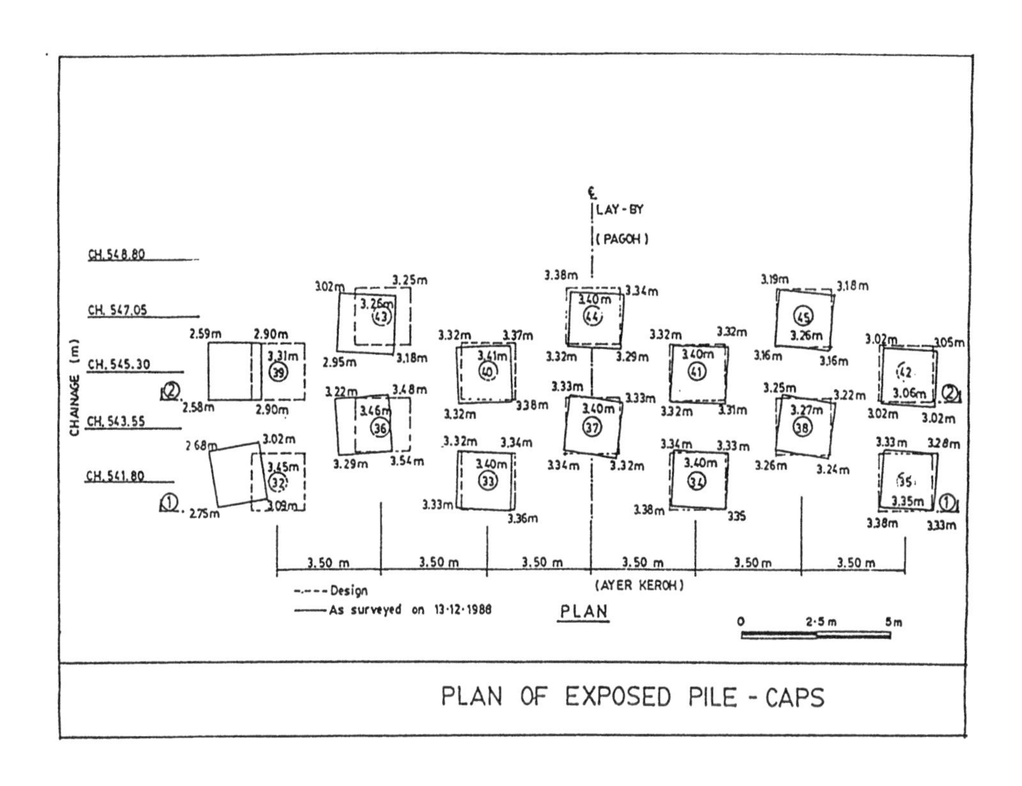

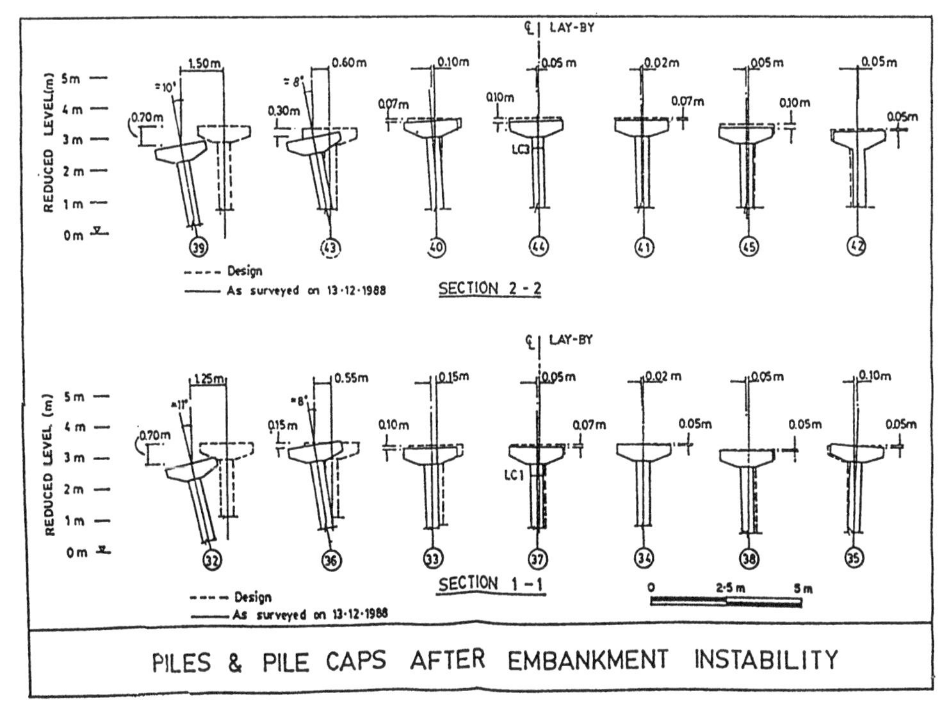

FIG. 9 PILE MOVEMENTS

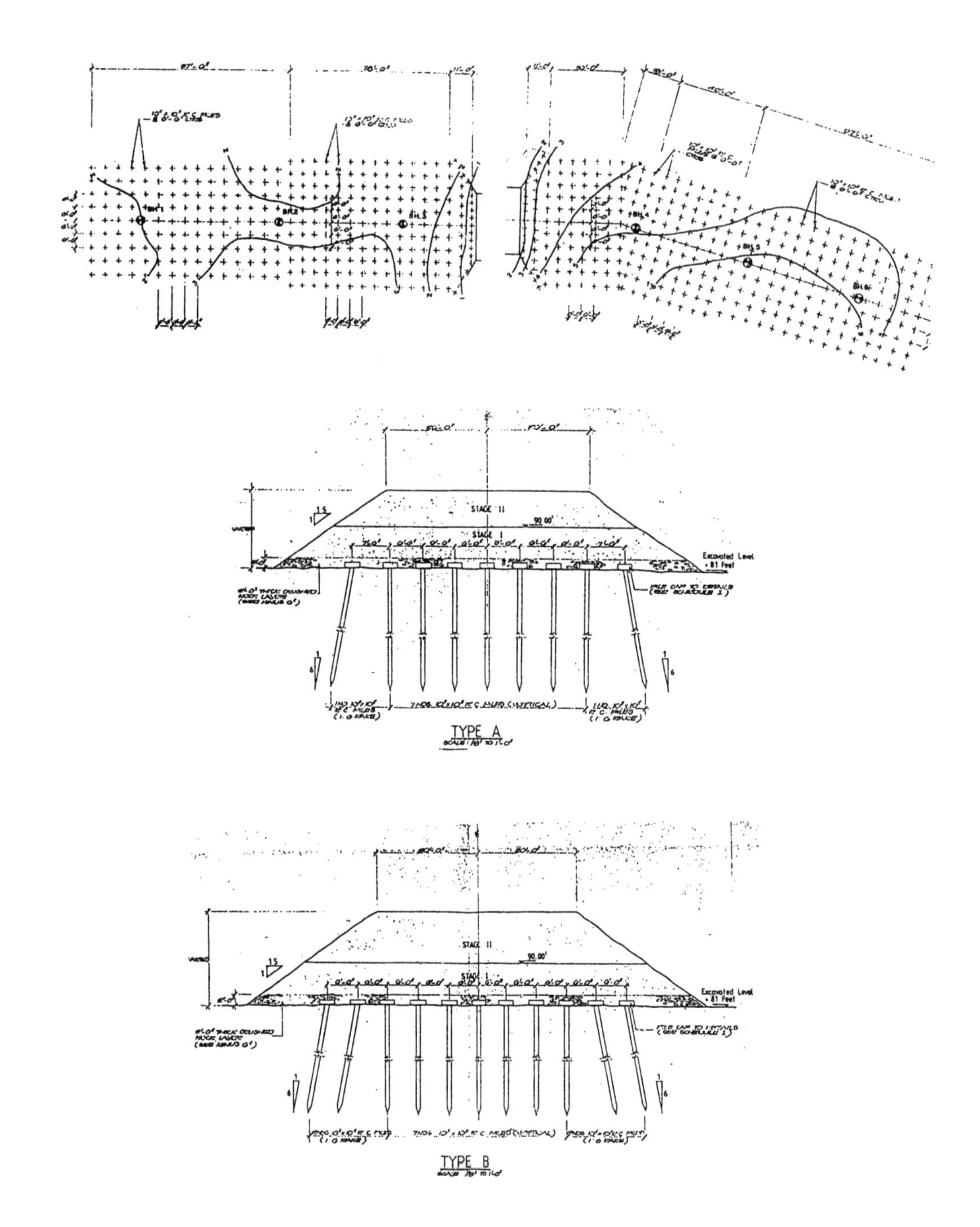

FIG. 10 GROUND MOVEMENTS

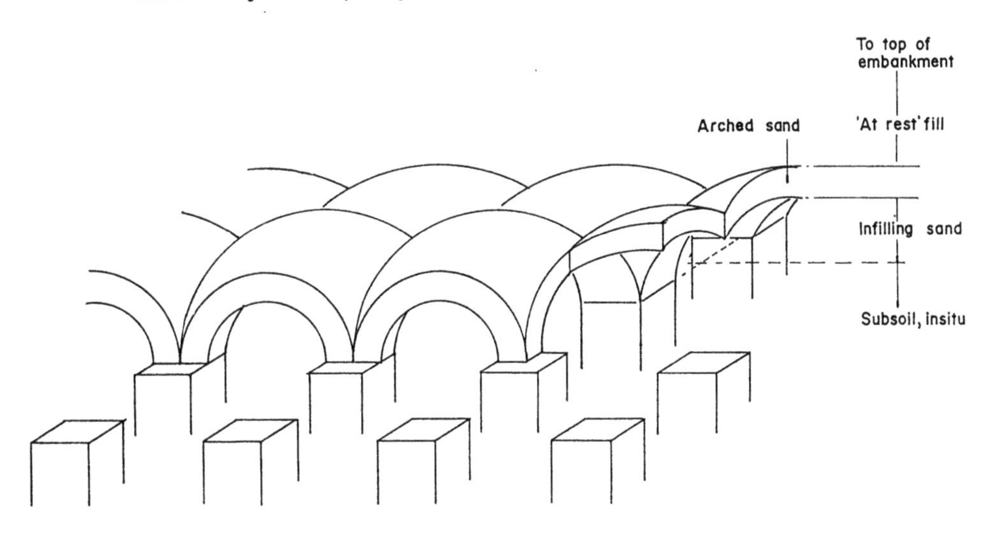

FIG. 11 VAULTS

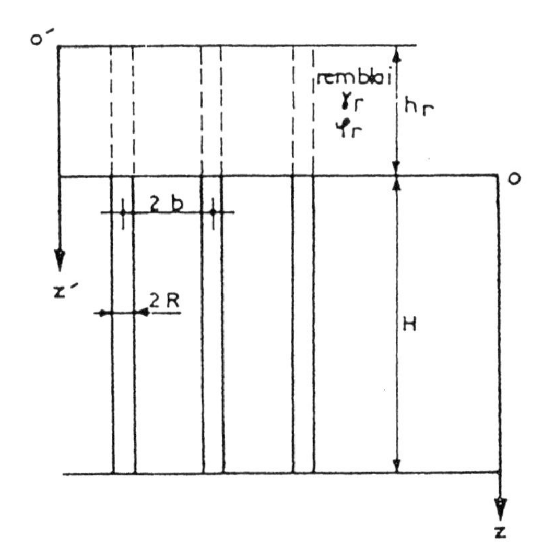

FIG. 12 GEOMETRY OF REINFORCED SOIL AND FILL

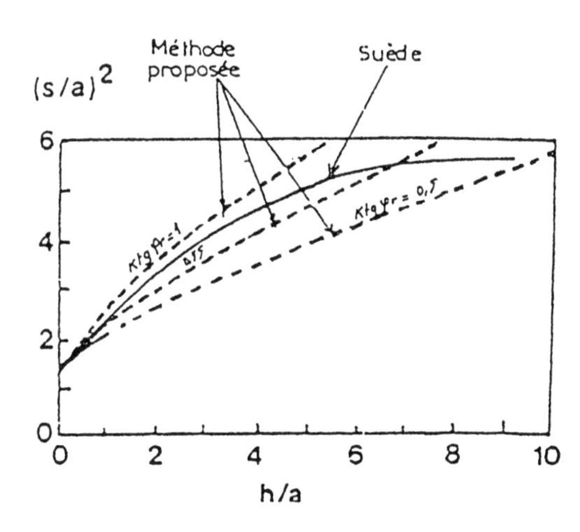

FIG. 13 GRAPH FOR PILES NETWORK DESIGN

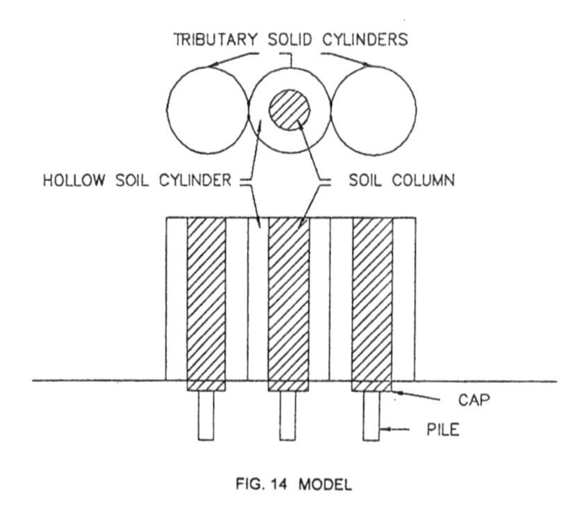

FIG. 14 MODEL

3.3.3.2. Stone Column Model Analyses

3.3.3.2.1. Basic Equations

The shaft resistance of the assumed column at slip can be deduced from the analyses of stone columns by Priebe (197?) on reinforcement of soil masses by stone columns. Priebe deduced the column reaction (X_f) at depth (h) for the friction case to be:-

$X_f = \gamma.A.f.(\exp(-h/f)-1) + \gamma.A.h$, where
γ = soil density
A = Area of tributary solid cylinder
$f = (A-A_c)/(2.\sqrt{(\pi.A_c)}.\tan\phi_s)$
A_c = Area of column

For the cohesion case the column reaction (X_c) at depth (h) is deduced to be:-

$X_c = 2.\sqrt{(\pi.A_c)}.c.h$, where
c = cohesion

The residual pressure (p_s) at the base of the cylinder at depth (h) for the friction case is:-

$p_s = \gamma.A.f.(1-\exp(-h/f))/(A-A_c)$

The pressure at the base of the cylinder of the embankment for the cohesion case is zero when there is no slip between the cylinder and the column.

3.3.3.2.2. Design

3.3.3.2.2.1. Methodology

A design method is proposed based on the analyses of the stone column model.

At slip between the soil column and the hollow soil cylinder, the total shaft resistance of the column is

FIG. 15 : PILE SUPPORTED EMBANKMENT 15m HIGH

equal to the column base reaction.

It is assumed in the proposed design method that the total ultimate shaft resistance must be able to just support the hollow soil cylinder on its own.

The requirement leads to the necessary expressions to limit the spacing of the piles to avoid downward movement of the hollow soil cylinder.

To avoid movement for the friction case:-

$A/A_c < (h/f).(1/(1-\exp(-h/f))$

For the cohesion case:-

$A/A_c < ((2c/\gamma).\sqrt{(\pi/A_c)}) + 1$

Both undrained (cohesion case) and drained (friction case) conditions are to be considered. However, the drained case is of primary importance as any loss of support of the embankment soil between pile caps is restrained by the consolidation process of the subsurface soil.

It has also to be ascertained that the column reactions (X_f and X_c) do not exceed the column strengths.

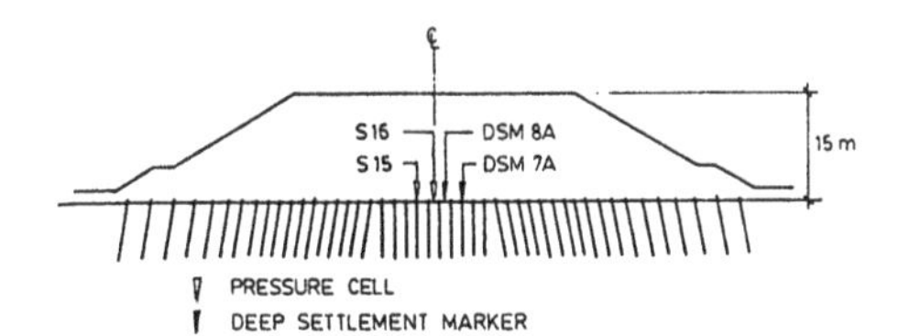

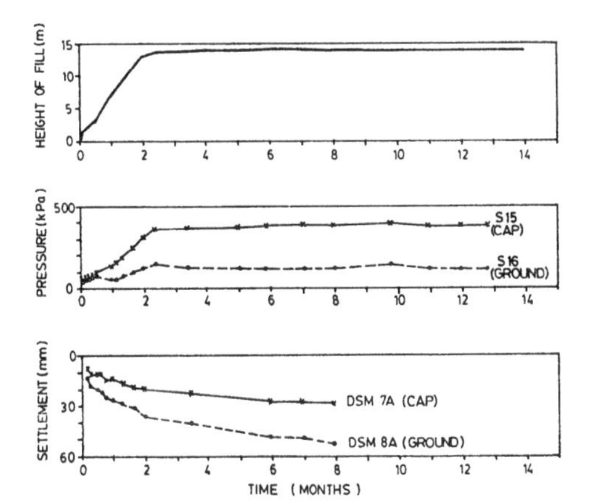

FIG. 16 PRESSURE AND SETTLEMENT AT PILE CAP LEVEL, EMBANKMENT AT 22 KM

3.3.3.2.2.2. Typical Solutions

Some actual cases calculated with the column model produced reasonable results. The cases included that described in Section 3.2.1. The results are compared with those deduced by the Swedish Method and are as shown in the table below.

The solutions are expressed in terms of two geometric factors the embedment ratio (h/d_c, where d_c is lateral dimension of pile cap) and spacing ratio s/d_c (where s is the centre to centre spacing of the caps)

CASES	d_c (m)	h (m)	h/d_c	s (m)	s/d_c	$(s/d_c)^2$
Case Ia						
As built	1.0	3.4	3.4	2.40	2.40	5.80
Swedish Method	1.0	3.4	3.4	2.07	2.07	4.30
Column Model	1.0 (1.2)	3.4	(2.83)	2.20	(1.83)	3.35
Case Ib						
As built	1.0	6	6.0	2.40	2.40	5.8
Swedish Method	1.0	6	6.0	2.32	2.32	5.4
Column Model	1.0 (1.2)	6	(5.0)	2.40	(2.0)	4.0
Case IIa						
As built	1.8	3.4	1.89	3.8	2.11	4.45
Swedish Method	1.8	3.4	1.89	3.3	1.84	3.40
Column Model	1.8	3.4	1.89	3.0	1.67	2.79
Case IIb						
As built	1.8	6	3.33	3.4	1.89	3.57
Swedish Method	1.8	6	3.33	3.7	2.05	4.20
Column Model	1.8	6	3.33	3.3	1.83	3.35

3.3.3.2.2.3. Interpretation of Results

The current design method is the Swedish Method which explains why in the above table, the as-built parameters are close to that of the Method.

When embankment ratio is greater than approximately 3.4, the column diameter in the column model is assumed to be enhanced by 20 per cent. This approach is consistent with the observations of the anchor pull-out tests and fits in with observations from field performance. The proposed method results in over conservative designs for high embedment ratios if the diameter enhancement factor is not applied.

In the case of the larger pile caps, the column model produced smaller values of cap spacing than suggested by the Swedish method especially for the lower embedment ratio cases. This result is of significance because the cap spacing has probably contributed towards the `mushroom' formation of

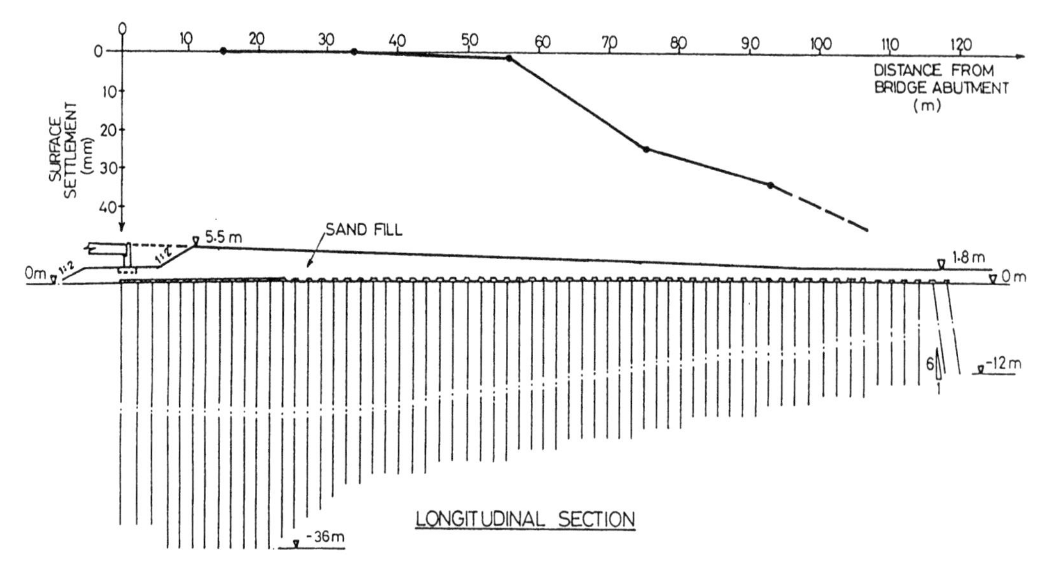

FIG. 17 SURFACE SETTLEMENT, SANTUBONG EMBANKMENT

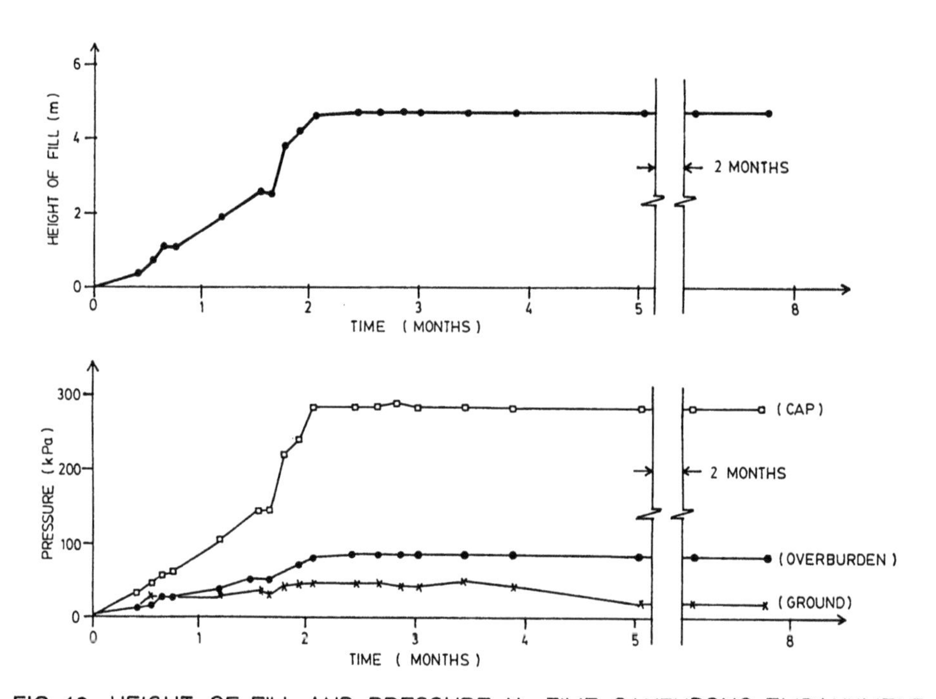

FIG. 18 HEIGHT OF FILL AND PRESSURE Vs TIME, SANTUBONG EMBANKMENT

the Penang Bridge Approach problem. It is a contradiction in terms that a lower embankment height requires a closer pile cap spacing. However, in practice, this result can be compensated by maintaining the same sizes of pile caps for the complete embankment, but the supporting piles can be made smaller in the low embankment areas when compared with that for the high embankment areas.

3.3.3.3. Further Development of Model

The advantage of the column model is its relative simplicity in concept and in analysis in contrast to the complexity of Hewlett & Randolph's vault model.

It is an improvement on the currently applied Swedish method in that it has possibilities of analysis in terms of geometry and materials and can overcome the problems arising from shallow embedment and the use of larger piles.

A further development will be to determine a relationship between embedment ratio and column diameter enhancement factor. The studies on pull-out of anchor plates (e.g. Rowe & Davis, 1982) will be applicable.

4. CASE HISTORIES

4.1 North-South Expressway

A total of 11 pile supported embankments of various heights were built in the Seremban-Air Hitam section of the Malaysian North-South Toll Expressway. The highest embankment was located at Chainage 22km with a height of 15m above original ground level. Fig. 15 shows a general view of this massive embankment that will be discussed below.

The embankment was built across a narrow valley on weak alluvial deposits, mainly soft clay, with a maximum depth of 14m. The fill materials used were residual soils. The piles were driven through the alluvium to bear on weathered shale and sandstone. These piles, which were made of high strength precast concrete, had a triangular cross-section with a side dimension of 300mm. A cross-section of the embankment showing the arrangement of piles is given in Fig. 16. The pile spacings were generally limited to 1.44m with a corresponding cap size of 0.6m x 0.6m. The caps were tied in the transverse direction except near the end connected to the bridge where 2-way ties were provided.

Fig. 16 shows typical data for the settlement of pile caps and for the soft ground between caps. During the construction period, the caps and the ground settled up to approximately 30mm and 50mm, respectively. The post construction settlements were small indicating satisfactory performance of the embankment. Fig. 16 also shows the extent of arching as indicated by the difference between the total stress measured on the pile cap and that on the adjacent original ground between caps.

4.2 Santubong Bridge

A pile supported embankment was built using sand fill on either side of the new high-level Santubong Bridge in Sarawak to serve as the approach from the surrounding low-lying swamp land. The subsurface profile consists of a top stratum of very soft organic silty clay up to a maximum depth of 22m, followed by a loose to medium dense sand 5m to 15m thick and this is underlain by weathered mudstone and shale.

A longitudinal section of the embankment together with the arrangement of piles is shown in Fig. 17. Except in the transition zone, the piles were driven through the weak soils to bear on the weathered mudstone and shale. Two sizes of triangular precast concrete piles were used with side dimensions of 186mm and 235mm. Pile spacing ranged from 1.8m to 2.3m whereas the corresponding cap size varied from 0.8m x 0.8m to 1.2m x 1.2m. The pile caps were tied together transversely in the transition zone, whereas in the high zone near the bridge two-way ties were provided.

A post-construction settlement profile at the road surface level 6 months after completion of the embankment is shown in Fig. 17. It can be seen that the settlement is negligible at the high end where the embankment meets the bridge. On the other hand, in the transition zone the settlement is increasing gradually towards the low end so as to be compatible with the settlement of the conventional embankment. Fig. 18 shows the variation of total stress measured on a typical pile cap and on the adjacent original ground between caps both during construction and thereafter. The results show that the total stress on the cap was many times that on the ground, indicating that arching had taken place.

5. FINDINGS AND CONCLUSIONS

5.1. Swedish Method - Applicable for Piles 300mm and below

The Swedish method of design was applied successfully when smaller piles of less than 300 mm sizes were used, provided some important design details were adhered to.

The details pertain to provisions for lateral earth loads in the way of lateral ties for the piles and raking piles at the embankment toe area.

5.2. Full-Scale Performance

Case histories on the performance of the embankment and piles in full scale problems are presented and exposed many of the problems that are to be expected.

5.3. Development of Stone Column Analogy

Design methods based on assumed mechnisms and various observed and theoritical mechanisms are discussed.

A stone column analogy model has been proposed for the design of pile supported embankments that can be applied to large supporting piles. It has the advantage over the Swedish Method in that it has analytical possibilities. The solutions obtained from the proposed method are consistent with observations from field performance of full-scale problems as well as laboratory models.

REFERENCES

Baskaran, S. & Sundaraj, K. . (1993). "Private communications on a pile supported embankment". Internal Consulting Engineers Report.

Bergdahl, U., Lingfors, R. & Nordstrand, P. (1979) "The Mechanics of Piled Embankment - a study including model tests". (Translated by Sehested, K.G., Swedish Geotechnical Institute).

Combarieu, O. (1988). "Amelioration des sol par inclusions rigides verticales-application a l'edification de remblais sur sols mediocres". Revue Franc. Geotech. no 44, pp. 55-79.

Hewlett, W.J. & Randolph, M.F. (1988). "Analysis of piled embankments". Ground Engineering. April 1988. pp 12-18.

Malaysian Highway Authority. (1989). "Trial embankments on Malaysian Marine Clays". Vol. 1. pp 13-1 to 22.

Priebe, H.J. (197?) "Design criteria for ground improvement by stone columns".

Rowe, R.K. & Davis, E.H. (1982) "The behaviour of anchor plates in clay". Geotechnique 32, No.1, pp. 9-23.

Swedish State Road Department. (1974). "Embankment supported on piles". Technical Division, Geotechnical Office.

Ting, W.H., Toh, C.T. & Chan, S.F. (1983) "Pile supported fill". Proc. Int. Symp. on Geot. Problems, Bangkok, Dec. 1983. A.A. Balkema. pp. 95-100.

Developments in Geotechnical Engineering, Balasubramaniam et al. (eds) © *1994 Balkema, Rotterdam, ISBN 90 5410 522 4*

Innovation in buried infrastructure

D.H. Shields
The University of Manitoba, Winnipeg, Man., Canada

ABSTRACT: Excavation to install and repair buried municipal services causes inconvenience and economic loss. Innovative alternatives are proposed which include bundling communications and power cables in common utility ducts and designing networks to benefit from no-dig technology.

INTRODUCTION

Geotechnical engineers will have to take the lead in revolutionizing the way in which services are routed underground in cities. These installations form a three-dimensional hodge podge of separate, intermingled cables and pipes. We inevitably run into these 'intertwined snakes' or 'bowls of spaghetti' when we drill test holes and, especially, when we excavate. Gas, water, sewer, electricity, and cable TV are just some of the services that are buried in the ground. Other buried infrastructure include traffic signal controls, telephone lines, pneumatic tubes, central heat and cooling pipes, and computer interconnections. Also buried, but less likely to be misplaced because they are so much bigger, are subway lines, train and automobile tunnels, and encapsulated rivers and streams.

Traditionally, geotechnical engineers have been little involved in the design and installation of small-diameter pipes and conduits. Our involvement has mainly been restricted to installations that are large and require appreciable depths of open cut, or require tunnelling. When we *are involved*, we limit that involvement to advice on excavation conditions (rock or soil? wet or dry?) and earth pressures on temporary or permanent structures. Sometimes we offer specialized advice, addressing, for example, whether there will be problems of shrinkage or swelling of the soil, whether the route is in permafrost or subject to seasonal frost, and whether the ground is corrosive. The major decisions of location, route, and depth of any buried service are left to others. The result of this "leaving to others" is anarchy, and an underground Gordian knot.

It seems that geotechnical engineers must take the initiative to bring reason and sanity to pipeline and conduit installations - particularly in urban settings - if for no other motive than that no one else is taking the lead. We have already assumed the initiative in other areas outside our traditional bread and butter, soil mechanics and foundation engineering. Many of us today gain our livelihood in the environmental field, others in mining, and still others in the design and maintenance of pavements.

HOW WE GOT WHERE WE ARE

In the case of most cities, suburbs and villages, public and private services were provided sequentially, on a hit-or-miss basis, and in a completely uncoordinated way. The town or city council decided that a piped water supply was in order. The pipes were installed. The increase in water supply overloaded cesspools and septic fields. Then it came time to replace privies or septic systems with sewerage systems, and a second set of pipes was put down, at least to some of the houses and businesses. Then piped gas was offered, and a third set of pipes needed. And so it progressed over the years. Most often, the gas company was separate from the electric company, which in turn was separate from the telephone company. Hence, there was little or no coordination in the routing of the pipes or of installation schedules. The same lack of coordination existed even within a single municipal government; the water department merrily went its own way, ignoring, because of tradition or inter-service rivalry, the waste (i.e. sewer) department.

It is also true that in new towns or subdivisions, underground installations end up in chaos, *even when the opportunity exists to bring order to disorder*. The local Department of Health rules that there must be a two meter spacing between a water line and a sewer line. The telephone company does not want to cooperate with the TV cable company because they are competitors. Storm sewer lines must operate by gravity (or so it is said), and their elevation is dictated by connections to older works. Finally, the wrong materials are chosen (ductile iron which soon corrodes is used for watermains), the systems must continually be modernized (fibre optic cables replace copper wires), or new installations made after a few years (sensors for traffic flows).

THE EFFECT ON ROADS

The existence of buried infrastructure affects roads and streets in many ways. Shallow pipes and cables compromise the thickness of the road structure itself. Catchbasins and subgrade drainage systems are planned with an eye to the locations of buried infrastructure, resulting in less than ideal design. Contractors do not have a free hand to shape and slope the subgrade in order to direct seepage water to drains. And once shaped, who is to say that the subgrade will remain that way when it is dug up and put back by a pipeline or cable repair crew? Even the choice of surfacing material for the roadway should--but all too often does not--take into account the fact that it will have to be penetrated on a regular basis in order to maintain underlying services or install new ones. (An argument for going back to cobble stones is that they can be operated like a zipper; opened up or closed as the need arises.)

The secondary effects of buried services on existing roads and streets, even when newly laid, are also of great concern. All too often we see evidence of subsidence where trenches have been backfilled poorly - either with contaminated or compressible material, or using little or no compaction. Or the surface may be stripped away, accommodating yet another excavation to expose a buried service. The subsequent surface patch is often made with little care or attention, resulting in poor ride quality and a much shorter service life for the road structure. (If it is not the roadway that is being dug up, it is the sidewalk, with an equal amount of inconvenience and poor workmanship.)

DO ALTERNATIVES EXIST?

Yes. The famed, century-old sewers of Paris provide ready access to many parts of the city. Communications cables and watermains have been affixed to the roof of the Paris tunnels, and are accessible at any time for upgrading. In other parts of the world, service tunnels interconnect buildings on university campuses and in research parks. But by and large, tunnels, with the possible exception of storm sewers, represent a negligible part of the buried infrastructure of cities.

Modern 'no-dig' methods provide an alternative to cut and cover. No-dig is now applied to buried infrastructure placement and repair. It is possible, for example, to put a slippery coating on the inside of badly corroded and pitted pipes; it is even possible to expand and, if necessary, destroy the existing pipes from the inside so that there is no subsequent loss in diameter after the new coating is in place. Problems arise, however, when there are side connections to the pipes - the new coating will close off the connection. It is difficult to go around small radius corners, and difficult, too, to inspect the finished product.

It is also possible to directionally drill underground horizontal holes into which new pipes and cables can be threaded. This method results in little or no disturbance to other buried infrastructure or to surface facilities such as roads. There are difficulties, though, in piloting the drill through and around nests of existing buried services. And there are problems of rights of way (land ownership) and in mapping the final locations.

ECONOMIC IMPLICATIONS

At first glance, one would think there would be economic advantages to cooperating in the installation and maintenance of buried services. This does not appear to be the case, however, since there is often little cooperation between the owners of the services. An obvious form of cooperation would be to schedule maintenance or upgrading so that two or three services are worked on at one time. One wonders whether the lack of cooperation is due simply to precedent and resistance to change (particularly as it applies to municipal or city-run departments) - or, as we will discuss below, due to a lack of appreciation of the true economic costs of go-it-alone activities.

The economic implications of anarchy in buried pipe and cable locations are considerable. The need to contact six, eight or ten separate departments or firms simply to drill a soil test hole is a small example. Excavations of any size are made much more complicated and costly when it becomes necessary to work through a maze of pipes and cables, provide temporary or permanent support for them, or relocate and reroute existing infrastructure. And if, by error or accident, a service is damaged, the result can be gas seeping into adjacent buildings and exploding, interruption of a million-dollar-a-minute television broadcast, or the flooding of basements and subway lines.

The economic implications of cutting a roadway in order to access buried services are now the subject of heated debate. Not only is there the cost of the excavation, backfill and resurfacing to consider, but there is the cost of the lower life expectancy of the reinstated road surface. More important, though, is the cost of interfering with traffic while the excavation remains open and the roadway barricaded. There is a direct cost to truck transport that is subject to rerouting and delay. There is a direct cost to hotels and shops that front onto a road that is cut, making these business inaccessible. Similar losses occur when businesses are bypassed because traffic is diverted or because on-street parking is prohibited. Attempts are now being made in some countries, England for example, to calculate the value of these losses *and to charge the cable company, the telephone company or the water department for these losses*. The value can be in the hundreds of thousands, if not the millions, of dollars. (An ancillary problem is the distribution of the funds to compensate the 'victims' equitably.)

THE BUNDLING OF SERVICES, AN ENTREPRENEURIAL OPPORTUNITY

Modern buildings are constructed with service ducts and shafts into which all municipal and private services are bundled. Why not a common underground duct or tunnel to house existing and future services under city streets (Figure 1)? We have the example of the Paris sewers, and, in the north of Canada, the example of utilidors, which are service boxes or ducts that run along the surface of the ground and which often serve a secondary purpose as sidewalks. Utilidors are placed on the surface because the ground is rock or is permanently frozen. Buried utiliducts need be only a meter or less in diameter in most instances, but they could be greatly expanded to include large storm sewers and, perhaps, provide people and goods movement as well.

Do utiliducts represent an opportunity for private enterprise of the kind adopted for railroads or canals? Could not a separate entity - a firm or a government department - install the buried duct and rent out space to whoever needs it (telephone company, gas company and so on)? Or, perhaps, the entity could install and own the fibre optic cable or pipe through which information or gas or water might flow? A starting point might be the realization that the ground under our roads, while public or common property, is owned by the community. Access to this ground must be controlled in future by the owners. The owners could, if they wished, authorize a single entity to manage this resource for them. We now sell or lease air rights above public property; why not sell or lease ground rights? Once an economic value is placed on the underground, and access to this ground restricted to those who purchase the right of access, we will have gone a long way towards encouraging cooperation through the realization that use (and abuse) of this limited resource is no longer free.

In order to facilitate the bundling of services in a utiliduct, many jurisdictions will have to change those existing regulations which restrict or forbid bundling. One such regulation is that enforced by some health departments, referred to earlier, which places a minimum spacing (often of two meters or more) between potable water line and sewer mains. The restriction is intended to reduce possible cross-contamination of the water by sewerage. Presumably, if the pipes were not buried separately in the ground but instead were exposed in a service duct, this spacing requirement need not apply; a leak in a sewer line would be obvious, and the line could be repaired. (There is no spacing requirement once the pipes enter a building and become exposed.) Another example of an outdated regulation is that imposed in cities where the ground freezes; the watermains must be located at depth, below the frost line. If watermains were bundled with sewer lines and electricity lines (both of which are sources of heat) in a (insulated?) duct, the possibility of freezing becomes remote (this is the experience with the surface utilidors in the Canadian arctic in temperatures as low as -40°C.)

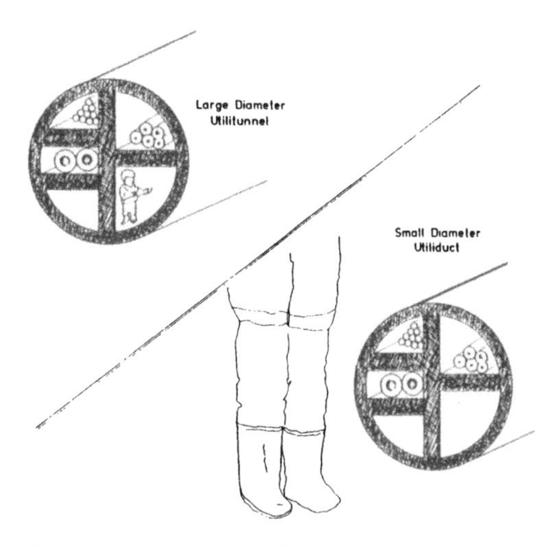

Fig.1: Bundled municipal services

It goes without saying, that the utiliducts in which the services are bundled must make provision for easy access by installers and maintenance workers. Person holes could be provided as necessary. Access would, of course, be on a restricted basis in order to prevent entry by unauthorized persons and also terrorists, and to ensure that the managers of the facility maintain control. A security and safety system equipped with alarms would ensure response to unauthorized entry, the escape of(toxic or explosive) gases or liquids, and temperatures that are dangerously high or low.

One of the original topics of the EUREKA research program of the European Common Market was the design of precast concrete units which could be buried in the ground and into which municipal services could be bundled. A number of technical problems must be overcome, including for example, the prevention of electrical interference which might disturb communications signals. But research needs to be broader than the design of utiliducts; there must be research into the jurisdictional and economic aspects of buried infrastructure bundling. Geotechnical engineers need not limit themselves to the soil mechanics aspects of utiliduct research, but should, as well, involve themselves in studies of the best way to own and manage this future resource.

WHY UNDER A ROAD? AND WHY IN THE GROUND?

A fundamental question is "Why bury services at all?" In the centre of most cities today there are parallel rows of underground basements that, if interconnected, would provide ready housing for pipes and cables. In many residential areas (at least in Canada), houses have basements. The basements are not contiguous, but could be joined readily by short sections of duct work (Figure 2). Infrastructure sewer, water, telephone and so on, housed in this way, would leave the roadway free. The road structure could then be designed without constraints, and, once built, would remain undisturbed by excavation. Commercial establishments and transportation companies would not be subjected to economic losses because of road closures. Again, questions of right-of-way, security, and access would have to be resolved. But these questions arise with buried services today (personhole covers welded shut along the routes of visiting dignitaries, for instance).

If it is necessary or desirable to bury services, why bury them under a road if there are boulevards or lawns which parallel the road? If the services were located along one side of the road only, the dwellings or buildings on the other side of the road would require cross-road connections in order to be serviced. Thought should be given in this case to having two utiliducts running under the boulevards, one on each side of the road (Figure 3).

When it is absolutely necessary to use the road right-of-way for buried services, the road surface itself should be included in the 'bundle'. It is possible to envision a utiliduct that runs along the road and, at the same time, stretches from one side to the other. Precast, hollow-core, concrete slabs of the kind used for floors in buildings might be one answer. The upper surface of the slabs could be the road surface. Even if the utiliduct were buried deeper and so could not be used to provide the road surface, the column of soil and road materials above the utiliduct might be managed by the body that manages the utiliduct facility itself.

THINKING EVEN BROADER

It is time we turned our attention, creatively, towards resolving the problems posed by other infrastructure services. One such question revolves around the disposal of solid waste. It is Middle Age thinking to tolerate garbage trucks and crews trundling through cities and towns, picking up solid wastes. We learned to dispose of human excrement in a fashion other than physical collection. Today, kitchen garburators grind up organic wastes and flush them down the drain. Once upon a time we disposed of paper wastes by burning them in the individual furnaces which were used to heat our homes or to heat water. We are no longer allowed to do this for environmental protection reasons or because we converted our solid fuel furnaces to oil or gas.

It should be possible to reduce the generation of solid waste, and hence reduce the need for waste collection. For example, liquid detergents, or soft drinks, or milk might be piped directly to each home, rather than each household buying these products in cans or bottles. Electronic data transmission and storage make it possible now to do away with paper--whether newspapers, letters or files. (Electronic forms of communication might replace another infrastructure anachronism: mail delivered door to door by persons.)

Fig. 2: Basement – housed utiliduct

However, at least for the foreseeable future, there will be *some* solid waste to dispose of. The way in which we collect, transport and finally get rid of this waste requires imagination. We no longer have coal delivery wagons traveling through towns and cities; energy supplies now arrive in the form of piped oil or gas, or as electricity by means of cable. Can we transform the solid waste disposal problem in a similarly advanced manner?

ACTION IN THE SHORT TERM

Geotechnical engineers can insist and ensure that service trenches be backfilled properly. The Canadian geotechnical community has sponsored a handbook of good practice for backfill as one of its activities of the past few years. Private industry came up with 'no-shrink backfill'--in essence, low strength concrete. Geotechnical firms offer inspection services, not only of the backfill but also of the quality of the road surface patch. Nevertheless, one key factor remains: the need for insistence on good workmanship by the owners of the ground--the municipalities. Geotechnical engineers can lead in making municipalities aware that they are the stewards of the common ground in our cities and villages, and that they are expected to manage this valuable resource. One effective argument might be to point out that the municipalities could gain revenue by selling temporary permits to service departments and firms that want to dig trenches through roads. In the application for a permit, a service organization would have to undertake to reinstate the ground and road in a fashion acceptable to the municipal engineer. Another positive argument is the acknowledgement of the economic cost of shorter pavement life and the political cost of potholed roads, both the result of lack of controls on digging out and replacing sections of the roads to access buried services.

Geotechnical engineers can use their expertise with materials to help limit the need to tear up a road. Poor research and little or no advance testing led to the premature adoption of ductile iron for watermains in many cities (in my city, Winnipeg, service lives as short as seven years were the result).

The use of water service pipes as the grounding medium for electrical services has led directly to premature corrosion of the metal pipes. One way around this problem is to use plastic pipe. Not only is plastic not an electrical conductor, but it is not subject to corrosion. Ground movements caused by frost heave and moisture changes also lead to pipe failure. Radical redesign of the appurtenances to piping (the fire hydrant standpipe that is jacked upwards by frost, for example) and specifying a different piping material (slippery plastic that frozen soil will not adhere to?) could lengthen service lives and reduce the need for excavation to replace broken mains. The National Research Council of Canada and the City of Calgary are experimenting with new, insulating backfill materials that reduce the requirement of depth of burial of watermains to counter the effects of frost.

Geotechnical engineers can enlarge the scope of their activities in no-dig underground pipe and cable installations. In addition to helping in the design of drill heads and drilling muds, thought should be given to layout of the particular pipeline or cable so that it better accommodates no-dig drilling. For example, instead of installing individual service lines running at right angles off a main supply line, it would be advantageous to radiate the service lines from a central point (Figure 4). Directional drilling in a number of directions from the central point could then cover six, eight or more needs, including the installation of the main to the next central point. It would be a simple matter to monitor the effluent from individual establishments, for compliance with environmental laws, if sewer lines were to be installed in this way; moreover, the central point could be used as a metering station for products such as gas and water.

Geotechnical engineers might also develop the non-destructive tests that are required to map the existence and location of buried services. A combination of ground penetrating radar, geophysics and tomography could be used to prepare three-dimensional computer graphics representations of the buried infrastructure at any location. The same graphics package might be extended to illustrate how the services could be bundled into a commonutiliduct with the least amount of disruption and cost.

Other areas in which geotechnical engineers can show leadership include the development of standards for buried infrastructure installations. One example of the need for a standard is the bedding of pipes. The present practice of bedding pipes in sand and gravel has proven to be counter-productive for sewer lines. The porous sand and gravel virtually ensures that water from the ground or from leaking neighbouring water pipes finds its way to weaknesses in the sewers. Infiltration water often accounts for a large portion the waste water that must be transported and treated by a sewer system. Standards are also required to ensure that buried infrastructure withstand earthquakes. All too often watermains and gas lines break as a result of earthquakes, leading to disastrous fires that cannot be contained because water is unavailable.

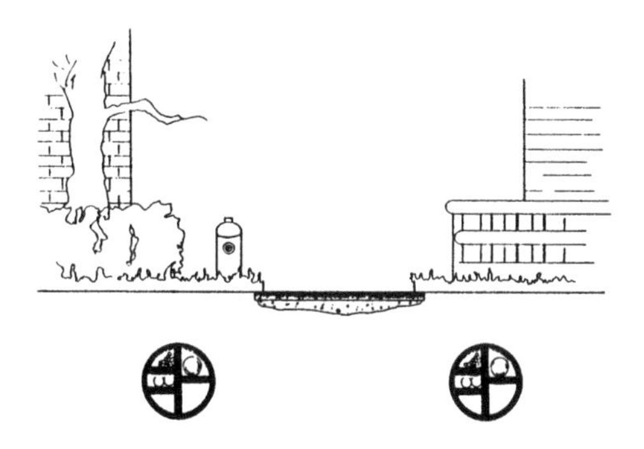

Fig. 3: Parallel services leaving roadway undisturbed

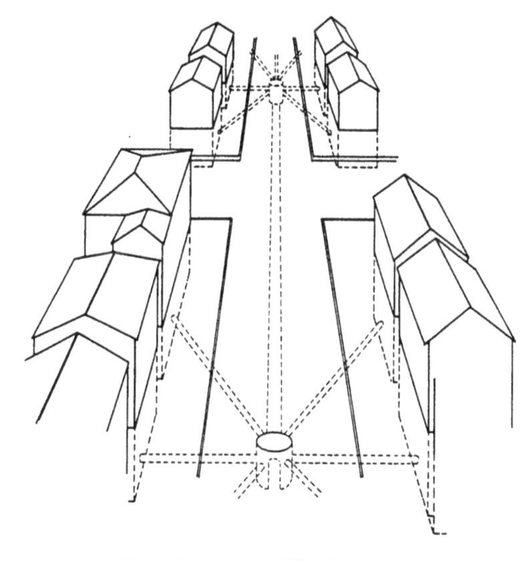

Fig. 4: Central access facilitating no-dig installation and repair

Another area open to geotechnical engineers is the application of geosynthetics to municipal infrastructure. The use of geomembrane wrappings might be one way to reduce water penetration into sewers, to reduce frost-jacking of personholes and fire hydrant standpipes, and to prevent corrosion. Geotextiles would help ensure that road base course gravels are not contaminated by poorly selected trench backfill materials. Geotextiles might also prevent the creation of craters and washouts when a watermain bursts.

Geotechnical engineers can also ameliorate the disruption caused by burying infrastructure by adopting the concept of narrow trenching. Much of the width of service trenches is taken up by shoring. If shoring were eliminated, there would be a dramatic reduction in the volume of material to be excavated and backfilled, and hence in the rate at which the work can proceed. One idea might be to apply slurry trench methods. A difficulty with the idea is that today's slurries are opaque, making it impossible to see what is going on in the trench.

The discovery of an inexpensive, non-toxic, clear, thixotropic gel for municipal trench work might revolutionize this industry. Work in the gel-filled trench might be carried out remotely by robotic devices or directly by scuba-diver personnel.

SUMMARY

The present chaos in underground infrastructure must be replaced by logic and order. Municipal councils need to be encouraged to exercise their right to manage the public right-of-way of our city streets and the subsoil that lies beneath them. One way might be for the municipal councils to devolve this right to a private or public enterprise, such as Infrastructure Enterprises Incorporated, which would, in turn, provide the services of a common carrier, much in the way that railroads, canals and pipelines are common carriers today. This hypothetical Infrastructure Enterprises Incorporated would, as a minimum, construct utiliducts. It could either rent out space in the utiliducts to water, sewer, gas and all other service departments or companies which would install and maintain their own pipes and cables, or Infrastructure Enterprises Incorporated could install cables and pipes itself and transport services for a fee. Infrastructure Enterprises Incorporated could also be required to provide and maintain the road surface, and provide such things as underground parking where warranted.

Geotechnical engineers have a vested interest in bringing order to underground infrastructure. We are the ones who drill through nests of pipes and cables, and it is we who supervise excavations through and adjacent to these same services. Our profession is all too aware of the liability and inefficiencies that these operations entail. Because of our interest, we should take the lead in convincing municipal councils to change their ways.

In the meantime, there are many opportunities for improving existing conditions using available geotechnical knowledge. These options range from improving no-dig practice to applying slurry trench technology. The problems of urban infrastructure are all around us; now is the time to think of solutions.

ACKNOWLEDGEMENTS

Mr. George Seadon, Director of the Institute for Research in Construction of the National Research Council of Canada, is responsible for encouraging the development of the ideas which make up this article. Special thanks are also due to LaVerne Palmer of the Institute, and to members of his Section, for the opportunity to debate and discuss the merits of the various alternatives which will, hopefully, revolutionize the design and construction of buried infrastructure.

Developments in Geotechnical Engineering, Balasubramaniam et al. (eds) © 1994 Balkema, Rotterdam, ISBN 90 5410 522 4

Warragamba Dam – A case study of environmental implications of the building of a hydraulic structure

Hans Bandler
Macquarie University, Sydney, Ryde, N.S.W., Australia

ABSTRACT: In 1960, a major dam for water supply to the City of Sydney in Australia was completed in the Warragamba River Valley. This is examined as a case study. Geographical features, the history of the area, including early proposals of use of the river are examined.

The question of submergence of the valley, social consequences, tree removal in the valley, anthropological aspects, recreational use of the area around the dam crest, protection of water quality, catchment management are important parts of environmental implications of the dam.

All these aspects have wide general application, they are of concern, must be carefully analysed and given consideration in concepts of hydraulic structures.

1 INTRODUCTION

The Hawkesbury River basin is very large area, covering 21,720 square kilometres. A major tributary of the Hawkesbury is the Warragamba River a branch of the Nepean River. A major dam was constructed across the Warragamba for water supply to the city of Sydney, in Australia. It was completed in 1960. Some major environmental implications resulting from the creation of this hydraulic structure are worthy of examination. They have wide general application and can be applicable to other such structures which need consideration in the initial concepts.

2 GEOGRAPHICAL FEATURES - USE OF THE RIVER

Warragamba applies to only a short length of river about 22 km to its confluence with the Nepean. The river flows through a 90 to 120 m deep sandstone gorge, which is about 300 m wide near the Nepean, widening to about twice that width upstream at the junction of its two main tributaries the Cox River and the Wollondilly River.

3 WARRAGAMBA DEVELOPMENT

The average annual rainfall over the catchment area is 825 mm. There were many other obvious details in favour of selection of the site. Some eight years were devoted to detailed investigations to assure the correct technical decisions were made before commencement of construction of the dam. The investigation continued during the period of ten years of construction. Technical analyses have continued ever since. Excavation commenced in 1950, first concrete was placed in 1953.

Warragamba is a straight concrete gravity wall of significant magnitude. At the time of its completion it was the second biggest concrete dam in the southern hemisphere.

In the investigations the geology of the area, the question of silting, drought flow and flooding, provision and transport of the materials for dam construction and many other technical details were scrutinised and determined for the final design.

4 WARRAGAMBA DAM

4.1 Principal Dimensions and Data

TABLE 1.

Storage capacity	2,057,000 Ml
Length of wall	350 m
Length of Spillway over crest	95 m
Concrete in wall	1,200,000 m^3
Maximum height of concrete	137 m
Greatest depth of water	104 m
Width of wall at base	104 m
Catchment area	9,013 km^2
Annual average rainfall	825 mm
Annual average evaporation	1,270 mm
Lake area at full supply level	75 km^2
Maximum length of lake	52 km
Length of foreshores	30x km

4.2 Environmental Implications to be Examined

Submergence of the Valley
Tree Removal in the Valley
Anthropological Aspects of Submergence
Recreational Use of the area around the dam crest
Protection of water quality
Catchment area - Pollution within the area
Effects on the valley below
Summary and Conclusions

4.2.1 Submergence of the Valley

Before the Dam was built, Burragorang Valley on the Wollondilly River, the widest valley of the tributaries, was a prosperous area. Not only did it have extensive orchard and dairy farming but it had a thriving tourist industry, it was truly popular.

When it became known that the construction of the Dam would cause flooding of the valley, the local inhabitants formed a Defence League in 1941, which asked for a Royal Commission to examine the need for Warragamba Dam. That request was never granted. Land acquisition proceeded fairly slowly with strict payment to the owners at current valuation price plus 10% displacement money.

Severe flooding occurred in July 1952 and again in February 1956 which induced farmer to sell more readily. Land acquisition was completed just before submergence occurred. A total of ten years elapsed from commencement of land acquisition to the time of submergence. With hindsight we can say, that this showed considerable foresight by the Water Board at the time.

By comparison, during about the same period, there were several occasions in U.S.A. were populations of the valleys, which were to be submerged by a man-made lake, to be created by the building of a dam on a major river, were given little time for evacuation and no assistance to resettle.

The question of removing and resettlement of people from valleys to be submerged by the building of dams, is to this day, a serious problem in many countries throughout the world, occasionally leading to grave upheaval. There are many cases of displacement of populations from fertile valleys.

The most recent disturbing example is the agreement to the proposal to build a huge dam across the Yangtze River in China, drowning the scenically beautiful "Three Gorges". At least one million people living in villages and on farms in the area to be submerged will be displaced.

4.2.2 Tree Removal in the Valley

An important work that would have been done before the dam was constructed was the removal of all trees from the area to be submerged. This is necessary because:

1) The drowning, rotting trees would affect the quality of the water

2) The logs and dead branches would be likely to drift downstream to the dam, where they might cause damage to the outlet works or obstruct flood gates

3) Dead and dying trees, logs and branches submerged along the shore, can present a serious hazard to lake access in case of fire or accidents and other emergency situations, for example in air disasters and safety investigations.

Up to 500 men were employed to cut down the timber on the slopes of the valley. This was not always easy because the slopes of the valley are in most parts very steep. A small sawmill was installed at Warragamba to deal with as much as possible of the timber which could be cut. It was used for the building of the numerous work sheds and offices, as well as for the barracks and houses at Warragamba Township which accommodated the engineers and many workers at the dam.

At its peak, about 3,500 people were housed at Warragamba Township. The timber was also available for scaffolding and general construction work at the Dam as well as for the construction of a naval dock yard in Sydney.

4.2.3 Anthropological Aspects of Submergence

It is important to realise that the period of Aborigines in Australia goes back over more than 40,000 years. The Hawkesbury basin was inhabited by quite a number of separate Aboriginal bands or family groups, their language was Dharuk.

The group in the Burragorang Valley was known as Gundungurra. During their time on, much activity was carried on in the valley which was later drowned by the lake, created by the dam. In this valley, there was a well known example of Aboriginal rock art, known as the "Hands on the Rock". It was on a separately free standing sandstone rock, about 6 meters long and 4 meters high. On it, hand stencils and boomerangs were engraved. This rock was submerged by the rising water, but could easily have been saved by placing it at a higher location and retaining it as an item of heritage from the Gundungurra tribe.

Much more safety operations of Aboriginal relics are now carried out at most dam constructions sites in Australia. Some remarkable Aboriginal relics are sometimes found. After some negotiations with the local, Aborigines they are recorded and where possible preserved by moving them to higher levels.

4.2.4 Recreational Use of the Area around the Dam Crest

Extensive garden and tree planting can enhance the surroundings of dam crests. Following termination of construction work on the dam, an area, about 1.6 km long by 0.6 km wide was landscaped and transformed for use as recreation area, . Over one thousand trees and shrubs were planted, roads were sealed and a picnic area were provided. There are extensive parking areas, children's play grounds,

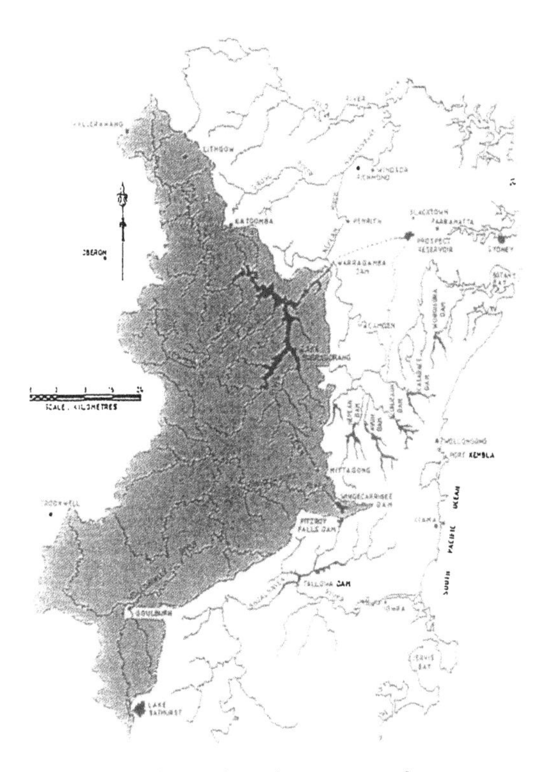

Fig. 1. Location and catchment area of Warragamba Dam.

barbecue facilities and hot water taps. A kiosk with a model of the dam open to the public was established. It is estimated that about 700,000 visitors per year make use of the facilities.

4.2.5 Lake Burragorang and Protection against Pollution

Lake Burragorang and its surroundings are very beautiful with towering sandstone walls and the gentler slopes of the Southern Tablelands.

However, to prevent pollution, the public is not permitted within 3 km of the waters edge except at the dam. A high fence surrounds the area and trespass is threatened with high penalties. This results in good quality water which does not need treatment apart from disinfection by chlorination. This has saved the Water Board and the rate payers considerable amounts of money.

There have been and still are a number of coal mines owned by Clutha Ltd. in the Valley. Originally coal washing arrangements were environmentally not too satisfactory, but this has been very strictly controlled in later years.

4.2.6 Catchment Area

The catchment area is shown in Figure 1. There is a great variety of topography but little variations in micro climate. Over 55 percent of the area is timbered, while there are distinct pastoral and agricultural areas in the north and in the south. Land use of the catchment area has not changed significantly in the period since introduction and completion of the dam in 1960. For administrative convenience the large Warragamba catchment area of 9,013 square kilometres has been divided into an inner and outer area. The inner area of 2,515 square kilometres is strictly controlled by the Water Board. By-laws restrict excessive urban development and govern farm management. There is close co-operation with the Soil Conservation Service of N.S.W to keep erosion to a minimum. No logging has been permitted within the catchment. The main concern is to minimise damage due to wildfire, especially within the fenced area. A single thunderstorm can cause the outbreak of several fires in separate locations. A special radio system is devoted to fire fighting only. Approximately 560 km of fire roads, not accessible to the public, have been constructed and are constantly maintained. To reduce the fuel load in the forest area occasional understorey burning is carried out under very strict control.

The Board claims that no significant siltation of the Lake has occurred. A large part of the catchment has been declared National Parks.

4.2.7 Effects on the Valley Downstream of the Dam

Flooding in the Hawkesbury River valley has always been a serious concern, ever since population centres have been established in the fertile flood plains, along the banks of the river. Floods occur most irregularly and are unpredictable. Official records of floods commenced in 1891.

Although the highest *officially recorded* flood occurred in 1900, it is known that a higher flood occurred in June 1867.

The Effect of Warragamba Dam was hoped to mitigate effect of floods in the Hawkesbury Valley. However, as the operation of the pondage behind the dam is directed essentially towards holding water for drinking purposes only, the mode of operation of the spillway gates is not entirely directed to control floods. The large storage capacity certainly diminishes the severity of small floods and reduces their peaks.

If operating normally a flood inflow of 8,500 cubic meters per second would be reduced to 6,660 cubic metres per second. The Warragamba Dam incorporates five spillway gates. A central 27 m wide 8 m high drum gate and four 12 m wide, 13 m high radial gates. Recent studies showed that a Possible Maximum Flood would overtop the dam several metres. Therefore a Flood Protection Program was instituted to ensure greater safety.

The Interim Flood Protection Measures entailed the raising of the dam by 5m. That entailed considerable concrete work and hydraulic lifting of the roadway bridges.

The central bridge weighs 360 tons and the other four bridges each weigh 170 tons. This increases the spillway capacity to about 60% of the Possible Maximum Flood.

The wall was also strengthened by 30 post tensioned anchors. Some post tensioned anchors were also provided in the side walls.

As a further Flood protection measure, it is proposed to construct a side spillway at some future date. This is still under investigation.

5 SUMMARY AND CONCLUSIONS

Where dams or similar structures, such as barrages are considered necessary, careful consideration must be given to technological aspects of positioning them, their size, shape and materials of construction. Once the location has been fixed, the dam, when built will create a man-made lake and bring about changes in the river basin upstream and downstream.

It is essential that the environmental implications in ecological, social, meteorological, anthropological, aesthetic and other aspects are given full consideration.

The submergence of the valley upstream of the structure causes swamps, agricultural, forest and grass land to disappear and be replaced by an expanse of water which may extend many hectares or even many square kilometres.

For the families who were occupants of the area, which will be submerged, and who gained their livelihood from the land, often for many generations, the submergence can be catastrophic. The flooding forces them away from their traditional land and homes. This enforced displacement, even where resettlement is provided, is distressing. Early action to advise these people and guarantee of adequate compensation is therefore paramount.

Where water storage is planned, especially if it is to provide drinking water, it is essential to clear the trees from the area to be submerged. This can be a major task. The cleared timber should not be wasted and no effort should be spared to ensure that the logged timber is used as far as possible.

Around the periphery of the stored water level, the retention and possibly planting of additional trees is advantageous. This can provide protection against erosion and enhancing the lake aesthetically. Maximum tree cover and sound forestry management throughout the catchment have important effects on run-off and micro-climate.

Recognition, retention and preservation of archaeological and anthropological relics is most important.

Recreational use of the area around the dam crest will have a long term influence on acceptance of the structure by the public.

Management of the introduced body of water requires considerable attention. Water quality and many local conditions will influence the solution.

To completely bar the public from the lake, the shore and a wide strip of land from the water's edge is an extreme form of management. The other extreme is to give general access and allow any and all sorts of water sports on the lake. The choice will require good judgement and must give environmentally sustainable results which finally will be acceptable to the public.

Once the crest area of the dam is available to the public, amenities for visitors are required.

Provision of toilets, tables and benches, playing fields, parking areas, etc. are important. Prevention of pollution of these areas are essential.

Dedication of some of the catchment area as national parks is advantageous for good water quality.

A change of flora and fauna in and around the introduced water is to be expected.

With the change from flowing water in the stream, to an almost stationary lake, different types of fish and marine population must be expected.

Many animals established in the catchment, indigenous or exotic need consideration. Special efforts to prevent introduction of feral cats, dogs, pigs and goats are important.

In forest areas, fire protection is essential. Construction of access trails and look-out towers is most desirable.

Flood mitigation is an important function of hydraulic structures. It is necessary that flood gates are consistently and correctly operated to protect the population downstream.

Architectural planning and shaping the dam itself, can result in an artistic structure which is impressive and pleasant to the viewer. Similarly, good landscaping, with colourful gardens and extensive tree planting, can enhance the surroundings of the dam.

Where dams are located in stark alpine surroundings, especially above the tree line, it is almost impossible to initiate special plantings. However, the grounds can be attractively contoured and cleared of debris. Simple facilities for shelter can be introduced, access roads and the whole structure fitted aesthetically into the landscape.

Public discussion can assist in the formulation and facilitation of acceptance of environmental policy for the dam, the lake and the shore line.

REFERENCES

Aird, W.V., 1961. The water Supply and Drainage of Sydney

Bandler, H., 1980. Environmental Implications of Warragamba, A Dam for Water Supply of Sydney, N.S.W., Australia, I.F.A.C. Symposium on Water and Related Land Resource Systems, May 1980, Cleveland, Ohio

Bandler, H.,1988. Catchment Management of a Dam for Water Supply, Review after a Quarter

Century, Sixth World Congress on Water Resources, International Water Resources Association IWRA, May-June 1988, Ottawa, Canada

Bandler, H.,1991. Case Studies of Two Major Works of Operational Prevention of Flood Damage, XXIV Congress of IAHR, Sept. 1991, Madrid, Spain

Nicol, T.B. 1964. Warragamba Dam, Journal of the Institution of Engineers, Australia, Vol. 36, Oct. November 1964, 239-262

Developments in Geotechnical Engineering, Balasubramaniam et al. (eds) © 1994 Balkema, Rotterdam, ISBN 90 5410 522 4

Experience of design of large dams on soft rocks in Pakistan

Amjad Agha
National Engineering Services Pakistan (Pvt) Limited, Lahore, Pakistan

ABSTRACT: The design of two large dams, Mangla and Kalabagh has given the profession a very useful insight on the behaviour and engineering characteristics of the soft rocks in general and sheared clays in particular. Mangla was designed when the concept of long-term stability of slopes underwent radical changes. The design of the project had to be modified during construction as it was realized that shearing in clays was far more extensive than what had been assumed during the design stage. Design parameters were revised and residual strength of clays was adopted instead of peak strength, owing to which major design changes had to be incorporated. The research conducted at Mangla on the properties of these over-consolidated clays is a landmark and has provided an important guidance for dealing with such foundations in the future projects. The experience of design, construction and monitoring during operation of Mangla dam and later the design of Kalabagh dam, afforded the profession an excellent opportunity for the study of these soft rocks. The special design features adopted for these large projects will also be of considerable help for the designers dealing with similar foundation conditions.

INTRODUCTION

A number of large dams have been designed and constructed in Pakistan on a variety of foundation materials including igneous and metamorphic rocks, deep alluvial filled valleys and fresh water deposits of the Siwalik system. Two of these large dams namely Mangla dam and proposed Kalabagh dam are earthfill embankments which have been designed on soft rock foundations. These rocks are fresh water deposits of Miocene age and have undergone considerable tectonic activity. These tectonic movements have developed shearing within the clays which has an important bearing on the engineering characteristics of these soft rocks.

The Siwaliks were deposited by streams rejuvenated by a major uplift of the Himalayas, initiated in the Miocene epoch. These sediments occupy the low outer most hills of the Himalayas along its whole length from the Indus to Brahamaputra. The Siwalik system comprises sandstones, clays and subordinate conglomerates ranging in thickness from 4500 to 5200 meters (Wadia, 1966). The removal of about 1300 meters of sediments rendered the existing Siwalik rocks heavily overconsolidated.

The engineering characteristics of the Siwaliks were first studied in detail during the design and construction of Mangla dam. In fact, major advancement in the understanding of engineering characteristics of these rocks developed during the construction of this project. The concept of long-term stability and the design on residual strength rather than peak strength underwent radical changes as elaborated by Prof. Skempton in his 4th Rankine lecture (Skempton, 1964). Major design modifications had to be incorporated during the construction stage.

Subsequently, the design of Kalabagh dam was undertaken with the experience gained at Mangla, and the shear zones were studied in even greater detail and further refinements in the design were made possible. Owing to the higher seismic risk at Kalabagh site and the presence of shallow dipping sheared clays, some innovative design had to be adopted.

Engineering characteristics of the soft rocks of Siwalik system and their impact on the design and construction of major civil engineering projects have been discussed in this paper in the light of experience gained during the design and construction and operation of Mangla dam and the design of the proposed Kalabagh dam.

SOFT ROCKS IN PAKISTAN

The Indo-Pakistan subcontinent may be divided into three main regions (Farah et al., 1979); the stable peninsula in the south, the rising mountain arcs in the north, and the Indo-gangetic plains in the middle. It has been postulated that in Tertiary era, the mountains began to rise and at the same time, they started pushing the foredeep against the stable peninsula, folding it and causing down warping. Subsequently, the foredeep was filled with the products of erosion, derived from the mountains and swept down by numerous rivers and streams. The foot hills of these rising mountains were formed by the sediment brought down by the Siwalik river system.

The Siwalik system contains various sedimentary rocks such as sandstones, clays and subordinate conglomerates. The term "clays" has been generalized for materials ranging from clays to sandy silts (based on the clay fraction) and varying in consistency from soft to very hard.

The bulk of the Siwalik formations, though lithologically similar to the materials constituting the recent alluvium of rivers are quite different in engineering character from them as the former are heavily overconsolidated, have undergone a long period of lithification and have been folded and faulted owing to tectonic forces.

The Siwaliks are spread over a large area in Pakistan as shown on Fig. 1.

STRUCTURAL DISCONTINUITIES

The Siwaliks lie on an unstable part of the earth crust which has undergone several episodes of tectonic activity, resulting in folding and faulting of the formations. Apart from numerous major faults in the region, there are two major regional faults, the Himalayas Thrust System running from Karakoram to Assam and the Balochistan Arc running north-south along the Suleiman Range.

The tectonic movements which have affected the Siwalik system have developed numerous structural discontinuities in the Siwalik formations particularly in the clay beds. These structural discontinuities play a major role in the engineering behaviour of soft rock formations. An important phenomenon is the development of shear zones which are a nightmare for the design engineer. There have been extensive studies of sheared clays and their implication on design of structures founded on them which are discussed later. The major types of structural discontinuities are discussed hereunder:

Random Fissures

Fissures, which are typical of most overconsolidated clays, are polished and occasionally slickensided. The shear strength of the clays along the fissures has reduced almost to the residual value.

Thrust Shear Joints

These joints, which have occurred as a result of beds adjusting to folding of the bedrock under compressive stresses, are typically found in conjugate sets. In clays, joint surfaces are polished and the shear strength along these surfaces has reduced to its residual value (Binnie & Partners, 1968). The shearing is generally concentrated in the finer grained beds and has not been detected in sandstones.

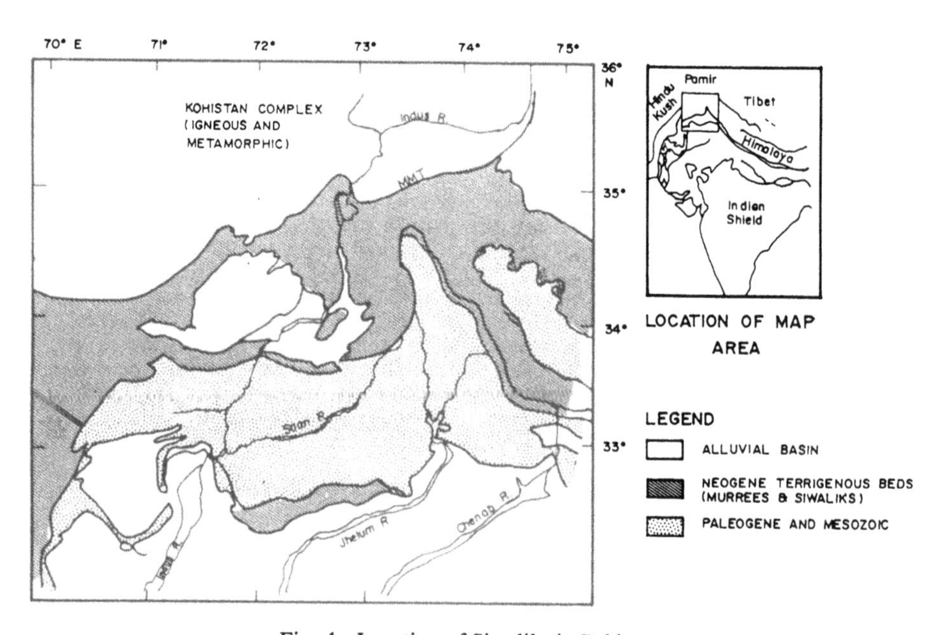

Fig. 1 Location of Siwaliks in Pakistan

Shear Zones

The slips along the bedding planes, which have occurred as a result of folding, are termed as shear zones. The thickness of the shear zones has been found to vary from a few millimeters to about one meter (Binnie & Partners, 1968). These zones appear to be continuous throughout the full extent of lens or bed. Like the other discontinuities, the shear strength along these zones has reduced to its residual value.

The shear zones in the clays of Siwalik formations were first detected during the geotechnical investigations carried out for the Mangla Dam Project. It was observed that some layers of the clays had more than one shear zone whereas the other layers were unsheared. On the average, out of every five clay beds, three were sheared. A typical shear zone in Siwalik clays is shown on Fig. 2.

ENGINEERING CHARACTERISTICS

As already mentioned, the tectonic movements in the past have resulted in folding and faulting of Siwalik bedrocks. The intensity of shearing increases with severity of folding. This shearing is concentrated in the fine grained beds and has been rarely detected in the sandstone beds.

The shearing in the fine grained beds, which range from thrust shear joints to fully developed bedding shear, reduce the shear strength of the bedrock often to residual or near residual values depending on the degree of shearing. Due to presence of the shear zones, it is essential to evaluate two sets of the shear strength parameters of the clays; one along the bedding planes and the second across the bedding planes.

Since the measurement of residual strength of soils in the laboratory is a sophisticated, costly and time consuming process, various researchers have attempted to correlate the residual shear strength of Siwalik clays with some basic clay properties such as liquid limit, plasticity index, and particularly, clay fraction. Most of these correlations give quite wide scatter of estimated shear strength parameters. However, the correlation of clay fraction (CF) with the residual angle of internal friction (Ø) is considered valid for engineering purposes.

Several laboratory techniques had been adopted for measurement of residual strength of clays. These basically comprise the following two approaches;

- Testing of specimens containing natural shear surfaces, and

- Testing techniques in which shear surfaces are created.

The first category includes reversal shear box and drained triaxial tests on samples with natural shear surfaces. The second category includes reversal shear box on remoulded and cut-plane samples and ring shear tests on remoulded samples. The reversal shear box and ring shear tests allow measurement of residual strengths at considerably large strains. The residual angle of internal friction has generally been measured to be 3.5 to 10 degrees lower than the peak value. The difference in peak and residual strengths is smaller at clay fraction less than about 25 percent and is larger at high clay fractions. The residual shear strength is generally evaluated from the angle of internal friction only, assuming the cohesion intercept as zero.

In order to investigate the shear strength of clays during seismic conditions, ring shear tests are performed at large strain rates. In case of testing for Kalabagh dam, a strain rates of about 700 mm/min was achieved.

Based on laboratory test data for Mangla dam and Kalabagh dam, Skempton (Skempton, 1985) has developed a correlation between $Ø_r$ and "CF" for Siwalik soft rocks having activity (PI/CF) ranging from 0.5 to 0.8. This correlation is plotted on Fig. 3 and generally shows a decrease in $Ø_r$ with increase in CF. The residual strength decreases significantly when CF increases from about 15 percent to 50 percent. For CF lower than 15 percent or greater than 50 percent $Ø_r$ is not very sensitive to changes in CF. The correlation has been further reviewed by Sheikh (Sheikh, 1986) by taking into account the effect of normal effective stress which shows that residual strength decreases with increase in normal effective stress. This correlation for the Siwaliks is shown on Fig. 4.

The elastic and shear moduli of Siwalik bedrocks have been determined on the basis of the laboratory and in-situ tests. These laboratory tests include unconfined compression, consolidation and consolidated drained triaxial tests. The in-situ tests include plate loading and self-boring pressuremeter tests. The average static and shear moduli are given in Table-1.

Dynamic moduli for the Siwalik bedrocks have been estimated from the shear wave velocities measured through the crosshole testing the average dynamic moduli are also given in Table-1.

The fines content in Siwalik sandstones varies from nil to about thirty five percent. The clay content in Siwalik clays varies from fifteen to over sixty percent. Generalized gradation envelopes of Siwalik soft rocks are shown on Fig. 5.

DESIGN OF DAMS

Design of two large dam projects; Mangla dam and Kalabagh dam, provided an excellent opportunity to study the engineering characteristics of Siwalik soft rocks. An account of experience gained from these two projects is given in the following sections.

MANGLA DAM PROJECT

Mangla dam completed in 1967 has been built on river Jhelum for providing water for irrigation and generation of hydro power. The main features of the project include three rolled earthfill embankments, two spillways, five power and irrigation tunnels and a power station. The Main embankment is 3350 m long at crest level and 116 m high from river bed level. One of the earthfilled embankments i.e., Jari dam is located about 24 km away from the main embankment. The project is located amid the low foothills of Himalayas. These foothills are composed of beds of the Upper Siwalik system of Miocene - Pleistocene age. The area is seismically unstable and earthquakes occur frequently. Most important system of thrust faults, known as Main Boundary Fault, passes within 50 km of Mangla Project.

At Mangla, the beds dip at about 12 degrees and at Jari the dip is about 45 degrees. The most obvious structural feature of the region is the change of strike on either bank of Jhelum river. On west of the river, the structural trend is north-east; on east of the river it is north-west, producing a sharp inverted V - pattern with an apex about 160 km north of the site. The regional geology of the area is shown on Fig. 6.

It is postulated that about 1200 to 1800 meters of Siwaliks have been eroded at the damsite rendering the formation in a highly over consolidated state.

It was during the detailed investigation at Jari (one of the three main embankments of Mangla) that a proper appreciation was first developed of the discontinuities in the Siwalik clays. Shear zones were later discovered in the foundations of all major structures in the project. These shear zones were not detected during earlier geotechnical investigations because of considerable difficulty in their identification. In many cases these shear zones are only a few millimeters thick and are almost always obscured by weathering in natural exposures. The shear zones are particularly missed in the core drilling, therefore wherever over-consolidated clays are found, extensive trenching and exposure are necessary to study them.

After the detection of the shear zones, extensive field and laboratory investigations were carried out to assess the engineering properties of sheared Siwalik clays. Multi-reversal direct shear tests were devised to determine the residual strength of shear zones at large strains. These tests were conducted on natural slip surfaces as well as on intact samples. In order to facilitate the evaluation of design parameters for clays, correlations of residual shear strength were developed with the index properties.

The contract design of the project was based on the design parameters evolved from the initial soil investigations. For the clays, the cohesion values ranged between 0 to 24 kPa and angle of internal friction ranged between 28 to 32 degrees. As a consequence of discovery of the shear zones the parallel to bedding shear strength of the clay beds was reduced to angle of internal friction 13 to 16 degrees with zero cohesion value. For design proposes, it was assumed that all clay beds contain a shear zone; that the shear zones are continuous; and that the clay within the shear zone has been reduced to the residual conditions. Thus the residual factor R was taken to be 100

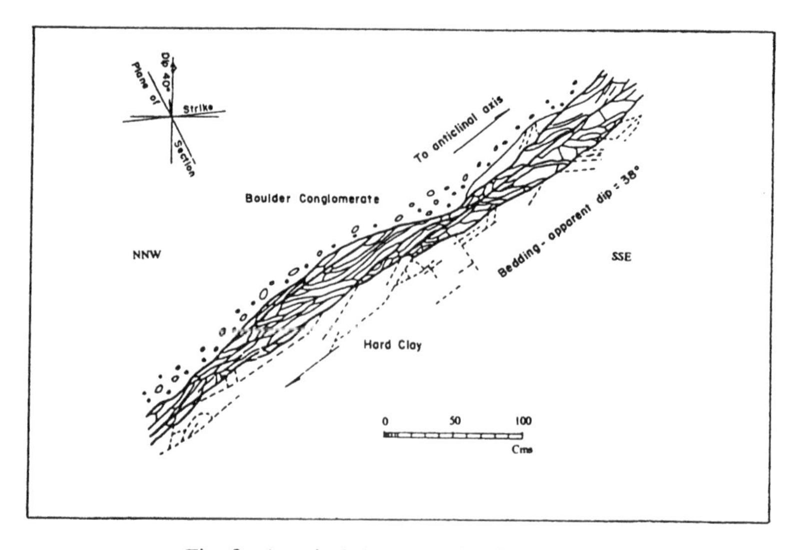

Fig. 2 A typical shear zone in Siwalik clays

percent for all potential failure surfaces parallel to bedding planes in clays.

The shear strength across the bedding was also reduced to take into account the presence of fissures and joints in the bedrock produced by tectonic activities. Depending upon the loading conditions and position of failure planes, the values of residual factor R were selected between 50 percent and 100 percent.

In the original design of the project, the heavily loaded structures were so planned that they would be founded on sandstones which were considered more competent in terms of their engineering characteristics. Accordingly, the spillway headworks and the powerhouse were founded on massive sandstone beds. However, most of the embankments had to be founded on clays. The discovery of shear zones in foundation bedrocks resulted in the following revision in design of various structures during their construction:

a. Construction of the Main dam was initiated prior to the modification of the design parameters. Toe weights were added on the upstream and downstream of the main dam to increase the stability to the required level.

b. Steel tunnel lining were introduced over that length of the tunnel in bedrock already not so lined in original design.

c. In the intake embankment, a blanket of rolled sandstone was placed over the sandstone bedrock on the upstream side to reduce the permeability.

d. In the intake embankment, a second drainage adit with wells was introduced on the upstream side. Some additional excavation of the bedrock was undertaken to remove some critical clay beds. An upstream toe weight was also added underwater.

e. The design parameters were modified prior to initiating construction of the Jari dam, and the dam and the rimworks were completely redesigned.

f. In the final design, the clay bedrock from the foundations of the Sukian Dam was excavated up to 15 meters below original ground level and backfilled with rolled sandstone.

g. Changes in the spillway design included addition of a low level adit with drainage wells, extra excavation to remove critical clay layers on both sides of approach channel, and installation of deep pumped wells to control porewater pressure under the side slopes of the two stilling basins and the lower chute.

h. The inclination of the chute of the emergency spillway was reduced from 1 on 1.75 to 1 on 2.25.

The design modifications resulted in an increase of about 22 percent in the estimated cost of civil works of the project. Similar design problems were faced during the construction of Gardiner Dam on the South Saskatchewan River in Canada founded on clay shale foundations. The design changes in this case resulted in an increase of about 17 percent from the original cost estimates.

Since its completion, Mangla dam project has functioned remarkably well. The project has been subjected to moderate earthquakes and high floods. In September 1992, during a very high flood, the spillway was operated very close to its design discharge. A detailed subsequent inspection of the project revealed that all the features of the project functioned well during and after the flood of record without any damage.

KALABAGH DAM PROJECT

Kalabagh dam is planned to be constructed in the Indus river about 190 km downstream of the existing Tarbela Dam. The dam will create a reservoir to provide about 9,300 million cubic meters of usable storage and the project will have an initial installed capacity of 2,400 MW of hydropower with eight units of 300 MW each. The ultimate power generation capacity will be 3,600 MW.

The project features include 81 meters high earthfill main embankment dam, auxiliary embankments, two spillways with combined discharge capacity of 6,400 cubic meters per second, four diversion conduits, eight power conduits, each 11 meters in diameter and a power station.

The main dam will have a zoned cross-section consisting of compacted materials from the excavations required for the spillways, power station and other features. Penstocks will be built in a 22 meters thick roller compacted concrete slab placed between the bedrock foundation and the compacted fill.

The experience gained during construction of Mangla dam regarding geological and geotechnical characteristics of the Siwalik rocks was utilized right from the site selection stage of the Kalabagh dam. An initially selected site for the project, which would have resulted in relatively shorter embankments was rejected due to the reason that the site was underlain by clays having about 1.5 meter thick shear zone. Treatment of such a zone would have resulted in high construction cost. An alternate site was thus selected which had comparatively lower degree of shearing of clay beds.

The selected damsite is located near the confluence of the Indus-Soan rivers about 16 km

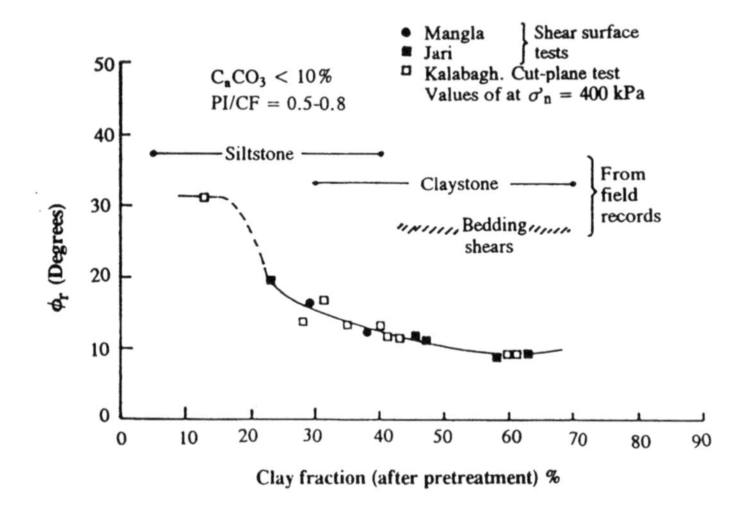

Fig. 3 CF correlation plot for Kalabagh (after Skempton, 1985)

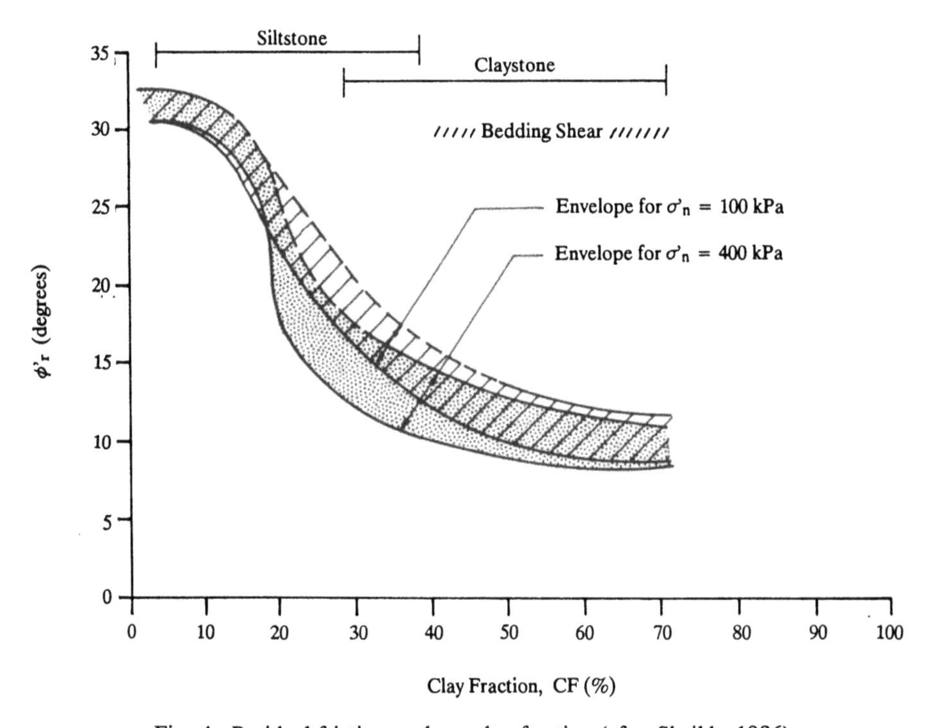

Fig. 4 Residual friction angle vs clay fraction (after Sheikh, 1986)

TABLE 1. Average moduli for sandstone and claystone bedrock (kips/sq.ft)

BEDROCK	LABORATORY						FIELD													
	UU or CU		UNCONFINED COMPRESSION		CD		PLATE LOADING								PRESSURE METER		CROSS HOLE	
	Eu	Gu	Eu	Gu	Ed	Gd	Ev1	Gv1	Evu	Gvu	Eh1	Gh1	Ehu	Ghu	Ghu	Gh1	Edy	Gdy
Sandstone	2500	850	4700	1600	3800	1400	13000	5400	15500	6500	16500	6800	20000	8300	10700	9500	106400	38000
					(u=0.32)		(u = 0.2)										(u=0.4)	
Pressure Range	0 to 12.5				0 to 12.5		0 to 69								Insitu 0 to 27		Insitu 0 to 27	
Claystone	2000	650	7900	2650	4200	1450	16000	5300	17000	5600	50500	16800	49200	16400	9600	6700	81000	30000
					(u = 0.43)		(u = 0.5)										(u = 0.35)	
Pressure Range	0 to 13.6				0 to 3.1		0 to 69								Insitu 0 to 27		Insitu 0 to 27	

Eu = Undrained elastic modulus
Ed = Drained elastic modulus
Gu = Shear modulus
u = Poission's ratio
Gd = Drained shear modulus
Eh1 = Elastic modulus (lateral) loading
Ehu = Elastic modulus (lateral) unloading

Ev1 = Elastic modulus (vertical) loading
Evu = Elastic modulus (vertical) unloading
Edy = Dynamic elastic modulus
Gdy = Dynamic shear modulus
Gv1 = Shear modulus (vertical) loading
Gh1 = Shear modulus (vertical) unloading
Gh1 = Shear modulus (lateral) loading
Ghu = Shear modulus (lateral) unloading

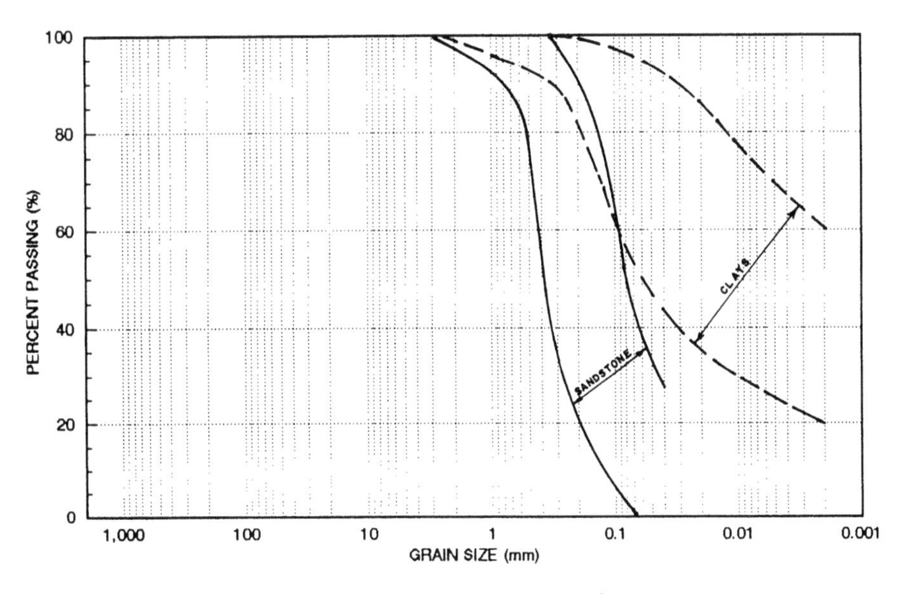

Fig. 5 Generalized gradation envelope of claystone

upstream of the town of Kalabagh. Indus river at the site flows in a narrow gravel filled channel between 15 to 30 meters high cliffs of gently dipping Siwalik bedrock covered on both sides by terraces of silt and gravel. Bedrock is exposed in the high ground above the terraces having a dip of about 4 degrees. The Siwalik rocks underlying the damsite area are the upper part of the Middle Siwalik Dhok Pathan formation of Middle Miocene age and consist of alternate beds or sandstone and clays with occasional lenses of gravel. It is postulated that about 1200 to 1500 meters of strata has been removed by erosion to reach the present ground level.

The tectonic activity in the geological past has caused folding resulting in shearing of generally the fine grained beds, i.e. clays and siltstone. The fine grained beds contain a variety of shear features, i.e. thrust joints, oblique shears, flattened shear lenses and fully developed bedding shear. The intensity of shearing increases with the clay fraction. Displacements have taken place along all the discontinuities to an extent sufficient to reduce the strength to residual value. This is evident from the polished or slickensided surfaces and from gouge material.

Keeping in view the experience gained at the Mangla dam regarding the shear strength parameters of the Siwalik rocks, the extent of shearing of the clays/siltstone beds was studied

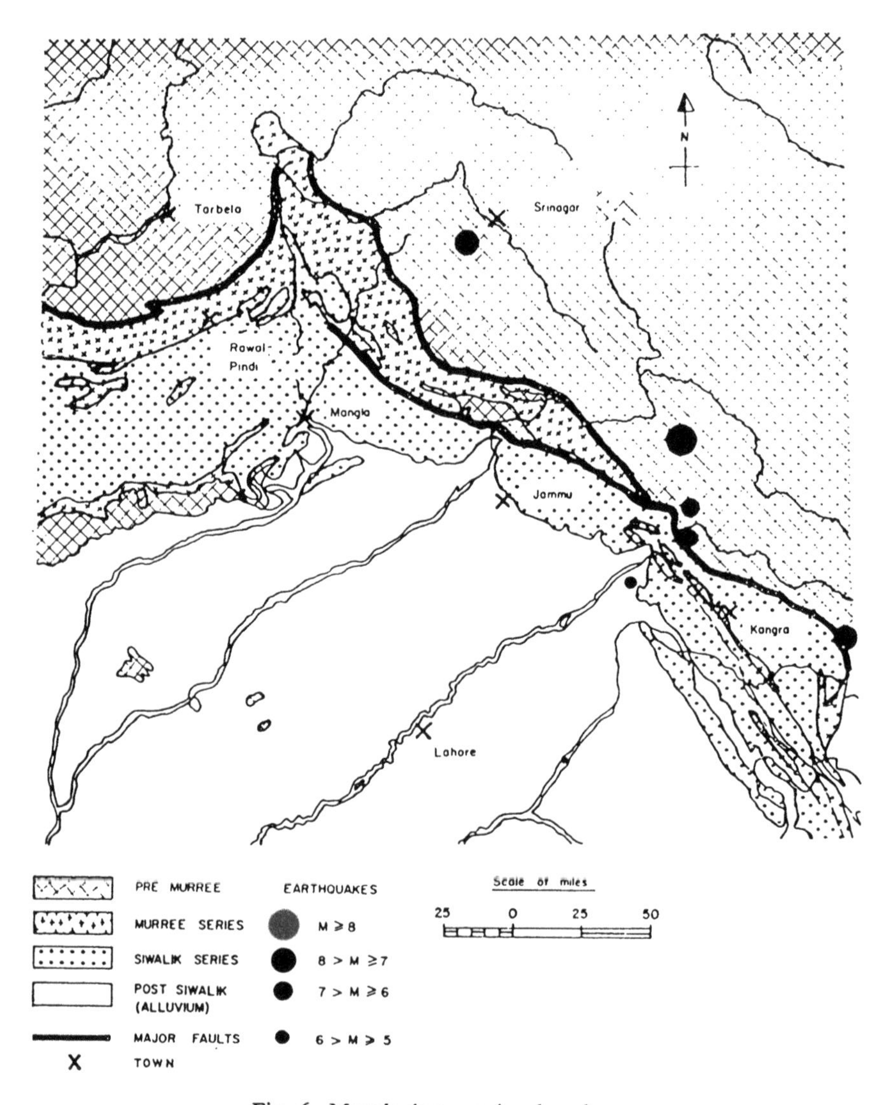

Fig. 6 Mangla dam - regional geology

through careful logging of exposures in excavated trenches. Representative samples were also taken from these trenches for index tests and shear tests. Shear tests conducted included multi-reversal shear box tests on cut-plane samples and natural slip surfaces and ring shear tests on remoulded samples. These tests provided an opportunity to study the effects of large strains on shear strength characteristics of the Siwalik bedrock. Based on the index properties and the results of series of shear tests, correlations were established between the clay fraction and the residual strength. Since Siwalik clays at Kalabagh are similar to those at Mangla/Jari, the residual strength tests of Mangla/Jari were also used in developing the correlation for Kalabagh. The plot so developed is shown in Fig. 3. The study of trench logs and index properties has shown that bedding shear occurs only in clays with an average clay fraction exceeding 45 percent or an average liquid limit greater than 50.

For carrying out engineering analyses and design of the Kalabagh dam the shear strength parallel to bedding was quantified as follows (Kalabagh Consultants, 1985):

$$\emptyset_D = \emptyset_r + i$$

where

$\emptyset_r$ = residual angle of shearing resistance as measured by multi-reversal shear-box and ring shear tests, and;

i = some conservative measure of the angle between a thrust joint and the bedding plane. This was determined from mapping of bedding shears, joints, discontinuities etc. in the clay beds as exposed in exploratory trenches and adits.

The shear strength across bedding plane was evaluated by the following two approaches.

i. Taking weighted average of shear strength of sandstone beds occurring within the clay beds and assuming an average clay fraction of the combined clay beds.

ii. Ignoring the shear strength of the sandstone beds occurring within the finer beds and assuming an average clay fraction of the clays/siltstone bed.

The design of shear strength in clays/siltstone beds along the bedding planes ranged between 10 and 13.5 degrees and across the bedding planes ranged between 13.5 and 25 degrees with zero cohesion in both cases.

The stress-strain relationship for samples of intact sandstone tested under drained conditions in triaxial compression indicated the material as brittle with a sharp post-peak loss of strength at small strains. The following shear strength parameters of sandstones were adopted for carrying out stability analyses for the embankments.

Condition	Cohesion (kPa)	Angle of Internal Friction (degrees)
Peak	62	55.5
Critical State	0	40

The design strength parameters established on the basis of degree of shearing and the residual strength of clay beds did not always give the required factors of safety during the stability analysis of the various embankments and the cut slopes involved in the project. Therefore, at such places special treatment measures would be required to achieve the desired level of safety. The more significant measures are listed below (Agha and Ahmed, 1990):

i. To achieve the desired level of stability of the main embankment dam, the sheared clay beds in certain areas under the downstream shoulder would be excavated upto 75 meters depth from the ground level and replaced with rolled sandstone fill.

ii. Under the auxiliary dam, replacement of sheared clay beds with rolled sandstone fill would be carried out upto 30 meters depth.

iii. In the power complex area, the power tunnels would be intersected by sheared clay beds and large displacement in shear zones in case of an earthquake could rupture the conduits. Huge excavations are now planned to remove sheared clays from the power complex area and replace power tunnels with conduits. A 22 meters thick roller compacted concrete (RCC) pad will accommodate the 11 meters diameter power conduits. The removal and replacement of sheared beds with RCC has also been planned in the power station and intake structure foundations (Fig. 7).

iv. At numerous other locations, permanent cut slopes expose the sheared clay beds causing instability of these slopes. At such locations, these exposed clay beds shall be cut back and replaced by rolled sandstone fill to improve the stability upto the desired levels.

Substantial costs will be incurred on the above mentioned ground improvements to achieve a safe design and to mitigate the potential problems, owing to the presence of shear zones in the Siwalik bedrock.

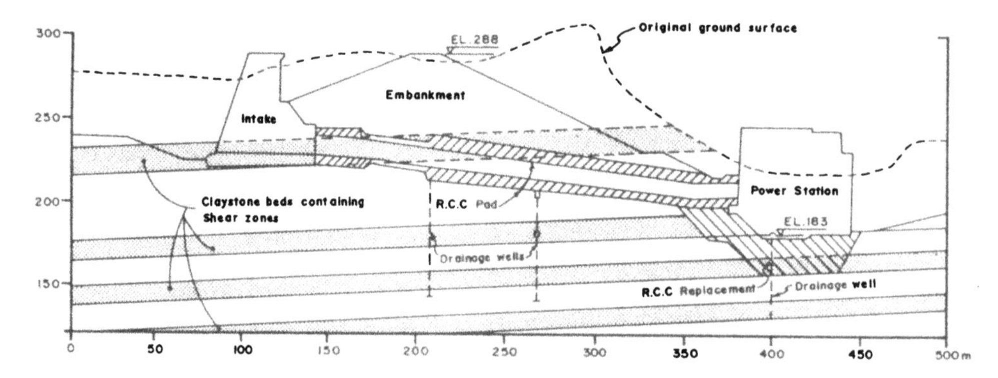

Fig. 7 Kalabagh dam power complex area treatment measures for sheared clays

CONCLUSIONS

The following conclusions are based on our experience of design of large dam projects on soft rocks:

- There is always a possibility of the existence of shear zones in soft rocks such as overconsolidated clays especially in tectonically active zones.
- Identification of shear zones is extremely difficult on weathered surface, unless you examine fresh surface of the bed inch by inch.
- The detection of shear zones requires careful logging of open-excavations as weak sheared clays are difficult to recover in core drilling.
- For clays containing bedding shear, residual strength should be used as parallel to bedding design strength. Across the bedding, strength may vary from residual to peak strength value, depending upon the intensity of shearing.
- Testing of thousands of samples has shown that the residual strength of clay can be correlated with clay fraction.
- Siwalik sandstones are non-cohesive and do not contain shearing planes.
- Core recovery in highly friable sandstone is problematic.
- Siltstones behave as clay and are assigned with residual strength values.
- The clay foundations after excavation have to be protected as they weather quickly.
- Sandstone develops stress release fissures, particularly on steep slopes or cliff-edges.
- Both sandstone and clays develop rebound.
- Sandstones are easily excavated by a heavy dozer, except hard lenses which sometimes need blasting.
- Clays have to be pulverized, watered and processed before using as an impervious fill.
- Water seeps through sandstone beds whereas clay beds act as impervious barrier. Owing to this, excess pore pressure conditions may develop in sandstones, therefore proper drainage should be provided in these beds.

ACKNOWLEDGEMENTS

The staff members of Geotechnical Division of National Engineering Services Pakistan (Pvt) Limited assisted me in preparation of this paper. Mr. Mansur Ahmed and Dr. Tahir Masood compiled the necessary information and offered comments and suggestions to give the final shape to this paper.

The manuscript was also reviewed and valuable comments were given by Mr. M. Saleem Sheikh and Ch. Altaf-ur-Rahman.

The author extends his appreciation to all his colleagues for their valuable contributions.

REFERENCES

Agha, A. and M. Ahmed (1990), "Experience of Geotechnical Engineering on Dams in Pakistan", Proc. of 1st Intl. Seminar on Soil Mechanics and Foundation Engineering of Iran, Nov. 1990, Iran.

Binnie & Partners (1968), "Mangla" A book on Mangla Dam Engineering.

Farah, A., S. Kees, and A. DeJong (1979), "Geodynamics of Pakistan", Geological Survey of Pakistan, 1979.

Kalabagh Consultants (1985), "Kalabagh Dam Project, Design Report-2, Dams and Embankments". December, 1985.

Sheikh M. Saleem (1986), "The Residual Strength of Soils and its Relevance to the Siwalik Clays in Pakistan", Thesis submitted to University of Engg. & Tech. Lahore in partial fulfillment of the requirements for the Degree of Master of Science in Civil Engg.

Skempton, A. W. (1964), "Long-term Stability of Clay Slopes", 4th Rankine Lecture, Geotechnique, Vol. 14, No. 2.

Skempton, A. W.(1985), "Residual Strength of Clays in Landslides, Folded Strata and the Laboratory", Geotechnique Vol. 35 No. 1, pp. 3-18.

Wadia, D.N. (1966), "Geology of India", Third Edition, English Language Book Society, London.

NOTATION

The following symbols are used in this paper:

CF clay fraction

$\varnothing$ angle of internal friction

$\varnothing_r$ angle of internal friction corresponding to residual strength

$\varnothing_D$ design value of angle of internal friction

PI plasticity index

R residual factor

Developments in Geotechnical Engineering, Balasubramaniam et al. (eds) © *1994 Balkema, Rotterdam, ISBN 90 5410 522 4*

Some remarks on the stability of towers

A. Desideri & C. Viggiani
University of Napoli Federico II, Italy

ABSTRACT: The stability of tall structures on a deformable subsoil is analysed referring to the inverted pendulum model. The condition of equilibrium and the stability of equilibrium are discussed, assuming either a linear soil (elastic half space or Winkler subgrade) or a non-linear strain hardening elasto-plastic restraint. The effect of creep and repeated loading cycles is briefly commented. Finally, the effects of some possible stabilising measures are analysed by means of the interaction loci of the bearing capacity. Some of the concepts developed are applied to the case history of the leaning tower of Pisa.

1 INTRODUCTION

According to Needlemann, the concept of stability is one of the most unstable concepts in the realm of Mechanics. Among the various different forms of stability, this paper is devoted to the stability of the equilibrium of tall structures, such as a tower or a smokestack, resting on deformable soils.

If a critical height, function of the stiffness of the subsoil and the geometry of the foundation, is attained, the structure reaches a state of neutral equilibrium and eventually fails by overturning even in the absence of an external moment (Froehlich, 1953; Gorbunov Posadov & Serebrajany, 1961; Habib & Puyo, 1970; Schultze 1973).

The real problem may be modelled as an inverted pendulum (Fig. 1), i.e. a weightless shaft of length h, with a concentrated mass of weight W at the top. The shaft is rigidly connected to a horizontal base that, in turn, is connected to a spring hinge capable of reacting to a rotation with a moment.

Let α be the angle of rotation of the shaft from the vertical position, and α' the angle of rotation of the base. The spring will react to a small rotation α' of the base with a moment $M_r = f(\alpha')$; a small rotation α of the shaft will produce an external moment $M_e = W\, h \sin\alpha \approx W\, h\, \alpha$. Being the shaft and the base rigidly connected, the two angles are linked by the relation:

$$\alpha' = \alpha - \alpha_o$$

where α_o represents an initial deviation from the verticality of the shaft due to factors such as constructional imperfections.

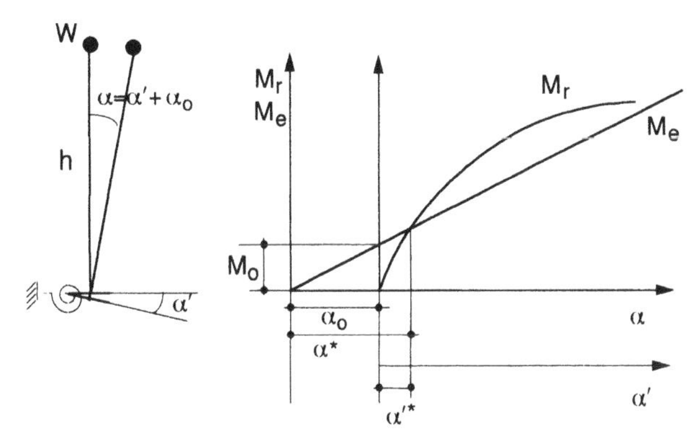

Fig. 1. The inverted pendulum as a model of the stability of the equilibrium of a tall structure

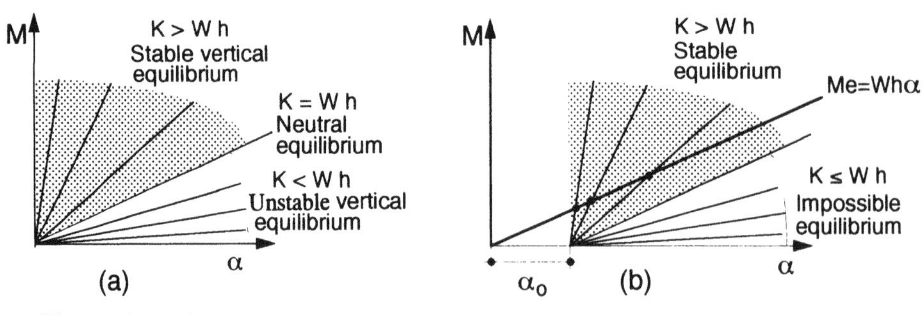

Fig. 2. Analysis of the stability for a linear foundation. a) vertical initial position ($\alpha_o = 0$); b) inclined initial position ($\alpha_o \neq 0$)

The equilibrium of the system is possible if a value α^* may be found such that:

$$W\,h\,\alpha^* = f(\alpha^* - \alpha_o) = f(\alpha'^*) \quad (1)$$

$$W\,h\,\alpha_o + W\,h\,\alpha'^* = f(\alpha'^*) \quad (1bis)$$

$$M_o + W\,h\,\alpha'^* = f(\alpha'^*) \quad (1ter)$$

In any configuration of equilibrium, characterized by a value α^*, the degree of safety may be characterized in a number of ways. If we refer to an existing structure, for which the values of W and h do not vary, the following factors of safety may be defined:

$$FS = \frac{M_{r,\,max}}{M_r(\alpha'^*)} \quad (2)$$

$$FS = \frac{\alpha'_{max}}{\alpha'^*} \quad (3)$$

$$FS = \frac{\left(\frac{\partial M_r}{\partial \alpha'}\right)_{\alpha'=\alpha^*}}{\left(\frac{\partial M_e}{\partial \alpha}\right)_{\alpha=\alpha^*}} \quad (4)$$

where α'_{max} and $M_{r,max}$ represent the maximum possible values of the rotation and reacting moment.

Eq. (2) and (3) allow an appraisal of the distance between the equilibrium configuration and the failure, respectively, in terms of moment and rotation. They express the degree of safety against failure.

For the problems where the external loads vary with the geometric configuration of the system, as the one under consideration, an assessment of the stability of equilibrium (eq. 4) is particularly important. If $\frac{\partial M_r}{\partial \alpha'}$ is smaller than $\frac{\partial M_e}{\partial \alpha}$, than the equilibrium is instable and a virtual increment $\delta\alpha$ of the inclination of the shaft leads to the collapse (the equilibrium is no more possible). If $\frac{\partial M_r}{\partial \alpha'}$ is equal to $\frac{\partial M_e}{\partial \alpha}$ the equilibrium is neutral, and a virtual increment $\delta\alpha$ of the inclination af the shaft leads to a new equilibrium configuration. If finally $\frac{\partial M_r}{\partial \alpha'}$ is larger than $\frac{\partial M_e}{\partial \alpha}$ the equilibrium is stable, and the unbalanced moment produced by a virtual increment $\delta\alpha$ brings the shaft back into the starting equilibrium configuration.

Once the function $M_r = f(\alpha')$ and the initial value of the inclination α_o are known, an analysis of the equilibrium and stability is possible.

Let us consider firstly an ideal subsoil reacting to the rotation as a linear spring: ($M_r = k_\alpha \alpha'$, with k_α = const.). Fig. 2 and Table 1 show that the equilibrium in a non vertical configuration is possible if and only if $k_\alpha \geq W\,h$. In particular, if $\alpha_o = 0$ and $k_\alpha = W\,h$, every configuration is a neutral equilibrium one; if on the contrary $\alpha_o > 0$, only one stable equilibrium configuration exists.

Let us now consider a non-linear spring, with a maximum value of the reacting moment M_r attained asymptotically. Fig. 3 and Table 2 show that, in a non vertical position, the equilibrium is possible if and only if $\alpha_o > 0$ and the value of $\left(\frac{\partial M_r}{\partial \alpha'}\right)_{\alpha'=0}$ is larger than W h. In this case the stability of the equilibrium depends on the value $\left(\frac{\partial M_r}{\partial \alpha'}\right)_{\alpha'=\alpha^*}$ as compared to W h.

Table 1. Analysis of the stability for a linear foundation

	$\alpha_o = 0$		$\alpha_o > 0$	
	Equilibrium	Stability	Equilibrium	Stability
$k_\alpha > Wh$	$\alpha = 0$	stable	$\alpha = \frac{\alpha_o K_\alpha}{(K_\alpha - Wh)}$	stable
$k_\alpha = Wh$	any α	neutral	impossible	-
$k_\alpha < Wh$	$\alpha = 0$	unstable	impossible	-

Table 2. Analysis of the stability for a non-linear foundation

	$\alpha_o = 0$		$\alpha_o > 0$	
	Equilibrium	Stability	Equilibrium	Stability
$\left(\frac{\partial M_r}{\partial \alpha'}\right)_{\alpha'=0} > Wh$	$\alpha = 0$	stable	possible if $\alpha \neq 0$	stable neutral unstable
$\left(\frac{\partial M_r}{\partial \alpha'}\right)_{\alpha'=0} = Wh$	$\alpha = 0$	neutral	impossible	-
$\left(\frac{\partial M_r}{\partial \alpha'}\right)_{\alpha'=0} < Wh$	$\alpha = 0$	unstable	impossible	-

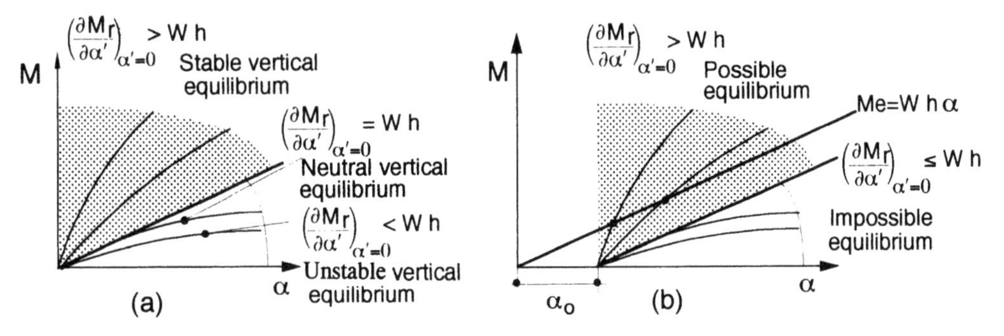

Fig. 3. Analysis of the stability for a non-linear foundation. a) vertical initial position ($\alpha_o = 0$); b) inclined initial position ($\alpha_o \neq 0$)

In order to assess the stability of the equilibrium of a tower inclined to the vertical of an angle α the function $M_r = f(\alpha')$ and the value of the initial inclination α_o must be known. It will be shown below that, with some simplifying assumptions, α_o and M_r can be obtained from the measurements of inclination and settlement, if available.

2 EQUILIBRIUM, STABILITY AND SAFETY FOR SOME SIMPLE MODELS

2.1 *Circular foundation on elastic half space*

For a circular foundation resting on a linearly elastic, homogeneous half space the rotation α' and the settlement ρ may be expressed as functions of the moment M and the normal load W as follows (Poulos & Davis, 1974):

$$\alpha' = \frac{M}{k_\alpha} \; ; \quad \rho = \frac{W}{k_\rho} \qquad (5)$$

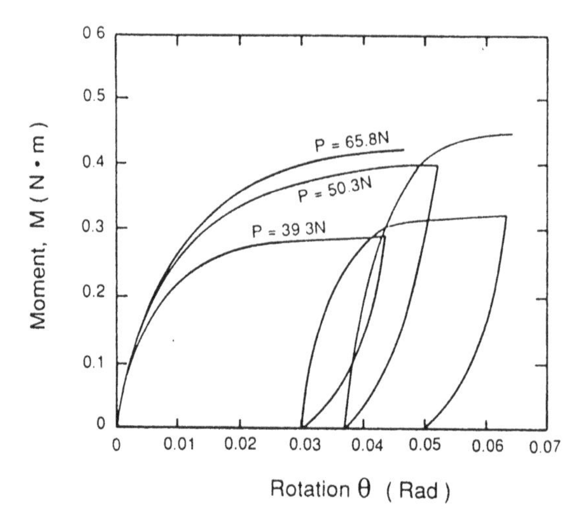

Fig. 4. Relationship between M_r and α' obtained by Cheney *et al.* (1991) by means of centrifuge tests

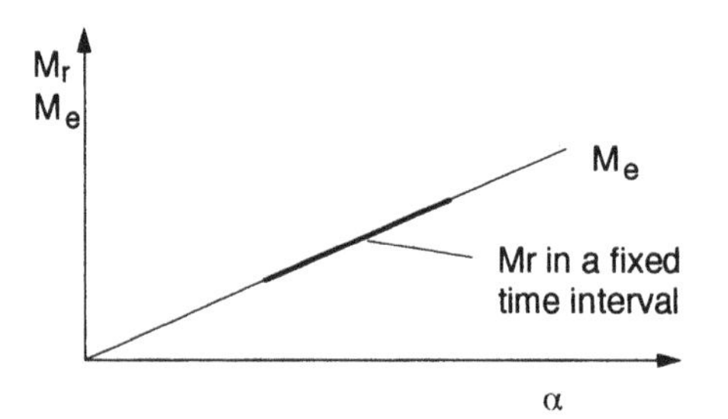

Fig. 5. Successive equilibrium configurations during an increase of the rotation with time

$$k_\alpha = \frac{ED^3}{6(1-\nu^2)} \tag{6}$$

$$k_\rho = \frac{ED}{1-\nu^2} \tag{7}$$

where E and ν are respectively Young modulus and Poisson ratio of the half space and D is the diameter of the circular foundation. By combining eqs. (5), (6) and (7) one obtains:

$$k_\alpha = \frac{W D^2}{6\,\rho} \tag{8}$$

$$\frac{\alpha'}{\rho} = \frac{6M}{W D^2} \tag{9}$$

A comparison between the calculated and observed values of the ratio $\frac{\alpha'}{\rho}$ allows an assessment of the significance of the assumed subsoil model. Eq. (9), furthermore, allows an evaluation of the rotation α' and hence, by comparison with the rotation of the tower, of the value α_o. The value of k_α given by eq. (8), compared with W h, allows finally an assessment of the stability of the equilibrium.

2.2 *Circular foundation on a Winkler subgrade*

For a circular foundation resting on a Winkler type subgrade, the rotation α' and the settlement ρ may be expressed as functions of the moment M and the normal load W as follows:

$$\alpha' = \frac{M}{k_\alpha} \; ; \quad \rho = \frac{W}{\Omega\, k_\rho} \tag{10}$$

$$k_\alpha = J\, k_\rho \tag{11}$$

where $\Omega = \frac{\pi D^2}{4}$ is the area of the foundation; $J = \frac{\pi D^4}{64}$ is its diametral moment of inertia and k_ρ is the coefficient of subgrade reaction of the Winkler springs. By combining eqs. (10) and (11), one obtains:

$$k_\alpha = \frac{W D^2}{16\,\rho} \tag{12}$$

$$\frac{\alpha'}{\rho} = \frac{16\,M}{W D^2} \tag{13}$$

As in the previous case, eq. (13) allow an

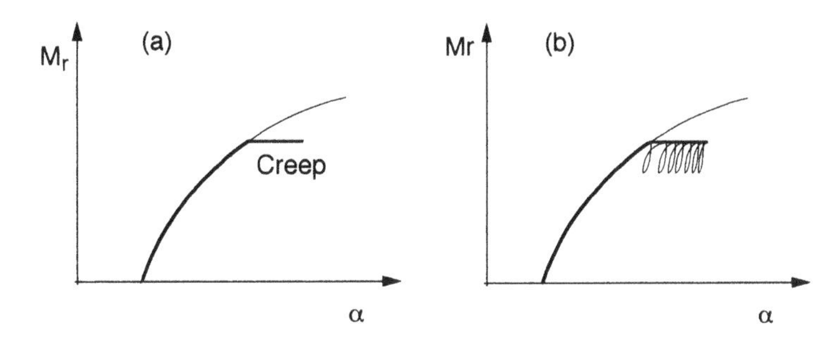

Fig. 6. Increase of the inclination due to: a) creep; b) small loading - unloading cycles

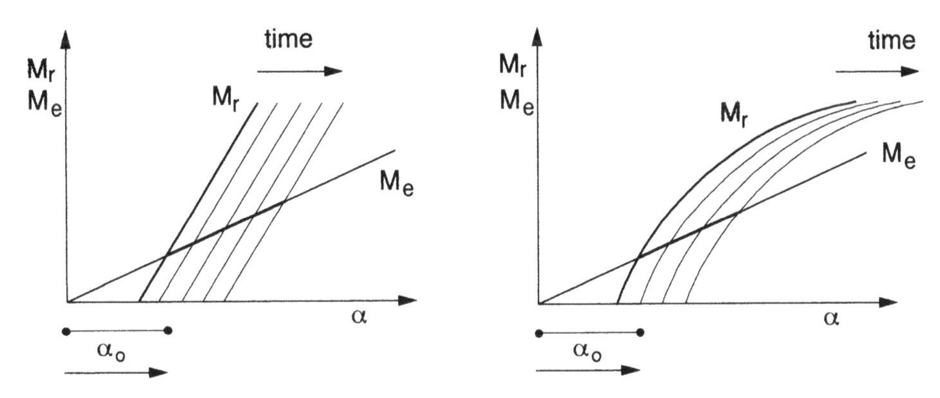

Fig. 7. Variation of α_o corresponding to successive equilibrium configurations

assessment of the significance of the model, and an evaluation of the initial rotation α_o. The value of k_α given by eq. (12), compared with the term W h, allows an assessment of the stability of the equilibrium.

2.3 *Non-linear restraint*

If we assume that the behaviour of the subsoil is strain hardening elasto-plastic, the relationships between M and α' and between W and ρ become non-linear and asymptotic. These relationships, however, cannot be derived only from the parameters of the model, because they depend also on the loading history.

Cheney *et al.* (1991) have obtained some examples of M_r-α' relations by centrifuge testing under different constant values of W. The observed behavior (fig. 4) is amenable to a representation by means of an exponential equation such as the one suggested by Lancellotta (1993):

$$M_r = A(1 - e^{-B\alpha'}) \qquad (14)$$

Different asymptotic expressions, as for instance an hyperbola, could be equally suited.

In order to judge the stability of the equilibrium, some of the unknown parameters have to be fixed. A possibility is to assume the value of $A = M_{r,max}$ and of the initial tangent $AB = \left(\frac{\partial M_r}{\partial \alpha'}\right)_{\alpha'=0}$, and then calculate the initial rotation α_o from the equilibrium configuration of the structure.

In this case, if α* and M* = Whα* are the equilibrium values, one obtains:

$$\alpha_o = \alpha^* + \frac{1}{B}\ln\left(1 - \frac{Wh\alpha^*}{A}\right)$$

The minimum possible value is $\alpha_o = 0$ (negative values are not possible). In this case, the equilibrium configuration is vertical ($\alpha_o = \alpha^* = 0$). As already pointed out in Table 2, the inequality:

$$\left(\frac{\partial M_r}{\partial \alpha'}\right)_{\alpha'=0} = AB > Wh$$

must hold. The maximum possible value is:

$$\alpha_{o,max} = \frac{A}{Wh} - \frac{1}{B}\left(1 + \ln\frac{AB}{Wh}\right)$$

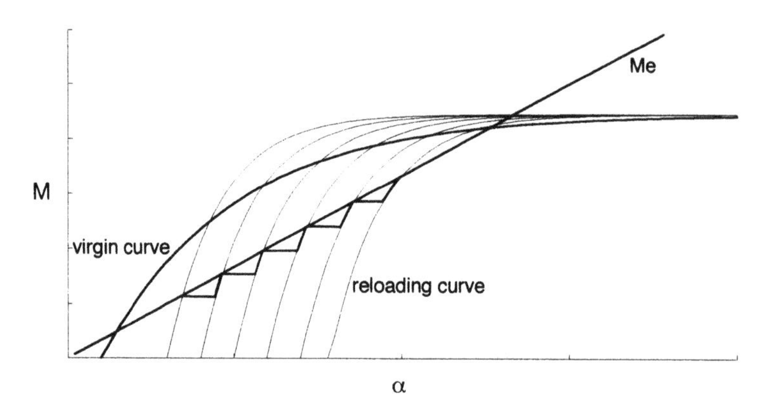

Fig. 8. Increase of the inclination due creep and/or plastic cumulated deformations with hardening

Such a value occurs when the line $M_e = Wh\alpha$ is tangent to the curve $M_r = f(\alpha)$.

Alternatively, the values of A and α_0 can be fixed and the value of B obtained from the equilibrium condition.

2.4 *Increase of the inclination with the time*

If an increase of the inclination in the time is observed, as it is often the case, the external moment and the reacting one must coincide over the range of values assumed by α. Accordingly, the overall relationship between M_r and α', at least in a range, must be linear (Fig.5).

To explain such a behaviour, different hypotheses can be formulated:

a. The relationship between M_r and α' is actually linear and $\alpha_0 = 0$. The laws of variation of M_r and M_e with α coincide; the equilibrium is (and has always been) neutral ($k_\alpha = Wh$)

b. The relation between M_r and α' does not coincide with that of the external moment. Other factors, such as creep deformation and/or small loading and unloading cycles, produce a gradually increasing inclination (Fig.6) making the superposition of M_r and M_e possible in the range considered.

In order to analyse these conditions, it might be assumed that the delayed deformation due to creep, or the cumulated one due to loading-unloading cycles, produces a simple translation of the relation between M_r and α', with no hardening. The final effect of these irreversible deformations, in this hypothesis, is an increase of α_0 with increasing time (Fig. 7).

It is also possible to suppose that plastic hardening does actually occur; in this case the equilibrium configuration belongs to a reloading curve (Veneziano, 1993). Such a curve may be represented too by means of an exponential expression similar to eq. (14). It must have the same asymptotic value ($M_{max} = A$) of the virgin one, but a higher value of the initial stiffness AB.

With increasing inclination, the point representing

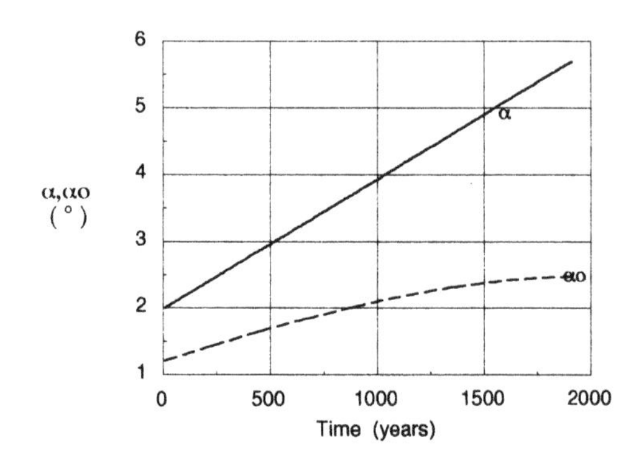

Fig. 9. Variation of α_0 corresponding to successive equilibrium configurations

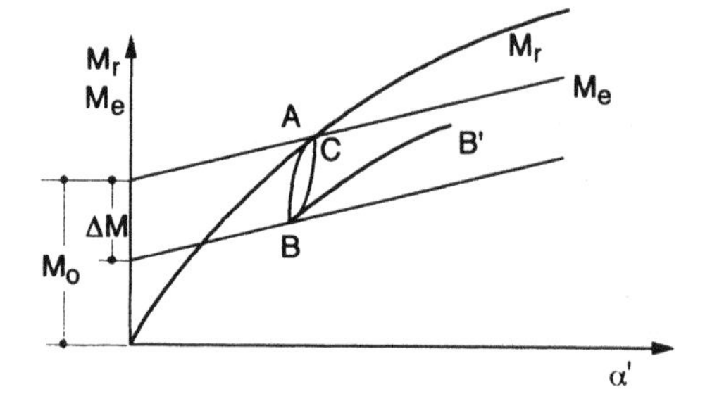

Fig. 10. Possible elasto-plastic paths of loading-unloading-reloading

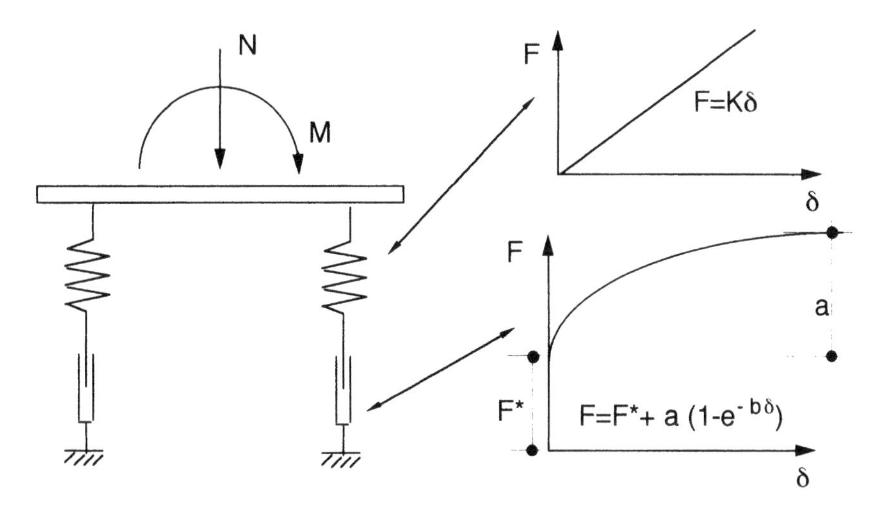

Fig. 11. Model of a beam on two elasto-plastic strain hardening supports

the equilibrium configuration moves on a succession of reloading curves that may be generated by successive translation of a same curve (Fig. 8). It is again possible to express the effects of creep or cumulated plastic deformations by a variation of α_0.

It is interesting to point out that, in the case of a non-linear law of the type expressed by eq. (14), increase of α_0 with a logarithmic law such as:

$$\alpha_0 = \alpha_{0,in} + c \ln(\frac{t}{t_i})$$

implies a nearly linear variation of the equilibrium value α^* with the time in the range where $M_r = M_c$. In Fig. 9, some compatible variations of α^* and α_0 are reported. It is also possible that a gradually decreasing factor, such as a logarithmic creep or a cyclic stabilization, could produce a constant effect.

From the point of view of the stability of the equilibrium, such a situation can be regarded as a succession of stable equilibrium configurations. In each of these configuration, i.e. for each value of α_0, the condition:

$$(\frac{\partial M_r}{\partial \alpha'})_{\alpha'=\alpha^*} > (\frac{\partial M_e}{\partial \alpha})_{\alpha=\alpha^*}$$

is then satisfied. Alternatively, the situation can be regarded as a situation of overall neutral equilibrium.

If the relationship between M_r and α' tends asymptotically to a maximum value, the safety factor defined by eq. (4) gradually decreases and approaches unity. As this value is attained, the equilibrium is locally neutral since:

$$(\frac{\partial M_r}{\partial \alpha'})_{\alpha'=\alpha^*} = (\frac{\partial M_e}{\partial \alpha})_{\alpha=\alpha^*}$$

The overall equilibrium, however, is instable since a further increase of α_0 makes the equilibrium impossible.

3 POSSIBLE STABILISING MEASURES

3.1 *Effect of the loading path*

To increase the safety of an existing tower, a reduction of the applied moment can be envisaged. To study the effects of such a measure, let us assume that the soil behaviour is strain hardening elasto-plastic. Let the relationship between M_r and α' be the one represented in Fig. 10, for a given value of the load W and for a given loading history. In these condition, a decrease ΔM of the external moment shifts the equilibrium point from A to B. If the moment increases again starting from the point B, apparently the resulting rotation of the structure should increase following the reloading path BC and then merge into the virgin curve. Being the reloading path much steeper than the virgin curve OA, a reduction of the value of the external moment seems to produce an improvement of the situation from the point of view of stability. In other words, in the configuration represented by the point B, the factor of safety defined by eq. (4) is larger than in A. It can be shown, however, that this is not necessarily the case and that in some instances a reloading path such as BB' is possible. The decrease of the moment would then not be beneficial to the stability of the structure, and could even be detrimental.

As a matter of fact, the relationship between M_r and α' does not represent all the applied forces and

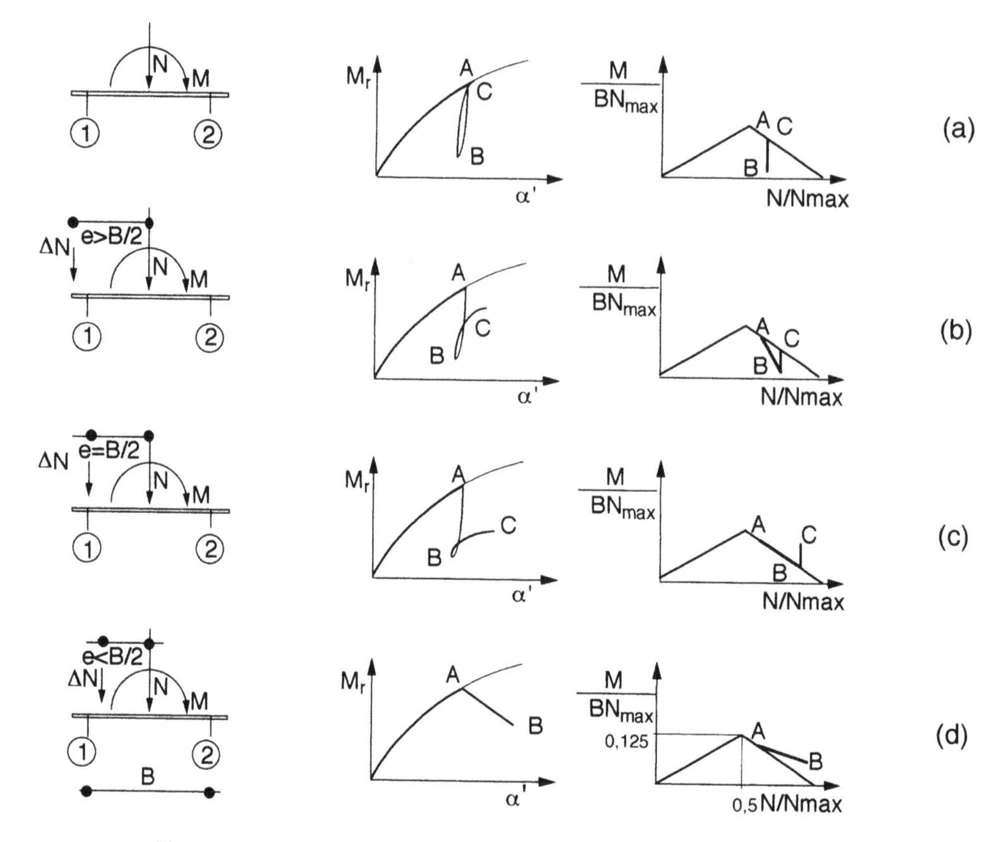

Fig. 12. Different loading-unloading-reloading paths on the model of fig. 11

the induced displacements; for instance, it does not consider the vertical load and the vertical displacement (settlement). The load-unload-reload relation (path ABC) is unique only if the unloading and reloading paths are the same and there are no variations of other parameters.

To elucidate this statement, let us consider the simple example of a beam resting on two elasto-plastic strain hardening supports (Fig. 11).
Let the initial condition be that of a centered load N that brings both supports into the plastic range ($F^* < \frac{N}{2} < F^* + a$). Starting from this condition, let us apply a pure moment first increasing, and then following an unloading-reloading cycle.

The resulting behaviour is that depicted in Fig. 12a. The virgin M_r- α' curve is non-linear along the path OA because the support 2 keeps deforming in the plastic range following the exponential relation reported in Fig. 11, while the support 1 is unloaded in the elastic range. Along the unloading path AB both the supports deform elastically; the unloading of the support 2 in the elastic range is equal to:

$$\Delta F_{2,unload} = \frac{\Delta M_{unload}}{B} .$$

If the moment increases again along the path BC, the transition from elastic to plastic behaviour occurs when the support 2 comes back to the initial state of load:

$$\Delta F_{2,reload} = \frac{\Delta M_{reload}}{B} = \Delta F_{2,unload}$$

$$\Delta M_{reload} = \Delta M_{unload}$$

Let us now consider a decrease of the applied moment, again starting at point A but obtained by applying a vertical force ΔN at an eccentricity e. The resulting behaviour may be shown to depend on the value of $e = \frac{\Delta M}{\Delta N}$ (Fig. 12).

If $e > \frac{B}{2}$, the behaviour is that reported in Fig. 12b. Along the path AB the support 1 deforms elastically, while the unloading of the support 2 in the elastic range is equal to $\Delta F_{2,unload} = \frac{\Delta M_{unload}}{B} - \frac{\Delta N}{2}$. If the moment increases again along the path BC, the transition from elastic to plastic behaviour occurs at the point C, for a value of M smaller than that in A. In C the support 2 comes back to its initial load; from the equality $\Delta F_{2,reload} = \Delta F_{2,unload}$, it follows that:

$$\frac{\Delta M_{reload}}{B} = \frac{\Delta M_{unload}}{B} - \frac{\Delta N}{2}$$

and hence:

$$\Delta M_{reload} < \Delta M_{unload}$$

If $e = \frac{B}{2}$ (Fig. 12c), the support 1 carries the whole load increment while the load acting on the support 2 does not change, and hence, the support 2 remains in the plastic range (path AB). On increasing again the moment (path BC), there is no elastic deformation at all.

If finally $e < \frac{B}{2}$ (Fig. 12d), the support 2 keeps deforming plastically along path AB; although the load increase is smaller in the support 2 than in the support 1, the displacement may be larger in 2 since the support 1 is in the elastic range. Accordingly, a decrease of M can produce an increase of the rotation α.

The yield loci and the loading paths in the plane $\frac{M}{BN_{max}}$, $\frac{N}{N_{max}}$ are also shown in Fig. 12. Such a representation appears very useful, since it allows an assessment of the expected behaviour associated to a given loading path.

3.2 *Interaction locus for a circular foundation*

The interaction locus of the bearing capacity is a useful tool also for analysing the response of a tall structure to variations of the load (vertical, moment and possibly horizontal).

For combined vertical and moment loading on a rectangular surface footing of width B and any length, the conventional bearing capacity theory (Meyerhof, 1953) gives:

- for a drained failure ($c = 0$; φ'):

$$\frac{M}{BN_{max}} = \frac{1}{2}\left(1 - \sqrt{\frac{N}{N_{max}}}\right)\frac{N}{N_{max}} \quad (15)$$

- for an undrained failure ($c = c_u$; $\varphi_u = 0$):

$$\frac{M}{BN_{max}} = \frac{1}{2}\left(1 - \frac{N}{N_{max}}\right)\frac{N}{N_{max}} \quad (16)$$

where N_{max} is the bearing capacity under a centered load ($e = 0$).

Dean *et al.* (1992) suggest the following empirical relation:

$$\frac{M}{BN_{max}} = \beta\left(1 - \frac{N}{N_{max}}\right)\frac{N}{N_{max}} \quad (17)$$

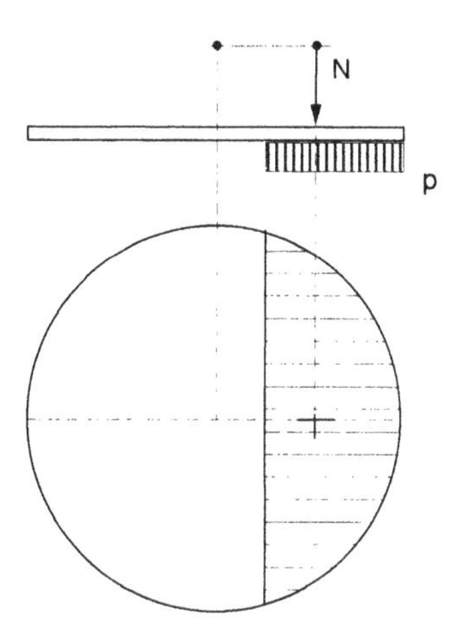

Fig. 13. Assumptions adopted to obtain the interaction loci for the circular and the ring shaped foundation

and claim that it fits the results of small scale model tests with $\beta = 0,35$. It may be observed that, with $\beta = 0,5$ which is less conservative, eq. (17) reduces to eq. (16).

For circular footings, Cheney *et al.* (1991) carried out some centrifuge tests; the best fit with the experimental findings is obtained by the relation:

$$\frac{M}{M_{max}} = 1,72\left\{1 - \left(\frac{N}{N_{max}}\right)^5\right\}\frac{N}{N_{max}}$$

that can be written:

$$\frac{M}{BN_{max}} = 0,19\left\{1 - \left(\frac{N}{N_{max}}\right)^5\right\}\frac{N}{N_{max}} \quad (18)$$

For the case of a circular footing, an interaction locus may be obtained on the following very simple assumptions (fig. 13):

- the unit bearing capacity p is costant (the same assumption underlies eq. 16);
- the reacting area is a circular segment.

By simply imposing the equilibrium, the curves reported in Fig. 14 are obtained. In the same figure the loci defined by eqs. (15) to (18) and that for a ring shaped footing are also reported. For this latter, a ratio of the outer to the inner radius $\frac{r}{r_o} = 4$ has been adopted.

According to Nova and Montrasio (1991), it may be assumed that the shape of the yield loci, internal to the limit domains reported in Fig. 14, remains the same as that of the domain (isotropic hardening).

An assessment of the effectiveness of a possible remedial measure may then be obtained by comparing the yield locus with the loading path that represents, in the plane $\frac{M}{BN_{max}}$, $\frac{N}{N_{max}}$, the examined intervention.

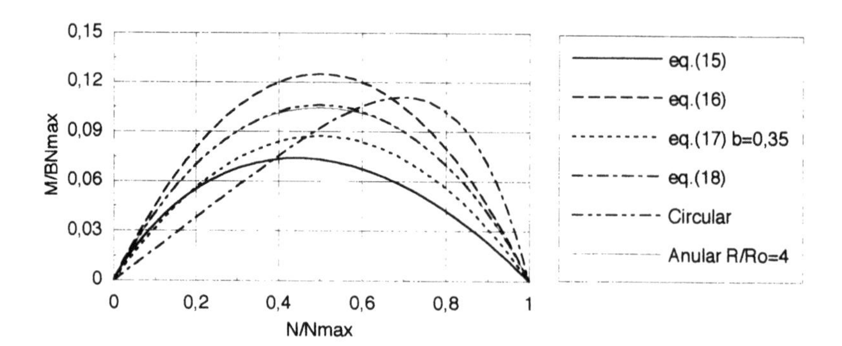

Fig. 14. M - N interaction loci

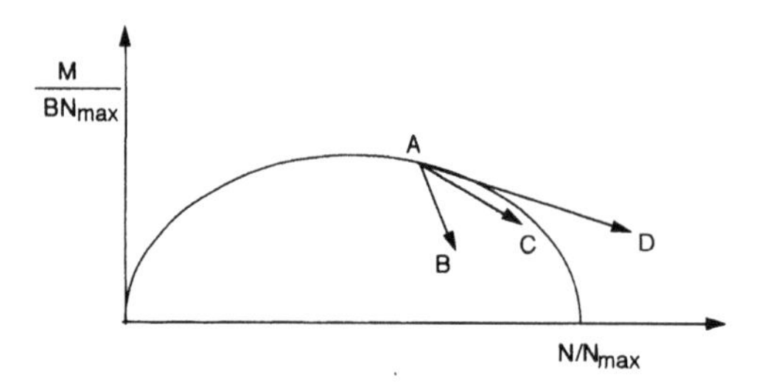

Fig. 15. Loading paths of some possible remedial measures

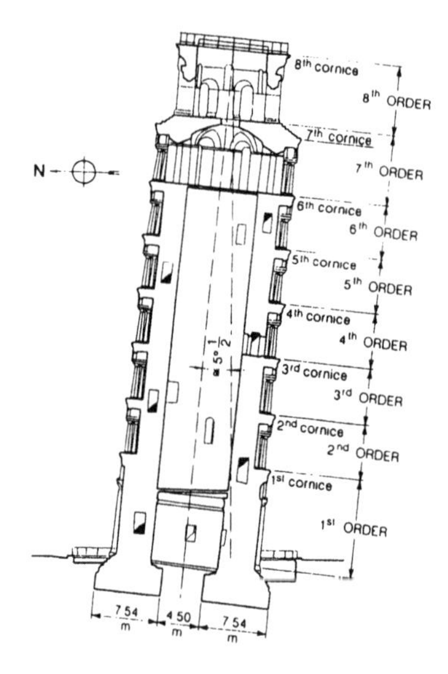

Fig. 16. Cross section of the Tower of Pisa in the plane of maximum inclination

The initial condition of the structure is assumed to be on the yield locus (point A in Fig. 15). If the loading path is directed inwards (path AB in Fig. 15), it can be shown that all the factors of safety defined by eqs. (2) to (4) increase, since the moment decreases (eq. 2), the rotation decreases (eq. 3) and the rotational rigidity increases (eq. 4).

On the contrary, an intervention implying a decrease of the moment with a loading path that is tangent to or directed outwards the yield locus (such as path AD in Fig. 15), is detrimental. In this case the factor of safety on the moment (eq. 2) increases. It is possible, however, that the rotation increases and the rigidity decreases due to plastic flow; accordingly, the factors of safety expressed by eqs. (3) and (4) decrease.

It is also possible that along some loading paths (such as AC in Fig. 15) the behaviour is initially elastic and eventually plastic depending on the size of the load increment.

4 A CASE HISTORY: THE TOWER OF PISA

A cross section of the famous leaning tower of Pisa is reported in Fig. 16. To assess the condition of equilibrium and stability of the monument, the relevant parameters are (AGI, 1991; Di Stefano & Viggiani, 1992):

$W = 141{,}8$ MN; $h = 22{,}5$ m; $e = 2{,}3$ m; $\alpha \approx 0{,}1$ rad $= 5°44'$; $D = 19{,}6$ m; $\rho = 2{,}5 \div 3$ m.

It is found that:

$$M = 141{,}8 x 2{,}3 = 326{,}1 \text{ MNm};$$

$$\frac{6M}{WD^2} = 0{,}036 \text{ m}^{-1}; \quad \frac{16M}{WD^2} = 0{,}096 \text{ m}^{-1}$$

Being $\frac{\alpha}{\rho} = 0{,}03 \div 0{,}04$ m^{-1}, and recalling eqs. (9) and (13), it seems that the elastic half space is more suited than the Winkler bed of independent spring to model the behaviour of the subsoil. Furthermore, by eq. (9), one obtains

$$\alpha' = \frac{6M}{WD^2}\,\rho = 0{,}036x(2{,}5\div3) \approx 0{,}1$$

$$\alpha_0 = \alpha - \alpha' \approx 0$$

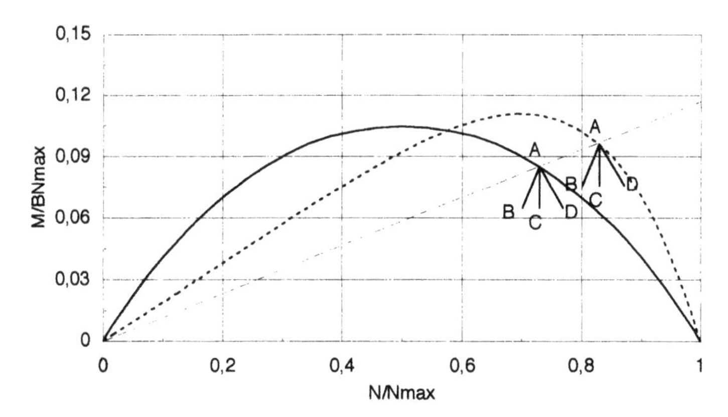

Fig. 17. Effects of possible stabilizing measures

for the elastic half space. On the contrary, the Winkler model leads to a negative, and then impossible, value of α_o.

It follows that (eq. 8):

$$k_\alpha = \frac{WD^2}{6\rho} = 3026 \div 3632 \text{ MNm}$$

The factor of safety against the instability of equilibrium is then (eq. 4):

$$FS = \frac{k_\alpha}{Wh} = 0{,}95 \div 1{,}14$$

In other words, according to this simple model the tower appears to be in an overall state of neutral equilibrium. This should explain the steady increase of inclination occurring since its construction (1173 - 1370) and documented by measurements in the last 170 years.

It may be interesting to observe (Burland &Viggiani, 1994) that irradiation by the sun during the day produces a differential deformation of the structure that, in turn, gives rise to a small cyclic variation of the moment and consequently to a small cyclic rotation of the foundation. A preliminary evaluation of the rotational stiffness k_α for these small cyclic movements leads to a value two or three times larger than the long term one. A detailed analysis of this topic is being performed by Veneziano (Probabilitas, 1992).

Recently, lead counterweights have been applied to the tower in an attempt to temporarily improve its safety (Burland *et al.*, 1993). The weight applied at present (november 1993) is 5,45 MN, with an eccentricity of 6,36 m, resulting in a stabilising moment of 34,7 MNm.

The effect has been a decrease of the inclination $\Delta\alpha = 27'' = 13\text{x}10^{-5}$ rad.

Accordingly, on this particular unloading path, the ratio between the applied moment and the rotation seems to correspond to a rotational stiffness $k_\alpha = \frac{34{,}7}{13} \text{x } 10^5 = 267000$ MNm; this value is almost 100 times larger than the overall long term value.

In the mean time an average vertical displacement of the base of the tower equal to 1,5 mm has been measured by precision levelling. Accordingly (eqs. 10 and 14):

$$\frac{\alpha'}{\rho} = 0{,}08 \text{ m}^{-1}; \quad \frac{6M}{WD^2} = 0{,}099 \text{ m}^{-1}; \quad \frac{16M}{WD^2} = 0{,}265 \text{ m}^{-1}$$

In this case, the observed value of the ratio between the increments of rotation and settlement is again close to the theoretical value for an elastic half space, and much smaller than of that predicted by the Winkler model.

Possible stabilising measures, that have been suggested, include: the temporary dismantling of the bell chamber at the top; the application of a stabilizing moment via a vertical eccentric force (weights or tensioned ground anchors); the decrease of rotation by inducing a controlled subsidence of the foundation soil. The first measure induces a decrease of both the vertical force and the moment; the second one, a decrease of the moment and an increase in the vertical force; the third one, a decrease of the moment with constant vertical force.

The effectiveness of such remedial measures may be assessed by analysing the related loading paths compared to the interaction (yield) locus of the foundation of the tower.

In Fig. 17 the interaction locus for a circular foundation obtained by Cheney *et al.* (1991) by centrifuge testing, and that developed by the writers for a ring shaped foundation similar to that of the leaning tower are reported. In the same figure the point A, representing the present condition of the tower, is obtained by intersecting each locus with the straight line through the origin with an inclination:

$$\frac{M}{ND} = \frac{e}{D} = \frac{2{,}3}{19{,}6} = 0{,}117.$$

Starting from the point A, the dismantling of the bell chamber is represented by a path such as AB; it would be surely beneficial for the safety of the tower. Unfortunately, such a measure is considered severely detrimental for the integrity of the monument.

The application of a vertical force is represented by a path such as AC in Fig. 17. Its effect may be either detrimental or beneficial depending on the inclination of the path, i.e. on the ratio between the applied moment and vertical load. In other words, the larger the eccentricity of the applied vertical force, the more beneficial is its effect. The inclination of the path AC reported in Fig. 17 is that pertaining to the lead counterweights applied as a temporary measure, as recalled above. Their effectiveness had been independently assessed by sophisticated finite element modelling; in any case, the eccentricity and the amount of the applied counterweight is such that the loading increment is totally internal to the yield locus in Fig. 17.

Finally, the reduction of the inclination by controlled subsidence is represented by the vertical path AD in Fig. 17; also this measure appears surely beneficial for the safety.

5 SUMMARY AND CONCLUSIONS

A tall structure, such as a tower or a stack, founded on a deformable soil, may experience a collapse due to a particular form of instability that is governed by the deformation rather than strength properties of the soil. The problem has been modelled as an inverted pendulum connected to some forms of linear or non linear deformable restraint.

The equilibrium and stability conditions are systematically reviewed, referring to some simple linear models (elastic half space and Winkler subgrade) and to a strain hardening elasto-plastic restraint. The effects of creep or cumulated plastic deformations due to cyclic loading are briefly discussed.

The safety of a tall structure threatened by instability may be improved by applying a stabilizing moment, which can be obtained in different ways. In order to assess the effectiveness of the possible stabilising measures, the use of interaction loci of the bearing capacity under combined vertical and moment loading is suggested. The procedure is based on the assumption that the behaviour of the foundation is strain hardening elasto plastic, and that the interaction locus acts as a yield locus.

Finally, the concepts developed are applied to the leaning tower of Pisa. It is shown that the tower is very nearly in a state of overall neutral equilibrium, and the effects of possible stabilising measures are briefly discussed.

REFERENCES

AGI 1991 The Leaning Tower of Pisa: Present Situation *X Europ. Conf. Soil Mech. Found. Eng.* Florence.

Burland J.B., Jamiolkowski M., Lancellotta R., Leonards G.M., Viggiani C. 1993 Leaning Tower of Pisa: what is going on. *ISSMFE News* 20, n. 2.

Burland J.B., Viggiani C. 1994 Osservazioni sul comportamento della Torre di Pisa *RIG* 28, n. 2.

Cheney J.A., Abghari A., Kutter B.L. 1991 Stability of Leaning Towers. *Journ. Geot. Eng., Proc. ASCE* 117: 297-318.

Dean E.T.R., James R.G., Schofield A.N., Tan F.S.C., Tsukamoto Y. 1992 The Bearing Capacity of Conical Footings on Sand in Relation to the Behaviour of Spudcan Footings of Jackups. *Proc. Wroth Mem. Symp. on Predictive Soil Mechanics* : 230-253. London: Thomas Telford.

Di Stefano R., Viggiani C. 1992. La torre di Pisa ed i problemi della sua conservazione. *Restauro* 152.

Froehlich O.K. 1953. On the Settling of Buildings Combined with Deviation from their Originally Vertical Position. *Proc. 3rd ICSMFE* 1:362-365. Zurich.

Gorbunov-Posadov M.I., Serebrajanyi R. V., 1961. Design of Structures upon Elastic Foundations. *Proc. 5th ICSMFE* 1: 643-648. Paris: Dunod.

Habib P., Puyo A., 1970. Stabilité des fondations des constructions de grande hauteur. *Annales IBTP* : 117-124.

Lancellotta R., 1993. Stability of a Rigid Column with Non-Linear Restraint. *Géotechnique* 43: 331-332.

Nova R., Montrasio L., 1991. Settlement of Shallow Foundations on Sand. *Géotechnique* 41:243-256.

Poulos H.G., Davis E.H., 1974. *Elastic Solutions for Soil and Rock Mechanics*. New York: Wiley.

Probabilitas, 1992. Statistical Analysis of rotations. *Unpublished Report.*

Schultze E., 1973. *Der schiefe Turm von Pisa.* Mitteil. Technische Hochschule Aachen.

Veneziano D., Van Dyck J., Koseki J., 1993. Simple Rheologic Models for the Leaning Tower of Pisa. *Unpublished Report.*

Developments in Geotechnical Engineering, Balasubramaniam et al. (eds) © 1994 Balkema, Rotterdam, ISBN 90 5410 522 4

Remedial treatment for Menjer concrete gravity dam (Indonesia)

Soedibyo
Civil Engineering Service Center, State Electricity Corporation of Indonesia (PLN), Jakarta, Indonesia

ABSTRACT: During construction of Menjer Concrete Gravity Dam for harnessing hydro-potential resource of Lake Menjer as well as Serayu River, there were some problems caused by poor rock condition and pervious layer beneath the dam foundation. Remedial measures to control the seepage water and uplift pressure to ensure the dam stability are very important. This paper describes counter measures already done to ensure the dam stability during construction of this dam.

1 INTRODUCTION

On Serayu River Basin there are several hydro - potential resources among others are : Garung Hydro Electric Power Plant (HEPP), Mrica HEPP (184,50 MW) already in operation, Tulis HEPP Project (13 MW), under construction, Maung HEPP Project (380 MW), under review design, Rawalo HEPP Project (12 MW), Gintung HEPP Project (23 MW) and Klawing (7 MW), under feasibility study stage. Garung HEPP project was started in 1961, had been completed and put into commercial operation since 1982.

A brief information concerning Garung HEPP is as follows :

- Part of the flow from the Klakah and Serayu River is diverted into the Menjer Lake through the Klakah Serayu tunnel (535 m length) and Serayu Menjer tunnel (2,075 m length).
- From this lake, water enters the intake conduit through to Menjer Dam to a long penstock (2,318 length) to rotate hydraulic turbines and generators in the powerhouse and then released to the Serayu River again, through tailraces, after generating the electricity.
- Menjer Dam : concrete gravity type, height 36 m, crest length 100.50 m, crest width 4.50 m, concrete volume 17,280 m^3.
- Spillway : open chute type, ungated, 6 m width and 119 m length.

2 GEOLOGICAL CONDITION AT THE DAM SITE AND ITS SURROUNDING AREA.

Since the dam is not so high, during the feasibility study and design stage, the geological investigations only consisted of 11 boreholes with a total length of 360 m, and rock shear test were not done. Soon after the excavation reached the soft soil, it was decided to carry out additional geological investigations of the lower part of the foundation consisting of 5 drillholes with water pressure test, rock shear test in an adit tunnel.
After all the additional investigations were completed, it was confirmed that the geological conditions are as follows.

2.1 The dam site and its surrounding area is composed of quarternary andesite volcanic products such as volcanic breccia, lapilli tuff and lava flow. The andesite is a hard and fresh rock, partly porous and auto brecciated.

2.2 The shore line of lake is composed of talus deposit, sandy clay, gravels and boulders. Since the dominant material is talus it is generally impervious.

2.3 The upper part of the dam area is composed of volcanic breccia, middle to weakly cemented deposit matrix, sandy clay, thickness about 10 m, with a coefficient of permeability k = 10^{-4} cm/s to 10^{-5} cm/s, only rarely 10^{-3} cm/s.

2.4 Underlain by volcanic breccia weakly cemented deposit partly very loose, thickness about 3 m to 7 m with a coefficient of permeability k = 10^{-3} cm/s to 7 x 10^{-4} cm/s and partly 10^{-2} cm/s. Which means it is very pervious.

2.5 Underlain by lapilli tuff weakly cemented deposit, partly very loose with a coefficient of permeability k = 10^{-3} cm/s, partly 10^{-4} cm/s with thickness about 20 m.

2.6 Underlain by andesite partly auto brecciated with a coefficient of permeability k = 10^{-3} cm/s with thickness about 15 m.

2.7 Underlain by andesite hard and fresh with many cracks with a coefficient of permeability k = 2 x 10^{-3} cm/s with thickness about 7 m.

2.8 Underlain by tuff of depth unknown.

3 CONSTRUCTION OF THE DAM.

During construction of the dam, there were some problems caused by the geological conditions which were rather poor compared with the anticipated during the design stage as well as the pervious layer beneath the dam, so it was necessary to prepare the design change. The chronological design changes are as follows:

Original Design, November 1977.

Crest width	:	7 m
Crest length	:	112 m
Height of dam	:	37 m
Dam base	:	31.82 m

After the excavation reached the lower part, it was seen that the geological conditions were rather poor, so it was decided to carry out the test adit with the rock shear test (May, 1979).

From the result of the rock shear test is was found that shear resistance and coefficient of internal friction were smaller than estimated during detailed design, so it was necessary to carry out additional investigations.

3.1 First Design Change (August, 1979).

While waiting for the additional investigation test results, the first design change was proposed as a tentative design. This was done to keep the schedule. The dam base was elongated and the extent of consolidation and curtain grouting was increased.

3.2 Second Design Change (September, 1979).

From the preliminary result of the additional investigation, the lower part of the dam foundation was not only found to be poor but also an artesian acquifer was found. So a second design change was proposed while waiting for the result of safety againts seepage.
Again the dam base was elongated, provided with two rows of keywedges, postponement of curtain, additional contact grouting, additional longitudinal joint with grouting, and the possibility of pipe cooling. The shear strength for calculation in the upper part of the foundation was taken as : = 52 ton/m² and lower part = 15 ton/m².

3.3 Third Design Change (February, 1980).

From this second design change, the stability of the dam itself was ensured, but there were still some difficulties remained concerning possible seepage which was estimated to flow through the foundation of the dam.

From the additional investigation, it was found that there were pervious layers underlying the dam and geological conditions were complicated.

The new investigation results, indicate that a very pervious zone, with a permeability coefficient of about 5 x 10^{-3} cm/s, is distributed to a depth of about 25 m below the dam base, and the zone below that has a coefficient of permeability of about 1 x 10^{-3} cm/s, and a thickness of more than 25 m. Two additional test boreholes were already finished but the other are still underway.

Based on the results of the investigation and also on the geological data already available at that time, those remedial measures against seepage were studied carefully by means of flow net analysis by the Finite Element Method (FEM) with the aid of electronic computers of various combinations.

There were three factors to be considered which were : seepage quantity ; uplift pressure ; and piping action through the foundation. The analysis was made based on two kinds of two dimensional flow net models, which were : horizontal section models and vertical section models.

As mentioned before, it was predicted that the distribution of seepage would not only through the dam foundation but also widely from both abutment areas.

Therefore, the section models of flow net were prepared to study the distribution of flow net and effect of curtain grouting and toe drainage.

The results of the analysis were as follows :

1. In order to ensure stability of the dam, a drainage system shall be provided at the toe of the dam to restrict uplift pressure within the assumed value in the stability calculation.

2. A drainage system consisting of a drainage tunnel parallel to the dam axis, 64 m in length, with drainage wells located in the tunnel was recommended.

3. The seepage amount was not expected to be reduced by increasing the curtain grouting which required additional cost.
Therefore, it was not recommended to increase curtain grouting beyond a certain limit. However, since the flow net analysis could not show if any direct seepage existed between the reservoir and the dam site, curtain grouting was considered to be an effective and inevitable precaution to close any such seepage passages.

4. For remedial measures against seepage, the lay-out of case 8 for horizontal sections model analysis was recommended. The estimated probable total seepage was 5 to 6 m^3/min.

For safety against piping, adequate filter materials were provided in the drainage wells.

A third design change was proposed with conditions:
The dam base was 54.36 m, provided with two rows of keywedges, cancelation of pipe cooling, introduction of drainage tunnel 64 m long with 14 drainage wells 250 mm diameter, about 12 m deep, at least 3 m into the pervious layer, and an additional 1,542 m curtain grouting.

3.4 Fourth Design Change (February, 1981).

Prior to the drilling of drainage wells in the drainage tunnel, three pilot holes were drilled to check the geological and seepage condition of the foundation.
In the course of drilling, a rather large quantity of spring water was observed at each hole from the pervious lapilli tuff layer lying below the impervious caprock with a depth of 3.30 m to 6.20 m from the tunnel invert. The estimated discharge was 0.308 m^3/min. For each hole, with the lake water elevation + 1170.00, a much larger quantity of seepage than that of the design estimation total seepage was 22 m^3/min. or about 4 times bigger.

According to recent observation record in the drainage wells, the water head at the well was 20 cm to 30 cm higher than the lake water level for several days. This fact suggests that the Menjer lake may not be the sole source of the water seepage but there is a possibility of other sources. Though it could not be traced definitely at that stage.

Conceivable other sources are the ground water in both banks or in the pervious andesite stratum which extends widely in the surrounding area, lying about 25 m below the dam foundation with exposed outcrops on the lake slope apart from the dam on both sides.

Four methods of remedial measures against increase of water seepage surface were considered as follows :

1. Blanketing on slope surface and bottom of Menjer lake in the foreground of the dam. Earth blanket for

below MOL, and gunite or wet masonry blanket for above MOL.

2. Extension of curtain grouting on both sides of the planned curtain grouting.

3. Provision of counterweight at the down stream side of the dam to cope with uplift water pressure, the penstock needed to be protected by a concrete culvert with the counterweight, consisting of rockfill, placed above.

4. Addition of drainage wells to relieve the uplift pressure, the seepage water amount of about 22 m^3/min. would have to be pumped up into the reservoir or for irrigation water use.

After careful studies, it was decided to use provision of counterweight at the downstream side of the dam to cope with uplift pressure as mentioned in point 3 above, the reasons were :

1. The method was effective, even if the main seepage was not directly from the lake.

2. The cheapest cost compared with the other methods.

3. This method has no risk of uncertainty from a construction point of view.

Fourth design change was proposed with counterweight as the most suitable solution. The top of counterweight elevation proposed was + 1187.000.

Dam base 54.36 m, provided with two rows of keywedges and possibly increased additional curtain grouting.

3.5 Fifth (Final) Design Change (November, 1981).

To check the right values of seepage amount, it was necessary to make a study after impounding.

Before the impounding test started, it was necessary to carry out 15 observation holes by drilling in the dam site, for observation of ground water levels. The observation holes, 3 located in the upstream of the dam and 12 in the downstream, drilled from the ground surface to the elevation equivalent to that of the dam base of the center block (elevation 1165.80).

The diameter of the holes was 73 mm with the depth ranging from 25 m to 50 m. The observation data especially those obtained during the impounding test, would be useful to show the distribution of seepage flow in that area.

3.5.1 Impounding Tests.

Before the Menjer Lake impounding, it is necessary to carry out the impounding test.

Impounding test were carried out with 5 objectives :

1. To clarify more accurately the relationship between water level of the lake and seepage amount of drainage wells.

2. To clarify the relationship between the lake water level and uplift pressure on the dam base and also the ground water level in the area surrounding the dam.

3. To estimate a more accurate value of the coefficient of permeability of the pervious layer under the dam.

4. To evaluate the effect of curtain grouting already provided for reducing seepage amount and uplift pressure on the dam base.

3.5.2 The Impounding Test Findings.

Findings regarding the seepage problem are summarized as follow :

1. A quite impervious natural blanket with its permeability coefficient in the order of 10^{-4} cm/s to 10^{-5} cm/s develops on the lake side slope.

2. Lapilli tuff underlying the dam is pervious as its permeability coefficient is 3 x 10^{-3} cm/s on average.

3. Some thin high pervious layers develop in the lapilli tuffzone, of which permeability coefficient was evaluated as 1 x 10^{-2} cm/s.

4. The artesian ground water level in the far right abutment of the dam was found quite high by the impounding test, so the seepage was considered rather larger from the right abutment than from the lake.

5. Anticipated seepage quantity by high water level of reservoir would be more or less 7.60 m^3/min.

6. The seepage amount of 10 m^3/-min. as applied in the design of counter measures is considered reasonable as a conservative value considering some unknown factors.

From the impounding tests, it was observed that :

1. The uncontrolled drainage through the drainage wells after filling the lake up to Full Supply Level was estimated at about 10 m^3/min.

2. Out of the total amount of 10 m^3/min., a discharge of 6 m^3/min., should be released from the drainage wells in order to control the ground water level in the downstream area to prevent unexpected water seepage.

3. By restricting the release from the wells to 6 m^3/min. the uplift pressure on the dam base would be controlled to 12 m at the downstream toe of the dam while it is 2.70 m in the case of free drainage and 25.20 m in the case of no drainage.

4. The counterweight could be made half the size of the previous design, which meant that the top elevation could be lowered by 5 m.

5. Additional curtain grouting was judged unnecessary since the effectiveness of the curtain grouting is not so significant, as far as cut off of the seepage is concerned.

The impounding was carried out in August and September 1981, until elevation + 1,179.50 was reached.

Based on the findings in this test, a counterweight as proposed on the fifth design change was constructed and completed by February 1982.

3.5.3 Design Change Data.

From the impounding test results, the seepage amount was estimated to be about 10.40 m^3/min. or about a half of that of the fourth design change. So the fifth design change was proposed.

The design was the same as the fourth design change except that the top elevation of the counterweight was lower.

Dam base 54.26 m, two keywedges, cancelled increase of additional grouting, top of counterweight elevation + 1182.000 with impervious blanket and estimated amount of water 10.40 m^3/min.

4 THE RESERVOIR IMPOUNDING

Soon after the fifth design change was approved, construction of the counterweight and other appurtenant work started and was completed in February 1982.

In April 1982, immediately after the completion of the drainage wells at the downstream toe of the dam, reservoir impounding started again. During impounding, it was found that the ground water levels at the observations holes were still higher than expected, in spite of the fact that the designed seepage quantity was released from the drainage wells.

Besides, a leakage was observed downstream of the reservoir, in a spring located in a brook in the village about 600 m on the right bank side from the dam and 300 from the edge of the reservoir.

The other springs were located about 1 km downstream of the dam along the Menjer river. The springs were seen in quite a wide area ranging about 40 m wide along the river and 10 m high along the slope of the right bank and also leakages were observed from original ground elevation around the dam area.

Leakage were found at the weep holes on the facing concrete between the counterweight and 3 block of the penstock foundation. After the cracks occured on the facing concrete, packed palm fibres and gravels were taken out from the weep holes in order to release uplift pressure between the concrete facing and the bedrock and then the confirmed seepage water gushed out from the weep holes. These weep holes were enlarged and filled up with gravel 3 mm - 15 mm to decrease discharge velocity and to prevent piping in the bedrock, additional weep holes were provided near the above weep holes from which comparatively large amount of water seepage come out. So it was decided to suspend the impounding when the elevation reached 1186.60 on July 1982.

As result of intensive review and study of the seepage problem, construction of 8 relief wells along the penstock line downstream of the dam and some improvement works were proposed.

Soon after the construction of the relief wells was completed, reimpounding started in April 1983 with rise in level ranging from 8.6 cm/day to 40.10 cm/day.

When the reservoir water level had reached elevation 1190.73 on May 30, 1983, ground water was observed flowing from the surface of the back filled soil between the drainage canal of the relief wells and the steep cut slope of volcanic breccia behind it.

The location of the spring was nearly 3 m from RW 6 toward RW 5 and was named SPGI for identification. Upon excavating the surface of the backfill material, minor boilings of sand were observed at the spot where water was rising up. However, it could be judged that it was washed sand by spring water from the backfill materials, and not from the bedrock. The spring was accompanied by small bubbles with a light smell of sulphur. From these findings, it was considered that the spring water was coming up from the artesian aquifer of lapilli tuff through the caprock of volcanic breccia, which is about 12 m thick in the vicinity of RW 6.

On the other hand this spring seemed to have appeared shortly after suppression of the discharge quantity from RW 6 so as not to exceed the allowable capacity of 1 m^3/min. measured on May 29, 1983.

In order to avoid the risk of piping action in the bedrock, the reservoir water level was lowered from elevation 1,190.735 and was kept constant at elevation 1,190 from June 1, 1983 to June 15, 1983 until the remedial measures were completed.

From remedial measures against the possibility of piping, the backfilled materials behind the drainage canal at the foot of the cutslope was replaced with well graded sand and gravels in three layers, 20 cm thick lower layer with particle size less than 15 mm, a 30 cm thick middle layer with 3 to 40 mm and a 30 cm thick upper layer with 15 to 40 mm. In addition, 50 mm diameter perforated PVC pipe was embedded in the middle layer, along the drainage canal, to dis-charge spring water. The area provided with filter protection was about 95 m x 1 m, at the right side of the penstock line of block 2 and block 3. This was done in the area behind the drainage canal between RW 3 and RW 6 and RW 2 as well.

In the afternoon of August 4, 1983, cracks appeared on the facing concrete along the penstock line between block 1 and block 2 and on the right side of block 3 and in the 14 m section downstream of block 3. Also, some cracks were found at the drainage canal of the relief wells which had been laid on the berm to lead the well discharge water to the pump house.

After finding these cracks, the gates at the Serayu intake (entrance of the Serayu Menjer tunnel) were immediately closed to lower the reservoir water level from elevation 1196.735 to elevation 1196.000 for safety and level was maintained until August 5, 1983.

Later on, some of the drains and the relief wells were opened a little to decrease uplift pressure acting on penstock line downstream of the dam. When the cracks occured, the weep holes on the concrete facing were almost clogged by palm fibres and gravels.

Therefore, it could be judged that the uplift pressure built up was due to the weep holes not working effectively. The concrete facing downstream of block 3 was found to have lifted by 2-3 cm, which was observed by opening a hole of 30 cm^2 in the concrete facing. The elevation of the penstock was checked on the section from the toe of the dam up to the culvert downstream of block 3, and it was found that the penstock was 1-2 cm lower than at the time of installation. This settlement was deemed due to the weight of water which filled the penstock and saddle piers.

In order to relieve uplift pressure acting on the concrete facing, the saddle piers and anchor blocks, additional weep holes and enlargement of the existing weep holes was carried out.

According to a further review of the core sample from OB-15, which is located about 20 m upstream of RW 6, and the boring record of RW 6, the caprock over the confined aquifer is thinnest in the part just downstream of BL 3 and seepage water could most easily rise up to the ground surface.

As a result, additional counterweight on the concrete facing on both sides of the penstock just downstream of block 3 was recommended. It was made by rockfill and wet masonry surfacing, approximately 160 m^3 or about 320 ton.

This construction work started on August 15, 1983 and was completed on September 10 except for some finishing work.

This reimpounding up to the Full Supply Level at elevation + 1,199.00 was completed on September 14, 1983.

5 CONCLUSIONS

1. Adequate investigation to check the geological condions as well as permeability, especially on the dam area are very important before construction to avoid problems.
2. Remedial measures and uplift control are very important to ensure the dam stability during construction as well as operation and maintenance.
3. During impounding of the reservoir, it is necessary to check the site condition, not only around the dam area, but also all the area where cracks and seepage might occur.
4. The role of geotechnical engineers are very important for construction of large dams so their participation should start from the beginning of the investigation.

REFENRENCES

Difficulties arising from inadequate investigation in Menjer Dam. Presented by Mr. Soedibyo in XIV ICOLD Congress in Rio de Janeiro.

Observation and monitoring on seepage during reimpounding in 1983. Nippon Koei Co, Ltd Tokyo.

Japanese National Committee on Large Dams.

Remedial measures for seepage and uplift control, prepared by Soedibyo in XV International Commission on Large Dams (ICOLD) in Laussanne, 1985.

Developments in Geotechnical Engineering, Balasubramaniam et al. (eds) © 1994 Balkema, Rotterdam, ISBN 90 5410 522 4

Efficient driving of precast concrete piles

A.F. van Weele & A.J.G. Schellingerhout
Institute of Foundation Verification-IFCO B.V., Gouda, Netherlands

ABSTRACT: Pile foundations fulfill a dominant role in foundation construction. Driven prefabricated concrete piles are in this respect used on a worldwide scale. The driving of such piles should be done in an efficient way in order to limit public nuisance, while at the same time care should be taken that the driving stresses do not surpass the strength of the pile. During driving many parameters can be deduced and by a clever use of the observed values it might ultimately be possible to predict the pile's static bearing capacity with an acceptable degree of accuracy. By making use of electronic data acquisition, fit for site use, very detailed information can be collected continuously. The cost thereof are small enough, not to affect competition. Until now, electronic data collection has been the domain of the specialists and mainly due to the limited scale of application, the degree of accuracy of such predictions is still too low to be used as the only source of information.

1 INTRODUCTION

The behaviour of prefabricated foundation piles during driving is influenced by the driving equipment and by the soil profile. The driving equipment is selected prior to the start of the work and the characteristics thereof are known during driving. The general soil profile is obtained in advance by means of site investigations so that this is known also. In soil that varies strongly, there may be substantial variations in pile behaviour and it is per definition uneconomical to have a so detailed investigation, that at the location of each and every single pile the soil profile becomes exactly known. If, however, the bearing capacity of the piles is, in general, known for the average soil profile, it would be very useful if local variations could be deduced from the driving behaviour of each single pile. If this would be possible and the bearing capacity of each pile could be more or less accurately assessed on the basis of the actual driving behaviour, a much better quality control of the completed pile foundation would be achievable, allowing the use of a reduced factor of safety, relative to the ultimate pile capacity. This would contribute to make foundation construction more economical.

A pile foundation of better quality and for a lower price comes within reach, if data collection could be made so simple, that it can be done by the piling crew as a standard routine for every pile and if the computerized data processing could lead to an on-line information for the piling operator. So, each pile could be driven to its own level, necessary to achieve a predetermined static bearing capacity. This would greatly improve the (soil mechanical) quality of the foundation by a more uniform load/settlement behaviour of the piles. Moreover, data collection would immediately signal pile damage, because the driving behaviour of a broken pile is very much different from that of an intact pile. Finally, this way of data collection per pile is more reliable and much more comprehensive than a skilled supervisor will ever be able to achieve. These data allow also a detailed interpretation afterwards, when queries might arise.

The step from a specialist activity to a daily routine for non-specialists is facilitated by the present day possibilities in the field of data collection and data-processing, using powerful, tailor made field-computers. IFCO has carried out a research program for the Dutch Association of Precast Concrete Pile Manufacturers (PREPAL) for developing a simple and reliable way of data collection. The result of the research is, that such a system is feasible and its application will not increase the cost of pile installation. It will replace the piling supervisor and makes integrity testing superfluous. PREPAL has ordered IFCO recently to equip 5 piling rigs with a professional version of the new data collection system, together with a field computer for on-line data processing, able to give detailed information to the operator.

First priority will be given to data collection. Experts from IFCO expect to need 1 year to compare the data to be collected with the 5 rigs, with the static bearing capacity of the driven piles, in order to refine the processing of the dynamic data to such an extent, that a value for the static bear-

Fig.1 Accelerometer, attached with a single M-8 bolt to the side of a 400 x 400 mm precast pile.

ing capacity of the piles will become available with adequate accuracy. The calibration to the actual static bearing capacity of the piles will, in general, be assessed by means of cone penetration tests (CPT's) and in particular also by static load tests to pile failure.

The main difference between the new system and already existing techniques, such as PDA in combination with Case method or Capwap analysis is, that the new system requires only a single, small and robust sensor and the data collection and interpretation is done automatically by the field computer. The final results are presented to the piling operator for immediate use. This does not leave any room for the application of a matching technique by a specialist. In order to compensate this "shortcoming", IFCO made it possible to use the total settlement, the rebound and thus, also the permanent settlement for each blow. These values are computed from the acceleration signals. In this way the computer gets adequate information to present an acceptable approximation of the soil resistance. PDA-analyses in combination with the Case- or Capwap-method are different as they r-equire more and also more complicated instrumentation (usually 2 sensors for strain and 2 sensors for acceleration). This makes the known methods unfit for an exclusive use by the piling crew on a daily routine basis, without an increase of piling cost.

2 DATA COLLECTION

Data are collected by an accelerometer, temporarily attached to the pileshaft, just below the piletop by means of a single M-8 bolt (Fig. 1). The obtained analog signal is transmitted via a cable to a 486 tailor-made field-computer, mounted in the operator's cabin, by which it is converted into a digital signal, stored in the memory and processed by the computer. Processing is able to supply the following information per blow:

1. time-interval between the blows in ms.
2. velocity of stress-wave through pile in m/s.
3. maximum compressive stress just below pile-top in MPa by multiplying the maximum velocity with the pile material impedance.
4. duration of contact between pile and hammer in ms.
5. efficiency of energy transfer to pile as a percentage, neglecting the reflected wave during hammer impact.
6. total penetration per blow in mm.
7. rebound of pile-top per blow in mm.
8. blow-count vs. penetration.
9. warning to replace the soft wood packing between driving cap and pile-top.

This amount of information is too much for practical routine, so that it has to be condensed to the most useful data only. The experience to be gained with the 5 rigs will help to assess, what will be essential and what can be left away. The intention is, that a single sheet of paper (maximum A-4) is used to present the relevant data per pile and that at the end of the job more general information will be presented for all driven piles together, such as real production time as a function of total time, variation in driving depth, same in driving resistance, etc.. It may be sufficient to collect data from every blow, but to process only every 5th blow. The sensor is equipped with 2-3

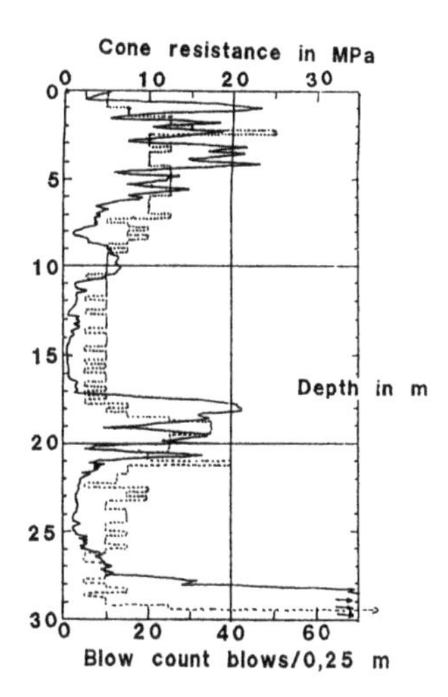

Fig.2 CPT and blow count diagram of a working pile, during the driving of which PDA- and PREPAL-measurements were caried out.

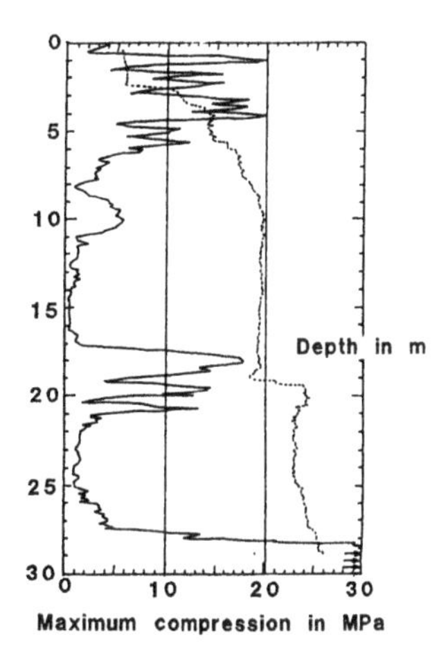

Fig.3 CPT vs. maximum average compression, observed in the pile during driving. No relation with static resistance.

m cable together with a connector. The piling machine has a cable fixed onto its leader mast and a connector at 1,50 m above ground. Normally, the connection between the sensor and the computer is made as soon as the pile has penetrated so much that the connectors have come within reach and can be mounted. It is, however, also possible to make use of a separate extension cable between the 2 standard connectors, so that the entire penetration can be monitored.

The data do not only serve the client with valuable information, but also the contractor with accurate details about daily working time, down time, driving time, etc. etc.. These data can be used in preparing future quotations for piling jobs under comparable circumstances. By applying different hammers under the same circumstances, the efficiency can be very accurately compared, leading to a better hammer-selection. Hammer efficiency will reveal the hammer quality. The system is also able to reveal the efficiency of alternative materials for the dolley or the packing in the driving cap.

Accurate data collection during piling as a daily routine at site level is now made possible and the computer will show to be a very powerful tool in the hands of piling contractors to improve their work in quality and in efficiency. It will, however, take time and much effort to exploit the new possibilities in full and to apply them in the best possible manner.

3 MAXIMUM COMPRESSIVE STRESSES

Precast, prestressed concrete piles were used for the foundation of piers, supporting a fly-over in Amsterdam. These piles had a section of 400 x

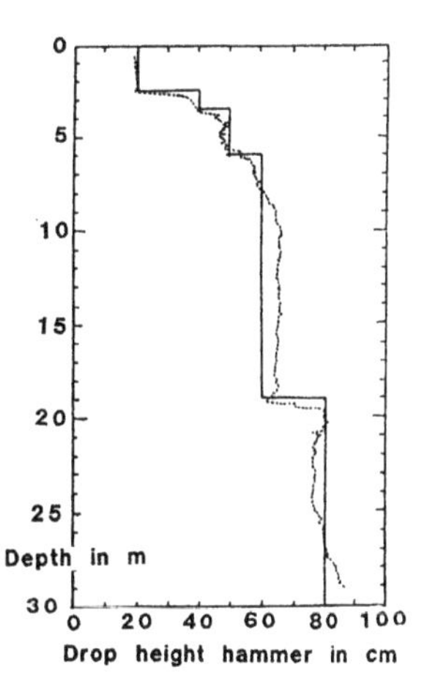

Fig. 4 Maximum average compression in pile vs. drop height of hammer. Excellent correlation.

400 mm² and a length of 29,5 m. The pile mass was 11,3 ton and they were driven with a hydraulically operated free-fall hammer with a mass of 10 ton. In Fig. 2, the cone-resistance at the pile location is presented together with the blow-count diagram. There is a good relation between the 2, indicating, that the blow-count could be a clue to the penetration resistance of the pile. Fig. 3 shows the cone-resistance as function of depth together with the maximum compressive stress in the pile-top, as obtained per blow. It is obvious, that there is no relation between the static soil resistance and the maximum force in the pile during driving. This

Fig. 5 Damage due to overstressing the pile material.

very important aspect is not yet widely understood, especially not among people who are directly involved in piling. Fig. 4 shows again the maximum compressive stress in the pile-top as a function of pile penetration, together now with the applied drop height of the hammer. It makes clear that there is a strict relation between both. This relation is so, because the speed with which the hammer hits the pile (in fact the driving cap hits the pile-top) determines the maximum compressive stress in the pile-top during the initial stages of the impact, together with the overall quality of the driving cap. This overall quality is determined by the mass of the cap and by the stiffness and damping properties of the dolley and the packing. The properties of the packing may change during its usage. The compressive stresses do not reflect, in any sense, the static soil resistance.
The actual velocity of the hammer at impact will have been smaller than the theoretical one, so that the true multiplication factor between the hitting velocity and the maximum compressive stress will be larger than indicated. The foregoing implies, that the piling operator is able to keep the driving stresses below a certain preset limit, determined by the quality of the pile-concrete. He can do this by limiting the drop height of the hammer. In The Netherlands, the usual pile-concrete has a characteristic strength of 52,5 MPa, meaning that the compressive strength of 150x150x150 mm test-cubes, stored for 28 days under water of 20° Centigrade, is equal to or in excess of 52,5 MPa for 95% of the tests. Experience has shown, that the average peak-compression, measured during driving at 3 pile diameters below the pile top, should remain below 40 MPa in order to limit pile damage to an acceptable degree.

A maximum compressive stress of 40 MPa requires for the Dutch conditions a hammer, of which its drop height is limited to such an extent that the hitting velocity is not more than 40/8 = 5 m/s. The hitting velocity = v = $\sqrt{2.gh}$, in which g = 10 m/s² and h = free fall height. This equation leads to: $h = v^2/20$. In order to limit the compressive stresses to 40 MPa, the drop height should be maximum: $5^2/20 = 1{,}25$ m, regardless the mass of the hammer, the mass of the pile and regardless the soil resistance to be overcome. In order to achieve a good penetration of the pile per blow, a heavy hammer is to be preferred above a light hammer. A heavy hammer has an impact that does not induce a larger compressive force than a light hammer, but the impact lasts longer than that by a light hammer, and it thus gives a larger pile penetration per blow. If concrete piles would be driven onto very hard soil, e.g. rock, the maximum compression could very well occur near the pile base in stead of near the pile top. Under such conditions, a different way of driving will result.

There is an upper bound to the weight of the hammer. If the hammer is too heavy, the impact lasts long enough to give the pile a large velocity, while upon deceleration of the hammer-velocity, the pile may loose contact with the hammer. Although this happens only during a split second, it is long enough to render the pile a free end so that the compression wave, reflected by the pile foot as a tensile wave, will be able to let the pile-top rebound freely. The efficiency of the energy transmission decreases as soon as this happens. It is to be preferred that the hammer remains in contact with the pile-top. In that case, the hammer mass will reduce rebound of the pile and the energy transfer will be more efficient. Efficient driving is obtained when the mass of the hammer is approx. 75% of the pile mass. The upper limit is approx. 100%, while the lower limit is 50 - 60%. A hammer that has a mass equal to or less than 50 % of that of the pile is, in fact, too light. Such a hammer should hit the pile-top very hard in order to achieve pile penetration and this hard hitting induces easily pile damage. See Fig. 5.
The values given are not applicable for concrete piles if they are driven onto rock.

The correct selection of the hammer mass has shown to be very essential in the daily piling practice. What happens in The Netherlands, and possibly elsewhere, is the following. Contractors have certain equipment available. The demand of the designers is for piles with increased bearing capacity, while at the same time piling contractors are often inclined to offer smaller numbers of heavier loaded piles as economical alternatives. Both developments result in the need to drive heavier piles with available equipment and thus relatively light hammers. This, in its turn, results more often in pile damage, until the demand for heavier piles is large enough to encourage contractors to invest in heavier hammers. But heavier hammers require usually also heavier piling equipment (crawler cranes) and this is a very costly matter. It is more logical to design pile foundations, bearing in mind the application of available piling hammers. By selecting the hammer first, the maximum pile mass can be set as 1,5 or maximum

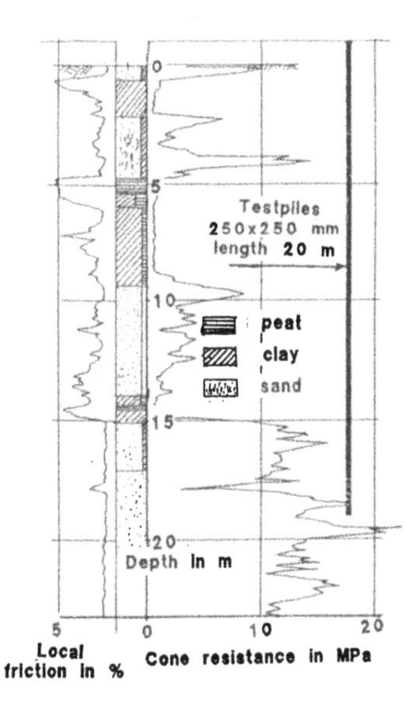

Fig. 6 Soil profile in Delft, where 4 identical testpiles were driven by different hammers.

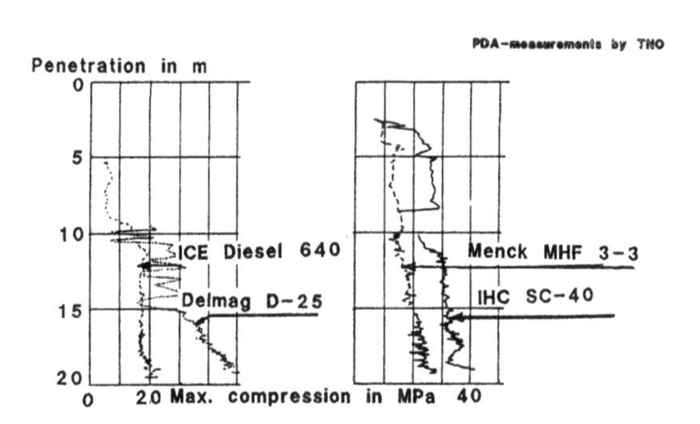

Fig. 7 Maximum pile compression vs. depth in a pile by using 4 different hammers.

2 times the hammer mass. The driving depth results from the local soil profile. Now, the selected pile mass in combination with the selected pile length gives the maximum pile cross-section that is to be applied. That section and the given length determine together the pile capacity, so that finally the number of piles results. Such an approach in the design stage of a pile foundation reflects the importance of aiming at an efficient execution of the piling work. *A good and efficient foundation design allows an uninterrupted and trouble-free execution, whereas difficulties on site during execution are the result of an inadequate foundation design.* This aspect is insufficiently understood by many designers.

The driving cap serves two important functions. The first one is to distribute the impact evenly over the pile section and to avoid as much as

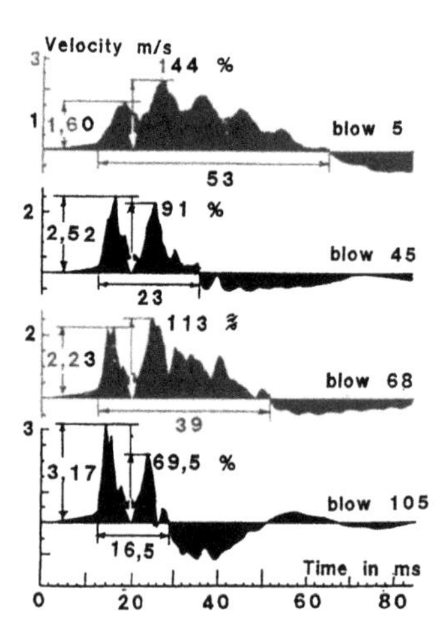

Fig. 8. 4 characteristic velocity/time diagrams at 4 different penetrations, with different soil resistance.

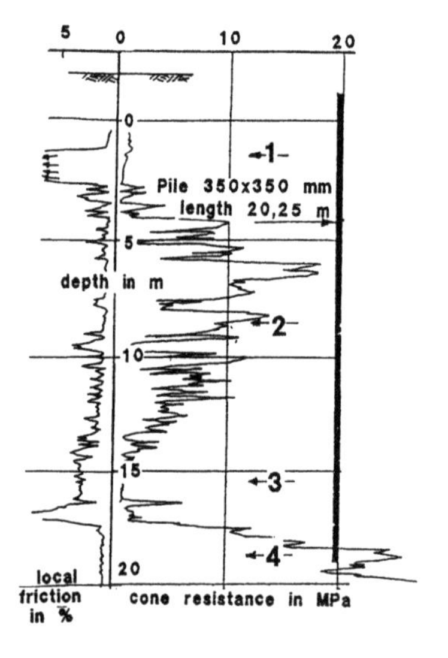

Fig.9 CPT diagram of piling site with approximated penetrations, where the 4 four diagrams of fig. 8 were collected.

possible local peak-stresses (i.e. directly above the main reinforcing bars). The second one is to decrease the peak force by extending the duration of the impact. IFCO's envisaged data processing system will be able to detect the very moment that a certain defined compressive stress in the piletop is reached and the system is thus able to warn the operator to replace the soft wood packing between the driving cap and the piletop. The operator will then have the choice to do this right away, or to proceed the driving with a lower drop height, until the driving cap is accessible from the ground, to facilitate its replacement.

4 SELECTION OF PILING HAMMERS

The maximum compressive stress in a pile during driving can be influenced by the speed with which the hammer (or more correctly: driving cap) hits the pile-top. Different hammers will also generate different stress/time relationships during the driving of a given pile. An interesting example was obtained by the driving of 4 identical prestressed concrete piles during the 4th International Stress-wave Conference held in 1992 in The Hague and Delft, The Netherlands. The piles were 20,00 m long and had a section of 250 x 250 mm². Each was driven in a very uniform soil profile to a penetration of 19,00 m (Fig. 6).

The hammers used were:

ICE, double acting diesel hammer DPH-640, Delmag diesel hammer D-25, Menck hydraulically operated free-fall hammer MHF 3-3 and IHC hydraulically operated free-fall hammer SC-40.

The ICE hammer needed the shortest time to reach penetration, followed by the D-25 and next both hydraulically operated free-fall hammers. The maximum compressive stresses observed by TNO in the piles during driving are presented in Fig. 7. The double acting ICE diesel hammer generated the lowest stresses in the pile, whereas the Delmag hammer gave maximum stresses as high as 45 MPa on average. Also, between both free-fall hammers, substantial differences in generated compressive stresses were observed. These results confirm that not the static soil resistance is influencing the stresses during driving but only and alone the dynamic performance of the selected hammer. It shows how important the hammer selection is and also the way the hammer is operated.

This last mentioned aspect is important for the hydraulically operated hammers, for which the operator selects the drop height. Especially during the initial stages of driving, when the pile is above ground over most of its length, driving should be done with a low drop height. In that stage damping is low due to minimal side friction and toe resistance. As a result, the stress-wave will reflect several times. With increasing pile penetration, the side friction will increase making the reflected wave smaller and smaller, while an increasing end-resistance will decrease the duration of the downward pile movement per blow. Refer to Fig. 8, in which 4 velocity/time diagrams are presented, recorded during the driving with a dieselhammer of a single precast concrete pile at increasing penetrations. The compressive stresses near the piletop are proportional to the velocity, until the first reflection reaches the pile section to which the sensor has been attached. The period, during which the piletop moved downwards, reduced from 53 ms for blow 5 to 16,5 ms for blow 105 when the pile-base had entered the bearing layer. The CPT near the location, where the pile was driven, is presented in Fig 9. The approximated pilefoot penetration at the 4 mentioned blows has been indicated in the diagram.

The more pronounced the impact-wave is, the more pronounced the reflected wave will be. Fast changes in velocity with time (steep slopes in the diagram) will help to induce tensile stresses. Tensile stresses are usually highest in the pile section, between 50 and 66% of the pile length below the top.

Diesel hammers are, in their behaviour, very different from free-fall hammers. The compression, taking place within the hammer-cylinder just prior to impact of the piston, firmly pre-stresses the system: hammer - driving cap - pile. This is a very favourable feature of the diesel hammer. The explosion is to happen a few ms after impact. This explosion increases the duration of the impact. Both pre-stressing and explosion reduce the development of tensile stresses in the pile during driving. The diesel hammer adjusts its hitting velocity automatically relative to the soil resistance. If the soil resistance is very low, the pile may even move already under compression, that the hammer does not hit the anvil hard enough to cause ignition of the fuel.

The major contribution to the pile's penetration comes from the impact, while a minor part is caused by the explosion. The larger the soil-resistance, the more the explosion will be used to lift the piston. The smaller the soil resistance however, the more the pile will sink during the explosion and the smaller will be the lift of the piston. Only when the pile meets near refusal, the hammer will jump highest and only then the maximum energy is delivered by the hammer to the pile. Driving the pile is done by the impact of the piston and not by the explosion. The manufacturers indication of rated hammer energy is misleading because the hammer may jump as high as 3 - 3.5 m height, whereas the impact velocity is not exceeding 5,5 m/s. The hammer speed is in last instance slowed down by the compression, so that not the real drop height is effective, but a substantially lower one. Sometimes when the hammer is overheated, the explosion advances in time and may even take place immediately before the impact. If that happens, the impact is counteracted by the explosion and the hammer is not at all effective anymore. Therefore, the s.c. rated energy is a value that should not be given too much weight.

5 PILE PENETRATION PER BLOW

By recording the acceleration of the pile as a function of time, the obtained values can be double integrated as a function of time in order to obtain pile penetration, also in time intervals of 0,1 ms. This routine is straight forward and seems

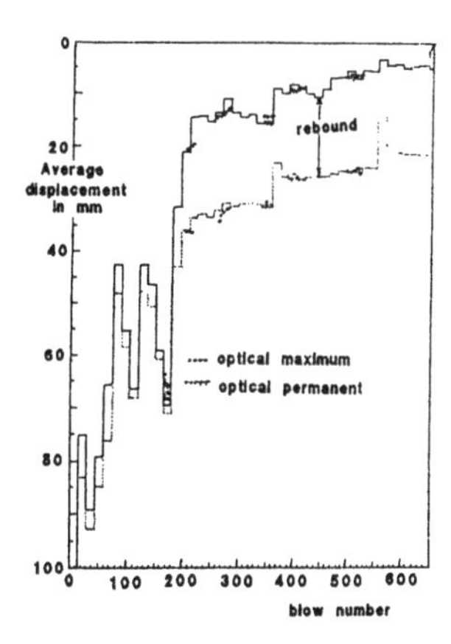

Fig.10 Calculated displacements of piletop, by double integration of accelerograms. Comparison with the same values, obtained by direct optical displacement measurements.

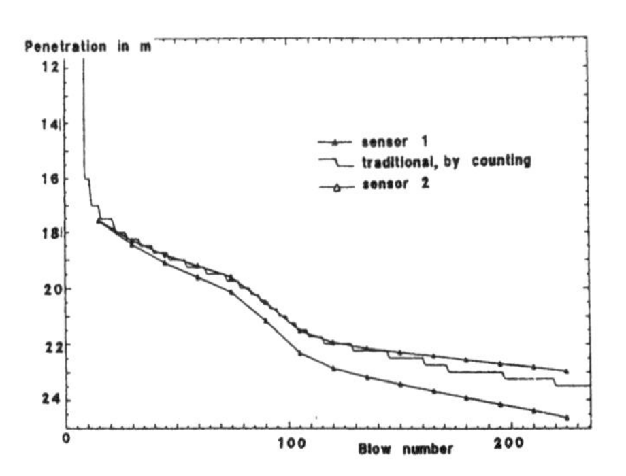

Fig.11 Blowcount diagrams as obtained via 2 independent accelerometers and obtained in the conventional way by counting.

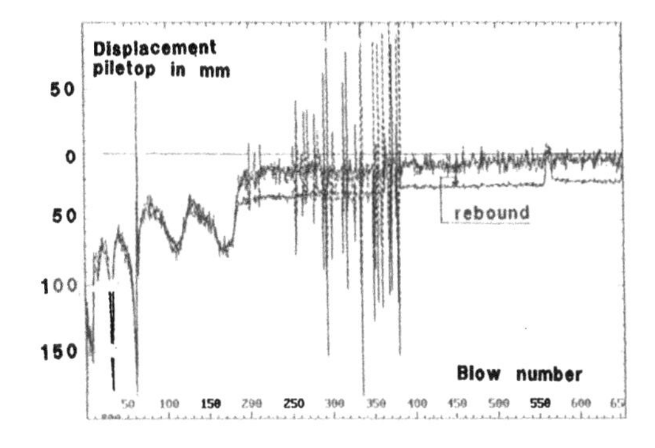

Fig.12 Calculated pilehead displacements per blow. Scatter in permanenent displacements relatively large. Much less for maximum displacements.

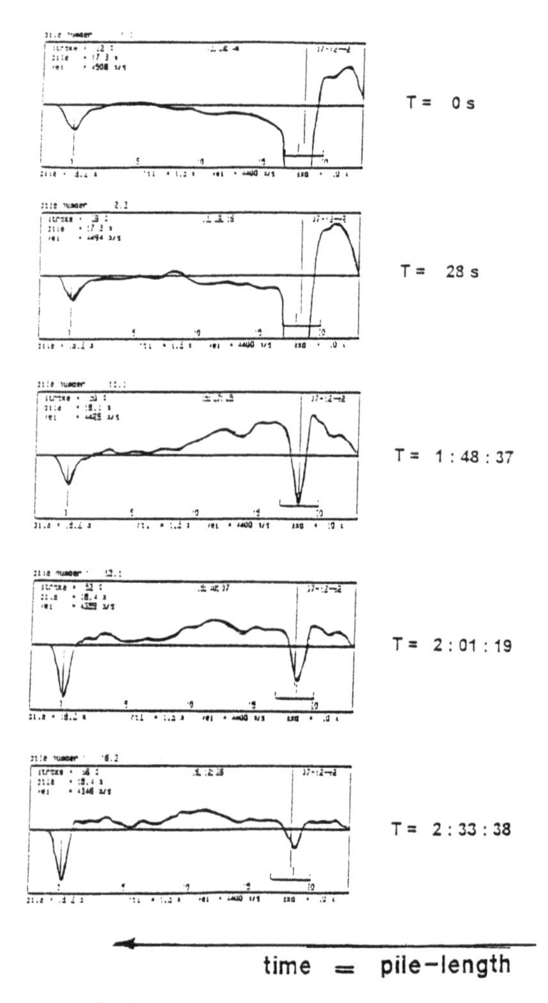

Fig.13 Pile Integrity tests on same pile at increasing time intervals after termination of the driving.

simple. The reality is, however, different. The problem is that, it may take approx. 200 ms before the pile-top is at rest after each blow and only then the permanent deflection is reached. The longer the period of time, the larger the imperfections in the output of the integration will be. Minor movements of the sensor at the very beginning, just prior to the impact, in stead of standing completely still is detrimental for a good end-result. The same applies after the end of the blow, when the sensor should come also to a complete standstill.

IFCO has been able to solve this problem with a computer model which calculates the most likely displacement versus time graph. The pile deflections can be assessed from the accelerations measured, so that the blow count diagram can be processed accurately enough. This has been shown by computing the pile movements and by comparing the results with those of direct deflection measurements at certain intervals. See Fig. 10.

Until now, this calculation procedure is only tested for concrete piles with settlements larger than 8 mm. In Fig. 11, two calculated blow count diagrams based on the integrated accelerations supplied by 2 different sensors are compared with the diagram observed in the conventional way by a supervisor. The results are promising but require still some further refinement. IFCO is however confident, that blow count diagrams obtained through data collection by a single accelerometer will become reliable enough for daily use in the very near future. By using the system on 5 piling machines, a further degree of refinement will be achievable. In Fig. 12, the computed average **total** deformation per 10 blows is presented for all blows of a driven pile, while also the calculated **permanent** deformations have been presented in the same manner. The distance between both lines represents the difference = the elastic rebound of the pile-top at the end of each blow.

Interesting is it to see, that the scatter in the total deformation is much smaller than that for the permanent penetration and, thus, also for the rebound. That is because the maximum deflection occurs soon after the impact, so that a relatively short integration leads to the result, whereas the permanent deflection is only obtained after a much longer integration and that result is of course less accurate. Nevertheless, the average result is accurate enough to serve the purpose. To be frank, the supervisor in the field is per definition not accurate either.

When the permanent deflection per blow is obtained, the computer is able to present the blow-count in any appreciable way. For instance, in a number of blows per 0,25 m penetration as is usual in The Netherlands in order to be able to compare with the cone-resistance observed with CPT's. But also as a continuous line of number of blows against total penetration, as is done in other countries, or as a settlement per blow, just as is done in Fig. 10 and 12. Careful study of Fig. 12 shows, that during easy driving the rebound is small and it shows an intention to increase when the permanent settlement decreases.

The fact that there is a relation between the blow-count diagram and the cone-resistance measured with CPT's, shows that frictional resistance during driving is low. This is confirmed by the set-up that occurs when driving is stopped. Set-up indicates that friction develops strongly after the driving has been stopped. It implies that the soil resistance opposing pile penetration during driving consists for a large part of end-resistance. End-resistance determines the pile elasticity as well as the soil elasticity below the pilefoot. So, the rebound could very well be some form of a yardstick to determine the static end-resistance. There is also another parameter that seems useful and that is the duration of pile penetration. The larger the soil resistance, the shorter and smaller the

Fig.14 Tensile cracks in shaft of reinforced concrete pile.

reflected wave lasts. An increasing part of the impulse is transferred into the bearing stratum, so that a smaller part of it is reflected and able to reach the very pile-top. The duration of the pile penetration is accurately measured and can be used to reveal a useful relation between soil resistance and pile behaviour.

The information collected with the PREPAL method is different from that collected with the conventional PDA-systems. The new system doesn't collect enough information to separate the compression wave from the reflected wave, but at the other hand, it renders the total deflection of the pile head, the permanent deflection and the rebound per blow, which form important information. The new system will be fully automatic and hopefully to all piles driven. Interference by an expert will ultimately not be required.

The set-up of friction with time after the termination of driving can be made visible by means of Integrity Testing. The reflection from the hammer blow is generally obscured by side friction. By testing the pile immediately after driving, a first echo is obtained when the set-up is still minimal. By repeating the test at time-intervals after driving, a good impression is obtained about the set-up in general. Some of the results are presented in Fig. 13. Re-driving a previously driven pile and observing the stress diagrams shows a similar effect in reverse. Friction is large in the beginning and decreases with the number of blows and the penetration achieved.

6 PILE DAMAGE

Giving blows with a heavy piece of steel onto a prefabricated concrete element compares with ordinary demolition. Driving needs, therefore, great care to prevent damage. A major cause of pile damage is overdriving by using a hammer that is too light compared to the pile mass. A free-fall hammer should have a drop height during the driving of concrete piles not exceeding 1,25 m. For diesel hammers, the actual drop height often reaches 3 m, but nevertheless, the hitting velocity remains below 5,5 m/s. Driving needs also a good matching driving cap in order to minimize eccentricity during driving.

Driving relatively heavy piles with a relatively light free-fall hammer through soft strata, using a heavy driving cap with a stiff packing, are conditions that all contribute towards the generation of pronounced tensile reflections, easily resulting in substantial tensile forces in the pile shaft. These forces may even be so large, that the tensile strength of the pile is surpassed. If that happens, the pile will be subject to tensile cracking (Fig. 14). If a cracked pile reaches a little later the bearing stratum or another hard layer making severe driving necessary, the resulting stress-changes will slowly deteriorate both sides of the concrete at all cracks. Prolonged hard driving will make, that the shaft will ultimately fail under shear near one of the cracks and the top section of the pile will move alongside the lower section. This will be immediately recognizable from the driving behaviour of the pile, that will differ strongly from that of intact piles.

In Fig. 15, the velocity/time diagrams of the very last 4 successive blows before termination of the driving of a precast concrete pile resulting in failure have been presented. The first of the succession shows still the normal pile reaction prior to failure. Blow 2 gives a small, but premature reflection in between the main impact and its normal reflection (between 16 and 20 ms) This makes it clear, that something noteworthy is happening. Blows 3 and 4 show that the pile has failed.
In both diagrams, the first reflection comes much sooner (after 4,5 ms in stead of 9 ms) and the pile sinks for a much too long period (36/37 ms in stead of 14 ms), indicating that the soil resistance has decreased abruptly and substantially. Pile failure happened very fast in this example, so fast that a timely intervention by the operator was made impossible even when he would have had the data displayed in real time on his monitor.

The integrity of the pile in the example was checked twice by means of a standard IT-system. See Fig. 16.
The reflection at the pile-top was observed at a nearby pile. This is the reflection that was used as a reference. The reflection in the middle was the one collected about 10 blows prior to the pile's failure, whereas the lowest diagram was obtained immediately after failure. The last mentioned diagram shows clearly that the pile failed at

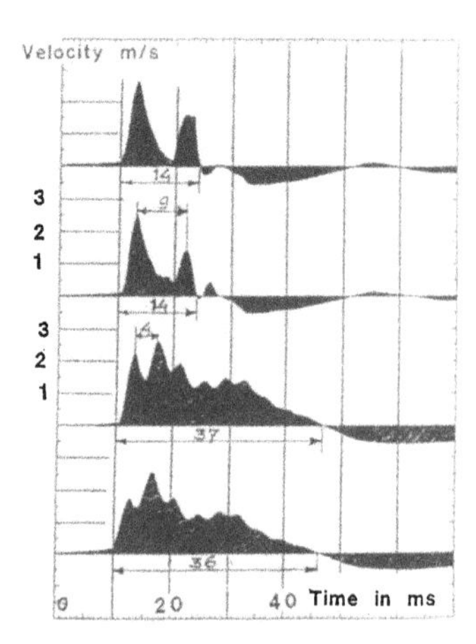

Fig.15 The last 4 succesive blows on a precast pile. The pileshaft sheared during the blow, shown in the second diagram from top.

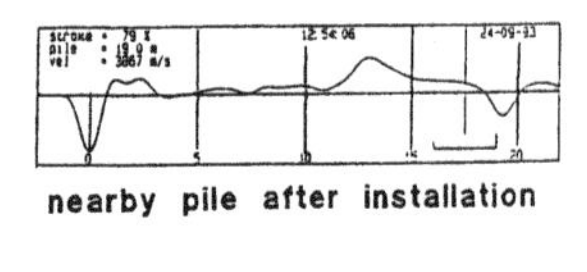

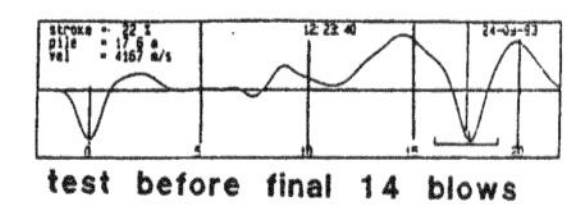

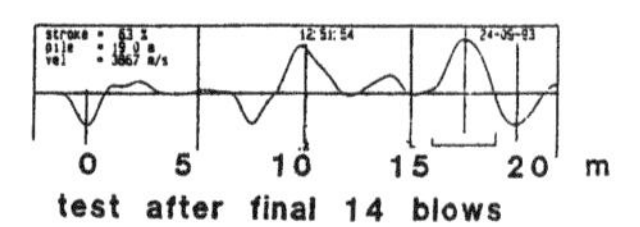

Fig.16 Pile integrity testing before and after failure of the pile.

approx. 7,5 m - top. By reconsidering the middle velocity/time diagram, it can be seen that the pile had already been damaged at the time of its testing (10 blows prior to failure). In that echo, a distinct reflection can be observed at 7,5 m - top, although the foot-reflection is still clear. This pile was already different from its neighbours, which should not be the case. Most probably, the prestressed pileshaft had at that moment seriously been overstressed, but was sufficiently intact to let the signal pass as well as its echo.

Fig. 17 shows a picture of another prestressed pile, that was on the verge of shear-failure at termination of the driving. All piles in the area were exposed by excavation and this revealed that one of the shafts had been overstressed seriously but both pile sections were still more or less in line with each other. Compressive forces could

Fig. 17 Exposed perstressed piles, which show seious damage due to overdriving. Use of too light hammer.

thus be transferred, as long as the pile would be supported laterally by the soil. The pile in the middle of the picture was in the final stage of damage, as both parts had sheared and moved along each other. The driving behaviour of the first pile did not indicate any anomaly because the blow-count diagram was not different from that of its (good) neighbours. The serious defect could thus not be deduced from its driving behaviour alone, not even after re-study of that behaviour, when the defect had been disclosed.

7 CONCLUSION

The foregoing makes it clear, that the driving of precast concrete piles is an art. Careful selection of the hammer and clear instructions to the crew how to use that hammer are essential to get the piles at a depth that is pre-determined. It is also essential to ensure that the condition of the piles after driving is still good enough to form a safe foundation. Present day possibilities, by means of electronic data collection and computerized processing, will greatly enhance our insight in the driving process. Such means will replace ultimately piling supervisors and enable operators to learn precisely how large the compressive stresses in precast concrete piles during driving are and to take adequate steps to prevent pile damage during the process of driving. By order of PREPAL, IFCO will equip in 1994 5 pile drivers with a specially developed data collection system combined with a field computer to present essential information on a real time basis to the piling operator. In this way, pile-driving will be improved and pile damage almost excluded. At the same time, it will enable the operator to drive the piles to a penetration, just enough to achieve a uniform, defined static bearing capacity, allowing the introduction of a smaller safety factor in the foundation design.

By collecting data and by driving piles with different hammers and different driving energies, by measuring pile behaviour before and after changing the soft wood packing in the driving cap and by comparing re-drives with the termination of the preceding driving, essential information will be collected. This information will be used to compare with the static bearing capacity of the piles, as derived from CPT's and static load tests so that on the long term, an acceptable degree of accuracy will be obtained in deriving the static bearing capacity on the basis of the actual dynamic driving behaviour. To establish such a general correlation is PREPAL and IFCO's final goal. The design and the construction of pile foundations will both very much benefit from it. A reliable and generally applicable correlation does not yet exist, notwithstanding the optimistic publications of many different specialists about this subject.

5 Soil-structure interactions

Developments in Geotechnical Engineering, Balasubramaniam et al. (eds) © 1994 Balkema, Rotterdam, ISBN 90 5410 522 4

Piles subjected to externally-imposed soil movements

H.G. Poulos

Coffey Partners International Pty Ltd & University of Sydney, N.S.W., Australia

ABSTRACT: There are several circumstances in which piles are subjected to loading by the movement of the surrounding soil past the pile. Such circumstances include:

1. the consolidation of soft clays around a pile, leading to negative friction forces;
2. the expansion or contraction of reactive clay around a pile, leading to either tensile or compressive additional forces;
3. the lateral movement of unstable or creeping slopes past piles, leading to additional bending moments and shears;
4. the lateral movement of soil during an earthquake, again inducing additional bending moment and shears.

The basis of analysing piles subjected to such movements is described briefly. The behaviour of piles subject to the above sources of soil movement are then described, and the implications for pile design are discussed.

1 INTRODUCTION

The majority of piles are designed to support "conventional" loadings, that is, loads which are applied directly to the pile head by a structure. However, there are many cases in which the piles must support "special" loadings, in particular, loads induced in the piles by the action of soil moving past the piles, vertically and/or horizontally.

This paper deals with the analysis and behaviour of piles which must withstand externally imposed soil movements under the four circumstances illustrated in Figure 1:

1. vertical soil movements due to consolidation of clay, which give rise to negative friction;
2. vertical soil movements due to swelling or shrinking of an expansive clay;
3. horizontal movements arising from slope movement or instability;
4. horizontal movements due to seismic effects.

In each case, the main design decisions are discussed, and some of the important characteristics of behaviour are described with reference to example cases.

2 BASIC ANALYTICAL APPROACHES

It is possible to develop a unified approach to the analysis of piles in a moving soil, using a boundary element approach. Such an approach has been described in detail for both vertical and horizontal response (e.g. Poulos and Davis, 1980), and involves the following broad steps:

1. discretisation of the pile into a number of elements;
2. development of equations for the displacement of the pile elements;
3. development of equations for the displacement of the soil adjacent to the pile elements; there are two components of soil displacement, one due to the pile soil interaction stresses, and the other due to the imposed "free field" soil displacements;
4. imposing compatibility of pile and soil displacements (while the soil remains in a non-failure state);
5. solution of the resulting compatibility equations, together with the appropriate equilibrium equations, to obtain the pile-soil stresses and displacements.

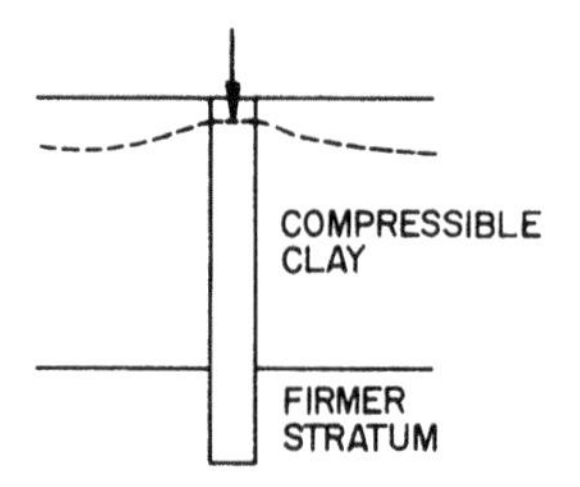

(a) PILE SUBJECTED TO NEGATIVE FRICTION

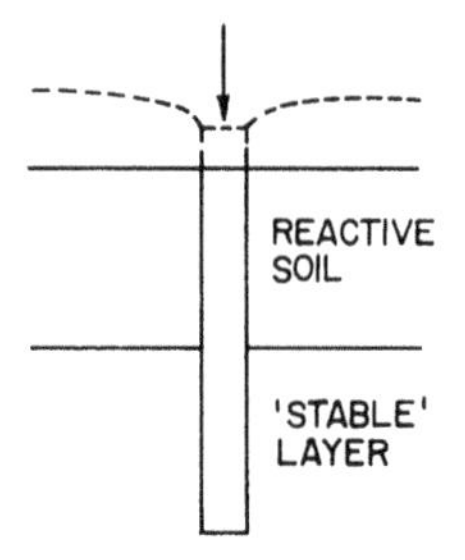

(b) PILE IN EXPANSIVE SOIL

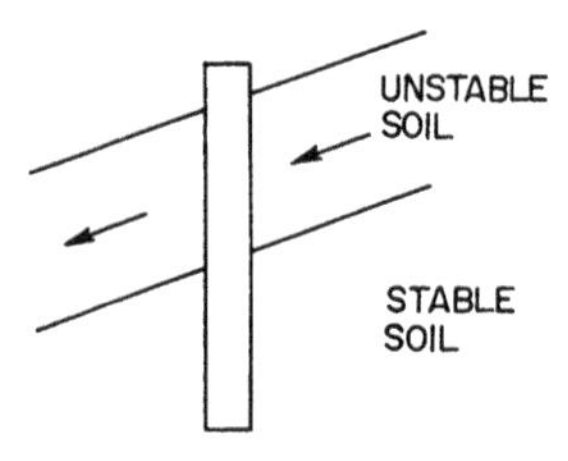

(c) PILE IN UNSTABLE SLOPE

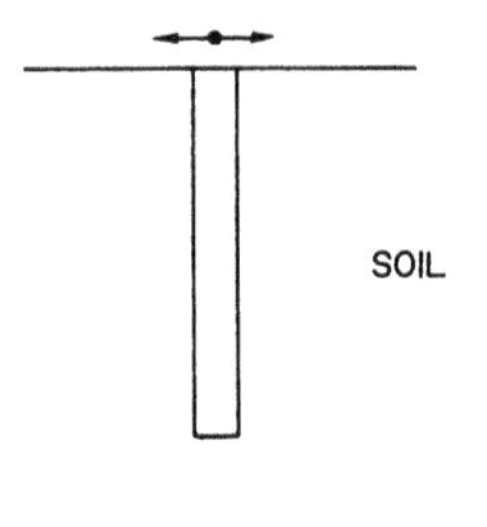

(d) PILE IN SOIL SUBJECTED TO SEISMIC EXCITATION

FIGURE 1 Examples of pile subjected to externally imposed soil movements

The most versatile and numerically tractable approach is to assume that the soil can be represented by an elastic continuum, but that there is a limit to the pile-soil stresses which can be developed. For vertical loading, these limiting stresses are the pile-soil shaft resistance (or skin friction) along the shaft, and the end-bearing resistance at the pile base. For lateral loading, the limiting lateral pile-soil pressures are specified.

It is possible then to account for the following aspects:

1. non linear pile-soil response;
2. non homogeneous soil;
3. group effects
4. sequential analysis, i.e. analysis of a pile subjected to a series of different types of loading.

A summary of the axial and lateral response analyses is given below.

2.1 *Axial Response*

The pile is divided into a series of elements, the vertical movement of which is related to the applied load, the pile-soil interaction stresses, the pile compressibility and the pile tip movement. The vertical movement of each supporting soil element depends on the pile-soil interaction stresses, the modulus or stiffness of the soil, and also on any free-field movements which may be imposed on the pile. To simulate real pile response more closely, allowance may be made for slip at the pile-soil interface, i.e. the pile-soil interaction stresses cannot exceed the limiting pile-soil skin friction.

By considering compatibility of the incremental pile and soil movements at each element, the following equation may be derived when pile-soil interface conditions remain in the non-failure state:

$$[I_s - A_D.F_E]\{\Delta p\} - \Delta\rho_b\{1\} = -\{\Delta S_e\} \quad (1)$$

where $[I_s]$ = matrix of soil influence coefficients

$[A_D]$ = summation matrix

$[F_E]$ = pile compression matrix

$\{\Delta p\}$ = vector of incremented pile-soil interaction stresses

$\Delta\rho_b$ = incremental displacement of pile tip

$\{\Delta S_e\}$ = vector of incremental free-field soil movements.

Details of the various matrices are given by Poulos (1989a).

At an element k at which the failure state has been reached, i.e., the pile-soil stress has reached the specified limiting value, the corresponding row in Equation 1 above is replaced by the condition:

$$\Delta p_k = 0 \qquad (2)$$

Equations 1 and 2 can be solved, together with the equilibrium equation, to obtain the incremental pile-soil stresses and pile tip displacements, from which the axial force and vertical displacement distributions within the pile may be computed.

The analysis of axial pile response therefore requires a knowledge of the pile modulus, the distribution of soil modulus and limiting pile-soil skin friction with depth, and the free-field vertical soil movements. This analysis has been implemented via a FORTRAN 77 computer program, PIES (Poulos 1989a).

2.2 *Lateral Response*

The lateral response analysis also relies on the use of a simplified boundary element analysis. In this case, the pile is modelled as a simple elastic beam, and the soil as an elastic continuum. The lateral displacement of each element of the pile can be related to the pile bending stiffness and the horizontal pile-soil interaction stresses. The lateral displacement of the corresponding soil elements are related to the soil modulus or stiffness, the pile-soil interaction stresses, and the free-field horizontal soil movements. A limiting lateral pile-soil stress can be specified so that local failure of the soil can be allowed for, thus allowing nonlinear response to be obtained.

By consideration of the compatibility of the horizontal movement of the pile and soil at each element, the following equation may be derived while the conditions of the pile-soil interface remain elastic:

$$[D + \frac{I^{-1}}{K_R n^4}] \{\Delta \rho\} = \frac{[I]^{-1}}{K_R n^4} \{\Delta \rho_e\}$$

where [D] = matrix of finite difference coefficients

$[I]^{-1}$ = inverted matrix of soil displacement factors

K_R = dimensionless pile flexibility factor

n = number of elements into which pile is divided

$\{\Delta \rho\}$ = incremental lateral pile-soil pressure

$\{\Delta \rho_e\}$ = incremental free-field lateral soil movement.

In addition, the horizontal and moment equilibrium equations, and the pile head and tip boundary conditions, may be expressed in terms of the displacements. The incremental pressure may then be evaluated from the equation of bending of the pile, and added to the existing pressures to obtain the overall pile-soil pressures. These values are compared with the specified limiting lateral pressures, and at those elements where the computed equation for that element is replaced by the pile bending equation which incorporates the condition that the lateral pressure increment is zero. The solution is then recycled until the computed lateral pressures nowhere exceed the limiting values.

In summary, the lateral response analysis requires a knowledge of the pile bending stiffness, the distributions of lateral soil modulus and limiting lateral pile-soil pressure with depth, and the free-field horizontal soil movements. A FORTRAN 77 computer program, ERCAP, has been developed to implement this analysis.

2.3 *Summary of Required Soil Parameters*

For axial response, the parameters required are:

Young's modulus of soil E_s
limiting skin friction f_s
end-bearing resistance f_b

For lateral response, the parameters required are:

Young's modulus of soil E_s
limiting lateral pile-soil pressure p_y

Assessment of these parameters is usually made on the basis of:

1. correlation with strength properties of soil;
2. correlation with insitu test data (e.g. SPT, CPT);
3. interpretation of appropriate pile load test data.

Further details are given by Fleming et al. (1985), Poulos (1989b) and Kulhawy and Mayne (1990).

3 NEGATIVE FRICTION

Much of the understanding of the behaviour of piles subjected to negative friction has come from consideration of end-bearing piles whose tip rests on or in rock or a very hard stratum. In such cases, the main design concern is the calculation of the additional force developed in the pile, and the consequent structural integrity of the pile.

In cases where the pile is founded in a stable, but still compressible layer, the negative friction problem becomes more complex, as both the force in the pile and the pile head settlement must be considered. In such cases, as stated by Van der Veen (1986):

"the design of a foundation on piles where negative skin friction will occur is a settlement of the pile problem."

A proper understanding of the negative friction problem in this case requires consideration of pile-soil interaction.

Using the computer program PIES, the hypothetical example in Figure 2 has been analyzed. The pile is a driven steel tube 20m long and 0.5m in diameter with a 10mm thick wall. The soil profile consists of a 12m layer of soft clay overlying a deep layer of dense sand. 3m of fill is assumed to have been placed on the surface of the clay, giving rise to the soil movement profile with depth shown in Figure 2. It is assumed that the skin friction is the same for both compressive and tensile loading, i.e. that positive and negative friction values are identical; it is believed that this is a realistic assumption for most cases.

A hyperbolic interface model has been adopted in the program PIES with the hyperbolic parameter R_f being taken as 0.5 for points along the shaft and 0.9 for the pile tip. The pile has been divided into 1m long elements for the numerical analysis.

Figure 3 shows the computed load-settlement behaviour of the pile, for values of the soil surface movement S_o varying from 0 to 300mm. It is found that the effect of the soil movement is to increase the movement of the pile, this effect becoming more pronounced as either S_o or the load P increases. The ultimate axial load capacity of the pile is about 3.17 MN, independent of the value of S_o. However, for larger values of S_o, this ultimate load will be developed only after very large values of settlement have been experienced.

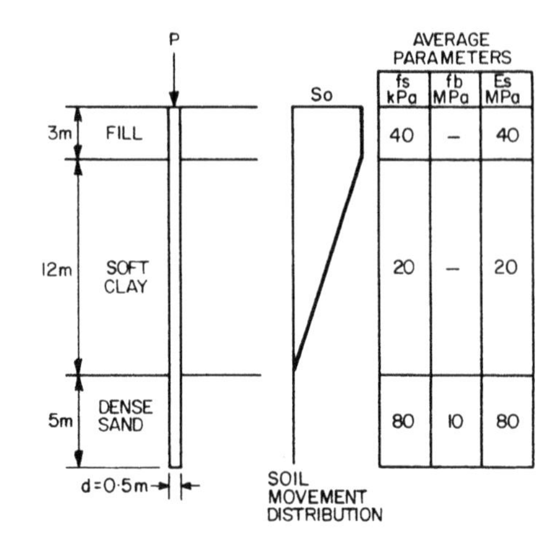

FIGURE 2 Hypothetical problem analysed

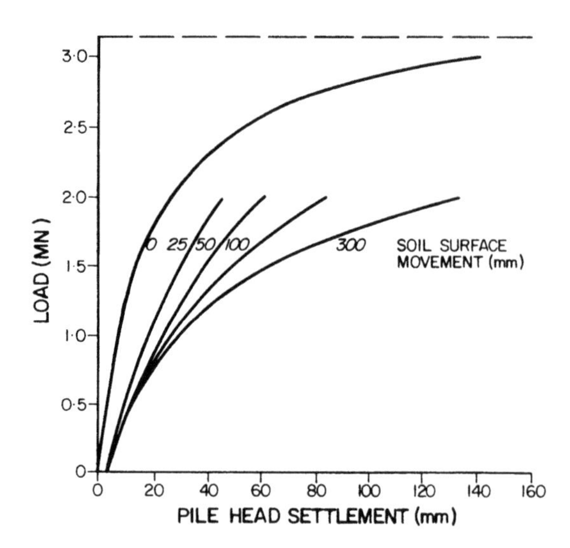

FIGURE 3 Influence of soil surface movement on load-settlement behaviour - uncoated pile

Analysis have also been carried out assuming that the pile is coated to a depth of 15m, and that the coating reduces the skin friction to 10% of the value for the uncoated pile. Figure 4 shows the load-settlement curves for the coated pile. The load capacity of the pile is reduced from 3.17 MN to 2.59 MN, but now the influence of the external soil movement is relatively small, and the increase in pile head movement due to negative friction is almost insignificant compared to the case of the uncoated pile. Also, the pile settlement behaviour

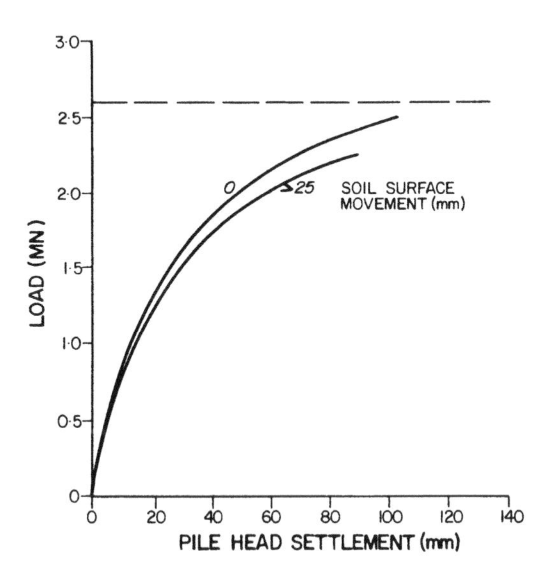

FIGURE 4 Influence of soil surface movement on load-settlement behaviour - coated pile

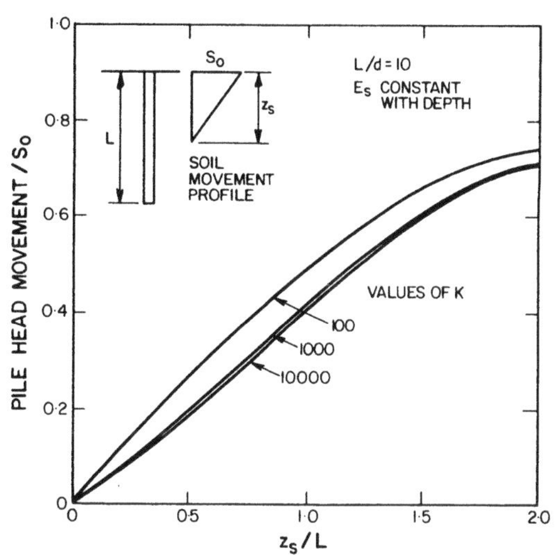

FIGURE 5 Elastic solutions of pile movement in expansive soil - uniform pile diameter

stabilises at small soil surface movements, generally less than 25mm.

The important points arising from this study are as follows:

1. the ultimate axial load capacity of a pile is not influenced by negative friction;
2. the load-settlement behaviour can be very significantly affected;
3. the greater the pile head load, the more a pile will settle under the influence of negative friction;
4. coating a pile can lead to a much improved performance, with higher allowable loads and smaller settlements than an uncoated pile.

4 PILES IN EXPANSIVE SOIL

In this case, the main design problems include:

1. estimation of the amount of movement of the pile;
2. estimation of the forces induced in the piles by soil movement.

The design of piles in an expansive soil is perhaps more difficult than for negative friction, because:

1. the ultimate skin friction is dependent on the moisture content (or soil suction) and is often difficult to assess;
2. the "free field" soil movements in an expansive soil are also difficult to estimate.

Theoretical solutions for pile movement are shown in Figure 5. The ratio of pile head movement to free field soil surface movement is plotted as a function of the relative depth of expansive soil. This ratio is relatively insensitive to the pile relative stiffness K (ratio of pile Young's modulus to soil Young's modulus). The solutions shown in Figure 5 assume elastic interface conditions. If pile-soil slip can occur, then the pile movement may be reduced. It is found that the pile movement and the maximum tensile stress in a pile both decrease if pile-soil slip occurs. It is also found that pile movement is relatively insensitive to pile diameter, suggesting that foundation movements may be suppressed by the use of relatively small-diameter piles.

A series of laboratory tests on model piles has been carried out by Challa and Poulos (1991) to investigate further the behaviour of a pile in an expansive soil and to provide data against which to compare the theory. Figure 6 shows comparisons between theoretical and measured pile head movements, for various depths of swelling, and reveals good agreement. These and other tests have indicated that the theory can give realistic estimates of relative pile movement and the distribution of magnitude of tension in the pile during the later stages of soil swelling, provided that appropriate values of skin friction and end

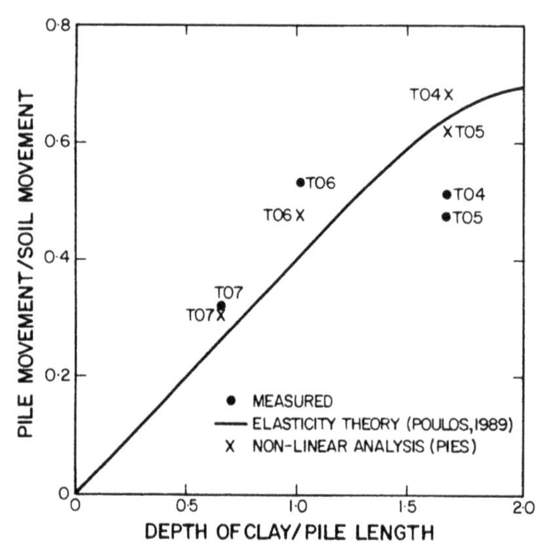

FIGURE 6 Comparison between theory and measurement - ratio of pile movement to soil surface movement at the end of each test

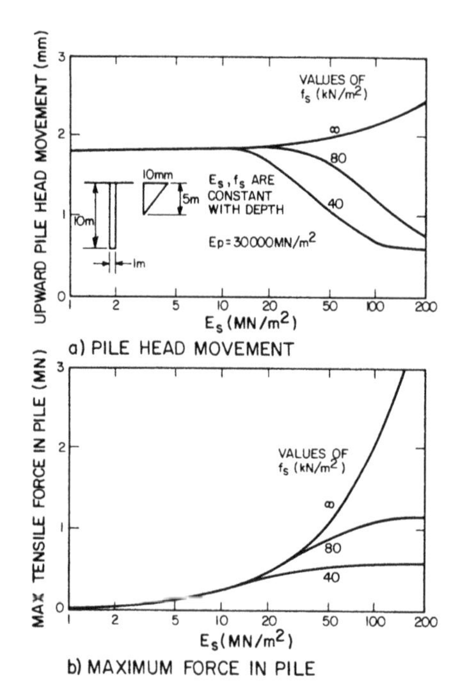

FIGURE 7 Example of Influence of soil parameters on behaviour of pile in expansive soil

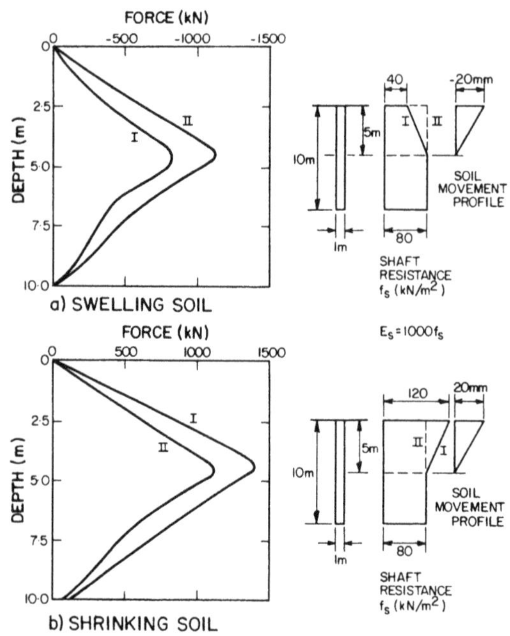

FIGURE 8 Differences between pile in swelling and shrinking soil

bearing resistance are selected.

The sensitivity of pile displacement and force calculations to soil modulus E_s and skin friction f_s is shown in Figure 7. For relatively soft soils (low E_s) the pile head movement is almost independent of E_s and f_s for a given soil movement, but for stiffer soils, the pile head movement depends on the value of f_s. The maximum tensile force tends to increase as E_s increases, and as f_s increases.

The effects of E_s and f_s changing because of changes in moisture content are shown in Figure 8. In a swelling soil, the decrease of f_s and E_s will reduce the tensile force developed (as compared with the initial values of f_s and E_s), while conversely, the increase in f_s and E_s in a shrinking soil will cause an increase in tensile force.

The amount of upward movement experienced by a pile in swelling soil will depend also on the applied axial load. A compressive load will tend to suppress upward movement, whereas a tensile load will tend to significantly increase the upward movement. Consequently, considerable caution must be exercised when designing anchor piles in expansive soils, as it is possible that excessive movement of the piles may occur under the combined effects of tensile loads and upward soil movements.

In pile groups, the forces and displacements induced in the piles are generally less than those in an isolated pile. Thus, ignoring group effects will be conservative in this case.

5 PILES SUBJECT TO LATERAL SOIL MOVEMENTS DUE TO SLOPE INSTABILITY

Figure 9 shows a simplified diagram of the problem. The main design issues are:

1. to determine the forces and bending moments developed in the pile by the movement of the unstable soil;
2. to estimate the increase in stability of the slope because of the presence of the piles.

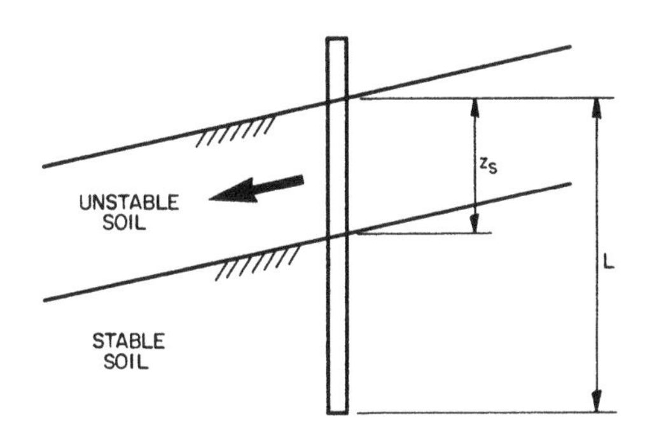

FIGURE 9 Schematic of pile in unstable soil problem

Theoretical analyses have revealed the existence of the modes of failure shown in Figure 10. These modes are:

1. the "Flow Mode" when the slide is shallow and the unstable soil becomes plastic and flows around the pile;
2. the "Short-Pile Mode" when the slide is relatively deep and the length of pile in the stable soil is relatively shallow; the sliding soil carries the pile through the stable soil layer, and full mobilisation of soil strength in the stable layer occurs;
3. the "Intermediate Mode" when the soil strength in both the unstable and stable soil is fully mobilised along the pile length;
4. "Long-Pile Failure": this occurs when the pile itself yields because the maximum bending moment reaches the yield moment of the pile section; this mode can be associated with any of the three modes of soil failure above.

For the case of a 0.75m diameter steel tube offshore pile in an unstable submarine slope environment, Figure 11 shows the theoretical solutions for shear force and bending moment in

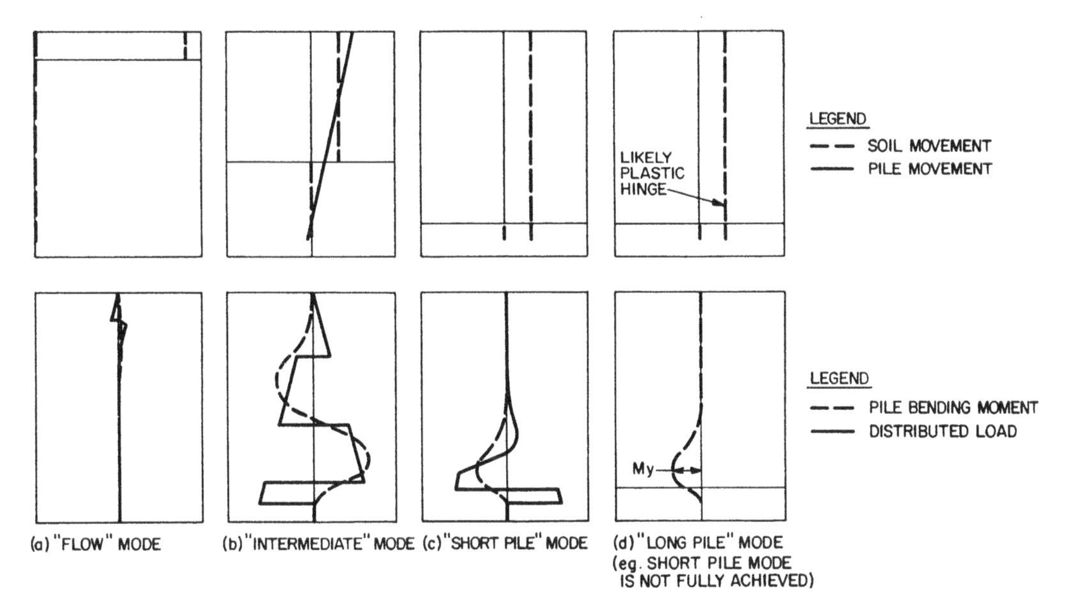

FIGURE 10 Typical distributions of deflection, bending moment and distributed load generated by the sliding soil

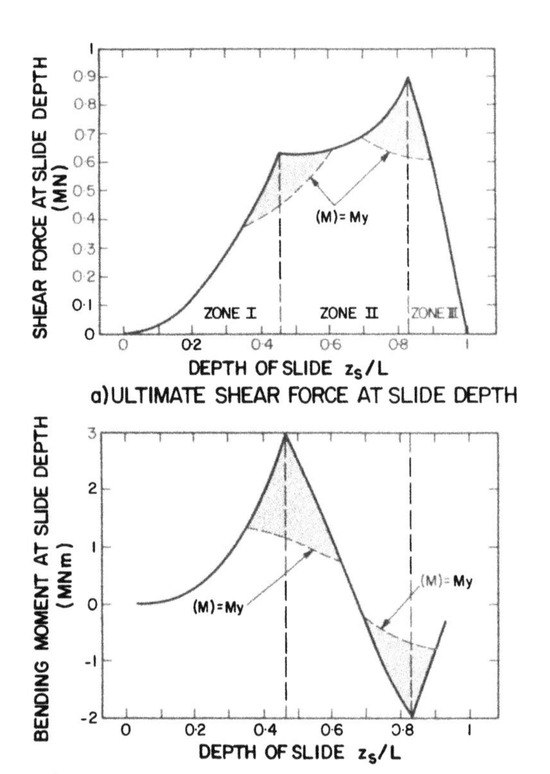

FIGURE 11 Shear force and bending moment in pile in unstable slope

the pile at the slide depth, as a function of the relative slide depth. Modes 1-3 above occur in Zones I,II and III respectively. The shaded zones indicate those slide depths for which the maximum moment induced in the pile reaches or exceeds the yield moment of the pile. For these slide depths, plastic hinges would form in the pile. Development of full soil failure in the clay requires soil movements ranging between about 50mm and 1.2m, depending on the slide depth. The largest soil movements are required when the slide depth is between 0.45 and 0.55 times the length, or between 0.75 and 0.8 times the length.

6 PILES SUBJECTED TO EARTHQUAKE-INDUCED HORIZONTAL SOIL MOVEMENTS

In this case, the design issues for the pile include consideration of the deflections and bending moments induced in the pile by the ground movements.

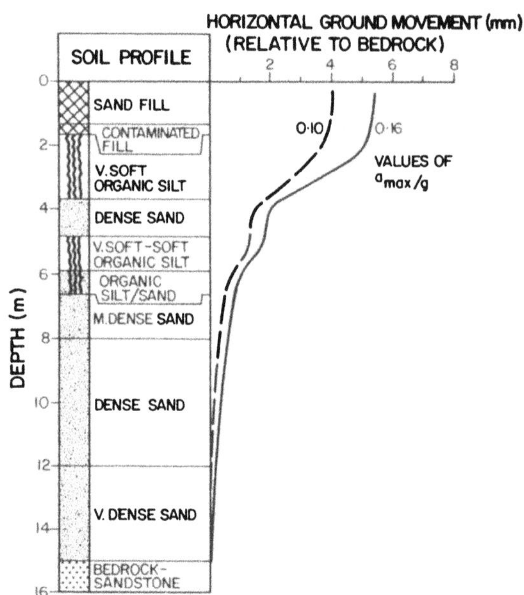

FIGURE 12 Computed maximum relative horizontal ground movement during earthquake

It has been found that, for piles subjected to earthquake-induced movements, inertia effects are often not significant. In such cases, the pile will behave more or less the same as if the soil movements had been applied statically.

Thus, the analysis of pile response can be carried out in two stages:

1. analysis of the free-field soil movements by means of a site response analysis, using a time-acceleration history of the earthquake at bedrock level as input;
2. analysis of the pile response to these soil movements (e.g. via the program ERCAP).

Figures 12 and 13 show the results of an analysis for 700mm diameter concrete piles in the Newcastle (Australia) area. The site consists of about 15m of silt and sand layers overlying sandstone bedrock. The estimated time-acceleration history of the 1989 Newcastle earthquake has been applied to the bedrock, and the site response analysed using the program ERLS (Poulos, 1991).

An earthquake duration of 6s has been assumed, and peak accelerations of 0.10g and 0.16g have been analysed. The dynamic soil parameters have been assessed from SPT data using the correlations of Ohta and Goto (1978). The maximum computed soil movements for each

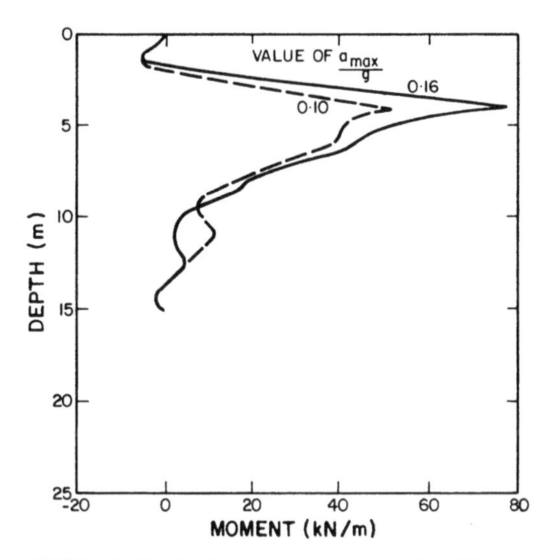

FIGURE 13 Earthquake response of 700mm R.C. pile

case are shown in Figure 12, and the corresponding bending moments in the pile are plotted in Figure 13. As would be expected, larger bending moments occur for the larger bedrock acceleration. The computed pile movements follow the soil movements closely.

Both the distribution and magnitude of bending moment and pile displacement are highly dependent on the assessed modulus and damping characteristics of the soil, and the time-acceleration history of the earthquake.

7 CONCLUSIONS

This paper has described briefly the analysis of piles subjected to externally-imposed soil movements and has attempted to demonstrate the importance of considering the pile-soil interaction if a proper understanding of the pile behaviour is to be developed.

For piles subjected to axial soil movements, it has been shown that significant additional forces can be developed in the pile by the soil movements, and that applied axial loading in the direction of the soil movement can lead to substantial increases in pile displacement.

For piles subjected to lateral soil movements, additional moments and shears are developed in the pile, and these may be large enough to cause structural failure. In the case of a pile within an unstable slope, several mechanisms of failure are possible. The initial mechanism depends largely on the extent of embedment of the pile into the stable zone, the depth of unstable soil, the relative flexibility of the pile, and the yield moment of the pile section.

REFERENCES

Challa, P.K. and Poulos, H.G. (1991). "Behaviour of Single Pile in Expansive Clay." *Geotech. Eng. Vol.* 22, No. 3 pp 189-216.

Fleming, W.G.K., Weltman, A.J., Randolph, M.F., and Elcon, W.K. (1985). "Piling Engineering." *New York: Surrey* University Press, Halsted Press.

Kulhawy, F.H., and Mayne, P.W. (1990). "Manual on Estimating Soil Properties for Foundation Design." *Report EL-6800, prepared for EPRI*, Cornell University, USA.

Ohta, Y., and Goto, N. (1978). "Empirical Shear Wave Velocity Equations Determined from Characteristic Soil Indexes." *Earthquake Eng. Struct. Dyn*, Vol. 6, No. 2 pp 167-187.

Poulos, H.G., and Davis, E.H. (1980). "Pile Foundation Analysis and Design." *John Wiley & Sons, New York.*

Poulos, H.G. (1989a). "PIES Users' Guide." *Centre for Geotech. Research*, University of Sydney.

Poulos, H.G. (1989b). "Pile Behaviour - Theory and Application." *Geotechnique, Vol.* 39, No. 3, pp 365-415.

Poulos, H.G. (1991). "Relationship Between Local Soil Conditions and Structural Damage in the 1989 Newcastle Earthquake." *Civ. Eng. Trans. IEAust, Vol* CE 33, No. 3, pp 181-188.

Van der Veen, C. (1986). "A General Formula to Determine the Allowable Pile Bearing Capacity in Case of Negative Skin Friction." *Proc. Int. Conf. on Deep Foundations*, Beijing, China Bldg. Industry, Vol. 1.

Developments in Geotechnical Engineering, Balasubramaniam et al. (eds) © 1994 Balkema, Rotterdam, ISBN 90 5410 522 4

Cost-effective use of friction piles in non-cohesive soils

Sven Hansbo
J&W Bygg & Anläggning AB, Lidingö & Chalmers University of Technology, Göteborg, Sweden

ABSTRACT: The number of piles installed in pile foundations on granular soil can be strongly reduced, simply by adopting another design philosophy than the conventional one. Thus, instead of assuming that the load is carried in its entirety by the piles, a more cost-effective and equally safe design is attained if the load bearing capacity of the pile cap is taken into account. The piles only serve the purpose of keeping settlement within permissible limits. Results of recent research on the behaviour of piled footings with the pile cap in contact with the soil are presented. Design principles based on these results are suggested.

1. INTRODUCTION

When piles were first utilised for foundation purposes the main object was certainly to ensure the long-term serviceability of the then buildings. Piles were installed in areas with bad soil conditions where by experience buildings would suffer damage due to settlement. The number of piles to be installed was adjusted on the basis of experience. Looking back on the number and types of piles utilised in early building history it is obvious that piles were installed only with the purpose of reducing settlement. Increasing safety requirements, forming the basis of building codes, have resulted in a design philosophy where the piles are assumed to carry the total load of the superstructure with no contribution whatsoever of the load carrying capacity of the pile cap itself, or of the raft. It is about time to learn from building history that a more cost-effective use of piles is possible than that achieved by conventional pile design of today.

A lot of research has been carried out, both theoretically and experimentally, to increase the understanding of the interaction phenomena taking place between the piles in a pile group, the pile cap (or the piled raft) and the soil.

The idea of using piles as settlement reducers was born anew in the seventies (Burland *et al.*, 1977). In the case of piled rafts on clay this design philosophy was introduced in connection with the construction of an integrated building complex covering an area of 53,000 m^2 on soft clay to depths of 50–90 m (Hansbo *et al.*, 1973). It was later on developed into a more advanced and refined design method for piled rafts on clay according to which the building load inducing stresses in excess of the preconsolidation pressure of the clay was carried by piles in a state of creep failure while the remaining part of the load was carried by contact stresses at the pile/raft interface (Hansbo & Källström, 1983; Hansbo & Jendeby, 1983; Hansbo, 1984, 1993). A similar approach was introduced in the UK by Burland (1986). The difference between a conventional pile foundation and a pile foundation according to the new principle with creep piles is illustrated for two neighbouring, equally heavy buildings constructed at a sight with very soft, normally consolidated clay to

Fig. 1. Two buildings in Gothenburg founded on friction piles, one according to conventional pile design (total pile length 24,000 m) and the other on 'creep piles' (total pile length 5,000 m). Normally consolidated clay to a depth of about 100 m.

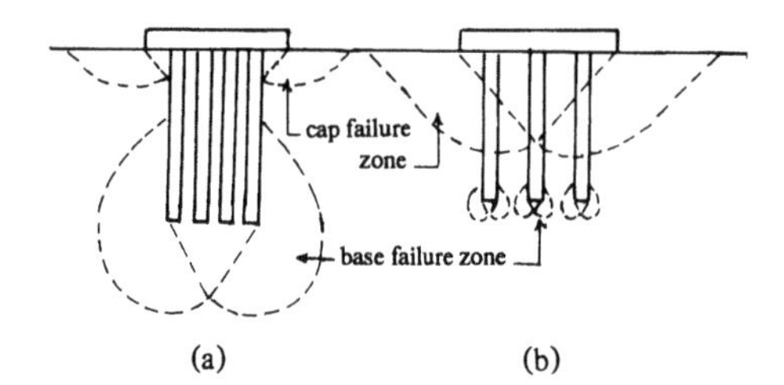

Fig. 2. Assumed failure zones at piled footings: (a) pier failure; (b) individual pile failure. (Kishida & Meyerhof, 1965)

great depth in Gothenburg, Sweden (Fig. 1).

The influence of the rigidity of the basement on the contact stress distribution and the distribution of pile loads can be taken into account by finite element analysis and the result has been followed up by settlement observations of buildings founded according to the creep pile principle (Svensson, 1993).

The design philosophy applied to foundations on clay has not yet been applied to foundations on non-cohesive soils. The compass of this problem is more far-reaching than in the case of cohesive soils which obey more closely the theory of elasticity. Pile installation may give rise to severe changes in soil properties with large variations inside the pile group which make a theoretical analysis extremely difficult.

A lot of research has been carried out, both theoretically and experimentally, to increase the understanding of the interaction phenomena taking place between the piles in a pile group in non-cohesive soil, the pile cap and the soil. Most of the experimental studies concern free-standing pile groups, *i.e.* cases where the pile cap is not in contact with the soil. A comprehensive literature survey of the studies performed has been presented by Phung (1993). According to Phung, only six cases of full-scale tests and two cases of large-scale tests with pile cap in direct contact with the soil (including those carried out by Phung himself) have been published. The conclusions drawn from these tests are not consistent. Moreover, the information obtained through the tests is insufficient.

In this paper, some important results arrived at in recent research on piled footings in non-cohesive soil will be reviewed. On the basis of these results, a new and cost-effective design principle will be presented.

2. PILE LOAD DISTRIBUTION

The installation of a pile group in non-cohesive soil gives rise to a change in soil properties and a consequential change in pile behaviour. Therefore, the individual pile in the group cannot be expected to behave in a way similar to that of a single pile installed under similar soil conditions.

If the soil were still homogeneous after pile installation, the frictional forces along the pile shafts would cause a settlement bowl underneath the pile cap with a tendency towards increasing load share among the outer piles and a decreasing load share among the inner piles in the pile group. In granular soil, however, pile installation—particularly in the case of displacement-type piles—generally causes changes in the deformation characteristics of the soil inside the pile group in such a way that the load distribution among the piles will change in the opposite direction to that mentioned. Thus, the piles in the centre of the pile group will generally be subjected to higher load than the outer piles.

3. BEARING CAPACITY

The ultimate load of a pile group with friction piles is generally different from the sum of the ultimate loads of the individual piles in the group.

According to Kishida & Meyerhof (1965) the total bearing capacity of the foundation and its surcharge effect on the point resistance of the piles in the group can be estimated either by considering the bearing capacity of the pile cap as a whole and the bearing capacity of the individual piles in the group (individual pile failure) or by considering only the contribution to the bearing capacity of the outer rim of the cap, outside the pile group (the pier), and the bearing capacity of the pile group as a whole (pier failure), Fig. 2.

In an extensive test series on bored pile/pile cap/soil interaction effects in sand, comprising 51 pile groups and 23 single piles, Liu *et al.* (1985) investigated the group effect on both pile groups with free space between pile cap and soil and pile groups with pile cap in direct contact with soil. The pile groups consisted of 2–15 piles, 125–330 mm in diameter, and with lengths of 8–23 times the pile diameter. Pile spacings were 2–6 times the pile diameter. From the results obtained they found no evidence of block failure and, therefore, propose that the analysis of pile group failure be based upon the bearing capacity of the cap as a whole in combination with individual pile failure.

The ultimate load Q_{grf} of a pile group is generally expressed as:

$$Q_{grf} = \eta n Q_{fs} \tag{1}$$

where η = group efficiency factor,
n = number of piles in the group,
Q_{fs} = ultimate load of a single pile under similar soil conditions as for the pile group.

A large number of investigations have been carried out for the purpose of determining the pile group efficiency factor under various soil conditions and pile group geometries. The investigations include both free-standing pile groups and piled footings with pile cap in direct contact with soil.

Table 1. Example of utimate total pile loads and ultimate shaft loads (in kN) observed by Ekström (1989). Free-standing square pile groups. Individual pile tests.

Pile spacing Initial I_D %	$3b_p$ 47	$4b_p$ 47	$6.5b_p$ 60	$3b_p$ 47	$4b_p$ 47	$6.5b_p$ 60
		Total			Shaft	
Single pile	8–10	11	24	3–6	6	10
Pile group, 5 piles:						
centre	18	21	33	13	14	20
corner	11–18	25–26	5–8	–	8–11	–
Pile group, 9 piles:						
centre	22	27	38	15	18	22
corner	14	25	23	6	12	9
Pile group, 13 piles:						
centre	–	35	36	–	19	21
corner	–	25	35	–	10	17
Pile group, 25 piles:						
centre	–	37	33	–	19	17
corner	–	25	25	–		7

Full-scale and large-scale tests on free-standing pile groups in loose to medium dense sand have resulted in efficiency factors $\eta > 1$ with a maximum ($\eta = 2$) at a pile spacing of about 2–3 times the pile diameter. For free-standing pile groups in dense sand, both $\eta > 1$ and $\eta \leq 1$ have been found. However, in the latter case only small-scale tests have been performed.

In reality, a piled foundation on friction piles is usually in direct contact with the soil. Consequently, the pile cap itself may contribute considerably to the bearing capacity of the piled foundation. In the case considered—non-cohesive soil—the combined action of the piles in the pile group and the piled footing itself has been shown to have a great influence on the group efficiency. The most influencial factors, besides soil conditions, are pile spacing S and the ratio of pile length l_p to width B of the pile cap.

Liu *et al.* (1985) proposed that the bearing capacity Q_{fpc} of a pile group with pile cap in direct contact with soil be expressed by the relation:

$$Q_{fpc} = n(\eta_s Q_{fs} + \eta_t Q_{ft}) + Q_{fc} \quad (2)$$

where $\eta_s = C_s G_s$ = group efficiency factor with reference to pile shaft,

$\eta_t = C_t G_t$ = group efficiency factor with reference to pile tip,

G_s and G_t = factors of influence of pile/soil interaction on shaft and tip bearing capacities of the individual piles in the group,

C_s and C_t = factors of influence of cap/pile/soil interaction on shaft and tip bearing capacities of the individual piles in the group,

Q_{fs} and Q_{ft} = shaft and tip bearing capacities of the individual piles in the group,

Q_{fc} = bearing capacity of the cap alone.

In the case of a pile group with pile cap in direct contact with the soil, the contribution to the bearing capacity due to the cap has two main causes: the bearing capacity of the cap itself and its surcharge effect on pile shaft friction. For pile lengths exceeding two times the width of the pile cap, the surcharge effect on the pile tip resistance can be neglected.

As pile installation in granular soil will always affect the properties of the soil in one way or the other, the load/settlement behaviour of the piled cap can be expected to differ from that of the unpiled cap. In order to take this into account, Phung (1993) modifed Eq.(2) and proposed the following relation:

$$Q_{fpc} = n(\eta_s Q_{fs} + \eta_t Q_{ft}) + C_c Q_{fc} \quad (3)$$

where C_c = factor of influence of cap/pile/soil interaction on the bearing capacity of the pile cap.

An illustrative effect of pile installation in granular soil was presented by Ekström (1989). A test series was performed on single piles on one hand and on individual piles in free-standing pile groups on the other. The number of piles in the group varied from 5 to 25 and the pile spacing from 3 to 6.5 times the pile width. The piles which were of hollow steel with square cross-section (width b_p = 60 mm; wall thickness = 5 mm) were driven to a depth of 3.3 m in sand. The influence of pile installation on the lateral earth pressure coefficient is exemplified in Fig. 3 and on the ultimate load of the

Table 2. Group efficiencies for bored piles according to Liu *et al.* (1985). Pile diameter D_p = 0.25 m.Loose silty sand. η_s and η_t include pile/soil interaction effect while η_G also includes action of pile cap in contact with soil.

S/D_p	l_p/D_p	Group	Shaft η_s	Tip η_t	Total η_G
3	8	3x3	0.36	1.44	1.64
3	13	3x3	1.09	1.51	1.69
3	18	3x3	1.16	1.49	1.51
3	18	3x3	1.42	0.91	1.36
3	23	3x3	1.16	1.12	1.15
2	18	3x3	0.98	0.70	1.21
4	18	3x3	1.11	0.93	1.46
6	18	3x3	0.82	1.06	2.23
3	18	1x4	1.11	1.10	1.49
3	18	2x4	0.88	1.51	1.40
3	18	4x4	1.03	1.45	1.19
3	18	2x2	1.20	1.22	1.60

Table 3. Efficiency factors at pile failure obtained by Phung (1993). Pile cap action included.

I_D %	S/b_p	Shaft η_s	Tip η_t	Total η
Free-standing group:				
38	4	2.6	2.0	2.4
67	6	3.2	0.8	1.1
62	8	2.0	1.0	1.2
Cap in contact with soil:				
38	4	3.2	3.0	3.1
67	6	4.4	0.7	1.3
62	8	4.4	1.4	2.0

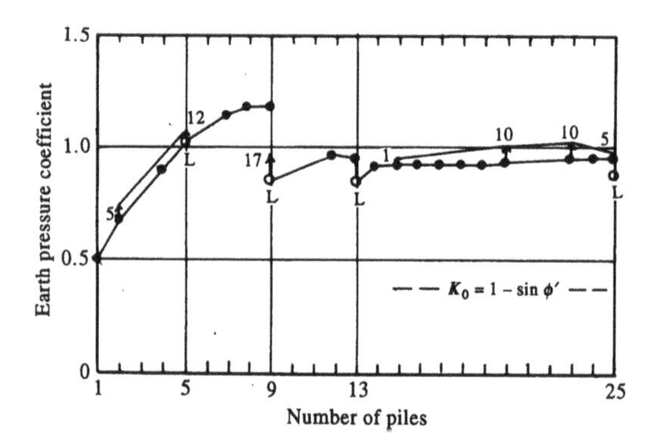

Fig. 3. Average lateral earth pressure coefficient with reference to the centre pile in pile group with pile spacing equal to $4b_p$, driven into sand with I_D = 47%. The letter L represents results after loading test. Arrows indicate change in lateral eart pressure with time, in days. (Ekström, 1989).

centre and corner piles in the various pile groups in Table 1.

As can be seen there is a strong influence on the soil properties and the load distribution among the piles in the group and, as was to be expected, the application of results of loading tests on single piles for the determination of pile group capacity, assuming η = 1, is a conservative approach. In his test series on the ultimate load of square pile groups comprising 5 piles, Ekström found the pile group efficiency factor η = 2 for pile spacing $4b_p$ and η = 1.4 for pile spacing $6.5b_p$.

From their tests on group effects on bored piles with pile cap in direct contact with soil, Liu *et al.* (1985) presented the group efficiency factors given in Table 2.

A wealth of information regarding the cap/pile/soil interaction problem is to be found in a doctoral thesis by Phung (1993). Phung performed a test series comprising cap without piles (spread footing), single pile, free-standing pile group (5 piles) and pile group (5 piles) with pile cap in direct contact with soil. The test arrangement is shown in Fig. 4. The piles were of the same type as those used by Ekström, *i.e.* hollow sand-covered steel pipes with square cross-section, 60 mm by 60 mm and 5 mm wall thickness. The pile length embedded in sand was 2.0 m. The load *vs.* settlement relationship for the single piles and the different pile groups at relative soil densities of 38, 62 and 67% obtained by Phung are summarised in Figs. 5–7. The

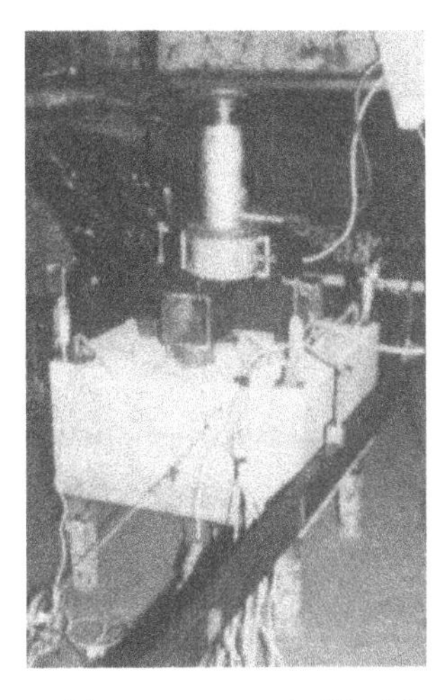

Fig. 4. Photo of test arrangement utilised by Phung for determination of pile group efficiency. The tests were perfomed in a sand pit. Sand influenced by the test was excavated and replaced by sand with desired density.

efficiency factor in respect of shaft resistance, tip resistance and total resistance thus obtained are presented in Table 3.

As could be expected, the group efficiency factor increases with decreasing relative density of the soil. It is also obvious that the surcharge induced by the pile cap, when in direct contact with the soil, has a very favourable effect on the bearing capacity of the pile group.

4. SETTLEMENT

Different proposals have been presented about how to define, under equal soil conditions, the settlement of pile groups in relation to the settlement of a single, free-standing pile—the so-called settlement ratio ζ.

A fairly large number of tests have been carried out to find the ζ value but the results obtained are often difficult to analyse both because of the settlement ratios being related to different factors of safety against pile failure and because of the settlement ratios not being defined. As settlement is very much dependent on the factor of safety applied, the results presented by different authors show great scattering and are often contradictory. Roughly speaking, results of loading tests on free-standing pile groups indicate that $\zeta > 1$ in dense sand while $\zeta < 1$ for driven piles in loose to medium dense sand. In the latter case, ζ has been found to vary from about 0.2 at $B_{gr}/D_p \approx 3$ (where B_{gr} is the

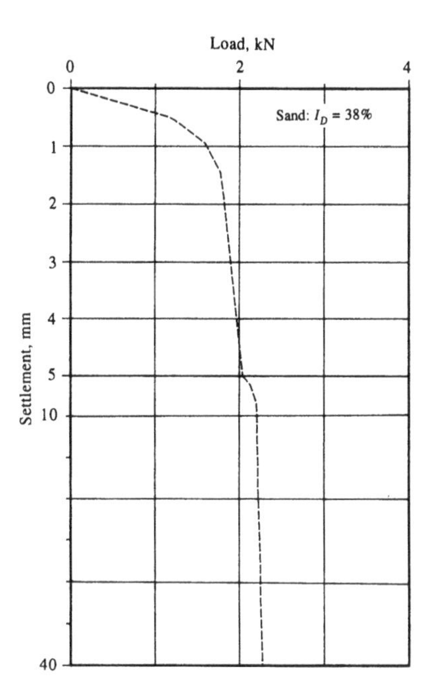

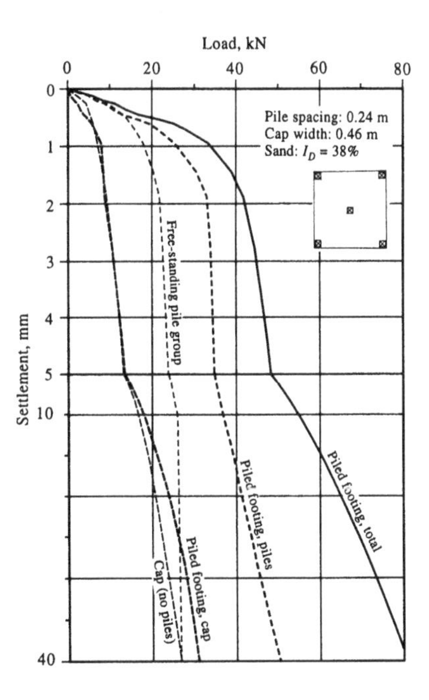

Fig. 5. Comparison of load/settlement relationship of cap (shallow footing), free-standing pile group and pile group with pile cap in direct contact with soil. Cap width = 0.46 m. Pile spacing = $4b_p$. Sand with I_D = 38% (After results presented by Phung, 1993)

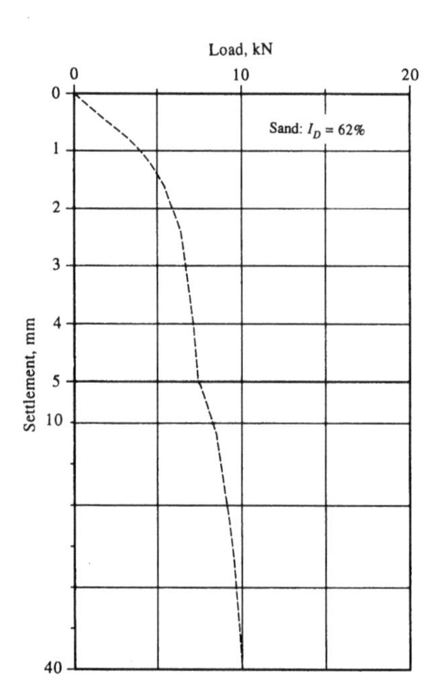

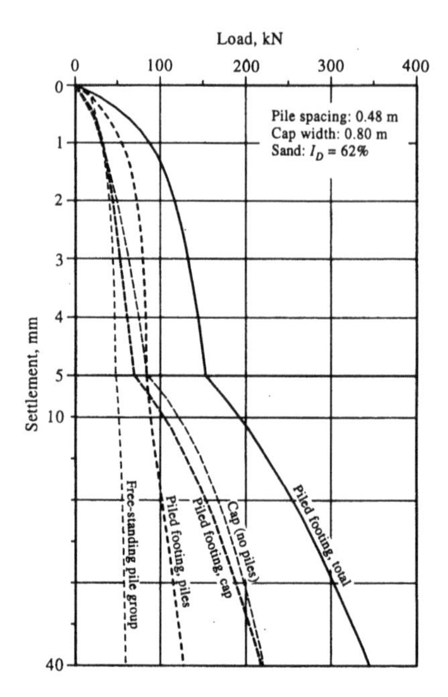

Fig. 6. Comparison of load/settlement relationship of single pile with that of pile cap alone, free-standing pile group, and pile group with pile cap in direct contact with soil. Cap width = 0.8 m. Pile spacing = $8b_p$. Sand with I_D = 62%. (After Phung, 1993).

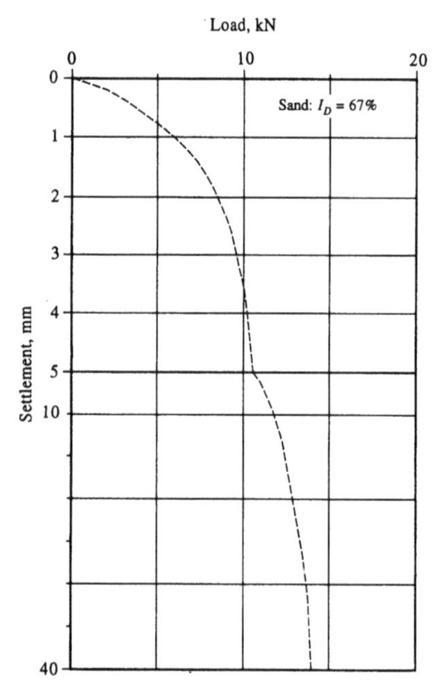

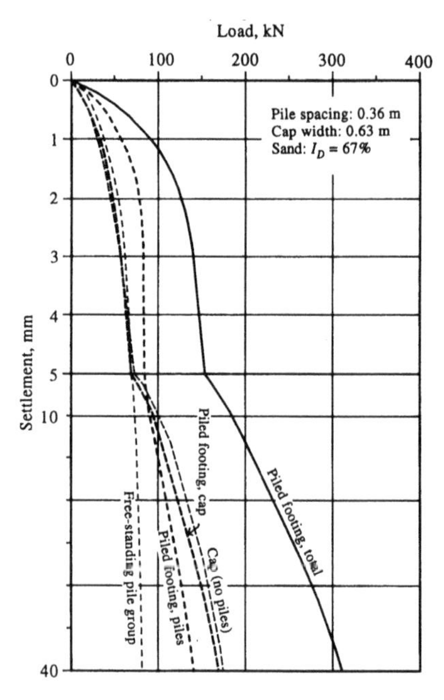

Fig. 7. Comparison of load/settlement relationship of single pile with that of pile cap alone, free-standing pile group, and pile group with pile cap in direct contact with soil. Cap width = 0.63 m. Pile spacing = $6b_p$. Sand with I_D = 67%. (After Phung, 1993).

width of the pile group) to about 0.7 at $B_{gr}/D_p \approx 10$. As for bored piles in loose sand, ζ has been found to vary from about 0.6 at $B_{gr}/D_p \approx 3$ to about 2 at $B_{gr}/D_p \approx 5$.

Among the proposed definitions of ζ it seems preferable to use either one of the following definitions:

- ζ is the ratio of the initial slope of the average pile load *vs.* settlement curve of the free-standing pile group to that of the single pile, or
- ζ is the ratio of the free-standing pile group settlement to single pile settlement at equal pile loads.

Based on the results of full-scale investigations, Vesic (1969) suggested that the ζ value be determined by the relation:

$$\zeta = \sqrt{B_{gr}/D_p} \qquad (4)$$

As in the case of the ultimate load of pile groups in sand the influence on pile group settlement of the surcharge exerted by the pile cap may be quite important. This is exemplified for the pile group installed in sand with relative density $I_D = 62\%$ (Fig. 8). According to the results obtained by Phung, the load is carried by the piles at small settlement and is then gradually transferred to the cap. Simultaneously, the lateral earth pressure against the pile shafts in the group increase considerably with increasing settlement which leads to a successive increase in the ultimate load of the piles. The additional load placed after pile failure is carried by the pile cap.

An important question to be considered in foundation analysis is that of estimating the long-term settlement—in the case of granular soil caused by creep. An example of the creep behaviour observed by Phung for the pile group installed in sand with $I_D = 62\%$ is given in Fig. 9. As can be seen the creep load of the unpiled footing and that carried by the piled cap itself in the pile group are more or less the same.

An important conclusion drawn by Phung from the results of his study is that the load/ settlement behaviour with reference to the part of the load carried by the cap itself in a piled footing is very similar to that of a corresponding shallow footing, irrespective of the relative density of the soil.

Phung relates the settlement of the cap to the ratio between the bearing capacity of the unpiled cap and that of the piled cap, defined as the relative cap capacity $\alpha = Q_{fcap}/Q_{ftot}$, where Q_{fcap} is the bearing capacity of the pile cap alone and Q_{ftot} is the total bearing capacity of the piled footing with cap in contact with soil. Based on the results of his study, the cap settlement increases approximately linearly from zero at $\alpha = 0$ (all of the load carried by the piles) to a value of 0.12 at $\alpha = 0.4$ and from then on an almost linearly up to unity when α reaches unity (shallow footing).

In practice the application of a ζ value for settlement analysis of a piled footing with pile cap in direct contact with soil is quite intricate. In the opinion of the Author, a more direct and easier approach is to study the settlement of the unpiled cap under the part of the load carried by

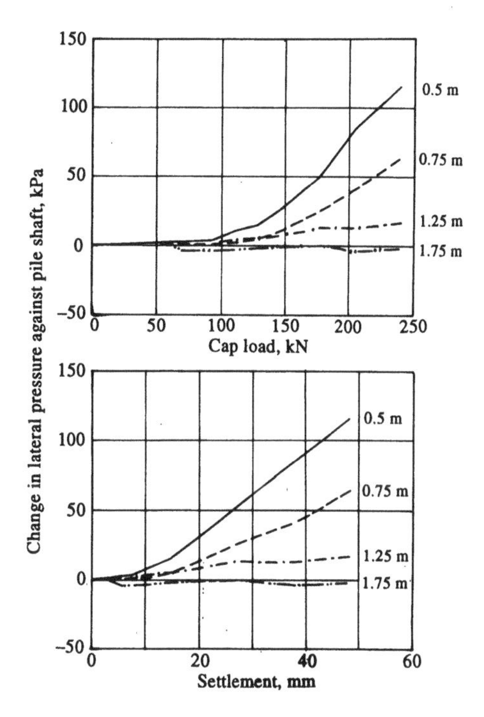

Fig. 8. Influence of lateral earth pressure against the shaft of the centre pile due to surcharge induced by pile cap. Pile group: 5 piles c/c $8b_p$. Pile cap 0.8 m square. Sand with $I_D = 62\%$. (After Phung, 1993).

contact pressure at the cap/soil interface. Thus, in reality it is more practical to determine the load which can be carried by an unpiled footing within the settlement limits given and then determine the number of piles $n\eta Q_f$ required to carry the total load than to apply the ζ value according to the principle given above.

In conclusion of the study performed by Phung (1993), it is obvious that a considerable decrease in the settlement can be expected due to the pile cap being in contact with the soil. This has been found to be true also in the case of bored piles for B_{gr}/D_p ratios between 2 and 5 (Garg, 1979).

5. PROPOSAL FOR DESIGN OF PILED FOOTINGS

According to conventional design, the load acting on a pile group, irrespective of whether the the piles are in cohesionless or cohesive soils, is assumed to be carried by the piles alone with a certain factor of safety against failure. This approach is rational when the piles are end bearing, or mainly end bearing, or when we have to deal with footings on normally consolidated clay. However, as previously mentioned it is used even in cases where the bearing capacity of the footing itself would be satisfactory, the reason being that settlements without piles are felt to be too large. In situations where the piles are friction piles for which the load/settlement relationship does not show a marked decrease after peak, this approach is quite conservative and unnecessarily expensive. In granular soil piles are generally used only when the soil is in a loose to medium dense state. Driving of piles in those soils will cause compaction and reorientation of the grains into a denser state which will improve the settlement characteristics of the soil (*cf.* compaction piling used for soil improvement). This favourable effect on the soil properties ought to be considered. Doubtless, a more cost-effective approach is first to investigate (on the basis of the existing soil properties) how much of the load can be carried by the pile cap without causing excessive settlement and then design the pile group to carry the remaining part of the load. The intricate problem of analysing the influence of pile/cap/soil interaction on the ultimate load of the piles in the pile group and on the settlement can be totally disregarded. Thus, the settlement obtained under the part of the load that is carried by contact pressure at the soil/cap interface can be assumed to be equal to the settlement obtained under an equally large load carried by a corresponding unpiled cap (shallow footing). This fact simplifies the design procedure. The number of piles required to limit the settlement can be determined on the basis of the failure load of the single pile and the settlement can be calculated as if the cap were a shallow footing carrying the load not taken by the piles.

The principle of pile group design can thus be summarised as follows:

• Determine the load Q_1 that can be placed on the unpiled cap without causing unacceptable settlement.

• The remainder of the load $Q_2 = Q - Q_1$ should be carried by settlement reducing piles. As the permissible settlement will be large enough for shaft resistance to be fully mobilised, the piles can be designed as friction piles in a state of creep failure.

• The settlement of the piled footing can be estimated at about the same value as the settlement of the unpiled footing under load Q_1.

This design procedure leads to a conservative design, the more conservative the looser the soil.

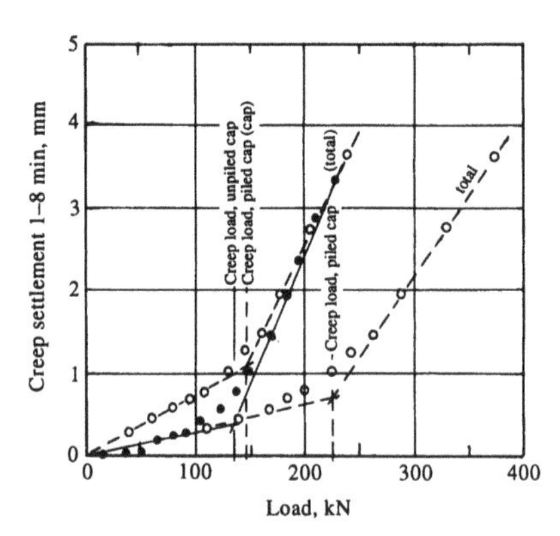

Fig. 9. Creep settlement *vs.* load during 1–8 minutes observed by Phung (1993). Unpiled cap and piled cap. Conditions equal to those given in Fig. 6. Open circles represent the creep *vs.* load carried by the piled cap (the cap itself and total) and filled circles that of the unpiled cap. The creep load corresponds to the point of intersection between the two straight parts of the creep curve.

CONCLUSIONS

The results of research on soil/structure interaction strongly supports the introduction of a new design philosophy concerning piled footings in non-cohesive soil. The purpose of pile installation can be considered as a means of reducing settlement of an unpiled footing to acceptable limits. The number of piles required by this design philosophy is strongly reduced in comparison with conventional design. On one hand the piles will only have to carry the load in access of the load that can be carried by the pile cap, considered as spread footing, without causing unacceptable settlement and, on the other hand, the piles can be designed in a state of creep failure.

The idea of using piles as settlement reducers was put forward by Burland *et al.* (1977) and the idea of designing piles in a state of creep failure, also for the purpose of reducing settlement, has been applied with great success on piled raft foundations on soft clay (see Hansbo, 1984). Up to now, the design of piled footings on granular soil takes no account of the load bearing capacity of the pile cap. It is about time that this design philosophy is abandoned and replaced by a new one in which the bearing capacity of the pile cap itself is considered.

REFERENCES

Burland, J. B., Broms, B. B. & De Mello, V. F. B., 1977. Behaviour of foundations and structures. Proc. 9th ICSMFE, Tokyo, Vol. 2, 495–546.

Burland, J. B., 1986. The value of field measurements in the design and construction of deep foundations. Proc. Int. Conf. on Deep Foundations, Beijing, Vol. 2, 177–187.

Ekström, J., 1989. A field study of model pile group behaviour in non-cohesive soils. Influence of compaction due to pile driving. Doctoral Thesis, Chalmers Un. of Technology, Gothenburg.

Garg, K. G., 1979. Bored pile groups under vertical loads in sand. J. Geot. Eng. Div., ASCE, Vol. 105, No. GT 8.

Hansbo, S., Hofmann, E. & Mosesson, J., 1973. Östra Nordstaden, Gothenburg. Experience concerning a difficult foundation problem and its unorthodox

solution. Proc. 8th ICSMFE, Moscow, Vol. 2, 105–110
Hansbo, S. & Jendeby, L., 1983. A case study of two alternative foundation principles: conventional friction piling and creep piling. Väg- och Vattenbyggaren, No. 7–8, 29–31.
Hansbo, S. & Källström, R., 1983. Creep piles—a cost-effective alternative to conventional piling. Väg- och Vattenbyggaren, No. 7–8, 23–27.
Hansbo, S., 1984. Foundations on friction creep piles in clay. Proc. 1st Int. Conf. on Case Histories in Geot. Eng., St Louis, Vol. 2, 913–922.
Hansbo, S., 1993. Interaction problems related to the installation of pile groups. Proc. 2nd Int. Geot. Seminar on Deep Found. on Bored and Auger Piles, Ghent, 59–66.
Jendeby, L., 1986. Friction piled foundations in soft clay—A study of load transfer and settlements. Doctoral Thesis, Chalmers Un. of Tech., Gothenburg.
Kishida, H. & Meyerhof, G. G., 1965. Bearing capacity of pile groups under eccentric loads in sand. Proc. 6th ICSMFE, Toronto, Vol. 2,.
Liu, J. L., Yan, Z. L. & Shang, K. P., 1985. Cap-pile-soil interaction of bored pile groups. Proc. 11th ICSMFE, San Francisco, Vol. 3, 1433–1466.
Phung, D. Long, 1993. Footings with settlement reducing piles in non-cohesive soil. Doctoral Thesis, Chalmers Un. of Tech., Gothenburg.
Svensson, P. L., 1993. Soil-structure interaction of foundations on soft clay—Experience during the last ten years. Proc. 10th ECSMFE, Florence, Vol. II, 583–586.
Vesic, A. S., 1969. Experiments with instrumented pile groups in sand. American Society for Testing Materials, Performance of Deep Foundations, ASTM STP 444, 172–222.

Developments in Geotechnical Engineering, Balasubramaniam et al. (eds) © 1994 Balkema, Rotterdam, ISBN 90 5410 522 4

Response of block foundations in vertical vibrations

S. Prakash
University of Missouri, Rolla, Mo., USA

B. M. Das & V. K. Puri
Southern Illinois University, Carbondale, Ill., USA

ABSTRACT: Block vibration tests were conducted on two test blocks measuring 3.0m x 1.5m x 0.7m and 1.5m x 0.75m x 0.70m, cast on level ground. The blocks were excited into vertical vibrations and the amplitudes of vibration at different frequencies of excitation were measured using acceleration transducers mounted on appropriate faces of the block. Dynamic shear modulus at this site was also determined by conducting in-situ tests. The natural frequencies and the vibration amplitudes of the test blocks were calculated by (i) the linear spring method, (ii) the elastic half space method and (iii) the impedance function method. A comparison was then made of the observed and computed natural frequencies and the vibration amplitudes of the blocks. The results of this comparison showed that for the cases of vertical vibrations, the natural frequencies in this case could be reasonably predicted by either of the methods used. The calculated and observed amplitudes, however, showed a wide variation.

1 INTRODUCTION

Rigid block type foundations are commonly used for supporting reciprocating machines. In order to ensure long term satisfactory performance of a foundation for a machine, its design should meet the criteria for both static and dynamic stability. The criteria for static design are: no shear failure in soil, and no excessive settlement of the footing. The criteria for dynamic stability require that the natural frequency of the foundation soil system should be far away from the operating frequency of the machine, and the amplitude of vibration under operating conditions should not exceed the specified limits. The vibrations produced due to operation of the machine should not be harmful to the people working in the vicinity of the machine, and to the adjacent structures. The design of the machine foundations is generally made either by the linear spring method (Barkan, 1962) or by the elastic half space method (Richart, Hall and Woods, 1970; Prakash and Puri, 1988; Puri and Das, 1993; Das, 1992). Recently, a new method has been proposed which utilizes the impedance-compliance functions for obtaining the stiffness and damping values for the foundation soil system ((Dobry and Gazettas, 1986), Dobry, Gazettas and Stokes, 1986), (Gazettas, 1991), and Gazettas and Stokes, 1991)). Very little data, however, are available on the calculated response of a machine foundation and its actual performance. In fact no attention is paid to the measurement of vibration amplitudes until some problem develops.

This paper presents the results of a comparison of the computed and observed response of two block foundations made of concrete. The test blocks were excited into vertical vibrations and their natural frequencies and amplitudes were measured. The pertinent soil properties were determined by conducting in-situ tests and oscillatory shear tests in the laboratory. The response of the two test blocks was then computed by the (i) linear weightless spring method, (ii) elastic half space method and (iii) the impedance function approach. A comparison was made of the observed and predicted response of the test blocks. The details of the tests conducted, data obtained, and comparison of the observed and computed response are discussed here.

2 TESTS CONDUCTED

2.1 *Block Vibration Tests:*

The block vibration tests were conducted on rigid concrete blocks. The sizes of the test blocks were 1.5m x 0.75m x 0.70m and 3.0m x 1.5m x 0.7m. The blocks were cast on level ground. Vertical vibration tests were conducted on each of these two blocks. These tests were conducted by exciting the block in the vertical vibrations with a mechanical oscillator. A speed controlled D.C. motor was used to operate the oscillator. The oscillator-motor assembly was mounted centrally on the top of the test block, and rigidly attached to it. The block was set into vertical vibrations by operating the mechanical oscillator. The vibrations of the block were measured with an acceleration transducer mounted on the top surface of the block and oriented to sense vertical vibrations. The output from the accelerometers was amplified and recorded. A schematic sketch of the test set up is

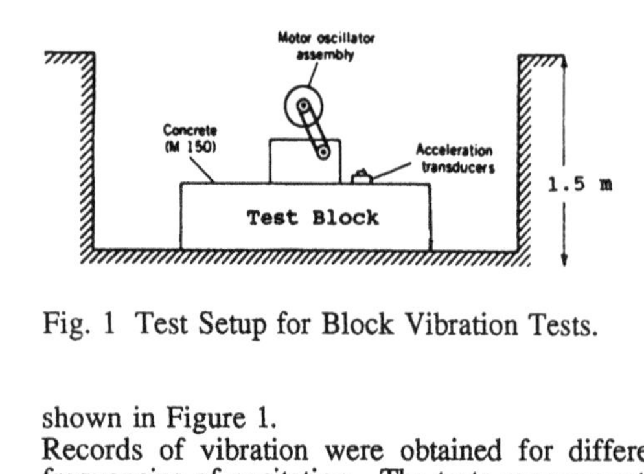

Fig. 1 Test Setup for Block Vibration Tests.

shown in Figure 1.

Records of vibration were obtained for different frequencies of excitation. The tests were repeated by changing 'θ' the angle of setting of the eccentric masses. From the recorded test data, the frequency and the corresponding amplitudes were determined. Values of observed natural frequencies and corresponding amplitudes are given in Tables 1 and 2.

2.2 *Tests for Determination of Dynamic Soil Properties*

The dynamic properties of the soil used in the analysis of machine foundation may be determined by laboratory and in-situ tests. These properties are affected by a number of factors which should be accounted for when selecting the design values. The most important factors which affect these properties are (1) the mean effective confining pressure, (2) the shear strain amplitude, and (3 density in the soil. A good discussion on these corrections has been presented by Das (1992), Prakash and Puri (1977, 1988), Nandakumaran et al (1977), Prakash and Puri (1981) and Indian Standard Code (IS 5249 - 1977).

In-situ soil investigations consisted of (1) wave propagation tests, (2) cyclic plate load tests and (3) standard penetration tests. From these tests, the values of dynamic shear modulus "G" were computed. From the uncorrected standard penetration (N) values shear wave velocity V_S at a particular depth was determined from equation (1) Imai (1977) and dynamic shear modulus "G" was computed from equation (2).

$$V_S = 91.0\ N^{0.337} \quad (1)$$

$$G = V_S^2 \times \rho \quad (2)$$

in which $\rho=\gamma/g$ = mass density of soil. Values of "G" from different tests were corrected for (1) effective confinement in each case and computed for an effective overburden pressure of $100 kN/m^2$ using equation (3).

$$\frac{G_1}{G_2} = \left(\frac{\bar{\sigma}_{v1}}{\bar{\sigma}_{v2}}\right)^{0.5} \quad (3)$$

in which G_1 = shear modulus at an effective overburden pressure of $\bar{\sigma}_{v1}$, and G_2 = shear modulus at an effective overburden pressure of $\bar{\sigma}_{v2}$.

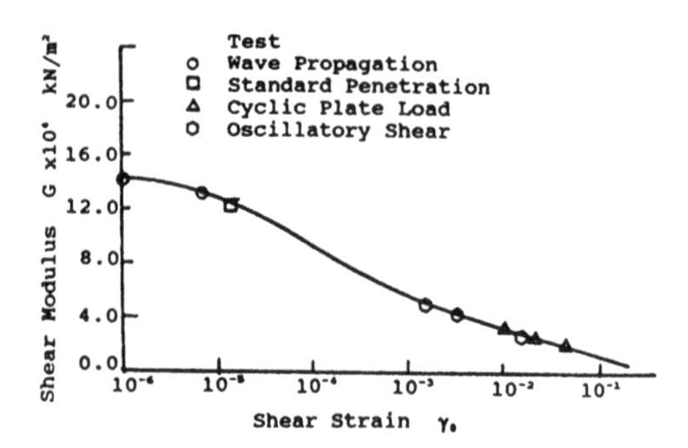

Fig. 2 Dynamic Shear Modulus versus Shear Strain.

The laboratory investigation consisted of oscillatory shear tests conducted on undisturbed representative samples. The tests were conducted using several different combinations of normal and shear loads. The values of dynamic shear modulus 'G' from the laboratory oscillatory shear tests were also computed for an effective overburden pressure of $100 kN/m^2$. Based on the results of in-situ and laboratory tests, a plot of dynamic shear modulus 'G' versus shear strain 'γ_θ' was obtained as shown in Figure 2.

3 PREDICTED RESPONSE OF THE TEST BLOCKS

The methods commonly used for the analysis and design of foundations for machines are (1) Barkan's method and (2) elastic half space method, and (3) impedance function method. In the Barkan's approach or the linear spring method (Barkan, 1962) the foundation soil system is represented as a spring-mass system. The spring stiffness due to the soil and mass of the foundation and supported equipment only are considered and inertia of the soil and damping are neglected. In the elastic half space approach the vibrating footing is treated as resting on the surface of an elastic, semi-infinite, homogenous, isotropic half space (Richart, 1962). The elasticity of the soil and the energy carried into the half space by waves travelling away from the vibrating footing (geometric damping) are thus accounted for and the response of such a system may be predicted using a mass-spring-dashpot model known as the half space analog (Richart and Whitman (1967) and Richart, Hall and Woods (1970)). The impedance function method (Gazettas, 1991) gives an approach for calculating soil spring and damping values which are frequency dependent.

The dynamic response of the foundation was computed using the above methods of analysis. The values of dynamic shear modulus were selected depending on the effective overburden pressure and the shear strain induced in the soil by the vibrating block. Effective overburden pressure at a depth equal to one half of the width of test block was used and was obtained from equation (4).

$$\bar{\sigma}_v = \sigma_{v1} + \sigma_{v2} \quad (4)$$

in which σ_{v1} = Overburden due to weight of soil
and σ_{v2} = Vertical stress intensity at a depth equal to ½ (width) due to superimposed load of machine and foundation and may be computed using Boussinesq theory.

3.1 *Barkan's Method*

Natural frequency of vertical vibrations ω_{nz} is given by

$$\omega_{nz} = \sqrt{\frac{C_u \cdot A}{m}} = \sqrt{\frac{K_z}{m}} \quad \text{.......... (5)}$$

and amplitude of vertical vibration A_z is given by

$$A_z = \frac{P_z}{m(\omega_{nz}^2 - \omega^2)} \quad \text{............ (6)}$$

in which,

m = mass of the foundation and machine
k_z = stiffness of vertical soil spring
ω = operating frequency, and
C_u = the coefficient of elastic uniform compression and is given by equation (7).

$$C_u = \frac{1.13 \times 2G(1+\nu)}{(1-\nu^2)} \cdot \frac{1}{\sqrt{A}} \quad \text{...... (7)}$$

in which ν = Poissons ratio (assumed 0.337).
The computed values of undamped natural frequencies and undamped amplitudes of vibration as obtained from the linear spring method, for the two test blocks, are given in Tables 1 and 2.

3.2 *Elastic Half Space Method*

The natural frequency of the foundation in vertical vibrations is computed by equation (5). The soil spring is computed as follows (Prakash and Puri, 1988) and Richart, Hall and Woods, 1970):

$$K_z = \frac{4Gr_0}{1-\nu} \quad \text{................... (8)}$$

in which, r_0 = Equivalent radius of the foundation and is obtained as follows:

For vertical vibrations or sliding

$$r_0 = \sqrt{\frac{A}{\pi}} \quad \text{...................... (9)}$$

The damped amplitude of vertical vibrations is given by

$$A_z = \frac{P_z}{k_z\left\{[1-(\omega/\omega_{nz})^2]^2 + (2\xi_z\omega/\omega_{nz})^2\right\}^{1/2}} \quad \text{............................... (10)}$$

Where,

$$\xi_z = \text{Damping ratio} = \frac{0.425}{\sqrt{B_z}} \quad \text{........ (11)}$$

and
B_z = Modified mass ratio

$$= \frac{(1-\nu)m}{4\rho r_0^3} \quad \text{................. (12)}$$

The computed values of undamped natural frequencies and damped amplitudes of vibration obtained from the elastic half space method for the two test blocks are given in Tables 1 and 2.

3.3 *Impedance Function Method*

The soil spring for surface foundations undergoing vibrations in the vertical direction is calculated from equation 13 (Gazettas, 1991):

$$k_z = K_z \cdot k_z^* \quad \text{...................... (13)}$$

in which, K = static stiffness, and k_z^* = dynamic stiffness coefficient.
Static stiffness K_z is obtained as:

$$K_z = \left[\frac{2GL}{1-\nu}\right](0.73+1.54X^{0.75}) \quad \text{..... (14)}$$

in which, L = Length of the foundations (as shown in Figure 3), and

$$X = \frac{A_b}{4L^2} \quad \text{....................... (15)}$$

where, A_b=Area of the foundation base.

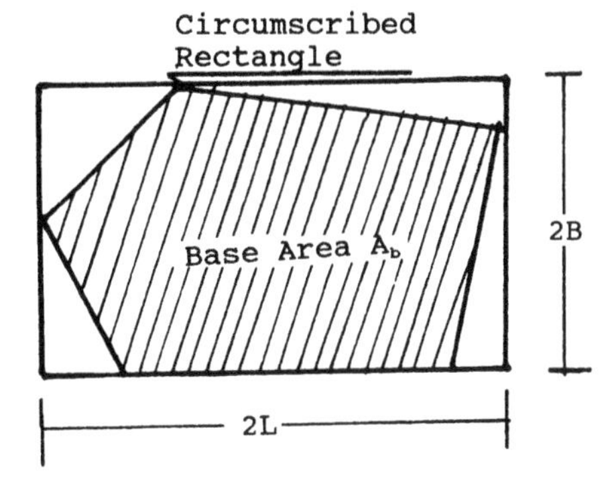

Fig. 3 Concept of Surface Foundation of Arbitrary Shape.

The values of dynamic stiffness coefficient k_z^* depend on the L/B ratio, poissons ratio ν and the dimensionless frequency factor 'a_o'. The dimensionless frequency factor is given by:

$$a_o = \frac{B\omega}{V_s} \quad \text{...................... (16)}$$

Table 1 Comparison of Observed and Computed Response of 1.5m x .75m x 0.70m Block in Vertical Vibrations

Angle of Eccentricity $\theta°$	Observed Data		Shear Strain γ_θ	Shear Modulus G kN/m^2	Coefficient of Elastic Uniform Compression C_u kN/m^3	Computed Data					
						Linear Spring Method		Elastic Half Space Analog Method		Impedance Function Method	
	Natural Frequency Hz	Amplitude Az mm				Natural Frequency Hz	Undamped Amplitude Az mm	Natural Frequency Hz	Damped Amplitude Az mm	Natural Frequency Hz	Damped Amplitude Az mm
25	38.5	0.031	4.13×10^{-5}	35,500	266,600	44.8	0.36	41.4	0.127	55.6	0.0761
50	37.5	0.069	9.2×10^{-5}	29,000	288,800	40.5	1.067	37.1	0.27	49.4	0.170
75	37.0	0.0975	1.3×10^{-4}	26,000	196,000	38.5	4.94	35.2	0.406	46.3	0.248
125	36.7	0.182	2.43×10^{-4}	22,100	166,500	35.3	3.23	33.1	0.531	42.6	0.250
180	35.0	0.240	3.3×10^{-4}	18,100	136,400	31.0	1.38	29.3	0.720	37.4	1.00

Table 2 Comparison of Observed and Computed Response of 3.0m x 1.5m x 0.70m Block in Vertical Vibrations

Angle of Eccentricity $\theta°$	Observed Data		Shear Strain γ_θ	Shear Modulus G kN/m^2	Coefficient of Elastic Uniform Compression C_u kN/m^3	Computed Data					
						Linear Spring Method		Elastic Half Space Analog Method		Impedance Function Method	
	Natural Frequency Hz	Amplitude Az mm				Natural Frequency Hz	Undamped Amplitude Az mm	Natural Frequency Hz	Damped Amplitude Az mm	Natural Frequency Hz	Damped Amplitude Az mm
25	36.0	0.006	4×10^{-6}	70,000	222,615	42.0	0.0954	42.0	0.0261	29.5	0.021
50	34.5	0.012	8×10^{-6}	67,500	214,660	41.2	0.157	41.3	0.0478	29.4	0.0381
75	37.0	0.018	1.2×10^{-5}	66,000	209,900	40.8	0.449	40.7	0.0852	29.05	0.051
100	37.0	0.024	1.6×10^{-5}	65,000	206,700	40.5	0.616	40.4	0.109	28.8	0.050
125	35.0	0.030	2.0×10^{-5}	61,000	193,900	39.1	0.566	39.2	0.119	27.9	0.0423
150	34.5	0.036	2.4×10^{-5}	60,000	190,800	38.8	0.578	38.8	0.127	27.7	0.025
180	33.5	0.053	3.5×10^{-7}	56,000	178,100	37.5	0.63	37.5	0.133	26.7	0.102

The values of k^*_z are plotted in Figure 4 as a function of a_o.

The radiation damping coefficient C_{z1} is obtained from equation (17).

$$C_{zl} = (\rho V_{La} A_b)\overline{C}_z \quad \ldots\ldots\ldots\ldots\ldots\ldots \quad (17)$$

in which,

$$V_{La} = \frac{3.4 V_S}{\pi(1-\nu)} \quad \ldots\ldots\ldots\ldots\ldots\ldots \quad (18)$$

and $\overline{C}_z$ = Dynamic damping coefficient.
The values of $\overline{C}_z$ are given in Figure 5 as a function of the dimensionless frequency factor a_o.

The total damping C_z in the system may then be obtained from equation (19).

$$C_z = C_{zl} + \frac{2k_z}{\omega}.\beta \quad \ldots\ldots\ldots\ldots\ldots\ldots \quad (19)$$

The values of 'ß' depend on the type of soil. Using the values of dynamic stiffness k_z and the total damping coefficient C_z, the dynamic response is calculated using the theory of vibrations. The calculated values of natural frequencies and damped amplitudes are given in Tables 1 and 2.

4 CONCLUSIONS AND DISCUSSION

1. It is observed from Tables 1 and 2, that the natural frequencies of the two test blocks calculated by using the linear spring method and the elastic half space method are generally of the same order. Also these calculated natural frequencies are in reasonable agreement with the observed natural frequencies. The calculated undamped natural frequencies are generally within 20% of the experimentally observed natural frequencies. The natural frequency of vertical vibrations calculated by the impedance function method are somewhat higher for the 1.5m x 0.75m high block as compared to the observed natural frequency. The calculated values of natural frequency by this method are somewhat smaller than the observed

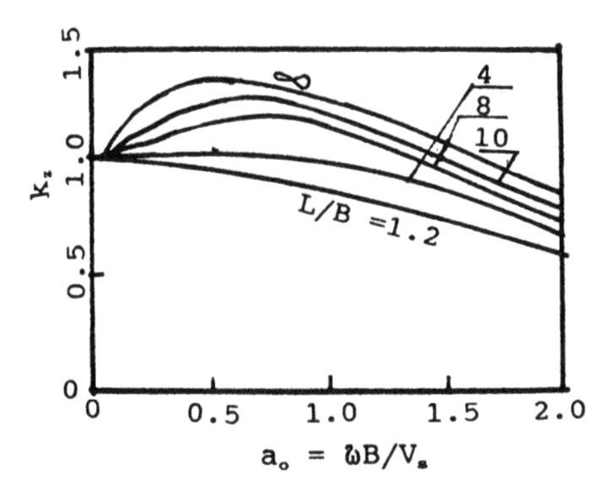

Fig. 4 Dynamic Stiffness Coefficient k^*_z versus a_o (ν=0.4). (Gazettas,1991)

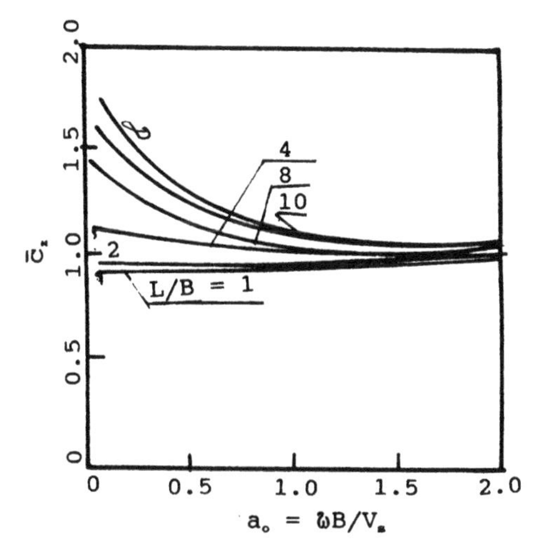

Fig. 5 Dynamic Damping Coefficient versus a_o (ν=0.4). (Gazettas,1991)

values for the 3.0m x 1.5m x 0.7m high block.

2. The undamped amplitudes of vertical vibration calculated by using the linear spring method are much higher than the observed amplitudes. This is to be expected since damping has been omitted in this method.

3. The damped vibration amplitudes as obtained from the elastic half space method are smaller than those calculated by the linear spring method. However, the amplitudes calculated by the elastic half space method also do not show any reasonable agreement with the observed amplitudes and are larger than the observed amplitudes.

4. The damped amplitudes calculated by the impedance function method for the two concrete blocks used in this study show a more reasonable agreement with the observed values. The limited data, however, does not justify a general conclusion.

REFERENCES

Barkan, D.D. (1962), "Dynamics of Bases and Foundations", McGraw Hill Book Company.

Dobry, R., and Gazettas, G., (1986), "Dynamic Response of Arbitrarily Shaped Foundations," ASCE, Journal of Geotechnical Engineering, Division, Vol. 112, No. 2, Feb., pp. 109-125.

Dobry, R., Gazettas, G., and Stokoe, K.H., (1986), "Dynamic Response of Arbitrarily Shaped Foundations: Experimental Verification," ASCE, Journal of Geotechnical Engineering Division, Vol. 112, No. 2, Feb., pp. 126-154.

Imai, T. (1977), "Velocities of p- and s- Waves in Sub-surface Layers of Ground in Japan", Proc IX: International Conference on Soil Mechanics and Foundation Engineering, Vol. 2, pp. 257-260.

IS: 5249-1977. Code of Practice for Determination of In-situ Dynamic Properties of Soil. Indian Standards Institute, New Delhi, First Revision.

Nandakumaran, P., Puri, V.K., Joshi, V.H. and Mukerjee, S. (1977), "Investigations for Evaluation of Dynamic Shear Modulus of Soils", Bulletin, Indian Society of Earthquake Technology, Vol.14, No.2, June, pp. 39-54.

Prakash, S. and Puri, V.K. (1977), Critical Evaluation of IS 5249-1969, Indian Standard Code of Practice for In-situ Dynamic Properties of Soil, Indian Geotechnical Journal, Vol. VIII, No. 1, January, pp. 43-56.

Gazettas, G., (1991), "Formulas and Charts for Impedances of Surface and Embedded Foundations," ASCE, Geotechnical Engineering Division, Vol. 117, No. 9, Sept., pp. 1363-1381.

Gazettas, G., and Stokoe, K.H., (1991), "Free Vibrations of Embedded Foundations: Theory versus Experiment," ASCE, Journal of Geotechnical Engineering Division, Vol. 117, No. 9, Sept., pp. 1382-1401.

Das, B.M. (1992), "Principles of Soil Dynamics", PWS Kent, Boston, MA.

Prakash, S. and Puri, V.K. (1981), "Dynamic Properties of Soils from In-situ Tests", Proc: ASCE, Journal of Geotechnical Engr. Div., Vol. 107, No. GT7, July, pp. 943-964.

Prakash, S. and Puri, V.K. (1981), "Observed and Predicted Response of a Machine Foundation", Proc X: International Conference on SM & FE Stockholm, Vol. 3, pp. 209-273.

Prakash, S. and Puri, V.K. (1988), "Foundations for Machines Analysis and Design", Wiley Series in Geotechnical Engineering, John Wiley, NY.

Puri, V.K. and Das, B.M. (1993), "Dynamic Response of Flock Foundation", Third International Conference on Case Histories in Geotechnical Engineering, St. Louis, June, Vol. III.

Richart, F.E. (1962), "Foundation Vibrations", Transactions: ASCE, Vol. 127, Part 1, pp. 863-898.

Richart, F.E., Hall, J.R. and Woods, R.D. (1970), "Vibrations of Soils and Foundations", Prentice Hall, Inc., New York, NY.

Richart, F.E. and Whitman, R.V. (1967a), "Design Procedures for Dynamically Loaded Foundations", Journal SMF, Proc: ASCE, Vol. 93, SM6, Nov., pp. 168-193.

Richart, F.E. and Whitman, R.V. (1967b), "Comparison of Footing Vibrations with Theory", Journal SMF, Proc: ASCE, Vol. 93, NOSM6, Nov., pp. 143-167.

Developments in Geotechnical Engineering, Balasubramaniam et al. (eds) © 1994 Balkema, Rotterdam, ISBN 90 5410 522 4

Prediction of movement and behavior of sheet pile wall for deep braced cut excavation in soft Bangkok clay

Wanchai Teparaksa
Department of Civil Engineering, Chulalongkorn University, Bangkok, Thailand

ABSTRACT: The sheet pile bracing system is the most common practical system for excavation work of basement and foundations in soft Bangkok clay. The sheet pile wall is defined as the flexible wall trends to show a large significant lateral deformation and induces ground settlement whenever the excavation system is carried out without a good quality control. Sheet pile bracing system with fully effective preloading or prestressing strut system was introduced and designed for basement and foundation excavation of 11.3 m. deep. The results shows that the preloading strut system trends to induce the lower wall deflection ratio against basal heave than those proposed by Mana & Clough (1981). The variation of measured ground settlement with distance is located in zone A and B of guide chart proposed by Peck (1969). The prediction of maximum wall deflection based on simplified method proposed by Wong & Broms (1989) agrees well with the field measurement at the final stage of excavation sequences.

1. INTRODUCTION

The construction of high-rise building and deep basement is increasing in recent years in Bangkok city, capital of Thailand. Basement floor is really needed in the center area of Bangkok due to limited land as well as to comply with the regulation of the Bangkok Metropolitan Authority (BMA) for car park area. As it is known, Bangkok subsoils consist of very thick soft dark marine clay with undrained shear strength of 1-1.5 t/sq.m. The sheet pile bracing system is a commonly used method by contractors in doing excavation works. Generally, the maximum excavation depth of sheet pile bracing system is approximately about 9-10m below ground surface. This limitation is due to the allowable lateral wall movement and vertical settlement of ground surface. This lateral and vertical movement will lead to induced damages to the nearby structures. For the excavation work of deeper than 10m, the rigid retaining structures such as diaphragm wall, secant pile wall, jet grouted gravity wall as well as the Berlin steel structure wall is more appropriate (Teparaksa, 1991 and 1992).

The Bai Yok II tower is the 90 storey building under construction and seems to be the highest highrise buiding in Bangkok city. The excavation for construction of basement and mat foundation is about 11.3 m. deep below ground surface which seems to be the deepest excavation for sheet pile bracing system. The flexible sheet pile bracing wall with fully effective preloading strut system, which was firstly introduced by the author, is adopted for the Bai Yok II excavation work project. The underground construction including mat foundation, basement, underground wall, and ground floor was completed at the end of February 1992.

This paper presents the installation of neccessary geotechnical instruments for detection of lateral and vertical displacement, as well as evaluation of displacement and soil parameters. The influence of factors effected to the lateral wall deflection of sheet pile bracing system for deep excavation is also presented and compared with field performance.

2. SUBSURFACE SOIL CONDITIONS

General Bangkok subsoil condition consists of soft marine clay (Figure 1) which is sensitive and has anisotropic and time dependent undrained stress-strain-strength properties and is a creep susceptable materials. These make the design and construction of deep excavations and filled embankments on soft clay difficult. The first medium dense to silty fine sand located at about 20 to 30 m., contributes to variations in skin friction and the mobilization of end bearing resistance of the pile foundation, upon using different piling construction methods. The problem is also similar for the dense and coarser second silty fine sand found at about 50 m. deep (Figure 1).

At the Bai Yok II project, four bored holes of about 80-100 m. deep below ground surface was carried out. The Bangkok clay consists of about 13.5 m. thick soft to medium marine clay. The undrained shear strength is in the order of 0.6 - 2.0 t/sq.m., water content is between 60 - 70%, liquid limit is about 70 - 96% and the unit weight is in the order of 1.55 - 1.62 t/cu.m. The typical soil conditions including undrained shear strength and SPT N-values of soft clay, stiff clay and first silty sand layer is presented in Figure 2.

3. GEOTECHNICAL INSTRUMENTATIONS AND EXCAVATED SYSTEM

The research site (Bai Yok II tower) is located on

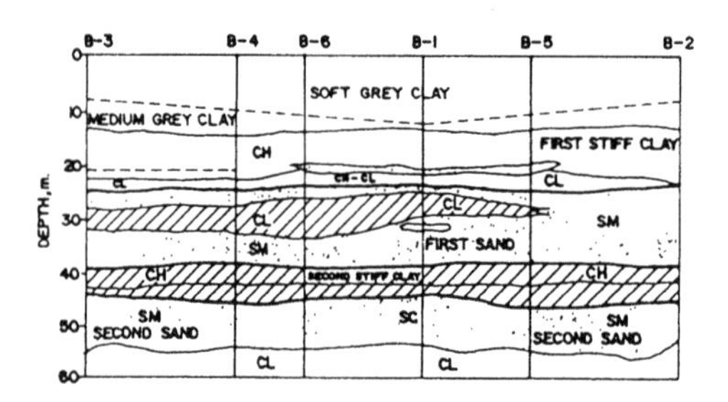

Figure 1 Typical profile of Bangkok subsoils

Rajaparop Road in the center of textile market of Bangkok, and the excavation area is about 50 x 90 m. The excavation area is surrounded with commercial buildings where the nearest building is the four storey commercial building located about 70 cm. behind the sheet pile wall. This narrow site causes a lot of troubles for earth hauling during excavation. The excavation sequences were arranged into two zones as zone 1 and zone 2 where the excavation depth was about 9.5 and 11.3 m., respectively. The excavation was firstly started at zone 1 to about 5 m. deep, then the second zone was followed. These excavation sequences was planned according to the supply of sheet piles which is shortest during period of construction. Zone 1 excavation is the 19 storey podium zone consisted of 2.5 m. thick mat foundation, while zone 2 is the 90 storey highrise zone consists of 5.0 m. thick mat foundation. Construction joint of about 1.5 m. wide was provided between podium and highrise building or between zone 1 and zone 2, where the inner sheet pile wall was installed along this boundary until the completion of the underground work. Figure 3 shows the excavation plan including sheet pile, strut, large bored piles, instrumentations, and the nearest 4 storey building at the corner of zone 2 excavation.

Four layers of bracing were designed for this excavation as shown in Figure 4. For zone 1 excavation (9.5 m. deep), only three strut bracings were used, while four strut levels were applied for zone 2 excavation (11.3 m. deep). Sheet pile wall type FSP IV of 18 m. long with stiffness (EI) of 8,106 tf-sq.m. per linear meter was designed for excavation

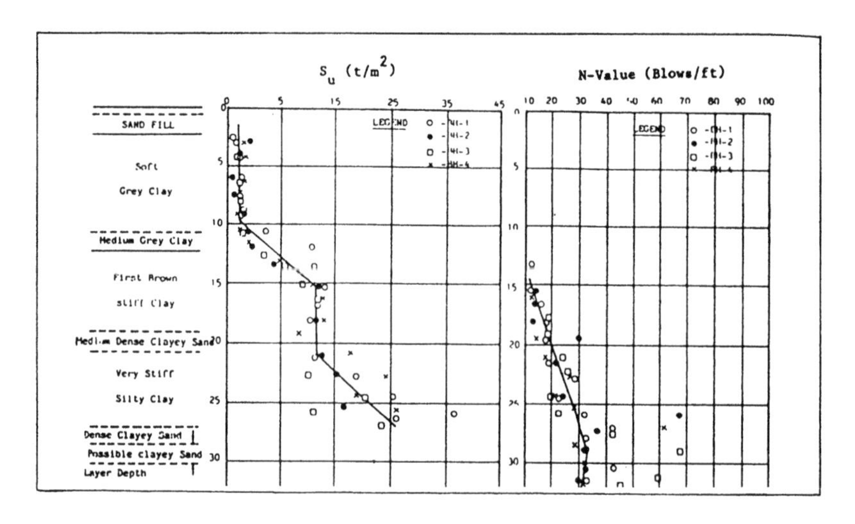

Figure 2 Typical soil properties against depth at research site

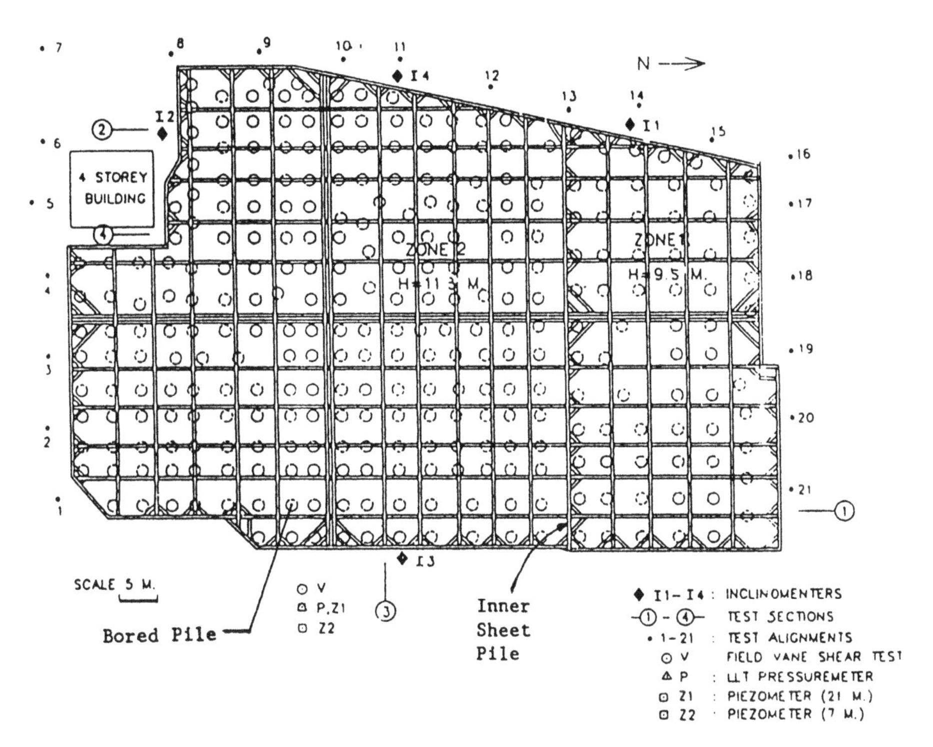

Figure 3 Excavation layout and location of geotechnical instrumentations

works. The section and properties of sheet pile was presented in Figure 5. The strut and wale of W 300 x 300 x 94 kg/m. section was used for the first bracing layer while W350 x 350 x 137 kg/m. section was for the 2nd, 3nd, and 4th bracing layer. King post of 21 m. long of W350x 350 x 137 kg/m section was arranged with spacing of 5.5 m. interval. The king post was designed to carry only the excavation machine and truck on the platform. The strut systems are designed to be slided on the support that is fixed on the side of the king post, except the first strut layer is designed to be fixed on the top of the king post to carry the loading of platform.

Preloading of strut or prestressing of strut was also adopted for excavation work. Preloading with two hydraulic jacks at both ends of a strut was applied immediately after completion of each strut installation. Preloading was applied at both end of strut at the same time with the same order of loading step. The preloading in each strut of all layers was applied about 70% of the apparent pressure diagram proposed by Terzaghi & Peck (1967) assuming a 6 t/sq.m. of uniform surcharge load on the ground surface behind the sheet pile wall.

The field vane shear test and pressuremeter test were carried out before excavation. The instrumentations

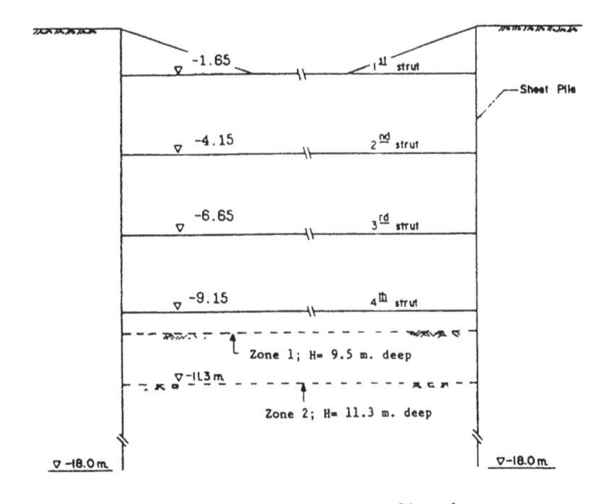

Figure 4 Typical cross section of bracing system

including 4 inclinometers, 2 piezometers and 21 surface settlement points were installed before excavation. The installation of geotechnical instrumentations followed the method proposed by Dunnicliff and Green (1988). Plan of instrumentation layout was shown in Figure 3. The inclinometers were installed in the steel casing that is welded to the sheet pile before driving. After driving

SHAPES

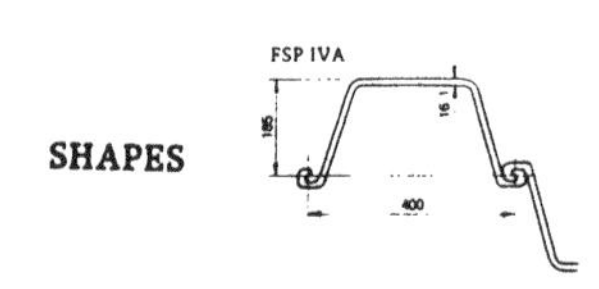

Section	Dimensions			Sectional Area	Weight		Moment of Inertia		Section Modulus	
	w	h	t	per pile	per pile	per wall width	per pile	per wall width	per pile	per wall width
	mm in	mm in	mm in	cm^2 in^2	kg/m lbs/ft	kg/m^2 lbs/ft^2	cm^4 in^4	cm^4/m in^4/ft	cm^3 in^3	cm^3/m in^3/ft
FSP IV	400 15 7	170 6 69	15 5 0 610	96 99 15 03	76 1 51 1	190 38 9	4.670 112	38.600 283	362 22 1	2 270 42 2

Figure 5 Typical section and properties of sheet pile type FSP IV

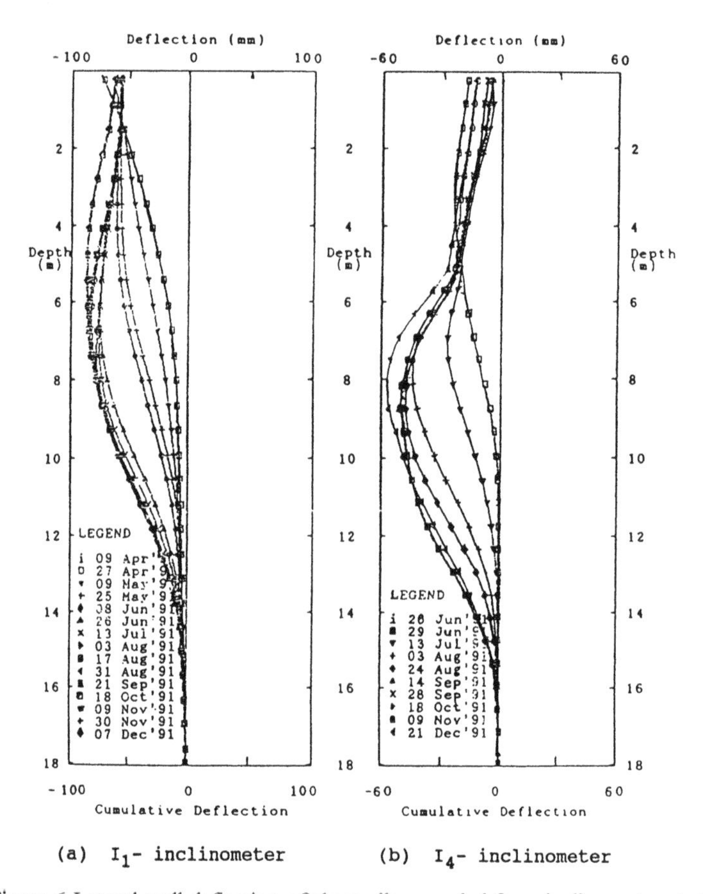

Figure 6 Lateral wall deflection of sheet pile recorded from inclinometer at all stages of excavation sequences

sheet pile fixed with steel casing, cement-benetonite slurry was filled in the casing then inclinometer tube was installed into the steel casing. The average coefficient of lateral earth pressure at rest, K_o, obtained from LLT pressuremeter test in soft clay is equal to 0.58. The piezometric level of soft clay at about 7 m. deep measured from piezometer varied between -0.44 to -0.55 m. However, in the first silty sand layer at about 21 m. deep, the piezometric level shows draw down phenomena at level -19.5 to -20.7 m.

The measurement of wall deflection during the

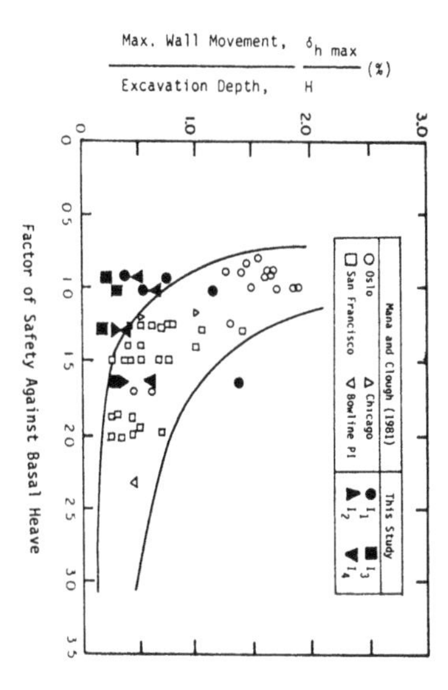

Figure 7 Relationship between maximum lateral wall deflection ratio and safety against basal heave

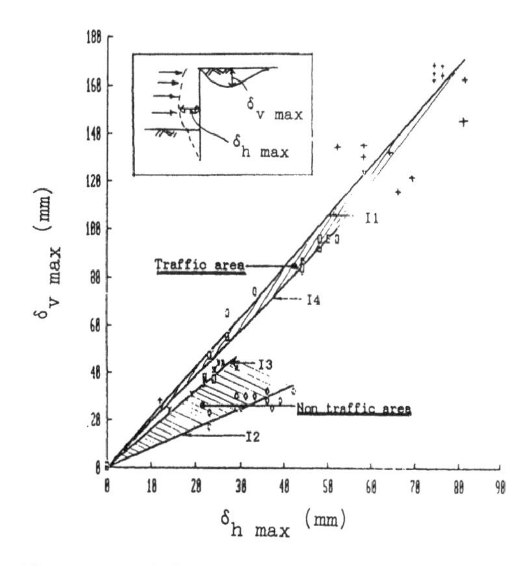

Figure 8 Relationship between maximum ground surface settlement and maximum lateral wall deflection for traffic and non-traffic area

excavation was recorded from the inclinometers I_1, I_2, I_3, and I_4. The inclinometer I_1 was measured at zone 1 excavation area (9.5 m. deep), while inclinometers I_2, I_3, and I_4 were at zone 2 excavation area (11.30 m. deep). The first step of excavation was free cantilever excavation of about 2.0 m. deep before bracing first strut without provision of berm beside the sheet pile wall. The wall movement recorded from I_1 inclinometer during cantiliver excavation was about 90 mm. (as shown in Figure 6(a)) which is very high and causes some damages on the pavement. The excavation depth of cantiliver step is rather close to the critical depth proposed by Peck (1969) and Clough and Davidson(1979). At I_1 inclinometer area , the truck traffic for earth hauling was passed behind the sheet pile wall during the cantiliver excavation, therefore, the excessive wall movement was induced. However, at I_4 inclinometer as shown in Figure 6(b), the measured wall movement was only in the order of 24 mm. This small deflection at I_4 inclinometer is due to the control of truck traffic for earth hauling by using platform.

The lateral wall deflection shows an inward bulging shape against excavation side, however, some order of wall deflection was forced back by applied preloading in the strut. The maximum lateral wall movement was induced at about 9 m. deep for final excavation of 11.3 m. deep or at elevation of approximately 80% of final depth as shown in Figure 6. The mode of sheet pile wall deflection was in the pattern of wall rotated about the bottom or fixed end type.

4. RELATIONSHIP OF MAXIMUM LATERAL WALL DEFLECTION AND POTENTIAL OF BASAL HEAVE

The factor of safety against the basal heave for the intermediate excavation stage at 4.5, 7.0, 9.5, and 11.3 m. deep was calculated based on the method proposed by Mana & Clough (1981). Figure 7 shows the relationship of the ratio between maximum lateral wall movement to the excavation depth and the factor of safety against basal heave. However, the maximum lateral wall deflection for first stage of excavation or cantilever stage was not included in this figure, because it is not the braced cut system. It can be seen that the measurement was within and below the boundary proposed by Mana & Clough (1981). This lower measurement is due to the effect of 70% preloading applied in the strut system.

5. RELATIONSHIP BETWEEN MAXIMUM SURFACE SETTLEMENT AND MAXIMUM LATERAL WALL MOVEMENTS

The maximum surface settlement was measured from 21 number of surface settlement plates. The measurements were recorded both for case of truck

Table 1 Soil parameter and condition for prediction of wall movement

Excavation Depth H(m)	$\sigma_v = \gamma H$ (t/m^2)	T (m)	E_s (t/m^2)	T/E_s	δH max/H (%)	
					Wong & BromS (1989)	Average field performance
4.5	7.18	16.5	1170	2.24	0.78	0.41
7.0	11.09	14.0	1320	1.69	0.59	0.51
9.5	15.02	11.5	1528	1.19	0.42	0.36
11.3	17.86	9.7	1737	0.88	0.31	0.31

Note : T is the thickness beneath excavation to firm layer ,
E_s is secant undrained modulus (from Pressuremeter test)

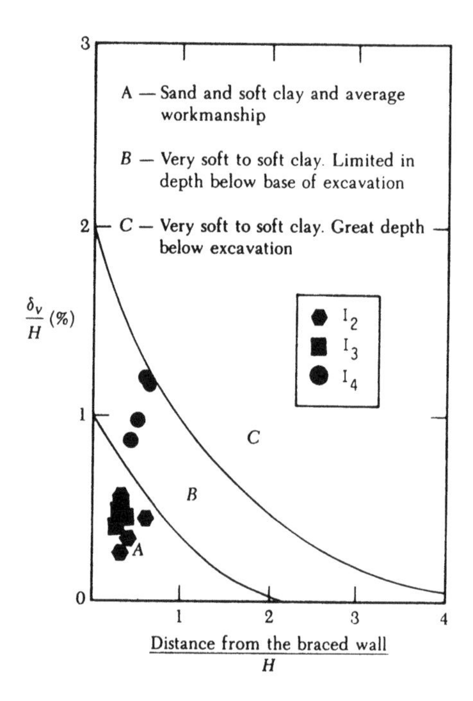

Figure 9 Variation of ground settlement with distance (after Peck, 1969)

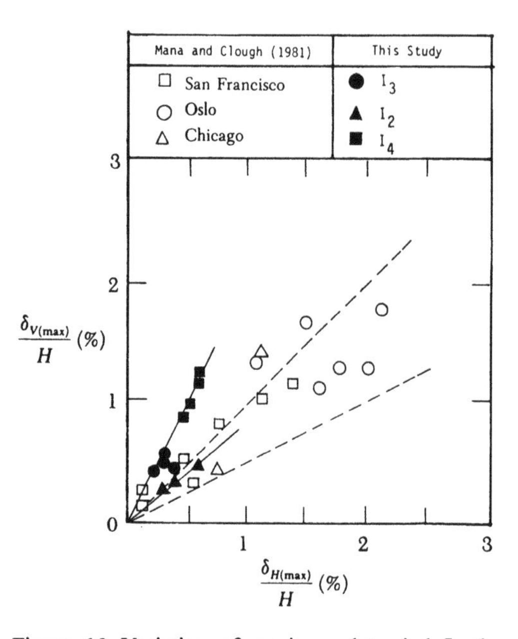

Figure 10 Variation of maximum lateral deflection ratio yield with maximum ground settlement ratio

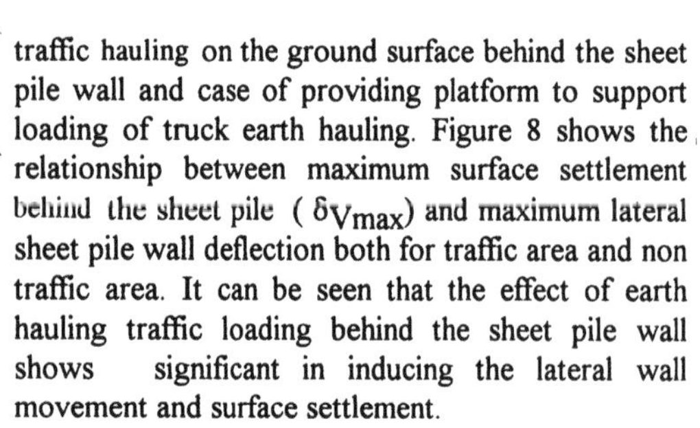
traffic hauling on the ground surface behind the sheet pile wall and case of providing platform to support loading of truck earth hauling. Figure 8 shows the relationship between maximum surface settlement behind the sheet pile ($\delta_{V max}$) and maximum lateral sheet pile wall deflection both for traffic area and non traffic area. It can be seen that the effect of earth hauling traffic loading behind the sheet pile wall shows significant in inducing the lateral wall movement and surface settlement.

The yielding of flexible wall will generally induce ground settlement (δ_V) around a braced cut. This is generally referred to as ground loss. Figure 9 shows the relationship between ground settlement and distance proposed by Peck (1969). and the field data obtained from this research. It can be seen that the field performance of sheet pile wall is located in zone A and zone B where the soil conditions should be soft clay and average workmanship or limited in depth below base of excavation. The performance of sheet pile bracing wall in soft Bangkok clay is normally fitted in the upper portion of zone C, however, the field performance in this research is rather excellent due to the effect of preloading system applied in the struts. Apart from the above relationship, the variation of maximum lateral deformation yield with maximum ground settlement based on this study and field performance proposed

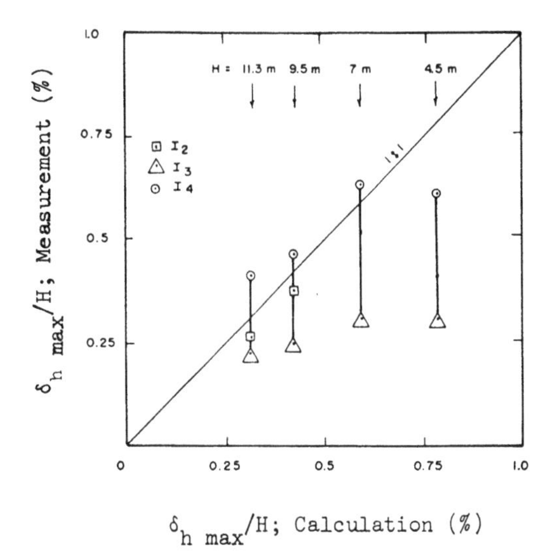

Figure 11 Comparison of maximum lateral wall deflection between prediction and field performance

by Mana & Clough (1981) is presented in Figure 10. It can be seen that this field performance shows the relationship of δ_{Vmax} = 1.25 to 2.0 δ_{Hmax}. This measured performance is quite different from those proposed by Mana & Clough (1981) due to the effect of preloading system.

6. COMPARISON OF PRIDICTION AND FIELD PERFORMANCE OF MAXIMUM LATERAL SHEET PILE WALL DEFLECTION

The prediction of maximum lateral sheet pile wall deflection (δ_{Hmax}) was carried out based on the simplified method proposed by Wong and Broms (1989). The prediction was based on the field measurement of excavation area at zone 2 or from the inclinometer number I_2, I_3, and I_4. The analysis was run without surcharge on the ground surface, because the earth hauling was strictly allowed only on the platform and not allowed behind the sheet pile wall in zone 2 excavation area. Table 1 summaizes the parameters, estimated maximum horizontal wall deflection and average field performance. The reference firm layer was assumed at about 21 m. deep in the first silty sand layer. Figure 11 compares the prediction of maximum horizontal wall movement with the field performance. It can be seen that the prediction of maximum lateral wall deflection agree well with the measurement at the final stage of excavation (H= 11.3m.), while at the intermediate stages of excavation depth (H = 4.5, 7.9 and 9.5m.), the predictions were higher than field performances.

7. CONCLUSIONS

Based on the measurement of field performance with full installation of the geotechnical instrumentations for deep excavation in soft Bangkok clay using sheet pile bracing system, it can be concluded that:

1. Mode of sheet pile wall deflection is defined as the rotation about the bottom or fixed end type.
2. Traffic loading behind the sheet pile wall shows significant influence on the maximum lateral wall movement.
3. The measured deflection ratio against basal heave is lower than those proposed by Mana & Clough (1981).
4. The variation of ground settlement with the distance is located in zone A and B of guide chart proposed by Peck (1969).
5. The relation between the maximum lateral yield of sheet pile and ground surface settlement is higher than those proposed by Mana & Clough (1989).
6. The prediction of maximum wall deflection based on simplified method proposed by Wong & Broms (1989) agree well with the field performance at the final stage of excavation.

REFERENCES

Clough, G.W., and Davison, R.R.(1979): Effect of construction on geotechnical performance, Proc. of Specialty Session III, 9th Int. Conf. on SMFE, Tokyo.

Dunnicliff , J. and Green G.W.(1988): Geotechnical instrumentation for monitoring field performance, John Wiley and Son.

Mana, A.I. and Clough, G.W.(1981): Prediction of movements for braced cut in clay, Jour. of Geot. Engg. Div., ASCE, Vol. 114, No. GT6.

Peck, R.B. (1969): Deep excavation and tunnelling in soft ground, State of The Art Report, 7th Int. Conf. on SMFE, Mexico, pp.225-290.

Teparaksa, W.(1991): Design of sheet pile bracing system for deep excavation in soft Bangkok clay, Seminar on Foundation Work and Underground Construction, Engineering Institute of Thailand (in Thai).

Teparaksa, W.(1992): Use of sheet pile bracing system for deep excavation in soft Bangkok clay, Jour. of Engineering Institute of Thailand for 48th year anniversary publication for highrise technology.

Terzaghi, K. and Peck, P.(1967): Soil Mechanics in

Engineering Practice, John Wiley and Son.
Wong, K.S., and Broms, B.B.(1989): Lateral wall deflection of braced excavation in clay, Jour. of Geotech. Engg. Div., ASCE, Vol. 115, No.6.

Developments in Geotechnical Engineering, Balasubramaniam et al. (eds) © 1994 Balkema, Rotterdam, ISBN 90 5410 522 4

Ultimate uplift capacity and shaft resistance of metal piles in clay

E.C. Shin, V.K. Puri & B.M. Das
Southern Illinois University, Carbondale, Ill., USA
Moon-Seup Shin
Kunsan National University, Korea

ABSTRACT: Laboratory model pile tests were performed to evaluate ultimate uplift capacities and critical relative displacements (CRD). The uplift capacities are significantly affected by pile surface conditions and the depth of pile embedment. The test results show that the CRD is not constant and is independent of the pile size , but is most closely related to the magnitude of the critical pile-soil adhesion along the pile shaft. An empirical relationship between these two quantities was established from the experimental results. This empirical equation was then verified by comparison with the reported results of conducting field tests on piles. A method has been proposed to calculate the value of adhesion coefficient for smooth metal piles in clay.

INTRODUCTION

Foundation piles are used to resist the uplift loads in offshore structures, transmission towers, and other elevated structures on land. In spite of the widespread use of pile foundations, little information is available about the ultimate uplift capacity and the magnitude of critical relative displacement along the pile shaft in clay. Most of the past researchers have discussed the uplift capacity of piles in sand.

The model tests for the uplift capacity of anchor piles in clay were conducted by Ali (1968) and Bhatnagar (1969). Recently, the ultimate uplift capacity of piles embedded in saturated clay was studied by Das and Seeley (1982). Das and Shin (1993) studied the effect of load inclination on the ultimate uplift capacity of rigid vertical metal piles in clay. An equation for estimating the adhesion coefficient along the interface of pipe pile with soil was presented by Das and Seeley (1982).

In practice, a certain amount of the relative displacement between the pile and the soil is needed to fully mobilize the frictional resistance along the pile shaft. This relative displacement is independent of pile size & length and ranges from 5.08 mm to 15.24 mm. The values of relative displacement are smaller for clay and larger for sand (Whitaker and Cooke; 1966, Vesic; 1967, 1969, 1977, Das; 1990, Davie et al.; 1993).

This paper presents results of pullout tests on piles in clay which were conducted to study their uplift behavior. The diameter of piles, embedment depth, roughness of pile surface, centrifuge acceleration levels and spinning times were varied to investigate their individual effects on the uplift capacities and critical relative displacements of the piles. The experimental test results of the net ultimate uplift capacities are compared with the results calculated by using the equation of Das and Seeley (1982). The equation suggested by Das and Seeley (1982) for estimating the adhesion coefficient is evaluated based on the results of 16 model tests on pipe piles.

LABORATORY TEST EQUIPMENT AND MODEL PILE TESTS

A total of 27 model pile tests were

performed in clay. Piles with smooth and rough surface were used in the investigation. The depths of embedment being 254 mm and 178 mm were used. Each test was repeated to obtain reproducible results. Piles with diameters of 4 mm, 5.6 mm, 7.1 mm, and 15.9 mm were used. Each pile was 457 mm in length, and had a horizontal hole in the top part of the pile in which the pullout pin can be inserted. All the model piles consisted of aluminum tubes and were sealed at the bottom.

The pullout tests on model piles were conducted in the Instron loading machine. Test data was automatically recorded on the built-in strip chart recorder of the machine. A schematic sketch of the test set-up is given in Figure 1.

The centrifuge equipment was employed to consolidate the clayey soil. The centrifuge used in this study is the Genesco G-Accelerator Model 1230-5, capable of generating angular velocities as high as 470 rpm.

The clayey soils used in this study were made by mixing Georgia Kaolin with Bonny silt. The proportion of Kaoline to silt was kept as 1:4 in order to produce a clay that consolidates in a reasonable period of time in the centrifuge, while still maintaining the general properties of clay. The engineering properties of this soil are shown in Table 1.

Bonny silt and Georgia Kaolin were first mixed with a small spatula in a dry condition and then mixed with 50% water by weight in an electric blender for approximately 30 minutes. The resulting mixture was carefully poured into the sample container which was placed on the swinging basket of the centrifuge. Since this slurry in its natural condition was not capable of holding a pile, the mixture was consolidated by spinning the soil in the container inside the centrifuge at 20g for 60 minutes prior to the installation of the pile. The model pile was then installed in the middle of the container through the hole of alignment panels. Three alignment panels were stacked up to make the length of the alignment hole approximately 127 mm. The centrifuge was spinned at 60g level for 80 minutes. Then, the spinning was stopped and the sample container taken out from the centrifuge and brought to the Instron machine. The sample container was placed on the crosshead of the Instron and the alignment panel was removed. By adjusting the location of the container the pile was aligned with the pull-out system. The pile was pulled out using displacement rate of 0.51 mm per minute. The displacement and the pullout resistance were both recorded automatically.

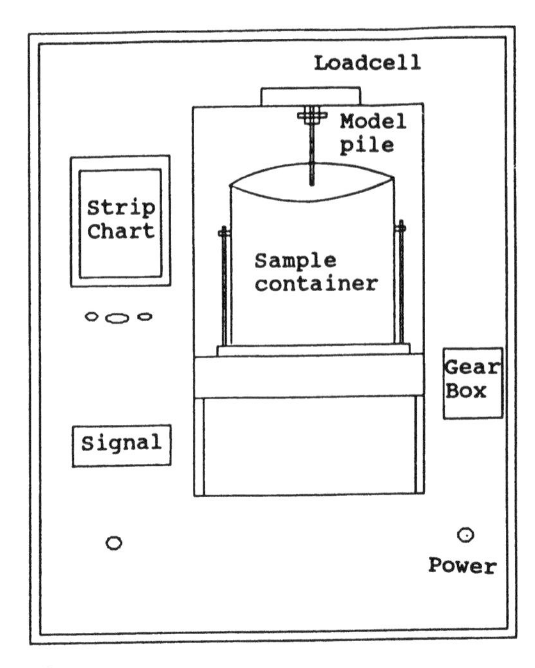

Fig. 1 Model test set-up with Instron machine.

Table 1. Physical properties of clay

Parameters	Quantity
Liquid Limit, LL	24.87 %
Plastic Limit, PL	18.63 %
Plasticity Index, PI	6.20 %
Specific Gravity, G_s	2.70
Activity, A	0.31
Undrained Shear strength, c_u	5.5 kN/m^2

TEST RESULTS AND ANALYSIS

The results of pullout tests on rough surface and smooth surface model piles are given in Tables 2 and 3, respectively. The data on critical displacement and peak shaft resistance for the model piles is given in Table 4. The results on uplift capacity of model piles were compared with the existing theories. The net ultimate uplift capacity of piles can be expressed as:

$$Q_{un} = Q_{ug} - W \qquad (1)$$

where Q_{un} = the net ultimate uplift capacity, Q_{ug} = the gross ultimate uplift capacity, and W = the effective weight of the pile. The net ultimate uplift capacity of a pile embedded in saturated clays can be estimated by using the equation below:

$$Q_{un} = L\alpha'\pi Dc_u \qquad (2)$$

where L = the embedded length of pile, α'= the adhesion coefficient at soil-pile interface, and D = the diameter of pile. From Equation (2), we get:

$$\alpha' = \frac{Q_{un}}{L\pi Dc_u} \qquad (3)$$

The coefficient of adhesion α' for $c_u \leq 27$ kN/m² may also be estimated as follows (Das and Seeley; 1982):

$$\alpha' = 0.715 - 0.0191\, c_u \qquad (4)$$

The undrained cohesion of clay used in this study is 5.5 kN/m² (Table 1) and is lower than 27 kN/m². Equation (4) can be used to estimate α'. The values of net ultimate uplift capacity of model piles calculated from Equation (2) are shown in Tables 2 and 3. A comparison of the calculated values of net ultimate pile capacity and adhesion coefficient with the observed values (Table 2) shows that for the case of model piles with rough surface used in this study, the values estimated from Das and Seeley (1982) are in good agreement with the observed values. For the case of smooth piles, the values calculated by Equation (4) are about 2.265 times higher than the experimental values. From this comparison, it may be concluded that the Equation (4) for calculating α' presented by Das and Seeley (1982) seems to be based on the experimental test results with rough pile surface condition. This equation needs modification to account for the pile surface roughness. Based on results of this study, and for smooth surface pipe pile, Equation (4) can be modified

Table 2. Net ultimate uplift capacity (Rough pile, Diameter = 7.1 mm, Length = 254 mm).

L/D	(a) Q_{un} (N)	(b) α' from Eq. (3)	(c) Q_{un} from Eq. (4) (N)	(d) α' from Eq. (4)	Ratio (c/a) or (b/d)
35.59	16.7	0.532	19.2	0.61	1.15
35.59	22.8	0.726	19.2	0.61	0.84
35.59	18.7	0.595	19.2	0.61	1.03
					Av. 1.01

Table 3. Net ultimate uplift capacity (Smooth pile, Diameter = 7.1 mm, Length = 254 mm).

L/D	(a) Q_{un} (N)	(b) α' from Eq. (3)	(c) Q_{un} from Eq. (4) (N)	(d) α' from Eq. (4)	Ratio (c/a) or (b/d)
64.10	4.7	0.268	10.6	0.61	2.26
45.66	5.9	0.240	14.9	0.61	2.53
35.59	8.9	0.283	19.2	0.61	2.16
16.00	20.2	0.288	42.6	0.61	2.11
					Av. 2.265

[a] Results shown here are the average values based on 4 tests.

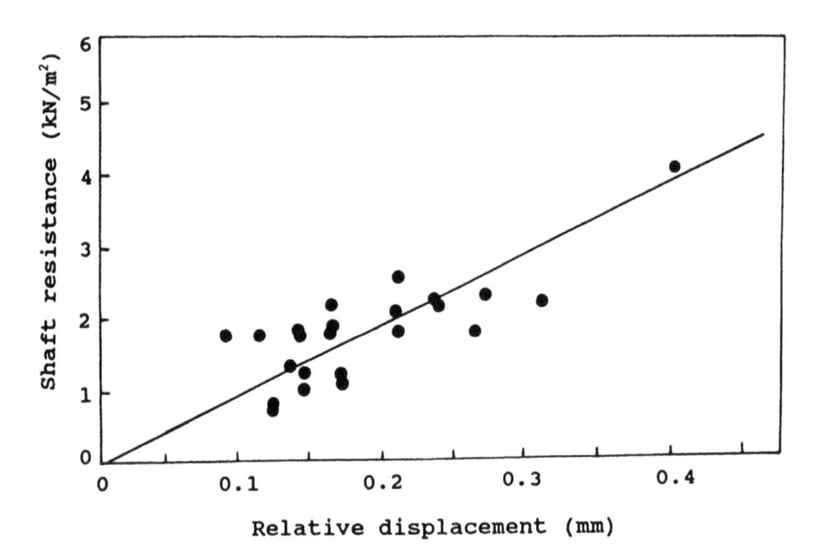

Fig. 2 Critical relative displacements versus shaft resistances.

as follows:

$$\alpha' = 0.316 - 0.0084\ c_u \qquad (5)$$

Previous investigators (Whitaker and Cooke; 1966, Vesic; 1967, 1969, 1977, Das; 1990, Davie et al.; 1993) have reported the amount of relative displacement needed to fully mobilize the skin resistance along the pile shaft irrespective of pile size in terms of its diameter and length. However, as noticed from Table 4, no definite relationship between the critical relative displacements and shaft resistances is observed for the model pile test of this study. The Critical Relative Displacement (CRD) is defined as the displacement at the point of maximum curvature in the pullout resistance versus relative displacement curve. From the experimental test results in Table 4, it ranges from 0.09 mm to 0.48 mm. It seems that in this case CRD is practically independent of pile length and diameter, but there is a some relationship between CRD and the critical adhesion. The data on critical relative displacements versus shaft resistances observed in this study were plotted in Figure 2. A linear regression analysis was performed on the data plotted in Figure 2 to obtain a regression line equation below:

$$Y(\mathrm{kN/m^2}) = 9.14\ X - 0.056 \qquad (6)$$

$$Y(\mathrm{kN/m^2}) \approx 9.14\ X \qquad (7)$$

Pile no. 16 in Table 5 (Whitaker and Cooke, 1966) has been used to verify the proposed empirical Equation (5). The uplift resistance versus displacement for this pile is shown in Figure 3. A critical point is chosen at displacement of 6.0 mm which has 1869 kN resistance (Figure 3). With this, the CRD (X) can be calculated as shown below by using Equation (7).

$$X(\mathrm{mm}) = \left(\frac{1}{9.14}\right)\left[\frac{1869\,(\mathrm{kN})}{2224\,(\mathrm{kN})}\right] 56.2 \left(\frac{\mathrm{kN}}{\mathrm{m^2}}\right)$$

$$\approx 5.2\ \mathrm{mm}$$

It may be noted that the computed value, 5.2 mm compares well with the existing theories and field data (Whitaker and Cooke; 1966, Vesic; 1967, 1969, 1977, Das; 1990, Davie et al.; 1993).

Table 6 shows the effect of the pile surface condition; smooth or rough. A rough surface was created by scrubbing a pile surface with coarse sand paper. It is seen from Table 6 that the pullout resistance ratio of rough to smooth piles in clay is about 2.16. It is reasonable to assume that piles in clay are less affected by the change of the surface condition than in frictional granular materials (Shin, 1987).

The measured pullout resistance in

Table 4. Critical relative displacement versus shaft resistance

L/D	X (mm)	Y (kN/m^2)	Remarks
64.10	0.11	1.59	L=254 mm Smooth piles 60g-80 min
64.10	0.17	1.59	
64.10	0.17	2.00	
64.10	0.24	2.00	
45.66	0.09	1.59	
45.66	0.14	1.59	
45.66	0.14	1.66	
45.66	0.17	1.72	
35.59	0.26	1.66	
35.59	0.27	2.14	
35.59	0.24	2.07	
35.59	0.21	1.93	
16.00	0.32	2.07	
16.00	0.21	1.66	
16.00	0.21	2.34	
16.00	0.14	1.17	
35.59	0.41	3.79	L=254 mm Rough piles 60g-80 min
35.59	0.36	4.96	
35.59	0.48	4.14	
24.91	0.17	1.10	L=178 mm Smooth piles 40g-40 min
24.91	0.17	1.10	
24.91	0.15	1.10	
24.91	0.17	0.97	
24.91	0.12	0.62	L=178 mm Smooth piles 40g-40 min
24.91	0.12	0.69	
24.91	0.17	1.10	L=178 mm Smooth piles 60g-80 min
24.91	0.15	0.90	

Table 7 shows that for piles embedded to a depth of 178 mm, its value is approximately 0.61 of the pullout resistance measured when it is embedded to a depth of 254 mm. Thus, pullout resistance is approximately proportional to embedment depth.

CONCLUSIONS

The results of laboratory model tests for short term uplift capacity and shaft resistance of metal piles in clay have been presented. Based upon the observed results, the following conclusions are drawn:

1. The observed values of ultimate uplift capacities are reasonably in good agreement with the existing theory for rough pile surface condition.
2. For smooth surface model piles, Equation (4) needs some modifications to match with the data observed during pullout tests. The modified equation for coefficient of adhesion ($c_u \leq 27$ kN/m^2) is

$$\alpha' = 0.316 - 0.0084\, c_u$$

3. Critical relative displacement (X) is most closely related to the critical pile-soil adhesion (Y). This relationship has been experimentally determined as

$$Y(kN/m^2) \approx 9.14\, X(mm)$$

4. Pile surface condition can affect the pile shaft resistance by

Table 5. Uplift test on anchor pile (Whitaker and Cooke, 1966)

Pile no.	Age at test (week)	D (m)	L (ft)	$W' + F_s$ (kN)	Uplift resist. (kN)	Shaft resist. (kN/m^2)
13	8	0.89	12.3	214	2082	55.4
17	4	0.94	12.3	240	2713	68.4
16	15	0.94	12.3	236	2224	56.2
15	28	0.94	12.3	238	2429	61.3
10	62	0.94	12.3	238	2509	63.6

Table 6. Pile surface effect (L = 254 mm)

L/D	(a) Rough pile shaft resist (kN/m^2)	(b) Smooth pile shaft resist (kN/m^2)	Ratio (a/b)
35.59	3.86	1.93	2.00
35.59	5.31	2.21	2.40
35.59	4.34	2.07	2.10
35.59	-	2.07	-
Av.	4.48	2.07	2.16

Table 7. Pile embedment effect for smooth surface piles

Pile dia. (mm)	(a) L=254 mm shaft resist (kN/m^2)	(b) L=177.8 mm shaft resist (kN/m^2)	Ratio (b/a)
7.14	1.93	1.31	0.68
7.14	2.21	1.24	0.56
7.14	2.07	1.31	0.63
7.14	2.07	1.17	0.57
Av.	2.07	1.26	0.61

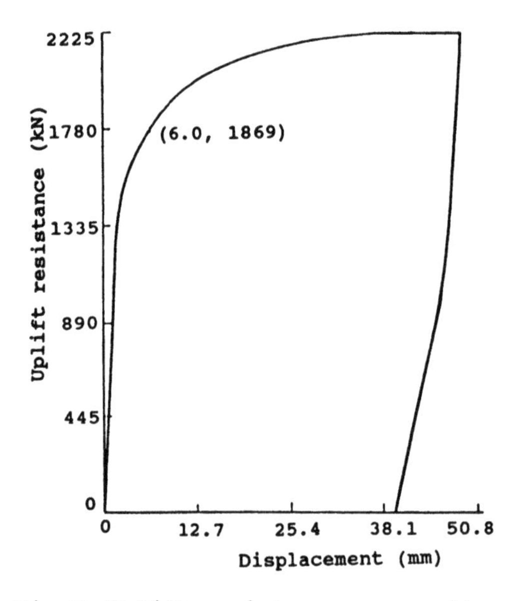

Fig.3 Uplift resistance versus displacement for pile no.16 (Whitaker and Cooke, 1966)

a factor 2.16 as observed during this study.

5. For the embedment length of model piles used in this study, the shaft resistance was found to be approximately proportional to the depth of pile embedment.

REFERENCES

Ali, M.S. (1968). "Pullout resistance of anchor plates and anchor piles in soft bentonite clay," M.S. Thesis presented to Duke University, Durham, N.C.

Bhatnagar, R.S. (1969). "Pullout resistance of anchors in silty clay," M.S. Thesis presented to Duke University, Durham, N.C.

Das, B.M. (1990). Principles of Foundation Engrg., Second Ed., PWS-KENT Publishing Co., Boston.

Das, B.M. and Seeley, G.R. (1982). "Uplift Capacity of Pipe Piles in Saturated Clay," Soils and Foundations, Vol.22, No.1, pp.91-94.

Das, B.M. and Shin, E.C. (1993). "Uplift Capacity of Rigid Vertical Piles in Clay Under Inclined Pull," J. of Offshore and Polar Engrg., Vol.3, No.3, pp.231-235.

Davie, J.R., Lewis, M.R., and Young, L.W. (1993). "Uplift Load Tests on Driven Piles," Proc. Third Int'al Conf. on Case Histories in Geotech. Engrg., St. Louis, Missouri, Vol.I, pp.97-103.

Shin, E.C. (1987). "On Pile-Soil Load Transfer," M.S. Thesis presented to the University of Colorado at Boulder, CO., USA.

Vesic, A.S. (1967). "A Study of Bearing Capacity of Deep

Foundations," Final Report, Project B-189, Engrg., Experimental Station, Georgia Institute of Tech., Atlanta, GA.

Vesic, A.S. (1969). "Load Transfer, Lateral Loads, and Group Action of Deep Foundations," Performance of Deep Foundations, ASTM, STP 444, pp.5-14.

Vesic, A.S. (1977). "Design of Pile Foundations," National Academy of Science, National Research Council, Highway Research Record 42, pp.9.

Whitaker, T. and Cooke, R.W. (1966). "Large Bored Piles," Proc. of the Symposium Organized by the Institution of Civil Engineers and the Reinforced Concrete Assoc. at the Institution of Civil Engineers, London, February, pp.6-49.

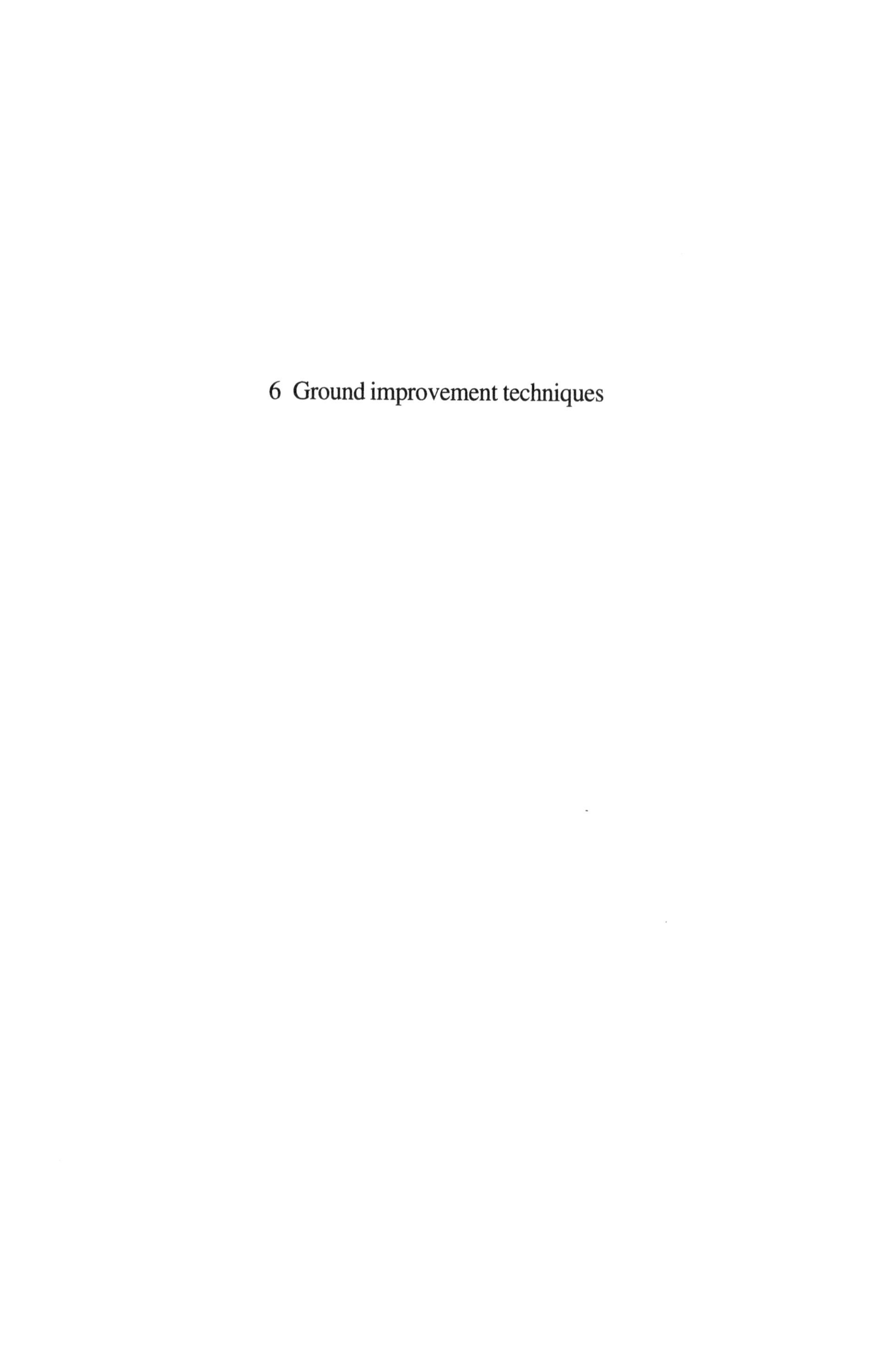

6 Ground improvement techniques

Developments in Geotechnical Engineering, Balasubramaniam et al. (eds) © 1994 Balkema, Rotterdam, ISBN 90 5410 522 4

Use of geotextiles in railway track rehabilitation

G. P. Raymond
Queen's University, Kingston, Ont., Canada

ABSTRACT: The work deals with various aspects that have been observed in the field to be desirable for longevity of railway track geotextiles. These involve first the geotextiles desirable physical properties, recommended specification requirements, and production control tests. Second are installation locations, transition zone requirements, depth of installation requirements, and an example of an actual installation. Also discussed is the beneficial effect of a geotextile in relation to frost susceptible underlying materials.

1 INTRODUCTION

Numerous case records of long term satisfactory installations of geotextiles in track rehabilitation are now being recorded (Raymond, 1985). These installations are generally at hard to maintain and drain areas such as grade crossings, track switches (Raymond 1986a), and the like. The fact that a geotextile is being used in track generally is indicative of a ballast support stability problem.

The two main requirements for adequate stability of the ballast support are first sufficient granular cover between the base of the crossties and the subgrade to prevent overstressing and thus failure of the subgrade, and second a filter blanket (generally granular) to prevent fine-grained particles from the subgrade penetrating the ballast (Raymond, 1985). A third little mentioned requirement is the stability of the subballast particularly after heavy rainfalls or during the thaw period if ice lensing has occurred.

Both (support subgrade or subballast) strength and the minimizing of particle migration are facilitated by good drainage. Such drainage provisions include (a) adequate side ditch draining to deal with surface water, (b) the lowering of the groundwater to increase the subgrade strength, and (c) the drainage of the track area to prevent water seeping into the subgrade bearing area. In the maintenance of subgrade or subballast stabilization work, whether using geotextiles or not, such drainage requirements are always the first and most essential item.

Clearly geotextiles should not be used in place of good drainage practice.

2 PHYSICAL PROPERTIES

Since facilitating drainage and preventing subgrade or subballast particle migration have been found to be the main basic functional requirements of satisfactory long term performance of geotextiles the geotextiles physical properties must be selected to achieve these requirements. The most important properties to achieve these requirements are (a) the hydraulic transmissibility, or in-plane permeability times thickness, must be large in comparison with the underlying soil (Gerry and Raymond, 1983b) (b) the pore size between the individual fibers of the geotextile, or Filtration Opening Size (FOS), must be small enough to prevent the soil particles in the underlying soil from migrating either through the geotextile or, more importantly, into the geotextile and causing excessive fouling and excessive loss of transmissibility (Gerry and Raymond, 1983a), (c) the ability to resist abrasion from large ballast particles or from impact during ballast placement (Costa and Raymond, 1983 and Vandine et al, 1982), (d) the ability to resist the chemical environment of a track bed from dripping freight tankers and weed killing chemicals, and (e) the ability to elongate over large ballast particles left on or protruding from the track bed after undercutting (Raymond, 1982, 1984, 1985).

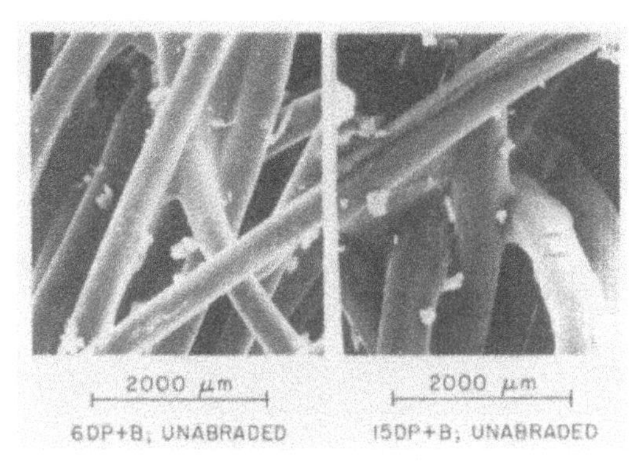

Figure 1 Resin bonding fibers to prevent separation when particles push into geotextile.

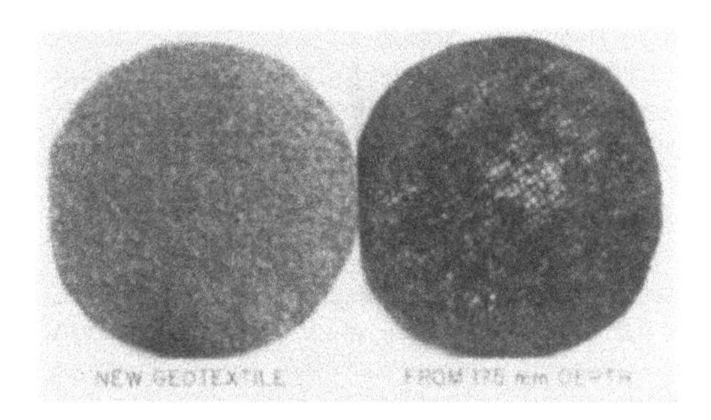

Figure 2 Comparison of new and used geotextile showing internal fiber wear.

3 SPECIFICATIONS FOR PROPERTIES

The above physical characteristics are generally achieved through the use of specific or performance requirements in a railroad specification. Such a specification, drafted by the writer, is in general use on Canadian National's Atlantic Region (Raymond, 1985) and has also been used by numerous other North American Railroads. The major requirements are that the geotextile be of nonwoven needle-punched construction. Woven and thin heat calendered geotextiles are not permitted since these have little to no ability to transmit water within their plane (Raymond, 1984). The nonwoven's fibers need to be mechanically interlocked for abrasion resistance so they are specified to have a minimum length of 100 mm (4 inch) and be needled with at least 80 needle penetrations/cm^2 (500 needle penetrations/in^2) during manufacture. In order to prevent particle penetration (due to repeated loading) a minimum of 5% by weight of acrylic or other non-water soluble resin that leaves the geotextile flexible is required. It is recommended that an upper limit of 20% by weight of resin be used otherwise the geotextile becomes too stiff to handle in the field. Resin treatment coats the fiber with about a 1 μm (one micron) abrasive layer and, as seen in Figure 1, bonds the individual fibers so they have resistance against a particle pushing the fibers apart and penetrating the geotextile. As seen in Figure 2 once penetrated the particle causes internal abrasion of the fiber matrix and thus a reduction of the in-plane transmissibility. A needle punched geotextile has a porosity of 85% to 90% so a thirty percent resin treatment, at most, lowers this to 80% affecting very little the geotextiles transmissibility.

Figure 3 Condition of undercut ballast surface

A nonwoven needlepunched geotextile has the ability of extendibility. This permits it to stretch around the ballast size particles such as those seen in Figure 3, left on the undercut surface or protruding from it, without cutting or puncture. Composite geotextiles of needled nonwoven fibers and a strong woven scrim, on the other hand, have no ability to extend around such particles and as seen in Figure 4 of an installed composite geotextile have always been observed to be cut and/or punctured. A one hundred percent non-woven is thus a requirement of the specification.

Along with being nonwoven the other factor which has been found essential for longevity has been a relatively heavy weight geotextile and the specification calls for a minimum final mass after resin treatment of 1050 g/m^2 (30 oz/yd^2). Such a mass is desirable so that the geotextile will have a service life at least equal to the overlying ballast. Thus the geotextile will not be the cause for the next cyclic undercutting of the track section.

Finally, and probably the most important requirements, are a fiber size of less than or equal to 0.67 tex (6 denier), and a fiber strength, after soaking for 24 h, of not less than 0.36 N/tex (4 g/denier). The fiber size is important since the FOS (in physical length units) is approximately equal to the fiber diameter divided by the fraction: one minus the geotextiles porosity. Smaller FOS values mean less particle penetration and thus less internal abrasion and less fouling. This allows the geotextile to maintain its in-plane transmissibility.

4 PRODUCTION QUALITY CONTROL

The writer's specification deals with the qualification testing of a geotextile for use in track and requires certain in-plane permeability, burst strength, FOS and other test criteria to be achieved for a geotextile to be accepted. Once accepted geotextiles should be monitored during production (manufacture). It is generally not normal for all the qualification tests to be performed at manufacture. Nor are qualification tests always the best method of controlling production.

The writer would recommend continued monitoring by an independent laboratory of the fiber mass and strength since most of the other physical properties except overall mass are dependent on these two values. Roll mass is generally evident during shipping if the geotextile has been given sufficient time to dry after resin treatment. Other manufacturing requirements are evident by the trained eye from visual inspection.

5 INSTALLATION LOCATIONS

When fouled ballast is removed from a track structure, it is important to achieve good drainage in the track shoulders, as illustrated in Figure 5, otherwise a canal effect occurs. Note the shoulder ballast should be removed and cleaned or replaced to a level below the freshly prepared track surface. If a geotextile is installed this would allow the geotextile's edges to act as a sink for water from the load bearing area of the track. This is of utmost importance since the subgrade's strength decreases in the presence of any excess water and the remaining fouled ballast tends to be frost susceptible resulting in considerable loss of strength during thaw. Thawing subballast and fouled ballast that are poorly drained allow segregation and pumping of their fines during the short thaw season.

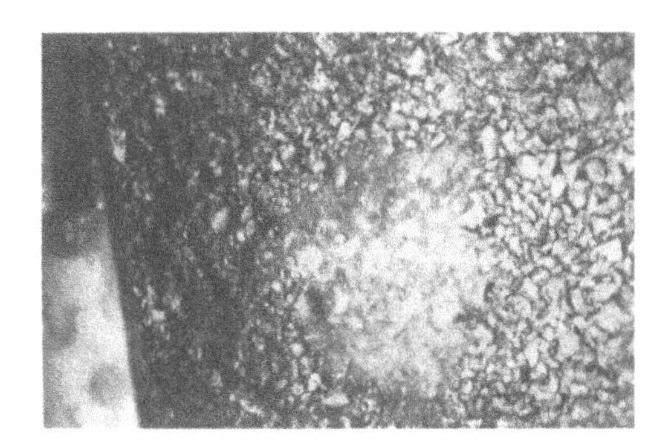

Figure 4 Ballast particles penetrating composite geotextile due to the woven's inability to elongate.

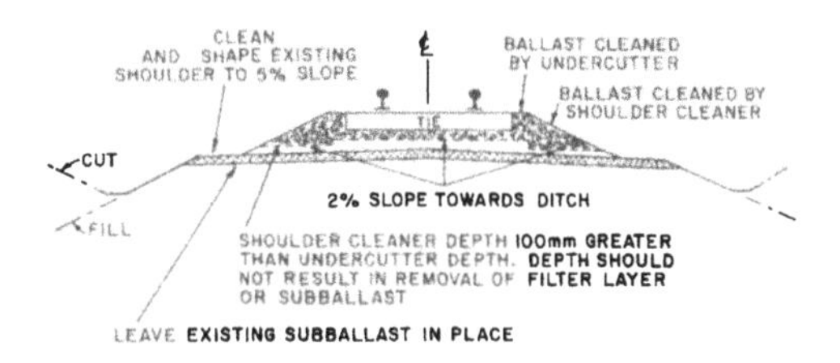

Figure 5 Recommended undercutting with removal of impermeable shoulder ballast.

In locations where the ground is flat and the undercut surface will be located below ground surface steps should be taken to (a) guard against the inflow of water from the ground surface and (b) to see that the water that falls on the track can drain from the load bearing area. Where geotextiles are used trench-drains should be provided on each side of the track, as shown in Figure 6, so that the geotextile will pump the water under repetitive loading into the trench-drains. Where a sink is not provided that is lower than the undercut load bearing area the geotextile will facilitate water flow into the load bearing area resulting in a worst condition than had it not been used at all. Note that geotextiles incorrectly installed do more harm than had they not been used. Correctly installed geotextiles, in the appropriate location, have been noted to be very beneficial. Correctly installed, they also contain the upward migrating fines and increase the bearing capacity of thawing soils that have become ice lensed during freezing.

At grade crossings and to a similar extent at turnouts a lack of adequate drainage along the crossing requires trench-drains or sub-drains similar

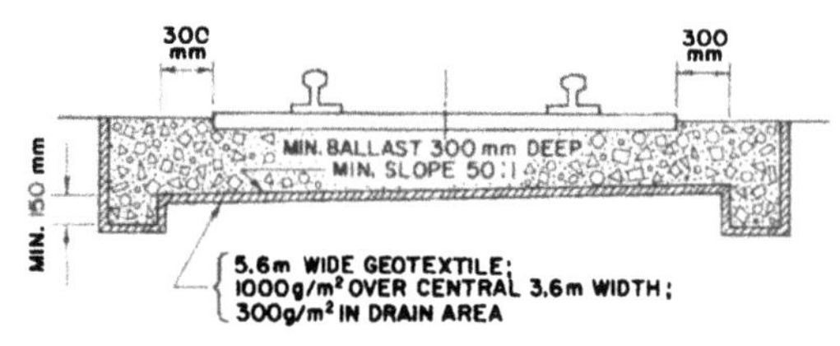

INSTALLATION OVER FLAT TERRAIN.

Figure 6 Recommended undercutting with french drains over flat terrain. Geotextile addition where required.

4 UNIT GEOTEXTILE GRADE CROSSING

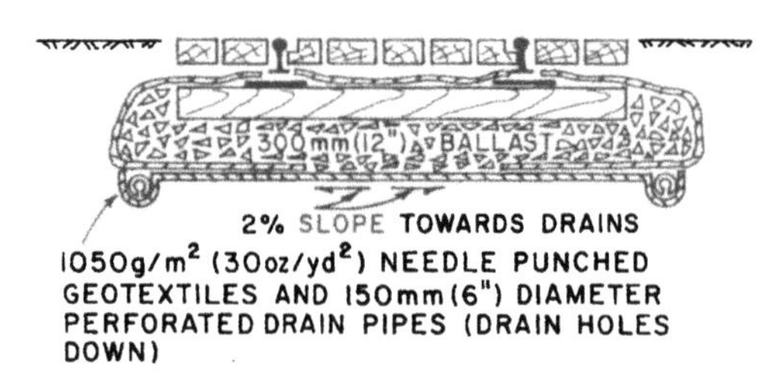

2% SLOPE TOWARDS DRAINS
1050g/m² (30oz/yd²) NEEDLE PUNCHED GEOTEXTILES AND 150mm (6") DIAMETER PERFORATED DRAIN PIPES (DRAIN HOLES DOWN)

Figure 7 Recommended grade crossing rehabilitation with encapsulating geotextiles to minimize fouling by winter salting and sanding.

BRIDGE ABUTMENT REHABILITATION

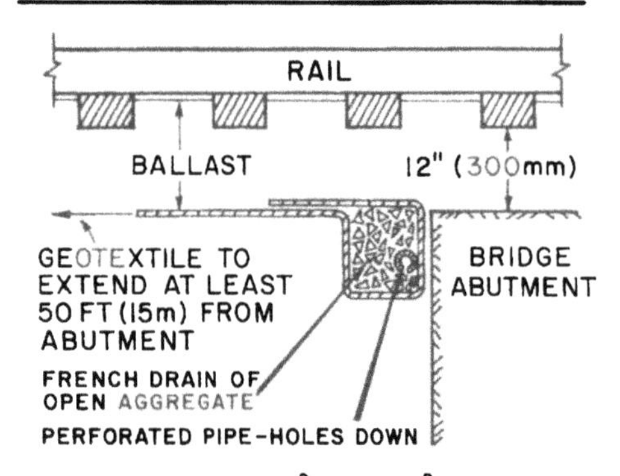

GEOTEXTILE: 30oz/yd² (1050g/m²) MINIMUM MASS PER UNIT AREA; NEEDLE PUNCHED; 2% MINIMUM CROSS-SLOPE FALLING TO SIDE DRAINS

Figure 8 Recommended bridge abutment rehabilitation using a geotextile.

to those for flat terrain. In cold climates where the highway is often subject to sanding geotextiles can be used to encapsulate the ballast as shown in Figure 7.

Figure 9 Transition zone between track not undercut and area with geotextile.

Another area where geotextiles are commonly used is at sudden transition zone areas such as bridge abutments. Here the sub-drain is not only used along the track but also at the abrupt transition from bridge to natural soil as shown in Figure 8.

6 GEOTEXTILE TRANSITION ZONES

The use of a geotextile correctly installed facilitates drainage and thus favours an increase in the subgrade stiffness due to a reduction of moisture. This stiffening effect occurs despite the theoretical argument that geotextiles are more compressible than soils and thus cause a reduction in track modulus. In order to allow as smooth a transition from the undercut track with geotextile to the unrehabilitated areas it has been found highly desirable to undercut some 6 to 9 metres beyond the geotextile laid ends as seen in Figure 9. This will decrease the suddenness of the track modulus change and reduce the associated dynamic traffic loadings. Such transition zones are important.

7 DEPTH OF INSTALLATION

One of the major contributing factors to a geotextile longevity in track has been found to be the depth at which it is placed below the base of the ties. Typical tamper tine settings are shown in Figure 10. These tines are inserted in the ballast to raise the track and/or compact the ballast.

Extensive excavations made to assess tamper (and other) damage has indicated that the tamper's tines vibrate the ballast to a depth of about 175 mm (7 in) below the base of the ties. This ballast particle movement is very abrasive to any geotextile such that an absolute minimum of 200 mm (8 in) of ballast is considered essential between the geotextile and base of the tie. This writer recommends a minimum of 250 mm (10 in) and a preference for 300 mm (12 in).

8 FROST SUSCEPTIBLE SUBGRADES

Geotextiles can be used to great effect over frost susceptible subgrades, subballast or dirty ballast. The mechanism of freezing and thawing damage is shown in Figure 11. As the snow gets ploughed from the track during winter it acts as an insulation layer to prevent thawing of the soil on the sides of the track. Thus, where poor drainage exists a thawed zone of ponded water results directly below the tracks. Where a good drainage surface has been provided the thaw of the ice lenses causes the water to escape as shown in Figure 12. As the water escapes upwards and as the ice melts the surface of the subballast will be left soft and its particles easily susceptible to (upward) migration. A geotextile installed to facilitate drainage will clearly accelerate consolidation of this loose soil and aid in preventing particle migration. The geotextile will also act to confine the granular soil aiding its bearing capacity. The important factor is to install the geotextile and/or undercut the track with drainage as shown in Figures 6, 7, 8 and 12.

Geotextiles may also be used to great effect in areas of plastic clay subgrade when used in conjunction with fine grain sands. The fine grain sand prevents the clay from migrating (pumping) upwards while the geotextile acts as a separator between the sand and the ballast (Raymond 1986c).

9 TURNOUT GEOTEXTILE INSTALLATION

In the Atlantic Region of the Canadian National Railways, a large number of turnout geotextiles have been installed since 1983. Best efficiency was obtained using a dedicated work gang. The geotextile is best delivered at the time of installation since vandals often remove any weatherproof packaging and, if soaked by overnight rain, the geotextile, with an 80% porosity, can increase dramatically in mass from about 340 kg (700 lb) to 1800 kg (4000 lb) making handling exceedingly difficult.

The first operation is to plough the shoulder ballast on the ditch sides of the track to a depth of 200 to 300 mm (8 to 12 in) below the ties. Then the switch undercutter, starting 9 m (30 ft) from the switch point, excavates a trench for its blade. The undercutter blade and trencher bucket then removed the ballast to a depth of 300 mm (12 in) below the track ties and deeper within the trench. When sufficient distance had been covered the process is repeated on the other side of the switch to a location well beyond the turnout's point. This is

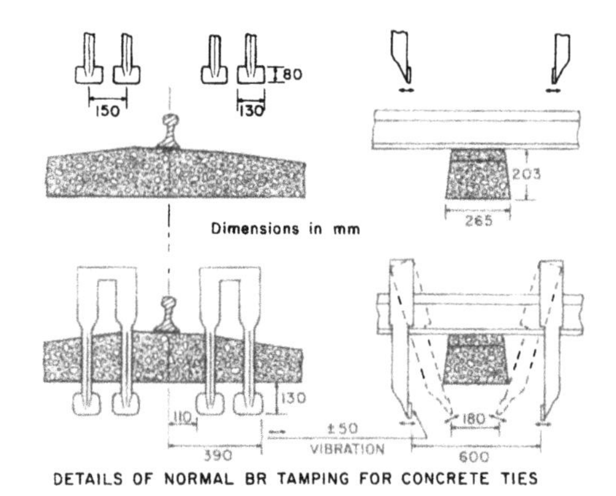

Figure 10 Typical tamper tine settings and movement during compaction.

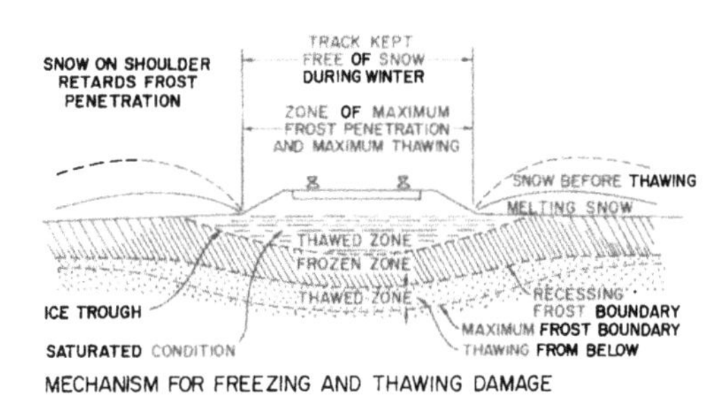

Figure 11 Typical concept of freezing and thawing mechanism for frost susceptible track support.

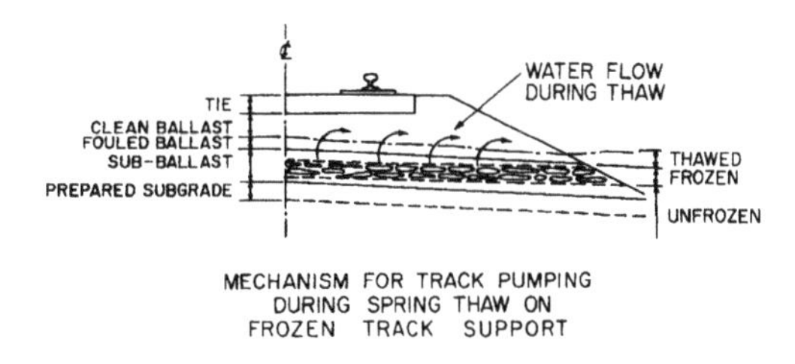

Figure 12 Mechanism for track support particle migration during spring thaw of frozen support.

TURNOUT UNDERCUT SEQUENCE

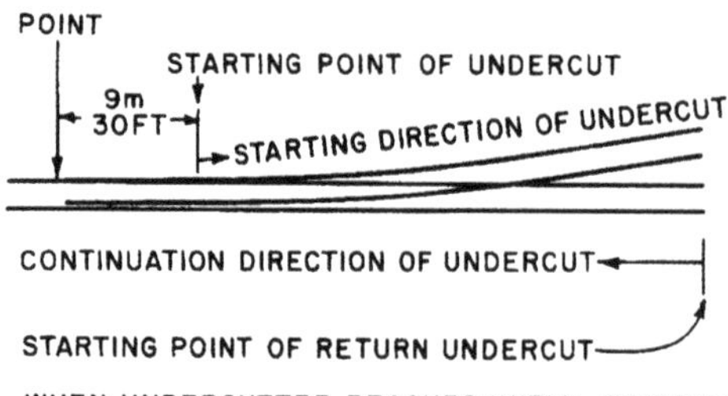

WHEN UNDERCUTTER REACHES INITIAL STARTING POINT ON ITS RETURN TRAVEL RAIL IS JACKED AND GEOTEXTILE (DOUBLED OVER) IS INSTALLED BELOW FROG

GEOTEXTILE EXTENDS FULL WIDTH AND DEPTH OF UNDERCUT TRENCH EXCAVATED BEYOND END OF TIES

Figure 13 Typical turnout undercutting sequence performed on CN's Atlantic Region.

TRACK JACKING ABOVE GEOTEXTILE

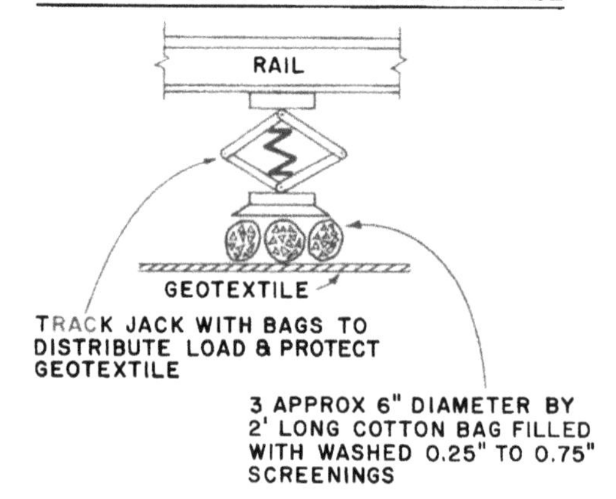

Figure 14 Method of protecting geotextiles from high imprint pressure of rail jack bases.

illustrated in Figure 13. Once completed the geotextile is layed out adjacent to the track and cut to the required length. The turnout point end of the geotextile is then lifted and pulled over to the wide-end portion of the turnout geotextile. The double thickness geotextile is then manually pulled under the track and aligned to its correct position under the wide-end of the turnout. The geotextile's edges now trail into the two ditches left on either side of the track by the bucket wheels. The upper layer of the geotextile is then fan folded towards the centre of the track.

Cotton bags, approximately 150 mm (6 in) in diameter, filled with fine-gravel-sized crushed aggregates, are then placed on the geotextile as spacers between the geotextile and the base of the ties. Two bags are placed at the end of every fourth to fifth tie so that when the rail jacks are removed the base of the ties are about 300 mmw(12 in) above the geotextile. Once removed, one set of rail jacks are repositioned at the fan fold end on three ballast bags placed side by side as shown in Figure 14. The fan fold is unfolded and the process repeated until the geotextile extends the full length of the turnout.

Once the geotextile had been fully extended and the ballast bags supporting the track are in place with the track jacks removed the turnout is ready to receive ballast. The ballast car is first unloaded into the side (undercutter trench) drains. Once the side drains are full the tracks are ballested until the ties are overflowing. The turnout is then ready to accept traffic. The final tamping and clean up may be left till the next day. In addition to the final clean up additional work involves the installation of perpendicular trenching and piping at both ends of the bucket made trenches on both sides of the track. This allows water to discharge from the drains to the side ditching as illustrated in Figure 15.

10 CONCLUSIONS

The work has reviewed and updated the use of geotextiles for railway track rehabilitation with particular emphasis on the geotextile developed for Canadian National and Canadian Pacific Railways. Assessment of the developed geotextile was established from excavations made on Canadian National's Atlantic Region. Also included in the work presented is a description of the method of installation followed by CN's Atlantic Region.

The established geotextile was made from one 100% needlepunched nonwoven fibres mechanically bound using not less than 80 penetrations/cm^2 (500 penetrations/in^2) and finally resin treated. A mass of 1050 g/m^2 (30 oz/yd^2) was found necessary to resist the harsh environment of a railway track bed. Good elongation was necessary for the geotextile to stretch around large particles left on the undercut surface. Strong fine fibres added to the abrasion resistance of the geotextile and maintained a low FOS essential for preventing geotextile fouling, internal wear and loss of transmissibility.

During installation attention to installing the geotextile to facilitate drainage from the load bearing area of the track was found to be of paramount importance.

A summary of the main points of the track specification is in the next section.

11 SPECIFICATION SUMMARY

The following is a summary of a railway track rehabilitation geotextile that has been on about 50 geotextile installations per year for over the last twelve years. Some of these have been exhumed and all were intact (Raymond, 1985). The ballast above the geotextile was clean. Track performance was satisfactory and judged to be highly cost effective. The geotextiles' durability has been excellent. The geotextile specifiction has also been used on other North American Company tracks with excellent and cost effective results (e.g. Raymond, 1991):

- Type:- Needle punched nonwoven with 88 penetrations/cm^2 (500 p/in^2) or greater.
- Fibre Size:- 0.7 tex (6 denier) or less.
- Fibre strength: 40 μN/tex (4 g/denier) or greater.
- Fibre polymer: Polyester.
- Yarn length: 100 mm (4in) or greater.
- Filtration Opening Size:- 75 μm or less.
- In-plane coeft. of permeability:- 50 μm/s or greater.
- Elongation:- > 60% to ASTM D-4632.
- Seams:- No longitudinal seams permitted.
- Colour:- Must not cause "Snow Blindness" during installation.
- Packaging:- Must be weatherproofed and clearly identified at both ends stating manufacturer, width, length, type of geotextile, date of manufacture.
- Wrapping:- 8 mil black polyethylene or similar.
- Abrasion Resistance:- 1050 g/m^2 geotextile must withstand 200 kPa on 102 mm burst sample after 5000 revolutions of H-18 stones each loaded with 1000 grams of rotary platform double head abrasor (ASTM D-3884).
- Width and length without seaming:- To be specified by client.
- Fibre bonding by resin treatment or similar:- 5% to 20% by weight low modulus acrylic resin or other suitable non water soluble resin that leaves the geotextile pliable.
- Mass:- 1050 g/m^2 or greater for track rehabilitation without the use of capping sand.

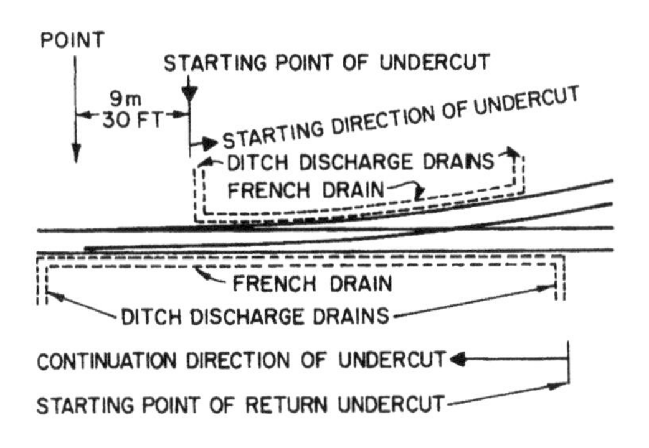

Figure 15 Illustration of positioning of perpendicular trenches to drain french drains.

12 ACKNOWLEDGEMENTS

The work reported forms part of general studies funded partially by Canadian Pacific Rail, Canadian National Rail, Via Rail, and Transport Canada Research and Development Centre through the Canadian Institute of Guided Ground Transport and partially by the Natural Sciences and Engineering Research Council of Canada.

13 REFERENCES

J.M.A. Costa and G.P. Raymond, 1983. "Burst Tester for Geotextile Bidirectional Strength", American Society for Testing Materials, Geotechnical Testing Journal, Sept. 1983, Vol. 6, pp 143-183.

B.S. Gerry and G.P. Raymond, 1983a. "Equivalent Opening Sizes for Geotextiles", American Society for Testing Materials, Geotechnical Testing Journal, June 1983, Vol. 6, pp 53-63.

B.S. Gerry and G.P. Raymond, 1983b. "In-Plane Permeability of Geotextiles", American Society for Testing Materials, Geotechnical Testing Journal, Dec. 1983, Vol. 6, pp 181-189.

G.P. Raymond, 1982. "Geotextiles for Railroad Bed Rehabilitation", Proceedings, Second International Conference on Geotextiles, Las Vegas, USA, August 1982, Vol. 2, pp 479-484.

G.P. Raymond, 1984. "Research on Geotextiles for Heavy Haul Railways", Canadian Geotechnical Journal, May 1984, Vol. 21, pp 259-276.

G.P. Raymond, 1985. "Subballast and Geotextiles for Railway Track Support Separation Applications", Canadian Institute of Guided Ground Transport, December 1985, Report No. 85-12.

G.P. Raymond, 1986a. "Installation Factors Affecting Performance of Railroad Geotextiles", Transportation Research Record, Transportation Board, Washington D.C., Record 1071, 1986, pp. 64- 71.

G.P. Raymond, 1986b. "Performance Assessment of a Railway Turnout Geotextile", Canadian Geotechnical Journal, Vol. 23, 1986, pp. 472-480.

G.P. Raymond, 1986c. "Geotextile Application for a Branch Line Upgrading", International Journal of Geotextiles and Geomembranes, Vol. 3, 1986, pp. 91- 104.

D. Vandine, S.E. Williams, G.P. Raymond, 1982. "An Evaluation of Abrasion Tests for Geotextiles", Proceedings, Second International Conference on Geotextiles, Las Vegas, USA, August 1982, Vol. 3, pp 811-816.

Developments in Geotechnical Engineering, Balasubramaniam et al. (eds) © *1994 Balkema, Rotterdam, ISBN 90 5410 522 4*

Japanese reclamation techniques for coastal and offshore areas with soft foundation

Y.Watari, N.Fukuda, S.Aung & T.Yamanouchi
Mizuno Institute of Technology, Hiroshima, Japan

ABSTRACT: Japan has a 400 years' history of shallow sea reclamation for new land. However, the application of Soil Mechanics Engineering to reclamation started just recently. In reclamation works, two types of fill-material, the pit sand and the dredged seabed soil, are generally employed. The water content of dredged soil used as fill-material is rather high due to the intrusion of seawater while dredging and as a result, an extremely soft ground is made-up. As such there arise the problems of insufficient bearing capacity and subsidence of the reclaimed ground. Some measures for ground improvement are required to solve these problems. This paper deals with the technology on reclamation, seabed dredging, revetment construction, the problems of reclaimed ground as viewed from Soil Mechanics Engineering and their solutions. These are discussed together with Japan's unique nature of technological development in the field of reclamation engineering.

1 INTRODUCTION

Reclamation works were most popularly carried out in the years from 1950 to 1960 when Japan achieved a high rate of economic growth. The aim of these reclamations is to use the new lands for large scale industrial projects. At present, the reclamation is limited to the minimum requirement for the peoples' welfare with due consideration of the environmental impact. Here, the shallow sea is economically feasible for reclamation; hence the shallow sea that extends toward far offshore area is usually selected for reclamation site. But, such shallow sea is mostly made up of soft foundation with a large deposit of alluvial clayey soil.

On the other hand, the method of using the soil dredged from seabed is considered the most economical. Recently, the clayey soil dredged from seabed is often used for reclamation as a means of surplus soil disposal. This often results in an extremely soft reclaimed land filled with soft dredged clay on soft foundation.

In this paper, the technology on soil behavior in coastal reclamation works will be discussed from both aspects of soft foundation and clayey fill-material soil.

2 RECLAMATION CONSTRUCTION TECHNIQUES

The materials used for reclamation are such types of soil as pit soil, seabed soil, construction leftover soil, industrial refuse, household waste, etc. The present practice is using the pit soil at about 80%, followed by the combined fill-material consisting of construction leftover soil, industrial refuse and household waste at 10% and the rest is mostly seabed dredged soil. Previously, the seabed dredged soil was first in volume in the utilization for reclamation works, and the percentage of the fill-material volume changes with the change of era (Watari 1991).

The extremely soft reclaimed ground is made up in case the seabed dredged clay is used as fill-material. In dredging soil from seabed, pump dredger is generally used for economical reason. Here, the explanation, limited to seabed dredging, is described as follows. The pump dredger dredges seabed soil and pumps it out together with seawater to the reclamation site by pipeline. During pumping the soil grains are unravelled and filled into the site with water content at 2000%~600%, forming an extremely soft foundation. The reclamation works of 1950s resulted in a number of new lands with such kind of extremely soft foundation. In those reclamation projects, the soil mechanics engineer played an active role mostly in the field of seabed soft foundation technology, but is hardly required for the fill-material technology except on special occasions. In such extremely soft foundation, it is quite difficult to attain a sufficient bearing capacity. Thus, the reclamation works of 1960s were carried out for the purpose of land development using pit soil as fill-material.

However, in the reclamation works carried out for the disposal of waste-soil, the soft clayey soil is used as it is without a proper soil selection. Recently, even in such kind of reclamation works,

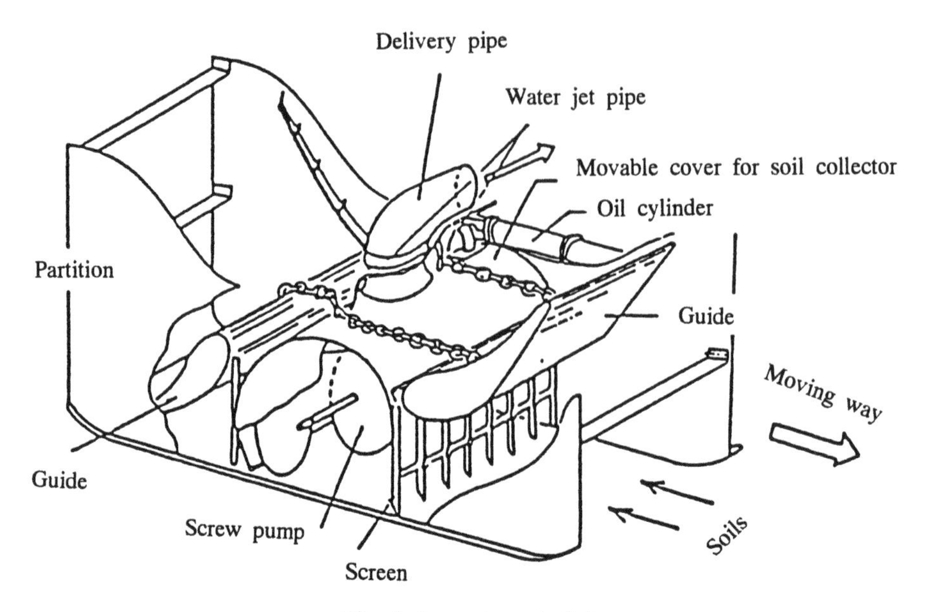

Fig. 1 Screw type dredging pump.

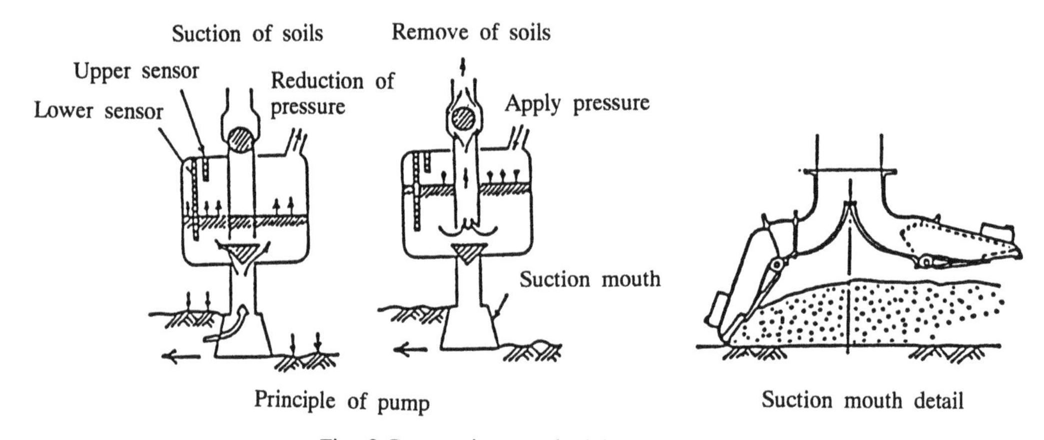

Fig. 2 Pneumatic type dredging pump.

a new method of soil cover using geotextile was developed, and as a result, it is possible to gain a high bearing capacity in the soft foundation. Hence, the reclamation with clayey soil has been reviewed favorably.

Mixing the seabed clay with seawater causes to increase the bulk density of the soil. Consequently, it is required to plan a greater volume of reclamation container, increasing the cost of construction. Hence, some new dredging techniques that reduce as much as possible the mixing of seawater with clay were developed around 1970s.

One method is to use the screw conveyer in the dredging mechanism (refer to Fig. 1) and the other is to make use of a pneumatic pump (refer to Fig. 2). In the latter method, the dredged seabed clay is sucked into the pump by vacuum pressure and conveyed to the delivery pipe by means of the pressurized air. Thus, the method enables the engineers to dredge the seabed clay without agitation causing the flow of muddy water. It is, however, difficult to completely shut out seawater from getting into the pump, and the dredged clay is mixed to 10~30% of its volume with seawater. These techniques were all developed with a precondition to get rid of the emission of muddy water and make it possible to achieve a high density dredging.

In the year 1989, a new dredging mechanism was developed with the main objective of high

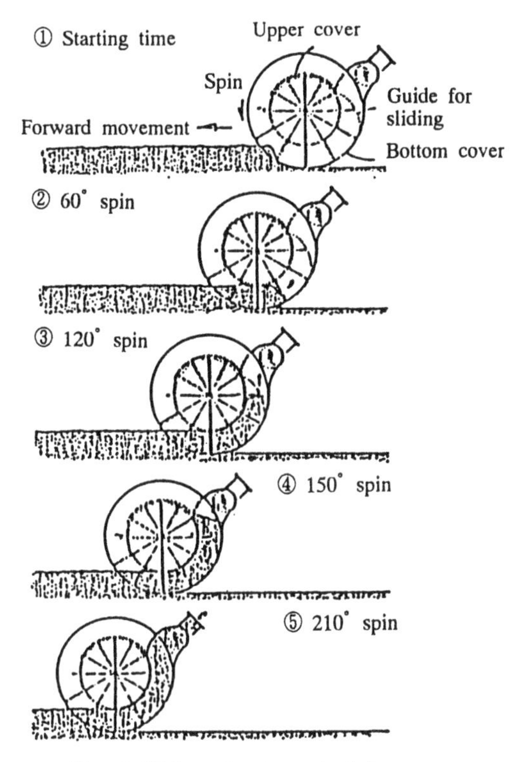

Fig. 3 Slide cutter type dredging pump.

density dredging. (Refer to Fig. 3). In this method, the seabed is dredged by the rotating bucket-type mud-collector machine equipped with numerous blades sliding at a speed synchronized with that of the dredger. The computer technology development enables the engineers to accomplish such a synchronized speed control for this new method. This method has a 0~10% seawater mixing.

These methods incorporate a minimum reduction of the muddy seawater as required from the environmental aspect and are employed throughout the entire country. However, when compared with the conventional pump-type dredger, the new method has a low capacity of dredging, hence the seawater mixed type of dredger is still being used for large scale reclamation projects as a main work force. And after construction works for reclamation are completed, the foundation improvement works usually follow.

3 CONSTRUCTION TECHNIQUE OF REVETMENT FOUNDATION

In case a revetment is constructed on the soft seabed foundation, the revetment load is dispersed by broadening the base area so that the structure remains stable. There are two ways of revetment construction, one is to carry out the construction works without foundation improvement, and the other is to conduct foundation improvement. The recent technique of revetment construction requires cost-effectiveness, the shortening of construction period and the minimization of construction material by adopting a small revetment section. Hence, the method of seabed foundation improvement is to be adopted for this purpose.

The seabed foundation improvement adopted in Japan typically consists of such methods as the seabed soil removal and replacement method, the sand drain method, the sand compaction method and the deep mixing method, and these methods are explained here. The historical background and major construction examples of the seabed foundation improvement are as shown in Fig. 4 (Suematsu et al. 1991). Until the year 1940, the seabed soil removal and replacement method of foundation treatment has been adopted in many cases. This is due to the fact that the sea at the reclamation site is shallow, the soft clayey seabed is rather thin and a plenty of replacement sand is available cheaply. However, the environmental problem due to turbidity of seawater, the problem of disposal of removed or dredged soil, and also the unfavorable nature of the conditions stated above finally lead to stop using the removal and replacement method in large scale reclamation projects, especially after the completion of Ougishima reclamation works in 1975.

The sand drain construction method was first adopted in Nagasaki Fishing Port project in 1952 (Offshore Construction Techniques Committee 1984). Afterwards, this method was adopted everywhere in Japan without limiting its application to port project. During fourteen years' period from 1970 to 1983 about 230,000 sand drains, each more than 20 m long, have been driven into seabed for foundation improvement in Japan (Suematsu et al. 1991).

During the above mentioned period, the foundation analysis technology, the instrumented management technology and the construction management technique have been considerably improved or developed. In the year 1987 when Kansai International Airport construction works started, about 1 million sand drain piles were driven into the seabed foundation of the revetment and reclamation site in a few-years time (Maeda 1989). The major objective of sand pile driving is to ensure the structural stability in the case of revetment construction, or to provide countermeasure against foundation subsidence in the case of reclamation works. In order to accomplish these objectives, in general, the design is conducted by computing the consolidation progress-time according to Barron's theory, and the on-site construction management is done on the basis of the computation results. In that there often arises a discrepancy between the assumptions made upon theoretical computation and the conditions that prevail at the site. Thus, how to determine the consolidation coefficient for the design is quite essential as it is directly related to the construction cost of the reclamation works. A

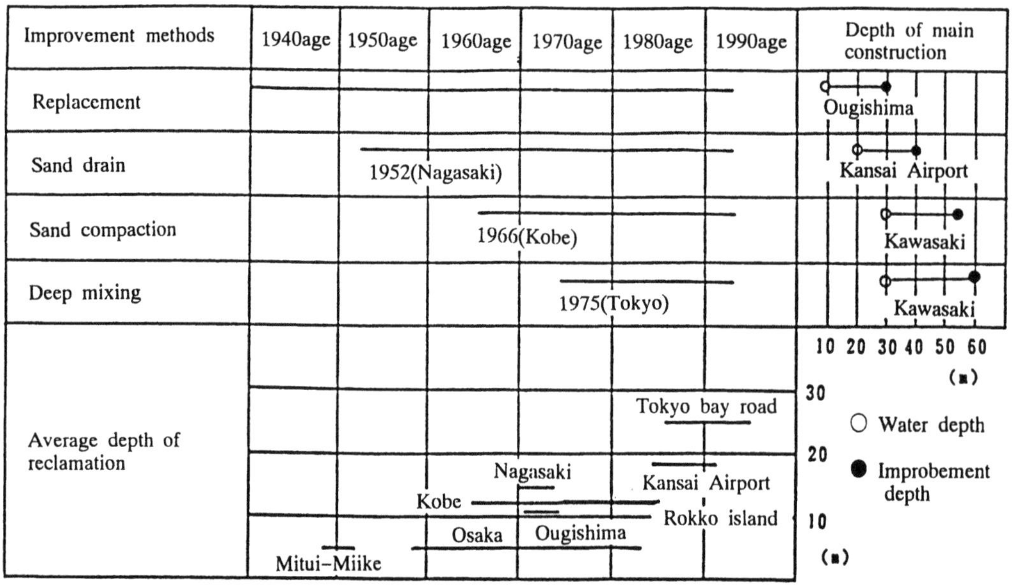

Improvement methods	1940age	1950age	1960age	1970age	1980age	1990age	Depth of main construction
Replacement							Ougishima
Sand drain		1952(Nagasaki)					Kansai Airport
Sand compaction			1966(Kobe)				Kawasaki
Deep mixing				1975(Tokyo)			Kawasaki
Average depth of reclamation	Mitui–Miike	Kobe, Osaka	Nagasaki, Ougishima	Kansai Airport, Rokko island	Tokyo bay road		10 20 30 40 50 60 (m); 30, 20, 10 (m); ○ Water depth; ● Improbement depth

Fig. 4 History of soil improvements.

general report on reclamation is that the consolidation coefficient in horizontal direction c_h is greater than that in vertical direction c_V,but the reversed results are also reported. Consequently, the design value is determined by backward calculation of consolidation coefficient using the actually observed initial subsidence value. The value of c_h is estimated and used afterwards. For the estimation of c_h from actually observed values, the method shown in Fig. 5 is proposed. This method is an application of the linear relationship between time factor T_h and $log(1-U_R)$ where U_R is the degree of consolidation (Monden 1963).

With the progress of consolidation, the strength of clayey soil increases. However, it is not easy to determine the increase rate at the design stage. Usually the method of deriving c_u using $c_u/p = tan\,\phi_{cu}$ from the shear resistance angle ϕ_{cu} obtained from unconsolidated undrained test is adopted.

When referred to Skempton, c_u/p is the function of plasticity index I_p and Mikasa has

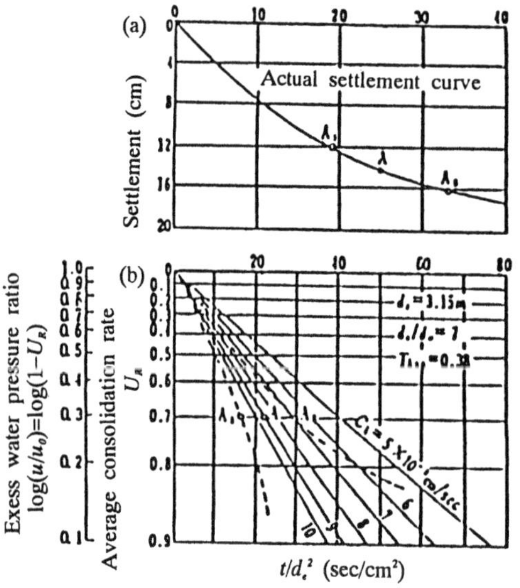

(a) Actual settlement curve; Settlement (cm)

(b) Excess water pressure ratio $log(u/u_0)=log(1-U_R)$; Average consolidation rate U_R; t/d_e^2 (sec/cm²)

Fig. 5 c_h estimation method using actual settlement curve.

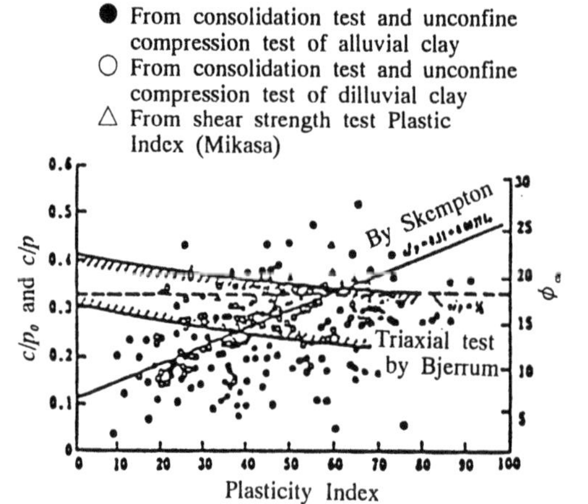

Fig. 6 Relation between c_u/p and plasticity index I_p.

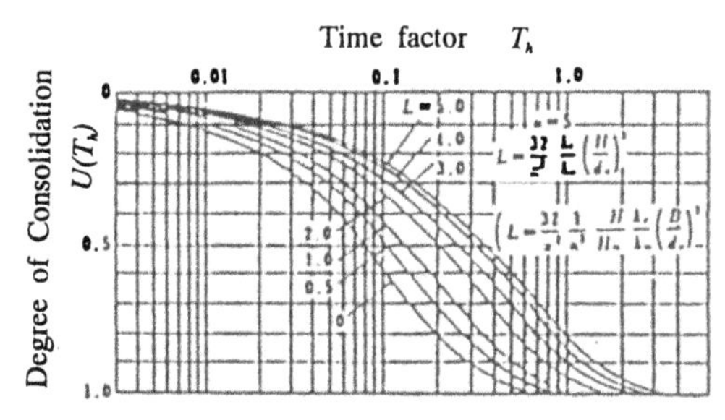

Fig. 7 Consolidation curve with well-resistance.
( ) with matt-resistance

prepared Fig. 6 showing the relationship between c_u/p and plasticity index I_p (Mikasa 1966). Obviously, there is a discrepancy between the value obtained from laboratory tests and that derived from stress distribution of the natural ground.

As indicated by Mikasa, owing to the fact that the relationship between stress and strain in the consolidation curve is not linear, the strength gain tends to be rather delayed as compared to the consolidation. Consequently, it is required to carry out construction management, adjusting the strength gain through instrumented measurement.

Yoshikuni (1974) has pointed out the fact that the permeability of the drain (discharge) depends not only upon the permeability coefficient of the drain material itself but on the drain length and the drain section area. The phenomenon is the so called well resistance. Similarly it is clarified from further research (Yoshikuni 1975) that the laying of sand layers results in the mat resistance. (Refer to Fig. 7). When very long drains and wide-spread sand mats are used, the drainage efficiency that is practically compatible to real situation can be realized.

The driving-type sand compaction construction method was first introduced to Japan in the year 1956. The method was lacking ease of construction and practically of no use. However, since the establishment of the vibration-type sand compaction method in 1960, the practical application of the method to on-land projects has started (Ishiwata 1974). And the method was applied to offshore construction works beginning in the year 1966. In 1970 the forced removal and replacement method with a high rate of replacement on soft clayey foundation started to enjoy popularity among engineers. Recently, the sand compaction method with a low rate of replacement was developed as a replenishment to the sand drain method and a rational approach to revetment construction was achieved (Okada et al. 1987, Takahashi et al. 1988). As to the application of the method to clayey foundation, Murayama has proposed in 1962 the basic theoretical concept of the present-day design method (Murayama 1962). Afterwards, numerous engineers and researchers conducted both laboratory and in-situ tests to establish the present-day design method.

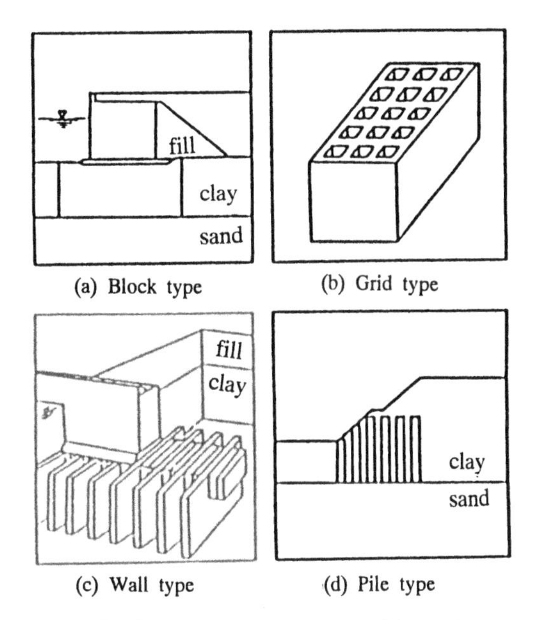

Fig. 8 Improvement pattern of DMM.

In the deep mixing method (DMM) developed in 1967 by Port and Harbor Research Institute, Ministry of Transport, the active lime and inert lime are used for foundation treatment as they have the water absorption and hardening qualities. These were practically applied to construction works in 1974 as DLM method (Deep Lime Mixing method). As these powder-form limes pose some problem, cement is used in CDM (Cement Deep Mixing) method developed in 1973. CDM has a long history of performance on both land and offshore areas. This method has two types of mixing; one is a mechanical method where forced mixing of soft foundation soil with cement takes place as a result of stirring rakes, and the other is a jet-type mixing where soft soil is mixed just in a similar way with chemical injection method. The former is the method almost used in reclamation works.

At the initial stages of development of the construction method, a usual practice is to conduct block-wise foundation improvement, but nowadays four kinds of improvement patterns such as block-type, grid-type, wall-type, pile-type are used, selecting a pattern according to the structure (Refer to Fig. 8).

4 CLAYEY SOIL RECLAIMED GROUND

The ground of the landmass formed by filling the reclamation site with dredged seabed clayey soil is in an extremely soft condition having 600% of water content at the initial stage, and in the later stages, the consolidation is in progress under its own weight. In such reclaimed ground, one problem is how to realize the consolidation settlement and the other is the estimation of fill-volume for reclamation. As to these two problems, the results obtained from research and development in Japan are described as follows.

In case the dredged seabed clayey soil mixed with seawater is dumped into the reclamation area, the volume changes of seabed soil take place as shown in Fig. 9, and it is required to predict the final volume of the seabed soil in terms of the percentage of the initial volume. Mikasa first conducted research on this problem, namely the consolidation problem of extremely soft foundation of the reclamation project in Osaka Bay (Mikasa 1963). In that project, one dimensional consolidation theory is formulated for the saturated clayey soil taking into consideration the effect of the clay's self weight and the changes that occur during consolidation such as clay's compressibility, permeability, consolidation load, clay layer thickness, etc. This is in other words Mikasa's proposition of theoretical concept considering the changes of the above factors as against the constant values taken in Terzaghi's theory. In so far as the clayey soil in the reclamation works is concerned, the changes of these factors are considerable, hence Terzaghi's theory is practically difficult in its application to an extremely soft reclamation ground. Thus, in Japan Mikasa's basic concept is adopted in dealing with such extremely soft reclamation ground.

In the reclamation projects of 1960s, the percentage of soil retained in reclamation site is considered to be 70% of the total volume of the dredged clayey soil that increases its void ratio with dredging. This is due to the fact that the small grain soil particles which are not required for reclamation are flowing out of the site. This pollution cannot be permitted as viewed from today's environmental requirements. Thus, Shiratori et al. (1982, 1983) conducted tests on water container model, a model of reclamation site, and reported the relationship between type of fill-material and concentration, the relationship between soil characteristics and formation of soil deposit, etc. Imai (1980) has pointed out the fact that the sediment time or sinking differs with the salt concentration and initial water content of the clayey fill-material, and Yano et al. (1984) has indicated that the fill-material deposited in a disintegrated condition, differing in qualities between seabed dredged clay and the deposited mass of reclamation area. Moreover, the testing methods have been developed to determine the consolidation constants of extremely soft clay at low loading condition from one dimensional sedimentation test and seepage consolidation test, and also by Umehara et al. (1980, 1982) from one dimensional sediment test and constant rate strain consolidation test. These are broadly used for consolidation analysis.

On the other hand, with the advance of computation technology that makes use of electronic computer, it was made possible to conduct consolidation analysis by FEM. Watari et al. (1986a, 1986b) performed prediction of fill-volume based on the above mentioned computation technology and the results of two dimensional sedimentation tests. The predicted results are checked with the results of on-site instrumented measurement.

5 SURFACE TREATMENT OF CLAYEY FILL-MATERIAL

The extremely soft foundation of a completed reclamation ground has no bearing capacity. Hence, both man and construction machine cannot be loaded on it. In the case of such an extremely soft ground, the surface layer improvement is performed to secure a staging ground for construction works as shown in Fig. 10 (Uehara et al. 1988). In this section, the surface hardening treatment method and the horizontal drain method are explained in detail.

The hardening agent used in the surface hardening treatment method is either cement type or lime type. Since the water content of the reclaimed foundation is very high, in general, the cement type hardening agent is used for many projects. But, it is inferior to the soil cover

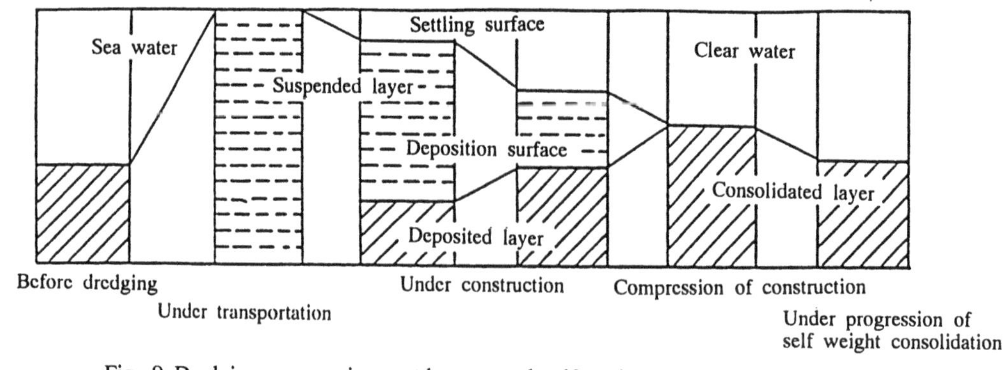

Fig. 9 Dredging, suspension, settlement and self-weight consolidation process.

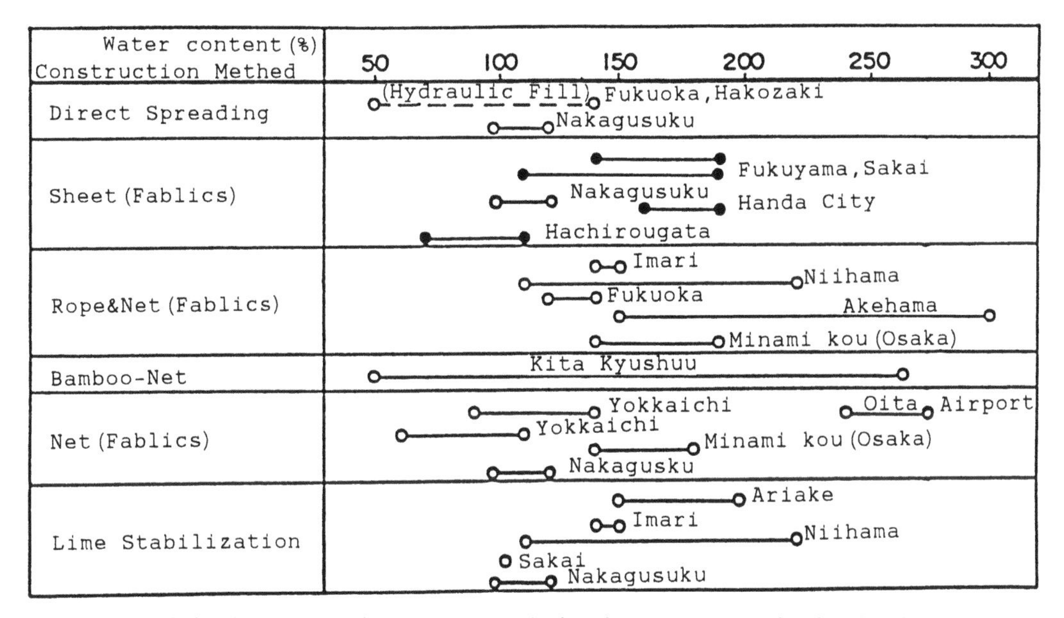

Fig. 10 Relationship between improvement method and water content of soft subsurface layers.
○ : Actual water content, ● : Estimated water content, – : Range of measurement.

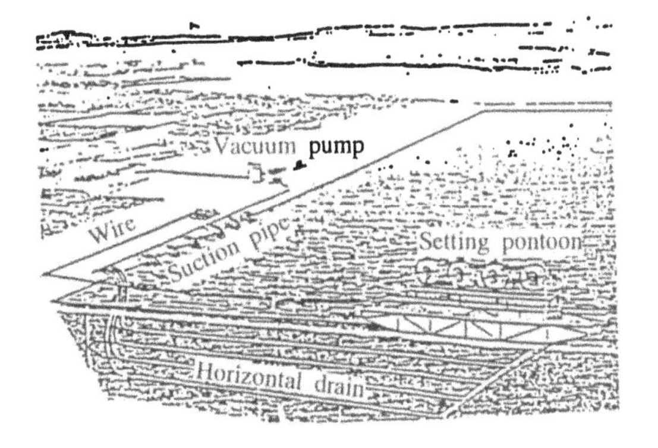

Fig. 11 Horizontal drain method.

method in economic viability, therefore it is used as a staging ground for the soil cover method. In the cement-type surface hardening method, the surface is hardened by mixing the clayey surface layer with cement milk in slurry form by means of a mixing machine. This kind of construction works has been carried out since the beginning of 1970.

The improved strength of the hardened mass and the improved size of the surface layer are quite important. As to the former, a symposium on testing method for stabilizing treatment soil was held in 1980 under the sponsorship of Japanese Society of Soil Mechanics and Foundation Engineering, and a draft report on standardization of testing methods was submitted for approval (JSSMFE 1980). In that report, the uniformity of terms and expressions are strictly enforced. On the other hand, as to the improved size of soil surface, Yoshida (1978) has proposed a design method based on the theory of limited length beam on an elastic ground.

The surface treatments of reclaimed ground are mostly carried out after the completion of reclamation works. The horizontal drain method is developed with a view to achieve shortening of construction time by means of execution of works in parallel with reclamation activities, and also to realize foundation improvement effect through water content reduction (Watari et al. 1987, 1993). This method includes horizontal insertion of drain-materials into the reclaimed clayey ground and aims at completing foundation improvement by accelerating the consolidation process during construction through the intensive drainage of excess pore water pressure and the parallel execution of vacuum consolidation process. The general outline of the surface treatment is shown in Fig. 11.

6 CONCLUSION

This paper presents the general outline of the overall reclamation technology mostly developed in Japan. Large scale reclamation works, such as New Kansai International Airport reclamation works, Haneda International Airport offshore extension works, etc. are now in progress in Japan. In these projects, there arise some new problems on soil mechanics and foundation engineering, and studies are being made as to the practical solution (through facts and figures) of the problems that have been ever encountered and widely discussed in Japan. From the above mentioned studies, the following points can be concluded.

1. Japan has a long history of coastal reclamation and accumulated technology and experience to successfully overcome the problems on soft foundation. Thus, Japan has the technological power to deal with coastal and offshore reclamation.

2. As to the dredging of seabed clay, the technology of prevention of seawater intrusion and preservation of environmental conditions has been developed.

3. The technology of wide application of sand drain method and sand compaction method according to objectives has been established. At the same time the rational foundation treatment method that can save sand as it is diminishing in quantity has been introduced by developing a directly in-situ hardening method for soft clayey foundation.

4. The prediction of required fill-volume for reclamation is made possible by realizing the phenomena of volume change in the fill-soil.

5. The surface treatment technology for reclaimed ground with extremely soft clayey soil has been established, and the method of reducing the water content in soft clayey ground during construction has been developed.

ACKNOWLEDGEMENT

The authors would like to express their heart-felt thanks to Dr. Nobuo Tajiri of Fukken Co., Ltd. for his collaboration in the preparation of this paper.

REFERENCES

Imai, G. 1980. Settlement behavior of clay suspension, *Soils and Foundations*, Vol. 20, No. 2, 61-77.

Ishiwata, T. 1974. Frontiers of Civil Engineering Technology Development, *Journal of JSCE*, Vol. 64, No. 1. (in Japanese)

Japan Port and Harbor Association, 1967. *Design Standards for Port and Harbor Structures*, 10-25. (in Japanese)

JSSMFE 1980. *Symposium on testing methods for stabilizing treatment soil.* (in Japanese)

Maeda, S. 1989. Future perspective and a case history of large scale man-made island construction in offshore area, *Proc. of JSCE*, No. 406/Ⅲ-1, 1-15. (in Japanese)

Mikasa, M. 1963. *Consolidation of soft clay*, Kashima Publishing Association, Tokyo, pp.126. (in Japanese)

Mikasa, M. 1966. Description methods for soil test's results, *11th Symposium of JSSMFE*. (in Japanese)

Monden, H. 1963. A new time fitting method for the settlement analysis of foundation on soft clay, *Memoir Fac. Eng. Hiroshima University*, Hiroshima, Japan, 2-1, pp.21.

Murayama, S. 1962. Some consideration on vibro-composer method for clayey soil, *Construction Mechanization*, Vol. 150. (in Japanese)

Offshore Construction Techniques Committee 1984. *Construction techniques of man-made island*, Japan Marine Development Association, Sankaido, Tokyo, 50-57. (in Japanese)

Okada, Y., T. Yagyu & I. Kouda 1987. In-situ destruction tests on foundation improved by sand compaction pile method, *Tsuchi-to-kiso*, Vol. 37, No. 8, 57-62. (in Japanese)

Shiratori, Y., H. Kato & S. Takeuchi 1982. Site investigation on settlement behavior of sand during reclamation (No.3), *Port and Harbor Technology Materials*, No. 420. (in Japanese)

Shiratori, Y., H. Kato & S. Takeuchi 1983. Site investigation on settlement behavior of sand during reclamation (No.4), *Port and Harbor Technology Materials*, No. 465. (in Japanese)

Suematsu, N. & H. Tsuboi 1991. Improvement technique of seabed foundation, *Tsuchi-to-kiso*, Vol. 39, No. 6, 77-86. (in Japanese)

Takahashi, K. & M. Shiomi 1988. In-situ tests on SCP method with low replacement rate, Maizuru Port, *Civil Engineering Technology*, Vol. 43, No. 10, 81-89. (in Japanese)

Uehara, H., M. Yoshizawa & S. Kohagura 1988. Experimental studies to improve the surface land reclaimed from the sea, *Proc. Int. Symp. Theory and Practice of Earth Reinforcement*, Fukuoka, Japan, A.A. Balkema Publishers, Rotterdam, 239-244.

Umehara, Y. & K. Zen, 1980. Constant rate strain consolidation for very soft clayey soils, *Soils and Foundations*, Vol. 20, No. 2, 79-95.

Umehara, Y. & K. Zen 1982. Consolidation and settlement characteristics of super soft harmful low quality soil, *6th Japan U.S. Specialist's Conference on treatment and disposal of harmful low-quality soil, Port and Harbor Engineering Report*, No. 87, 143-166. (in Japanese)

Watari, Y., H. Shinsha, K. Hayashi & H. Aboshi 1986. Forecasting of dredged soil volume based on self-wait consolidation analysis of clayey reclamation soil, *Research Report of Civil Engineering Department Hiroshima University*, Hiroshima, Japan, Vol. 34, No. 2, 159-173. (in Japanese)

Watari, Y., H. Shinsha, K. Hayashi & H. Aboshi 1986. Settlement and disintegrated sedimentation of dredged clayey soil, *Research Report of Civil Engineering Department Hiroshima University*, Hiroshima, Japan, Vol. 34, No. 2, 175-187. (in Japanese)

Watari, Y. & H. Shinsya 1987. Improvement of reclaimed cohesive soil ground by vacuum consolidation, *Proc. Int. Symposium on Geosynthetics -Goetextiles and Geomembranes-*, JCIGS, Kyoto, Japan. 125-133.

Watari, Y. 1991. Reclamation techniques of man-made island construction, *Tsuchi-to-kiso*, Vol. 39, No. 6, 109-114. (in Japanese)

Watari, Y., Y. Higuchi & H. Shinsya 1993. An

improvement method on extremely soft clayey ground using horizontal board drains, *8th Geotextiles Symposium of JCIGS*, Tokyo, Japan, 101–110. (in Japanese)

Yano, Y., K. Tsuruya & T.Yamanouchi 1984. Disintegrated sedimentation of dredged soil, *Tsuchi-to-kiso*, Vol. 32, No. 5, 23–28. (in Japanese)

Yoshida, N. 1978. Coefficient of subgrade reaction on very soft ground and modulus of deformation on improved ground, *Proc. 33rd JSCE Annual Conf.*, Vol. Ⅲ, 306–307. (in Japanese)

Yoshikuni, H. & H. Nakanodo 1974. Consolidation of soils by vertical drain wells with finite permeability, *Soils and Foundations*, Vol. 14, No. 2, 35–46.

Yoshikuni, H. & H. Nakanodo 1975. Consolidation of a clay cylinder with external radial drainage, *Soils and Foundations*, Vol. 15, No. 1, 12–27.

Developments in Geotechnical Engineering, Balasubramaniam et al. (eds) © 1994 Balkema, Rotterdam, ISBN 90 5410 522 4

The 'reinforced soil' in the future of geotechnics

F. Lizzi
Italsonda S.p.A., Naples, Italy

ABSTRACT

Geotechnics, like any other Engineering branch, needs appropriate parameters for the design of its structures. Such parameters, collected by means of sophisticated instruments and duly confirmed by wide field experience, have been and are successfully used for design in homogeneous soils (including hard or soft rock, provided they are continuous).

Unfortunately, reality does not always offer homogeneous soils to the designer. Several materials fall in the "grey" area between hard soils and soft rocks; the negative feature of such materials is the irregularity of the structure; the usual geotechnical parameters belonging to the different components, sometimes chaotically mixed, have no practical meaning for the design.

A completely different approach is necessary: the "Soil Reinforcement" could be a possible answer to the problem of giving some sort of " homogeneity" to an heterogeneous mass, for the solution of geotechnical problems.

1 INTRODUCTION

When Geotechnics, thanks to the pioneering work of Terzaghi, entered the field of consolidated Science, the attention was focused on the soil, as the main support of any man-made construction.

But natural soils not always showed sufficient bearing capacity for day by day increasing loads; some artificial improvement such as compaction, short timber piles, grouting, etc. were generally limited to the surface soil; the main problem was partially, but not completely, solved.

2 THE AGE OF PILING

The introduction of concrete piles in Geotechnique at the beginning of this century was, without any doubt, a turning point.

The problem of the meticulous knowledge of the parameters of the subsoil, in homogeneous as well as in heterogeneous materials, was completely overshadowed by the reality offered by data collected through direct load tests.

At any rate, Theory took advantage of this invaluable field confirmation, and considerable progress has been made in understanding the engineering behaviour as well as developing design and construction methods for geologic materials that clearly fall into the field of either soil or rock mechanics. For other heterogeneous materials, lacking convenient parameters, the load tests remained the only base for design.

As a matter of fact, the soil as direct foundation of the constructions was practically forgotten.

The success of piling pushed continuously for better performance, particularly toward the increase of the load bearing capacities.

Recent decades were characterized by a veritable rush to Large Diameter Piles (called Shafts in USA). On the other hand, for such piles some very serious perplexities arose; among others:

- exaggerated cost of load tests; limited number of them;

- in defect of such an important field contribution, the design based on theories derived by the long experience on medium diameter piles was not completely satisfactory for the larger diameters;

- technological serious difficulties of the drilling in presence, in the subsoil of obstructions, boulders, old masonries etc.

- risk of poor workmanship in the presence of a limited number of piles entrusted with thousand tons loads.

3 GEOTECHNICS GOES BACK TO THE SOIL

Due to the above difficulties, there was, recently, in Geotechnics a certain tendency to a revival of the long forgotten study of the soil, as the main support of Man's construction. A new actuality arose for the Terzaghi's Theories.

4 THE "PILED RAFTS"

Modern heavy constructions were characterized, as regards their foundations, by a great number of piles, collected by large R. C. rafts.

Some direct field tests clearly demonstrated that a not negligible part of the load entrusted to such foundations is carried by the soil below the raft, according to some sort of collaboration with the load carried by the piles.

The new developed theory of "Piled Rafts" has entered the Geotechnics in a moment that can be considered crucial for this Science.

Although the "Piled Raft" theory has demonstrated very interesting aspects for the design, some negative points prevent a large generalization of the system.

For instance:

- A direct contact Raft/Soil is essential. On the other hand the Rafts, as shallow structures, are in contact with a generally poor quality surface soil, disturbed by the excavation, and exposed to water infiltration etc.

- The collaboration is based on the compatibility of the settlements (of the piles and of the rafts); the assessment of the curve, loads- settlements for the raft, is not so reliable as the corresponding chart for the piles; acceptable for homogeneous soils it has no practical meaning for heterogeneous soils.

5 WHY NOT "REINFORCE" THE SOIL?

The success of the Concrete Constructions has been due to the introduction of"reinforcement".

Some basic features made possible a steel/concrete collaboration; the essential one was the "adherence" between the two elements.

And now, for Geotechnics, the question is: is there any possibility to "reinforce" the soil in order to obtain a unitary new stronger mixed element?

The question involves not only the

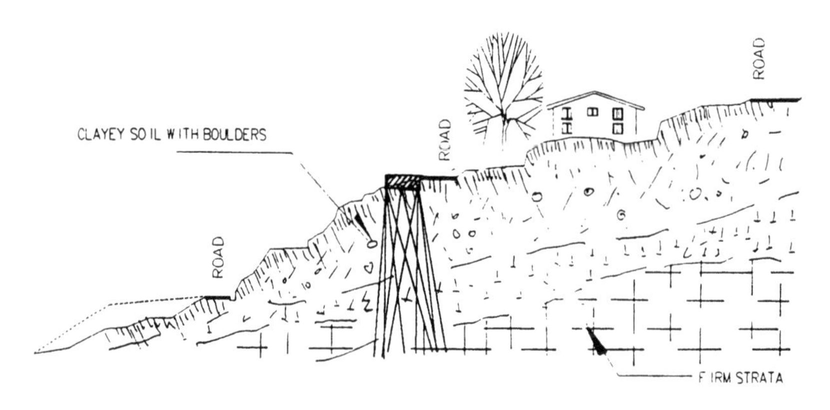

Fig. 1 Reticulated pali radice structure for stabilizing a landslides in loose soils.

Foundation problems but any other structure (for instance, retaining walls) in which the resistance of the soil is engaged.

6 INTUITION, EXPERIENCE AND THEORY

As in any other human research activity, also in Geotechnics the first step is reserved to intuition. The experience (that is, the facts) confirms or not the intuition. The theory tries to codify the results of the above steps.

7 THE FIRST EXAMPLES OF IN-SITU "REINFORCED SOIL"

The first examples of in-situ "reinforced soil" go back to 40 years ago when the Author introduced the "root piles" ("pali radice") and the "reticulated root piles" structures.

One of these well known examples, about a case of landslide prevention, is indicated in Fig. 1.

The "reticulated root piles" (herein after called R R P) is intended to create some sort of reinforced soil/pile "gravity retaining wall" where the soil supplies the essential resisting force (the gravity), whereas the piles are intended to encompass the soil and supply some additional resistance to the tensile and shear forces acting on the wall. The size and the structural resistance of these piles have a minor importance.

The efficacy of such a mixed structure is based on several requisites briefly recalled below:

- The reinforcement is entrusted to slender piles, vertical or at a rake, whose main peculiarity is *full adherence to the soil all along their shafts* (just like the steel reinforcement in R.C.). The original "root piles" fully satisfy such an essential requisite, whereas steel micropiles are not suitable for soil reinforcement.
- The unity of the mixed soil/piles mass is based on some sort of "network" effect among the piles, in the limits of not very large spacing. Of course there is not a physical contact among the piles but the soil/pile, and conversely the pile/soil, interactions are intended to supply some sort of virtual "knot" effect.
- The technology of drilling and casting of such piles must be a very smooth one, in order not to disturb the natural equilibrium of the soil.

8 THE "IN-SITU SOIL REINFORCEMENT" AND THE "TERRE ARMÉE"

About 20 years after the introduction of "Reticulated root piles" in France, the "Terre Armée was created by Vidal, and largely diffused.

There is only a formal analogy between

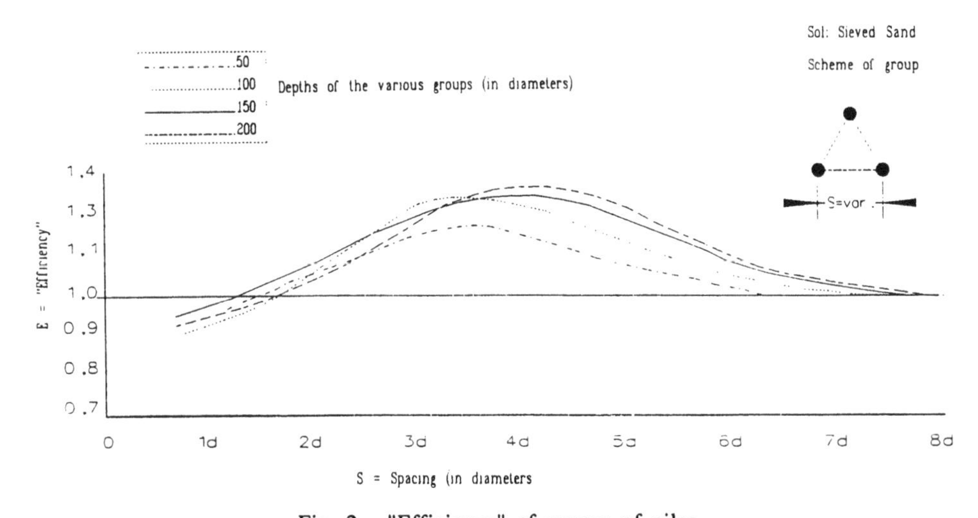

Fig. 2 "Efficiency" of groups of piles

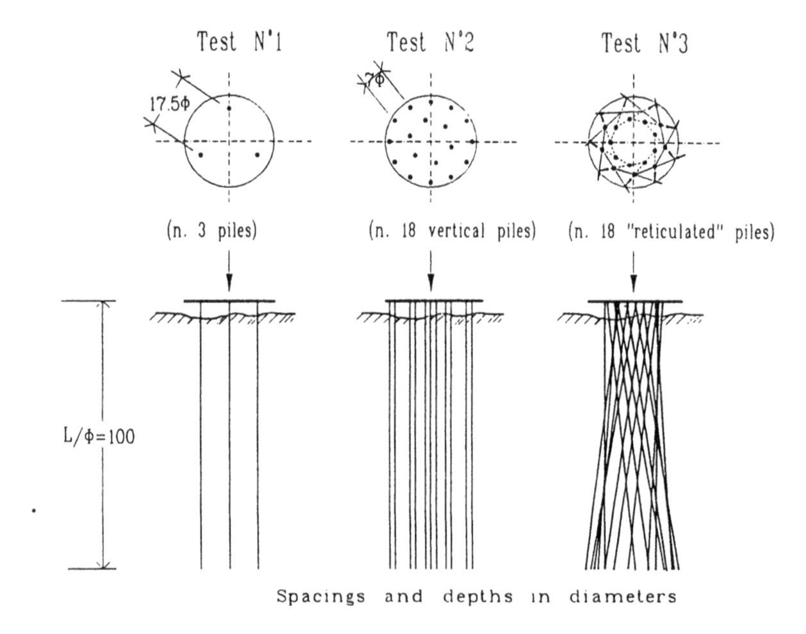

Fig. 3 Schemes of the tests on individual piles and group of piles

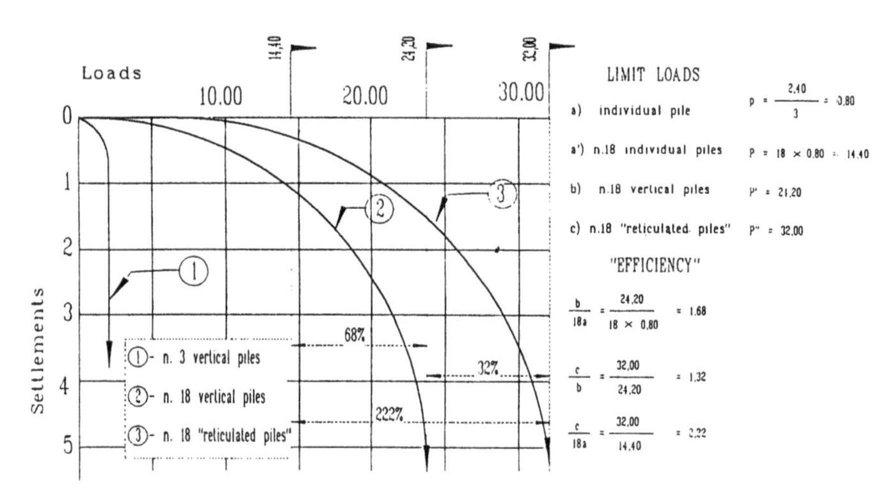

Fig. 4 Results of the tests of fig. 3

the two structures which are substantially completely different. In fact, the "soil reinforcement" above described is carried out in natural in-situ soil, without any excavation, whereas "Terre Armée" is a structure where the soil, excavated and selected, is used as a construction material.

9 FIELD TESTS AND MODEL TESTS

Field data about "Reinforced Soil" structures actually carried out are, for the time being, the basic elements for future applications. It is to be noted that in the majority of cases the stresses in the piles are very small (and very difficult to be detected) on account of the prevailing presence of the "gravity" of the soil.

On the other hand, the interaction of the piles in a "R.R.P." network has been the subject of several tests, at least in homogeneous soils, as specified below.

10 THE "EFFICIENCY" IN A "RRP" STRUCTURE

In the United States, where it was first used in Geotechnics, the term "efficiency" is generally synonimous to reduced load bearing capacity for a group of piles, compared with the sum of the load bearing capacity of the individual piles. But it is to be noted that this reduction is probably true when dealing with "driven" piles (current in the United states but not so in Europe) and, perhaps, for cast-in-situ piles drilled with old percussions technologies.

On the contrary, when piles are introduced in the subsoil without any substantial disturbing action the "efficiency" reduction is doubtful.

As for the "root piles", inserted in a RRP, the situation is, for instance, completely different.

After 40 years of experience it can be stated that "root piles" at a close distance, increase (rather than reduce) their global load bearing capacity. In order to check and possibly quantify this essential feature, (confirmed at any rate by the results of real cases), several experimental model tests were carried out; for instance:

10.1 Model tests on RRP

In Fig. 2 on a chart Spacing/Efficiency are collected the results of load tests carried out on groups of three piles according to different spacings and different depths. Notwithstanding the limited number of piles, the "efficiency" reached values of the order of 1.2 1.30; the mutual influence of the piles was evident up to seven diameters spacing.

Another experiment is indicated in Figs.3 and 4: The load bearing capacity of independent piles (largely spaced) is compared to the load bearing capacity of vertical piles at a closer distance (7 diameters) and to the load bearing capacity of "reticulated" piles; the "efficiency" of the Test no.2 is 1.68, raising to 2.22 for the "reticulated pattern" of the Test n.3.

That is: the soil encompassed by the piles supplies an extra load bearing capacity of 68% for the vertical pile group and 122% for the "reticulated" scheme. Tests carried out in several other homogeneous soils, as well as in heterogeneous soils and in full scale, always showed substantial, although different, increases in the "efficiency".

11 RRP IN FOUNDATION PROBLEMS

The positive aspects of a "Reinforced soil" solution is very clearly demonstrated by the Foundation of a heavy construction. In both Figs. 5 and 6 the foundation is assumed to be formed by "root piles" that is, by slender piles very tightly adherent to the soil, all along their shaft.

11.1 The classical scheme

In Fig. 5 the scheme, according to the classic pattern, is indicated (Fig. 5a); vertical piles, sufficiently spaced in order to exclude any possible interference between them. The load Q is supported by n piles, engaged in the limits of their safe load Ps

$$Q = n\,Ps$$

in accordance with the load/settlement chart of a load test for a single pile (Fig. 5b). For the sake of generality the raft is assumed to be of infinite stiffness and detached from the soil; consequently no direct contact between the raft and the surface soil is to be expected. Briefly, the load Q is supported exclusively by the piles, from the surface down (Fig. 5c).

11.2 The "reinforced soil" scheme

In Fig. 6 the same problem is entrusted to a "reinforced soil" block whose scheme is sketched in Fig. 6a. The load Q, in this case is supported by the combined action of soil and piles in the "reinforced" bulb.

In Fig. 6b in addition to the previous load settlement chart of the single pile, the corresponding chart for a pile inserted in a "network" is indicated. The Safe load and the Limit load are substantially increased;

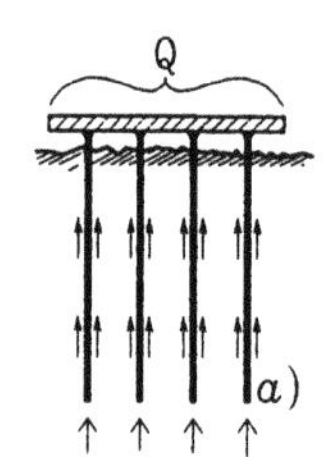

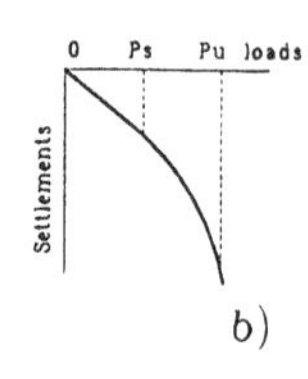

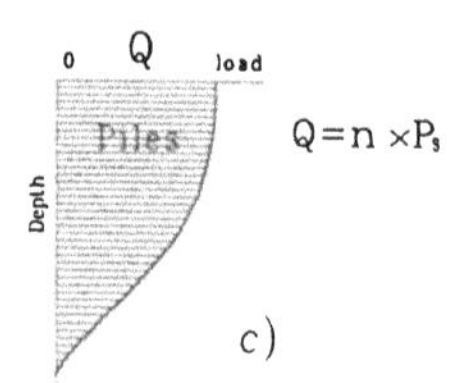

Fig. 5 - Vertical spaced piles (no group effect)

accordingly, the number of piles can be reduced from n to n' where n'<n

$$Q = n' P_s^*$$

where P_s^* is the safe load for the pile in the "network"with $P_s^* > P_s$.

The total load, along the depth of the structure, is supported partially by the soil and partially by the piles, according to some "amplification factor" (like in RC) which depends from the mechanical characteristics of the soil checked against the structural features of the pile, layer by layer.

To be noted:

- near the surface, before the intervention of the soil, the load is supported exclusively by the piles; there is a local extra stress for the piles, easily supported by them because their cross section near the surface is generally bigger than the cross section along the current shaft below.
- the scheme does not need very long piles; "root piles" generally show their full load bearing capacity in the limit of 30 meters approx., even in soft soils. This is peculiarity which turns to be strongly in favour of the "Reinforced Soil".

11.3 The design of a "Reinforced Soil" foundation

With reference to Figs. 5 and 6, the problem of the stability of the foundation has been analized primarily from the point of view of the piling for which, as a consequence of a "network effect", the load bearing capacity of the piles is increased. A substantial saving in the number of piles is at hand.

Anyway, it is interesting to note that the problem can be examined even from a different point of view; that is, as a case of "Reinforced Soil".

The stability of the bulb of "Reinforced Soil", as sketched in Fig. 6, is a problem of *internal stability* (resistance of the bulb) and *external stability* (the breaking of the block into the soil as unit).

As for the *external stability* we have:

$$Q = Q_b + S \times s$$

where: - Q_b is the ultimate bearing capacity of the base of the bulb;

- S is the side surface of the bulb;
- s is the average unit shearing resistance of the side soil.

As a matter of fact the more appropriate and safe geometry of the bulb is fixed by the designer without any difficulty.

The shearing resistance of the soil, s, at least as a rough average, can be evaluated through the data of the soil investigation.

The bearing capacity of the base of the block Q_h can be evaluated according to the old methods suggested by Terzaghi and others. The *internal stability* is entrusted to the network of piles; their number, as above calculated, is to be considered, generally, sufficient for keeping together the different layers of soil.

A more accurate calculation needs a full knowledge of the different parameters of the soil layers but in the majority of cases such uncertainty is to be overcome by a convenient Safety Factor.

Summarizing, for a sound design approach to a "Reinforced Soil" Foundation structure the following data must be gathered:

- A convenient, though not necessarily very accurate soil investigation.
- Diagrams loads/settlements for a single pile and, possibly, for a group of at least 3 piles.
- Some penetrometer tests.

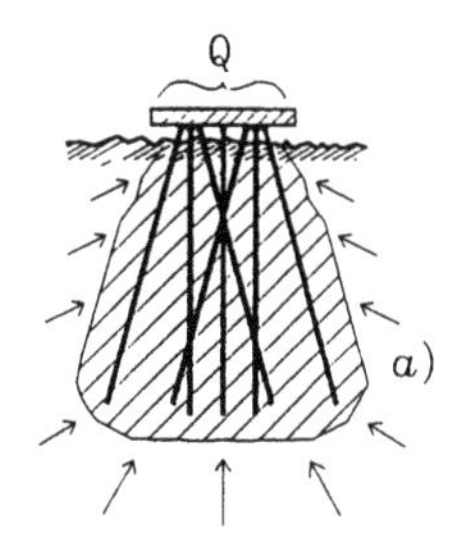

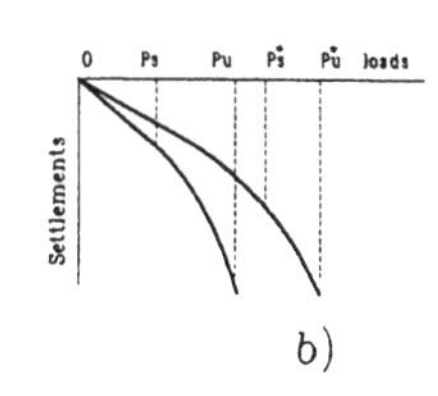

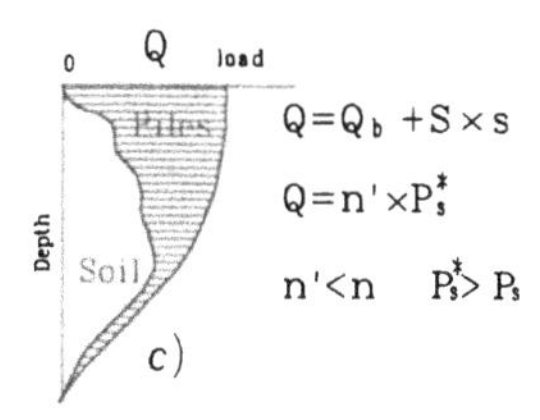

Fig. 6 "Reinforced Soil" bulb

- Evaluation of the average unit shearing resistance, s, of the soil.
- Evaluation of the base resistance of the block.

It is worthwhile stressing that in a foundation of "Reinforced soil", compared with a corresponding Large Diameter Piles foundation, the saving is a consequence of the following factors:

- a substantial part of the load is taken by the soil, thus reducing the load on the piles;
- large diameter piles are, generally very long, in order to reach very competent layers; in Reinforced Soil the reinforcing elements (the piles) do not need to be very long. A 30 meter depth for a bulb of "reinforced soil" is to be considered a maximum, even in medium soils.
- the Safety Factors for a structure such as a Large Diameter Pile Foundation, based on independent elements heavily charged, need to be higher than for a structure like the "reinforced bulb" based on a plurality of elements.

11.4 Something more about the mechanical characteristics of a "Reinforced Soil" foundation

Thc limit of the bearing capacity of a bulb of "reinforced soil" is reached when it fails as a whole by sinking, into the ground, provided it has a sufficient *internal stability* for not being disintegrated before.

The extra resistance is supplied by the piles, acting as a reinforcement; once the reinforcement is sufficient to this purpose, a greater number of piles is needless, because it doesn't increase the load bearing capacity.

At any rate, the rake imposed to the piles allows an increase in the dimensions of the bulb without increasing the number of piles. Therefore, as stated before, there is a problem of *external stability* as well as a problem of *internal stability.*

In conclusion, the design of the "reinforced soil" Foundation may be carried out as follows:

- definition of the geometry (size, depth) of the bulb, as a unit resisting structure sunk into the soil. This study is based on the geotechnical characteristics of the soil surrounding the bulb;
- convenient density in order to increase the resistance of the assumed block into the soil, for its full depth;
- check of the number of piles necessary for the above reinforcement against the number of piles necessary for the load bearing capacity as a "network" of piles. This second approach is based on the load tests, on a single as well as on group of piles. Generally this second value of the density is higher than the previous one; therefore the design carried out on the "network" hypothesis, can be sufficient for the purpose.

12 OTHER EXAMPLES OF "SOIL REINFORCEMENT"

According to the present level of technology and costs, the "Reinforced Soil" may find a place essentially in heterogeneous soils where other Geotechnical structures cannot be used. But there are several cases even in homogeneous soils, where conventional structures do not offer the same advantages of the "Soil Reinforcement"

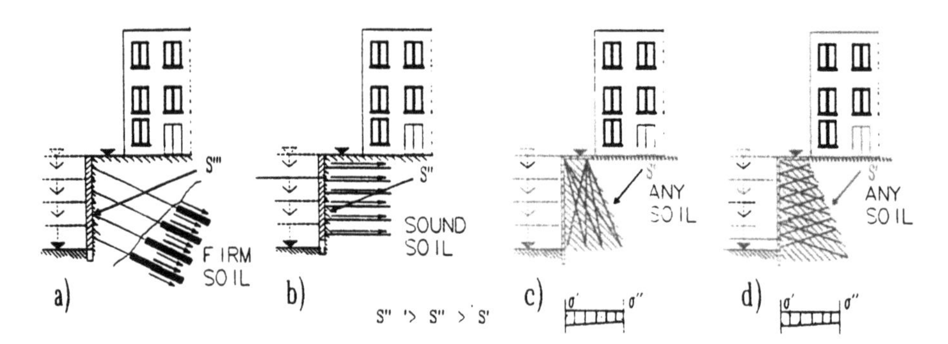

Fig. 7 Earth retaining walls; Different schemes

12.1 Retaining walls in areas to be excavated

In Fig. 7, four currently used schemes are indicated:

- 7a - Prestressed anchors
- 7b - Soil nailing
- 7c - Reticulated Root Piles (Soil Reinforcement).
- 7d -Reticulated Root Piles (Soil Reinforcement Alternate Solution)

Neglecting other differences it can be stated that schemes 7a and 7b consist of a front slab supported by elements introduced into the soil and entrusted by a pull resistance.

The static scheme of 7c is completely different; the network of piles has the main purpose of encompassing the block of soil, which forms a veritable "gravity" wall, where the weight of the soil is the main supporting element. The earth pressure is not acting against the front slab but on the back of the "gravity" wall.

Since the "reticulated" pattern has a great ductility its scheme may be adapted to any situation. In Fig. 7d, an alternate solution to the 7c scheme is indicated.

At any rate, as a matter of fact, for the Schemes 7a and 7b there are several design approaches taking, probably, advantage of the monodirectional pattern of the anchors (or nails). This is not the same for the RRP on account of the more intricate pattern, although this scheme offers undoubtedly a more appropriate physical approach to the problem. It is to be hoped that in future this veritable example of "Soil Reinforcement" could be better and more deeply analized.

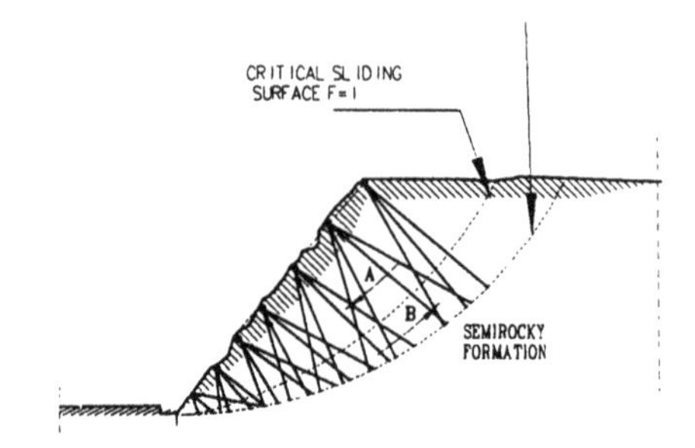

Fig. 8 "Reticulated pali radice" structure for landslide prevention in steep semirocky formation.

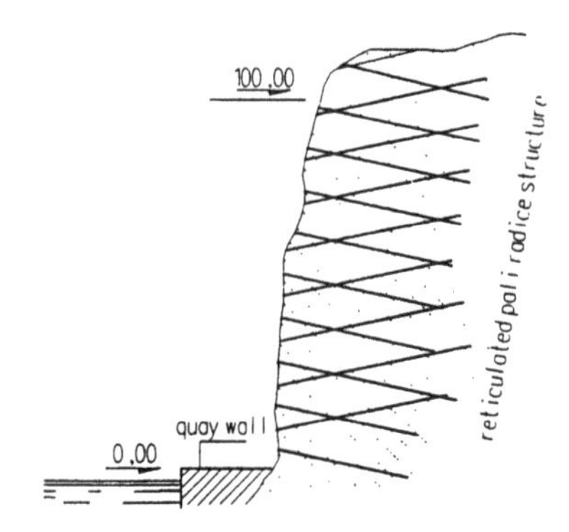

Fig. 9 Reticulated pali radice for consolidation of rocky formation.

12.2 Landslide prevention

In addition to the case illustrated in figure 1 there are several cases in which the "Soil Reinforcement", introduced by RRP network proves to be very effective. In Fig. 8, the reinforcement of a steep semirocky formation is indicated; the purpose of the

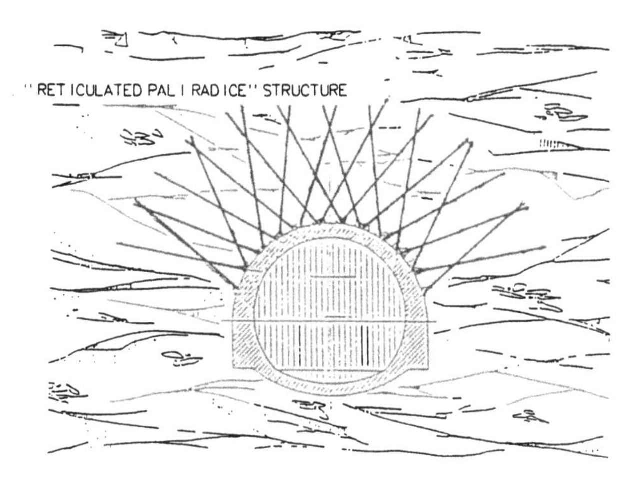

Fig. 10 Scheme of reinforcement of a tunnel in a loose clayey formation.

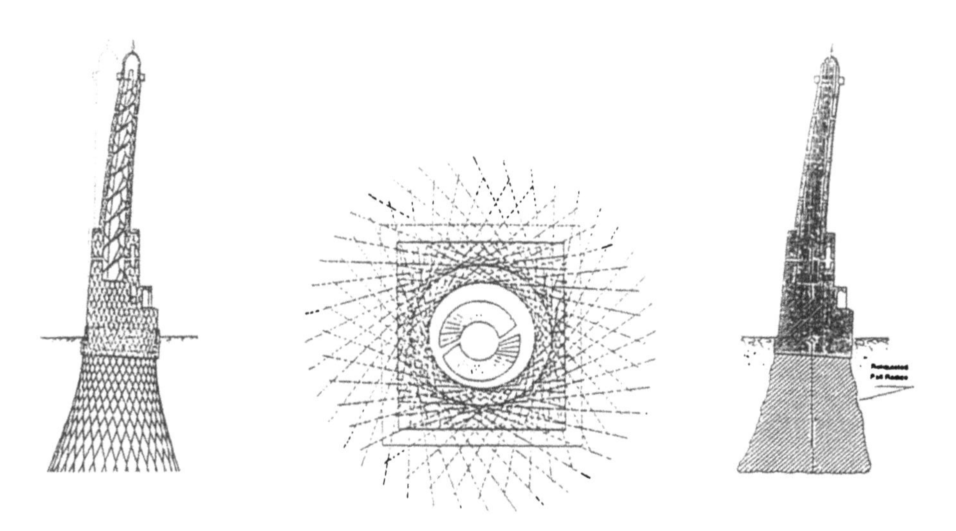

Fig. 11 The Al-Hadba leaning minaret Mosul (Iraq). Scheme of the under pinning by means of pali radice.

network is to connect the different parts of the slope (mixed soil and rock) in order to form a unitary solid mass laid on a lower, less steep, level; for environmental reasons the piles are connected by some small R.C. beams, very easily covered by the vegetation.

12.3 Consolidation of fractured rocky formation

This is (Fig. 9) a problem very frequent in tourist areas where ostensive protective measures against the collapse of the rock (as for instance R.C. anchors slabs) cannot be accepted for environmental reasons.

The three-dimensional R.R.P. network, introduced as a systematic reinforcement of the rocky mass leads to some sort of reinforced masonry wall without any need of a front R. C. slab.

12.4 "Soil Reinforcement" for the stability of tunnels

In Fig. 10, the case of a reinforcement

carried out for the stability of a series of tunnels, in critical static conditions, is illustrated. The structures were immerged in a chaotic clayish formation (flysch) which produced disequilibrated pressure on the lining, with the risk of final collapse.

A R.R.P. reinforcement was able to create, even in very loose soils, some sort of a reinforced arch sufficient to take and distribute uniformly the external thrust.

Similar reinforcements can be produced during the construction of new tunnels for the consolidation of the soil to be excavated.

It is to be noted that, presently, the current practice in the consolidation of the soil to be crossed by a tunnel is to create some sort of a continuous shell formed by jet-grouting, or grouted pipes, all along the external line of the lining.

It is the Author's opinion that such a substantially thin structural layer does not reduce the earth thrust, and on the other hand is able to support it only at a reduced extent. An upper arch of "reinforced soil", although not continuous, but of convenient thickness, whenever possible, would probably prove in several cases to be more effective.

12.5 "Soil reinforcement" for underpinning of tall buildings (towers)

A "root piles" underpinning very frequently used for tall buildings or towers in critical static conditions, realizes in the subsoil a veritable bulb of "reinforced soil" gravimetrically associated with the construction. The center of gravity of the system drops, ensuring in this way a total security to the stability of such a delicate structure (Fig. 11).

13. CONCLUSIONS

The diffusion of "piling" in foundation has overshadowed the long era of the soil as the main support of Man's constructions. No doubt, "piling" has restricted the foundation problem to the load bearing capacity of piles, ignoring any other possible contribution for the soil, declassed to "indirect" support of the piles.

It is the Author's opinion that it is time to reinstate Soil "duly reinforced" as the most appropriate foundation, and for extending the "Reinforced Soil" to several other geotechnical problems.

Geotechnics is a very large field, open to reflexion, intuition and perhaps to imagination, long before any theoretical approach. "Reinforced soil" is probably a promising new branch of Geotechnics.

The purpose of this paper is to raise interest in the above matter, and to stimulate appropriate studies in order to mark, for the "in-Situ Reinforced Soil", some appreciable progress.

REFERENCES

- F. LIZZI (1950/52) - First patents on Root Piles and Reticulated Root Piles, on behalf of the Italian Firm FONDEDILE, Naples, Italy
- F. LIZZI (1964) - "Root Pattern piles underpinning" - Symposium on Bearing Capacity of Piles, Roorkee, India.
- F.LIZZI (1977) - "Practical Engineering in Structurally Complex formations", International Symposium on the "Geotechnics of structurally complex formations" Capri, Italy.
- F. LIZZI(1978) - "Reticulated Root Piles" to correct landslides. A.S.C.E. Convention, Chicago (USA)
- F. LIZZI e CARNEVALE G. (1979) - "Les réseaux de pieux racine pour la consolidation des sols". Colloque international sur le renforcement des sols. Paris, France.
- F. LIZZI (1983) - The "Reticulated Root Piles" for the improvement of soil resistance, VIII ECSMFE, Helsinki, Finland.
- SCHLOSSER F., JURAN J. et al. Several publications on "Terre Armée" and "Soil nailing"
- THORNBURN S. & HUTCHISON J.F. (1985) "Underpinning" Surrey University Press, London, UK.
- BRANDL H. (1988) - "The interaction between soil and groups of small diameter bored piles". I International Geotechnical Seminar on Deep Foundation on bored and auger piles - Ghent.
- NASSEF SOLIMAN & GEORGE MUNFAKH (1988) - "Foundations on drilled and grouted minipiles. A case history". I

International Geotechnical Seminar on Deep Foundation on bored and auger piles, Ghent.
- MUNFAK G. A. (1989) - "Soil Reinforcement. A tale of three walls".XII ICSMFE, Rio de Janeiro.
- PLUMELLE C. (1989) - "Full scale experiments on three soil nailed walls". XII ICSMFE, Rio de Janeiro.
- LIZZI F. (1989) - "Anchors and Root Piles, Similarities and Differences" General Report, Session 12 - XII ICSMFE, Rio de Janeiro.
- GUDEHUS G. (1991) - "Displacements and soil-structure interaction of earth retaining structure". General Report Session 4 - X ECSMFE Florence, Italy.
- Van IMPE W.F. (1991) - "Deformations of deep foundations" X ECSMFE Florence, Italy.
- LIZZI F. (1991) "Root Piles (pali radice) as Soil Reinforcement for foundation problems", Civil Enginering European Courses, COMETT, Paris, 1-3 October.
- SCHLOSSER F., UNTERREINER P. and PLUMELLE C. (1992) - French Research Program, CLOUTERRE on Soil Nailing. Proceedings of the ASCE Conference, New Orleans, Luisiana, USA.
- THORNBURN S. & LITTLEJOHN G.S. (1993)j, "Underpinning and Retention" Blackie Academic Professional - London, N.Y., Tokyo,Melbourne, Madras.
- HANSBO S. (1993) - "Interaction problems related to the installation of pile groups". Deep Foundations on Bored and Auger Piles BAP II. Ghent.
- POULOS H.G. (1993) - "Settlement prediction for bored pile groups" Deep Foundations on Bored and Auger Piles BAP II. Ghent.
- RANDOLPH M.F. & CLANCY P. (1993) - "Efficient design of piles raft". Deep Foundations on Bored and Auger Piles BAP II. Ghent.
- CARVALHO D. & CINTRA J.C.A. (1993) - "Aspects of the bearing capacity of root piles in some Brazilian soils." Deep Foundations on Bored and Auger Piles BAP II. Ghent.

Developments in Geotechnical Engineering, Balasubramaniam et al. (eds) © 1994 Balkema, Rotterdam, ISBN 90 5410 522 4

Engineering of ground in lowlands

Norihiko Miura
Saga University, Japan
Madhira R. Madhav
Saga University, Japan & Indian Institute of Technology, Kanpur, India

ABSTRACT: Saga plain is a typical lowland reclaimed from and affected by the high tides of the Ariake Sea. The soft, compressible and very sensitive Ariake clay, 15 to 30 m thick, gives rise to many difficult geotechnical problems due to its low bearing capacity, very large total and differential settlements and inadequate safety of earth structures. Various ground improvement techniques are practiced to modify the ground. Examples of preloading in conjunction with sandpacked and plastic board drains, deep mixing methods and piled-geogrid reinforced granular fill rafts, are presented.

1 INTRODUCTION

Saga Plain, a typical lowland, is located adjacent to the Ariake sea. Reclamation of this part of the country from the sea, has been going on for more than one thousand years (Fig.1). The soil at this location, called Ariake clay, is a soft sensitive marine clay which has many interesting and unique characteristics (Miura et al. 1988). Hanzawa et al. (1990) report in detail the various physical, engineering and mechanical properties of the Ariake clay. Only a brief summary of the properties is given here to highlight the geotechnical challenges.

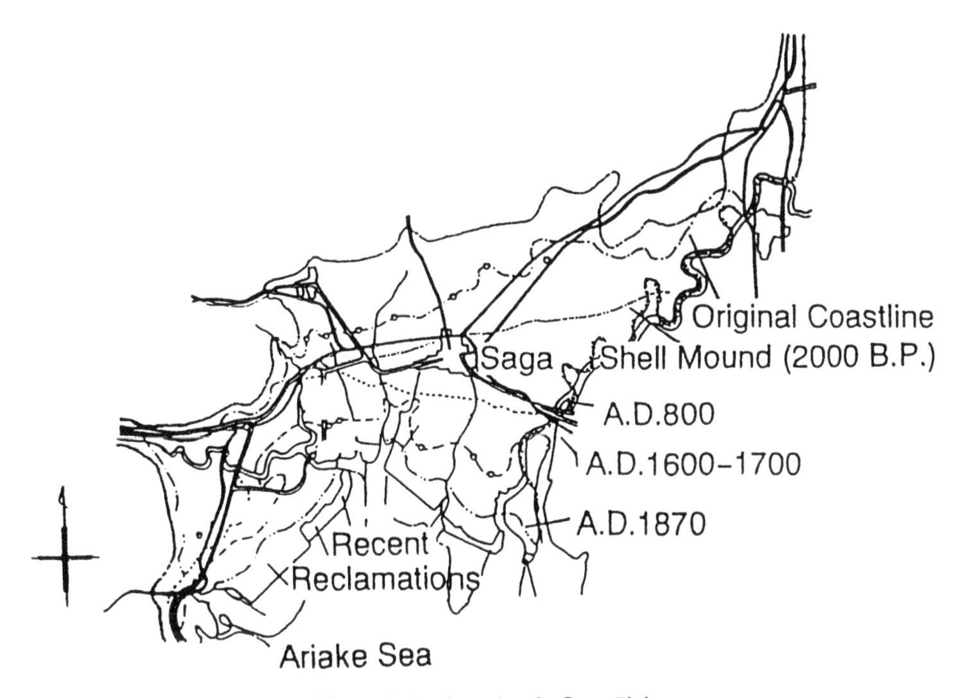

Figure 1. Reclamation in Saga Plain.

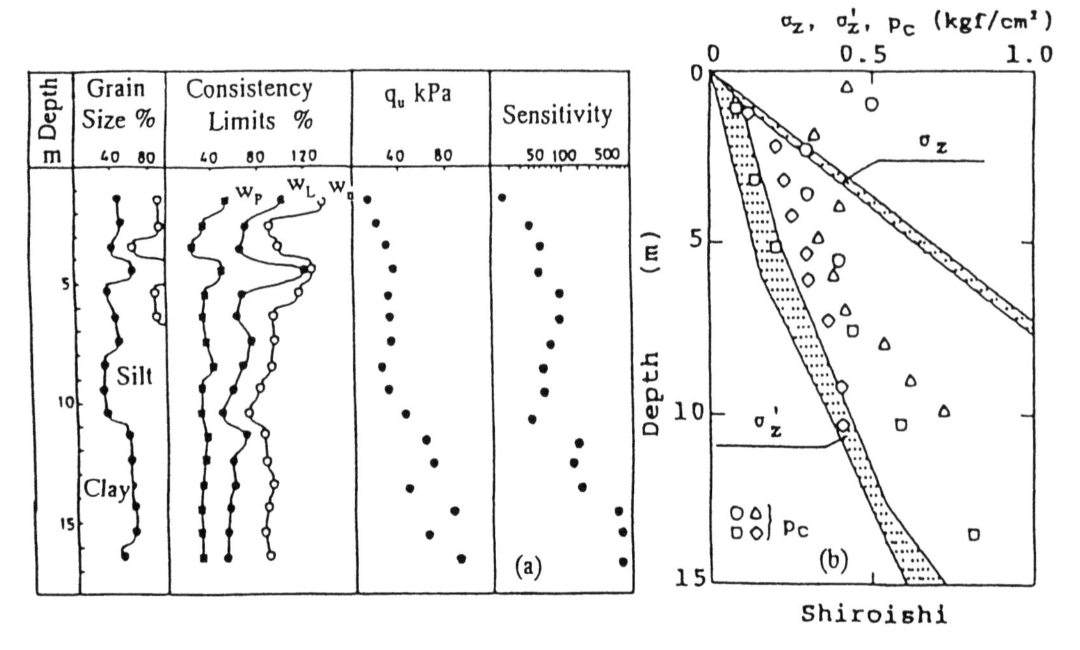

Figure 2. Physical and engineering properties of Ariake clay.

The physical and engineering properties of Ariake clay from a typical site are depicted in Figure 2. The clay deposit extending to depths of 15 to 30 m at various locations, is usually divided into about 12 to 15 m thick upper clay layer and a lower clay layer. The lower layer was formed under brackish conditions during the period 10,000 to 6000 B.P. when the sea level was rising rapidly (ABRG 1965). The upper clay layer was deposited under marine conditions. The ground water level is close (0 ~ 0.5 m) to the ground level.

The liquid limit and plasticity index values for the upper clay (Fig.2a) range between 60 to 120% and 25 to 80%, respectively. The liquidity index is greater than one, values as high as 1.5 to 2.0 are not uncommon. The undrained strength for the clay increases from a low (< 5 kPa) to about 40 kPa at a depth of 12 m. The strength increase in the lower clay is much more rapid than the increase in the upper clay. The sensitivity of the clay ranges (Fig.2a) usually between 12 and 100, but sometimes could be more than 100 (Onitsuka 1988) in the interior where reclamation was carried out long time back and saline porewater leached.

Smectite is the principal clay mineral of the Ariake clay (Ohtsubo et al. 1988). Kaolinite, illite and vermiculite also occur in smaller proportion. The cation exchange capacity (CEC) of the clay ranges between 27.0 and 40.0, the specific surface area is about 400 to 450 m^2/g and the salt content ranges between 0.44 to 1.08 g/l.

The preconsolidation stress, σ'_c, is predicted more accurately from constant rate of consolidation test (Fig. 2b) and indicates the over consolidation ratio (OCR) to be 1.5 to 3.0. The soil is in a lightly over consolidated state probably because of cementation and subsequent leaching. Consequent to structural breakdown, the compression index, C_c, is 2.6 to 4.0. The void ratio – log effective stress curves flatten at higher stresses and the C_c values range between 1.15 and 1.60.

2 SUBSIDENCE

Below the soft Ariake clay layer there exists a sandy aquifer from which groundwater had been pumped out in the past. Due to excessive pumping, sea water has intruded into the aquifer and contaminated the groundwater, the chlorine content exceeding the permissible limit of 200 ppm. Figure 3 depicts the time – settlement curves at two locations in the Saga Plain. The total subsidence observed for the period 1972 to 1985, is about 30 cm in Saga and 80 cm in Shiroishi. The relative elevations of the ground and groundwater levels over this period, depicted in Figure 4, show a close inter-relationship.

3 ENGINEERING OF GROUND

Design and construction of traditional foundations, either shallow or deep, for structures on soft, compressible and sensitive Ariake clay, is very exp–

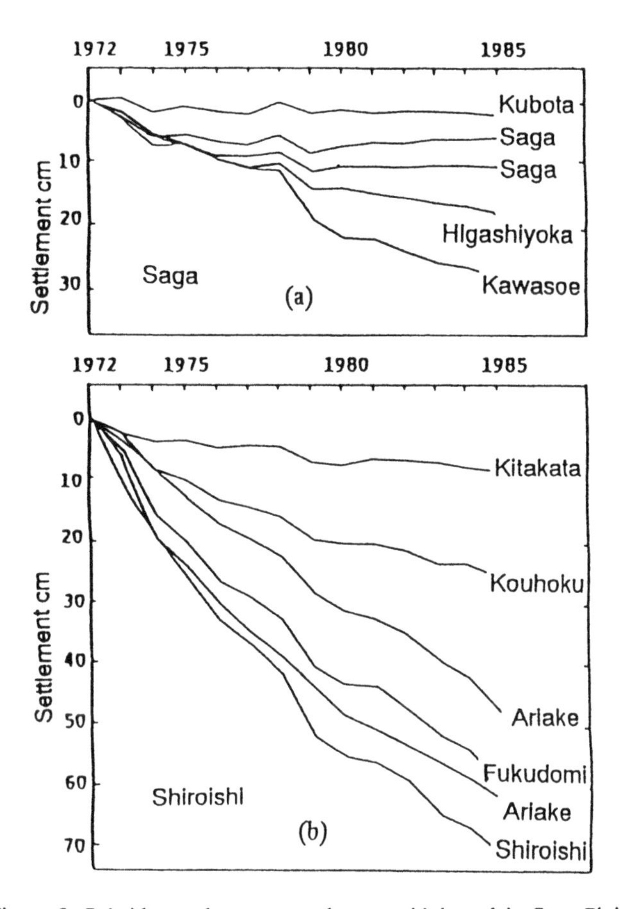

Figure 3. Subsidence due to groundwater withdrawal in Saga Plain.

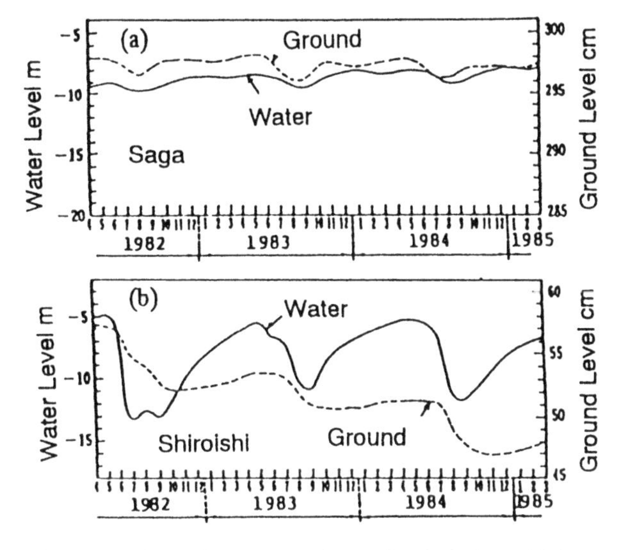

Figure 4. Ground and groundwater levels in Saga Plain.

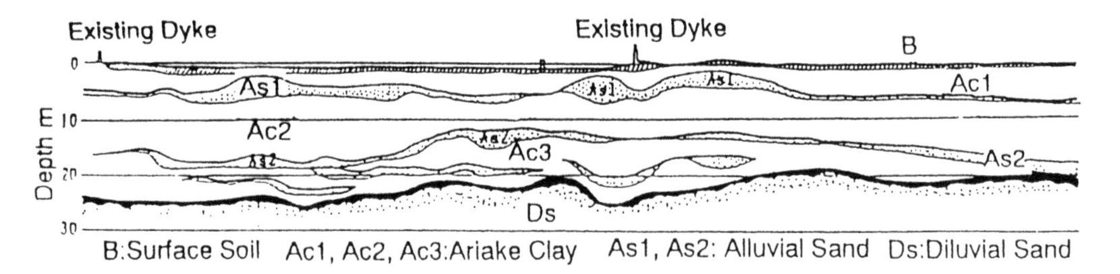

Figure 5. Soil strata at Saga airport site.

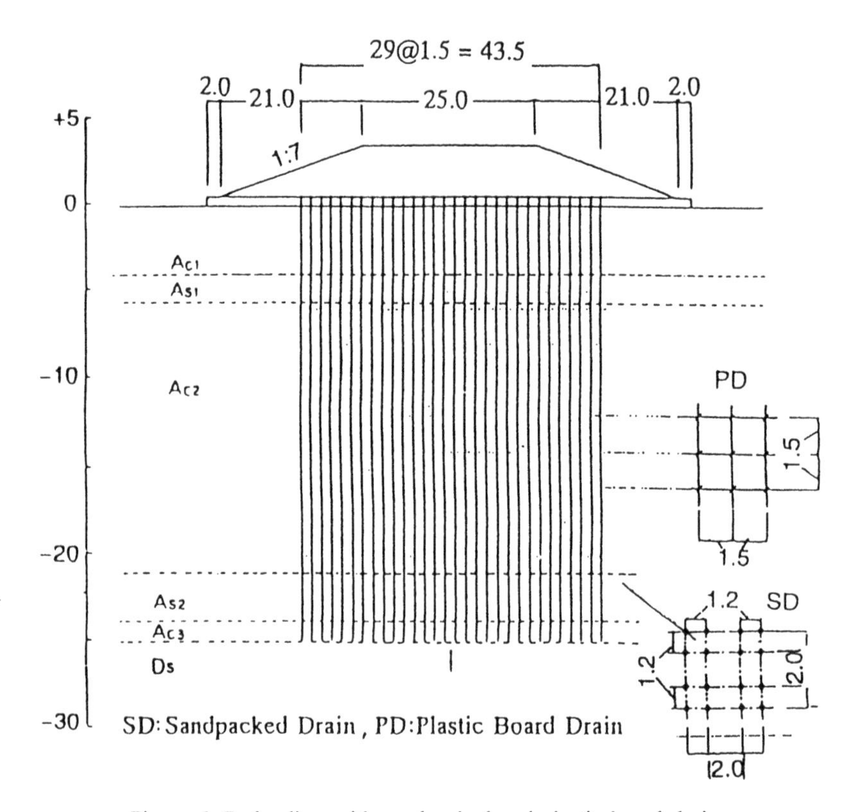

Figure 6. Preloading with sandpacked and plastic board drains.

ensive. It is found imperative to improve the ground to achieve the required standards of performance of the structures founded on it. The most preferred choices (Kawakami 1992) for the engineering of Ariake clay are:
(i) preloading,
(ii) vertical drains - sand compaction and plastic board drains,
(iii) deep mixing methods - High Pressure Mixing and Dry Jet Mixing (DJM), and
(iv) Geogrid reinforced foundation beds over floating piles. Examples of the application of these techniques, are presented in the following sections.

3.1 *Preloading and vertical drains*

Saga airport covering an area of 110 ha and having a runway of 2000 m is being built on reclaimed ground near Saga. The existing ground level which is presently at $-0.5 \sim +0.4$ m is being raised to the level of about $+1.5 \sim 1.7$ m above mean sea level. From the profile of the site shown in Figure 5, it can be noted that the thickness of the Ariake clay ranges

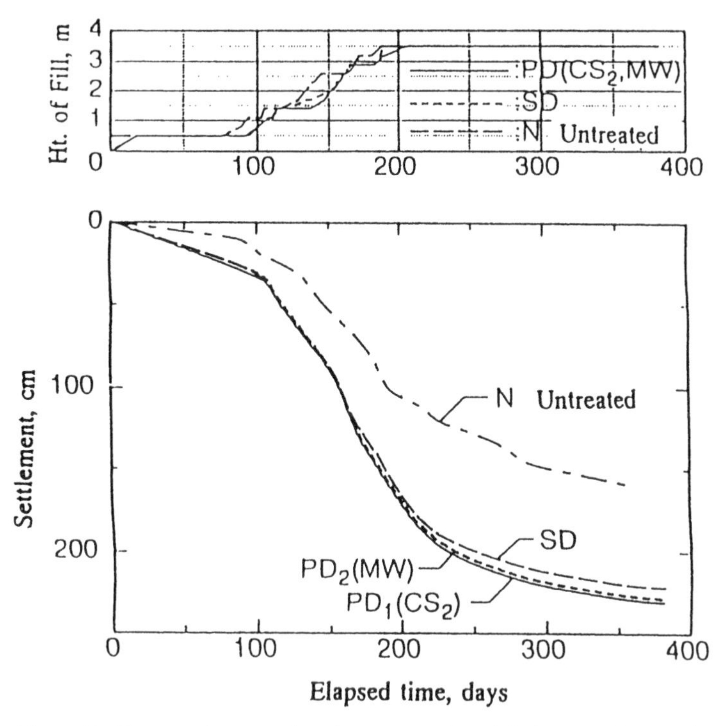

Figure 7. Time–settlement plots for untreated and drain treated ground.

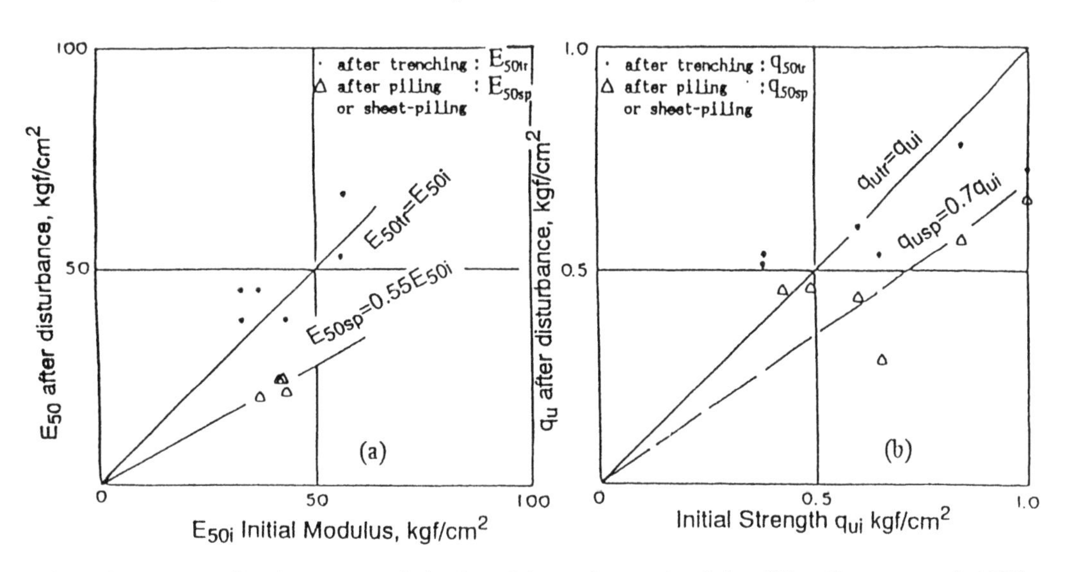

Figure 8. Effect of disturbance on undrained modulus and strength of clay (After Hanzawa et al. 1990).

between 20 to 25 m. The clay is intercalated with two or three layers of sand. A dense diluvial sand stratum underlying the clay permits free drainage from the bottom surface as well.

To evaluate the efficacies of different types of drains, a trial preloading, was carried out. A 3 m high embankment of top width of 25 m and side slopes of 7(H):1(V) was constructed on 0.5 m thick sand blanket (Fig.6). The arrangement of sand packed (SD) and plastic board (PD) drains is also depicted in Figure 6. The average length of the drains is 22.5 m. The sand packed drains were installed four at a time in a square pattern at a spacing of 1.2 m and the spacing between each set being 2.0 m. The plastic board drains were also installed in a square pattern but at a spacing of 1.5 m. Settlement gauges at various depths, total (surface) settlement gauges and piezometers were installed to monitor the progress of the consolidation of the ground.

The time–settlement plots of the untreated and treated ground are presented in Figure 7. By the end of the construction period of 200 days, the untreated ground settled by 110 cm while the treated ground settled by nearly 170 cm and achieved about 80 % degree of consolidation. Thus, it is very obvious that the vertical drains have definitely accelerated the rate of consolidation, the PD functioning slightly better than the SD. However, it is disconcerting to note that the total settlements were larger in the drain treated area compared to those of the untreated ground. The excess settlement of the drain treated ground is possibly due to the disturbance and remolding caused by the drain installation. Kawakami's (1992) data (Fig.8) indicated that after sheet piling, the modulus of deformation and the undrained strength of the Ariake clay reduced to 55 % and 70 % of their original undisturbed values respectively. Larger settlements of treated ground in comparison with those of untreated ground were also reported by Balasubramaniam et al. (1980), Akagi (1981) and Crawford et al. (1992).

Mechanistically, this larger settlement of the treated ground results from the changes in the initial stress state and compressibility due to the disturbance and remolding caused during the installation of the drains. Sensitive clays like Ariake clay, possess a relatively high preconsolidation or yield stress (σ'_y)(Fig.9). When the surcharge stress ($\Delta\sigma$) is applied to the undisturbed ground, the initial state point A moves through the yield stress on to the steep post yield compression line to point C. However, in the case the soil is disturbed and/or remolded, the yield stress effect is destroyed, the initial stress state point shifts to point D and the path for consolidation is represented by points D to E. The distance between A to D corresponds to the pore pressure set up due to drain installation. Thus since the final equilibrium stress state for undisturbed and disturbed zones (around the drains) is the same, compression is more in the latter case due to the reduction in the effective stress and destruction of the yield stress. The total settlement of the drain treated ground consisting of disturbed and undisturbed zones of soft sensitive clays exceeds that of the untreated ground by 20 to 30 %.

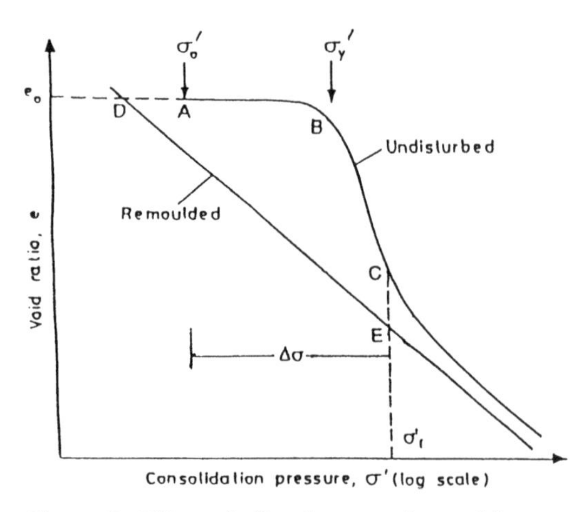

Figure 9. Effect of disturbance and remolding on settlements due to drain installation.

3.2 *Deep mixing methods*

Deep mixing methods have become very common to improve the mechanical properties of soft sensitive clays like Ariake clay. They are classified into two categories (Miura et al. 1987) as:

(a) Slurry Jet Mixing Methods, and
(b) Mechanical Mixing methods (e.g. Dry Jet Mixing or DJM Method).

In the former, cement or quicklime with or without admixtures, is injected into the ground in slurry form and mixed by rotating wings or blades. In the DJM method, cement or quicklime in powder form is introduced into the soil with the aid of compressed air mechanically with the clay. Since no water is added to the soil in the DJM method, the improvement effects on the strength of Ariake clay are more than those in which slurry is used. If quicklime is used instead of cement, the heat of hydration further reduces the water content of the soil and contribute to higher strength gain. For clays of high organic content (>8%), cement is preferred over quicklime.

3.3 *Piled geogrid reinforced granular fill*

The foundations for sluiceways across highway embankments founded on soft clays is another interesting problem. The conventional method of construction is to found the sluiceway on end bearing piles. However, such a practice has led to (i) the formation of a cavity beneath the sluiceway and (ii) cracking of the embankment near the shoulders. The embankment load causes the soft soil to settle while the piled foundation prevents the downward movement of the sluiceway, thus creating a void. The cracking of the embankment near its shoulders arises from the large differential settlement.

Geogrid reinforced granular fill, if used as a raft over piles, has the stiffness to transfer the loads to the piles and at the same time, has the flexibility to deform conforming to shape of the ground after settlement. In addition, the end bearing piles are replaced by floating piles to make the foundation consistent and compatible with the soft soil and the sluiceway deformations.

Figure 10 depicts two design alternatives provided for sluiceway foundations in Saga (Miura et al. 1992). Design alternative A (Fig.10a) uses timber piles of length 8 m at a spacing of 1.5 m in a square grid pattern. The raft on top of the piles consisted of 1.5 m thick granular fill in which two layers of geogrid were provided as reinforcement. Initial camber of 30 cm at the center was provided. Design alternative B (Fig.10b) has piles by the Dry Jet Mixing Method (DJM) in which the in situ soil was mixed with cement introduced in dry powder form. The DJM piles are 1.0 m in diameter and 7.5 m long. They were spaced at 1.6 m in a square arrangement. The improvement ratio, defined as the area covered by the DJM piles to that of the treated ground, is 30 %. A reinforced

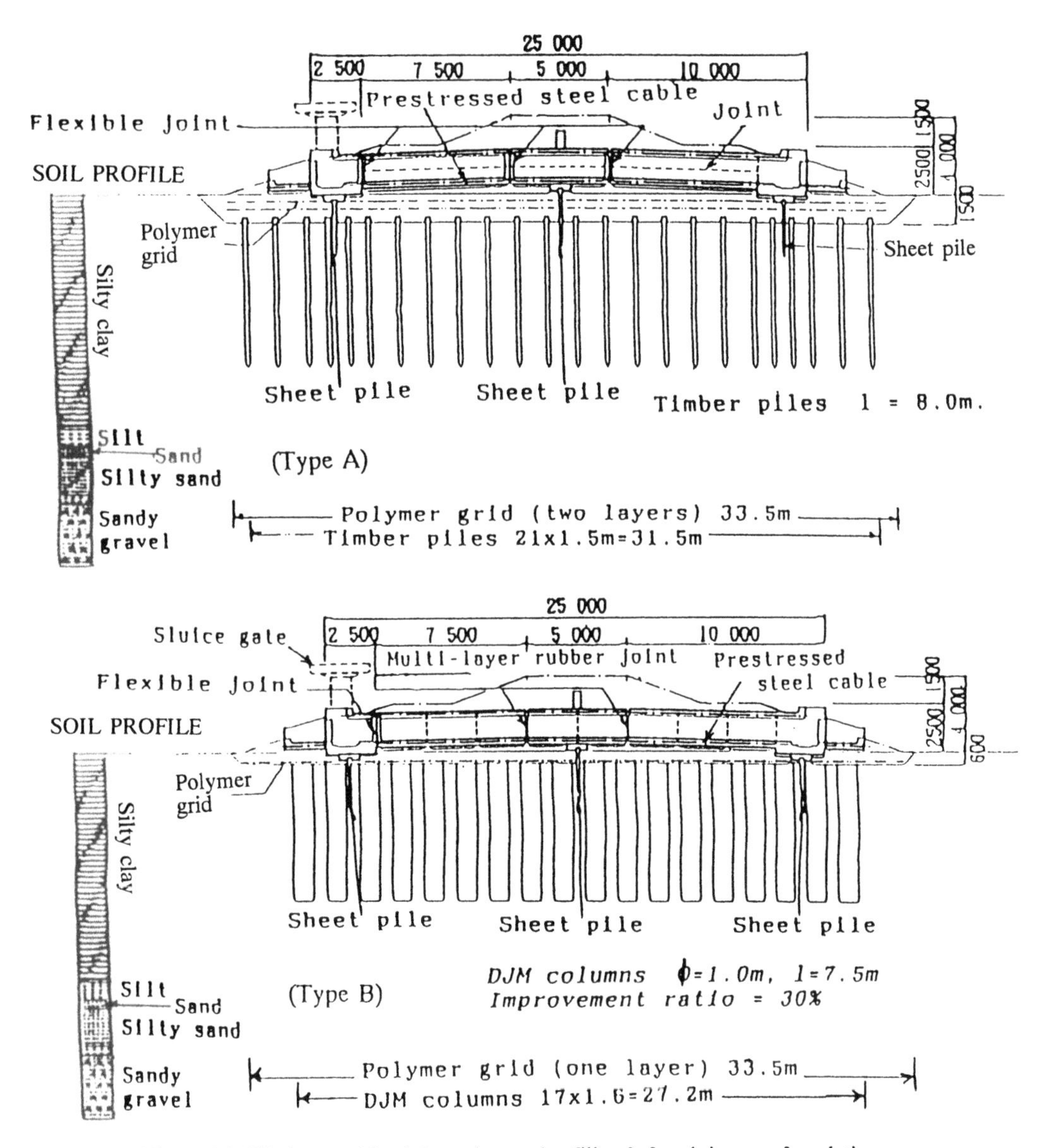

Figure 10. Piled-geogrid reinforced granular fill raft for sluiceway foundations.

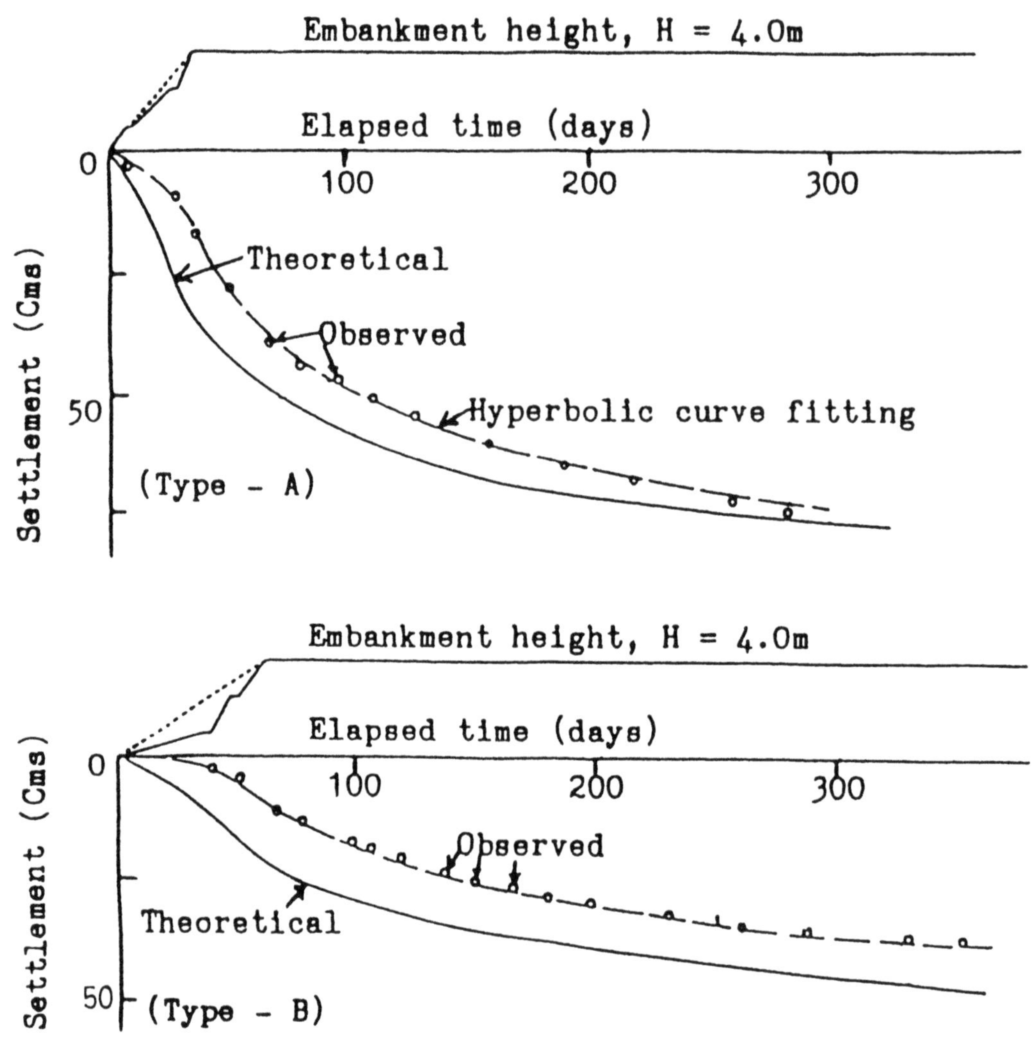

Figure 11. Time-settlement plots for sluiceway foundations.

granular layer 600 mm thick with a single layer of polymer grid was laid over the DJM Piles. The camber provided is only 20 cm at the center in this case.

The sluiceways were made flexible by constructing in the form of blocks and connecting them by flexible and multilayer rubber joints respectively for the two designs. Settlement of the untreated ground due to 4 m high embankment, is estimated to be 190 cm. The estimated settlements for design alternatives A and B are 80 cm and 50 cm, respectively. The settlements measured at the end of one year (Fig.11) are about 72 cm and 36 cm for the two alternatives. Fitting a hyperbolic plot to the observed time – settlement curves, the final settlements are estimated to become 91 cm and 61 cm which are about 20 % larger than the predicted ones. The sluiceways appear to deform and take the profile of the ground underneath after settlement (Fig.12). Thus, geogrid reinforced granular fill rafts function as a flexible plate transmitting loads to the piles and at the same time deforming with the ground so that part of the load is transmitted to the soil directly.

4 CONCLUDING REMARKS

Saga Plain, a typical lowland, consists of Ariake clay which is a soft, sensitive clay which has unique characteristics. The liquidity index is greater than one, values as high as 1.5 to 2.0 are not uncommon. The undrained strength for the clay increases from a low (< 5 kPa) to about 40 kPa at a depth of 12 m. The sensitivity of the clay ranges usually between 12

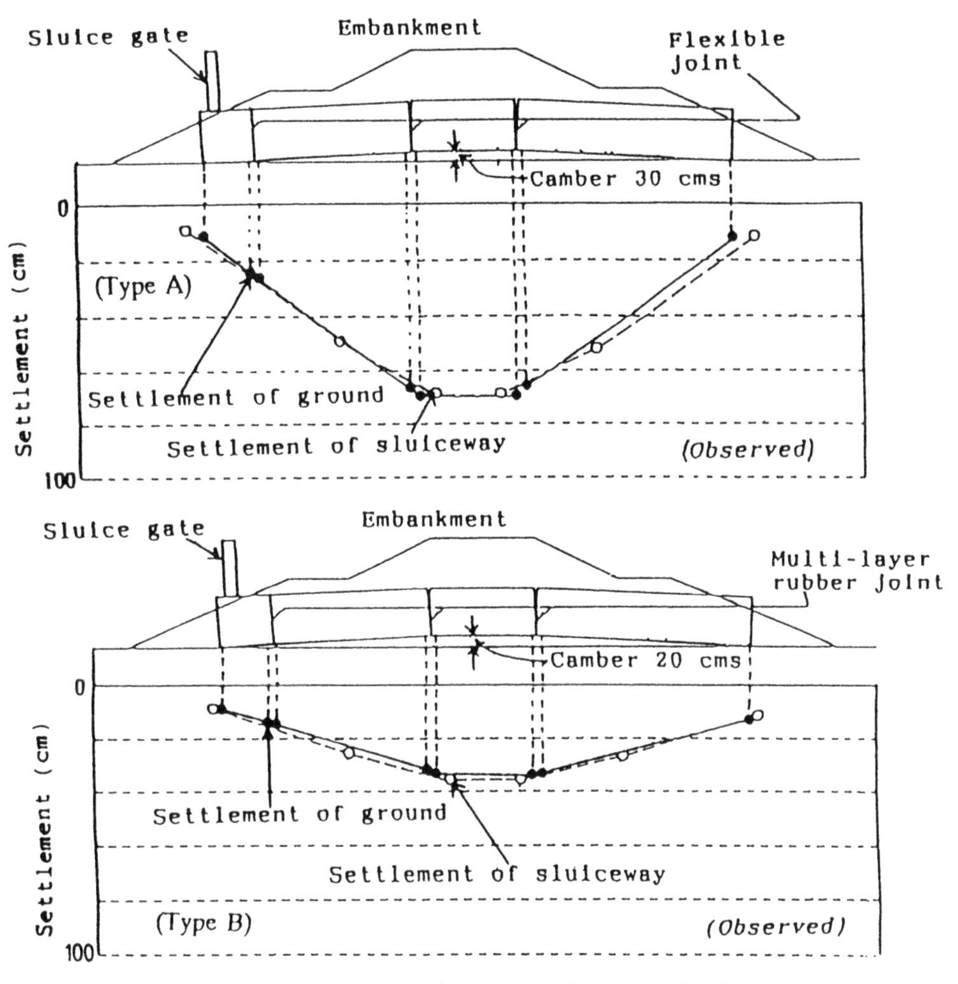

Figure 12. Deformations of sluiceways and the ground underneath.

and 100 but sometimes could be as high as 100 - 200 inland where reclamation was carried out long time back and saline porewater leached. Due to excessive pumping, sea water has intruded into the aquifer and the total subsidence observed for the period 1972 to 1985, ranges between 30 to 80 cm. The engineering of the Ariake clay is carried through:

(i) preloading,
(ii) vertical drains - sand compaction and plastic board drains,
(iii) deep mixing methods - High Pressure Mixing and Dry Jet Mixing (DJM), and
(iv) Geogrid reinforced foundation beds over floating piles.

Examples of the application of these techniques, are presented in the paper. It is shown that with proper care, appropriate ground modification techniques can be practiced and satisfactory to good performance of geotechnical structures are ensured.

REFERENCES

Ariake Bay Research Group (ABRG) 1965. *Quaternary system of the Ariake and Shiroishi Bay areas, with special reference to the Ariake clay.* The Association for the Geologic Collaboration in Japan (in Japanese with English summary).

Akagi, T. 1981. Effect of mandrel-driven sand drains on soft clay. *Proc. 10th ICOSMFE, Stockholm*. Vol.1, pp.581-584.

Balasubramaniam, A.S., R.P. Brenner, G.U. Mallawaaratchy & S. Kuvujitjaru 1980. Performance of sand drains in a test embankment in soft Bangkok clay. *Proc. 6th SEA Conf. on Soil Engrg., Taipei*. pp.447-468.

Crawford, C.B., R.J. Fannin, L.J. deBoer & C.B. Kern 1992. Experiences with prefabricated vertical drains at Vernon, B.C. *Can.Geotech. J.* 29: 67-79.

Hanzawa, H., T. Fukaya & K. Suzuki 1990.

Evaluation of engineering properties for an Ariake Clay. *Soils and Foundations* 30 (4):11–24.
Kawakami, Y. 1992. Improvements of soft clay near the Rokkaku river in Saga plain. *Proc. Int. Symp. on Problems of Lowland Development, Saga*. pp.231–238.
Miura, N., D.T. Bergado, A. Sakai & R. Nakamura 1987. Improvements of soft marine clay by special admixtures using dry and wet jet mixing methods. . *Proc. 9th SEAGC, Bangkok*. Vol.2, pp.8–35~8–46.
Miura, N., Y. Taesri, A. Sakai & M. Tanaka 1988. Land subsidence and its influence to geotechnical aspects in Saga Plain. *Proc. Int. Seminar on Shallow Sea and Low Land, Saga*. pp.151–158.
Miura, N., M.R. Madhav, Y. Kawakami, M. Hayakawa & J. Kaneko 1992. Performance of a new system of flexible sluiceway with floating foundation in highly compressible ground. *Proc. Int. Symp. on Prediction versus Performance in Geotech. Engrg., Bangkok*, Nov.–Dec. 1992, pp.15–28.
Onitsuka, K. 1988. Mechanical properties of very sensitive Ariake clay. *Proc. Int. Seminar on Shallow Sea and Low Land, Saga*. pp.159–167.
Ohtsubo, M., M. Takayama & E. Egashira 1988. Mineralogy, chemistry and geotechnical properties of Ariake marine clays. *Proc. Int. Seminar on Shallow Sea and Low Land, Saga*. pp.145–150.

Developments in Geotechnical Engineering, Balasubramaniam et al. (eds) © 1994 Balkema, Rotterdam, ISBN 90 5410 522 4

Performance of reinforced embankment on soft Bangkok clay with high strength geotextile reinforcement

D.T. Bergado & P.V. Long
School of Civil Engineering, Asian Institute of Technology, Bangkok, Thailand

C.H. Lee, K.H. Loke & G.Werner
Polyfelt Geosynthetics Co. Ltd, Kuala Lumpur, Malaysia & Linz, Austria

ABSTRACT: In order to study the improvement of embankment stability on soft ground, two test embankments were constructed to failure. One test embankment was reinforced with high-strength, non-woven geotextile as base reinforcement. The reinforcement consists of one layer of high-strength geotextile placed directly on the natural ground surface at the bottom of the embankment fill. For comparison, another full-scale, unreinforced embankment using the same fill material as that of the reinforced one was also constructed to failure at the adjacent site. The height at failure of the reinforced embankment was 6 m while that of the unreinforced embankment was 4 m. This paper presents the instrumentation program, construction procedure, monitored data, and stability analysis of the test embankments. The study indicated that the high strength, non-woven geotextile as base reinforcement, can increase considerably the ultimate height of embankments on soft clay. It was also shown that the rupture occurred at large deformation of foundation subsoils.

1 INTRODUCTION

The use of polymer materials such as non-woven and woven geotextiles, polymer grids, prefabricated vertical drains, etc. is now quite widespread for applications in infrastructure constructions on soft Bangkok clay. Both Tensar and Tenax geogrids were used as reinforcements of embankments on soft Bangkok clay in the Second Bangkok Express System (SBES). The prefabricated vertical drains have been used at the Klong 19 to Saraburi railway project of the State Railway of Thailand (SRT). Non-woven and woven geotextiles have been used as separator and for filtration at the bank stabilization of Mekong River and at the reclamation in Laem Chabang and Map Ta Put Deep Sea Ports, respectively. Non-woven geotextiles and geomembranes are being used for lining of artificial ponds and lakes for golf courses and for lining of waste containment projects. However, the use of high strength, non-woven geotextile as reinforcement of highway embankments is not yet widespread. Thus, this study was conducted to investigate the advantages of using geotextile reinforcement on embankments on soft ground. One embankment was reinforced by one layer of high strength, non-woven geotextile (PEC 200) with tensile strengths of 200 kN/m. Another unreinforced embankment was constructed nearby as control embankment. The research embankment project has been supported by Polyfelt Geosynthetics of Austria.

2 TEST LOCATION AND SOIL PROFILE

The test embankments were located in the campus of Asian Institute of Technology (AIT). Layout of the test embankments is given in Fig. 1 including the canal excavation. The locations of field vane tests and boreholes are also shown.

The general soil profile and soil properties of the test site are presented in Fig. 2. The uppermost 12 m can be divided into 4 layers. The weathered crust consisting of heavily overconsolidated reddish-brown clay forms the uppermost 2 m. This layer is underlain by a soft, grayish clay down to about 8 m depth. The medium stiff clay with silt seams and fine sand lenses is found at the depth of 8 to 10.5 m depth. Below this layer is the stiff clay layer. The undrained shear strength profile measured by field vane shear tests is also presented in Fig. 2.

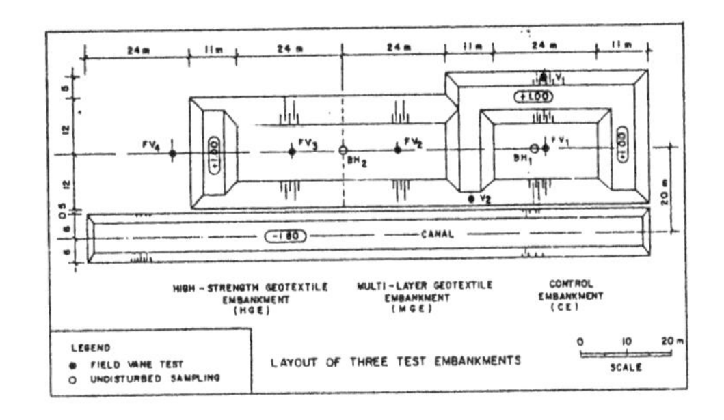

Fig. 1 Layout of the Test Embankments Including Locations of Field Vane Tests and Boreholes

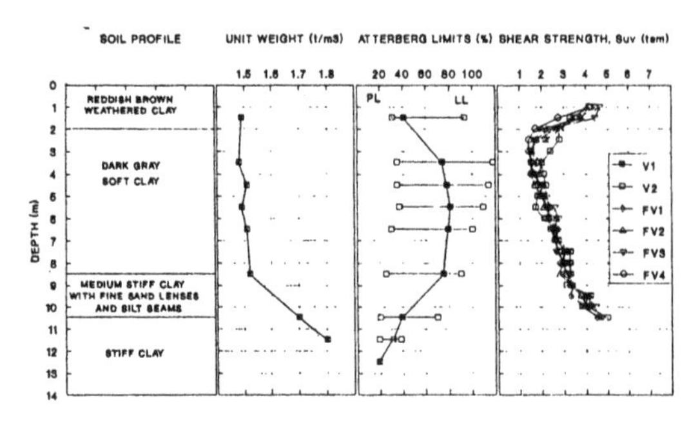

Fig. 2 Soil Profile and Soil Properties at the Site

3 INSTRUMENTATION PROGRAM

Foundation instrumentation of both embankments is given in Figs. 3. The instruments consisted of settlement gages, piezometers, inclinometer and earth pressure cells.

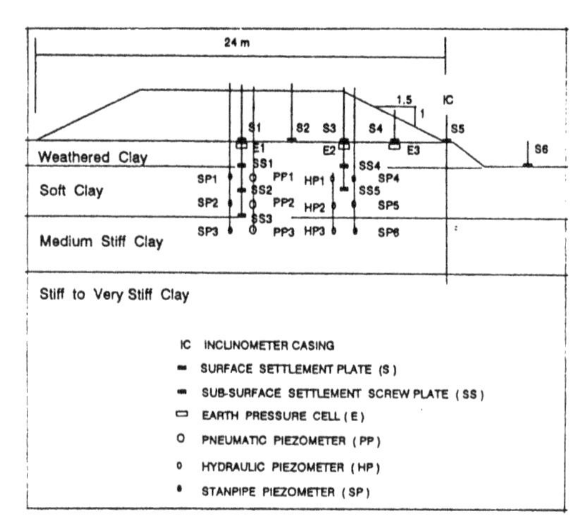

Fig. 3 Foundation Instrumentation of the Test Embankments

3.1 *Settlement measurement*

Settlements of ground surface and subsoils were measured by means of AIT surface settlement plate and subsurface screw plate, respectively. The settlements were measured at the depths 0 m, 2 m, 4 m, and 6 m from the original ground surface (Fig. 3). The surface settlement plate consisted of a 16 mm diameter steel rod connected to a 40 cm x 40 cm base plate and protected by a 19 mm diameter casing of GI pipe. The subsurface settlement screw plate consisted of a 250 mm diameter steel screw plate connected to a 16 mm diameter steel rod. Settlements were measured daily by a precise levelling machine.

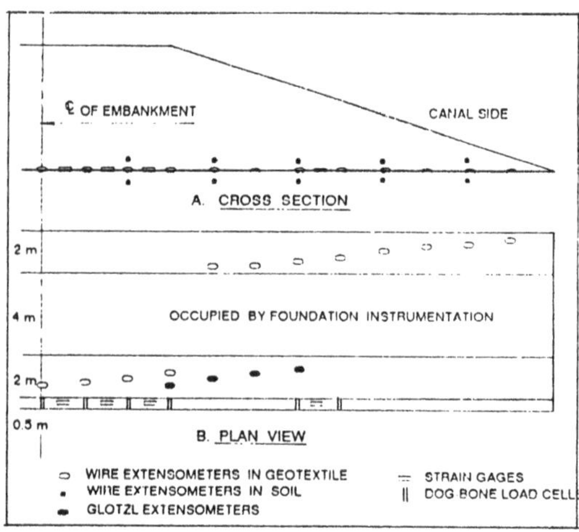

Fig. 4 Geotextile Instrumentation of HGE Embankment

3.2 *Pore pressure measurement*

Pore pressures were monitored at depths of 3 m, 5 m, and 7 m from the ground surface (Fig. 3). To minimize the uncertainty of pore pressure measurement, three types of piezometers were used, namely: closed-type hydraulic piezometer, open standpipe piezometer, and pneumatic piezometer.

3.3 *Total earth pressure measurement*

Pneumatic earth pressure cells (Sinco model # S1452) were located at the base of each embankment to measure the vertical contact pressures. The earth pressure cell is 230 mm diameter and 11 mm thick, made of stainless steel, filled with liquid (ethylene glycol), and is attached to pneumatic transducer.

3.4 *Lateral displacement measurement*

The lateral movements of the subsoil were measured using digitilt inclinometer. One inclinometer was installed to the depth of stiff clay layer at the toe of each embankment. Inclinometer casings of Sinco model # 51100100 were used. The casings have 69.8 mm outer diameter, and 58.9 mm inner diameter.

The instrumentation program for geotextile consisted of wire extensometers, Glotzl extensometers, strain gages and dog bone load cells. The side view and layout of geotextile instrumentation are shown in Fig. 4. Summary of these measurements are follows:

3.5 *Strain gages*

Strain gages of the type EP-08-40 CBY-120, manufactured by Micro Measurements Division,were used. These gages have a nominal resistance of 120 ohms, and are 10 cm long. Strain gages were located on 0.5 m wide strip of geotextile, at 1 m spacing, and were in between the dog-bone load cell points. Two strain gages were installed at one point of strain measurement, and were connected in series to be a sensor of double resistance (Fig. 5).

3.6 *Dog-Bone load cells*

Dog-bone load cells were used to measure tensile forces in geotextile. Manufactured in AIT, the dog-bone cells consisted of a stainless steel bolt installed with strain gages, clamps connected to geotextile, and an outer plastic tube for construction protection. The tensile forces were measured indirectly using strain gage principle. In the field, the load cells were installed in pairs, and were connected to geotextile through clamps (Fig. 6).

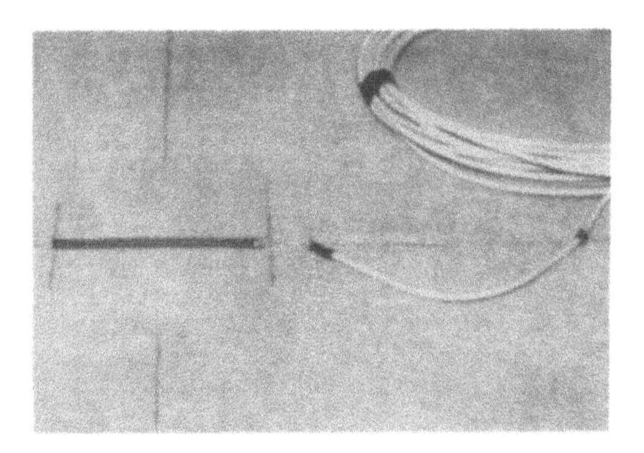

Fig. 5 Installation of Strain Gages in Geotextile

Fig. 6 Arrangement of Strain Gages and Dog-bone Load Cells in the Field

3.7 *Wire extensometers*

Wire extensometers were used to measure the displacements of geotextile and surrounding soils as well. The extensometer consists of an inner 2 mm diameter high-strength stainless cable and an outer flexible PVC tube. The PVC tube, has 4 mm inner and 6 mm outer diameters, enabled the free moving of the inner wire. One end of the wire was fixed at the measured point as shown in Fig. 7. The other end was connected to a counter weight of about 0.5 kg through a pulley system at the readout board.

3.8 *Glotzl extensometers*

The other type of extensometer installed on geotextile was the precise Glotzl type extensometer. Glotzl extensometer consists of a 10.7 mm diameter inner rod made of glass fibre, an outer PVC tube with 12 mm inner diameter and 16 mm outer diameter. One end of the inner rod was attached in geotextile at the measured location as given in Fig. 8. The opposite end of the outer tube was fixed at

Fig. 7 Arrangement of Wire Extensometers in the Field

Fig. 8 Arrangement of Glotzl Extensometers in the Field

the readout board. Relative displacements of the inner rod and the fixed end of the outer tube yielded the relative displacements in the geotextile.

4 CONSTRUCTION OF EMBANKMENTS TO FAILURE

The layout of the test embankments is presented in Fig. 1 including the canal excavation. Before construction of the embankments, the ground surface was excavated to about 20 cm to 25 cm depth and levelled. The canal of 2 m deep and 7.5 m wide at the bottom was excavated along the pre-determined failure side of all embankments. The ground levelling and canal excavation were carried out in 10 days and finished on January 8, 1993.

The first layer of control embankment was placed on January 8, 1993. The embankment was constructed in layers with compaction lift thickness of about 33 cm. The main equipment used for embankment construction consisted of one D4 bulldozer, one backhoe and one vibrating roller. Compaction at the area nearby the instruments were performed by a Walker hand compactor. Density and moisture content of each layer were controlled by field nuclear tests and sand cone tests. The dry density of 1.7 t/m^3 and moisture content of 9 % were maintained. The embankment reached the height of 3 m without any signals of significant deformations. The height of 4 m was reached in the afternoon of February 4, 1993. In the early morning of February 5, 1993 a crack of about 5 mm wide and 3 m long was observed near the settlement plate S6 in the canal. The width of the crack opened quickly to about 1 cm wide, and the embankment completely failed as shown in Fig. 9. The cross-section of failure surface is given in Fig. 10.

Fig. 9 Photograph Showing the Failure of CE Embankment

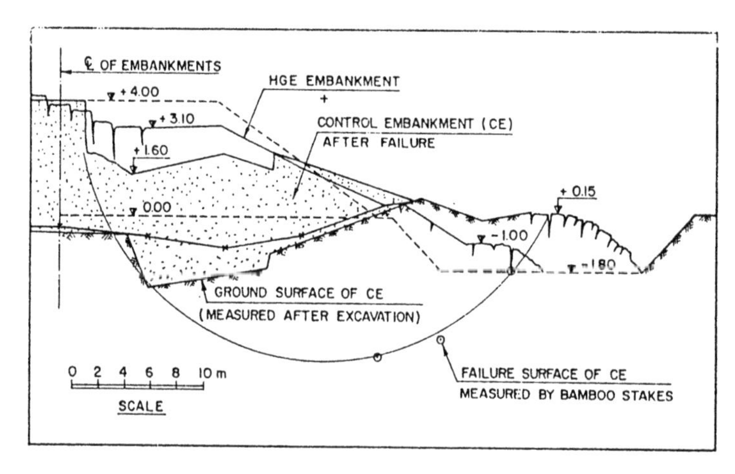

Fig. 10 Cross-Sections of CE Embankment after Failure and Induced Failure of HGE Embankment

Two months after the failure of the control embankment, two reinforced embankments were constructed adjacent to the unreinforced one. One embankment was reinforced by 4 layers of low-strength, non-woven geotextile (MGE). The other was reinforced by one layer of high-strength, woven-nonwoven geotextile (HGE). Both embankments were built at the same time and the same rate of filling. Construction procedure and quality control were kept to be the same as that of the control embankment. At 3 P.M. of May 28, 1993 when the two embankments reached 3.75 m high, one small crack of about 5 mm width was observed near the surface settlement plate S_6 in the canal. The crack developed into two larger cracks on the next day. Construction of embankments were continued to reach 4.2 m high on May 30, 1993. Failure of MGE embankment and induced failure of HGE embankment occurred at the night time of May 30, 93. The cross-section of HGE embankment measured on May 31, 93 is given in Fig 10 together with the failure section of the control embankment. After this event, all instruments of HGE embankment were still working except the inclinometer, strain gages and dog-bone load cells. The measured data indicated that the geotextile in HGE embankment was not yet broken. The HGE embankment construction was then continued starting on June 19, 1993. The HGE embankment reached 6 m height on June 30, 93 and failed at the night time of that day. The side view photograph and cross-section of embankment after failure are presented in Fig. 11 and Fig. 12, respectively.

It is noted that there were 35 rainy days during construction of MGE and HGE embankments. Consequently, the average field moisture content of MGE and HGE measured at the end of construction

Fig. 11 Photograph Showing the Failure of HGE Embankment at 6 m High

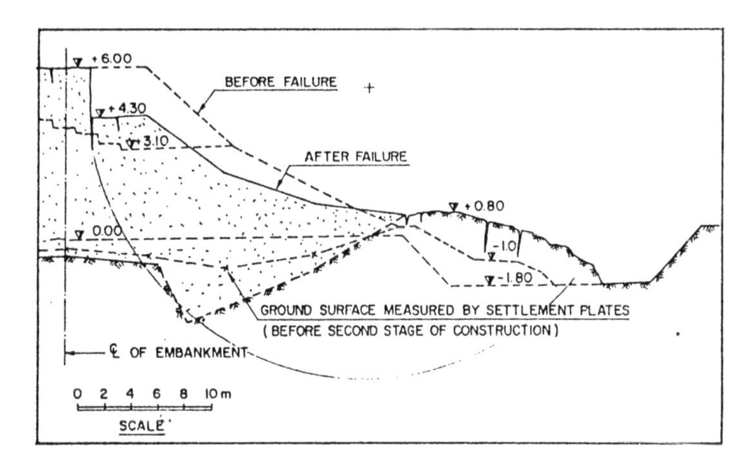

Fig. 12 Cross-Section of HGE Embankment after Failure at 6 m High

was about 13%, and was about 4% higher than that of the control embankment resulting in higher surcharge loading.

5 MEASURED DATA

The main results obtained during construction of the CE and HGE embankments are given in Fig. 13 to Fig. 17. The measured results are summarized as follows.

5.1 *Lateral displacement*

Lateral displacements of two embankments are plotted together in Fig. 13. Up to the height of 3 m, the displacements in both embankments are nearly the same, and the maximum displacement were observed at the depth of 3 m. The maximum horizontal displacement of CE embankment at the day before failure February 4, 93 is 17 cm, and occurred at the ground surface. The displacement of ground surface of HGE embankment at 4.2 m height is 6 cm smaller that of CE embankment on the day before failure (H = 4m). However, the maximum displacement of HGE embankment still occurred at the depth of 2.5 m with the magnitude of 19 cm, and was about 3 cm larger than that of CE embankment. This event can be explained by (a) the vertical load acting on the ground surface of HGE embankment was larger than that of CE embankment because of the extra 0.2 m height and the increase of total unit weight due to rainfall during construction of HGE, and (b) the reinforcement has no significant effects to the depth of the soft clay layer.

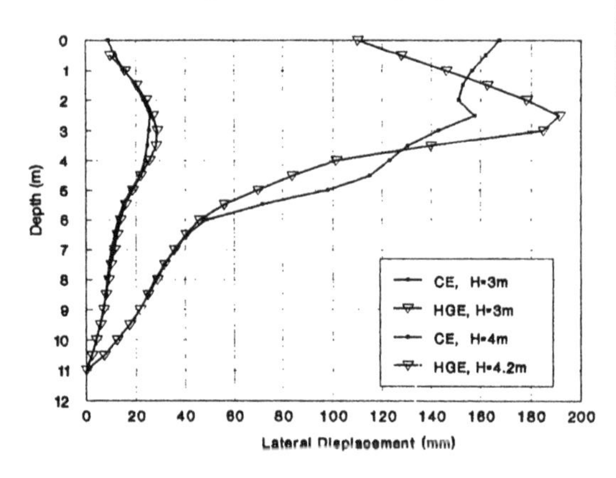

Fig. 13 Lateral Displacements of CE and HGE Embankments

5.2 *Excess pore pressures*

The excess pore pressures measured at the depth of 3 m, 5 m, and 7 m of CE and HGE embankment foundations are plotted versus the embankment height in Figs. 14 and 15, respectively. The excess pore pressure in HGE embankment foundation were larger than that of CE embankment at the same embankment height, e.g, at the height of 4 m the excess pore pressures of 6 tsm and 6.5 tsm were observed in CE and HGE embankment foundations, respectively. This can be explained by the effects of rainfall during HGE embankment construction.

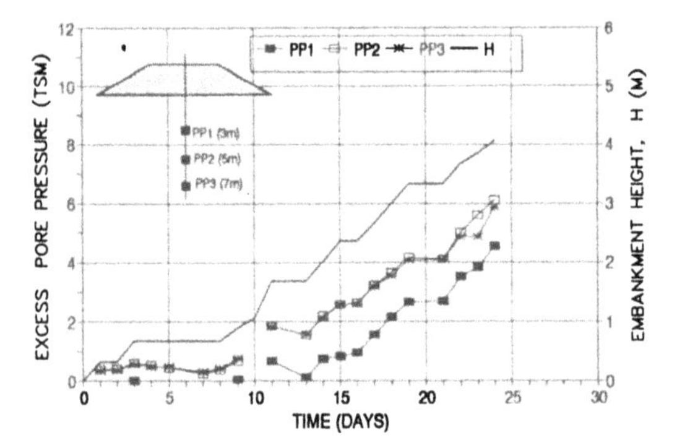

Fig. 14 Excess Pore Pressures in CE Embankment Foundation

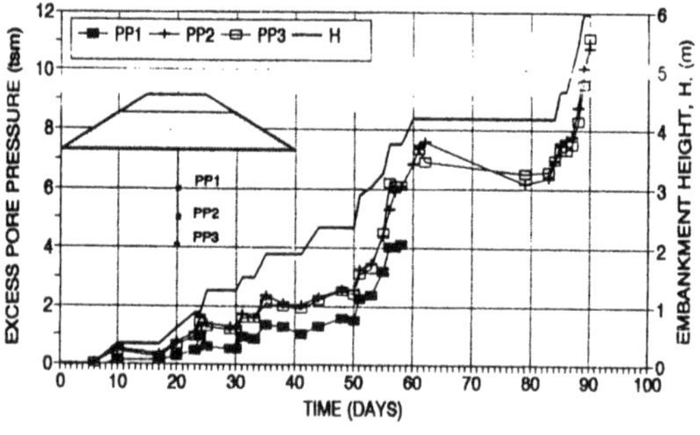

Fig. 15 Excess Pore Pressures in HGE Embankment Foundation

5.3 *Geotextile displacements*

The displacements of geotextile measured by wire extensometers are presented in Fig. 16a,b. At 4 m high, the maximum displacement of geotextile at the location near the toe of the embankment is about 10 cm which is corresponding to lateral displacement of ground surface measured by inclinometer at 4 m high (Fig. 13). Before the failure at 6 m high, the maximum displacement of geotextile was about 60 cm.

5.4 *Strain measurement*

Strain in geotextile measured by wire extensometers,

Glotzl extensometers, and strain gages are plotted together in Figs. 17a,b. There were no significant strains in geotextile at the embankment height of lower than 3.5 m. The strain of about 2 % were obtained at the height of 4.2 m. After the induced failure, the strain in geotextile was about 8 %. Increase of the embankment height, the maximum strain of 12 % was measured before embankment failure at 6 m high. The distribution of strains in geotextile does not correspond to the expected one with the classical design method. In fact, the maximum strain did not occur at the center of embankment and some compression was measured under the slope. Similar behavior was also reported by Delmas et al (1992). The validity of strain measurements can be found by the comparable results obtained from wire extensometers, Glotzl extensometers, and strain gages as shown in Fig. 17a,b.

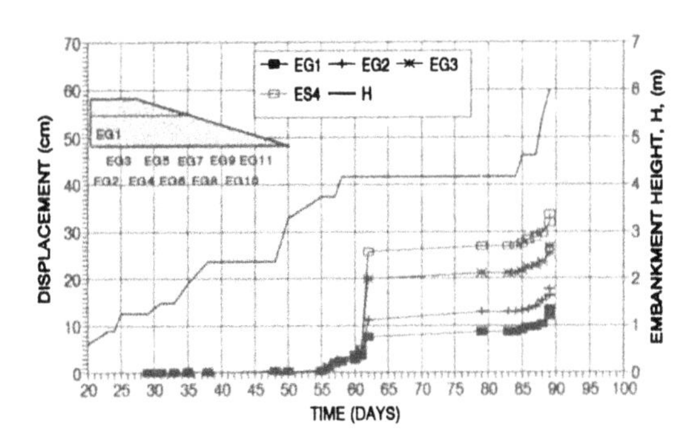

Fig. 16a Displacements of Geotextile Measured by Wire Extensometers EG1 to EG4

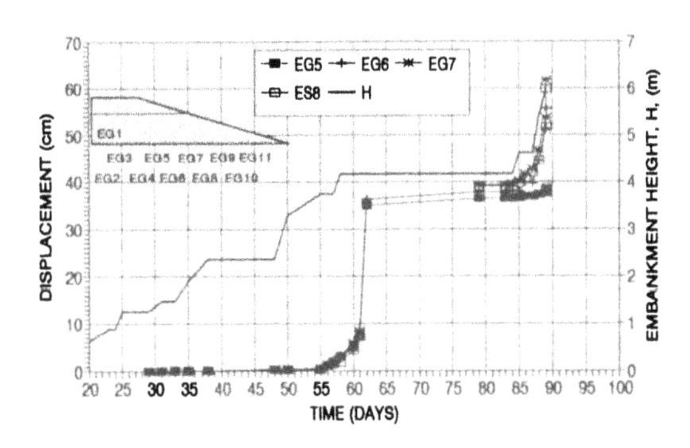

Fig. 16b Displacements of Geotextile Measured by Wire Extensometers EG5 to EG8

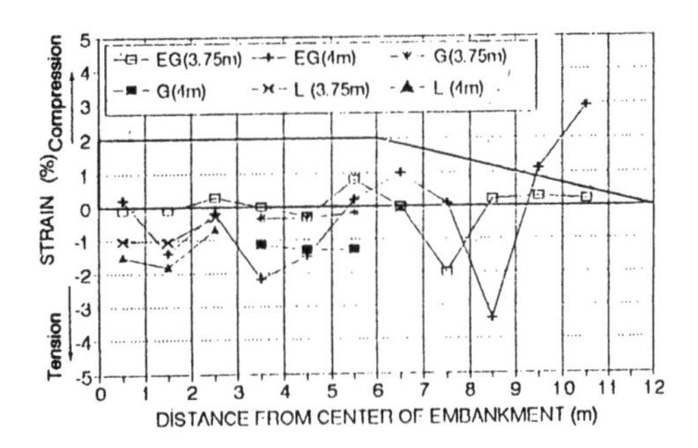

Fig. 17a Strains in Geotextile at Embankment Heights of 3.75 m and 4 m

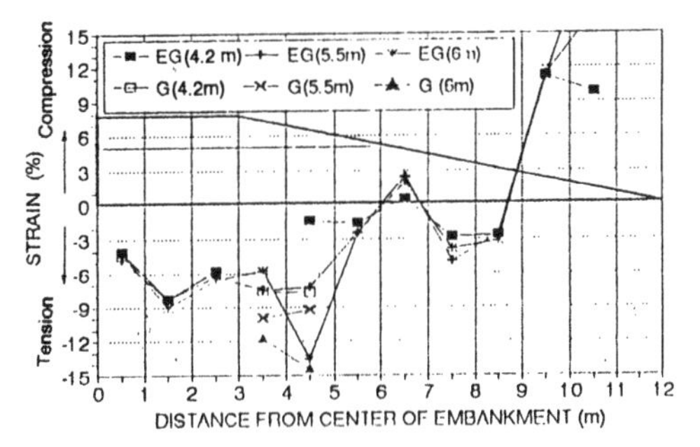

Fig. 17b Strains in Geotextile at Embankment Heights of 4.2 m, 5.5 m and 6 m

6 SLOPE STABILITY ANALYSIS

Limit analyses with circular slip surfaces have been commonly used in conventional design of geotextile reinforced embankments. In this type of analysis, two main assumptions have to be made: (a) the mobilized tensile force in geotextile at limit state, and (b) the orientation of the reinforcement which is often presented by the inclination factor, I_f. The mobilized tensile force may be taken from 2% to 10 % (Bonaparte & Christopher, 1987). The values of I_f ranges from 0 to 1 corresponding to horizontal and tangential direction of the reinforcement, respectively. The tensile force has been considered to act horizontally by some investigators (Fowler, 1982; Jewell, 1982; Ingold, 1983; Milligan & La Rochella, 1984), tangentially to the failure surface by the others (Binquet & Lee, 1975; Quast, 1983; Delmas et al, 1992) and between the above twc directions (Huisman, 1987).

The actual failure surfaces of the test embankments presented in this study can be fitted well by the circular surfaces as shown in Figs. 10 and 11. Hence, Bishop's simplified method using SB-SLOPE and STABL6 softwares are used for stability analyses of these unreinforced and reinforced embankments, respectively.

6.1 *Shear strength of backfill material*

The locally silty sand was used as backfill material in these test embankments. The shear strength parameters of backfill material used in this analysis were determined by large direct shear tests. The values of ϕ = 30° and c = 15 kPa which correspond to the moisture content of 9 % were applied for the case of the control embankment. In the case of the reinforced embankment (HGE), the measured moisture content at the field at the end of construction was 13 % due to the rainfall as mentioned in previous section. Corresponding to this moisture content, the values of ϕ = 30° and c = 10 kPa were obtained.

6.2 *Results of analyses*

The results of back analyses indicated that in the case of the control embankment, using the field vane shear strength with the correction factor μ = 0.8 for the soft clay layer, the factor of safety (FS) equal to 1 was obtained.

In the case of reinforced embankment (HGE), the maximum strain measured in geotextile at 4.2 m high (before the first failure) was 3.5% which corresponds to the mobilized tensile force of 60 kN/m. Using this value of tensile force, the FS of 1 and 1.08 were obtained for the case of the inclination factor, I_f, of 0 and 1, respectively. The results imply that the reinforced embankment was in the stable equilibrium state during mobilization process of tensile strain in the geotextile. For comparison, taking the same properties of backfill sand as used in HGE embankment, the factor of safety of unreinforced embankment must be 0.94 and 1.0 corresponding to the height of 4.2 m and 3.7 m, respectively. The results indicated that at low strain level of less than 3.5%, the reinforcement increased about 10% of factor of safety or 0.5 m height of the unreinforced embankment.

At the height of 6 m of HGE embankment, using the tensile force of 200 kN/m which is corresponding to the measured strain of 13% together with the residual strength of foundation soils, the calculated factor of safety were 1.05 for I_f = 1. The factor of safety of 0.65 was obtained for the equivalent embankment without reinforcements. Thus, the geotextile increased 60 % the factor of safety or at least 2 m in the ultimate height of the unreinforced embankment.

7 CONCLUSIONS

1) The use of high-strength geotextile as base reinforcement can increase considerably the ultimate height of embankment on soft clay (up to 2 m more). The rupture of geotextile occurred at large deformation of foundation soil (average strain of 5%).
2) At stress level lower than the limit state of unreinforced embankment, the strains in geotextile are controlled mainly by the lateral displacement of the weathered crust beneath the reinforcement. At low stress level, there were no differences in lateral displacements between the foundations of CE and HGE embankments. Increasing the embankment height, the base reinforcement decreased the lateral movement of the ground. However, it seems that the geotextile has no significant effects on the lateral movements of the soft clay layer below the weathered crust.
3) The high-strength geotextile reinforcement can be used effectively on the soft soils that can sustain large deformation (average strain of 5%) during construction.

8 REFERENCES

Binquet, J. & Lee, K. L. (1975). Bearing capacity analysis of reinforced earth slabs. *J. Geotech. Engng Div., ASCE*, 101(12), pp 531-534.
Bonaparte, R. & Christopher, B. R. (1987). Design and construction of reinforced embankments over weak foundation, *Transportation Research Record 1153*, pp. 26-39.
Delmas, Ph., Queyroi, D., Quaresma, M., De Saint Amand & Puech, A. (1992). Failure of an experimental embankment on soft soil reinforced with geotextile: Guiche. *Geotextiles, Geomembranes and Related Products*, Den-Hoedt (ed), Balkema, Rotterdam.
Fowler, J. (1982). Theoretical design considerations for fabric reinforced embankments. *Second International Conference on Geotextiles*, Las Vegas, Vol. 3, pp. 665-670.
Huisman, M.G.H. (1987). Design guideline for reinforced embankments on soft soils using Stabilenka reinforcing mats. *Enka Technical Report*, Arnhem.
Ingold, T.S. (1983). Some factors in design of geotextile reinforced embankments.*Proc. of the Eighth European Conference on Soil Mechanics and Foundation Engineering*, Helsinki, Vol. 2, pp. 503-508.
Jewell, R.A. (1982). A limit equilibrium design method for reinforced embankments. *Second International Conference on Geotextiles, Las Vegas*. Vol. 3. pp. 671-676.
Milligan, V. & La Rochelle, P. (1984). Design methods for embankment over weak soils. *Polymer Grid Reinforcement*. Thomas Telford, London, pp. 95-102.
Quast, P. (1983). Polyester reinforcing fabric mats for the improvement of the embankment stability. *Proc. of the Eighth European Conference on Soil Mechanics and Foundation Engineering*, Helsinki, Vol. 2, pp. 531-534.

Developments in Geotechnical Engineering, Balasubramaniam et al. (eds) © 1994 Balkema, Rotterdam, ISBN 90 5410 522 4

Author index